P9-CRF-408

Sustainable Energy

Sustainable Energy

Choosing Among Options

Jefferson W. Tester, Elisabeth M. Drake, Michael W. Golay,
Michael J. Driscoll, and William A. Peters

The MIT Press
Cambridge, Massachusetts
London, England

MIT Press books may be purchased at special quantity discounts for business or sales promotional use. For information, please email special_sales@mitpress.mit.edu or write to Special Sales Department, The MIT Press, 5 Cambridge Center, Cambridge, MA 02142.

This book was set in Times New Roman by Susan Robson and was printed and bound in the United States of America.

Manuscript/Copy Editor: Scott Howe
Book Design and Layout: Susan Robson
Electronic Typesetting: Susan Robson
Proofreading/Copyediting: Susan Palumbo
Figures and Artwork: Joe Pozerycki, Jr. and Ed Sirvio of Area D and Susan Robson

Library of Congress Cataloging-in-Publication Data

Sustainable energy : choosing among options / Jefferson W. Tester ... [et al.].
p. cm.
Includes bibliographical references and index.
ISBN 978-0-262-20153-7 (hc.: alk. paper)
1. Renewable energy sources. I. Tester, Jefferson W.

TJ808.S85 2005
333.79′4—dc22

2005041652

10 9 8 7 6 5 4

Contents

Preface

This is a book about **energy**. The resources that supply energy are essential to human survival. The intense exploitation of these resources has enabled the development of highly technological societies in the world's wealthier countries. But with the benefits of energy comes a dilemma. How can we keep providing humankind with energy-derived advantages without damaging the environment, affecting societal stability, or threatening the well-being of future generations? The resolution of this dilemma involves finding sustainable energy sources. Many experts view our energy use pathways as unsustainable for the long-term future, if present growth trends continue. A better future might depend strongly on how individuals and institutions choose amongst diverse and potentially contradictory technical, sociological, geopolitical, and environmental options, and how we set priorities for changing our present courses.

This book presents an overview of the development of energy technology, energy resources, and energy technologies available today, and of some emerging technologies for the future. But the issues surrounding the future role of energy extend far beyond technology, since energy use is integral to many activities, including our quality of life, our commerce, our mobility, and the stability of our social institutions. Our energy use impacts the environment in ways that may be irreversible—especially as the human footprint on the planet expands due to population increases. Population growth and economic growth lead to increased demands on natural resources and greater individual energy consumption. Moreover, the uneven geographic distribution of primary resources creates significant geopolitical consequences. Taken together, these factors pose substantial challenges to achieving a sustainable future.

The goals of this book are to clearly present the tradeoffs inherent in defining sustainability, to study technology and technology-intensive policy options, and to provide a framework for assessing solution options. The approach is quantitative, though not encyclopedic. We examine available and future technologies in the context of their environmental strengths and weaknesses, their technical and economic viability, and their ability to keep pace with evolving public and regulatory expectations for the sustainable use of the planet's resources.

About a decade ago, the authors of this book came together from a varied set of technological disciplines and experiences to develop a graduate-level course that would examine energy from a broad perspective and provide quantitative approaches for evaluating future tradeoffs in a context of "sustainability." We developed and taught the course as an interactive team and brought in guest lecturers to cover aspects of environmental and societal issues that were outside of our immediate expertise. The students also came from diverse technical (engineering, natural science, political science, and public policy) and cultural backgrounds (both US and international). These students participated actively in class discussions and in the examination of case studies.

This book evolved from this experience. It is organized as a text, with illustrative examples, homework problems, references for further reading, and links to some interesting Web sites. Our presentation is structured to be helpful to both energy experts and to inquisitive non-specialists—in government, industry, foundations, other non-profit organizations, and the public. The value of this book lies beyond the course work, and we believe that anyone seeking a better understanding of energy in a framework of enduring social and environmental stewardship will find this book of interest and use.

The first six chapters of the book examine the broader aspects of energy use from viewpoints of sustainability, resource availability, technical performance, environmental effects, economics, and a systems perspective. These chapters provide essential "tools" for making informed energy choices. Chapters 7 through 15 review the technology, environmental impacts, and economics of each of the main energy sources that are likely to be part of a future energy portfolio. Chapter 16 addresses the important topic of energy storage, transmission, and distribution. Chapter 17 examines the electric power sector. The next three chapters present an overview of the role of energy in the three main end use sectors—transportation, industry, and buildings. Chapter 21 looks at the challenges of integrating all the pieces of the energy puzzle into a framework that will promote understanding of present and future pathways, as well as the role of uncertainties and vested interests. The final chapter returns to the dilemma, and challenges us to act in an informed manner.

The difficulty of implementing major change must not be lightly dismissed. Energy-intensive services and "luxuries" are largely taken as entitlements in the wealthier countries. People in poorer countries need and aspire to the improved socioeconomic conditions that energy can facilitate. Further, many aspects of environmental protection require energy for their implementation. Most nations jealously guard their right to ignite, invigorate, or revitalize economic prosperity by all suitable means, including use of low-cost energy if appropriate. As populous poorer countries industrialize to promote a better life for their people and achieve a semblance of parity with the wealthier countries, the demand for increased energy services will be enormous, even with moderate rates of growth, increasing the adverse impacts of global energy use. Moreover, these developments increase the difficulty of making sound energy decisions in the face of great uncertainty.

Overlying these factors, however, is a welcome and growing awareness around the world that energy and its social and environmental consequences affect the whole planet. More and more, people are realizing that nations and individuals must discover, implement, and share the means to provide energy-related benefits more efficiently while preserving the earth's resources and ecosystems for future generations. This process of discovery, implementation, and cooperation can be considered the quest for sustainable energy.

But what is the end product of this quest? An objective of this book is to provide readers with the background and methodologies to answer this question for themselves—to develop their own conceptualizations of sustainable energy. To launch our thinking, we have chosen this operational definition:

> ***Sustainable energy: a living harmony between the equitable availability of energy services to all people and the preservation of the earth for future generations.***

In our view, pursuit of this harmony is worthwhile, but the journey will be long and arduous, with challenges, uncertainties, and some mistakes that, we hope, will teach valuable lessons and not be repeated. This book provides some navigational aids to those who wish to be active participants in the journey. Our hope is that it will give readers a better understanding of the strengths and weaknesses of different technology and technology-intensive policy options for transitioning to a more sustainable future—and that it will better equip readers to think critically and find their own sustainable energy pathways, while promoting local, regional, national, and global responsibility for a sustainable future.

Readers interested in learning more about the topics discussed in this book should visit our Web site:

http://web.mit.edu/10.391j/www

This site features supplementary material, updates, and errata.

Acknowledgments

The authors are grateful to a large number of people who have helped us develop our thinking about energy and sustainability. Our collaboration began with discussions at the MIT Energy Laboratory in 1996. We concluded that there was a need for an upper-level elective course that quantitatively presented energy in a broad framework and addressed today's sustainability challenges. We were inspired by two graduate-level, one-semester courses in the MIT School of Engineering. One was a popular course on energy in context taught by the late Prof. David Rose of the MIT Nuclear Engineering Department. The other, taught by Prof. Jack Howard and the late Prof. Hoyt Hottel of the MIT Chemical Engineering Department, was on new energy technology. We decided to collaborate on a new course that built on the work of these three energy scholars, while updating and extending several topics to address the new societal and environmental challenges that are central to today's use of energy.

Under the aegis of the MIT Energy Laboratory, first the course and then this book were developed. The Energy Lab also provided partial support for the production of this book. Additional support came from the MIT Dean of Engineering, Prof. Tom Magnanti, the V. Kann Rasmussen Foundation, and the Alliance for Global Sustainability (a collaborative research and educational partnership between MIT, Eidgenössische Technische Hochschule (ETH) – Switzerland, and the University of Tokyo), as well as from several academic departments and programs at MIT that were interested in having the elective offered. In 2001, the Energy Lab merged with the Center for Environmental Initiatives at MIT to become the Laboratory for Energy and the Environment, under the leadership of Prof. David Marks, who continues to support our activities.

Frequent guest lecturers in our course have been:

From MIT: Prof. Morris Adelman (petroleum resources), Stephen Connors (electric sector), Dr. Denny Ellerman (energy economics), Prof. Jeffrey Freidberg (nuclear fusion power), Prof. Ralph Gakenheimer (mobility), Prof. Leon Glicksman (eco-buildings), Howard Herzog (carbon management), Prof. John Heywood (automotive technology), Prof. Jack Howard (fossil energy), Prof. Henry Jacoby (integrated models for assessing global change), Dr. Edward Kern (renewable energy), Prof. David Marks (dam construction), Dr. David Reiner (international policy), Prof. Peter Stone (climate interactions), Dr. Malcom Weiss (transportation life cycle analysis), Prof. Donald Sadoway (battery technology), Dr. Joanne Kaufmann (geopolitical issues), and Prof. David Gordon Wilson (combustion turbines).

From elsewhere: Admiral Richard Truly and Dr. Stan Bull of the National Renewable Energy Laboratory; Prof. John McGowan of the University of Massachusetts – Amherst; Prof. Alexander Gorlov of Northeastern University; Dr. Jamie Chapman from OEM Development Corp.; Prof. Fraser Russell of the University of Delaware; and Prof. David Pimentel of Cornell University.

Many of these people provided valuable material that we have incorporated into the book with their permission and review. Over the years that we have been teaching the course, our teaching assistants have been most helpful in providing constructive comments, preparation of materials, problem sets, and course documentation. While the contributions of all our teaching assistants were vital to this endeavor, the following, in particular, made noteworthy contributions to the book: Dr. Murray Height, chemical engineering, and Tim Prestero, ocean engineering; Jason Ploeger, chemical engineering, and Chris Handwerk, nuclear engineering. We have also received valuable comments and feedback from a number of our students.

In addition, we are grateful to Profs. Mujid Kazimi and Neil Todreas for a critical review of the chapter on nuclear energy; Prof. János Béer for his review of the fossil fuels chapter; Michael Pacheco, Richard Bain, Richard DeBlasio, Robert Thresher, and other experts at the National Renewable Energy Laboratory for their review of chapters 9 through 16 (renewable energy topics and storage/distribution); John ("Ted") Mock and Robert Potter for their comments on the geothermal energy chapter; and Prof. John Sterman for a helpful critique of Chapters 6 and 21 (energy systems). Howard Herzog provided valuable information on carbon sequestration.

Many others have helped us better comprehend energy and environmental issues. Time and space prevent us from acknowledging all of these colleagues individually, but we do wish to thank Prof. David White, Founding Director of the MIT Energy Laboratory, Prof. Jack Longwell, Prof. Adel Sarofim, and Prof. William Thilly for leading multi-disciplinary research on energy technology and policy, fuels technology, pollution control, and environmental health effects of energy utilization.

In early 2003, a draft version of this book was used at MIT and in classes at Cambridge University with useful feedback. Prof. Ben Ebenhack of the University of Rochester, Enrico Borgarello and Giovanni Pacolini from ENI Technology, and Admiral Richard Truly and his staff at the National Renewable Energy Laboratory were also kind enough to review the manuscript, as well as provide many useful resources, documents, and illustrations.

Gwen Wilcox (assistant to Prof. Tester) and copy editor Scott Howe provided invaluable assistance in the final preparation of the manuscript. The final production of camera-ready copy, including design, layout, and electronic typesetting, was the work of Susan Robson, who did a masterful job in integrating all the rough materials into a polished product. Ed Sirvio and Joe Pozerycki, Jr. of Area D, Inc. adapted figures from other sources and produced many original illustrations. Susan Palumbo, Stephanie

Dalquist, and many students in our 2004 Sustainable Energy class provided essential proofreading. Jocelyn DeWitt and Bonnie Murphy provided considerable assistance on early drafts of the book.

Sustainable Energy—The Engine of Sustainable Development 1

In Xanadu did Kubla Khan
A stately pleasure-dome decree:
Where Alph, the sacred river, ran
Through caverns measureless to man
Down to a sunless sea.
So twice five miles of fertile ground
With walls and towers were girdled round:
And there were gardens bright with sinuous rills,
Where blossomed many an incense-bearing tree;
And here were forests ancient as the hills,
Enfolding sunny spots of greenery.

—Samuel Taylor Coleridge, *Kubla Khan*

1.1 Sustainable Energy: The Engine of Sustainable Development

Energy is one of the essential needs of a functioning society. The scale of its use is closely associated with its capabilities and the quality of life that its members experience. Worldwide, great disparities are evident among nations in their levels of energy use, prosperity, health, political power, and demands upon the world's resources. During recent decades, concern has grown that it is unwise for the poorest nations to be relegated to miserable poverty and for the richest ones to be able to command such large shares of the resource pie. The latter concern has been especially acute in regard to the United States, which has the world's largest economy (with about 5% of the world's population consuming approximately 25% of the world's energy production), supporting its preeminence culturally, politically, and militarily since the end of the Cold War.

Of the six-plus billion inhabitants on earth today, more than two billion are living below the poverty line as defined by the United Nations—equivalent to about 2US$ per day PPE (purchasing power equivalent). For these very poor, the basic needs of food and shelter are a daily challenge. Most scavenge biomass for any energy needs, leading to depletion of what little vegetation is locally available. Access to water is another challenge, and men, women, and children spend long hours hauling water, food, and fuel. For these people, health problems abound, and mortality rates are high.

To move out of abject poverty, access to some sort of income is necessary. This need is closely tied to the availability of energy—which makes local businesses and the transportation of commercial goods to markets possible. Energy development provides as well for health clinics, for lighting, and for basic services. The UN and the World Bank have launched initiatives to reduce world poverty. Energy development is a major aspect of their development plans.

Even developing countries that have populations well above the poverty line are seeking to improve living standards through more industrialization. Looking at per capita energy consumption is a good way to understand the disparities between the richest and poorest nations. Figure 1.1 shows the variation for typical countries.

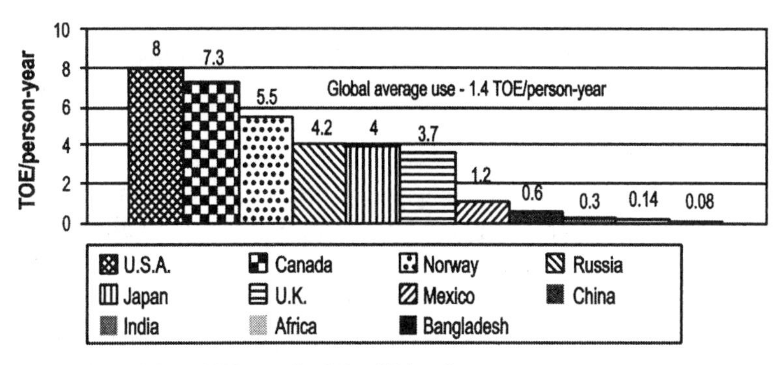

Figure 1.1. 1999 per capita average commercial energy use for selected countries.

The average US citizen uses 100 times more commercial energy than the average person in Bangladesh. If today's energy use were distributed equally over the world population, each person would use about 1.4 tons of oil equivalent (TOE)/year.[1]

World population is still increasing; it has tripled since the late 1930s. In 1700, world population was 600 million. More people and more development suggest even higher future needs for energy. The most available and affordable sources of energy in today's economic structure are fossil fuels (about 85% of all commercial energy is derived from them). Efficiency improvements and new technologies are part of the solution. Still, society may have to make some major changes if we wish to address the challenges of eradicating abject poverty and stabilizing greenhouse gas (GHG) emissions.

Such concerns have led to the creation of political movements pressing for changes in our demands upon the environment. These pressures have ranged from criticism of individual practices (e.g., growth in the heavy vehicle portion of the passenger automobile economy) to demands that the overall structure and economic basis of societies be changed (e.g., criticism of global trade markets, with the attendant suffering of those who lose within them). Many pressures are focused on energy, in terms of its uses, supply technologies, and efficiencies. These concerns have often been expressed in demands for less use, greater end-use efficiencies, and more reliance on solar and geothermally powered technologies, rather than fossil and nuclear energy, which extract their fuels from finite earth resources.

In its most extreme guise, sustainable energy is that which can be provided without change to the earth's biosphere. However, no such form of energy supply exists. All require some form of land use, with attendant disruption of the associated ecosystems,

1 One ton of oil equivalent (TOE) contains about 40 million Btus or 43 trillion joules of energy (see Table 2.1).

extraction, which can be disruptive for fossil fuels, and less so for nuclear ones. Ultimately, all of these extracted materials reenter the biosphere as wastes, where their sequestration practices are at least as important as their masses in determining the accompanying ecological disruption.

In this book, we treat energy technologies as being sustainable if their net effects upon the biosphere do not significantly degrade its capabilities for supporting existing species in their current abundance and diversity. This definition is inherently conservative and favorable to the status quo. It reflects our ignorance in assessing the quality of alternative ecosystems (some of which might be preferable to what exists now) and in understanding our effects upon them. However, as a response to human practices, which in many instances appear to be harmful, it is an improvement. We hope that, in time, our ecological knowledge will improve to the point that such conservatism will not be the best that we can do.

The objectives in the previous discussion demand definition (perhaps quantitatively) before a useful system for decision-making will exist. Such definitions are made via our political systems—which is another way of saying that they will be the outcomes of struggles among interest groups. This recognition is important, because, while the alternatives in achieving sustainability are technological, the social directions served by their use, and the criteria for what is acceptable, will always remain social. Thus, sustainable energy must be concerned not only with energy and environmental technologies, but also with the social and political factors that impact human lifestyles.

The patterns of energy use worldwide reflect the distribution of wealth among nations. Of the earth's approximately six billion people:

- Roughly 20% live in the wealthy, industrialized countries of western Europe, North America, and Japan. For example, the specific energy consumption rate in the US is 350 million Btu/capita annually; in western Europe and Japan, it is about half of that amount.
- An approximately equally sized group lives within rapidly industrializing South Korea, Taiwan, Malaysia, Thailand, South Africa, and China; these account for much of the growth in energy consumption. The specific energy consumption rate for China is 30 million Btu/capita with a wide variation among the population on an individual basis. More affluent urban populations are approaching energy consumption levels of Europe and Japan, while many rural populations remain at energy consumption levels typical of undeveloped countries. This group also includes countries such as Chile, Mexico, and Brazil, which are in a middle state of development but have only slowly growing economies.
- The rest of the world is largely primitive economically and is developing slowly, although many countries in this group have wealthy elite, small regions of prosperity, and more rapid development (e.g., Karala and the

region near Bangalore in India, which has specific energy consumption of 15 million Btu/capita annually).

Growth of energy demand reflects the pace of economic development. The first group accounts for most of the consumption of energy (and other resources) worldwide, and also, because of its wealth, is the group most able to protect the environment from disruption that occurs as a consequence of its consumption.

The second group is making the transition to wealth, has great disparities of individual and national abundance, and largely does not invest in environmental protection (e.g., because of regional air pollution, when flying over China it is common to be unable to see features on the ground distinctly).

The third group consists of a population subsisting largely outside the cash economy. These countries meet most of their energy needs via biomass, using their own muscles and those of their animals, and burning foraged vegetable matter. Energy-use data for these countries reflect primarily consumption within the cash economy, which is atypical of the situation of most of its members.

Overall, the available data indicate the great majority of energy consumption (80%) occurring in the first group of countries but with slow growth, most of the remainder occurring in the second group of countries (15%), and the remainder occurring in the last group. Investments in the energy supply infrastructure also reflect these trends.

In a typical developed country, such as the US, electricity production accounts for about 25% of total energy consumption (note that electricity is an intermediate energy product that is made from primary energy sources), with the remainder of energy needs met by direct fossil fuel consumption. Use of geothermal energy and the renewable technologies are negligible. Coal is used almost exclusively for electricity production (accounting for about half of the fuels used to make electricity). Remaining needs, especially in transportation, are met mainly via petroleum consumption, with natural gas being about as important as petroleum in the industrial, residential, and commercial sectors. Specific data for the US, France, and Japan are given in Table 1.1. In some special cases, national policies have affected the energy technology mix. For example, most French and Belgian electricity is from nuclear energy, and in Denmark about 20% of electricity is obtained from wind.

In partially developed countries (second group), reliance on petroleum for all needs is typically greater and fractional use of electricity is less than with the fully developed nations (reflecting the strong correlation between gross per capita domestic product and per capita electricity consumption). The partially developed countries include many that have some of the characteristics of industrialized countries (e.g., growing motorization and electrification) but retain substantial peasant populations. Those that are growing rapidly are also in the process of integrating this group increasingly into the cash economy.

Table 1.1. 2001 Primary Energy Consumption by Fuel for Selected Countries and Secondary Electricity Generation (million TOE*)

Country	Oil	Gas	Coal	Nuclear	Hydro	Total
USA	896	578	546	183	48	**2251**
Japan	248	71	103	73	20	**515**
France	96	38	12	96	18	**258**
Germany	132	75	85	39	6	**336**
UK	77	87	40	20	2	**226**
Belg/Lux	32	13	8	11	1	**64**
Netherlands	44	35	9	1	—	**89**
China	232	25	519	4	54	**834**
India	97	25	173	4	16	**314**
WORLD	**3517**	**2220**	**2243**	**601**	**585**	**9165**

Source: BP, 2001.

*Electricity is converted to TOE based on fossil fuel primary energy and 38% conversion efficiency. Since nuclear and hydro energy generate electricity at much higher efficiencies than 38%, the fractions of primary energy contribution are skewed when everything is put on a TOE basis.

The third group of countries is neither wealthy nor growing rapidly economically. These countries demand relatively few of the earth's resources. However, locally, their pressure on the environment can be heavy, due to their intense use of scavenged and harvested biomass. In these societies, the economic surplus needed to protect the environment in association with human activities is least, as is the protection. A major question for wealthy nations concerns how to persuade and help these nations protect their own environments, and, thereby, the global environment. The main feasible means for accomplishing this appear to be various forms of financial and social aid (i.e., education and health programs), and the creation of attractive technologies for their use. Past experience indicates that both of these avenues are difficult to implement successfully and may be inadequate.

Within the third group of countries, energy needs are met using indigenous sources to the extent feasible (e.g., coal in India and China, geothermal in El Salvador); most other needs within the cash economy are satisfied using petroleum. Data concerning consumption outside the cash economy are questionable, but it is evident that biomass that can be harvested by the affected individuals is the predominant energy source.

All three groups of countries aspire to more prosperity and an improved quality of life for their citizens. Such development will require expansion of energy services and infrastructure world wide. Can this be achieved in a sustainable manner so that our environment, social structures, and economy remain a viable legacy for future generations?

Energy production or utilization is often intertwined with consumption of other precious natural resources, such as minerals, forests, water, food, and land. Further, the everyday use of energy can damage human health and the earth's ecosystems over wide length and time scales (1 m to 10,000 km; 10 s to 100 yr). Yet in the developed countries, the availability of stable supplies of energy at manageable prices has propelled economic development and enfranchised most of the populace with mobility and a host of lifestyle benefits that were unimaginable a century ago. Developing countries are now dramatically expanding their energy use to extend economic prosperity within their borders. Because of growing populations and expanded development, world energy use over the next century may grow four-fold even with substantial improvements in conservation and end-use efficiency. At the same time, continued dominance of the world's energy by fossil fuels (about 85% of global consumption in 1998) is expected to be challenged, not by the red herring of scarcity, but by concern that emissions of fossil-derived CO_2 and fugitive CH_4 to the atmosphere will cause serious global climate modification.

In the energy sector, sustainability ideas confront a phalanx of trans-global forces—pervasive human impacts, current and aspirant economic well-being, ecological degradation, geopolitical equity—that are conspiring to create an energy-prosperity-environmental dilemma of epic proportions. That dilemma is how to maintain and extend energy-derived benefits for present and future generations while shepherding the planet's natural resources. Scholars differ on how to resolve this dilemma, but many support the precepts that there are multiple answers, that solutions change with time and circumstances, and that virtually every technology and policy option entails tradeoffs that must be carefully valued.

For a more sustainable energy future we need to develop a rich set of energy technology and technology-intensive policy options. These options include increased efficiency of energy production and use, reduced consumption, a new generation of renewable energy technologies, nuclear options that can win and retain public acceptance, and means to use fossil fuels in a climate-friendly way. If fossil fuel prices rise to include costs of carbon management, consumers may also modify their consumption patterns. Environmental and ethical concerns may also contribute to new attitudes about unconstrained economic growth patterns. Sustainability concepts provide a framework to focus the evaluation of energy technology and policy options and their tradeoffs and to guide decision-making on energy futures. The key to getting these concepts right is to develop a solid understanding of the multi-faceted technological, geopolitical, sociological, and economic impacts of energy use and abuse.

Sustainability is necessarily subjective because it reflects human value—the relative importance stakeholders assign to the activity to be sustained, to the perceived benefits of that activity, and to other values "traded off" to sustain the activity in question. Many sustainability tradeoffs are an inevitable result of tension between the benefits derived and the adverse consequences of the activity that provides those benefits. Examples of these tensions include: time horizons for reform, i.e., taking the "best" now versus preserving it for future generations; individual versus national versus global interests; and expansion of economic opportunity versus stewardship of resources.

Energy affects all people, is technologically intensive, and evokes diverse, value-laden, and controversial perspectives on what technologies and policies are appropriate in various contexts. Diverse views on what *sustainable energy* really means are thus inevitable. To catalyze meaningful discussion, we propose the following definition:

> **sustainable energy**: a dynamic harmony between the equitable availability of energy-intensive goods and services to all people and the preservation of the earth for future generations.

This definition is similar to the well-known one developed by Bruntland (1987)—"development that meets the needs of the present without compromising the ability of future generations to meet their own needs."

Given the past trajectory and current status of planet Earth, sustainable energy so defined is a *state of global being yet to be realized*. Some would proclaim this condition too optimistic to be possible, or, if attainable at all, inevitably unstable. We reject the idea that a condition of sustainable energy as here defined is impossible, but we recognize that to remain effective this condition is unstable in one particular sense. The specific attributes of a state of sustainability will inevitably change with time, owing to technologic and sociopolitical forces. To survive, sustainable energy must eventuate a state of *dynamic immutability*, in which desired energy-derived benefits are available without term, owing to technologic and policy measures that evolve, both rapidly and substantially, in response to evolving human and environmental needs. Engines of production transformation are technological innovation, reliable scientific understanding of how technology affects the earth and its occupants, and effective communication of technology risks and opportunities with and among stakeholders.

If sustainable energy is yet to be realized, we must consider how to get there. We see no easy or unique roadmaps, but rather a series of pathways that will be dominated by gradual transformations of institutions and individuals. This transition toward more sustainable energy use can be likened to a mission-oriented expedition, initially deliberate but sooner or later fragmented by various excursions, some rewarding, others frustrating. Because energy impacts virtually every aspect of human endeavor, this journey will attract or indenture a wide range of participants. These travelers, be their purpose "business" or "tourism," will reap more from this journey if they regularly update their understanding of mapped and unexplored territories.

This text is a travel guide for sustainable energy pilgrims. Our goal is to provide information and analysis tools for a more productive and pleasant passage. It is assumed that each traveler will wish to select his or her own major routes and side-trips and to make his or her own decisions on what to explore.

The supply and use of energy to improve human well-being has many attributes and consequences, some transparent, others opaque, some beneficial, others hostile. Thus energy exemplifies the types of tradeoffs that are the foundation of the sustainability approach. Throughout this text, we seek to identify and understand these energy tradeoffs (i.e., the human and ecological costs, benefits, and limitations in using energy to propel economic prosperity while preserving the environment and other irreplaceable resources). We will explore such questions as: How much energy is there? Is energy a thing? Is "work" the only type of energy, or are there many different forms? If so, are all of these forms equally useful to people? Can most or all energy, regardless of its form, be employed for productive purposes or must some energy always be wasted? If waste of energy is inevitable, how great are the losses and do they depend upon what energy processes or equipment are utilized? How do energy-related actions harm the ecological well-being of present and future generations? Are there countermeasures to prevent or remediate energy-related environmental degradation? How can we measure and fairly value the beneficial and detrimental effects of energy? Subsequent chapters will respond to these questions with credible technical information and with discussion of technological uncertainties, information voids, and alternative solution pathways when well-established information bases or broadly-based sociopolitical consensus are elusive.

1.2 Defining Energy—Scientific and Engineering Foundations

We need an understanding of energy suitable for quantitative study of sustainability. We must comprehend what energy *does* and what energy *is*. Webster's Dictionary (1986) defines "energy" as:

> **1:** vigorous exertion of power **2a:**. the capacity of acting or being active <intellectual ~> **b:** dynamic quality <narrative ~>; **3:** the capacity for doing work
> **4:** usable power (as heat or electricity); *also*: the resources for producing such power.

The implications are that "energy" embodies animated and possibly productive physical or mental activity—presumably by humans, animals, machines, nature, "electricity," etc. Definition (3) reprises what apparently was the first use of the term "energy" as "ability to do work," by Thomas Young in his 1805 Bakerian lecture to the Royal Society (Levenspiel, 1996).

Experientially, we think of "work" as arduous physical or mental exertion. To comprehend sustainable energy, we need to refine our understanding of four key concepts: "energy," "work," "heat," and "power."

Centuries of observations show that a certain quantity remains constant during physical, chemical, and biological changes. This conserved or "immutable" quantity is energy. First, consider some of the consequences if energy is conserved. One is that we cannot get rid of it. Let us divide up the entire universe into specific regions with well-defined boundaries and other particular characteristics. We define each such region as a *system*. Often we divide the universe into just two systems: the region we wish to analyze in some detail (System 1, e.g., the combustion chamber of an automobile engine), and all the rest of the universe (System 2), which we define as the *surroundings*. What is important with a system is not the absolute energy content of the universe, a system or even a set of systems, but rather the *change* in the energy content of particular systems within the universe and their interactions with their surroundings during the course of that change. In energy analysis, it is important to know, for particular circumstances, the total amount of energy a system can give to or take from its surroundings *and* what fraction of that exchanged energy can be converted to useful purposes such as the motion of an automobile or the generation of electricity.

The First and Second Laws of Thermodynamics (see Chapter 3) provide us with the theory to answer these questions quantitatively, assuming that we have the necessary data to implement the tools of thermodynamics for practical calculations. It is often important to know how rapidly energy can be generated within, assimilated by, or released from one or more systems (e.g., how fast can the chemical energy of a fuel be converted to the kinetic (motive) energy of an automobile or the thrust that propels a rocket). To address these questions, we need to draw on thermodynamics and on the disciplines of chemical kinetics, physical transport, and fluid dynamics to describe the rates of chemical reactions and of the exchange of heat, material, and momentum within and between single and multiphase media. Sections 3.4 and 3.5 introduce chemical kinetics and heat transmission in the context of sustainable energy.

The position or motion of matter causes energy to exhibit diverse forms. Many are readily observed (e.g., as changes in pressure [stress fields], volume, temperature, surface area, and electromagnetic properties) (Table 1.2).[2] Heat is a familiar form of energy, defined formally below. Thermodynamics shows that heat and all forms of energy are related to mechanical work, such as the raising or lowering of a weight in a gravitational field (Tester and Modell, 1997).[3]

A *closed* thermodynamic system is completely surrounded by movable boundaries (walls) permeable to heat but not matter (e.g., a vertical cylinder filled with a gas and covered with a piston that can be moved up and down). By adding weights to the piston, we can compress the gas and store energy in analogy to pushing on a coiled spring. This addition of weights is an example of *work performed by the surroundings* on our system. The resulting downward movement of the piston is *work obtained by the system* from its surroundings. Regardless of how meticulously we add the weights, the amount of work taken up by the system is always less than the work done on the system by its

2 These effects are practical scale manifestations of microscopic changes in matter, e.g., molecular vibrations and rotations, motion of electrons, spinning of nuclei, formation and scission of chemical and nuclear bonds. Thus, fundamental studies of molecular and nuclear phenomena provide crucial insights on how to make energy more sustainable.

3 Establishing the mechanical equivalence of heat is a seminal triumph of early thermodynamics scholars, i.e., Rumford, Mayer, Joule, and Carnot (Carnot, 1824).

surroundings by an amount of energy exactly equal to the heat gained by the system. This is a rudimentary expression of the Second Law of Thermodynamics. Thus *heat* is a form of energy. In the piston example, it arises from wasted or lost work. Rigorously, heat is defined as a mode of energy transfer to or from a system by virtue of contact with another system at higher or lower temperature (Levenspiel, 1996; Denbigh, 1981).[4] *Work* is defined as any mode of energy transfer, other than heat, that changes the energy of a system (e.g., by a chemical reaction, raising or lowering a weight, turning an electrical generator). *Power* is the rate of energy exchange between two systems. It has units of energy per time and may represent a flow of work, heat, or both.

Table 1.2. Changes in System Properties That Produce or Consume Work[a]

Category of Work	Responsible Physical Process	Energy-Related Example
"Generalized Work"	(A Force) *timesüR* *RÇ* (A Spatial Displacement)	NA[b]
Pressure-volume[c]	Volume change caused by force per unit area (pressure).	Movement of piston in internal combustion engine.
Surface deformation	Surface area change caused by surface tension.	Small (stationary) droplet of liquid fuel suspended in a quiescent fluid (assumes a spherical shape).
Transport of ionized (electrically charged) material	Movement of charged matter caused by an electric field.	"Electrostatic" precipitation of particulate pollutants in stack gas.
Frictional	Movement of solids in surface contact.	Generation of waste heat by unlubricated moving parts in machinery.
Stress-strain	Deformation (strain) of a material caused by a force per unit area (stress).	Pumping of a viscous (highly frictional) liquid through a pipe.

[a]Mathematical treatments are given by Tester and Modell (1997).

[b]Not applicable.

[c]Often called "*PdV*" work.

The formal thermodynamic statement of the law of conservation of energy is the First Law of Thermodynamics. To apply this law, we first state it mathematically. For a closed system, we have:

$$\Delta E = Q + W \quad \text{[closed system]} \tag{1-1}$$

4 In Section 3.5, we will consider the main mechanisms by which heat can be exchanged between two systems at different temperatures.

where ΔE is the change in the energy content of the system, Q is the amount of heat transferred *to* the system *from* its surroundings, and W is the amount of work done *on* the system *by* its surroundings. (Many textbooks write Equation (1-1) as $\Delta E = Q - W$ because they define W as the amount of work done by the system on its surroundings. Both expressions are correct provided work is defined consistently throughout.) In many practical energy problems, the system is not closed but "open," i.e., matter can flow inward, outward, or in both directions, across the system boundaries. Equation (1-1) must then be modified to account for this transport of matter and for any other processes that change the system energy content (see specialized thermodynamics texts such as Tester and Modell, 1997, Levenspiel, 1996, Smith and Van Ness, 1949, Weber and Meissner, 1939).

For practical calculations, we can rewrite Equation (1-1) as:

$$\Delta E = \Delta U + \Delta E_p + \Delta E_k = Q + W_{sh} - W_{PV} \quad \text{[closed system]} \tag{1-2}$$

This expression tells us that a change in the energy content, E, of a closed system can be divided into changes in the internal energy, U, potential energy, E_p, and kinetic energy, E_k, of the system. The internal energy can be changed by modifying the system temperature, changing its phase (e.g., solid to liquid), by chemical reaction (i.e., changing its molecular architecture), or by changing its atomic structure (e.g., by fragmenting [fission] or coalescing [fusion] nuclear particles; Chapter 8). The potential energy is changed by shifting the system location in a force field (e.g., gravitational, electrical, magnetic). The kinetic energy is varied by increasing or decreasing the system velocity (Levenspiel, 1996, and Feynman et al., 1963). Equation (1-2) disaggregates work into W_{PV}, which we call "*PV*" work, and W_{sh}, i.e., "shaft work." We define shaft work as any work other than *PV* work—it may involve rotation of a shaft, but it may also include electrical work and other forms. *PV* work arises from the fact that every system, however small, has some volume. Recalling our *gedanken* (thought experiment) of compressing a confined gas by adding weights to a movable piston, we conclude that a change in system volume changes the system's potential energy. Any system at equilibrium (i.e., at a fixed temperature, pressure, and composition) has a constant volume. To attain that volume, the system had to push its surroundings out of the way to make room for itself. The work done by the system to reach a volume, V', by shoving back a pressure p, is *PV* work and is given by:

$$W_{PV} = \int_{o}^{V'} P dV \tag{1-3}$$

It is often convenient to combine a system's internal energy, U, with the energy it has by virtue of its volume, V, at a pressure, P. The resulting thermodynamic quantity is the *enthalpy*, H, defined mathematically as:

$$H = U + PV \tag{1-4}$$

Note that P is the pressure *in* the system. In some situations, it may also be the pressure of the surroundings (e.g., the pressure of the atmosphere) but not always. As with other forms of energy, we are interested in the change in enthalpy when a system changes from a state 1 to a state 2:

$$\Delta H = H_2 - H_1 = \Delta U + \Delta(PV) \tag{1-5}$$

$$= (U_2 + P_2V_2) - (U_1 + P_1V_1) \tag{1-6}$$

Equations (1-4) through (1-6) apply to closed and open systems. If a closed system undergoes a change in energy but remains at constant volume, there is no PV work, and Equation (1-2) reduces to:

$$\Delta U + \Delta E_p + \Delta E_k = Q + W_{sh} \quad \text{[closed system, constant volume]} \tag{1-7}$$

A second case is when pressure remains constant so that from Equation (1-2):

$$\Delta U + \Delta E_p + \Delta E_k = Q + W_{sh} - (P_2V_2 - P_1V_1) \tag{1-8}$$

i.e.,

$$(U_2 + P_2V_2) - (U_1 + P_1V_1) + \Delta E_p + \Delta E_k = Q + W_{sh} \tag{1-9}$$

or from Equation (1-6):

$$\Delta H + \Delta E_p + \Delta E_k = Q + W_{sh} \quad \text{[closed system, constant pressure]} \tag{1-10}$$

For a change in the energy content of a closed system at constant pressure, we can write the law of energy conservation directly in terms of a change in the system's enthalpy, potential energy, and kinetic energy (Levenspiel, 1996). Further, if a closed system is at rest or moves at a constant velocity, and any motion in a force field does not modify the potential energy (e.g., the system remains at the same height in a gravitational field), then the enthalpy change accounts for all the change in energy brought about by addition or removal of heat and shaft work.

Now, can we know the *absolute* energy content of a substance? The answer, provided by Einstein's theory of special relativity (1905), is of great scientific and practical importance. Einstein discovered that the total energy content, E_{tot}, of a quantity of matter of mass, m, equals the product of m with the square of the speed of light, c, i.e.:

$$E_{tot} = mc^2 \tag{1-11}$$

Special relativity also teaches the speed of light in a given medium remains constant, and an object's mass increases with its velocity according to:

$$m(v) = \frac{m_o}{\sqrt{1 - \frac{v^2}{c^2}}} \tag{1-12}$$

In Equation (1-12), m_o is the "rest mass," or the mass of the object standing still ($v = 0$).[5] By expanding the square root term in Equation (1-12) in a Taylor series in (v^2/c^2), we see that m doesn't change very much from m_o unless the velocity approaches the speed of light:

$$m(v) = m_o\left[1 + \frac{1}{2}\frac{v^2}{c^2} + \frac{3}{8}\frac{v^4}{c^4} + \frac{5}{16}\frac{v^6}{c^6} + \ldots\right] \tag{1-13}$$

We can combine Equations (1-2), (1-11), and (1-12) to write:

$$E_{tot} = U + E_p + E_k = \mathrm{mc}^2 = \frac{m_o c^2}{\sqrt{1 - \frac{v^2}{c^2}}} \tag{1-14}$$

The deltas of Equation (1-2) have disappeared because now we are dealing with absolute amounts of energy, not differences in the energy content of two systems. From Equation (1-14), we find the expression for the absolute energy content of a body at rest, i.e., where $v = 0$ and thus $E_k = 0$:

$$E_{tot,\,rest} = U + E_p = m_o c^2 \quad \text{[body at rest]} \tag{1-15}$$

Thus, for the more general case of a body in motion from Equations (1-14) and (1-15):

$$E_{tot} = U + E_p + E_k = \frac{E_{tot,\,rest}}{\sqrt{1 - \frac{v^2}{c^2}}} \qquad \text{[body in motion]} \tag{1-16}$$

We now use Equation (1-13) to eliminate m from Equation (1-14):

$$E_{tot} = m_o c^2\left[1 + \frac{1}{2}\frac{v^2}{c^2} + \frac{3}{8}\frac{v^4}{c^4} + \frac{5}{16}\frac{v^6}{c^6} + \ldots\right] \tag{1-17}$$

$$= m_o c^2 + \frac{m_o v^2}{2}\left[1 + \frac{3}{4}\frac{v^2}{c^2} + \frac{5}{8}\frac{v^4}{c^4} + \ldots\right] \tag{1-18}$$

Using Equation (1-15) to replace $m_o c^2$, Equation (1-18) becomes:

$$E_{tot} = U + E_p + \frac{m_o v^2}{2}\left[1 + \frac{3}{4}\frac{v^2}{c^2} + \frac{5}{8}\frac{v^4}{c^4} + \ldots\right] \tag{1-19}$$

From Equation (1-14) or (1-16) this requires that:

$$E_k = \frac{m_o v^2}{2}\left[1 + \frac{3}{4}\frac{v^2}{c^2} + \frac{5}{8}\frac{v^4}{c^4} + \ldots\right] \tag{1-20}$$

---

5 Equation (1-12) is derived in physics textbooks, e.g., Feynman et al. (1963).

Equation (1-20) tells us that, except at system velocities approaching the speed of light, the kinetic energy of the system is almost the same as that computed from the non-relativistic formula of classical mechanics using the rest mass m_o, i.e.,

$$E_k = \frac{1}{2} m_o v^2 \tag{1-21}$$

Changes in system energy content are virtually always the crucial issue in sustainable energy calculations. Nevertheless, the theory of special relativity (Equations (1-11) through (1-20)) elucidates foundational concepts that are essential to a thorough understanding of energy engineering. In particular, relativity theory teaches us how to calculate the *absolute* energy content of a quantity of matter at rest (Equation (1-15)) and in motion (Equation (1-16)), and that the results differ. More important, the theory provides us with quantitative rules to tell us when these differences are consequential for practical applications (i.e., for objects moving at high velocity). Thus, relativity reveals that the mass of a substance increases as the velocity of that substance gets larger (Equation (1-12)). A practical consequence is that above a certain velocity, the classical formula for the kinetic energy of an object in motion (Equation (1-21)) will need to be replaced by Equation (1-20). When this becomes important depends upon the degree of accuracy required, but from Equation (1-20) it is clear that the non-relativistic approach (Equation (1-21)) increasingly underestimates the true kinetic energy as the velocity more and more closely approaches the speed of light.

An important result of special relativity is that energy and matter are different manifestations of the same thing (Pitzer, 1923), or *energy and matter are interchangeable* (Equation (1-11)). In writing Equation (1-14), we modified our earlier statement of the "law" of energy conservation (Equation (1-2)) to account for this fact. Equation (1-14) tells us that for a specified quantity of matter the sum of its internal, potential, and kinetic energy is constant and equal to mc^2. A practical consequence is that destruction of a small amount of matter can produce enormous amounts of energy (note the huge magnitude of c^2 in Equation (1-14)). Thus, if we could convert just 1 gram of matter to energy, we would produce 9×10^{13} joules, i.e., 8.5×10^{10} Btu or roughly the energy equivalent of 15,000 barrels of oil!

In practice, it is difficult to produce useful energy in this way, especially under controllable conditions. This is because fundamental building blocks of the atomic nucleus must be ripped apart (fission) or coalesced (fusion). Strong nuclear forces (for fission roughly 10,000 times those of chemical bonds, even larger for fusion) must be overcome. This is why nuclear fission or fusion can produce so much more energy from the same amount of material than can a chemical reaction. Today, controlled nuclear fission generates over 20% of the electricity consumed in the US and over 75% of that produced in France (Chapter 8). Nuclear fusion could become a commercial source of electricity, but this may require several more decades of research and development (Chapter 8).

For non-nuclear energy processes, the gain or loss of matter is far too small to be consequential. For example, burning 1 US gallon (3.78 liters) of gasoline in a typical spark ignition automobile piston engine releases about 140,000 Btu (1.48×10^8 joules) of energy, which, depending on the vehicle and driving conditions, provides 10-30 miles of travel. From Equation (1-11), the change in mass associated with release of this much energy is 1.64 μg, versus the 3,175 g in one gallon of gasoline (i.e., the relative mass decrease is $< 6 \times 10^{-10}$ g/g). This is negligible for practical purposes and would be difficult to detect.

In light of Equation (1-14), we can think of all our practical energy "sources" (Figure 1.2) as reservoirs "pre-stocked" with energy from elsewhere. In the language of thermodynamics, these reservoirs are "systems." Humans obtain useful energy by changing the relative stability of these systems. For example, we could open the sluice gate on a dam to release water that would gain kinetic energy by converting potential energy owing to a drop in the height of the water (Chapter 12). We could expose a fuel like natural gas to a high temperature and enough oxygen that chemical equilibrium demands a new state in which the carbon and hydrogen atoms of the gas respectively become CO_2 and H_2O, in the process releasing about 1,000 Btu for each standard cubic foot of gas reacted/burned (Chapter 7).

Figure 1.2 shows the origin of our familiar energy sources in a range of biochemical, geological, cosmic, and nuclear processes of diverse time scales. For example, only a single growing season of a few weeks is long enough for photosynthesis to convert solar energy plus CO_2 and H_2O to biomass, whereas millions of years are required for the bio-geochemical production of coal from plant and animal debris. Within the sun, continuous fusion of hydrogen nuclei consumes matter to produce energy, a minute fraction of which (about 5×10^{-10}, Feynman et al., 1963) is transmitted to the earth by thermal and other forms of electromagnetic radiation, such as solar energy. Interestingly, the earth's primary source of renewable energy is nuclear energy! In fact, Figure 1.2 shows that all the common forms of energy utilized on earth can be traced to ongoing or defunct cosmic processes (tides from lunar cycles, geothermal magma from the survival of deep subterranean molten material owing to the slow cooling and resolidification of the earth's inner core), or to nuclear phenomena in the heavens (the sun) or within the earth (geothermal heat created by subterranean nuclear decay).

As we quantify amounts and rates of transfer of various types of energy, it becomes immediately apparent that there are many different systems of units in common use. The metric system provides a good scientific set of units, but many still use British units. Further, each industry has developed convenient units to describe stocks and flows of energy—barrels of oil, standard cubic feet of natural gas, tons of coal, megawatts of electricity, etc. Energy practitioners need to become proficient in converting from one set of units to another. Conversion factors are provided in the Appendix to aid our readers.

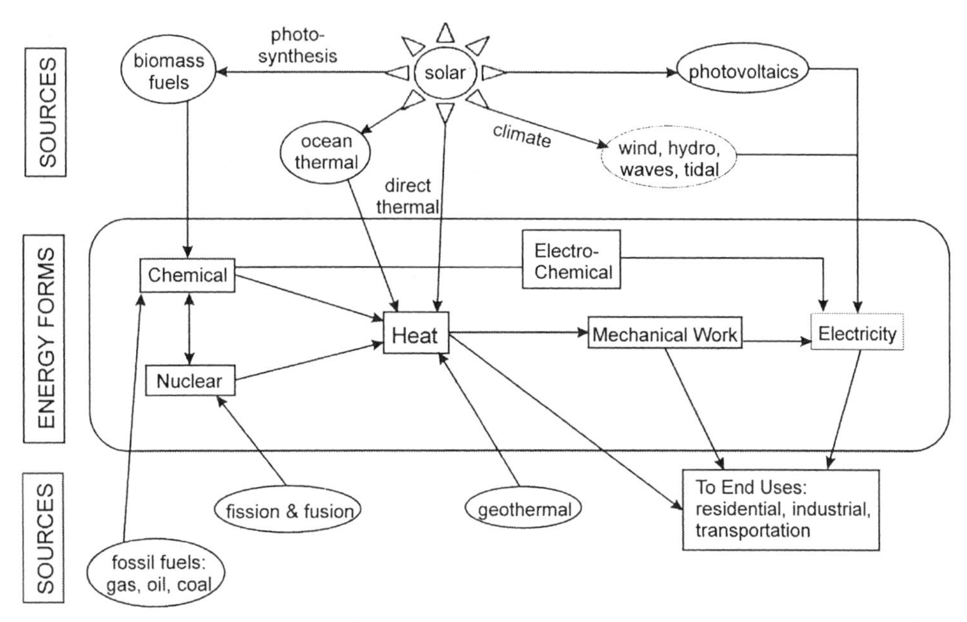

Figure 1.2. Energy sources and conversion processes.

A final caution is that electricity and hydrogen are energy "carriers" that are made from primary energy sources like fossil fuel or renewable energy. To calculate the primary energy consumption associated with producing a unit of electricity, for example, one must also divide the amount of electricity by the efficiency of the actual primary energy conversion process. To differentiate the different forms of energy, sometimes *e* (electrical) and *t* or *th* (thermal) are used, often subscripted.

1.3 Aspects of Energy Production and Consumption

In understanding energy needs and the technologies for supplying them, it is important to have an idea of the energy requirements of specific activities. Until the start of the Industrial Revolution in the 18th century, almost all energy needs were met by renewable sources (i.e., using biomass for heat and muscular energy; wind). Then, machines of increasing power and efficiency were devised to permit much greater power to be generated from the burning of fossil fuels. In the second half of the 20th century, these power sources were complemented by devices using much more energetic nuclear reactions.

In considering energy evolution since the Industrial Revolution, Table 1.3 (Levenspiel, 1996) compares the power dissipated by living creatures in various activities with the power consumed in some modern energy conversion machines. Note that even arduous human labor musters only as much power as that expended in a single 100W incandescent light bulb. Even the 10-fold higher power of a draft horse is barely 1% of the power needed to operate a compact car (25 miles per gallon of gasoline). Thus, driving a compact car is like having 1,000 people working hard to propel the vehicle and driver. If these workers were paid $8.00 per hour (somewhat above the US federal minimum wage in 1999), it would cost $8,000 an hour to operate this car. Some employers reimburse business use of private automobiles at 30-40 cents per mile to offset fuel and other operating costs, plus vehicle depreciation. If the car were driven at 60 mph, the highest reimbursement for an hour's travel would be 60 miles x $0.40/mile = $24.00 or only 0.3% of the $8,000 cost of human propulsion. If gasoline costs $1.30 per gallon, the corresponding fuel cost for gasoline power is [60 miles/hr] × [1 gal/25 miles] x [$1.30/gal] = $3.12 or about 5.2 cents per mile. Clearly, it would be economically (and mechanically) impossible to achieve the speed, service, and convenience of a modern gasoline-powered car by using human or animal power. This example shows the tremendous amplification of human labor made possible by the availability of high-quality energy at tractable costs.

Moreover, for many applications, human or animal labor cannot match the efficiencies and power densities of modern energy conversion technologies. Table 1.4 (Levenspiel, 1996) shows that an "average" human needs about 4,000 kJ/day of energy just to survive and about five times that (i.e., 20,000 kJ/day) to sustain intense exercise or hard labor. Let's estimate an efficiency for human conversion of energy by assuming that all the food intake above the subsistence level is consumed to produce work in an 8-hour shift:

Human Output (Hard Work) = 0.1 kW (Table 1.3) × 8 hr x 3,600 s/hr × 1 (kJ/s)/kW
= 2,880 kJ vs (20,000 - 4,000) kJ input

Percent Efficiency = [2,880 kJ/16,000 kJ] × 100 = 18%

This seems like a good efficiency for a person because many modern coal-to-electricity plants with SO_2 scrubbers have an efficiency of 37%, and many spark-ignition automobiles, at best, convert about 18% of the energy of their gasoline to useful power at the vehicle wheels (Chapters 7, 18). But the "fuel" to provide this energy is food, and the human body is only about 10% efficient in converting food to fat and muscle tissue that can in turn be converted to useful work (Levenspiel, 1996). Thus, the efficiency in using the human "engine" in going from food like grain to useful work is not 18% but 0.1 × 18, or 1.8%. Moreover, if the human eats lots of beef rather than cereal, the net efficiency from grain to human work drops 10-fold to 0.18% because the animal first converts grain or grass to animal fat and muscle at 10% efficiency. Using a bicycle, however, maximizes the efficiency of the human "engine." Bicycles use human power to transport people and light goods relatively efficiently and have added sustainability appeal in dense population areas where motor vehicles contribute to traffic, parking congestion, and air pollution.

Table 1.3. Power Expended in a Sampling of Activities

Power Producers and Users	Power Involved
Lifting a mosquito at 1 cm/s A fly doing one pushup	1 erg/s = 10^{-7} W = 10^{-10} kW
Cricket chirps	10^{-3} W = 10^{-6} kW
Pumping human heart	1.5 W = 1.5×10^{-3} kW
Burning match	10 W = 10^{-2} kW
Electrical output of a 1 m^2 solar cell 10% efficiency	100 W = 0.1 kW
Bright lightbulb	100 W = 0.1 kW
Human hard at work	0.1 kW
Draft horse	1 kW
Portable floor heater	1.5 kW
Compact automobile	100 kW
Queen Elizabeth (giant ocean liner)	200,000 kW
Boeing 747 passenger jet, cruising	250,000 kW
One large coal-fired power plant	1×10^6 kW = 1 GW of electricity
Niagara Falls, hydroelectric plant	2×10^6 kW = 2 GW of electricity
Space Shuttle Orbiter (3 engines) Plus its 2 solid booster rockets at take off	14×10^6 kW = 14 GW
All electric power plants worldwide	2×10^9 kW = 2,000 GW
U.S. automobiles, if all used at the same time (150 million)	15×10^9 kW = 15,000 GW
Humankind's total use in 2005	1.1×10^{10} kW = 1.1×10^4 GW = 400 Q/yr
SUV[a], 15 mpg at 60 mph	160 kW

Sources: Levenspiel (1996) and EIA (1998).
[a]Sport utility vehicle.

Table 1.4. Energy Requirements for Various Human Activities

Activity	Energy Needed Per Day	
	kJ	kcal[a]
Minimum to "survive," i.e., to just maintain vital cellular activity	4,000	1,000
"White collar" work, full-time/steady	8,000[b]	2,000
Bicycle riding, jogging, construction work	20,000	5,000

Adapted from Levenspiel (1996).

[a]Approximately. Note that here 1 kcal = 1 Calorie, as seen in US food labels, also called a "large" calorie or a kilogram calorie.

[b]This assumes 24 hours of work at this rate.

The limited productivity of human and animal labor has serious geopolitical ramifications. Nations lacking modern infrastructures for energy conversion and utilization face competitive disadvantages, not only in the manufacture of energy-intensive goods but in all commerce dependent upon reliable supplies of high-quality energy (e.g., transport, safe habitat, information technologies). Nations that possess up-to-date energy infrastructures need access to assured supplies of energy at manageable costs to sustain and grow their economies. Continuous technological improvements in efficiency and demand-side management will temper but not eliminate growth in energy consumption. Developed nations without indigenous energy resources can trade products of their economies for energy, at least in peace time. Developing nations lacking indigenous energy face a Catch-22 situation because energy drives the economic productivity that creates value tradeable for importable energy.

The most important fuel worldwide is petroleum. It is extracted in many countries and consumed primarily in the industrialized countries. An international trade exists from several unindustrialized producing countries, especially those of the Middle East (Figure 1.3). Because this commerce is essential for international economic vigor, the wealthy nations have demonstrated a willingness to use strong measures (including the Gulf War of 1991) to keep it stable. Since many of the producer countries have experienced political instability, or are in politically unstable regions, accomplishing this can be difficult.

Currently, international trade in energy (largely fossil mineral fuels) is colossal. Figures 1.3 and 1.4, respectively, show the magnitude and direction of international petroleum and natural gas trades in 2002. Also shown are the regional divisions of fossil energy exporters and importers, the former dominated by countries in the Middle East, Africa, Central and South America, Indonesia, and the former Soviet Union, the latter by Japan, the US, and western Europe. (The large transfers of petroleum from the Middle East to the general area of Malaysia reflect petroleum transport to Singapore for refining.)

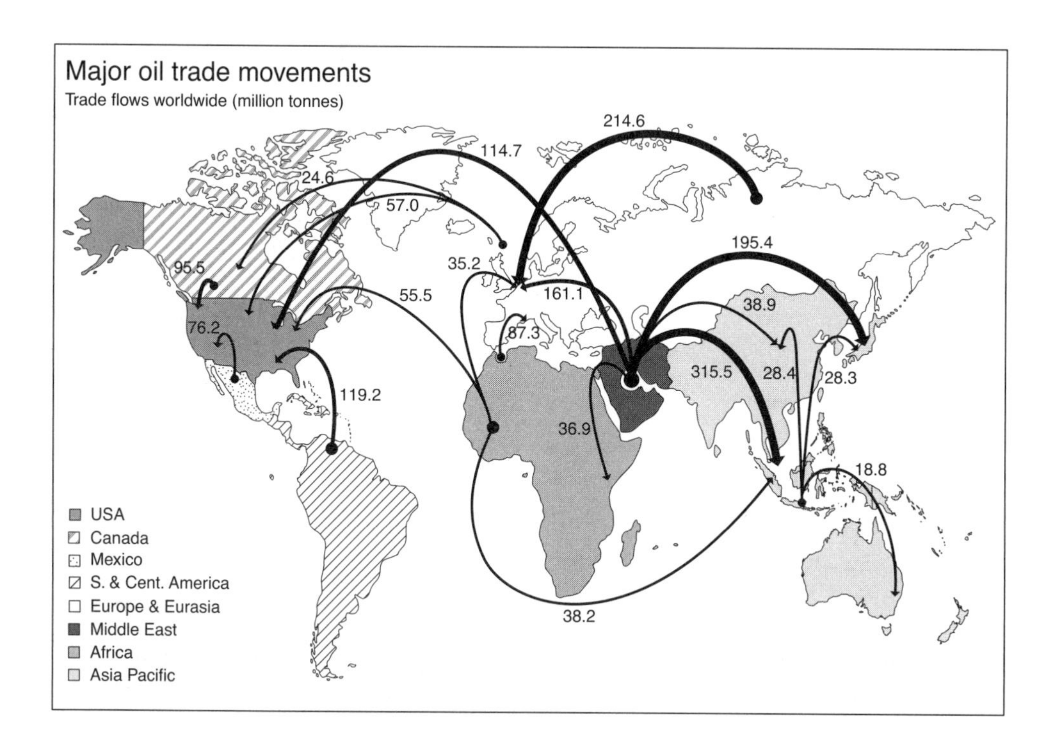

Figure 1.3. Major global trade movements of crude oil in 2002. Source: British Petroleum, (2003): www.bp.com. Reprinted with permission of British Petroleum.

Another important disruption of the petroleum market occurred with the 1973 Organization of Petroleum Exporting Countries (OPEC) shutoff of oil shipments to the US and western Europe in reaction to perceived favoritism of Israeli over Arab interests in the Middle East.

Many of the same countries that are rich in petroleum also have abundant natural gas supplies. However, a corresponding international commerce in natural gas has not emerged due to the greater difficulty and expense of transporting natural gas. Thus, natural gas found at sites far from population centers, especially if they are separated by water, has gone largely unexploited.

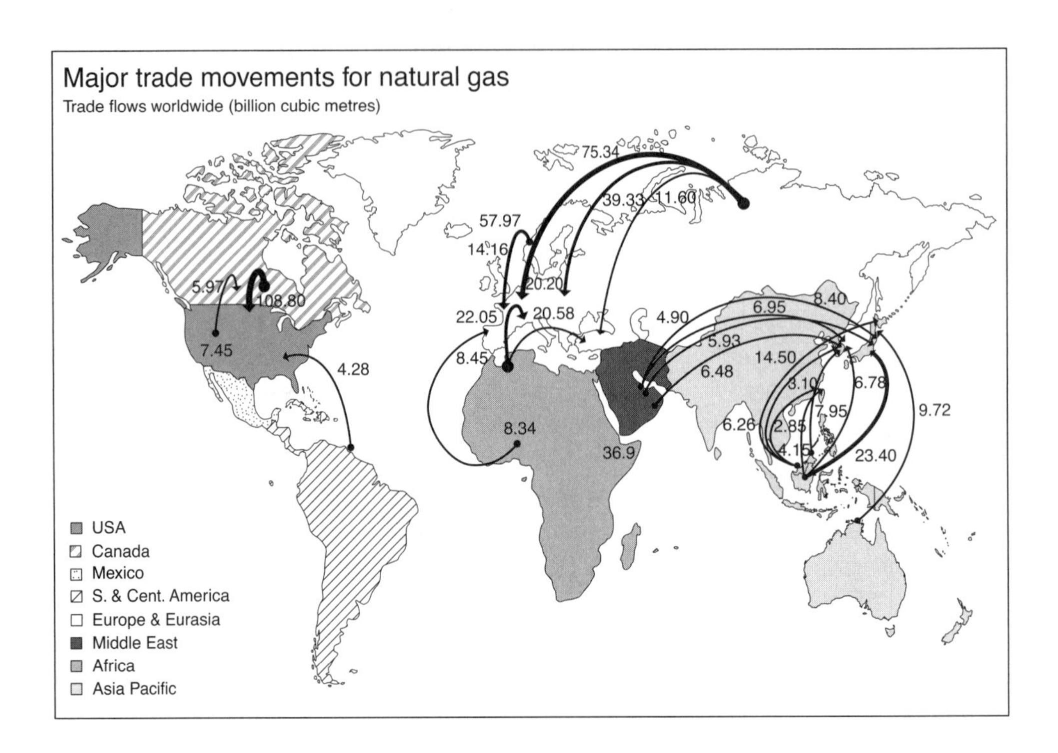

Figure 1.4. Major global trade movements of natural gas in 2002. Source: British Petroleum, (2003): www.bp.com. Reprinted with permission of British Petroleum.

The principal means of transporting natural gas is via pipeline. Important systems for this link are the fields of the western US, serving eastern and western population centers, and the gas fields of Russia and the North Sea, which supply users in western Europe. Liquefied natural gas is transported in ships more expensively (by about an order of magnitude) and less abundantly, thereby permitting some remote sources (e.g., Indonesia, Venezuela, and Algeria) to supply industrialized markets. Both modes of transport are likely to become more important, reflecting the convenience of natural gas for consumers and the relatively low environmental effects of its production.

Similarly, an international trade exists in coal. However, it is less important than those of oil and natural gas. This reflects the environmentally detrimental nature of its production and use, which effectively restricts its consumption primarily to production of electricity and, to a lesser degree, metallurgical coke manufacture, at least in the industrialized countries.

Figure 1.5 shows historical data and growth projections for world population from 1900 to 2100, while Figure 1.6 depicts the contributions of major sources to world energy consumption from1850 to 1995. The dramatic increases in world energy use since World War II reflect the revival of war-torn economies, expansion of the middle class, and exponential growth in world population. The intense coupling of population and energy use is further illustrated by the correlations between per capita gross domestic product (GDP) or gross national product (GNP) and per capita energy use shown for various countries in Figure 1.7. At least for the US, per capita GDP is even more strongly correlated with per capita electricity use, showing the importance of electricity over other forms of energy in driving wealth creation in the US. Figure 1.7 also reveals another important geopolitical issue in sustainability, namely the broad range in aggregate energy efficiency of different countries. Note that the slope of a line connecting the origin with the data point for any country has units [(GJ/person)/(gross national product (GNP)/person)], i.e., GJ/GNP, and is thus an aggregate measure of the amount of energy consumed in each country to produce a unit of GNP.

Composite efficiency metrics are useful, but policy and technology decisions are better informed by dissecting these indices to determine how energy is used in various sectors and then adjusting for patterns of industrial activity, mobility infrastructure, lifestyle, geography, and climate. For example, the lower aggregate energy efficiency of the US as compared to Japan might be explained by the lower population density of the US leading to larger living spaces and longer travel distances per journey. Figure 1.8 shows energy consumption, energy consumption per capita, and energy per unit of GDP for the US from 1949 to 2001. Note that energy/GDP has declined significantly over this period. The causes include sectoral shifts to less energy-intensive industries, as well as improved energy efficiency in manufacturing and the piston engine automobile. The ratio [energy per capita/energy per GDP], i.e., [GDP/capita] is a plausible aggregate measure of economic well-being or quality of life. Estimating this quantity as the ratio of Curve II to Curve III in Figure 1.8 shows that, by this metric, the US average standard of living has increased substantially from 1950.

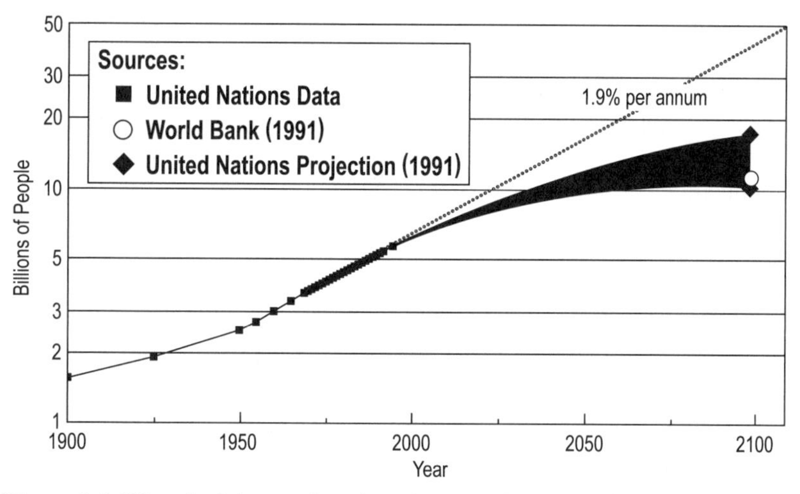

Figure 1.5. Historical data and projected growth in worldwide population (1900 to 2100).

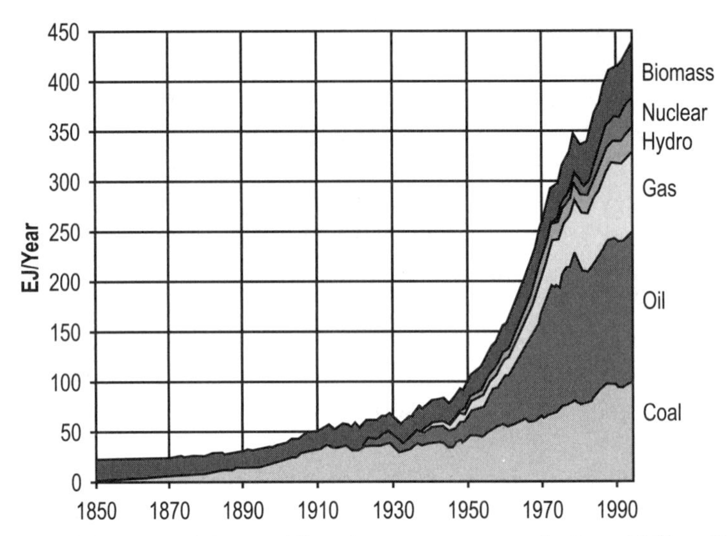

Figure 1.6. Sources of the world's primary energy supply from 1850 to 1995. Source: WEC (1995).

The message of this section is that energy is intimately commingled with the economic vitality and social progress of individual nations and of the world. Since the Industrial Revolution, energy has been a "critical enabler" of economic productivity. The availability of energy at tractable (although not necessarily competition-driven) prices has extended economic and social benefits to countless humans in the democratic countries of the world, which in pre-18th century, non-mechanized societies were reserved for a privileged few. Further, since the 1860s, sales of energy itself have been a "great provider" of economic opportunity. Global daily trade in crude oil is on the order of $1 billion (Moore, 1999). The addition of natural gas and coal probably triples this cash flow. The market value of electricity produced in the US is some $0.5 billion per day (Moore, 1999). The major geopolitical impact is the transfer of huge amounts of wealth to individuals, industrial companies, geographic regions, whole nations, and governments, which has created new global financial powers. Further, governments harvest astonishing revenues by taxing wholesale and retail transactions in energy (Chapter 7).

1.4 National and Global Patterns of Energy Supply and Utilization

In 1997, the world used almost 450 quads of energy. The sheer magnitude of this amount of energy is astonishing. If it were entirely petroleum, it would be enough material to cover an area of about 15,600 square miles (i.e., about the same area as the Netherlands and almost twice the area of Massachusetts) with a 1-foot deep layer of oil. Of course the amount of energy consumed and how that energy is used varies by country and region. Figures 1.7 and 1.9 respectively show the per capita and total energy consumption for various countries in 1992. Clearly, there are wide differences among nations, with developed countries accounting for large portions of the world's total annual consumption. However, these countries also account for much of the world's economy (Table 1.5). The correlation between economic prowess and energy consumption further illustrates the daunting challenges of how to make sustainability practical, how to equitably redress apparent imbalances in economic prosperity enabled by energy, and how to extend energy and energy-derived services to constituencies previously under-represented in the energy marketplace while continuing to meet the energy needs of established energy consumers.

Figures 1.10 and 1.11 respectively show historical records of energy production by source for the world (1970-2000) and for the US (1949-2002). The figures also show how both regions have been dominated by fossil fuels. Figure 1.12 shows detailed patterns of energy supply and consumption for the US for 1997 (EIA, 2002). Based on this figure, end-use consumption can be divided into four major sectors—transportation, residential, commercial, and industrial—that, respectively, consume about 28%, 21%, 18%, and 33% of total US energy. Total consumption in 2001 was about 97 quads, which was about 1/5 the world's total energy use. Fossil fuels accounted for about 85% of this amount.

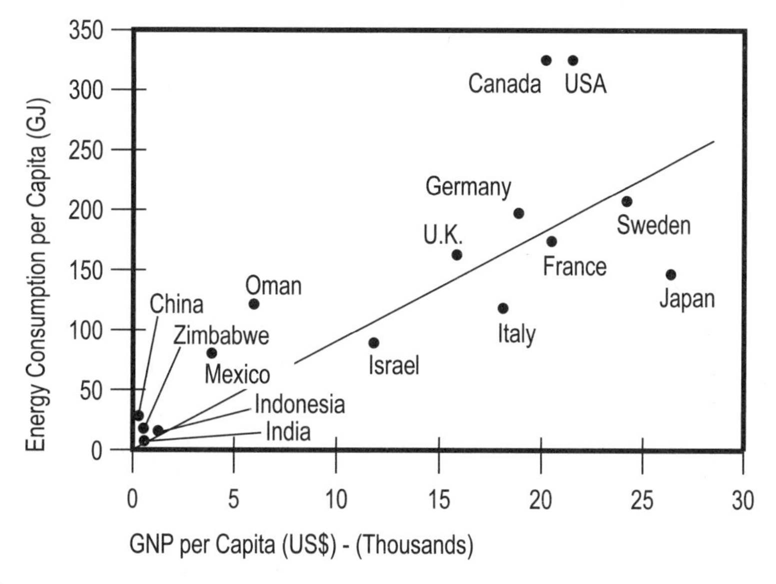

Figure 1.7. Comparison of energy use per capita per year versus GNP per capita for various countries. The line is only to show a general trend. 1 GJ = 10^9 Joules. Source: The World Bank (1992).

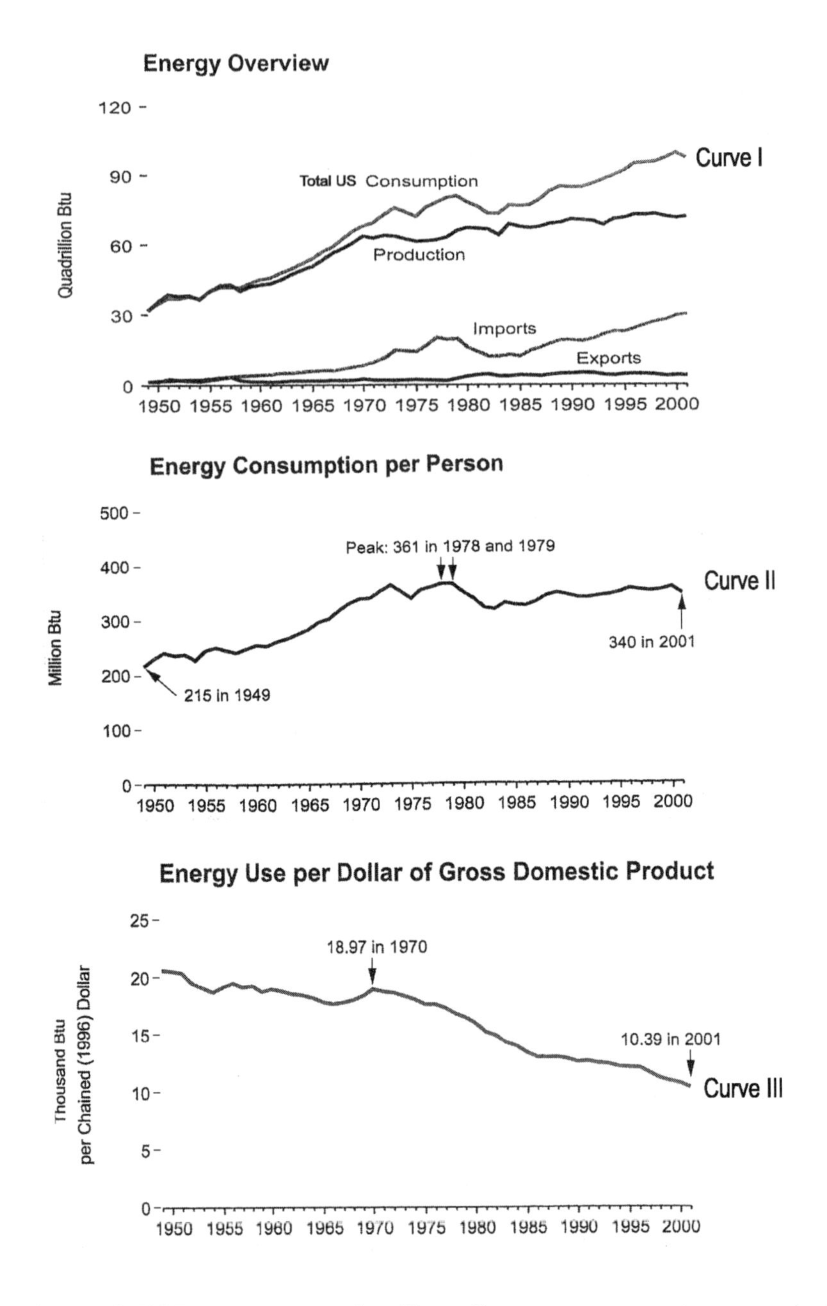

Figure 1.8. US Energy consumption (Curve I), energy consumption per capita (Curve II), and US energy consumption per dollar of Gross Domestic Product (Curve III), for the period 1949 to 2001. Note that some analysts take the ratio of Curve II to Curve III, i.e., $ of GDP/capita, as an index of aggregate economic well-being. Source: EIA (1998).

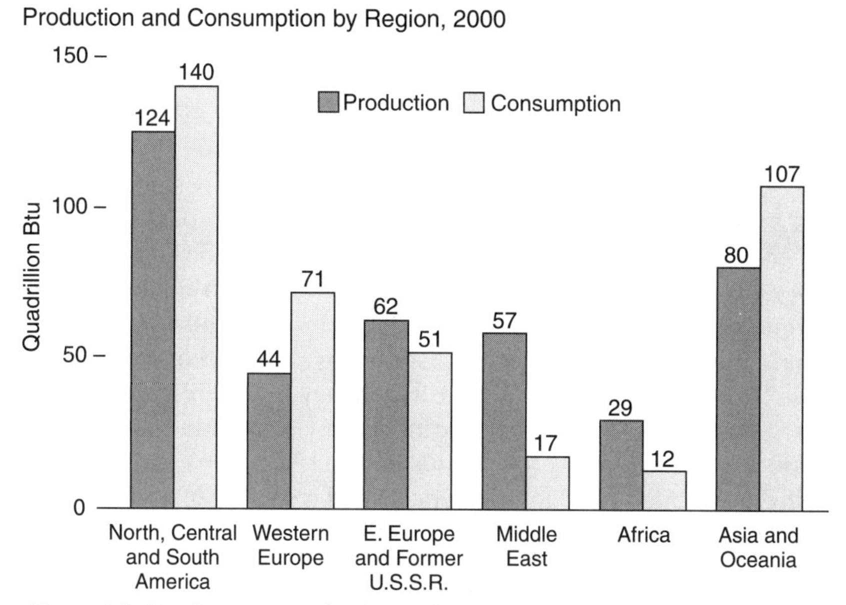

Figure 1.9. Total energy production and consumption in 2000 by region. Source: BP (2001).

Table 1.5. Percentage shares of world population, world GDP, and world commercial energy consumption for selected countries

Country	% of World Population 2001[a]	% of World GDP 2002[b]	% of World Energy Consumption 2002[c]
United States	4.6%	32%	24%
Japan	2.0%	12%	5%
France	0.9%	4%	3%
Germany	1.4%	6%	4%
United Kingdom	1.0%	5%	2%
China	20%	4%	11%
India	17%	2%	4%

[a]World Population 2001 was 6.2 billion. Country data from The Institute of Memetic Research. www.futuresedge.org/World_Population_Issues/

[b]World GDP 2002 was $32 trillion dollars. Country data from World Development Indicators database, World Bank, July 2003. www.worldbank.org/data/

[c]World primary energy consumption 2002 was 9.4 billion tonnes of oil equivalent. Country data from BP. Statistical Review of World Energy 2003. www.bp.com/centres/energy/index.asp

Interestingly, the US exports almost 5 quads of energy. About half of this energy is coal. The balance is a mix of electricity, metallurgical coke, and other fossil fuels and their derivatives. However, US energy imports of over 21 quads of petroleum and petroleum products, plus about 3.7 quads of natural gas, coal, electricity, and metallurgical coke more than offset these outflows. Renewable and nuclear energy, respectively, account for about 7.6% and 7.1% of US energy consumption.

Figure 1.13 shows that hydro (55%) and biofuels (38%) accounted for most of the renewable energy contributions in 1997. Most of the hydro and some of the biomass was used for electric power generation. Some biofuel was also employed for blending of ethanol into gasoline for motor transport. The hydro and geothermal figures include some imports from Canada and Mexico respectively. The role of wood in residential heating is not known, so the biofuels contribution may be underestimated.

Use of renewable technologies may expand in the US with time, although growth in market share may be slow owing to environmental resistance to new large hydropower dams, limited availability of major high grade wind resources in acceptable locations, and appreciable cost disadvantages for several other forms of renewables except in some niche markets. This prognosis could change if there are technological breakthroughs, or if concerns over global climate change result in mandatory reductions in fossil CO_2 emissions.

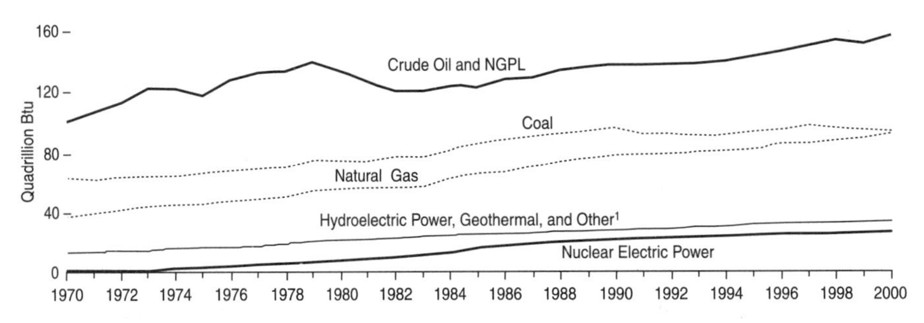

[1]Net electricity generation from wood, waste, solar, and wind. Data for the United States also include other renewable energy.
Note: NGPL is natural gas plant liquids.

Figure 1.10. World primary energy production by source for the period (1970-2000). Source: EIA (1998).

The outlook for nuclear energy is also uncertain. Economic factors and the desires for energy supply security and diversification of energy supplies may propel continued expansion of nuclear electric power production in Asian rim countries like China, Taiwan, and South Korea. Concerns over fossil carbon emissions may improve the acceptability of nuclear energy in other countries, but, barring a dramatic shift in US policy, it seems unlikely that new nuclear plants will be constructed in the US for several decades. However, it is difficult to envision how, over the longer term, the US can adopt a low-carbon energy society without a substantial contribution from nuclear energy.

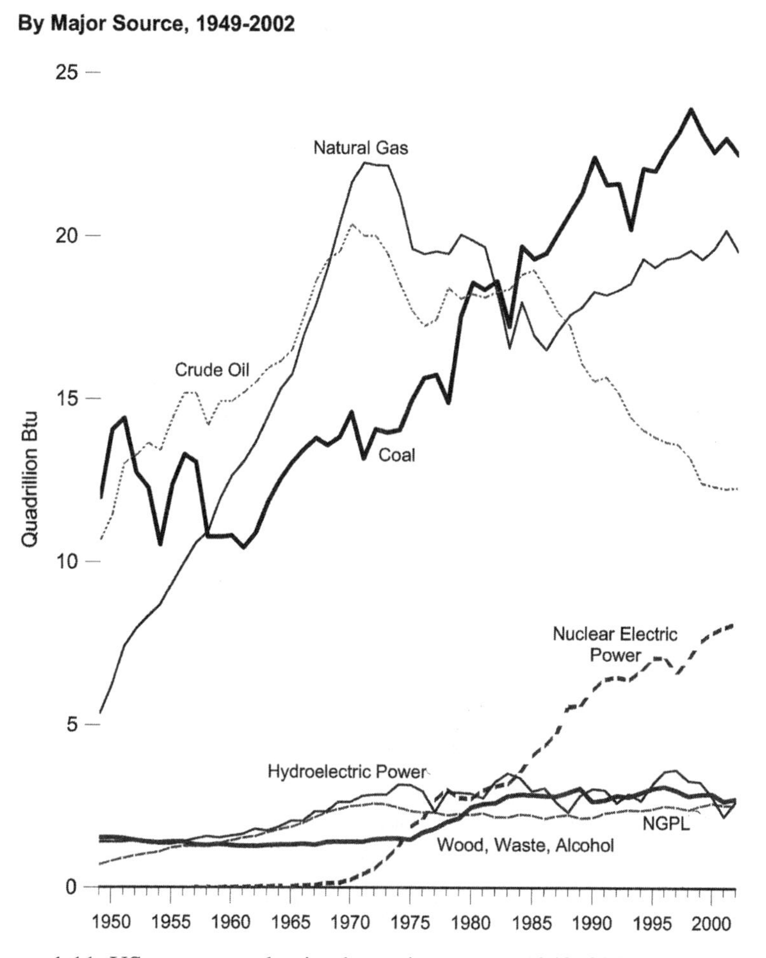

Figure 1.11. US energy production by major source, 1949–2002. Source EIA (2003).

Figure 1.14 provides historical data for and US Department of Energy (DOE) projections of US nuclear generating capacity from 1960 to 2055. This figure shows that without a major re-licensing program, US nuclear capacity will fall to about 1/4 of year 2000 levels by 2025. Even with 20-year re-licensing of 3/4 of the current plants, US nuclear capacity falls to about 1/5 that of year 2000 by 2045. Given that construction of a 1,000 MW nuclear power plant in the US can take 7-10 years, versus less than 5 years elsewhere, the real message of Figure 1.14 is that if the US contemplates, even as a backstop or insurance measure, a useful contribution from nuclear-generated electricity to reduce fossil-derived CO_2 emissions, serious planning for new capacity should begin within the next decade.

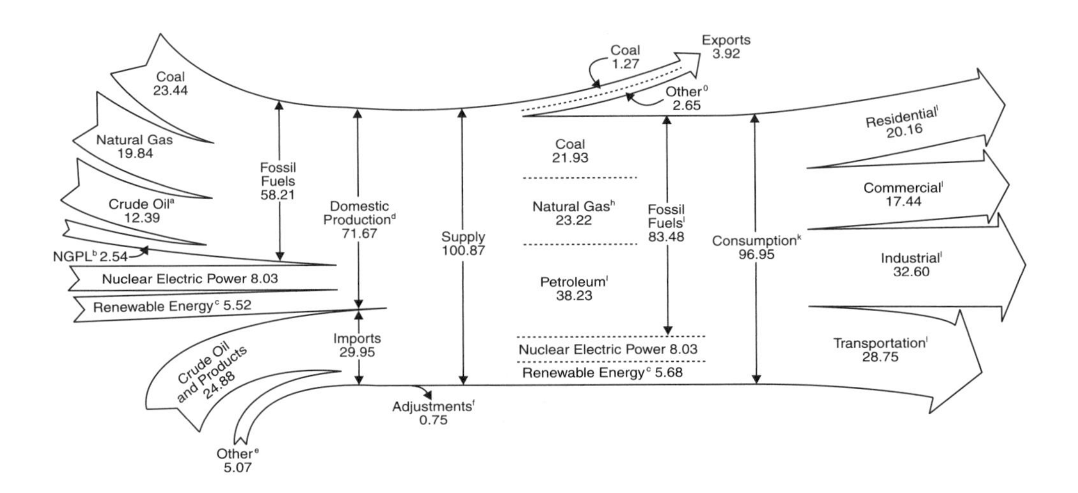

[a]Includes lease condensate.

[b]Natural gas plants liquid.

[c]Conventional hydroelectric power, wood, waste, ethanol blended into motor gasoline, geothermal, solar, and wind.

[d]Includes -0.09 quadrillion Btu hydroelectric pumped storage.

[e]Natural gas, coal, coal coke, and electricity.

[f]Stock changes, losses, gains, miscellaneous blending components, and unaccounted-for supply.

[g]Crude oil, petroleum products, natural gas, electricity, and coal coke.

[h]Includes supplemental gaseous fluids.

[i]Petroleum products, including natural gas plant liquids.

[j]Includes, in quadrillion Btu, 0.04 coal coke, net imports and 0.05 electricity net imports from fossil fuels.

[k]Includes, in quadrillion Btu, 0.09 hydroelectric pumped storage and -0.15 ethanol blended into motor gasoline, which is accounted for in both fossil fuels and renewable energy but counted only once in total consumption.

[l]Primary consumption, electricity retail sales, and electrical system energy losses, which are allocated to the end-use sectors in proportion to each sector's share of total electricity retail sales.

Notes: Data are preliminary. Totals may not equal sum of components due to independent rounding.

Figure 1.12. Pattern of energy supply and utilization in the US for calendar year 2001. Units are quadrillion (i.e., 10^{15}) Btus. Source: EIA (2002).

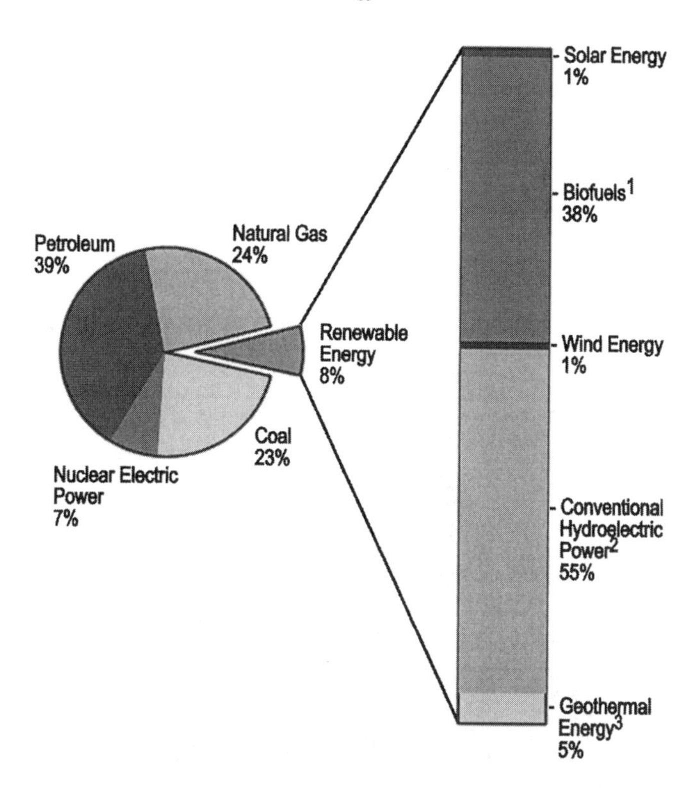

[1] Wood, wood waste, wood liquors, peat, railroad ties, wood sludge, spent sulfite liquors, agricultural waste, straw, tires, fish oils, tall oil, sludge waste, waste alcohol, municipal solid waste, landfill gasses, other waste, and ethanol blended into motor gasoline.
[2] Includes electricity net imports from Canada that are derived from hydroelectric power.
[3] Included electricity imports from Mexico that are derived from geothermal energy.

Figure 1.13. Contributions of various renewables and other major sources to US energy consumption in 1997. Source: EIA (1998).

Figure 1.15 provides a detailed breakdown of US production and utilization of electricity in 2001. The dominance of electric utility generation by fossil fuels (71%) is clearly shown in the upper part of the figure, especially coal (51%), although nuclear (21%) and hydro (9%) also make significant contributions. Biomass, wind, solar, geothermal, and other renewables have little US impact (0.4%), but their contributions are more significant in other countries (Chapters 9 through 15). "Conversion losses" in Figure 1.15 are largely energy lost as waste heat during the conversion of heat to mechanical work that rotates the shafts of electrical generators (Chapter 3). At present, these losses are about 67% for generating equipment installed in the US. Modern gas-fired, gas turbine-steam turbine combined-cycle (GTSTCC) generators (Chapter 7) reduce this loss to about 40% by using waste heat from the gas turbine to raise steam that generates additional electricity. Figure 1.15 also shows that the non-utility power producers contributed about 8.4% of the retail sales of electricity to end users. Data are not available to gauge the amount of heat wasted during electricity generation by these producers. Because many use efficient natural gas-fired GTSTCC generation equipment, we can assume that their waste heat losses are closer to the 40% noted above for these power cycles.

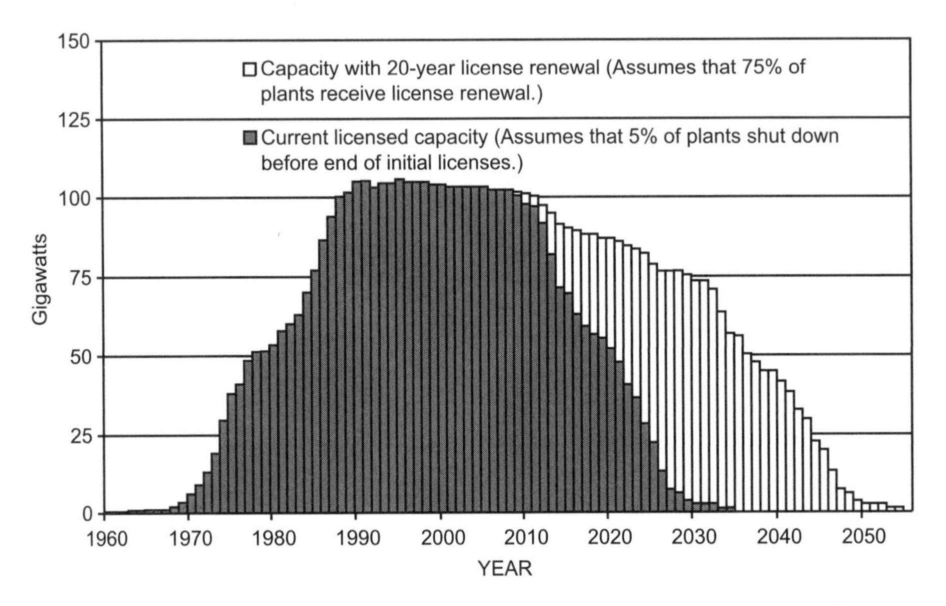

Figure 1.14. Historical and projected US nuclear electric generating capacity 1960-2055. Source: PCAST (1997).

1.5 Environmental Effects of Energy—Gaining Understanding

In Chapter 4, we consider environmental impacts of energy supply and utilization. We begin our study of these matters here because perseverant environmental stewardship must be a cornerstone of any scenario for sustainable energy. In fact, environmental considerations affect virtually all aspects of energy decision-making by government, the private sector, and increasingly, by consumers. Sustainability advocates and architects must resolve a vexing question: how can humankind equitably provide energy-derived benefits to a growing world population without degrading the environment or exhausting resources for which there are no apparent substitutes?

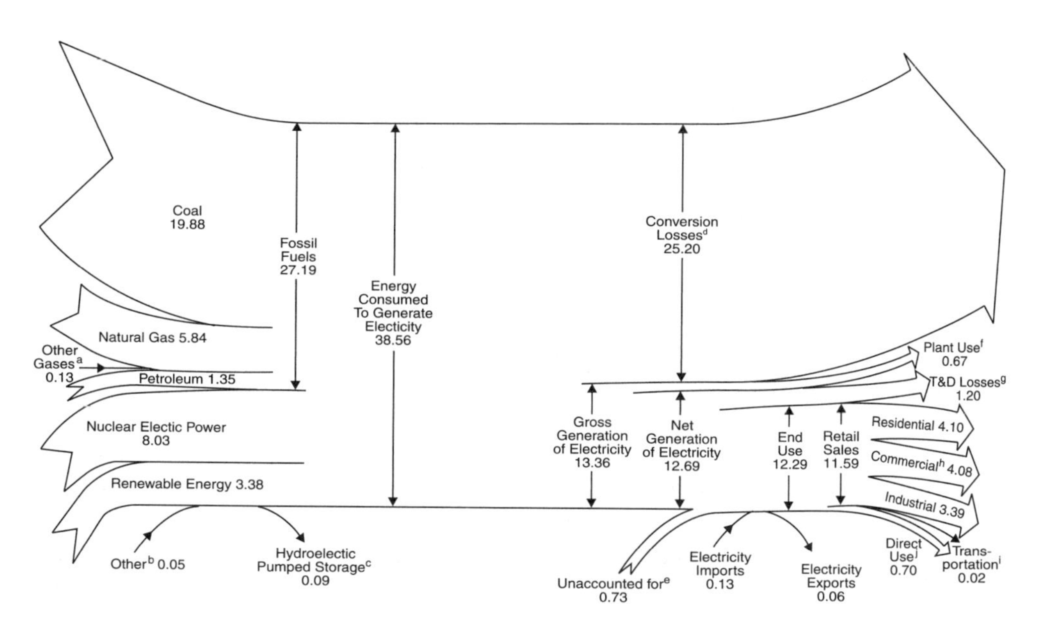

[a]Blast furnace gas, propane gas, and other manufactured waste gases derived from fossil fuels.

[b]Batteries, chemicals, hydrogen, pitch, purchased steam, sulfur, and miscellaneous technologies.

[c]Pumped storage facility production minus energy used for pumping.

[d]Approximately two-thirds of all energy used to generate electricity.

[e]Data collection frame differences and non-sampling error.

[f]Electric energy used in the operation of power plants, estimated as 5 percent of gross generation.

[g]Transmission and distribution losses, estimated as 9 percent of gross generation.

[h]Commercial retail sales plus approximately 95 percent of "Other" retail sales.

[i]Approximately 5 percent of "Other" retail sales.

[j]Commercial and industrial facility use of onsite net electricity generation.

Note: Totals may not equal sum of components due to independent rounding.

Figure 1.15. Sources and end use of electricity in the US for 2001. Units are quadrillion (i.e., 10^{15}) Btu. Source: EIA (2002).

Figure 1.16 displays some attributes of this *energy-prosperity-environmental dilemma*. Note that adverse impacts from energy include a vast legacy of prior damage, current pollution, and the real prospect of continued ecosystem degradation for decades or more to come. Table 1.6 lists some of the environmental and other hazards from a selection of renewable and non-renewable energy supply technologies.[6] Note that energy-derived pollutants include gases, liquids, solids, or mixed phases, and that they may adversely impact a host of environmental media and ecosystems. Further,

6 Links between energy and the environment are not new. In 1776, British physician Percival Pott attributed the high incidence of scrotal cancer among London chimney sweeps to their regular exposure to combustion-derived soot (NRC, 1972). Supposedly when the King of Denmark learned of Pott's findings, he ordered all Danish chimney sweeps to take daily baths, thereby promulgating one of the earliest energy-related environmental regulations.

energy-derived pollutants may act over a wide range of length and time scales (Table 1.7). On the other hand, as Figure 1.16 shows, energy-driven economic progress benefits humankind by preserving and extending prosperity *and* improving the environment through redress of past damage and prevention of future pollution.

Human activities have economic consequences that are not always reflected in the prices of related goods and services. Sometimes these "hidden costs" or *externalities* are so minute as to be inconsequential and not to justify their internalization or inclusion in the costs of an activity. Contrastingly, many scholars believe that the supply and use of energy is accompanied by significant expenses that are not captured in present-day energy prices. Many of these externalities are associated with adverse environmental impacts of energy, including the costs for health care, lost productivity, air and water pollution, aesthetic damage, destruction of open space, remediation of wastes, etc. (Table 1.7). Chapters 5 and 8 describe modern efforts to quantify the hidden costs of energy, with particular attention to the supply of energy from nuclear fission. What is interesting about attempts at internalizing external environmental costs is not that it is difficult to do, but that reasonable people make the attempt. This is because the basic knowledge needed to quantify such effects at a useful level of precision are often absent, and a reliable calculus for homogenizing a set of disparate external costs into a common basis (e.g., money) does not yet exist. Thus, the qualitative concept of internalization is appealing but difficult to implement in the absence of complementary mechanisms to induce the affected parties to compromise their differences in determining acceptable practices.

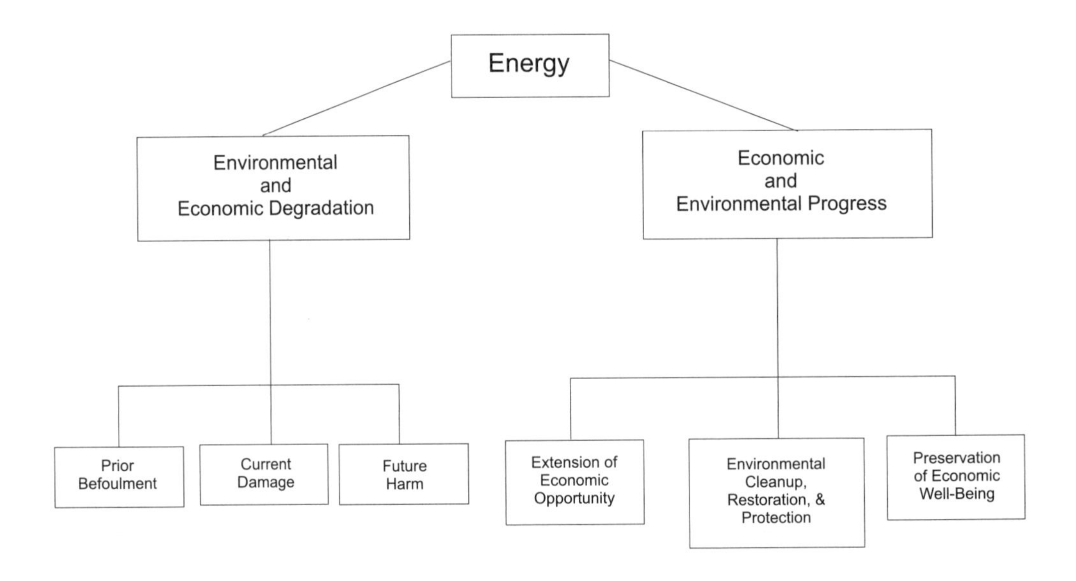

Figure 1.16. Schematic illustration of the energy-prosperity-environmental dilemma.

For example, any attempt to assign monetary value to loss of human health or life necessitates highly subjective judgments and risks alienating stakeholders. Actuarial data on life expectancies, insurance policy values, and lost productivity costs would seem to be a logical source of information for costing externalities. However, the appeal of this method quickly fades when faced with the need to apply its statistically based conclusions to a particular individual. Another vexing problem is the lack of unambiguous data relating environmental risk to specific causes. For example, in the case of human health effects of energy-derived pollutants, there are few cases where human morbidity or mortality can be unequivocally linked to specific toxic substances. To reliably "close the loop" between an adverse impact in a human and an energy-related cause of that impact, we need to know a variety of things: what agent caused the human impact; whether that agent was formed within the human body by transformation of some exogenous substance; whether multiple agents were involved within the body; if undesired transformations of exogenous substances occurred because the body's natural defense mechanisms were disrupted; whether the human was exposed to environmental agents that themselves, or via their transformation products (metabolites) within the body, caused adverse health consequences; whether such agents were produced directly as emissions, effluents, or by-products of some energy technology (e.g., air emissions from an automobile); what the identities and amounts of such agents emitted were; whether harmful agents generated from energy-derived effluents were attributable to chemical or physical changes in ambient air or water. These are just some of the questions that toxicologists ask as they seek to determine whether environmental agents contribute to adverse health effects in humans, and as they work with experts from other disciplines to learn whether energy or other technologies are responsible for the generation of such agents or their progenitors (W.G. Thilly, personal communication).

Another complication is that energy-related environmental effects can occur over diverse length and time scales—from millimeters to 10,000 kilometers—and from seconds to more than 100 years (Table 1.7). Thus, an environmental insult that leads to cancer or an inheritable birth defect might go undetected for several decades. Similarly, CO_2 emissions in one location may ultimately upset the climate halfway around the world 40 years later. The presence of diverse length and time scales also makes it difficult to develop an integrated mathematical simulation of all the suspected environmental agents and their unwanted impacts on humans and the planet. This is because any such model must accommodate wide variations in spatial and time coordinates, an historically difficult problem in systems evaluation (Chapter 6).

Table 1.6. Examples of Environmental and Other Hazards of Various Energy Supply Technologies (None is free of adverse effects)

Fuel/ Phase	Coal	Petroleum	Natural Gas	Nuclear	Hydro
Extraction	Mining Accidents Lung Damage	Drilling Spills (off-shore)	Drilling	Mining Accidents Lung Damage	Construction
Refining	Refuse Piles	Water Pollution	—	Milling Tails	—
Transportation	Collision	Spills	Pipeline Explosion	—	—
On-Site:					
Thermal	High Efficiency	High Efficiency	High Efficiency	Low Efficiency	—
Air	Particulates SO_2, NO_x	SO_2, NO_x	NO_x	Low Radiation	—
Water	Water Treatment Chemicals	Water Treatment Chemicals	Water Treatment Chemicals	Water Treatment Chemicals	Destroys Prior Ecosystems
Aesthetic	Large Plant Transmission Lines	Large Plant Transmission Lines	Large Plant Transmission Lines	Small Plant Transmission Lines	Small Plant Transmission Lines
Wastes	Ash, Slag	Ash	—	Spent Fuel Transportation Reprocessing Waste Storage	Fish Killed
Special Problems	—	—	—	—	Population, Agricultural Displacement
Major Accident	Mining	Oil Spill	Pipeline Explosion	Reactor Cooling Failure Nuclear Weapons Proliferation	Dam Failure

Table 1.6. Examples of Environmental and Other Hazards of Various Energy Supply Technologies (None is free of adverse effects) (continued)

Solar Terrestrial Photovoltaic	Solar Power Tower	Solar Satellite Photovoltaic	Nuclear Fusion	Geothermal	Wind
Mining Accidents	—	Mining Accidents	H^2, Li Production	—	—
—	—	—	—	—	—
—	—	—	—	—	—
				—	—
Low Efficiency Ecosystem Change	Ecosystem Change	Genetic Change	—		
—	—	—	—	H_2S	
Water Treatment Chemicals	Water Treatment Chemicals	Water Treatment Chemicals	Tritium in Cooling Water	Brine in Streams	
Poor Large Area	Poor Large Area	? Large Area (Antenna)	Large Area Plant	Large Area	Locally visible
Spent Cells	—	—	Irradiated Structural Material	Cool Brine	—
Construction Accidents	—	Vulnerability in Wartime	Occupa- tional Radiation Doses	—	Siting Structural Failure
Fire	—	Intense Microwave Beam	Tritium Release	—	—

Table 1.7. Approximate Length and Time Scales for Selected Known and Potential Environmental Effects of Energy Production and Utilization

Local	0.001-10 km e.g., Air Pollutants – Acute Respiratory Episodes: < 1 day – Lung Cancer: 10-50 years – Mutagenicity: 1 – 5 generations
Regional	100-500 km e.g., Acid Rain – Forest and Aquifer Damage: 1 – 20 years e.g., Particulate Pollution
Global	5,000-25,000 km e.g., Climate Modification – Sea–level Rise } 30 – 100 years – Desertification } or more

A further challenge is that pollutant inventories are often the result of a complex interplay of multiple physical and/or chemical factors, so that seemingly intuitive pollution control measures can be horribly flawed. The point is illustrated by ozone (O_3), a human respiratory irritant and smog precursor in the troposphere (0–10 km above the earth's surface).[7] Tropospheric O_3 is formed by sunlight-induced chemical reactions of two combustion-derived pollutants, nitrogen oxides, NO_x, and volatile organic compounds, VOCs. It is intuitive to reason that tropospheric O_3 pollution can be reduced by lowering emissions of NO_x or of hydrocarbons. Figure 1.17 shows that matters are not so simple. It plots, as a series of contours, atmospheric O_3 concentrations for an ambient air-shed as affected by corresponding ambient air concentrations of NO_x and VOCs. The figure is based on simulations of chemistry along moving parcels of air in Atlanta, and is a simulated version of one shown by Seinfeld and Pandis (1998) from Jeffries and Crouse (1990). The figure shows that reduction of NO_x emissions does not guarantee lower tropospheric ozone and may even be counterproductive. Depending on the prevailing VOC inventories, lowering NO_x may have no effect or even increase ambient O_3 concentrations! The reason is that the mechanisms for O_3 formation and survival are complex and, in general, are not linearly proportional to ambient concentrations of NO_x. The key message from Figure 1.17 is that effective regulation of energy-derived pollutants must be based on reliable quantitative understanding not only of pollutant origins, but also of their transport and transformation in air, water, and other media. Similar complexities and needs for care in formulating policies are encountered all over the energy-environment landscape as further explored by Kammen and Hassenzahl (1999).

In summary, every important energy-environment question is uncertain to a substantial degree because the models and data for comprehensive understanding are usually unavailable and will remain so. Coupling such uncertainties to the differing interests and values that affect the parties involved, it is almost unavoidable that serious disagreements will arise among them. In other words, treating such questions as purely technical ones, for which true answers can be found, is usually unrealistic. Rather, it makes greater sense to use analyses, including treatments of uncertainty, to characterize the plausible range of potential energy-environmental outcomes. To date, this last step

7 In the stratosphere, 10-50 km above the earth, ozone becomes a strong friend to humans by filtering out harmful UV radiation from the sun. Depletion of stratospheric O_3 by anthropogenic chlorofluorocarbons (CFCs) exemplifies an adverse environmental impact of human technology that could have culminated in a human catastrophe (Chapter 4). However, good science discovered the problem and led to an exemplary model of science-intensive public policy, namely the international decision to phase out use of CFCs in what has become known as the Montreal Protocol. Professors Mario Molina and Sherwood Rowland earned the 1995 Nobel Prize in Chemistry for their discovery that reactions with CFCs threatened to deplete stratospheric O_3.

of policy formulation has been largely ignored in the US and elsewhere. Not surprisingly, such ignorance has resulted in impasses and waste in the resolution of energy-environment questions.

The energy-prosperity-environmental dilemma is intensely intertwined with issues of fairness and equitable treatment of all stakeholders. These matters command high visibility with voters and consumers, as well as with politicians, advocacy groups, and commercial enterprises. Part of the dilemma is that there are diverse and seriously held convictions as to what energy technologies and policies constitute fair treatment of affected constituencies. Consensus becomes especially elusive when there are issues of risk to ecosystems and/or humans. Limitations in risk assessment, risk analysis, or risk communication (Chapter 6) can damage public confidence so badly that it may be virtually impossible to win acceptance for sustainable energy reforms.

Such impasses may be aggravated when the parties involved exaggerate the merits of their case and the weaknesses of those of their opponents. Such distortions are common in political power struggles. It should not be a surprise that they occur when such struggles occur under an energy-environment guise.

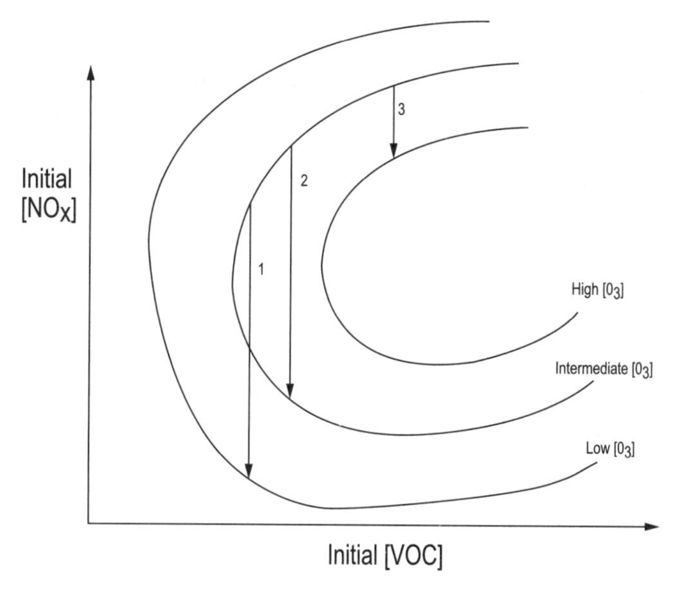

Figure 1.17. Schematic illustration of how changes in ambient concentrations of nitrogen oxides [NO_x] may impact levels of ozone [O_3] pollution in an urban airshed with high photochemical activity. Note that reducing NO_x inventories, depending on the ambient concentrations of volatile organic compounds [VOC], may decrease (Case 1), have no effect (Case 2), or increase ozone pollution (Case 3). Source: Seinfeld and Pandis (1998) citing Jeffries and Crouse (1990).

There are various motivations for public sector involvement in these issues. In developed countries, governments seek to protect people seen as especially vulnerable (e.g., lower-income people, children, the elderly, the homeless) from impacts of energy development, energy pollution, or loss of essential energy services. Objectives are to prevent regions occupied by these stakeholders from being unfairly selected as sites for energy or environmental cleanup facilities and to guard the financially destitute against termination of energy services. Developing countries may wish to use the benefits of energy to accelerate economic progress and improve the social well-being of their citizenry. Thus they may, in the near term, assign lower priority to environmental concerns in order to accelerate industrialization and infrastructure expansion by means of lower cost but potentially higher polluting energy resources. In many cases (e.g., until relatively recently in the Asian rim nations), this need has been driven by a combination of growing populations and dramatic annual increases in economic output. Poorer countries understandably ask why they should temper what they perceive to be essential growth in their own economies in the name of environmental protection, while already developed nations with dramatically higher standards of living do little to abate their own consumption of energy. Many lesser developed countries (LDC) and non-LDC experts believe developed nations that already enjoy the economic benefits of high per capita energy use should shoulder a "proportionate" share of the financial costs of providing for global environmental stewardship. To some, this means devising programs that curtail energy-related environmental damage in the developed world and assist LDCs in developing their own clean energy resources. Technology transfer, educational programs, and financial assistance are proposed as means to the latter end. One prominent global environmental thinker, Dr. Maurice Strong,[8] has suggested that developed countries formalize these goals by committing to spend a fixed percentage of their GDPs to improve environmental quality in developing nations.

Energy and environmental concerns ultimately boil down to two basic questions: who pays for pollution and when? Sustainable energy advocates respond that *everyone* pays for pollution *all the time*, whether or not the consequences of that pollution are immediately visible. For them, sustainable development seeks to confederate all stakeholders. Their mission is to devise technological and policy options to preserve and improve the earth's environment while protecting and extending socioeconomic progress throughout the planet.

A "confederation" does not mean that sustainability approaches will eliminate controversy or perfectly satisfy all expectations of every stakeholder. To the contrary, tradeoffs in the world's energy future are inevitable. The goal of sustainable energy is

8 Dr. Maurice Strong served as the Under Secretary General of the United Nations and Special Advisor to the President of the World Bank. Formerly Chairman of Ontario Hydro and senior executive in several other Canadian companies, he now sits on many advisory boards. He served as Secretary-General of the 1992 UN Earth Summit.

to understand these tradeoffs and then elucidate workable technological and policy responses that will create more environmentally sound *and* consumer-friendly energy futures. Part of the sustainability process must be science-intensive, inclusionary discourse to unite potentially adversarial stakeholders and build lasting covenants for environmental and economic survival. As illustrated above for tropospheric ozone pollution, such engagements will only succeed if they are based on solid scientific and engineering foundations. Furthermore, sustainability goals will evolve as more is learned about tradeoffs and as new policy and technology innovations are discovered. Thus, research on scientific, engineering, and policy issues is essential to assuring the timely availability of information that can overcome the energy-prosperity-environmental dilemma.

1.6 Confronting the Energy-Prosperity-Environmental Dilemma

Sustainability and Alternative Proposals

The magnitudes of world daily energy consumption, financial expenditures for energy, and international trade in energy are colossal. This textbook introduces sustainable energy as a plausible response to a technology-intensive problem that is expected to vex humankind for decades to come. We can headline that problem as *the energy-prosperity-environmental dilemma*. Figure 1.16 captures the essence of this dilemma in schematic form. The thesis of this book is that persistent application of sustainable development concepts to energy and its environmental impacts is a viable solution to this dilemma. Realization of this solution will be challenging because current and future stakeholders embrace diverse and frequently contradictory sociological, geopolitical, technological, and environmental objectives. Subsequent chapters examine tradeoffs created by these diverse requirements, together with technology and technology-intensive policy options, that can bring about sufficient consensus to create a more sustainable energy future. However, there are alternative proposals for managing or combating adverse environmental and resource consumption impacts of energy utilization. Chapters 2 through 6 provide background information and methodologies to assess the claims and prospective performance of sustainable energy and other approaches to the world's energy future.

Table 1.8 (Allenby, 1999) summarizes four approaches to societal governance of technology and their probable implications for population growth and economic expansion. Each approach provides a significantly different philosophy for addressing the energy-prosperity-environmental dilemma. Whether any of these philosophies can be translated into an efficacious operating strategy remains to be seen. Nevertheless, there is merit in examining the tenets of all four approaches in Table 1.9, noting that they differ substantially from each other. At one end of the spectrum, we have *continuation of the status quo*, a largely *laissez faire* methodology punctuated occasionally by high visibility interventions, such as the chlorofluorocarbon (CFC) ban. At the other extreme stands *radical ecology*, a deeply penetrating technological revisionism in which most of society would return to low-technology methods for providing goods and services. Since the 1960s, mini-movements to this end have surfaced from time to time in the US (e.g., in the form of individuals and families seeking solace from urban pressures by taking up residence and vocations in rural areas). Less extreme is so-called *deep ecology* based on use of "appropriate technology," including low-technology approaches where possible.

Table 1.8. Proposed Responses to Societal Concerns Over Adverse Impacts of Industrial Activity[a]

Response Strategy	Effect on Technology	Implications
Radical Ecology	Return to low technology	Unmanaged population crash: economic, technological, and cultural disruption
Deep Ecology	Appropriate technology, "low tech" where possible	Lower population, substantial adjustments to economic, technological, and cultural status quo
Industrial Ecology	Reliance upon technological evolution within environmental constraints: no basis for "low tech" unless environmentally preferable[b]	Moderately higher population, substantial adjustments to economic, technological, and cultural status quo
Continuation of Status Quo (*Laissez Faire*)	Ad hoc adoption of specific mandates (e.g., CFC ban): little effect upon overall trends	Unmanaged population crash: economic, technological, and cultural disruption

Source: Allenby (1999).

[a]These strategies and consequences are also plausible outcomes of societal reactions to the energy-prosperity-environmental dilemma (see text and Figure 1.16).

[b]Or as a better match to ambient socio-economics.

Table 1.8 projects painful consequences from both the *status quo* and *radical ecology* approaches (i.e., economic and social calamities including population explosions or collapses) and associates *deep ecology* with some decline in population as well as appreciable mutations in current economic, technological, and cultural norms. As presented in Table 1.8, *industrial ecology* shares appreciable common ground with the concepts of sustainable energy as presented in this book. However, sustainability leaves open the depth, direction, and time scales of "substantial adjustments to economic, technological, and cultural status quo," preferring instead to emphasize judicious use of technology and technology-intensive policy measures to both preserve and extend the economic opportunities enabled by energy.

Industrial ecology forecasts substantial technological and economic changes. Historically, technology has advanced by continuous evolutionary improvements, interdicted at certain defining moments by revolutionary changes (e.g., steam engines replacing human and animal power, semiconductors replacing vacuum tubes, information technology). Thus, given sufficient time, the technology prognoses of *industrial ecology* seem likely to prove correct. Less certain however, is whether over one or two inter-generational time scales (e.g., in the next two to five decades) the pace of technological change in the energy, environmental, and closely related sectors (e.g., automotive and electric power) will become more revolutionary than evolutionary, thereby unleashing dramatic transformations in industrial practice and consumer behavior.

The prospect of "substantial adjustments" to the economic status quo (Table 1.8) would be welcomed by some but certainly not all sustainable energy proponents. Serious adherents range from advocates of appreciable governmental intervention to die-hard free marketeers. Appropriate missions for government and for non-governmental institutions in stimulating economic growth and protecting the environment will continue to be debated in democratic societies. The existence of different perspectives on the depth and means of economic adjustment needed to achieve meaningful progress on sustainability should be recognized as a positive force that can enrich, strengthen, and diversify the pathways to effective reforms. Diversity is a strength—not a weakness—of the sustainable energy credo.

The extent to which sustainability ideals will penetrate industry and society more broadly is not known, but there are several encouraging signs. Many global industrial companies, without governmental regulations, have adopted industrial ecology and sustainability thinking as part of their business practices (Allenby, 1999, Graedel and Allenby, 1995, Schmidheiny and Zorraquin, 1996, Schulze, 1996, Socolow et al., 1999). In a 1997 address that attracted major attention from environmental groups, Sir John Browne, the chairman of British Petroleum, committed his company, now one of the world's three largest producers and marketers of fossil energy products, to major development of non-fossil renewable energy sources over the next several decades. Sustainable development is gaining industrial favor as a means not only to improve corporate images but also to retain and expand market share and to increase economic profitability.

Academic institutions and governmental laboratories in the US and abroad are mounting serious educational, research, and development programs to infuse sustainability concepts into the formal classroom training of new generations of scientists and engineers and into the design and implementation of emerging technologies and policy measures for supply and utilization of energy. Many of these initiatives have attracted financial support from industry. Some programs (e.g., the U.S. Program for a New Generation of Vehicles [PNGV]) feature industrial cost-sharing and hands-on partnering with industrial companies. These and many other examples suggest that there is reason for optimism that the benefits of sustainability approaches to the world's energy future will find increasing acceptance as the rational norm among industrial and governmental decision makers.

Each reader of this book must draw his or her own conclusions regarding whether the sustainability approach offers the best prospect for obtaining and maintaining global environmental stewardship and energy-enabled social progress. In reaching that decision, all of us are advised to consider possible obstacles to constructive change and ponder what unexpected events might modify the outlook for sustainability and alternative approaches. Thus, we posit that most sustainability advocates have embraced certain tacit assumptions about the world's future:

- that the apparent international trend toward increased democratization will continue
- that the globalization of commerce will not be seriously attenuated
- that protracted (exceeding one year), large scale, and, in particular, transcontinental military conflicts are unlikely, but local and regional conflicts will continue to plague the planet and will include flareups in areas containing or proximate to major deposits of fossil and other energy resources
- that cataclysmic economic disruptions among major economic giants such as Germany, Japan, and the US will not occur
- that within three decades certain developing nations such as China will hold economic power comparable to or eclipsing that of at least some, and possibly most, current economic superpowers

If reality proves to be substantially different from any of these assumptions, the cultural, social, technological, or economic incentives for timely adoption of sustainability concepts could change dramatically. If sustainability is to succeed, it must have broad-based public acceptance.

In particular, sustainability proponents must consider:

- **Practical political reality:** How to economically preserve the services and lifestyles in developed nations that historically have been enabled by high per capita fossil energy consumption.
- **Globalization of commerce and more democratic ideals:** How to equitably extend to developing nations and nations in transition energy-related goods and services, especially in light of elevated rates of economic and population growth.
- **Environmental stewardship:** How to prevent and/or redress environmental damage, known and potential, from the supply and utilization of energy.
- **Constructive engagement:** How to earn and maintain a consensus for sustainable energy throughout the global community, in light of diverse and potentially conflicting individual, local, national, and regional priorities for environmental responsibility, economic progress, and stewardship of natural resources.

A major pitfall in responding to these challenges is failure to impose a systems view (see Chapters 6 and 21). The beneficial and adverse consequences of energy utilization arise from many interacting processes that change with time and location. The energy-prosperity-environmental dilemma is a prototypical large complex system. It exhibits diverse length and time scales for the supply of energy (Figure 1.2), and for ecosystem damage (Table 1.7), sociopolitical transformations, and technological innovation. Further, this system is inextricably connected to related complex systems, such as resource management (e.g., water, minerals, fertile land), micro- and macro-economic growth, and geopolitical stability. These realities complicate analysis and decision-making on energy and its impacts, but they must be accounted for in devising and implementing sustainable energy strategies.

1.7 Mathematical Representations of Sustainability

If sustainability could be objectively measured and described quantitatively, analysts would have a powerful tool for prioritizing and defending sustainable energy strategies. However, sustainability is a cloth woven of objective and subjective fibers. Potent "hot button" issues in the subjectivity category are how to value human health and human life, open space, and species diversity. Methodologies to assign economic costs to these values have been proposed (Chapters 5 and 6). Many object to such metrics on moral grounds or seriously question the underlying assumptions for value assignments, especially in light of appreciable uncertainties or gaps in crucial data. Even putatively objective measures, like the mass of pollutant a technology emits per unit of useful energy provided (e.g., lbs of SO_2 per kWh of electricity), are vulnerable to subjectivity in that different evaluators will give more or less weight to a given attribute. Further, the weighting factors may well change with time as more is learned about human health and the adverse impacts of energy-related pollutants. Thus "metricators" face a complex stew of parametric uncertainty, imperfect miscibility of objectivity and subjectivity, and transient weighting factors. Consequently, a meaningful mathematical representation of sustainability will presumably need to be more robust than a single number or even a list of numbers. Such a representation should nevertheless be subject to well-established rules for mathematical manipulation so that metrics for various options can be easily compared, contrasted, and, where appropriate, reduced or combined. Given these requirements, we propose that tensors, which disaggregate and openly reveal anisotropy in real quantities (Jeffreys, 1961, and Margenau and Murphy, 1956), may provide a powerful mathematical apparatus for quantitatively describing and comparing the sustainability potency of diverse options. To the best of our knowledge, the mathematical analysis and data fitting to test this hypothesis have never been performed.

Some measurement approaches assume that the sustainability of a technologically intensive activity reflects:

(a) how many people use the technology,

(b) the role of the technology in the economy, and

(c) some measure of the resource consumption or environmental degradation caused by the technology.

Translating this hypothesis into useful formulas for comparing sustainability impacts is more challenging. One algorithm assumes that population (*P*) is a good surrogate for (a), but that the metrics implied by (b) and (c) should be expressed mathematically, not as absolute quantities, but rather as relative impacts on the population and economic activity respectively. Thus, useful simulants for (b) and (c) are per capita gross domestic product (*GDP*/*P*), and energy consumption per GDP (*E*/*GDP*), giving the following equation for the sustainability impact, S_I:

$$S_I = (P) \times (GDP/P) \times (E/GDP) \tag{1-22}$$

Some analysts interpret (*GDP*/*P*) and (*E*/*GDP*) as aggregate measures of standard of living and energy intensity (i.e., reciprocal energy efficiency). Multiplication of the three factors on the right hand side (RHS) of Equation (1-22) suggests the seemingly trivial result that $S_I = E$ or that total world energy consumption is a measure of sustainability, suggesting that energy use only benefits sustainability. This result alone might satisfy some sustainability "metricators," but Equation (1-22) offers much more. It can reveal the potency of specific adverse impacts simply by appending its RHS with one further multiplicative factor depicting the intensity of that impact per unit of energy used, e.g., the amount of carbon dioxide CO_2 emitted per Btu (CO_2/E):

$$S_I = (P) \times (GDP/P) \times (E/GDP) \times (CO_2/E) \tag{1-23}$$

Multiplying the RHS terms of Equation (1-23) equates S_I with total CO_2 emissions.[9]

Equation (1-23) can be modified in several ways (e.g., to allow for human actions that mitigate undesired impacts of energy). For example, CO_2 emissions can be combated by intentionally removing CO_2 from the atmosphere (e.g., by planting more trees; Chapter 10) or by capture and sequestration of anthropogenic CO_2 emissions (Chapter 7). Then, Equation (1-23) would become:

$$S_I = (P) \times (GDP/P) \times (E/GDP) \times (CO_2/E) - (CO_2)_{sq} \tag{1-24}$$

where $(CO_2)_{sq}$ is the amount of CO_2 sequestered. Another CO_2 mitigation approach would be to use some form of solar energy (e.g., tides, wind, photovoltaic electricity) to displace energy that was previously supplied from fossil fuels. This would reduce the

9 When used to quantify release rates of CO_2 to the atmosphere, Equation (1-23) is commonly referred to as the Kaya Equation, in recognition of Professor Yoichi Kaya.

carbon intensity (i.e., the last term on the right hand side of Equation (1-23)). Equation (1-23) could be further modified to account for the association of energy with a diverse array of impacts known or suspected to be harmful (e.g., air pollution, water consumption, degradation of wildlife habitat, global climate modification) and for the distinct probability that different stakeholders will weight the sustainability importance of each impact differently. Then, for a set of independent impacts, Equation (1-23) would become:

$$S_I = [(P) \times (GDP/P) \times (E/GDP)] \sum_{i=1}^{n} W_i(t) \left[A_i(E)/E\right] \tag{1-25}$$

where A_i (E) is the i_{th} particular impact related to energy, $W_i(t)$ is the weighting factor assigned to the i_{th} impact at a particular time, t, and n is the total number of impacts to be considered. The limitations of Equations (1-22) through (1-25) are that they assume the various terms are independent of one another, which is often incorrect. For example, living standards can cause birth rates to decline, in some cases below the 2.2 children per couple needed to sustain zero population growth. Similarly, advances in living standard may give rise to great ability to purchase or develop more efficient and less polluting energy generation/utilization technologies, or substitutes like effective transportation infrastructure, that will decrease the adverse impacts of energy. We will revisit these factors when we examine sustainable energy from a systems perspective in Chapters 6 and 21.

1.8 The Rest of This Book

This book was written for diverse audiences. It is a textbook for graduate or senior undergraduate courses in engineering, public policy, or environmental science, that are concerned with energy *and* its technological, socioeconomic, and geopolitical ramifications. Thus, the sections and chapters are crafted to provide an orderly study of sustainable energy. However, this book is also a sustainable energy reference work, designed to serve the needs of inquisitive non-specialists for a useful introduction, as well as to augment the understanding of serious cognoscenti.

To serve these different audiences, this book looks at the topic of sustainable energy in two ways. First, we consider sustainable energy as a complex system that, in the broader context, is subservient to the "supersystem" of sustainable development. Then, we candidly discuss the prospects for sustainable energy to have practical global impacts in the 21st century.

References

Allenby, B.R. 1999. *Industrial Ecology: Policy Framework and Implementation*. Upper Saddle River, NJ: Prentice Hall.

Ausubel, J.H. and H.D. Langford, eds. 1997. *Technological Trajectories and the Human Environment*. Washington, DC: National Academy Press.

British Petroleum. 2001. *Statistical Review of World Energy 2000*. London, See: www.bp.com/centres/energy/index.asp

British Petroleum. 2003. *Statistical Review of World Energy 2002*. London, See: www.bp.com/centres/energy/index.asp

Bruntland, H.G. 1987. *Our Common Future*, WCED. New York: Oxford University Press.

Carnot, S. 1824. *Reflections on the Motive Power of Fire*. Reprinted 1992. Magnolia, MA: Peter Smith Publishers.

Denbigh, K.G. 1981. *The Principles of Chemical Equilibrium*. New York: Cambridge University Press.

Einstein, A. 1905. Special Theory of Relativity. German Yearbook of Physics.

Energy Information Agency, US Dept. of Energy, 2003. Annual Energy Review 2002. Washington, DC. See: www.cia.doe.gov/emeu/aer/contents.html

Graedel, T.E. and B.R. Allenby. 1995. *Industrial Ecology*. Englewood Cliffs, NJ: Prentice Hall.

Energy Information Agency, US Dept. of Energy, 1998. *Annual Energy Review 1998*. Washington, DC. See: www.eia.doe.gov/emeu/aer/contents.html

Energy Information Agency, US Dept. of Energy, 2002. *Annual Energy Review 2001*. Washington, DC. See: www.eia.doe.gov/emeu/aer/contents.html

Feynman, R.P. et al. 1963. *The Feynman Lectures on Physics*. Boston: Addison-Wesley Publishing Company.

Institute of Memetic Research. See www.futuresedge.org/world_population_issues/

Jeffreys, H. 1961. *Cartesian Tensors*. New York: Cambridge University Press.

Jeffries, H.E. and R. Crouse. 1990. "Scientific and technical issues related to the application of incremental reactivity." Department of Environmental Sciences and Engineering, University of North Carolina, Chapel Hill, NC.

Kammen, D.M. and D.M. Hassenzahl. 1999. *Should We Risk It? Exploring Environmental, Health, and Technological Problem Solving*. Princeton, NJ: Princeton University Press.

Levenspiel, O. 1996. *Understanding Engineering Thermo*. Upper Saddle River, NJ: Prentice-Hall.

Margenau, H., and G. M. Murphy. 1956. *The Mathematics of Physics and Chemistry*. Second Edition, New York:Van Nostrand.

Moore, J.F. 1999. *Report of the Visiting Committee of the MIT Energy Laboratory*. (by J. Moore, chair). Cambridge, MA: MIT Energy Laboratory

National Research Council. 1972. *Biological Effects of Atmospheric Pollutants*. Washington, D.C.: National Academy of Sciences.

New York Times. 2000. *World Almanac 2000*. New York.

Pitzer, K.S. 1923. *Thermodynamics: McGraw Hill Series in Advanced Chemistry*. New York: McGraw Hill.

President's Committee of Advisors on Science and Technology. 1997. *Federal Energy Research and Development for the Challenges of the Twenty-First Century*. Washington, DC.

Schmidheiny, S. and F.J.L. Zorraquin. 1996. *Financing Change: The Financial Community, Eco-Efficiency, and Sustainable Development*. Cambridge, MA: MIT Press.

Schulze, P.C., ed. 1996. *Engineering Within Ecological Constraints*. Washington, DC: National Academy Press.

Seinfeld, J.H. and S.N. Pandis. 1998. *Atmospheric Chemistry and Physics*. New York: John Wiley and Sons.

Smith, J.M. and H.C. Van Ness. 1949. *Introduction to Chemical Engineering Thermodynamics*, Third Edition. New York: McGraw-Hill.

Socolow, R., C. Andrews, F. Berkhout, and V. Thomas. 1999. *Industrial Ecology and Global Change*. Cambridge, Great Britain: Cambridge University Press.

Tester, J.W. and M. Modell. 1997. *Thermodynamics and Its Applications*. Upper Saddle River, NJ: Prentice Hall.

Weber, H.C. and H.P. Meissner. 1939. *Thermodynamics for Chemical Engineers*. New York: John Wiley & Sons.

Webster, Noah. 1986. New Universal Unabridged Dictionary. New York: Simon and Schuster.

World Bank. 2003. See World Development Indicators database at www.worldbank.org/data/

World Bank. 1992. See on-line databases at: www.worldbank.org/data/databytopic/environment.html

World Energy Council and International Institute for Applied Systems Analysis. 1995. *Global Energy Perspectives to 2050 and Beyond*. London: WEC.

Estimation and Evaluation of Energy Resources 2

As to petroleum: "The dominant view of a fixed mineral stock implies that a unit produced today means one less in the future. As mankind approaches the limit, it must exert ever more effort per unit recovered. This concept is false, whether stated as common sense or as elegant theory" (Adelman, 1997).

2.1 Units of Measurement: Energy and Power

Over time, personal preferences and application sectors such as science, engineering, wholesaling, retailing, international trade, and public policy have led to the development of many approaches for measuring units of energy. Table 2.1 and the Appendix display many of these common energy units and their mathematical conversion factors. Common scientific and engineering units, e.g., joules (J), calories (cal), and British thermal units (Btu), have precise definitions. For example, 1.0 Btu is the amount of energy required to raise the temperature of 1.0 pound of water from 39.1 to 40.1°F when the water is at normal atmospheric pressure. 1.0 J is defined as the work done when a force of 1.0 newton moves an object 1.0 meter in the direction of the force; a newton is the force that accelerates a mass of 1.0 kg at 1.0 m/s^2 in the direction of the force.

Table 2.1 also presents units that are widely used in energy trade and commerce. Examples are the mass or volume of a particular fuel—barrels (bbls) of oil, thousands of (standard) cubic feet (MCF) of natural gas, and tonnes[1] or tons of coal. To convert these units to J or Btu, we multiply the stated volume or mass of fuel by a representative *heat content* per unit volume or unit mass of that particular fuel. The chemical composition of fossil mineral fuels (oil, natural gas, coal) and of biomass, especially municipal solid waste, is complex and variable with fuel source and extent of purification prior to utilization. Thus, the heat contents per unit mass or volume of these fuels also vary. For example, the heat content of 1.0 cubic foot of natural gas can range from 950 to 1,200 Btu.[2]

Nevertheless, for many commercial and engineering calculations where a preliminary but quantitative estimate of energy resources is needed, sufficient accuracy is obtained by using an average heat content for the fuel (1,000 Btu per SCF of gas, 5.7×10^6 Btu per bbl of oil). We sometimes call these estimates *orientation* or *scoping* calculations. An example would be to estimate how much energy is contained in a newly discovered petroleum deposit stated to contain 1 billion barrels of recoverable oil (see Section 2.4.4). Here it is accurate to use the average energy content of 1 barrel of oil, which leads us to infer that this oil discovery represents about 5.5 quads [i.e., $(1 \times 10^9$ bbl$) \times (5.5 \times 10^6$ Btu/bbl$) \times (1 \times 10^{-15}$ quads/Btu)]. But we must be cautious. The amount of useful energy

1 Tonnes are metric tons and equal 1,000 kg, i.e., 2,205 lbs; 1 ton = 2,000 lbs.
2 This is the *higher heating value* of the fuel (see Chapter 7).

obtainable from this quantity of fuel as productive work, even assuming the entire billion barrels is produced, will depend on how the oil is processed and utilized, and will always be less than the total energy content of the fuel (Chapter 3).

Despite the diverse nomenclature, magnitudes, and measurement systems, it is important to be familiar with commonly used energy units and their quantitative equivalents. Table 2.1 provides several useful conversion factors; the row for joules shows that 1.0 J = 9.48×10^{-4} Btu and 0.239 cal. In the long run, time will be saved by memorizing common energy equivalents, including the number of Btus and joules in a quad, a gallon of gasoline, a barrel of oil, a pound of coal, and 1,000 standard cubic feet of natural gas. Remember to distinguish energy content from energy production rate or power (Section 1.2). Two common units of power are kilowatts (kW) and horsepower (hp), which respectively equal 1,000 and 745.7 J/s, i.e., 3,413 and 2,545 Btu/hr (Table 2.1) (i.e., 1.0 watt = 1.0 J/s).

Table 2.1. Various Energy Units and Conversion Factors

	Btus	quads	calories	kWh	MWy
Btus	1	10^{-15}	252	2.93×10^{-4}	3.35×10^{-11}
quads	10^{15}	1	2.52×10^{17}	2.93×10^{11}	3.35×10^{4}
calories	3.97×10^{-3}	3.97×10^{-18}	1	1.16×10^{-6}	1.33×10^{-13}
kWh	3,413	3.41×10^{-12}	8.60×10^{5}	1	1.14×10^{-7}
MWy	2.99×10^{10}	2.99×10^{-5}	7.53×10^{12}	8.76×10^{6}	1
bbls oil	5.50×10^{6}	5.50×10^{-9}	1.38×10^{9}	1,612	1.84×10^{-4}
tonnes oil	4.04×10^{7}	4.04×10^{-8}	1.02×10^{10}	1.18×10^{4}	1.35×10^{-3}
kg coal	2.78×10^{4}	2.78×10^{-11}	7×10^{6}	8.14	9.29×10^{-7}
tonnes coal	2.78×10^{7}	2.78×10^{-8}	7×10^{9}	8,139	9.29×10^{-4}
MCF gas	10^{6}	10^{-9}	2.52×10^{8}	293	3.35×10^{-5}
joules	9.48×10^{-4}	9.48×10^{-19}	0.239	2.78×10^{-7}	3.17×10^{-14}
EJ	9.48×10^{14}	0.948	2.39×10^{17}	2.78×10^{11}	3.17×10^{4}

To convert from the first column units to other units, multiply by the factors shown in the appropriate row (e.g., 1 Btu = 252 calories).

Key: MWy—Megawatt-years; bbls—barrels; tonnes—metric tons = 1,000 kg = 2,204.6 lbs.; MCF—thousand cubic feet; EJ—exajoules.

Assumed calorific values: oil—10,180 cal/g; coal—7,000 cal/g; gas—1,000 Btu/ft^3 at standard conditions.

Table 2.1. Various Energy Units and Conversion Factors (continued)

	bbls oil equivalent	tonnes oil equivalent	kg coal equivalent	tonnes coal equivalent	MCF gas equivalent	joules	EJ
Btus	1.82×10^{-7}	2.48×10^{-8}	3.6×10^{-5}	3.6×10^{-8}	10^{-5}	1,055	1.06×10^{-15}
quads	1.82×10^{8}	2.48×10^{7}	3.6×10^{10}	3.6×10^{7}	10^{9}	1.06×10^{18}	1.06
calories	7.21×10^{-10}	9.82×10^{-11}	1.43×10^{-7}	1.43×10^{-10}	3.97×10^{-9}	4.19	4.19×10^{-18}
kWh	6.20×10^{-4}	8.45×10^{-5}	0.123	1.23×10^{-4}	3.41×10^{-3}	3.6×10^{6}	3.6×10^{-12}
MWy	3,435	740	1.08×10^{6}	1076	2.99×10^{4}	3.15×10^{13}	3.15×10^{-3}
bbls oil	1	0.136	198	0.198	5.50	5.80×10^{9}	5.80×10^{-9}
tonnes oil	7.35	1	1,455	1.45	40.4	4.26×10^{10}	4.26×10^{-8}
kg coal	5.05×10^{-3}	6.88×10^{-4}	1	0.001	0.0278	2.93×10^{7}	2.93×10^{-11}
tonnes coal	1	0.688	1,000	1	27.8	2.93×10^{10}	2.93×10^{-8}
MCF gas	0.182	0.0248	36	0.036	1	1.06×10^{9}	1.06×10^{-9}
joules	1.72×10^{-10}	9.48×10^{-19}	3.41×10^{-8}	3.41×10^{-11}	9.48×10^{-10}	1	10^{-18}
EJ	1.72×10^{8}	2.35×10^{7}	3.41×10^{10}	3.41×10^{7}	9.48×10^{8}	10^{18}	1

To convert from the first column units to other units, multiply by the factors shown in the appropriate row (e.g., 1 Btu = 252 calories).

Key: MWy—Megawatt-years; bbls—barrels; tonnes—metric tons = 1,000 kg = 2,204.6 lb; MCF—thousand cubic feet; EJ—exajoules.

Assumed calorific values; oil—10,180 cal/g; coal—7,000 cal/g; gas—1,000 Btu/ft^3 at standard conditions.

2.2 Comparison of Different Forms of Energy

Informative comparisons of different *forms* of energy (Figure 1.2) are more challenging than converting energy units. From a sustainability perspective, the quality of various types of energy is determined by scientific, technological, economic, and environmental factors, as well as by inputs such as user preferences and behavior. Sustainability views energy as a means (sometimes, but not always, the only practical means) to accomplishing a purpose that benefits humans and the earth's ecosystem. Thus, compatibility with prescribed tasks suggests one framework for comparing and even prioritizing different categories of energy. Depending on the desired energy service, where that service is needed, and the budget of the service providers and users, one or another form of energy may be preferable. In remote, off-grid locations in less well-developed regions, low-tech energy sources for heating, cooking, and electricity may better accommodate user needs than high-tech electricity sources, such as diesel generators fueled by imported and costly premium fuels, or expensive solar voltaic systems that require energy storage equipment. Examples of low-tech systems are simply designed compact stoves suited to a range of indigenous forms of energy like dung, and anaerobic digesters that produce *biogas* (a roughly 50/50 by volume mixture of methane and CO_2 with a volumetric heating value about half that of natural gas (see

Chapter 10), while recycling animal waste for fertilizer (Larson, 1993)). Ecologically sensitive regions, even those rich in indigenous petroleum, may benefit from substantive diversification of their energy sources. For example, development of indigenous solar, wind, geothermal, and natural gas resources could allow remote communities in Alaska to reduce their dependence on liquid petroleum fuels that are expensive to transport. Useful mechanical energy can in turn be obtained in a variety of ways (Figure 1.2), such as from a change in potential energy of water in a hydro plant or by using fossil, geothermal, nuclear, or solar heat to power a heat engine (Section 3.2).

Primary sources of energy can be converted to electricity which serves as an energy carrier. Electricity is an especially versatile form of energy in that it can directly provide heat by resistive (joule) heating, heating and cooling by powering a heat pump, or mechanical work (with an electric motor). Electricity can also drive chemical reactions (via electrochemical cells and plasma processing (see Chapters 17, 19)). On the other hand, electrical energy does not occur naturally in forms suited for practical applications.[3] It must be generated from other energy forms (Figure 1.2), and for many applications, it must be transmitted and/or stored before use. All three operations are expensive and inefficient (Chapter 17). Two forms of energy that can be converted to electricity are chemical energy via fuel cells (Chapter 7) or electrochemical batteries (Chapters 16, 17), and mechanical energy via rotation of electrical generators (Chapters 7, 17).

Hydrogen is also an energy carrier which can be made from primary energy or from electricity. Because hydrogen can be stored more efficiently than electrons in a battery with present technology, it is being considered as a future energy carrier. However, much more R&D will be required before its practical use can be justified.

Forms of energy can also be compared in terms of their technical suitability for a particular task, or the efficiency at which a given energy type can be converted to another form better matched to the job. The classic example is the conversion of heat to useful mechanical energy (Chapter 3). All energy conversion steps have inefficiencies that squander some or a great deal of the inputted energy (Chapter 3). Table 2.2 shows typical conversion efficiencies for various energy transformations of practical interest. Energy is wasted by imperfections in equipment, materials, and fuels, but ultimately energy is wasted because of inherent limitations in the conversion of heat to work. (We analyze these limitations quantitatively in Chapter 3.)

---

3 It is estimated that each day 40,000 thunderstorms worldwide result in 100 lightning strikes per second that deliver roughly 700 megawatts (MWe) of electric power to the earth's surface (Feynman et al., 1964). This is roughly the rate at which electricity is consumed by 3/4 million "typical" US residential customers. Despite Benjamin Franklin's kite experiments and much research since, no one has figured out how to harness lightning to supply energy to consumers. Even if natural electricity could be collected, its wide dispersal would make it commercially impractical. Indeed, lightning strikes are often the nemesis of electricity generation and transmission hardware (Chapters 16, 17).

Other technical issues include whether the available form of energy will flow to the location where it is needed, and if so, whether it will do so rapidly enough to be practical. A single 1.5 volt battery will not power a 3.0 volt camera flash, whereas two such batteries connected in series will. A fuel that burns in air with a maximum (*adiabatic*, Chapter 7) flame temperature of 1,400°C will not release its energy at sufficiently high temperature to heat a cauldron of molten glass at 1,600°C. It would not matter how much of this fuel we had; it still could not meet our performance requirements and would be an unsuitable energy form for this task. It could, however, perform splendidly in a job requiring heat at 1,000°C, because it would provide a substantial temperature difference (*driving force,* Sections 3.3 and 3.5) of 400°C for energy transfer.[4] A branch of thermodynamics known as "exergy" or "availability" analysis provides techniques for quantitative analysis of the compatibility of energy generated at a given temperature with particular performance needs (see Chapter 3; Levenspiel, 1996; Tester and Modell, 1997; and Kenney, 1984). Quantitative study of how rapidly or lethargically energy flows between venues is the province of heat transfer (Section 3.5).

2.3 The Energy Lifecycle

Several steps in series are the norm in transforming a natural source of energy, such as mineral fuel, uranium, geothermal heat, or solar radiation, to a useful end product or service such as mechanical work or heat at a usable temperature. This sequence can be thought of as *the energy lifecycle*, the steps (Figure 1.2) of which generally include:

- exploration for and discovery of the primary source
- production or harvesting the energy
- preparation, transport, and/or storage
- further processing, purification, and conversion
- utilization
- recovery, destruction or decontamination, or storage of by-products and/or wastes.

The overall efficiency of providing the desired energy-intensive product or service is the product of the efficiency of each sequential stage.

4 Hottel (1983) describes a quantitative procedure for comparing the relative ability of different fuels, including synthetic gaseous and liquid fuels from coal, to supply heat for practical purposes. His method defines and computes the *relative thermal value* (RTV) of the fuel, which he defines as the value for use in furnaces, of a Btu in that fuel compared to the value of a Btu in a standard fuel, such as gaseous methane.

Table 2.2. Efficiencies of Common Anthropogenic and Natural Energy Conversions[a,c]

Conversion	Energies[b]	Efficiencies
Large electricity generators	$m \rightarrow e$	98 – 99
Large power plant boilers	$c \rightarrow t$	90 – 98
Large electric motors	$e \rightarrow m$	90 – 97
Best home natural gas furnaces	$c \rightarrow t$	90 – 96
Dry-cell batteries	$c \rightarrow e$	85 – 95
Human lactation	$c \rightarrow c$	75 – 85
Overshot waterwheels	$m \rightarrow m$	60 – 85
Small electric motors	$e \rightarrow m$	60 – 75
Best bacterial growth	$c \rightarrow c$	50 – 65
Glycolysis maxima	$c \rightarrow c$	50 – 60
Large steam turbines	$t \rightarrow m$	40 – 45
Improved wood stoves	$c \rightarrow t$	25 – 45
Large gas turbines	$c \rightarrow m$	35 – 40
Diesel engines	$c \rightarrow m$	30 – 35
Mammalian postnatal growth	$c \rightarrow c$	30 – 35
Best photovoltaic cells	$r \rightarrow e$	20 – 30
Best large steam engines	$c \rightarrow m$	20 – 25
Internal combustion engines	$c \rightarrow m$	15 – 25
High-pressure sodium lamps	$e \rightarrow r$	15 – 20
Mammalian muscles	$c \rightarrow m$	15 – 20
Milk production	$c \rightarrow c$	15 – 20
Pregnancy	$c \rightarrow c$	10 – 20
Broiler production	$c \rightarrow c$	10 – 15
Traditional stoves	$c \rightarrow t$	10 – 15
Fluorescent lights	$e \rightarrow r$	10 – 12
Beef production	$c \rightarrow c$	5 – 10
Steam locomotives	$c \rightarrow m$	3 – 6
Peak field photosynthesis	$r \rightarrow c$	4 – 5
Incandescent light bulbs	$e \rightarrow r$	2 – 5
Paraffin candles	$c \rightarrow r$	1 – 2
Most productive ecosystems	$r \rightarrow c$	1 – 2
Global photosynthetic mean	$r \rightarrow c$	0.3

[a]All ranges are the first-law efficiencies in percentages.
[b]Energy labels: c = chemical, e = electrical, m = mechanical (kinetic), r = radiant (electromagnetic, solar), t = thermal
[c]Adapted from Smil (1998).

All real-world processes are imperfect (Section 3.2) and waste some energy (Table 2.2), and the overall energy utilization efficiency can be quite low. Figure 2.1 illustrates one of the most common energy lifecycles—transformation of the chemical energy of a mineral fuel to visible light using an incandescent electric bulb. There are different ideas on where to draw the boundaries for lifecycle analysis (Allenby, 1999). One approach includes activities well upstream and downstream of the primary energy generation source. In the light bulb example, the analysis would account for the energy and environmental impacts of manufacturing and ultimately disposing of the (spent) light bulb, of building the electric power station, of mining and delivering the coal, and of disposing of the ash and other wastes from the generation plant. A lifecycle view resonates with our core proposition that energy provides useful products or services, but typically does so with environmental penalties that may or may not be known contemporaneously.

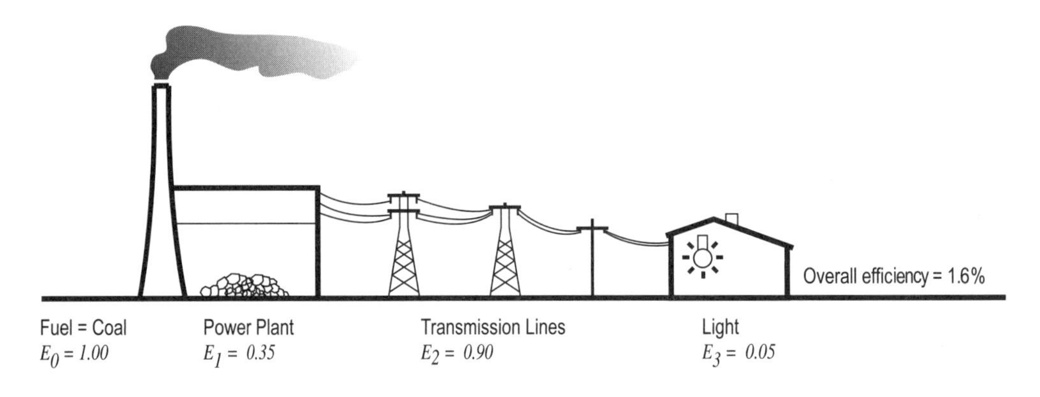

Figure 2.1. Multiple steps and associated efficiencies in converting the chemical energy of a fuel to energy as visible light for illumination.

To identify the substantive tradeoffs among sustainable energy options, we need an understanding of current and prospective consequences for the earth and its inhabitants of each option under consideration. However, the larger the boundaries of the lifecycle, the more difficult it becomes to define precise information requirements and to obtain putatively responsive data. Thus, flexibility should guide the selection of spatial and temporal boundaries for lifecycle analyses aimed at informing sustainability decision-making.

Rational compartmentalization of energy systems into manageable size lifecycles allows for informative "cradle-to-grave" analyses without the need to wait on further advances in information technology or additional data. Thus, for the light bulb of Figure

2.1, the lifecycles of the power plant itself, including its equipment, and of the end use device (the light bulb), could be disaggregated from the lifecycle of the fuel and the emissions and other effluents from its combustion and ash disposal.

We expect that spatially and temporally larger lifecycles will be susceptible to analysis as energy and environmental databases improve, as more is learned about shorter and longer term ecological impacts of various energy technologies and policies, and as more is learned about time scales for implementing alternative means of supplying energy services. Lifecycle thinking provides an operational framework for codifying the forms and impacts of energy at various stages of its discovery, production, transmission, and utilization. This framework also serves as an accounting sheet for valuing the tradeoffs between benefits and adverse consequences, and as a flowsheet to elucidate synergisms within the energy-environmental system (Chapters 6 and 21).

A lifecycle analysis for each energy service analyzed will start with identification of the primary energy source and progress through the appropriate steps for energy conversion and transportation up to the final device(s) providing the energy service. However, a broader lifecycle issue is whether the global and local cumulative use of the primary source is depleting the resource base in a way that will create unwanted future shortages. We examine the primary commercial fuels in use today in the order of their global economic importance.

Petroleum. Petroleum is the world's most important fuel. Used everywhere, it provides the greatest flexibility in utilization and ease of transportation. Petroleum can be typically produced from conventional resources with relatively small environmental effects, and it is available in a large international market. The overall market is somewhat free, constrained in part by the production component, by the OPEC cartel, and by different levels of taxation. These non-idealities have successfully raised prices, particularly during the "oil shocks" of 1973 and 1980. Vastly different and often times substantial levels of taxation exist among different countries.

The transportation sector of every country relies on petroleum to a major extent. Additionally, every consuming sector of any national economy relies substantially on oil, particularly poorer countries. Petroleum is of intermediate carbon intensity and environmental effect in its production.

Table 2.3 shows national proven petroleum reserve estimates for major world regions (see Sections 2.4.2 and 2.4.4 for further discussion). Many oil-producing nations are poor countries that sell to rich countries, as illustrated in Figure 1.3. The typical wealthy country is poor in terms of fossil fuel resources, including petroleum. Major exceptions include the US and Canada. However, much more of the conventional petroleum and gas produced from on-shore, lower-48 resources in the US come from stocks comparable in quality to those of major export nations.

The ratio of estimated proven reserves divided by production rates for different regions of the world for the various fossil fuels is also shown in Table 2.3. For petroleum and natural gas, values of the order of decades are typical. In the Middle East, however, values of the order of a century are seen. However, these ratios must be viewed as highly tentative and conservative because advances in technology continue to open up new stocks for development every year (Table 2.4).

Natural gas. The resource and depletion situation of natural gas is similar to that of petroleum (see Table 2.3). Natural gas is abundant, but there is great difficulty in moving it from remote sites (which often hold huge resources) to consumers. Natural gas is transported as a gas in pipelines or as a cryogenic liquid in ships, trucks, and trains. Due to the need to cool natural gas in order to liquefy it, the cost of liquefied natural gas (LNG) is typically an order of magnitude greater than that of the gas itself at the remote wellhead.

Because it has the lowest carbon intensity and produces the least environmental effects in production and consumption, natural gas is the preferred fossil fuel for stationary combustors. It is used primarily in the stationary consuming sectors of the wealthy countries. Its transportation use has been confined to politically popular niches, such as bus fleet propulsion in wealthy countries. The capital costs of pipeline and LNG facilities are sufficient barriers to deny it to most poor countries and deter its export from most poor producing countries (where it is often found in conjunction with petroleum). In such situations, natural gas is often either burned (flared) as a waste or reinjected into the ground. The international flows of natural gas are shown in Figure 1.4.

Table 2.3. Proven Reserves and Ratio of Proven Reserves to Production for Major World Regions (BP, 2003)

	Petroleum		Natural Gas		Coal	
Region	**2002 Proved Reserves (10^9 bbls)**	**R/P (years)**	**2002 Proved Reserves (10^{12} SCF)**	**R/P (years)**	**2002 Proved Reserves (10^9 tonnes)**	**R/P (years)**
North America	49.9	10.3	252.4	9.4	257.8	240
S. and Cent. America	98.6	42.0	250.2	68.8	21.8	404
Europe and Eurasia	97.5	17.0	2155.8	58.9	355.4	306
Middle East	685.6	92.0	1979.7	>100	1.7	>500
Africa	77.4	27.3	418.1	88.9	55.3	247
Asia Pacific	38.7	13.7	445.3	41.8	292.5	126
World	**1047.7**	**40.6**	**5501.5**	**60.7**	**984.5**	**204**

Table 2.4. World Production and Additions to Reserves of Petroleum, 1944–1993[a]

	1944	1945–1960	1961–1970	1971–1980	1981–1993	(Total) 1944–1993
OPEC						
– Cumulative production	–	26	55	103	100	284
– Gross reserve additions	–	219	251	128	434	1,032
– Reserves at end	22	215	412	436	770	770
Non-OPEC						
– Cumulative production	–	51	64	102	190	407
– Gross reserve additions	–	98	187	114	207	607
– Reserves at end	29	76	200	212	229	229
Total World						
– Cumulative production	–	77	119	205	289	690
– Gross reserve additions	–	318	439	242	640	1,639
– Reserves at end	51	291	611	648	999	999

[a]Data are given in millions of barrels
Source: Adelman (1997).

In the US, natural gas use has grown during recent decades. This is driven by its low cost, low pollution, and convenience. Today, natural gas–fueled combined-cycle gas turbine units are the technology of choice for new electric power generation in the US and elsewhere. As attractive on-shore deposits of conventional gas within the lower 48 states become used, demand will shift toward Canadian and Alaskan supplies, for which new pipelines will be needed. Growth of imported LNG supplies can also be anticipated, as well as calls in some quarters for more off-shore exploration around the lower 48.

For both petroleum and natural gas produced within the US, the ratio of assured resources to current demand has been equal to approximately 8-10 years for the past two decades. These data reflect increases in assured resource estimates roughly equal to annual production principally via expansion and reevaluation of the producible amounts of gas in these fields.

Coal. Of the established fossil fuels (petroleum, gas, coal), coal is the most carbon intensive and environmentally challenging to produce and use. Coal is used primarily for electricity production within industrialized countries, where the economies of centralized large-scale consumption permit it to be used at low cost and in compliance with regulations on air and other emissions. In some less wealthy countries, it also plays major roles in domestic and commercial heating (e.g., Central Europe), and in some poor countries (e.g., India and China), it continues to power heavy industry (as was common in the US a century ago). Without substantial emission control, coal-fired plants can cause substantial air pollution. The trend worldwide is toward less reliance upon coal in the industrial and residential/commercial sectors and more concentrated use to produce electricity.

Coal is important as a long-term fuel, as its ratio of resources to production or demand rates is substantially greater than with other fossil fuels (Table 2.3). This is important because as petroleum becomes more depleted, the attractiveness of coal (and also oil shale and tar sands, which are little exploited currently) as sources of syncrude and syngas feedstocks will likely grow. The estimated proven coal reserves of major regions of the world are shown in Table 2.3. Coal stocks are large both absolutely and relative to current demand, with identified supplies typically projected to be adequate for centuries, assuming reasonable growth in demand. Because of this expectation of long-term production, the state of knowledge of coal resources is less accurate than with petroleum and natural gas, as the incentives to refine current understanding are not as large.

Major coal-producing countries include the US, China, India, Russia, Australia, and several Central European countries. An international trade exists in coal, but its low price (relative to petroleum and natural gas) makes it more attractive to consume near the point of production, so that the costs of transportation do not render it less attractive to use. Important international suppliers tend to be nations with small economies and/or small heavy industry sectors (e.g., Australia and Poland).

In the US, coal consumption is almost entirely for electricity production. Because much of the coal used today is obtained from low-sulfur deposits in the west (e.g., Montana and Wyoming) but is consumed in the populous eastern states, coal transport by rail has become more important, with associated traffic disruption and accidents also growing.

Formerly, the important sources of US coal were the Appalachian states and Illinois, via underground and small-scale surface mining. As resistance to use of coal from these sources has grown (due to its higher sulfur content, associated surface ecosystem disruption and stream pollution, as well as uncontrolled underground fires and subsidence of abandoned mines), coal production has shifted to the west. In the latter sites, the coal seams are thicker and closer to the surface, justifying the use of larger scale mechanized mining. Federal law requires restoration of surface mined sites.

Uranium. The primary mineral used as a nuclear fission fuel is uranium, but thorium could also serve. Because of the slow growth of nuclear power worldwide, demand for a second fuel source has not yet arisen, but thorium is available as a potential complement to uranium. Uranium is found in many countries, but it is produced in only a few of these countries. This is the result of the low price of uranium for the past two decades, which has reduced investor incentives for investment in uranium mining and production. For example, during the early Cold War years, the US was the world's most important uranium producer. Today, most domestically consumed uranium is supplied from abroad even though abundant domestic deposits remain.

Identified worldwide, conventional uranium resources can meet current demand for at least a century and even more if estimated additional resources are also taken into account. These large numbers are vastly different from those obtained while nuclear power was growing rapidly during the 1970s, and when the scale of identified resources was smaller (serious exploration for uranium only started after World War II). At that time, the US Atomic Energy Commission estimated that the assured reserves of US uranium would be consumed by 1986. However, the reality has been very different with reduced growth and increased discoveries of deposits. This expectation was used as the basis of the US national breeder reactor program, which was stopped in 1980.

Important producer countries include Australia, Canada, Russia, Namibia, and Niger (Neff, 1984). Uranium is a low-cost commodity with abundant supplies. It is easily transported, reflecting its low level of penetrating radiation and ultra-high energy density. Because of these factors, the incentives of countries to develop technologies to permit more efficient use of the fuel are economically weak. However, they can pursue such policies as a means of reducing their vulnerability to interruptions of supplies (as can arise during wartime) and as a way of covertly developing a nuclear weapons program. Many countries will pursue the latter policies, but not based on economic arguments. This is important, because several countries (Iran, North Korea, South Africa) have offered fuel efficiency as their justification for constructing full closed-fuel cycles (see Chapter 8). This, but not the construction of nuclear power plants, is almost always a sure sign of the intention to develop a nuclear weapons capability.

The future prospect is one of abundant uranium, unless nuclear power should become used on a much larger scale than it is currently (e.g., in substitution for fossil fuel consumption). If this were to happen, the time scale for consumption of the existing identified resources would become decades rather than centuries. In that situation, exploration for additional resources and development of reprocessing technologies, which are less vulnerable to nuclear weapons proliferation than those in current use, would be expected.

Renewable energy technologies. Renewable energy technologies are typically taken as different versions of solar energy, as well as geothermal energy (Chapters 9, 10, and 13). Renewables are regenerated over short time scales compared to fossil and nuclear fuels (other than those from breeder reactors). Solar energy is viewed as "free," but the capital "equipment" (collectors, land, etc.) and operating costs (human labor, fertilizer, and water for biomass, etc.) are not.

Table 9.2 shows global resource bases for various forms of renewable energy. While the total amount of solar energy falling on the earth is large, the amount of solar energy reaching the earth's surface varies between 0 and 1,000 W/m^2 depending on latitude, diurnal and seasonal cycles, sun angle, cloud cover, and other factors.

Biomass is a widely used non-commercial source of energy and can match energy demands in regions of low population density. Biomass converts solar energy to a biofuel at an efficiency of up to 1-2%. When population energy needs exceed local biomass supplies, problems of desertification can occur (e.g., sub-Saharan Africa).

Hydropower is the largest of the renewable energy sources because solar evaporation of water ends up as rain over large watersheds where terrain may allow development of a dam and a reservoir. This also provides an intrinsic energy storage system which makes hydropower competitive with other commercial energy sources where it is available. It can also be used as an additional energy storage system since excess energy can be used to pump water into the reservoir for later reclamation of most of the energy.

Wind also can be considered a concentrator of solar energy, since heating and cooling of land and sea masses generate weather systems. Wind resources vary widely, but wind energy power generation with modern turbines can be cost competitive in high quality wind regions.

Solar thermal and solar photovoltaics are becoming more cost competitive, especially in regions with high insolation. But both solar and wind energy are variable and either require an energy storage device or a back-up alternate energy supply. Nevertheless, when they are generating energy they are reducing the total demand for fossil energy. At present, the solar-based renewables are a small part of the energy supply, but their importance is growing.

Finally, geothermal energy (generated by radioactive decay in the earth's interior) is used in several locations worldwide where hot steam is generated near the earth's surface. This energy is renewable as long as the drawdown does not exceed the regeneration. This and more futuristic ideas about heat mining deep in the earth are discussed in Chapter 11.

2.4 Estimation and Valuation of Fossil Mineral Fuels, Especially Petroleum

2.4.1 Asking the right questions and avoiding the unanswerable ones

As of 2000, fossil fuels (primarily petroleum, natural gas, and coal) supplied about 85% of the world's commercial energy and commanded daily trade flows of 2 billion US$. Fossil fuels will most likely dominate world energy supplies for at least the next three decades. By definition, fossil fuels are "exhaustible" in that they are replenished only over geological periods of time—a few million to many millions of years. These epochs are huge compared to the time scales for socio-economic activities that consume fossil fuels (Chapters 7, 9). Industry, academe, and governments in major energy-producing and energy-consuming nations have invested substantial time and resources to forecast when world supplies of fossil fuels, particularly oil and natural gas, will run out. In this section, we consider some of the geological, economic, and political factors that must be taken into account to make sense of surveys and forecasts of fossil mineral fuel resources.

In mineral fuel economics, seemingly logical reasoning can produce wrong conclusions. Simply stated, nobody knows when the US or any other part of the world will run out of oil or any other fossil fuel. This is partially because depletion, in the sense of consuming all of a resource that can be produced, does not happen. Rather, long before depletion would occur, the cost of extracting additional fuel from the earth will have become so high that investments to increase supply will be unprofitable and alternative forms of energy will be selected.

To estimate future energy supplies, it is essential to learn as much as possible about the cost of new stock (Adelman, 1997). Recognizing this crucial piece of understanding, Adelman and his co-workers (e.g., Adelman, 1999, 1998, 1997, 1995, 1993, and Lynch, 1998) have dramatically advanced our understanding of mineral fuel economics with particular concentration on the example of petroleum. New knowledge in this field has not come easily. Progress has depended upon tracking the behavior of mineral fuel prices and detecting and accounting for effects of markets that, at different times, have been impacted not only by supply and demand but by vendor collusion to sell less and charge more (Adelman, 1997). Data on world oil prices have only existed since 1947. Up to 1959, there was little public information on oil transactions between genuinely unattached sellers and buyers. Before 1947, "sales" of crude oil occurred only as transfers within integrated companies and their affiliates, and US oil prices became controlled by cartels in oil-producing countries. The US began importing large amounts of oil in 1948, but import volumes were restricted even after 1970. US price controls then further meddled with competitive market forces for roughly another decade. US and world oil prices did not actually converge until 1981.

2.4.2 Perspectives from mineral geology

In discussing fuel and other mineral resources, it is important to distinguish between:

- Resources - The amount of fuel that is estimated (at a stated confidence level) to be potentially recoverable in a geographic region under current economic and technological conditions; and
- Reserves - The amount of fuel that has been identified (at a much higher confidence level) in a geographical region by organizations in a position to produce it for the market under current economic and technological conditions.

Reserves are always much less than resources, and both are formulated with particular market conditions in mind. (See also the discussion in Section 2.4.4.) Geologists use scientific knowledge of the earth, its components, and how the earth changes over long time scales (tens of thousands to multiple millions of years) to guide the discovery of regions of the planet that may be developed for economically profitable recovery of mineral resources, including fuels. Geologists explore for zones of the earth with geological attributes that indicate the presence of minerals, fossil fuels, or metal precursors (ores), in sufficient quantity and quality to justify further investigation to assess economic potential. Geologists identify regions that contain mineral deposits in quantities large enough to be of scientific or practical interest. Geologists, however, do not find reserves. Reserves are made by investments in exploration and recovery (Section 2.4.3) (Adelman, 1997).

By identifying and sizing geographical areas showing greater probability of containing commercially interesting deposits of minerals, geologists provide information to help guide decisions on whether to invest in exploration and recovery operations to exploit these deposits so that some portion of them could become reserves (Weeks, 1969). If two regions of similar geology are 10-fold different in size, exploration and recovery of the larger region is 10 times more likely to result in a profitable return on investment. However, geological exploration alone cannot guarantee economically successful exploitation of mineral deposits, nor is it intended to.

2.4.3 Two interpretations of hydrocarbon fuel economics

There are two different perspectives on the consumption of fossil mineral fuels and its effects on fuel prices and fuel scarcity. One proposition holds that because fossil (and other mineral) fuels are "exhaustible," they exist in the earth in a *fixed stock*. Therefore, as this stock is used up, it will become harder to recover additional fuel from whatever stock remains in the ground. Therefore, the recovery cost and resultant price to users will increase. The underlying economic theory of these lines of reasoning was codified mathematically by Hotelling (1931). Since then, many geologists, politicians, energy developers, and economists have embraced this thinking. Specialists and non-specialists alike find these ideas logical, intuitive, and appealing.

In the last decade or so, several applications of the *fixed stock scarcity theory* have led to the conclusion that supplies of "low cost" oil have already peaked or soon will peak, and that, within a few decades or less, oil scarcity will give rise to major price escalations, political upheaval, and oil supply blackmail (e.g., Al-Jarri and Startzman, 1997, Calder, 1996, Campbell and Laherrère, 1998, Campbell, 1997a,b, MacKenzie, 1996, and Romm and Curtis, 1996).[5] MacKenzie (1996) and Campbell (1997a) project that scarcity will doom "cheap" oil and precipitate serious economic upheaval because of dramatically higher energy prices. Calder (1996) foresees oil insufficiency leading to political conflicts as energy-hungry nations, especially in the Asian Rim, scramble for diminishing supplies. Romm and Curtis (1996) presage a greater threat from resurgent power among the oil exporters. Each of these projections should be weighed carefully in composing technology and policy strategies for a more sustainable energy future.

An alternative line of thinking reveals two shortcomings in applying classical mineral depletion theory to petroleum. First, the tenet of a known fixed stock of mineral fuel is flawed. Paraphrasing Adelman (1995): humans will never know the earth's ultimate treasure chest of fossil fuels—well before this bounty is depleted, alternatives will have taken over, either because further fossil fuel production will become far too expensive or because a better and lower cost energy source will become feasible. Second, *when markets are not subverted,* the economical availability of fossil fuel reserves depends upon a perpetual tension between depletion and the extension of knowledge, including new technology. The classical view of depletion theory does not account for advances in technology that will liberate, for profitable exploitation, stocks of fuel that previously

5 We use quotes for "low cost" because this term must be carefully defined (Section 2.5).

were unknown or judged unattractive to develop. This reinforces the argument that ultimate reserves of fossil fuels in the earth and the time to run out of these reserves cannot be known. The amount of reserves appropriate for economical recovery depends upon future technology and no human knows what technology will be available in the years or indeed even the months ahead.[6]

The history of mineral resource knowledge to date favors the latter thinking, where, for the great majority of minerals, current prices are lower and estimated resources are greater than ever. This is due primarily to innovations in extraction technologies rather than in consumption of identified deposits. Whether this pattern will persist remains to be seen; however, it inspires caution in expecting supplies to run out soon.

Contrasting the two views of depletion, Adelman (1997) states: "The dominant view, of a fixed mineral stock, implies that a unit produced today means one less in the future. As mankind approaches the limit, it must exert ever more effort per unit recovered. This concept is false, whether stated as common sense or as elegant theory. Under competition, the price results from endless struggle between depletion and increasing knowledge. But sellers may try to control the market in order to offer less and charge more. The political results may feed back upon market behavior. These factors—depletion, knowledge, monopoly, and politics—must be analyzed separately before being put together to capture a slice of a changing history."

The far better fit of non-classical interpretation of mineral depletion economics to oil is supported by:

(1) a roughly 65-year historical record of growing, not declining, world petroleum reserves, achieved while consumption rates have steadily grown

(2) a long-term trend (1900–1996) of stable or declining petroleum and natural gas prices

Conventional depletion arguments express the temporal history of mineral fuel production rates in terms of so-called Hubbert curves named for M. King Hubbert of the US Geological Survey (Hubbert, 1956). Figures 2.2 and 2.3, respectively, provide Hubbert curves for US production of petroleum and natural gas together with actual data. Any suggestion that the appreciable excess of actual production rates over the predictions since 1960 is only temporary is undermined by several other observations. Figure 2.4 compares historical data with seven 1989–1991 forecasts of non-OPEC (Organization of Petroleum Exporting Countries) oil production. Trend lines drawn through these data exhibit Hubbert-like behavior especially after recognizing the relatively short time scales compared to Figures 2.2 and 2.3. However, Figure 2.4 also

6 For example, in the mineral fuel sector, advances in exploration and recovery techniques for remote and geographically difficult regions have emancipated huge new stocks of petroleum and natural gas.

shows that the actual production rates from 1994–1998 have dramatically exceeded even the most optimistic forecast. Moreover, the actual data display increasing yearly production rates, whereas at least four of the predictions show the opposite.

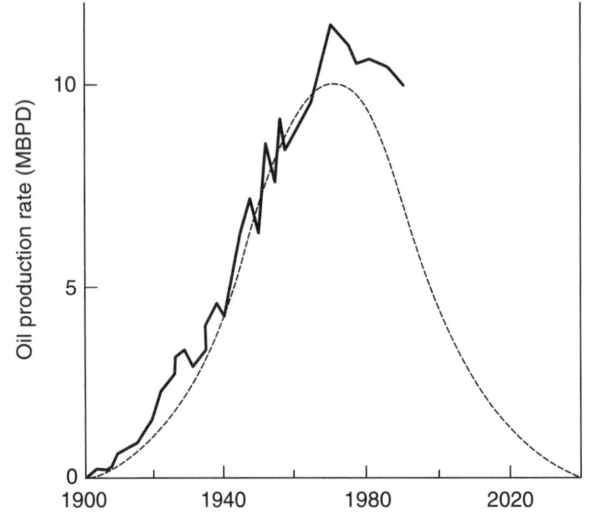

Figure 2.2. Comparison of an estimated Hubbert production curve for crude oil in the US with actual crude oil production data for the period 1900–1983.

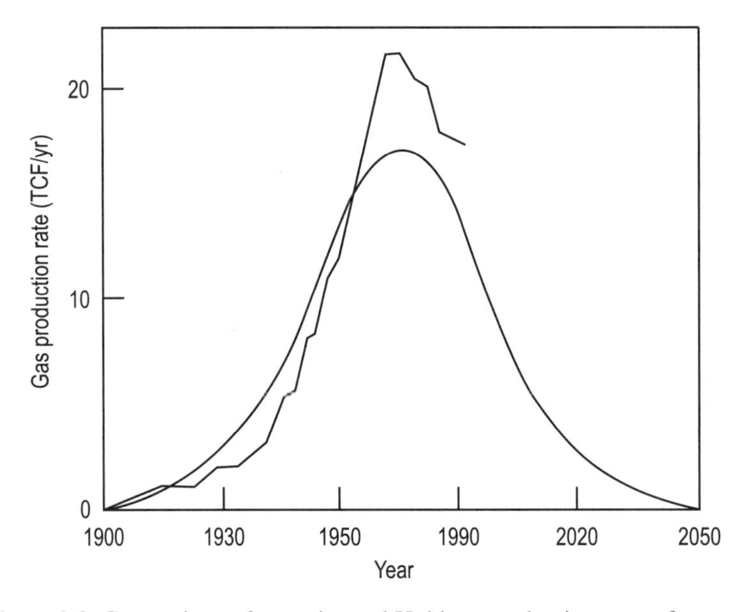

Figure 2.3. Comparison of an estimated Hubbert production curve for natural gas in the US with actual natural gas production data for the period 1900–1983.

Hubbert curves (Figures 2.2, 2.3) plot the *production rate*, $P(t)$ of petroleum or natural gas as affected by time, t. It is common for petroleum production rates to increase more rapidly before maximizing at $P(m)$ (at time m) than they decrease after this peak. Thus, Hubbert curves should not in general be symmetric about the peak production time (Ellerman, 2005). However for some calculations it is convenient to *approximate* Hubbert curves by a mathematical function with this symmetry, e.g., by a Gaussian probability density function (PDF) (Zelen and Severo, 1972), even though this function overestimates the rate of depletion at long times (Socolow, 2005):

$$\frac{P(t)}{P(m)} = e^{\left(\frac{-(t-m)^2}{2\sigma^2}\right)} \tag{2-1}$$

where 2σ is the width of the curve between its two points of inflection. The cumulative amount of the resource produced, P_{cum}, is the product of production rate in each time interval of production, summed over all production time intervals, i.e., P_{cum} is the time integral of the Hubbert curve:

$$P_{cum} = \int_{-\infty}^{+\infty} P(t)dt = P(m)\int_{-\infty}^{+\infty} e^{\left(\frac{-(i-m)^2}{2\sigma^2}\right)} dt = (2\pi)^{1/2}\,\sigma P(m) \tag{2-2}$$

where it is recognized that there is no relevant error in extending the lower limit of the integral from 0 to $-\infty$, and we have used the known definite integral:

$$\int_{-\infty}^{+\infty} e^{\frac{-(x-b)^2}{2q^2}} dx = (2\pi)^{1/2}\, q \tag{2-3}$$

where b is a constant, to evaluate the integral in Equation (2-2). $P(m)$ and σ can be evaluated by best fitting Equation (2-1) to resource depletion data. For a given Hubbert curve these two quantities are fixed and thus Equation (2-2) tells us that P_{cum} is also fixed. Selecting a specific Hubbert curve means signing on to advance knowledge of exactly how much of the mineral resource exists in the earth. History shows this knowledge is unavailable to humans. The only ways out of this dilemma are to chop off the Hubbert curve early, that is, while depletion is still on-going, or to make the Hubbert curve time-dependent. Either amendment overrides the fixed-stock hypothesis, but with that demise must also go the conclusion of an inevitable escalation in energy prices brought on by increasing scarcity. In the early termination fix, the forecaster can never be sure that the depletion curve won't change slope.

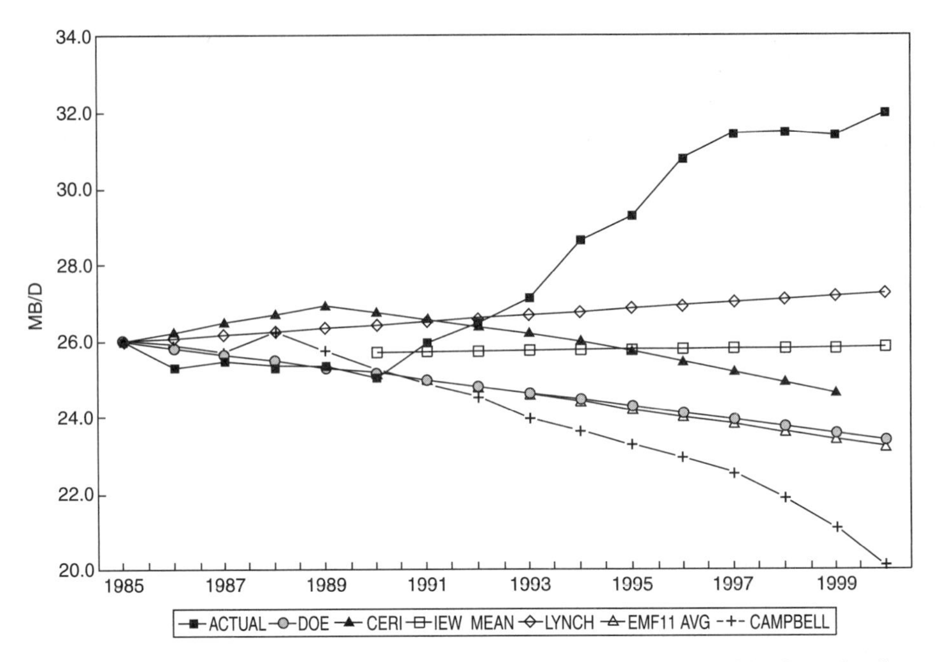

Figure 2.4. Comparison of various 1989-1991 forecasts of non-OPEC crude oil production, with actual production. Source: Lynch (1998).

Figure 2.5 illustrates this possibility by matching two different but equally plausible Hubbert curves to data on annual rates of US oil production for the period 1900–1990, adapted from Figure 2.2. The apparent leveling off of production rates could well be temporary. Then Curve II would represent the existing data quite well and do a much better job than Curve I in capturing the overall production rate history. Note that Curve II doubles the peak production rate and defers the onset of declining yearly production from m to $(m + 1.5\sigma)$, i.e., by 30 years since $\sigma = 20$ for Curve I. Using Equation (2-2) to calculate the cumulative production from the two Hubbert curves:

$$\text{Curve I} \quad P_{cum}(I) = (2\pi)^{1/2}\,\sigma\,P(m) \tag{2-4}$$

$$\text{Curve II} \quad P_{cum}(II) = (2\pi)^{1/2}\,(1.8\sigma)\,P(m') = 3.6(2\pi)^{1/2}\,\sigma\,P(m) \tag{2-5}$$

we see that Model II gives rise to 3.6 times as much oil.

Figure 2.4 confirms that real oil production rates can accelerate after a period of repose. This figure compares actual data for non-OPEC crude oil production with several predictions that exhibit Hubbert-like behavior. Lynch (1998) notes that some forecasters of pending oil "scarcity" in effect adopt the second fix-up. They employ what amounts to time-dependent (and expandable) Hubbert curves by regularly extending their predicted times for petroleum production to peak (i.e., they increase the quantity m in Equation (2-2)). A time-stationary Hubbert curve cannot account for undiscovered innovations in technology that will emancipate previously unknown or unexploitable reserves. Neither can it account for changes in energy policy that do the same thing, such as by allowing exploration and development of publicly owned or other lands where investment has been prohibited.

For example, an estimated 90% of Alaska's sedimentary basins have potential to provide commercially interesting quantities of natural gas and/or oil but have never seen a drill bit. Numerous offshore regions are embargoed even to exploration. Policy measures and enabling technologies that balance environmentally motivated restrictions on exploration and development for mineral fuels with the need for economical energy services would be useful contributions to sustainable energy thinking.

The fixed mineral fuel stock tenet is further challenged by the data of Table 2.4 showing that additions to world petroleum reserves from 1944–1993 have grown dramatically, far outpacing consumption despite prodigious global economic expansion during the same period. It might be suggested that Table 2.3 reflects fortuitous discoveries and that no major reserves remain to be exploited. However, this would be at odds with significant growth in oil production by non-traditional players. Figure 2.6 documents 50 years (1946–1996) of almost unimaginable growth in petroleum production by producers who are neither part of OPEC, the US, or the former Soviet Union. Today, these producers account for about 35% of all the world's oil production.

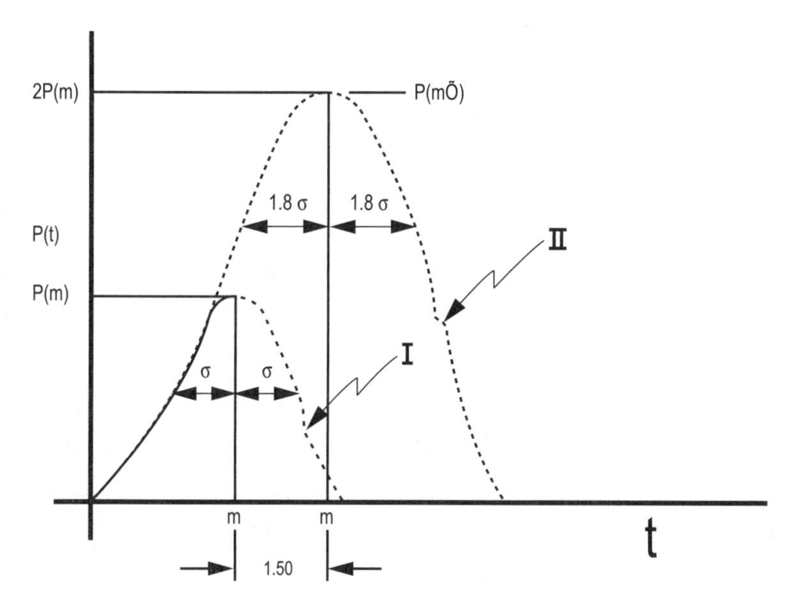

Figure 2.5. Two estimated crude oil production (Hubbert) curves (I and II) consistent with the actual crude oil production data of Figure 2.2 (shown here as the smooth curve). As discussed in the text, curves I and II imply the existence of a known, fixed stock of crude oil (see Equation (2-2)).

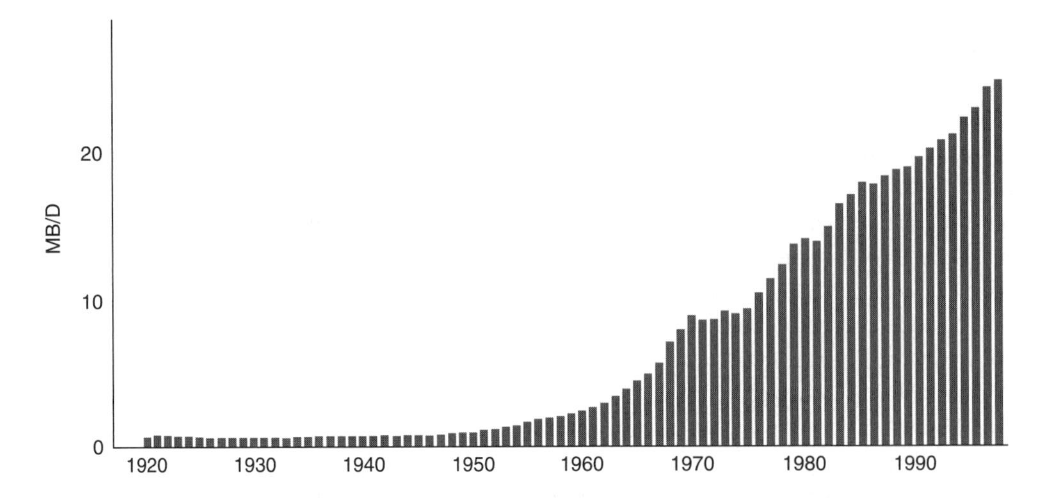

Figure 2.6. Production of crude oil for 1920–1997 by countries other than the US, OPEC members, or states of the former Soviet Union. Source: Lynch (1998).

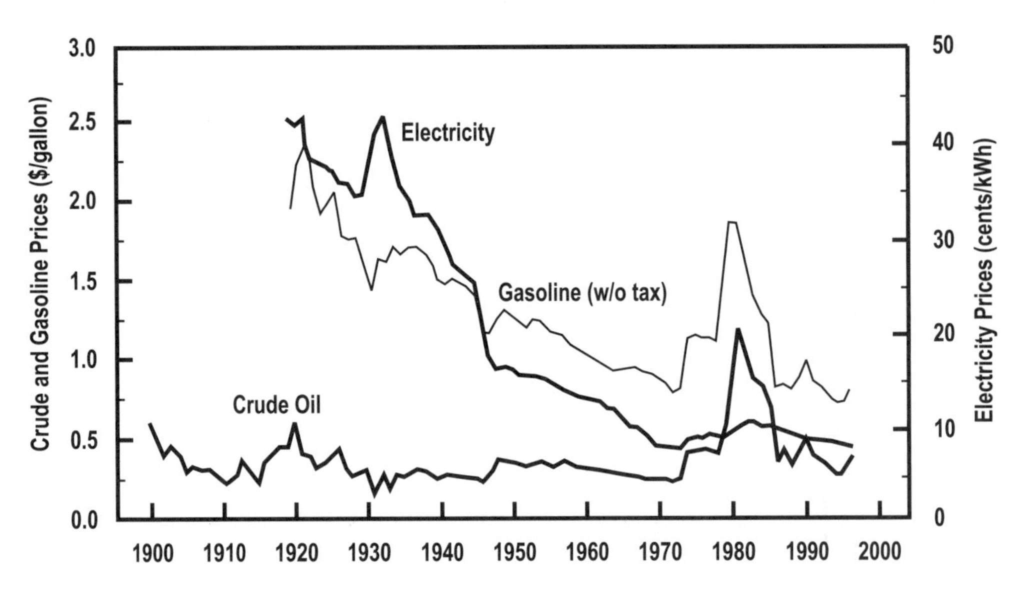

Figure 2.7. Energy prices for the period 1900–1996, expressed in 1996 dollars. Source: M.C. Lynch, pers. comm.

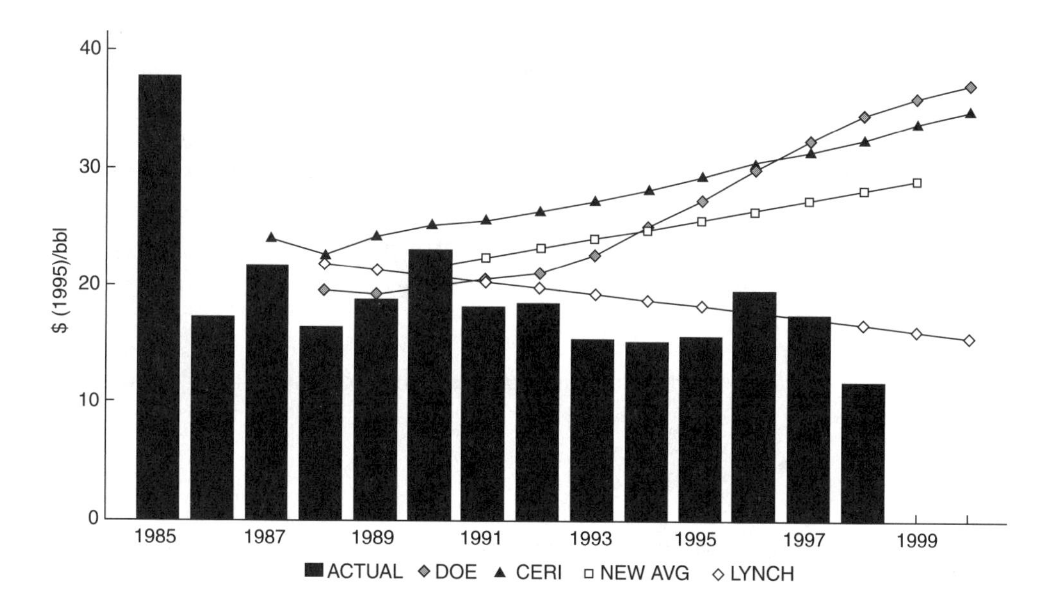

Figure 2.8. Comparison of various 1989 forecasts of crude oil prices for the period 1987–2000, with actual price data for 1985–1998. Source: Lynch (1998).

Historical data on crude oil prices reveal trends contrary to expectations from conventional mineral depletion economics. Figure 2.7 shows that, while there have been short-term episodes of oil price increases, over the long term (1900–1996), energy prices have been stable or have declined appreciably. Figure 2.8 shows more recent variations in world petroleum prices, as well as several forecasts. Again, the overall stability or downward shift, punctuated by upswings of short duration, is evident. The Hubbert curve concept is formulated on the basis of stable markets. The fluctuations observed in actual mineral production and discovery rates reflect political events, economic booms and busts, and similar situations. To the extent that such fluctuations are not corrected in the fitting of a curve, they will contribute uncertainties to the extrapolations made with a particular fitted curve.

2.4.4 Categories of reserves

Geological prospecting does not create reserves, but rather identifies regions of the earth more or less likely to contain "deposits" of oil or other mineral fuels. Deposits are quantities in place, *a portion of which* may be suitable for commercial development. Reserves are created from deposits by investment to construct and operate facilities (e.g., wells, supporting equipment) to produce some portion of the deposited fuel for practical use. How much fuel is recovered will be determined by the rate of fuel production and the cumulative operating time over which fuel production is profitable. Production rates for petroleum and natural gas almost always decline with time owing to declining well pressure and increased influxes of ground water. Technological interventions, such as well linking and re-pressurization by injecting CO_2, can counteract such effects for a time. The duration of profitability will be determined by prevailing energy prices and by advances in technology. It is useful to subdivide reserves as follows:

Proved developed reserves. This is a reservoir engineering concept defined by the accumulated production $R(\tau)$ from a source (e.g., a single oil or gas well, or several wells linked) over its useful operating lifetime, τ. The magnitude of these reserves can be estimated by integrating the well production rate $Q(t)$ over τ. For an assumed exponential decline in yearly production:

$$Q(t) = Q_0 e^{-at} \tag{2-6}$$

where Q_0 is the initial production rate and a is a fitting parameter (see below), $R(\tau)$ becomes:

$$R(\tau) = Q_0 \int_0^{\tau} e^{-at}\, dt = \frac{Q_0}{a}\left[1 - e^{-a\tau}\right] \tag{2-7}$$

In theoretical studies, τ would be taken as infinity, so that from Equation (2-7):

$$a = Q_0/R(\tau) \quad \left[\lim \quad \tau \to \infty\right] \tag{2-8}$$

However much owners and their heirs might wish it, τ is not infinity but some finite number that has typically ranged from a few years to a few tens of years. The useful lifetime is the period over which investment to initiate and then maintain production continues to be justifiable based on an acceptable rate of return on investment. When this is no longer the case, production is said to have *reached the economic limit*. This occurs when revenue from sales of incremental output of fuel can no longer keep pace with production costs. Production then ceases even though there may be gargantuan stocks of oil still in the well. This doesn't matter. Those stocks are deposits but they are not proved reserves because they cannot be economically developed under prevailing financial realities.

If we can estimate the quantity a, we can still use Equation (2-6) to predict a depletion schedule for a real-world well. On economic grounds, it is reasonable to assume that the ratio of the final to initial production rates ($Q(\tau)/Q_0$) is proportional to the ratio of initial production rate to total reserves ($Q_0/R(\tau)$). This shows that the higher the value of the ratio ($Q_0/R(\tau)$), the higher the fixed annual expenditures to maintain production, and the higher the value of $Q(\tau)$ at which sales no longer cover the yearly costs of continued production (Adelman, 1997). With this assumption, it can be shown that:

$$a = [Q_0/R(\tau)] - [Q_0/R(\tau)]^2, \quad [\tau \text{ finite}]. \tag{2-9}$$

Thus, from an estimate of the initial magnitude of the resource and the first year's rate of production, one can use Equations (2-6) and (2-9) to predict the annual production rate at any time t, as well as the time t_d for the yearly production rate to decline to any fraction $[Q(t_d)/Q_0]$ of the initial rate of drawdown. In summary, proved developed reserves are inventories that turn over, typically in periods of years.

Proved undeveloped reserves. These are defined as portions of the fuel stocks known to exist but not yet subject to production. They are the economic equivalent of *options on an underlying asset*, or the reserve that could be developed, for example, if new technology enabled economically profitable production.

Probable, possible, speculative, and "undiscovered" reserves. Estimates of the magnitude of each of these classes of reserves appear in the mineral geology and economics literature. These estimates are projections of future cumulative demand and supply of the particular form of mineral fuel, allowing for expansion of knowledge (Adelman, 1997). However, no human knows these future quantities. These four categories of reserves are best considered as *information* estimated at steadily decreasing confidence levels. They should not be thought of as assets, or even options, and certainly not as inventories. Despite the oxymoronic nomenclature, knowledge of "undiscovered" reserves is useful information because it sheds some light on the likelihood that commercially attractive quantities of a mineral fuel could be discovered by exploring a particular region. For example, if exploration identifies two more or less equally promising sedimentary regions, (A) and (B), and Region (A) is 10 times the size of Region (B), then (A) is a 10 times better place to go hunting for the mineral of interest (Adelman, 1998).

Unconventional reserves and resources. These are reserves that lie between what is known today and what may be known in the not-too-distant future that would enable previously unrecoverable resources to be economically produced. For example, before 1950, offshore oil was known to exist but, owing to inadequate production technology, it was considered impossible to develop. Since 1950, development of offshore oil has become commonplace. Technological advances have pushed back barriers. In 1998, for example, drilling to depths of three miles was possible. It was only possible to drill to depths of less than 100 feet when offshore production first began.

2.4.5 Forecasting mineral fuel prices and supplies

The track record on longer term forecasting (> two decades) of fossil fuel prices or supplies leaves much to be desired. Lynch has a better record than many, but even he reports (1998) that most organizations no longer engage in energy forecasting, at least publicly, and that much of the petroleum industry now uses scenario analyses to assist in their long-range planning. To be sure, there are some early warning signs of possible future scarcity, at least from a particular production source (e.g., if the cost of adding to reserves shows steady increases with time). However, no two deposits of fuel are identical. Even rather short-term projections of oil production rates (e.g., see Figure 2.4) or prices (Figure 2.8) have turned out to be remarkably too pessimistic.

Table 2.4 shows world petroleum production and additions to reserves for the period 1944–1993. Despite dramatic growth in total world oil consumption (implied by line 7 in Table 2.4), petroleum reserves have more than kept pace. Similarly, Figure 2.6 shows the startling expansion of annual production from non-traditional oil producers. Suppose one had attempted to forecast 1970 reserves or production using the 1944 data of Table 2.4, or a linear or even modestly exponential increase fitted to the 1945–1955 data of Figure 2.6. Even a confirmed optimist would have significantly under-predicted the actual results and probably been tempted to warn of ominous petroleum scarcities in the 1970s. During the 1970s, as in the 1960s and in the 1980s, total world additions to petroleum reserves soundly trounced total world petroleum production (compare lines 7 and 8, Table 2.4), and true shortfalls of oil never existed. However, major oil price spikes appeared twice in the 1970s (Figure 2.7). Some attributed these price hikes to oil scarcity, but we now know that they arose from panic buying brought on by the fear—not the reality (Table 2.4)—of an oil shortage (Section 2.5).

Of the diverse categories of reserves described in Section 2.4.4, only those proved and developed contribute meaningful estimates of future fuel supplies. By "meaningful," we mean forecasts based on engineering understanding of the production behavior of oil and gas wells (e.g., typical production rates and drawdown patterns) (Equations (2-6)–(2-9)). The other classes of reserves constitute a mixed bag of information (Adelman, 1998), because they are based on fragmentary knowledge and reflect various degrees of uncertainty. This information can, at best, provide only approximate estimates of what quantities of mineral fuels might be recoverable. Lewis G. Weeks, the first chief geologist at Standard Oil of New Jersery (now Exxon-Mobil), pointed out that estimates are "ordinal not cardinal," which is to say they are useful as indications of where to look to produce oil that may be in place (Adelman, 1998). But these estimates are of limited value in projecting new supplies, because even if we know how much is in the ground we do not know what fraction of it will ultimately be developed.

The ability of new technology to emancipate fuel sources previously judged unproducible should not be underestimated. Advances in know-how for producing oil offshore at increasing water depths were noted above in our discussion of unconventional reserves. Bitumens and ultra-heavy oils are two further examples of "unconventional" resources that have only recently entered the market. These resources came into account in fossil fuel resource estimates of cumulative production levels of several billion barrels and potential longer term outputs of many tens or even hundreds of billions of barrels.

This case illustrates an important aspect of consumption of most mineral resources. Where one form of the resource becomes increasingly expensive due to depletion, alternative forms or other mineral substitutes can become attractive enough to enter markets in which they were not previously participating. For fuels, many such alternatives exist, and we can expect such substitution to happen for many centuries. Coupled with this is the phenomenon that improvements in technology have typically driven down the cost of most minerals even as the earth's unproduced supplies of minerals from known stocks has decreased. Thus, in constant dollars, the cost of a mineral resource is typically lower now than at any time in the past.

An example is provided by petroleum. Technologies for finding and extracting oil have improved greatly over the past 30 years. The use of advanced computing, well pressurization, horizontal drilling, deep sea drilling, and ice technologies for roads and pads in cold regions are all examples of areas of high innovation. As a result, the typical fraction of oil recovered from a well has grown from 30 to 40%.

Whether this will continue to be true for many fuels is uncertain, as substitutes may not be found or technologies may not be improved (McCabe, 1998). We know that specific mineral deposits eventually become sufficiently depleted so that further production at available returns on investment is no longer economical and extraction eventually ceases (e.g., mine towns in the western US, anthracite mines and oil fields in Pennsylvania). However, the persistence and apparent generality of the trends discussed above create confidence that the pattern of technological advances opening up new producible stocks will continue into the future.

So, the problem to worry about is not *when* we will run out of fuels. Rather, it is *what will happen should we continue to use them*, especially fossil fuels, should their consumption be the cause of global warming. Today, fossil fuels are abundant and inexpensive to produce, and they will be so tomorrow. The consequence is that alternative energy technologies find it difficult to compete in an unfettered market. Thus, if it is desirable to restrict use of fossil fuels, non-market mechanisms (e.g., prohibitions, taxes) must be used. So far, such mechanisms have been able to influence consumption rates (e.g., the effects of high fossil fuel taxes in many European countries in promoting greater use of small automobiles and of rail transportation), but they have been unable to drive fossil fuels out of use. If it is truly desired to restrict the use of fossil fuels to much lower than current levels, stronger efforts will be required, and energy will likely become much more expensive for consumers. This would, in turn, strongly impact all sectors of the economy, necessitating careful analysis of prospective tradeoffs before enactment of fossil-restricting policies.

The experience of the 20th century was that the real cost of most mineral resources, including fuels, declined steadily where free markets were allowed to operate. Such free markets did not include pricing factors recognizing the value of preserving resources for future generations. When high prices were experienced, as with petroleum, it was the result of restrictions in the market—usually via taxes or cartel actions (which led to high oil prices during the 1970s and 1980s). During the latter episodes, the prices of other fuels also rose, reflecting less competition from oil for their markets. Should fossil fuel prices be increased in the future, we can expect short-term increases in the prices of alternative energy technologies. The effects of innovation and increased scale of use, however, could temper these increases.

Despite the foregoing discussion, some believe that supplies of fuels are finite, and that we shall eventually run out of them. Also, considering that our consumption rates typically grow steadily, it is argued that this may happen sooner rather than later. These points of view, as discussed above, pertain to a time in the future that is unknown. Other resource utilization issues are important for people living today and later in the 21st century. All of the currently used fuels have ample substitutes and are also available in lower grade deposits that are being marginally exploited.

A crucial question related to sustainability is how to extract these fuels and other minerals in ways that do not seriously harm the environment. In many cases, it appears that such harm can be avoided at reasonable costs and effort. Examples include the use of longwall rather than room-and-pillar mining (which, in mining coal, substantially avoids the problems of underground fires, acid seepage into aquifers, and uneven surface subsidence) and the use of ice rather than gravel as a construction material in cold environments (thereby reducing the amount of tundra that is permanently disturbed by oil and gas drilling operations). In other situations, an environmental price must be paid in order to extract some deposits. Examples include surface mining in arid regions, where restoration of stable, pre-existing ecosystems may not be possible, and building roads and pipelines for transport of minerals to markets (which involves some permanent alteration of the earth's surface, and can make previously remote regions accessible to the ravages of civilization). In these cases, the values of the societies involved will determine the balance of protection accorded economic and pristine environmental amenities. However, experience to date indicates that much of the mineral extraction needed to supply current energy demands can be performed with relatively small environmental disturbances compared to what is experienced when protection of the environment is not required and least-cost production is permitted.

Issues of US energy self-reliance and worldwide stable energy markets. At various times during the 20th century, energy markets were disrupted by wars, embargoes, and cartel actions (particularly by OPEC). This has led to demands within the US for "energy independence" and has increased the importance of large fuel-supplying countries—particularly those of the Middle East, where the bulk of the world's low-cost oil supplies is contained. The political instability of this region contributes substantially to the insecurity of the most important fuel powering the world's economy. The desire to stabilize global oil markets was surely a factor in the formation of a large international coalition to expel Iraqi armed forces from Kuwait following Iraq's 1990 invasion. Stabilization of oil markets was plausibly a factor in other conflicts including the 2003 war in Iraq. To the extent that such concerns alone motivate specific international military expenditures, such expenses can be viewed as effective subsidies in providing the affected fuels to the market. However, multiple factors (e.g., treaty obligations) typically motivate military spending, so it is not straightforward to disentangle energy resource protection from other driving forces. As the world's greatest military power, the US has played a larger role in these activities than any other country.

Such military activities have received large public support in the US. However, the support for being self-reliant in energy supply has been weaker and less consistent. This may be because the US has been largely self-reliant in most fuels other than petroleum (and, increasingly, natural gas), yet that this cannot continue to be the case in the future. The reason is that US consumption of conventional oil and natural gas has depleted known onshore deposits in the lower 48 states to a far greater extent than other fuel-supplying countries have exhausted their more easily producible deposits. Consequently, the ability to increase production from these US sources is weak, especially because less expensive foreign supplies are available.

Currently, more than half of US oil is imported, and this fraction is growing. Despite this situation, actions to increase self-reliance have been largely ineffective. The possibilities for action include reducing consumption (in reality, petroleum and natural gas consumption have continued to grow, partially through greater use of high-fuel-demand, on-road vehicles); creation of technologies that would substitute coal (which remains abundant in terms of current demand rates), nuclear, and renewable technology use for those relying on these two fossil fuels; establishment of a strategic petroleum reserve, and creation of taxes and regulations that would discourage fossil fuel use. While substantial nuclear and coal-fired base load electric power generation has prevented higher petroleum imports, none of the above actions has so far reversed the overall growth in petroleum and gas imports. Thus, the prospect for the future is for one of increasing reliance upon imported fluid fuels and increasing vulnerability to political instability in the supplying countries.

2.4.6 Geopolitical factors and energy supply "crises"

Military conflicts and politically motivated embargoes have disrupted oil shipments from producer to user nations on several occasions during the last 60 years. During World War II, strangulation of enemy supply lines was a major strategy of Germany, Japan, and the Allies. Strategically consequential oil embargos include that of Great Britain, the Netherlands, and the US against Japan in 1941, and those against the US and other western nations by Middle Eastern countries in 1967 and 1973 and by Iran in 1979. A comprehensive discussion of the causes and geopolitical ramifications of these and other energy-related international conflicts is beyond our scope. However, some consideration of how such politically-induced disruptions in petroleum shipments impact energy prices and economic growth, as well as strategies for effective countermeasures, merits our attention.

Lynch (1998) defines an *oil crisis* as a supply disruption severe enough to cause oil prices to rise to the point of causing global economic damage. Thus, the issue is not whether geopolitical unrest causes oil "scarcity," but whether the resulting temporary loss of oil supplies gives rise to serious economic dislocations.[7] The logical defense

7 "Temporary" is a relative term. World War II destabilized supplies for about six years. UN sanctions

strategy is to minimize the economic injury of supply disruptions (Lynch, 1998) by interrupting the chain of events that would lead to economic upheaval. Plausible actions are to prevent or promptly squash the supply disruption, attenuate the expected short-term rise in energy prices, or offset the economic dislocation brought on by those price increases. Logical countermeasures are to develop and maintain strategic stockpiles of fuel to offset lost imports during embargo periods and to switch to other fuels and energy alternatives. For example, natural gas and coal can be used to offset petroleum consumption in electric power generation. However, in many nations, including the US, relatively little oil is currently used in the electric sector (Chapter 17), so fuel-switching opportunities are limited. Expanded use of renewable forms of energy is another strategy, but if the motivation is to mitigate or prevent economic hardship, these renewables must be economically competitive with *pre-embargo fossil fuel prices* (Section 2.4.5).

Many scholars believe that low oil prices undermine sustainable development by removing financial incentives to reduce demand and diversify supply technologies. Despite significant volatility, oil prices have exhibited an overall downward trend since the first OPEC price shock of 1973 (Figure 2.8), and further longer term declines are projected. However, these prices may not fall to the true price under competition, which may be $5/bbl, long term. Thus, in the present discussion, "low" means small compared to the price at which alternative energy sources could compete seriously, i.e., achieve appreciable market penetration, without subsidies.

It has been argued that low oil prices dissuade private and even public sector investments to stimulate adoption of renewable energy sources. The gap between existing oil prices and the price(s) at which various alternatives could compete economically is so large that most energy consumers, at least so far, show little interest in the more expensive alternatives. In order to assess the commercial prospects of promising alternative energy technologies, substantial investment in development is needed. If results prove favorable, testing at scales large enough to evaluate the performance of individual components and of the ability of those components to work together in harmony, such as integrated process systems evaluation, are needed. If selling prices are projected to be non-competitive with petroleum and other low cost forms of energy, investors see poor prospects for reasonable returns. It is therefore highly probable that they will either deploy their capital elsewhere or demand premium rates of return to compensate for high perceived risks of alternative energy projects. Even if alternative energy technologies are able to attract sufficient interest and investment to be commercialized, they require higher selling prices to offset higher borrowing costs.

---

against business-as-usual shipments by Iraq lasted from 1991 at least through 2004. Many other politically driven disruptions in oil shipments have been far shorter.

Many argue that another "flaw" of low prices for oil and other traditional sources of energy is that these forms of energy have costs which are not reflected in their selling prices. It is further proposed that if these costs were included in the selling prices of oil, alternative forms of energy would be more economically competitive. These costs, typically referred to as *externalities*, refer to government actions that effectively subsidize the particular form of energy, and to environmental, economic, and health damage caused by the energy technology throughout its lifecycle of development, construction, and decommissioning. Identification of externalities and determination of their effects on energy costs and of how to account for externalities in energy pricing are complex subjects beyond our current scope. Among the major challenges are how to assign monetary values to human morbidity and mortality and to the earth's ecosystems; how to disaggregate government actions to quantify their value in supporting particular technologies or industries; and how to unequivocally link adverse human health effects and other damage to particular industrial causes. Chapter 5 provides additional information on accounting for externalities in energy pricing, with particular reference to the production of electricity.

Price differentials between petroleum and various alternatives, both renewables and synthetic fuels from coal, oil shale, and tar sands, are currently substantial (Chapters 7, 9–15). For example, for synthetic liquid transportation fuels from coal or biomass to compete with their existing counterparts from petroleum, oil prices would need to reach $30 to over $60/bbl. Past experience with oil price volatility discourages hope that non-governmental actions can sustain such prices long enough for alternative technologies to improve and penetrate markets sufficiently to cause substantial change. Low prices also promote increased consumption as growing SUV purchases show.

Another contention is that low oil prices hurt traditional energy producers, such as petroleum suppliers, by curtailing exploration for new deposits. This line of reasoning is challenged by historical data. Despite the "ultra-low" pre-1972 oil prices, huge deposits were discovered on the North Slope of Alaska, the North Sea, and southeastern Mexico (Adelman, 1999). Returns on investment for exploration and development were adequate for these regions, better in Venezuela, and better still in the Persian Gulf, where in 1972 ARAMCO anticipated an almost 3-fold increase in output (from 7 to over 20 million bbl/day) over the next 10 years at no higher costs. Its eventual participation in monopoly pricing may have been one of the factors that cost ARAMCO most of this hoped-for increase in export volume.

An alternative perspective sees low oil prices as good for consumers and a dynamo for sustainable development by enabling economic growth and lasting stewardship of all the earth's natural resources. Low-cost energy is an engine of economic growth, and global economic progress is essential to broad-based acceptance of sustainable energy concepts. Less expensive oil can help maintain the economies of developed countries and supply economical energy to developing nations that need it to play economic catch-up. Many of these nations are among those least able to purchase more expensive non-fossil alternatives. Sustainable energy will not truly succeed if major segments of the earth's population remain hungry or economically destitute. As independent sovereign nations, developing countries will not undertake crucial economic, industrial, and sociopolitical reforms to establish and maintain environmentally sustainable energy practices at the expense of economic advancement for their people. Likewise, developed countries will not risk military intervention in these regions to enforce environmental codes, nor (at least in the absence of a major war) are they likely to impose politically suicidal limits on the freewheeling energy consumption of their own populace. Yet some developing countries are committing to investing in cleaner and more efficient energy use as a prudent longer-term development strategy. Present investment in modern appropriate technologies offers a better pathway to industrialization.

Sustainable development implies a commitment to sound management of all the earth's natural resources for present and future generations. The development, environmental protection, and preservation of food, water, arable land, minerals, forests, and recreational venues will be energy-intensive operations. Low-cost energy will make it easier to apply sustainability principles to all of these resources. At present, industrial nations expend substantial amounts of energy to clean up past legacies of hazardous wastes and prevent current industrial and transportation technologies from emitting adverse pollutants. For the future, energy (and monetary) expenditures for environmental protection are certain to increase as a percentage of their respective totals in both developed and developing nations. The availability of low-cost petroleum-derived energy and related products will lower the cost of these environmental improvements and increase the probability of their broader penetration. Of course it must be recognized that to maintain a position of respect in a sustainable energy future, all energy technologies, including petroleum, must be available without adverse environmental impacts from their production and use.

2.5 Lessons for Sustainable Development

There are many drivers for sustainable energy but scarcity of oil and other fossil fuels is not one of them. Continued domination of world energy supplies by fossil fuels may be challenged by concerns over potential climate modification attributed to their emissions of CO_2 and fugitive methane, but sustained major shortages of oil, coal, or natural gas seem unlikely in the next four decades. Policy countermeasures such as carbon taxes could increase mineral fuel selling prices to the point of seriously harming economic sectors that depend on low-cost fossil fuels for their survival. On the other hand, the prospect of such policy interventions has already stimulated efforts to develop more efficient and non-fossil energy alternatives. Voluntary measures to curtail atmospheric CO_2 emissions could lead to increased consumption of non-fossil-derived electrical energy, but appreciable advances in technology would be required for major decarbonization of the transportation sector.

It is unclear whether sustained persistence of low oil prices is good or bad for sustainable development. Some experts believe that low prices harm conservation efforts and are a significant disincentive to consumers to adopt alternative energy technologies, especially renewables. However, low oil prices are a proven source of economic vitality for developed and developing nations. Some argue that low oil prices are a means to protect the environment and achieve the diverse goals of sustainable development, including stewardship of all the earth's natural resources—water, minerals, forests, arable land, open space, recreational areas, and energy—while maintaining and extending economic opportunity.

2.6 Summary and Conclusions

We began this chapter by discussing units to measure energy and power for applications in trade and commerce, scientific research, and engineering. We noted that energy can exist in various forms of different efficacy and economic value to end users. Thinking of energy in terms of its lifecycle, or from its exploration and discovery to final use, giving due heed to the health and environmental consequences of each step, proved useful in framing the costs and benefits of energy to humans. We then discussed how economists and earth scientists categorize and value energy resources. We carved out small slices of the history, politics, and economics of the volatility in oil prices, concentrating on this mineral fuel because of its importance to a broad range of sustainability questions and because its economics have been thoroughly studied by scholars.

Major lessons for sustainable development are that scarcity of fossil mineral fuels is unlikely in the foreseeable future and, thus, is not a driver for sustainable energy, although adverse environmental impacts of using fossil fuels may be. Low oil prices will impact sustainable development by affecting market penetration of conservation measures and alternative energy technologies, and by enabling economic growth and environmental preservation, as well as sustainable development of food, water, forests, minerals, and other natural resources. However, if important sustainability goals are not recognized in energy markets, a true transition will be accordingly delayed.

References

Adelman, M.A. 1999. "Oil Prices: Volatility and Long-term Trends." CEEPR Research Note, #CR-1. Cambridge, MA: MIT Center for Energy and Environmental Policy Research (August).

Adelman, M.A. 1998. Oral presentation on oil prices and reserves. Cambridge, MA: MIT

Adelman, M.A. 1997. "My Education in Mineral (Especially Oil) Economies." *Annual Review of Energy and the Environment* 22: 13-46.

Adelman, M.A. 1995. *The Genie Out of the Bottle: World Oil Since 1970.* Cambridge, MA: MIT Press.

Adelman, M.A. 1993. *The Economics of Petroleum Supply Papers by M.A. Adelman, 1962-1993*. Cambridge, MA: MIT Press.

Al-Jarri, A.S. and R.A. Startzman. 1997. "Worldwide Petroleum-Liquid Supply and Demand." *Journal of Petroleum Technology*: 1329-1338.

Allenby, B.R. 1999. *Industrial Technology: Policy Framework and Implementation*. Upper Saddle River, NJ: Prentice-Hall.

British Petroleum. 2003. *Statistical Review of World Energy 2002*. London. See www.bp.com/centres/energy/index.asp

Calder, K.E. 1996. "Asia's Empty Tank." *Foreign Affairs* 75(2): 55-69.

Campbell, C.J. 1997a. *The Coming Oil Crisis*. Brentwood, England: Multi-Science Publishing.

Campbell, C.J. 1997b. "Depletion Patterns Show Change Due for Production of Conventional Oil." *Oil and Gas Journal*: 33-37.

Campbell, C.J. and J.H. Laherrère. 1998. "The End of Cheap Oil." *Scientific American* 278(3) (March): 78-83.

Ellerman, A.D. 2005. Private communication.

Feynman, R.P., R.B. Leighton, and M. Sands. 1964. *The Feynman Lectures on Physics, Volumes I, II and III*. Reading, MA: Addison-Wesley.

Graedel, T.E. and B.R. Allenby. 1995. *Industrial Ecology*. Upper Saddle River, NJ: Prentice-Hall.

Hottel, H.C. 1983. "The Relative Thermal Value of Tomorrow's Fuels." *Ind. Eng. Chem. Fundam.* 22: 271-276.

Hotelling, H. 1931. "The Economics of Exhaustible Resources." *J. Polit. Econ.* 39: 137-175.

Hubbert, M.K. 1956. "Nuclear Energy and Fossil Fuels." *Drilling and Production Practice*: 17.

Kenney, W.F. 1984. *Energy Conservation in the Process Industries*. New York: Academic Press.

Larson, E.D. 1993. "Technology for Electricity and Fuels from Biomass." *Annual Review of Energy and the Environment* 18: 567-630.

Levenspiel, O. 1996. *Understanding Engineering Thermo*. Upper Saddle River, NJ: Prentice-Hall.

Lynch, M.C. 1998. "Oil Scarcity, Oil Crises, and Alternative Energies: Don't Be Fooled Again." Manama, Bahrain: Paper presented at the 7th International Energy Conference and Exhibition, ENERGEX '98 (March).

MacKenzie, J.J. 1996. "Heading Off the Permanent Oil Crisis." *Issues in Science and Technology* 12(4): 48-54.

McCabe, P.J. 1998. "Energy Resources - Cornucopia or Empty Barrel?" *American Association of Petroleum Geologists Bulletin* 82(11): 2110-2134.

Neff, T. 1984. *The International Uranium Market.* Cambridge: Ballinger Publ. Co.

Romm, J.J. and C.B. Curtis. 1996. "Mideast Oil Forever." *Atlantic Monthly*: 57-74 (April).

Smil, V. 1998. *Energies: An Illustrated Guide to the Biosphere and Civilization.* Cambridge, MA: MIT Press.

Socolow, R. 2005. Private communication.

Tester, J.W. and M. Modell. 1997. *Thermodynamics and Its Applications.* Third Edition. Upper Saddle River, NJ: Prentice-Hall.

Weeks, L.G. 1969. "Offshore Petroleum Development and Resources." *J. Petroleum Technology* 21: 377-385.

3 *Technical Performance: Allowability, Efficiency, Production Rates*

The comprehension of the laws which govern any material system is greatly facilitated by considering the energy and entropy of the system in the various states of which it is capable.
—Josiah Willard Gibbs (1876)

3.1 Relation to Sustainability

To achieve sustainable energy, we must make informed choices among competing policies and technologies. Ideally, options will be selected because their behavior fulfills enough expectations of enough stakeholders to create a broad consensus. Besides the obvious requirement of technical feasibility, economic viability has historically been the nonnegotiable requirement for broadly based adoption of a given energy technology. Sustainability calls for energy technologies to fulfill additional performance standards related to the longer-term public good. These standards include preservation of natural resources, extension of economic opportunity, and protection of the environment. In this chapter, we will consider basic analytical tools for gauging the technical performance of energy conversion technologies. We will focus on machines that transform heat into useful work, but the principles discussed apply to all modes of energy conversion.

Questions 1 and 2 in technical evaluation are: (1) Will the technology work? In other words, is it consistent with the fundamental laws of science? And (2) if so, how well does it work? Thermodynamics provides powerful tools for answering these questions. Its methods allow us to screen out impossible ideas, compelling as they may seem to their inventors or proponents, and to forecast the technical performance of ideas that are scientifically possible. Consider the seemingly trivial example that water will not spontaneously flow uphill. Thus, a hydroelectric plant requiring water to independently travel against gravity is doomed to failure. Yet we also know that in certain locations, natural forces, (e.g., strong tides) will cause the effluent of a river or inlet to reverse direction so that twice each day water does, for awhile, flow uphill, back upstream and against gravity. Tidal flows are harnessed for electric power generation in France and Nova Scotia, and development of several other sites has been considered (Chapter 14). Different physics govern the behavior of water in these two flow situations. Water flowing spontaneously against the force of gravity violates a basic law of science and has no possibility whatsoever of occurring. On the other hand, tidal power, although perhaps not economically competitive for broad-based applications, is completely feasible technically. It violates no scientific law because an external force, the gravitational force of the moon, intervenes to counteract the "downhill" force of the earth's gravity.

Practically speaking, "how well" a technology works boils down to determining how proficiently the technology performs a task beneficial to humans. Tasks beneficial to today's societies include rotating a turbine; transporting people, goods, or information; generating electricity; and climate conditioning a building. Thermodynamics helps us answer the "how well" question by predicting the best that the specific technology could possibly do. "Best" can be expressed using several metrics of thermodynamic performance. For instance, it may be the maximum amount of useful work an energy conversion machine can produce from a given amount of energy inputted.

For a machine that converts heat energy to useful mechanical energy (i.e., a *heat engine*), a useful thermodynamic coefficient of best performance is the *ideal thermal efficiency*. This is the ratio of the maximum work that can be achieved for specified operating conditions and equipment design to the amount of heat energy put into the machine. It turns out that this is an idealized quantity in that it assumes flawless operation of the machine, and as such, provides a theoretical target for comparing and improving actual processes.

If you are a newcomer to thermodynamics, note carefully the implications of this concept. There are limits on the ability of even a flawless heat engine to convert heat into useful work. No amount of technical innovation or political intervention will further improve the machine's performance beyond this ideal. Because practical machines are not flawless, they fall short of this ideal. Thermodynamic analysis is useful because it helps define opportunities for overcoming process imperfections by benchmarking the inherent limitations that can never be surmounted. Many specialized textbooks detail thermodynamic fundamentals and applications of thermodynamics energy conversion processes (e.g., Tester and Modell, 1997, Levenspiel, 1996, Heywood, 1988, Wilson, 1984, Smith and Van Ness, 1975). Section 3.2 introduces foundational concepts for thermodynamic analysis of energy systems.

Thermodynamics alone will not tell us if an energy conversion concept is technically feasible. To construct and operate practical-scale technologies for converting and using energy, we must have materials able to withstand any locally punishing effects of the technology. For example, depending on the process, we may need materials that are resistant to corrosion, erosion, ionizing radiation, vibration, temperature extremes (hot or cold), among others.

Further, to be practical, a process or technology must provide energy, or an energy-intensive product like steel, aluminum, or glass, sufficiently rapidly to match the demand for that energy or product. Failure in this department can have serious practical consequences, such as the inability of a supplier to keep pace with customer demands for goods or services. More specifically, slow rates of energy conversion mean lower available power. This can compromise a host of practical functions—some a matter of life and death. Examples are aircraft takeoff and course adjustment, safe entry of a motor vehicle into a high-speed, traffic-congested highway, and stable supply of electricity even when industrial or residential customers accelerate their demand. But where do rate limitations in energy technologies come from? The answer is that chemical reactions that produce energy or various processes that transmit heat or material may be too slow. Causes include heat source temperatures that are too low or converter devices that are too small. Causes also include a local shortfall or excess of some crucial chemical reactant, the failure of materials, inadequate design of mixing devices, and numerous other imperfections. There are many specialized textbooks and monographs that provide detailed analyses of the kinetics of chemical reactions (e.g., Fogler, 1992, Levenspiel, 1972), of the rates of heat, mass and momentum transport in media of interest to energy conversion and utilization (e.g., Deen, 1998, Cussler, 1997, Levenspiel, 1984, Denn, 1980, Whitaker 1977, Bird et al., 1960, McAdams, 1954), and of situations in which rates of chemical reactions and of physical transport are inter-connected (Rosner, 1986, Frank-Kamenetskii, 1969). Methods for analysis of (uncoupled) chemical kinetic and heat transfer effects in energy conversion technologies are introduced in Sections 3.4 and 3.5, respectively.

Continuing advances in numerical analysis and molecular theory, data availability, and computational hardware are making mathematical modeling and simulation increasingly powerful tools for design, performance analysis, control, and optimization of practical energy conversion devices and systems. Simulation can eliminate tedious testing and focus expensive measurement programs on the most crucial issues. For hazardous materials or processes, computer experiments eliminate the human and equipment risks of actual testing programs. Nonetheless, any new energy conversion technology that shows promise after thorough thermodynamic, materials, kinetic, and economic analysis must eventually be tested experimentally. Before a reliable decision can be made on the technical and profitability prospects of a commercial venture, the technology must be tested at "appropriate scale." This means at a size large enough to provide reliable information for the technical design and economic evaluation of a full-scale plant. Especially important is to choose a scale large enough to probe the complexity of a full-scale process, which often requires synchronous subsystem interactions to operate. The value, timing, and funding of pilot and demonstration scale energy projects have been the subject of animated policy debates in the US (see Chapters 6, 13, and 21 for further discussion).

3.2 An Introduction to Methods of Thermodynamic Analysis

In this section, we have drawn heavily on material from the textbooks of Levenspiel (1996) and Tester and Modell (1997). As an introduction, you may wish to review our earlier discussion of thermodynamic states and the First Law of Thermodynamics in Section 1.2.

3.2.1 Allowability, efficiency, and the Second Law

The Second Law of Thermodynamics determines what physical and chemical changes can and can't happen spontaneously in the real world. Further, it provides recipes for calculating the maximum amount of useful work obtainable from an idealized energy conversion or utilization process. Useful work emerges only when a system undergoes a change in its thermodynamic state (i.e., pressure, temperature, and composition). The "working currency" for Second Law analysis is the entropy S, a thermodynamic property that depends only on the state of the system. Some scholars like to visualize entropy as a thermodynamic measure of a system's disorder or inventory of random information. The magnitude and the sign of the entropy change ΔS

experienced by a system as it traverses from one thermodynamic state to another tells us a great deal about energy conversion processes. In particular, the Second Law teaches that a change in the state of an isolated system is allowed only if its entropy does not decrease.[1] If a system is not isolated (i.e., if it is able to communicate with its surroundings by accepting or giving up heat or work, or both), then the corresponding Second Law rule is that a change in the state of the *system* is allowed only if the combined total of the entropy of the system *plus* that of its surroundings does not decrease. Mathematically, we can write these two rules as

$$\Delta S_{system} \geq 0 \qquad \text{for an isolated system} \qquad (3\text{-}1)$$

$$\Delta S_{system} + \Delta S_{surroundings} \geq 0 \qquad \text{for a non-isolated system} \qquad (3\text{-}2)$$

For practical Second Law calculations, we need a mathematical rule for computing ΔS. For an isolated system undergoing a change from one fixed state (state-1) to another (state-2), ΔS is given by:

$$\Delta S = S_2 - S_1 = \int_{state-1}^{state-2} \delta Q_{rev} / T_{system} \qquad (3\text{-}3)$$

where T_{system} is the absolute temperature of the system and Q_{rev} is an idealized heat uptake or release by the system (i.e., the heat that *would be* exchanged when all mechanical energy changes occur perfectly (i.e., reversibly)) and where T_{system} is the same at every location within the system. The prefix δ signifies that Q_{rev} is a path-dependent quantity.

The performance of real energy conversion machines is compromised by various dissipative processes that can never be completely eliminated (e.g., mechanical friction, heat leakage, etc.). To take detailed account of these dissipative effects would make entropy computations horrendously unwieldy. Happily, applications-oriented entropy calculations can always get around this complication by utilizing the fact that ΔS only depends on its end states (Equation (3-3)). This is done by assuming a path from state 1 to 2 whereby all relevant changes of state are *reversible* (i.e., they occur so that both the system *and* its surroundings *could* be restored to their original condition). A reversible process is a thermodynamic idealization that can never be achieved in practice. In a reversible process, there are no dissipative processes, e.g., no mechanical friction and wasted heat. Nevertheless, the assumption of *thermodynamic reversibility* is a useful approximation. It not only simplifies performance analysis but also represents the best possible case of what could be achieved if the energy conversion machine was totally defect-free.

1 Because a change may be too slow to detect before our patience runs out, it is more precise to say that the change is not prohibited if the entropy does not decrease.

In an *irreversible* process, it is impossible to restore both the system *and* its surroundings to their original state, no matter what manipulations we try. After an irreversible process, we can restore a system to its original state but, in doing so, the surroundings will not be able to resume their original state. In particular, they will have gained or lost entropy. All real-world processes are irreversible. To test your understanding of the concepts of entropy and irreversibility, what does the last sentence imply regarding the entropy of the universe?[2]

If your main interest is the public policy aspects of sustainable energy, you may be thinking that entropy is yet another abstract mathematical concept cooked up by engineers to torment social scientists and others not formally trained in engineering thermodynamics. If you fall into this category, you may be tempted to skip over Section 3.2. Don't be hasty. Even a policymaker claiming no technical expertise needs to be confident that new and innovative policy decisions do not violate the fundamental laws of science, including thermodynamics. You don't need to be an expert on "thermo" to be an effective policy analyst. But you will want to understand enough about the Second Law so that you can ask informative questions when proponents try to persuade you of the merits or limitations of policy reforms that depend on technology.

3.2.2 More about entropy

All energy conversion devices function by changing the thermodynamic state of some substance. The *heat engine* is an important energy conversion machine that follows a path that absorbs heat from a suitable source and then converts a portion of that heat into useful work. The sun (Chapter 13), chemical potential energy (Chapters 7, 10), a geothermal reservoir (Chapter 11), and nuclear decay (Chapter 8) are important heat sources for sustainable energy (Figure 1.2). In a heat engine, the substance undergoing a state change is the *working fluid.* With rare exceptions, practical working fluids are a gas, a liquid, or gas-liquid mixtures. In heat engines, typical working fluid state changes are heating, cooling, vaporization, condensation, expansion, and compression. Second Law analysis is based upon calculating the sign and magnitude of the entropy changes that accompany these state changes (i.e., on making practical use of Equation (3-3)). But wait a minute. It turns out that there are an infinite number of reversible and irreversible pathways for transforming a system from one thermodynamic state to another. Further, the quantity δQ_{rev} in Equation (3-3) depends upon the pathway taken. Thus, how do we know which pathway to choose in applying Equation (3-3) to real-world energy converters?

---

2 One of the patriarchs of thermodynamics, Rudolph J. E. Clausius (1865), provides an answer:
"Die Energie der Welt ist konstant.
Die Entropie der Welt strebt einem Maximum zu."
[The Energy of the World is constant. The Entropy of the World strives toward a Maximum.]

Here is where we begin to obtain a glimmer of the value of basing our analysis on entropy. Recall that ΔS does *not* depend on the path between two states, only on the states themselves (thermodynamicists call Q_{rev} a *path function* and S a *state function*). As a result, we choose the most convenient path for evaluating ΔS – a reversible one. Thermodynamics texts show how to compute ΔS for various state changes involving all sorts of working fluids. Sometimes it is possible to obtain a simple mathematical expression for ΔS. This requires a mathematical formula that relates the pressure, P, of a substance to its temperature, T, volume, V, and composition, x_i, where x_i is the mole fraction of a pure species, i, in the working fluid. Such a formula is called an *equation of state* (*EOS*). EOSs range from simple to remarkably elaborate expressions depending on the chemical and physical heterogeneity of the substance, the complexity of its thermodynamic behavior, and the ranges of P, T, V, and x_i of interest. The simplest EOS for ideal gas or vapor is:

$$PV = nRT \quad \text{or} \quad P = (n/V)RT \tag{3-4}$$

where n is the number of moles of the gas or vapor and R is the universal gas constant and T is in absolute temperature units (K = °C + 273.15). Equation (3-4) is called the *ideal gas equation of state*. Ideal gas molecules behave as infinitesimally small particles that exert no forces on each other. The volume of the gas increases (decreases) directly (inversely) in proportion to an increase in the gas temperature (pressure). The quantity (n/V) is the molar density of the vapor. Reid et al. (1987), Tester and Modell (1997) and various engineering handbooks provide EOSs for a variety of more complicated working fluids of practical interest.

We illustrate an entropy change calculation for expansion of the working fluid, one of the state changes that typically occurs in a heat engine. Let 20 moles of a gas be compressed to 10 times atmospheric pressure in a sealed container and then brought to room temperature (25°C = 298 K). If the container is opened under an external pressure of 1 atm, it is intuitive that the gas will expand and that we can harness some of the force of the expanding gas to do some productive mechanical work, like pushing back a piston or rotating a turbine. Let's check our intuition by calculating the entropy change for expansion of our gas. We begin with an appropriate expression for the First Law of Thermodynamics (see also Equation (1-2)):

$$\Delta U + \Delta E_{PE} + \Delta E_{KE} = Q - W_{sh} - W_{pV} \tag{3-5}$$

Next, we must specify how the expansion is carried out and how the gas behaves thermodynamically when there is a change in its state. Let's assume ideal gas behavior (Equation (3-4)) and expand reversibly and isothermally. At 25°C (and up to much higher temperatures) most gases cool when expanded and heat up when compressed; if we were actually performing this expansion, we would design our apparatus to add just the right amount of heat to keep the gas temperature constant. We neglect any change in the potential or kinetic energy of the system, i.e., $\Delta E_{PE} = \Delta E_{KE} = 0$. This is equivalent to assuming that the energy of our gas is not affected by gravity or any other external forces including a minuscule change in vessel volume arising from release of the gas pressure. For an isothermal change of state $\Delta U = 0$ for an ideal gas, and $Q = Q_{rev}$ for a reversible process, Equation (3-5) reduces to:

$$Q = Q_{rev} = W_{rev} = W_{sh} + W_{PV} = \int_{state-1}^{state-2} PdV \tag{3-6}$$

Because differentiation and integration are inverse mathematical operations, Equation (3-6) requires that:

$$dQ_{rev} = PdV \tag{3-7}$$

To compute ΔS, use Equation (3-7) to replace dQ_{rev} in Equation (3-3):

$$\Delta S = \int_{state-1}^{state-2} PdV/T_{system} \tag{3-8}$$

In this integral P is the system pressure, which decreases throughout the expansion. P equals the pressure of the surroundings *only* at the end of the decompression, i.e., when $P = 1$ atm. Thus, P cannot be removed from the integral in Equation (3-8). We assume that the gas is ideal, and we can use Equation (3-4) as an EOS to express P in terms of V, giving:

$$\Delta S = \int_{state-1}^{state-2} \left[\frac{nRT}{VT}\right] dV = nR \int_{state-1}^{state-2} \frac{dV}{V} = nR \ln\left[\frac{V_2}{V_1}\right] = -nR \ln\left[\frac{P_2}{P_1}\right] \tag{3-9}$$

where we have dropped the subscript to denote that T is the system temperature, and we have used Equation (3-4) again to replace V_2/V_1 by P_1/P_2. Substituting $P_1 = 10$ atm, $P_2 = 1$ atm, and $n = 20$ moles in Equation (3-9) we obtain:

$$\Delta S = -(20 \text{ moles}) \times (1.987 \text{cal/mole–K}) \times \ln[1.0 \text{ atm}/10 \text{ atm}] = 91.5 \text{ cal/K} \tag{3-10}$$

Note that ΔS has units of [energy/temperature]. Confirming our intuition, ΔS for this change of state is positive, i.e., the gas spontaneously begins to decompress once the container is opened. The corresponding reversible work done by the expanding gas on its surroundings is found from Equation (3-6), again using Equation (3-4) to replace P by a function of V:

$$W_{rev} = \int_{state-1}^{state-2} PdV = \int_{state-1}^{state-2} \left[\frac{nRT}{V}\right] dV = nRT \int_{V_1}^{V_2} dV/V \tag{3-11}$$

i.e.,

$$W_{rev} = nRT \ln\left[\frac{V_2}{V_1}\right] = -nRT \ln\left[\frac{P_2}{P_1}\right] \tag{3-12}$$

Substituting numerical values in Equation (3-12) we find:

$$W_{rev} = -(20 \text{ moles}) (1.987 \text{ cal/mole–K}) (298 \text{ K}) \ln[1.0 \text{ atm}/10 \text{ atm}]$$

$$W_{rev} = 27{,}300 \text{ cal}$$

Again our expectation is met in that the expansion did give rise to work. However, even in a reversible expansion, not all of this work can be harnessed to turn a shaft or other useful mechanical chores. Some work, W_{PV}, had to be expended to push back the atmosphere. This work is:

$$W_{PV} = \int P_{ext}\, dV = P_{ext}(V_2 - V_1) \tag{3-13}$$

where P_{ext} is the external atmospheric pressure = 1 atm. Using Equation (3-4) to evaluate V_1 and V_2 we find:

$$W_{PV} = P_{ext}\, nRT[1/P_2 - 1/P_1] \tag{3-14}$$

$$W_{PV} = (1 \text{ atm}) (20 \text{ moles}) (1.987 \text{ cal/mole–K}) (298 \text{ K}) [1/(1 \text{ atm}) - 1/(10 \text{ atm})]$$

$$W_{PV} = 10{,}700 \text{ cal}$$

Therefore, the net useful work, W_{sh} is:

$$W_{sh} = W_{rev} - W_{PV} = 27{,}300 \text{ cal} - 10{,}700 \text{ cal} = 16{,}600 \text{ cal} \tag{3-15}$$

Remember that because we chose the expansion to be reversible, W_{rev} is the maximum work obtainable by expanding an ideal gas in the stated manner.

Desired attributes of a working fluid are high heat capacity and heat of vaporization, as well as chemical stability, compatibility with materials of construction, safe handling and use, and low cost. Fluids meeting these requirements do not exhibit ideal gas behavior, making entropy change calculations appreciably more difficult than in the examples we just presented. Fortunately, with empirically fit, high-precision *PVTN* equations of state (EOS) and correlations for ideal-gas–state heat-capacity, calculation of state points in thermodynamic cycles is straightforward.

The best known working fluids are water and steam, used for over 200 years in a variety of heat engines, including modern electric utility boilers. More esoteric working fluids include pentane, ammonia, and chloropentafluoroethane. These fluids have been employed during the last two decades to convert geothermal heat (Chapter 11) and solar heat (Chapter 13) to mechanical work in Rankine cycles because of their relatively high vapor densities at condensing conditions. Thermodynamics texts (e.g., Tester and Modell, 1997, Levenspiel, 1996) and many engineering handbooks (e.g., Perry et al., 1997) detail methods for entropy calculations on real-world working fluids.

Sometimes entropy calculations are simplified by using diagrams that graph two or more thermodynamic quantities. Two charts especially useful in analysis of heat-to-work conversion processes are the temperature-specific entropy (*T-s*) diagram (Figure 3.1) and the specific enthalpy-specific entropy (*h-s*) or Mollier diagram (Figure 3.2). Recall how we defined enthalpy in Equation (1-4) of Section 1.2. *Specific* thermodynamic quantities refer to the amount per unit mass or unit number of moles. To illustrate use of Figures 3.1 and 3.2, let's again calculate how much useful (shaft) work can be obtained by expanding a gas from a high pressure, P_1, to a lower pressure, P_2, which we will take to be atmospheric pressure. Now, let the expansion occur *adiabatically*, i.e., so that no heat is exchanged between the gas and its surroundings. This is an idealization, because there are no perfect thermal insulators and some heat will always flow between the gas and the vessel used to confine it. Nevertheless, the adiabatic assumption is one of two reasonable limiting case approximations to describe the expansion and compression of gases in heat engines. The other limiting case, also an idealization, is the isothermal approximation described above. Further, let the adiabatic expansion be reversible so that the change in heat content of the gas (our working fluid), $\delta Q = \delta Q_{rev}$. Because no heat flows, $Q = 0$ and $\Delta S = 0$. Further, we will no longer require that the gas be ideal, only that we have its T-s and h-s charts[3] represented by Figures 3.1 and 3.2, respectively, for a pure, single component substance. We expect the gas to cool on expansion and therefore that the final temperature T_2 will be lower than T_1. Because $\Delta S = s_2(T_2, P_2) - s_1(T_1, P_1) = 0$, we can calculate the temperature change from Figure 3-1 by measuring the vertical distance between the lines of constant P_1 and P_2 at the specific entropy of the working fluid in state 1 (or state 2). Normally, T_1, the working fluid temperature at P_1, is specified.

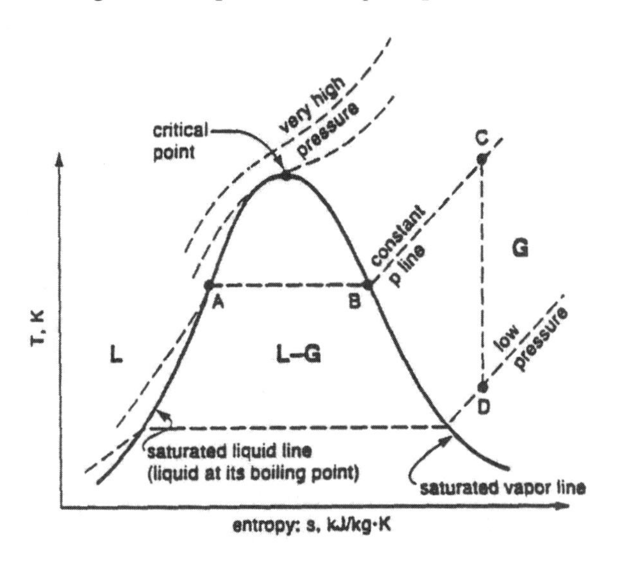

Figure 3.1. Example of a temperature-specific entropy diagram for a pure substance. Point C corresponds to state 1 and Point D to state 2. Source: Levenspiel (1996).

3 Of course, the charts must cover the ranges of operating conditions of interest to our analysis. Normally, diagrams like Figures 3.1 and 3.2 do not extend to low temperatures where liquid working fluids solidify. Even when suspended in gases or liquids, solids are difficult to pump around in heat engines and are not used as working "fluids."

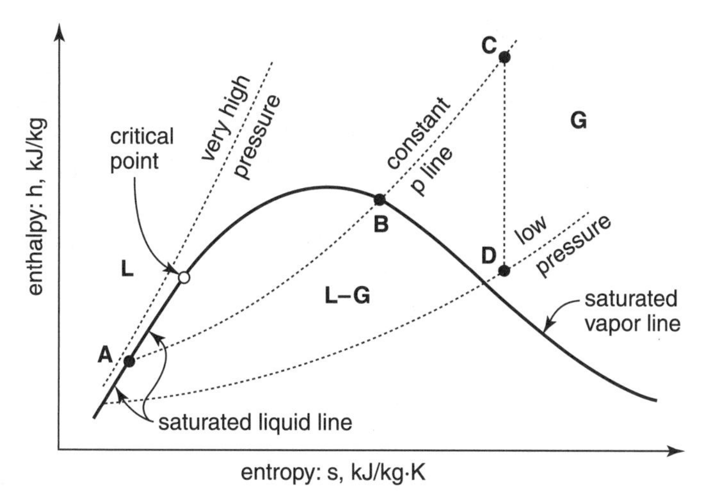

Figure 3.2. Example of a specific enthalpy-specific entropy (Mollier) diagram for a pure substance. Point C corresponds to state 1 and Point D to state 2. Source: Levenspiel (1996).

Thus, from Figure 3.1 we can immediately find s_1 at state point C = s_2 at state point D where a line perpendicular to the T axis at T_1 intersects the curve of constant P_1. The value for T_1 then follows immediately by dropping a line from this point perpendicular to the s axis to intersect the line of constant P_2 (Point D). Let the initial conditions be those of the isothermal expansion calculations illustrated in Equations (3-5) through (3-15), i.e., $P_1 = 10$ atm, $T_1 = 25°C$, and 20 moles of gas which we further assume to be air. We must specify what gas we are using because T-s and h-s behavior depends on the gas. Using the appropriate T-s chart for air, we would find that $P_2 = 1$ atm and T_2= -119°C, i.e., the gas is appreciably cooled. Because $Q_{rev} = 0$, it can be shown from the First Law of Thermodynamics that the reversible work provided by this expansion, W_{rev}, equals $n\Delta h$ (n, the number of moles of working fluid is needed because Δh is the specific enthalpy change for one mole of material). Again, we are neglecting any change in kinetic and potential energy. Because $\Delta s = 0$, we can calculate Δh directly from Figure 3.2 by measuring the vertical distance (i.e., line CD) between the two lines for P_1 and P_2 at the point s_1. Note that we already determined s_1 using Figure 3.1. Using the h-s chart for air and the above conditions, we would find that W_{rev} = 14,270 cal. Knowing P_1 and T_1 (initial conditions), P_2 (given), and T_2 (calculated above from the T-s chart), we can then use an appropriate EOS to calculate the work to push back the atmosphere, W_{PV} in an equation analogous to Equations (3-13) and (3-14):

$$W_{PV} = p_{ext}[V_2 - V_1] = P_{ext}\, nR[T_2/P_2 - T_1/P_1] = 4{,}950 \text{ cal} \qquad (3\text{-}15a)$$

so that the net work, W_{sh}, would be 14,270 - 4,950 = 9,320 cal.

We can also use a T-s diagram to determine the entropy change when a working fluid undergoes a change in phase (e.g., from liquid to vapor and vice versa). These processes, evaporation and condensation respectively, are crucial steps in many practical heat engines. For a pure substance at a constant pressure, a phase change occurs at a constant temperature. Thus, we can obtain Δs for evaporation or condensation at a prescribed pressure. This is done by measuring the horizontal distance between the curve for saturated liquid in equilibrium with its vapor and the curve for saturated vapor in equilibrium with its liquid (for example, line AB in Figure 3.1). Figure 3.1 also shows that the entropy change for vaporization of a pure substance can change appreciably with pressure (i.e., the distance A-B changes depending on what line of constant pressure pertains to our system). We will see shortly that a *T-s* diagram conveniently summarizes heat and work flows in energy conversion cycles.

3.2.3 Analysis of ideal (Carnot) heat engines

We now define a *heat engine* more precisely as any device or machine that converts heat into work by a *cyclic process*. A cyclic process is a set of changes in the thermodynamic state (i.e., the temperature, pressure, and composition) of a working fluid that culminates in restoration of the fluid to the thermodynamic state from which the changes began. Not surprisingly, heat engines extract the energy they need to function by heating their working fluid. To complete the cyclic process, at least one other operation must occur in the engine (i.e., cooling, compression, expansion, evaporation, or condensation of the working fluid).

Curiosity or pragmatism suggests an obvious question: what is the maximum amount of work obtainable when a particular engine absorbs a specific amount of heat? Carnot (ca.1811) provided the answer by logical reasoning. Carnot's analysis shows that for two heat reservoirs at different temperatures ($T_1 > T_2$) there are only four possible situations (of eight imaginable) by which heat Q, and work W, can be exchanged with a reversible heat engine (Figure 3.3). In Figure 3.3, Case (a) is a Carnot heat engine and Case (b) is a Carnot Heat Pump. The other two cases not shown, although technically feasible, are of little practical interest because they, respectively, expend work and heat unnecessarily. Carnot showed that the maximum efficiency η_c of a heat engine is given by:

$$\eta_c = \frac{W}{Q_1} = \frac{Q_1 - Q_2}{Q_1} = \frac{T_1 - T_2}{T_1} \tag{3-16}$$

The second entry in Equation (3-16) is intuitive in that it states that the maximum efficiency of work production must be the fraction of heat extracted from the higher temperature reservoir Q_1, that ends up as work W. The third entry derives from the obvious energy balance that $Q_1 = W + Q_2$, while the last entry arises from the direct proportionality between reservoir temperatures in Carnot heat engines and the amount of heat flowing to or from those reservoirs. This relationship was proved by Kelvin.

We also wish to know the best possible performance attainable from a heat pump, which is a Carnot heat engine operated in reverse (Figure 3.3b). Carnot's methods again provide the answer. We define "*summer*" and "*winter*" *coefficients of performance*, $(COP)_s$ and $(COP)_w$, respectively. $(COP)_s$ pertains to use of the heat pump as a refrigerator or air conditioner to remove unwanted heat (Q_2, Figure 3.3b). $(COP)_w$, is applied when the heat pump is used as a heater to supply heat (Q_1, Figure 3.3b) e.g., to warm a building. The resulting expressions are:

$$(COP)_s = \frac{Q_2}{W} = \frac{T_2}{T_1 - T_2} \quad (3\text{-}17)$$

and

$$(COP)_w = \frac{Q_1}{W} = \frac{T_1}{T_1 - T_2} \quad (3\text{-}18)$$

The last term in Equations (3-17) and (3-18) reflects the energy balance (from the First Law of Thermodynamics) $W = Q_1 - Q_2$ and use of Kelvin's relation to replace Q_1 and Q_2 by T_1 and T_2, respectively.

Carnot proved three important theorems about heat engines: (1) all reversible heat engines operating between the same two temperatures, T_1 and T_2, must have the same efficiency; (2) reversible heat engines have the highest efficiency, for the same reservoir temperatures T_1 and T_2; and (3) for the same high temperature T_1, the reversible heat engine that has the larger temperature difference between T_1 and T_2, has the higher efficiency (i.e., it produces more work).

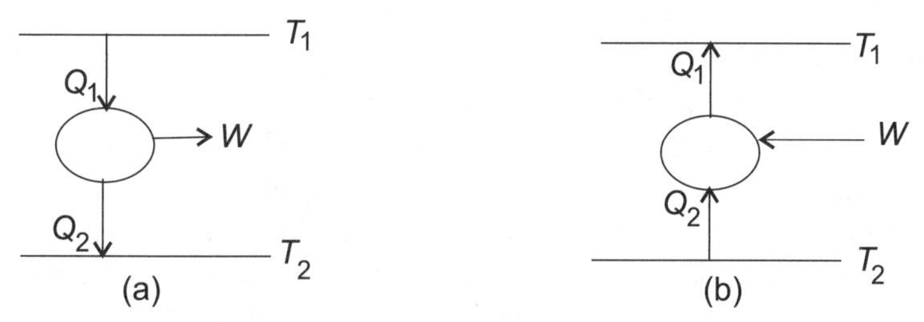

Figure 3.3. Two combinations of continuous heat and work flows between a high temperature (T_1) and a low temperature (T_2) heat reservoir, allowed by the First and Second Laws of Thermodynamics. The ellipse represents a device that can produce work from heat or utilize work to move heat from a lower to a higher temperature. For reversible systems (see text), Case (a) is a Carnot heat engine and Case (b) is a Carnot heat pump. Adapted from Levenspiel (1996).

Figure 3.4 shows what a typical Carnot cycle would look like on a P-V and a T-s diagram. One complete cycle of engine operation entails four sequential changes of state of the working fluid: (1) Step AB: by adding heat Q_{AB} isothermally (T = constant = T_1), do some work W_{AB} by expanding the gas from P_A to P_B; (2) Step BC: do some more work W_{BC} by reversibly and adiabatically ($Q_{BC} = 0$) expanding the gas from P_B to P_C. Because $Q_{BC} = 0$ and the expansion (in this case)[4] cools the gas, the gas temperature falls to a lower value T_2; (3) Step CD: provide work W_{CD} to the engine to remove heat isothermally (now T = constant = T_2) thereby compressing the gas from P_C to P_D; and (4) Step DA: input some more work W_{DA} to the engine to further compress the gas adiabatically ($Q_{CD} = 0$) and reversibly until T rises to T_1. Thermodynamic equations for computing the work in each step are given in Table 3.1. W_{BC} and W_{DA} are equal in magnitude and opposite in sign. Thus, the cycle efficiency is the work derived from Step AB minus that expended in Step CD, divided by Q_{AB} the amount of heat added to the machine, i.e.:

$$\eta = \frac{W_{AB} - W_{CD}}{Q_{AB}} \tag{3-19}$$

i.e., from Table 3.1:

$$\eta = \frac{R(T_1 - T_2)\ln\left(\frac{P_A}{P_B}\right)}{RT_1\ln\left(\frac{P_A}{P_B}\right)} = \frac{T_1 - T_2}{T_1} \tag{3-20}$$

as expected from Equation (3-16).

3.2.4 Analysis of real world (irreversible) heat engines

Basic principles. We now consider the thermodynamics of practical, or irreversible, heat engines. Although Carnot cycles are idealizations, they are useful for analysis of real-world machines. To begin, we must recognize that all practical engines end up wasting some of the inputted heat that in a perfect engine would be converted to useful (shaft) work. Simply stated, in the real world, there will be defects in every operating stage of the engine (i.e., in steps AB, BC, CD, and DA in Figure 3.4 and Table 3.1). But why? Neither Step AB nor CD can be perfectly isothermal because all of the working fluid cannot be instantaneously heated to T_1 or instantaneously cooled to T_2. Likewise, steps BC and DA can never be truly adiabatic because there are no perfect thermal insulators. In fact, many factors degrade the performance of actual heat engines, including: (1) heat losses when transferring heat from the high temperature reservoir to the machine and from the machine to the cold reservoir; (2) leakage of heat from the

4 Up to a certain temperature (the *Joule-Thompson inversion temperature*), all gases and vapors cool when expanded adiabatically. Above this temperature, gases *heat up on expansion*. These forces are not captured in the ideal gas EOS of Equation (3-4). Most gases and vapors of interest in sustainable energy have high *Joule-Thompson inversion temperatures* and cool when expanded adiabatically at room temperature and even much higher temperatures. This fact is practically utilized in many of today's refrigeration systems. Molecular hydrogen (H_2) is a notable exception. It heats up when expanded adiabatically at temperatures above -80°C. This effect, plus the low ignition energy and broad flammability limits of H_2 in air (Chapter 7), mean that even H_2 stored or pipelined at low temperatures (even down to -110°F) poses a risk of igniting spontaneously if it is rapidly decompressed (e.g., owing to rupture of a containment vessel) and mixed with air. This is an important safety issue for storage, transport, and utilization of H_2 and clearly of concern should H_2 become a major energy carrier (Chapter 16).

working fluid to the engine hardware and then to the surroundings; (3) pressure losses from friction between the working fluid and engine parts; (4) generation of waste heat from mechanical friction between parts;[5] (5) expansion or contraction of working fluids when they flow past or through various mechanical obstructions such as valves and nozzles; (6) mixing of different components or different phases of a working fluid; (7) chemical reactions (e.g., thermal degradation of the working fluid, corrosion of the machine parts).

We can analyze the behavior of real heat engines using the methods we learned for the ideal Carnot cycle. Operating *irreversibilities* cause practical heat engines to have *T-s* diagrams (Figure 3.4d) that are more complex than those of a reversible cycle (Figure 3.4b). Because of these performance flaws, a smaller amount of useful work is obtained from an actual heat engine than from a reversible engine operating between the same temperature limits. Mathematically:

$$W_{produced}^{heat engine} < W_{rev} \tag{3-21}$$

$$W_{required}^{heat pump} > W_{rev} \tag{3-22}$$

Even though the Carnot heat engine is "ideal," this does not mean that it has a heat-to-work efficiency of 100%. Its "perfect" efficiency is only that allowed by Equation (3-16). Nonetheless, the Carnot efficiency continuously increases as the upper reservoir temperature increases. Therefore, it is reasonable to hypothesize that we could improve the efficiency of a practical heat engine by also increasing its upper operating temperature. For example, we could increase the temperature at which the working fluid expands (Step AB in Figure 3.4). It turns out that this works in practice.

Carrying this idea further: can't we just keep making practical heat engines that operate at higher and higher temperatures so that we can realize better and better efficiencies? The idea is appealing but, at some point, other limitations appear. For example, above a certain operating temperature, all materials of construction will fail and all working fluids will undergo chemical degradation. Both of these processes drain away energy and, ultimately, would physically destroy the engine. Cost is another practical factor. Suppose we could develop chemically inert working fluids and construction materials extraordinarily resistant to high-temperature degradation. We would still need to worry about financing the manufacture and safe operation of practical devices to take advantage of these improvements.

5 In Chapter 18, we discuss an interesting tradeoff with serious sustainability overtones—how to optimize use of lubricants in automotive engines to reduce friction and needless waste of energy while avoiding particulate and other pollutants traceable to lube oil.

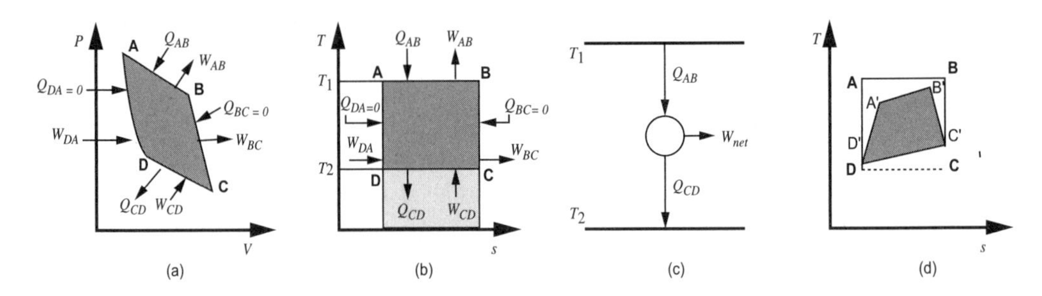

Figure 3.4. (a) pressure-specific volume (*P-V*) and (b) temperature-specific entropy (*T-s*) diagrams for operation of a Carnot heat engine (c); (d) *T-s* diagram for an irreversible (i.e., actual heat) engine. In (a), (b), and (d) the area of the shaded region ABCD or A′B′C′D′ is equal to the net useful work (W_{net}) obtained from each heat engine. In (b) the wasted heat (Q_{CD}) is given by the area of the rectangle below ABCD formed by the line segment DC and the x-axis below. Similarly in (d), the wasted heat ($Q_{C'D'}$) is given by the quadrangle area formed by D′C′ and the x-axis below. The lower efficiency of the real world engine is clear from the smaller W_{net} and larger $Q_{C'D'}$ areas of (d) versus Q_{CD} of (b). Table 3.1 provides thermodynamic equations for calculating the four work terms in Part (b) for the case where the working fluid is one mole of an ideal gas. Adapted from Levenspiel (1996).

Table 3.1. Thermodynamic Equations for Computing the Work Generated or Consumed in the Four Steps of the Carnot Heat Engine Shown in Figure 3.4 (c)[a]

Step AB (isothermal expansion): $W_{AB} = Q_{AB} = RT_1 \ln\left(\frac{P_A}{P_B}\right)$

Step BC (reversible adiabatic expansion): $W_{BC} = C_v\,(T_2 - T_1)$

Step CD (isothermal compression)[b]: $W_{CD} = Q_{CD} = RT_2 \ln\left(\frac{P_C}{P_D}\right) = -RT_2 \ln\left(\frac{P_A}{P_B}\right)$

Step DA (reversible adiabatic compression): $W_{DA} = C_v\,(T_1 - T_2)$

[a]The equations assume reversible changes of state for one mole of an ideal gas of constant heat capacity at constant pressure.

[b]The last entry is obtained from the P,T relationship for reversible adiabatic expansion and compression of an ideal gas, i.e., in this case $\left(\frac{P_B}{P_C}\right) = (T_1/T_2)^{\frac{\kappa}{\kappa-1}} = \left(\frac{P_A}{P_D}\right)$, where κ is the ratio of the gas specific heats at constant pressure and constant volume. Thus, $\left(\frac{P_C}{P_D}\right) = \left(\frac{P_B}{P_A}\right)$.

Costs, materials challenges, and inherent efficiencies notwithstanding, we know from experience that practical heat engines run well enough to be useful to humans. Thermodynamics plays a central role in enabling engine operability and efficiency improvements. To illustrate, let's look at a few practical engines in more detail. We define two additional performance indices for practical heat engines: (1) the cycle efficiency, η_{cycle}, which is the net useful work produced divided by the heat input to the machine; and (2) the thermodynamic utilization efficiency, η_u, which compares the net work (or power) produced by the machine to the maximum possible work (or power) that could be produced under fully reversible conditions by the corresponding Carnot heat engine:

$$\eta_{cycle} \equiv \frac{W_{net}}{Q_{in}} \tag{3-23}$$

$$\eta_u \equiv \frac{W_{net}}{W_{Carnot}} \tag{3-24}$$

η_{cycle} is the actual efficiency of our real world engine. η_u tells us how well this engine is doing compared to the best it would do if it performed perfectly as a reversible Carnot engine. To test your understanding of these ideas, see if you can prove the following, using Equation (3-16):

$$\eta_{cycle} = \eta_u \, \eta_{Carnot} = \eta_u \left(\frac{T_1 - T_2}{T_1} \right) \tag{3-25}$$

Equation (3-25) tells us that a high absolute efficiency requires large values for both η_{carnot} and η_u.

Stationary power cycle systems. We concentrate on three distinct heat engine cycles and a hybrid of two of them. All four cycles feature prominently in modern energy conversion hardware and two are used in generating electricity from alternative energy sources. The Rankine cycle is widely used in large central plants that generate electricity from fossil fuel combustion (Chapter 7) or nuclear fission (Chapter 8). It is also applied in producing electricity from alternative heat sources, such as solar (Chapter 13), geothermal (Chapter 11) and biomass (Chapter 10). Rankine cycles can reach an impressive scale (e.g., one coal-fired electric utility boiler can generate 350 MW of electric power). A single large nuclear-fueled Rankine boiler occupies less "firebox" volume than a 350 MW coal-fired unit and produces over four times as much electric power—enough capacity to meet the electricity needs of about 1.3 million people.

Figure 3.5 shows typical hardware found in a Rankine cycle. Each of the six cycle steps (Figure 3.6)—compression of a liquid working fluid (AB), preheating (BC) and vaporization (CD) of that fluid in a boiler, raising the temperature of the resulting vapor (superheating) (DE), production of work by expansion of the superheated vapor in a turbine (EF), and conversion of the vapor to liquid in a condenser (FA)—have one or more inefficiencies that squander energy and thereby downgrade the Rankine cycle performance versus that of the ideal Carnot cycle. To see this better, let's compare these cycles side-by-side on a *T-s* diagram (Figure 3.6). The Carnot cycle performs some work while absorbing heat (Step B′E) at constant temperature (T_1), and consumes some work while rejecting heat (Step FA) at constant temperature (T_2). Real heat engines typically use an inexpensive working fluid like steam. In this case, the two steps analogous to B′E and FA in the reversible engine occur with the fluid in the vapor-liquid region (i.e., by boiling and condensing the fluid) (Steps CD and FA, Figure 3.6). However, a practical problem sets in. Turbine and compressor equipment which are used to extract work from or add work to fluids do not operate well on vapor-liquid mixtures owing to erosion and other damage caused by the collapse of gas bubbles or formation of liquid droplets that are created when the pressure suddenly changes (e.g., cavitation). To avoid these problems, Rankine cycles are designed so that their expansion (work extraction) and compression steps occur while the working fluid exists as one phase (i.e. as vapor and liquid, respectively). This adds steps and complicates the cycle (Figure 3.6).

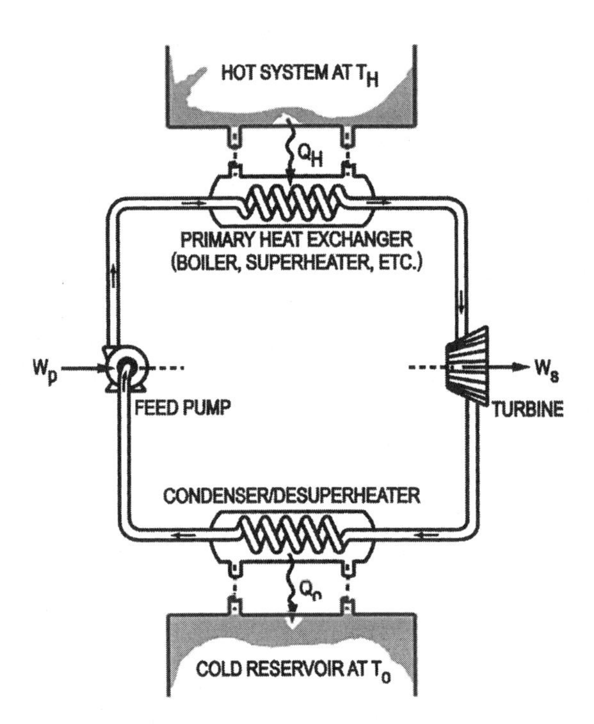

Figure 3.5. Schematic of a typical Rankine heat engine cycle. Source: Tester and Modell (1997). Reprinted with permission of Pearson Education, Inc.

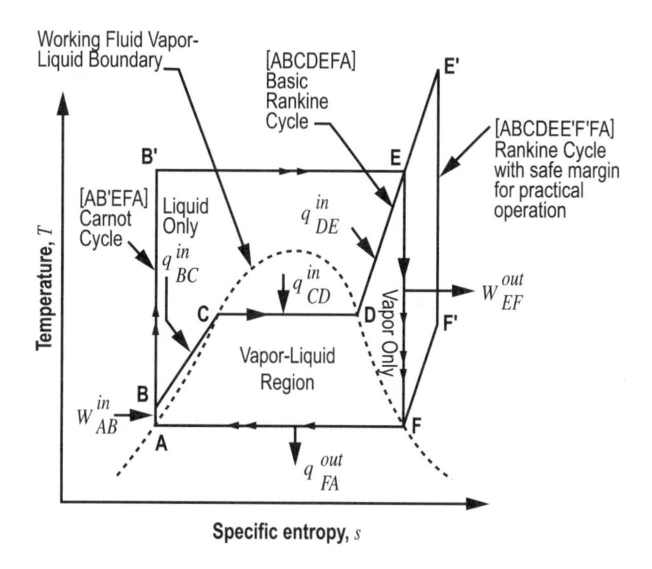

Figure 3.6. Temperature-specific entropy diagram for a Rankine power cycle (ABCDEFA), its Carnot idealization (AB′EFA), and a Rankine cycle designed to prevent condensation of working fluid vapor to liquid during expansion in the work recovery turbine (ABCDEE′F′FA). The dashed curve defines coexistence of working fluid vapor and liquid.

In Figure 3.6, the bell-shaped dashed curve separates liquid only, vapor only, and coexisting liquid and vapor regions. The areas AB′EF and ABCDEF, respectively, denote the useful work provided by the Carnot and the Rankine cycle. The lower efficiency of the practical Rankine cycle is evident from its smaller area on the T-s chart. We can relate this efficiency, $\eta_{Rankine}$, to changes in the enthalpy, h_i, of the working fluid by writing an energy balance for the cycle using Figure 3-6:

$$\eta_{Rankine} = \frac{q_{in} - q_{out}}{q_{in}} = \frac{q_{BC}^{in} + q_{CD}^{in} + q_{DE}^{in} - q_{FA}^{out}}{q_{BC}^{in} + q_{CD}^{in} + q_{DE}^{in}} = \frac{(h_E - h_B) - (h_F - h_A)}{(h_E - h_B)} \tag{3-26}$$

i.e.,

$$\eta_{Rankine} = \frac{(h_E - h_F)}{(h_E - h_A)} \tag{3-27}$$

where we have neglected the small amount of work needed to compress the liquid (i.e., we have assumed $h_B = h_A$). A further complication is that practical engines typically operate with a margin of safety to avoid turbine and compressor damage. This is commonly done by maintaining the working fluid entirely as vapor throughout the expansion step and entirely as liquid for all of the compression stage. The former would incur efficiency penalties by extending the cycle into the EE′F′F region (Figure 3.6).

The extra steps are heating to higher temperature before turbine expansion, terminating the expansion at a higher temperature, $T_{F'}$, and adding an additional cooling step (F′F) after expansion to desuperheat the vapor to its condensing temperature.

Rankine cycles have an important practical restriction. The durability of construction materials sets an upper limit on the operating temperature for Rankine boilers (tubes). Flame temperatures (Chapter 7) from combustion of common fossil and biomass fuels in air are typically between 1,000-1,500°C, whereas boiler tube materials cannot tolerate high-pressure steam at more than about 600°C. This large temperature difference (or ΔT as engineers frequently call it) between the hot combustion gases and the colder primary working fluid is marvelous for speeding up heat transfer from the flame to the outside of the boiler tubes (Section 3.5), but degrades appreciable work-producing potential. In response, engineers have developed another cycle to directly utilize more of the high-temperature heat energy of fuel combustion. This is the Brayton cycle, which powers many stationary electric-power-generation turbines fueled with natural gas and oil.

The steps in the Brayton cycle are: (1) compression of air which is then mixed with liquid or gaseous fuel in a combustion chamber; (2) addition of heat to the pressurized mixture of air and fuel by burning the fuel, essentially at constant pressure; (3) expansion of the hot combustion gases through the turbine to generate useful work. The major combustion products are CO_2 and H_2O, plus N_2 and some oxygen not consumed by combustion; there are also traces of pollutants like NO_x (Chapter 5); and (4) rejection to the environment of heat remaining in the turbine exhaust gases after they exit the turbine.

An introductory analysis is possible by approximating the four steps of the Brayton cycle (Figure 3.7) as adiabatic reversible compression (AB), isobaric heating (BC), adiabatic reversible expansion (CD), and isobaric cooling (DA). For each step, the working fluid is in the vapor phase but in contrast to the Rankine cycle, there are no boiling or evaporation steps. There are no boiler tubes and no need to worry about limitations on materials compatibility with steam at the high temperatures and pressures of the Rankine cycle. Another advantage of the Brayton cycle is that it generates heat (q_{BC}, Figure 3.7) directly within the engine by combustion of the fuel. This eliminates the inefficiencies of transferring heat from an exogenous source through the engine walls. Ultimately, the Brayton cycle is one type of *internal combustion engine*.

The Brayton cycle does have materials challenges. The higher the inlet temperature of the working fluid tolerable by the turbine blades, the better the efficiency of the Brayton cycle (Equations (3-16) and (3-25)). If the steps in the Brayton cycle are approximated as described above (Figure 3.7) and ideal gas behavior is assumed, it can be shown that the heat-to-work efficiency of this cycle is given by:

$$\eta = \frac{T_B - T_A}{T_B} = 1 - \left(\frac{P_A}{P_B}\right)^{\frac{\kappa-1}{\kappa}} \tag{3-28}$$

where κ is the ratio of heat capacities (C_p/C_v). Innovations in materials, or turbine-blade cooling that enable Brayton cycles to operate at higher temperatures, are good examples of *sustainable energy technologies* as they lead to more efficient use of fuels. Further, because the Brayton compression step involves a gas instead of a liquid, much more compressor work is required because a much larger volume of material must be handled (gases have a much higher volume per unit mass than do liquids). Consequently, Brayton cycle compressors must have high efficiencies. As we might expect, a high cycle efficiency also requires that the turbine efficiency be high. Modern Brayton compressor and turbine efficiencies now approach 90%, leading to total cycle efficiencies of 28% or better. Brayton cycles are used in stationary turbine engines (single and combined cycle, see below) for electric power generation and in jet engines and other mobile turbine engines, such as those found in buses and heavier duty trucks. The book by Wilson (1984) is an excellent source of detailed quantitative information on Brayton cycles and gas turbine engines.

Brayton efficiencies are still far from the Carnot limit because of a disadvantage when this cycle is operated in isolation. The turbine exhaust gases are still at a high temperature (relative to ambient (see end of Step D; Figure 3.7)). These hot gases still have considerable work-producing potential, much of which is surrendered to the surroundings as heat, q_{DA}^{out} during the cooling step, DA.

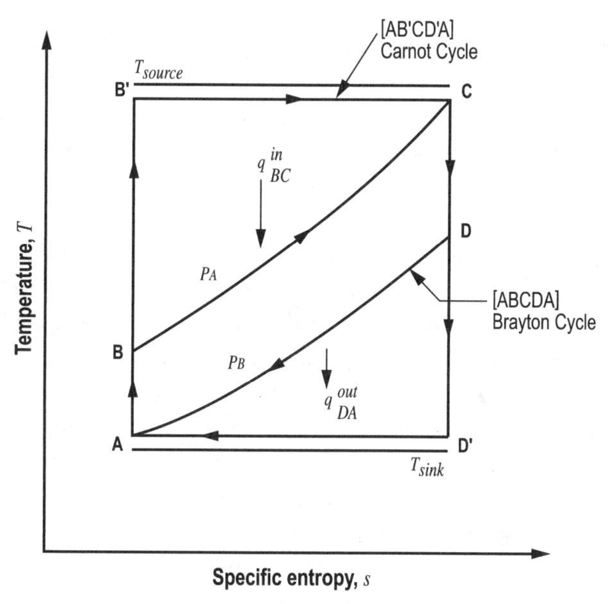

Figure 3.7. Temperature entropy (T–s) diagram for a Brayton power cycle (ABCDA) and the Carnot cycle idealization (ABB′CDD′A) operating between fixed source and sink temperatures, T_{source} and T_{sink}. Note the absence of a vapor-liquid coexistence curve because the working fluid remains in the vapor phase throughout the cycle.

Another type of cycle, the *integrated or combined (power) cycle,* has been developed to recover additional work from heat rejected by a power cycle at elevated temperatures. One of the most common combined power cycles thermally joins a Brayton and a Rankine cycle. In this approach, the Brayton turbine exhaust heat is used to heat the working fluid of the Rankine cycle. Thus, the waste heat of the Brayton (*topping*) cycle becomes the heat source of its companion (*bottoming*) Rankine cycle. Figure 3.8 shows a *T-s* diagram for a combined Brayton-Rankine cycle with combustion gases and water as the working fluids. A higher overall efficiency results from the combined cycle. This is easily seen in Figure 3.8, recalling that the useful work from a power cycle is the internal area circumscribed by the cycle steps in the *T-s* diagram. State-of-the-art combined power cycles consist of a natural gas-fueled topping (Brayton) cycle linked to a steam (Rankine) bottoming cycle. In modern electric power plants, these combined cycles provide total efficiencies of 60% or more, with very low NO_x emissions (Chapter 7). Because the efficiency of converting fuel (chemical) energy to cycle input heat exceeds 98%, the corresponding fuel-to-electricity efficiencies (i.e., electrical energy generated as a percentage of fuel energy inputted) for these cycles are close to 60% or better.

Heat rejected by a power cycle may also be used for space heating or generation of steam for process applications. This is referred to as *cogeneration* or *cogen.* It is also a form of combined cycle—rigorously a *combined energy cycle* rather than a combined power cycle. Typical cogen applications include the use of waste heat from Rankine or Brayton electric power generation cycles for district heating and heating institutional buildings, such as those at universities and hospitals. There have also been proposals to use smaller-scale, gas-fired heat engines to supply electricity and residential heating to apartment buildings and single-family homes (Chapter 20) and in other distributed energy applications (Chapter 16).

Internal combustion engines. So far, we have not considered the heat-to-work cycles that move most of the people and freight in modern-day road transport. Virtually all automobiles, trucks, and buses are powered by internal combustion (IC) piston engines that operate, to a reasonable first approximation, according to the Otto (spark ignition, SI) or diesel (compression ignition, CI) power cycles. The SI engine fueled with gasoline is used in most cars, Sport Utility Vehicles (SUVs), and some light-duty trucks. The CI (diesel) engine is used in most buses and heavy-duty trucks, as well as some light-duty trucks and a considerable number of cars, especially outside North America. Advances in alternative means of moving people and goods (public transport, information technology, alternative on-the-road vehicle propulsion technologies) are expected over the long term to temper and eventually displace IC engines in human mobility (Chapter 18). But for the foreseeable future, SI and CI piston engines will continue to play a dominant role in the transportation sector.

The impact of this forecast on sustainability issues may not be as severe as some might perceive. This is because over the last 30 years there have been continuous evolutionary advances in the technology to improve fuel efficiency and reduce pollution. Further, there are encouraging prospects for additional progress on SI and CI engine technologies and on devising cleaner fuels for these engines (Chapters 7, 18, and 23).

Here, we set out some basic tools for analyzing SI and CI power cycles. To understand the thermodynamics of the SI-internal combustion engine power cycle, we approximate it as an ideal Otto cycle. A chart that maps out changes in the pressure and temperature of the working fluid within the engine (i.e., a *P-V* diagram; Figure 3.9a), amplifies the standard *T-s* diagram (Figure 3.9b). The Otto cycle steps are: (1) a mixture of air and vaporized fuel is admitted to an engine cylinder at constant pressure by drawing back a piston (ED) where (2) they are adiabatically compressed by inward movement of the piston (DA) to give a hot, high pressure, fuel vapor-air mixture, which (3) is then *ignited* by discharge of an electrical spark to begin rapid oxidation of the fuel (*combustion*) (Chapter 7). Rapid release of heat by combustion raises the working fluid temperature and pressure so quickly that the fluid undergoes essentially constant volume pressurization in the cylinder (AB); (4) this hot, high-pressure mixture of combustion products and vitiated air then undergoes adiabatic expansion (BC) to provide useful work by pushing back the engine piston; (5) some of the working fluid, having divested much of its power-producing capacity, is exhausted from the cylinder at constant volume by opening the cylinder exhaust valve and allowing the gas, still at a pressure higher than ambient, to flow out (CD); and (6) to complete the cycle, the remaining exhaust gases are ejected from the cylinder at constant pressure by an inward stroke of the piston (DE′).

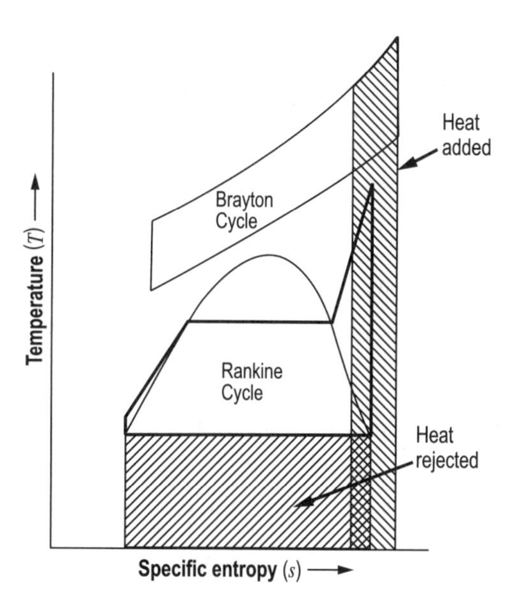

Figure 3.8. Temperature-specific entropy *T-s* diagram for a combined power cycle. Heat from the higher temperature exhaust of the Brayton (topping) cycle (e.g., a gas turbine) enters the lower temperature steam Rankine (bottoming) cycle.

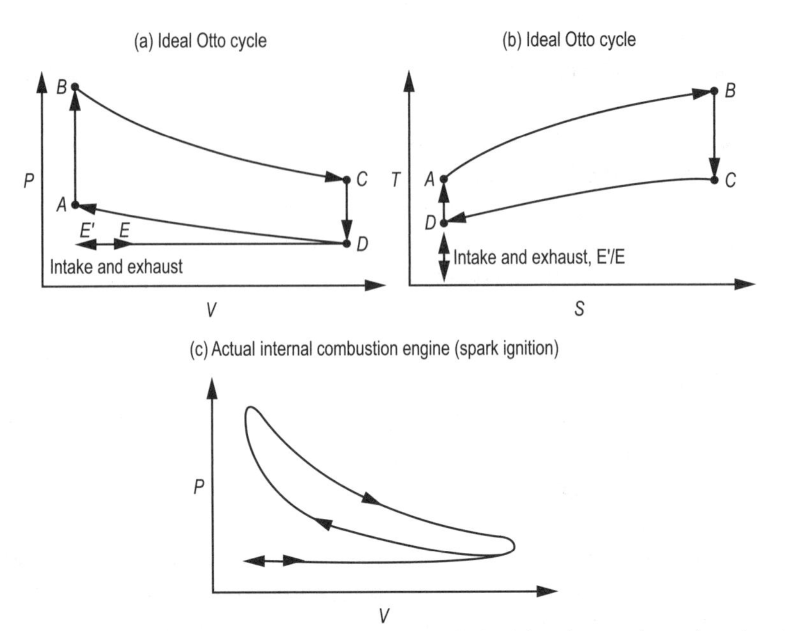

Figure 3.9. Power cycle diagrams for the spark ignition internal combustion (SI-IC) piston engine (i.e., the common automobile "gasoline" engine): (a) *P-V* and (b) *T-s* diagrams for the Otto cycle approximation, and (c) *P-V* diagram for an actual engine. Source: Tester and Modell (1997). Reprinted with permission of Pearson Education, Inc.

The Otto approximation does not reveal the full thermodynamic complexity of an actual SI-IC engine. To see this, compare the *P-V* diagrams for the ideal Otto cycle (Figure 3.9a) and for an actual SI engine (Figure 3.9c). Irreversibilities occur in the compression and expansion steps because of heat losses and mechanical friction between the pistons and cylinders. The Otto approximation is also compromised by imperfections in the combustion process, in the transfer of heat and of the working fluid within and beyond the engine cylinders, and by the non-ideal thermodynamic behavior of the working fluid mixture. Nevertheless, the Otto cycle analysis illustrates important features of SI-IC engine behavior, including how operating parameters impact engine efficiency.

By applying the First and Second Laws and assuming ideal gas behavior for the working fluid, the efficiency of the Otto engine is given by:

$$\eta_{Otto} = 1 - \left[\frac{1}{r_c}\right]^{\kappa-1} \tag{3-29}$$

where

$$r_c \equiv \left[\frac{V_C}{V_B}\right] = \left[\frac{P_B}{P_C}\right]^{\frac{1}{\kappa}}$$

is the *compression ratio*. Equation (3-29) tells us that, for the SI engine, efficiency increases with increasing compression ratio, but *also* with increasing ratio of working fluid specific heat at constant pressure to that at constant volume, $\kappa \equiv C_p/C_v$ (note that $\kappa>1$). Next time you're haggling about price and gas mileage during the purchase of a car, you may be tempted to ask about the compression ratio of the vehicle's engine *and* the specific heat ratio of its working fluid! The salesperson may tell you that he or she can't offer you very much leeway on κ, because the working fluid composition, largely air depleted of O_2 plus CO_2 and H_2O from gasoline combustion, won't change much, even with gasoline octane rating. There may be some choice with respect to engine type and therefore, some modest difference in r_c, but it is likely to be small.

The thermodynamics of the CI-IC (diesel) engine are a little more complicated, but we dare not leave out the diesel, given its widespread use in transportation. Advanced diesel engines also offer promise for increasing road vehicle efficiencies (i.e., passenger miles per gallon). A key hurdle is the development of particulate emissions control technologies compatible with current and emerging fuel formulations (Chapters 7, 18). Figure 3.10 shows *P-V* and *T-s* charts for the ideal diesel cycle, a good first approximation for thermodynamic analysis of the CI-IC power cycle. The diesel cycle steps are: (1) air is drawn into the engine cylinder from the surroundings, by extending the piston (E′D); (2) the inlet valve closes and the air is compressed reversibly and adiabatically by inward movement of the piston, which heats up the air (DA); (3) diesel fuel (as fine droplets of liquid) is injected into this hot, high-pressure air where it combusts. Diesel fuel, however, burns more slowly than gasoline in the SI engine. Consequently, the burning diesel fuel releases its heat more slowly, which causes the working fluid to expand isobarically and push back the piston until injection of the fuel stops (AB); (4) the hot, high-pressure products of combustion then undergo reversible adiabatic expansion, generating useful work by pushing the piston farther back (BC); (5) some of the working fluid is then expelled from the cylinder at constant volume by opening an exhaust valve on the cylinder and allowing the high-pressure gas to expand (CD); and (6) to complete the cycle, the remaining exhaust gases are ejected to the surroundings from the cylinder at constant pressure by an inward stroke of the piston (DE).

Application of the First and Second Laws provides a mathematical formula for the efficiency of the ideal diesel power cycle in terms of two ratios, one for compression r_c, and one for expansion r_e, i.e.:

$$\text{Ideal diesel compression ratio: } r_c = V_D/V_A = [P_A/P_D]^{1/\kappa} \quad (3\text{-}30)$$

$$\text{Ideal diesel expansion ratio: } r_e = V_C/V_B = [P_B/P_C]^{1/\kappa} \quad (3\text{-}31)$$

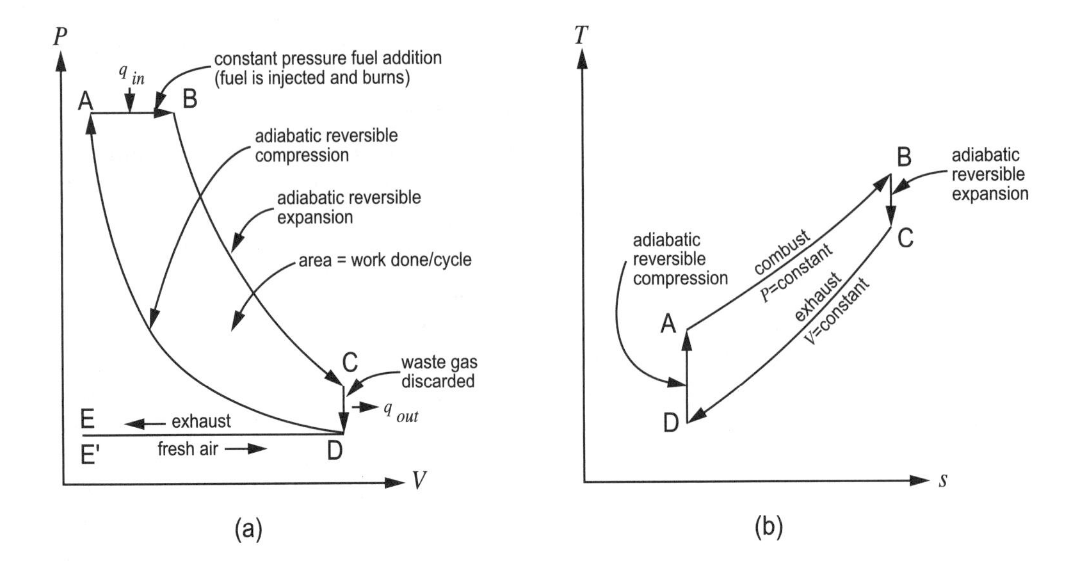

Figure 3.10. (a) *P-V* and (b) *T-s* diagrams for the diesel approximation of the compression ignition, internal combustion piston engine (i.e., the common automative diesel engine) power cycle. Source: Levenspiel (1996).

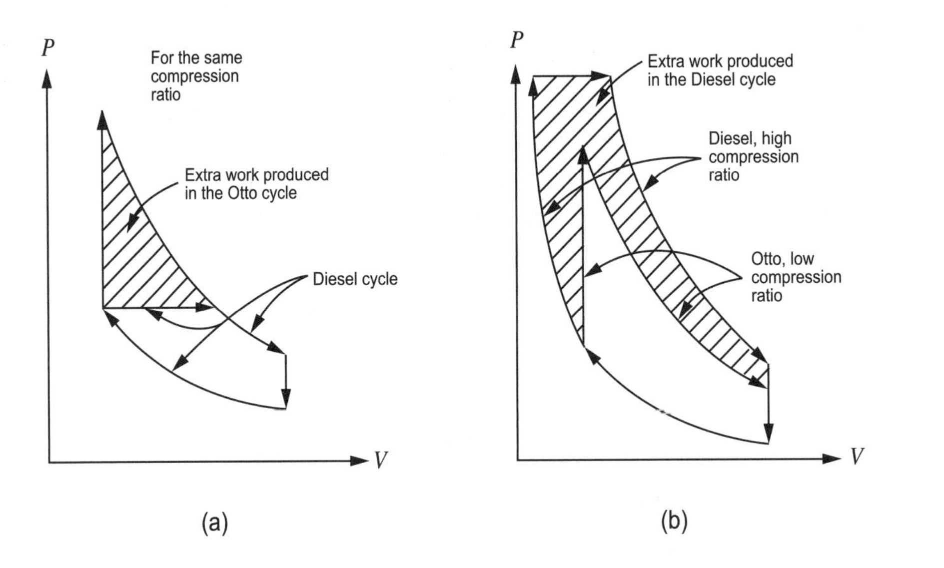

Figure 3.11. Comparison of Otto and diesel power cycles on the same *P-V* diagram where: (a) both engines have the same compression ratio; (b) the diesel has a higher compression ratio, usual in modern diesel engines. Source: Levenspiel (1996).

The pressure ratio terms in Equations (3-30) and (3-31) assume that the working fluid is an ideal gas of constant heat capacity. The ideal diesel efficiency, η_D, is given by:

$$\eta_D = 1 - \frac{1}{\kappa}\frac{\left(\frac{1}{r_e}\right)^{\kappa} - \left(\frac{1}{r_c}\right)^{\kappa}}{\frac{1}{r_e} - \frac{1}{r_c}} \tag{3-32}$$

Equation (3-32) is a good deal more complicated than Equation (3-28) for the ideal Otto cycle. So which engine is more efficient for a passenger car or light-duty truck? Levenspiel (1996) compares the two cycles (Figure 3.11). The "winner" in terms of efficiency depends on whether or not the engines have the same compression ratio. Figure 3.11a shows that if they do, the Otto cycle engine comes out ahead. Typical compression ratios are 8 to 12 for SI engines and 12 to 24 for CI engines (Heywood, 1988).[6] Figure 3.11b shows that because of its higher compression ratio, the diesel cycle efficiency can substantially exceed that of the Otto engine. This is one of the reasons advanced diesels show promise for improving motor vehicle fuel economy. From a sustainability perspective, automotive engines must fulfill multiple performance requirements. For example, the higher compression ratios that improve efficiency must be attainable in commercial-scale engines with acceptably low emissions of NO_x or fine particulate matter (Chapters 4, 18, 21). It is likely that advanced diesel engines will overcome this emissions challenge. So, for on-the-road vehicles, keep your eye on the advanced diesel.

Virtually every analyst agrees that energy efficiency is a cornerstone of sustainable development. A key aspect of achieving high efficiency is understanding how to use thermodynamic analysis to estimate the theoretical and practical efficiencies of heat engines. These are the practical machines that dominate the conversion of primary energy to two of its forms most useful to people—mechanical work and electricity. Although much progress was made in the 20th century, scientists and engineers continue to improve the efficiency of practical heat engines and bring them closer to their reversible limits. Figure 3.12 compares efficiencies from various practical heat engine

6 The lower SI engine efficiency arises in part from fuel properties. In an SI engine cylinder, a compressed parcel of fuel (gasoline), premixed with air, burns rapidly when a flame (ignited by the spark) moves steadily across the cylinder chamber, pressurizing the chamber contents (Figure 3.9a, AB). Excessive pressure buildup may preheat some fuel so much that it ignites before the flame reaches it. The result is premature and rapid burning of enough fuel to generate intense high-frequency pressure oscillations within the cylinder, audible as sharp metallic engine noise (*knock*). Modern gasoline formulations will knock at compression ratios above 8-12. In diesel engines, combustion occurs more slowly when fuel droplets encounter preheated air within the engine cylinder and burn as many tiny flamelets (Figure 3.10, AB). Compression ratios are no longer constrained by premature ignition of fuel (knocking) (J.B. Heywood, pers. comm., 1988).

cycles with different heat source temperatures to their corresponding theoretical best case (Carnot) efficiencies. The figure shows encouraging present-day performance as well as considerable opportunity for further progress *at all source temperatures.*

Heat pumps. Finally, let's look at a Rankine cycle in reverse also known as a *heat pump*. Figure 3.13 shows a hardware schematic and a *T-s* diagram for a heat pump that might be used to warm a building by extracting heat from below the frost line during winter. The cycle steps are: (1) liquid working fluid, returned from the region needing heat, is cooled adiabatically by passing it through an expander (AB); (2) vaporization of this liquid then gobbles up heat Q_{in}, from the ground or other source (BC); (3) by providing shaft work W_{in}, the cycle compresses and heats this vapor (CD); heat Q_{out} is then deposited at the target location by three processes, first, (4), by cooling the working fluid vapor (DE), then, (5), by condensing this vapor to liquid (EF), and finally, (6), by cooling the liquid (FA). For practical heat pumps of this type, typical COPs (Equation (3-18)), i.e., Q_{out}/W_{in} in Figure 3.13, range from 4 to 6.

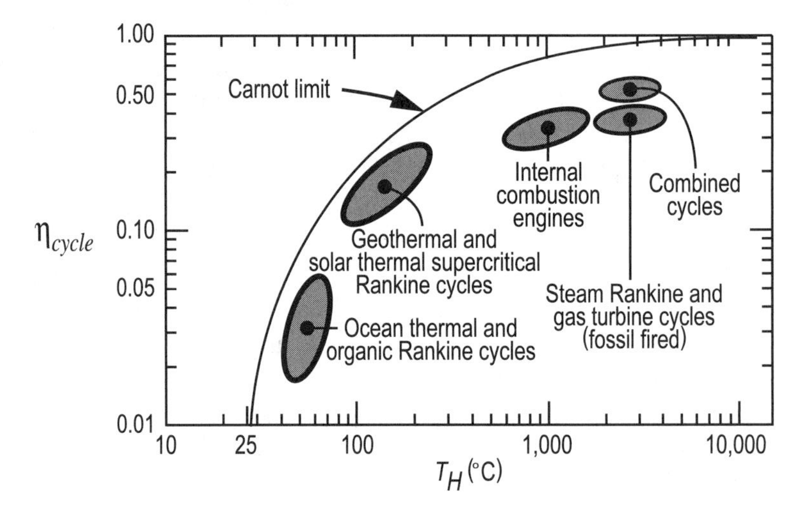

Figure 3.12. Practical efficiencies for heat-to-work conversion in various heat engine cycles of interest in sustainable energy applications and corresponding Carnot limits, as affected by heat source temperature. Heat is assumed to be rejected at average ambient conditions, i.e., 25°C and 1 bar pressure. Source: Tester and Modell (1997). Reprinted with permission of Pearson Education, Inc.

3.3 The Importance of Rate Processes in Energy Conversion

Thermodynamic analysis allows us to keep a balance sheet on where energy goes in various physical and chemical changes (First Law) and to determine the maximum efficiency of converting heat to useful work under various circumstances (Second Law). Thermodynamics cannot tell us how slowly or rapidly energy is generated or exchanged

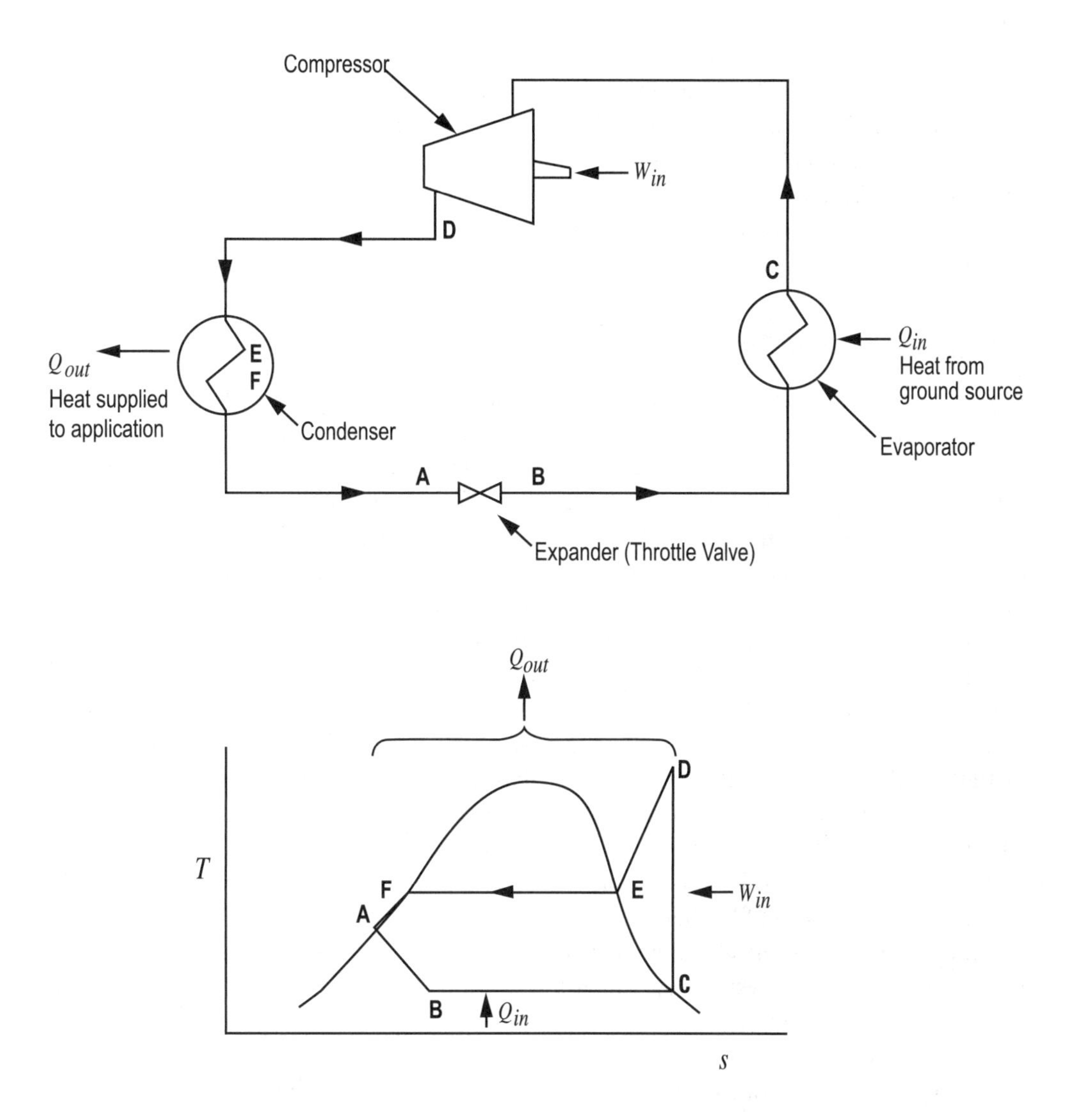

Figure 3.13. Hardware schematic and *T-s* diagram for a typical ground source heat pump. Note that this heat engine is effectively a Rankine power cycle (Figure 3.6) operated in reverse.

within or between systems. Knowledge of rates of energy production and transfer is needed to design and safely operate technologies for the efficient generation, utilization, and conservation of energy. In most cases, the benefits provided by energy depend on the ability of some technology to deliver that energy at a certain rate. One example is building climate conditioning where heating (or cooling) must be achieved rapidly enough to offset heat transfer to or from the surroundings. Safely accelerating a motor vehicle to merge into traffic and ramping up electric power generation to keep pace with sudden demand surges are two more examples. The quantitative analysis of the rates of energy flows require characterization of chemical kinetics and physical transport of mass, heat, and momentum. These disciplines study the kinetics of the production, transmission, and depletion of energy owing to chemical reactions and physical transport including fluid mixing. Now, we will briefly introduce two rate phenomena relevant to a first study of sustainable energy: heat release by chemical reactions and heat transmission by physical mechanisms.

3.4 Chemical Rate Processes

Chemical kinetics play a crucial role in chemical, physical, biological, and geological phenomena important to sustainable energy. They are involved in the growth of biomass; the production of fuel gas (methane) by decay of wastes; the formation and transformation of mineral fuels in the earth; and the generation and modification of pollutants in soils, sediments, aquifers, the atmosphere, and in various energy conversion processes, like combustion (Chapters 4, 7, 10). Focusing on heat release, chemical kinetics affect how rapidly a fuel like biomass, coal, or natural gas can release its chemical energy by oxidation. In some combustors, the oxidation rates are so fast that the entire energy content of the fuel is liberated essentially "instantaneously." By this we mean that the rates of exothermic heat release by breaking chemical bonds are very fast compared to other relevant rate processes in the combustion equipment. Examples of these "other" rates are those for fuel-oxidant (e.g., air) mixing, for heating fuel and oxidant to a temperature high enough to sustain combustion, and for compressing the fuel-air working fluid in an SI-IC engine (Section 3.2.4). In these cases, the overall heat release rate is governed by mixing or other physical transport processes, such as heat transmission (Section 3.5). There are important situations, however, where the rates of chemical reactions themselves determine how fast the fuel burns. In this text, we can only introduce the analysis of chemically-controlled heat release rates. More detailed treatments are found in the combustion literature, e.g., Glassman, 1996, Fristrom, 1995, and Lewis and von Elbe, 1987.

A simplified model of fuel combustion combines fuel and oxidant:

$$fuel + oxidant \rightarrow products + heat \tag{3-33}$$

To make the example more tangible, consider the combustion of methane, the dominant component of natural gas, using molecular oxygen as the oxidant:

$$CH_4 + 2O_2 \rightarrow CO_2 + 2H_2O \tag{3-34}$$

An overall average (sometimes called a *global* or *engineering kinetics*) model for the chemical rate of burning of fuel, according to Equation (3-33), is:

$$\dot{m} = -d[fuel]/dt = k(T)[fuel]^a\ [oxidant]^b \tag{3-35}$$

where $\dot{m}$ is the fuel burning rate, e.g., in moles per second, $[\,i\,]$ denotes (typically molar) concentration of species i (the units are moles per unit volume of reacting mixture), and a and b are experimentally-determined parameters that describe how the observed reaction rate depends on the concentration of the fuel and the oxidant, respectively. In Equation (3-35), the rate is said to be of order a in fuel concentration and of order b in oxidant concentration. The minus sign signifies that fuel is being consumed rather than being formed. The quantity $k(T)$ is the *rate constant*, sometimes referred to in British literature as the *reaction velocity*. One commonly used form of $k(T)$ is the *Arrhenius rate expression*:

$$k(T) = k_o \exp[-E_a/RT] \tag{3-36}$$

where k_o is the pre-exponential factor, E_a the activation energy, and R the universal gas constant. The absolute temperature, T, is expressed in degrees Kelvin (K) or Rankine (°R). The Arrhenius rate constant is empirical, but both of its parameters have a molecular interpretation: k_o is related to how many times the molecules collide per unit time; E_a has been shown to be equal to the difference between the average energy of those molecules that actually react and the (smaller) average energy of the entire collection of molecules. Equation (3-36) has been found to adequately correlate the apparent average (*global*) rate constant for a wide range of chemical reactions. Equations (3-36) and (3-35) state that the rate of a chemical reaction rises exponentially with increasing temperature. Thus, even a modest increase in temperature can dramatically increase the rate of a chemical reaction. This consequence becomes more apparent by differentiating Equation (3-36) with respect to T, to determine the relative change in k caused by a relative change in T:

$$d(\ln k)/d(\ln T) = E_a/RT \tag{3-37}$$

Many reactions important in sustainable energy, from corrosion of steel to oxidation of fuels, have high activation energies, so that even at high temperatures $E_a >> RT$. For example, despite the elevated temperatures (ca. 1,300°C), a fractional increase in the temperature within a pulverized coal combustor (Chapter 7) would cause about a 15-fold higher fractional increase in the rate of pyrolysis (thermal decomposition) of the finely divided coal particles.

We need to describe heat release rates $\dot{Q}$. If Equation (3-35) is valid, then:

$$\dot{Q} = \dot{m}\,\Delta H_{ox} \tag{3-38}$$

where ΔH_{ox} is the exothermal heat of oxidation of the fuel. Note that, in general, ΔH_{ox} depends on the reaction temperature, but for engineering calculations, it can be adequately represented by the heating value of the fuel (Chapter 7).

Chapter 8 will draw a loose analogy between heat generation by chemical and nuclear processes, the former by scission and formation of chemical bonds in a fossil fuel, the latter by fission and fusion of nuclear bonds within the core structure of the atom itself. Further, Chapter 8 shows that to understand nuclear energy as a source of useful heat for power cycles, we need to learn about the rates of nuclear reactions. In chemical reactions, nuclear reactions, and other rate phenomena, it often proves convenient to express the kinetics in terms of some characteristic time related to a readily visualized amount of change. One such time is the *half-life*, defined as the minimum amount of time that must elapse so that a fixed amount of material is depleted to one half that amount. The consumption mechanism may be nuclear decay (radioactivity), chemical reaction, or any rate process. Let's derive an expression for the half-life of the following simple first order chemical reaction:

$$\mathrm{A} \rightarrow \mathrm{B} \tag{3-39}$$

The rate of disappearance of substance A, $\dot{m}_A$, using Equation (3-35) with $a = 1$ and $b = 0$ is,

$$\dot{m}_A = d[A]/dt = -k[A] \tag{3-40}$$

and,

$$d[A]/[A] = d\ln[A] = -kdt \tag{3-41}$$

Note that we have assumed the reaction is first order in the concentration of A and that it does not depend upon the concentration of any other species, like an oxidant. Equation (3-41) is a linear first-order differential equation and has the same mathematical form as the rate equation for radioactive decay. To solve this equation, we need one initial condition. This requirement is met if we know the initial concentration, [A(t=0)]. The solution for Equation (3-41) is then:

$$\left[\frac{[A(t)]}{\left[A(t{=}0)\right]}\right] = \exp(-kt) \tag{3-42}$$

Equation (3-42) tells us that [A] is decaying exponentially with time, which also allows us to obtain a quantitative expression for the half life of a substance:

$$\left[\frac{\left[A\left(t_{1/2}\right)\right]}{\left[A(t{=}0)\right]}\right] = \frac{1}{2} = \exp(-kt_{1/2}) \tag{3-43}$$

giving

$$t_{1/2} = \frac{-\ln(0.5)}{k} \approx \frac{0.69}{k} \tag{3-44}$$

If we know the rate constant for any first-order rate process (chemical, nuclear, physical transport, or whatever) that follows a functional relationship given by Equation (3-40), we can immediately calculate its half-life. Likewise, if we have data on the half life of such a rate process, we can immediately calculate its rate constant. Half-lives for chemical reactions relevant to sustainability range from microseconds for combustion and thermal explosion processes to tens of years for atmospheric reactions that deplete the greenhouse gases CO_2 and CH_4. To test your understanding of these concepts, derive a mathematical expression that will tell you how the half-life of a first-order chemical reaction depends upon temperature. Half-lives for radioactive decay range from microseconds to tens of millions of years depending on the radionuclide (Chapter 8).

Because of their conceptual and mathematical simplicity, chemical kinetic equations like Equations (3-35) through (3-37), and (3-39) through (3-44) are especially appealing for exploratory or scaling calculations. Often, appreciable physical and/or chemical complexity is embedded in these rudimentary chemical models. The good news is that this allows us to make an informative head start on the quantitative analysis of more complicated systems.

Nonetheless, more complex approaches are often needed, as in the treatment of pollutant formation in full-scale combustors (Chapter 7) and the analysis of the transport and transformation of pollutants in the ambient environment (Chapter 4). To get a sense of this complexity, consider again Equation (3-34). In actuality, the reaction partners may include not only the fuel (CH_4) and one oxidant-like molecular oxygen (O_2), but also transient (short lifetime, chemically reactive) species, such as different fragments of the fuel molecule (e.g., CH_3^-., the methyl radical), and other species capable of functioning as potent oxidants under the right chemical circumstances (e.g., the hydroxyl radical OH^-). Individual reaction steps may affect each others progress or be affected by non-chemical processes like mixing or the physical transport of heat or material.

Even when a single reaction rate expression like Equation (3-35) describes the mechanistic chemistry with sufficient reliability, there can be other complications. The reaction kinetics may also vary with the total pressure of the system because changes in pressure can change the rate constant k, i.e., $k = k(T,P)$. Further, rates of many chemical reactions can be accelerated or slowed down by other materials (*catalysts*). These may be exogenous agents intentionally added to a process or machine to change (generally speed up) the rate of an important chemical reaction. They may also be products of the chemical reaction itself. Importantly, the materials of construction of the machinery or vessels used to carry out chemical reactions to destroy pollutants or accelerate energy production may act as catalysts. Sometimes their catalytic effects are beneficial. Other times endogenous materials catalyze the formation of unwanted by-products or inhibit the rates of desired reactions.

Methods are available to deal with these complications. Typically, the necessary modeling power for equipment design or performance monitoring can only be realized by computer simulation of processes and process sub-systems (e.g., the flame in a gas turbine, a swarm of individual coal particles burning in an electric utility furnace). At this stage, the important points to remember are that: (1) chemical rates may sometimes dictate how fast we can recover heat from a power cycle; (2) Equations like (3-35) through (3-37) and (3-39) through (3-44) (or similar engineering kinetics models) are often useful for initial screening calculations to determine chemical reaction rates; (3)

for many practical technologies relevant to sustainable energy, the governing chemistry is complicated and intricately intertwined with physical processes. In these cases, reliable analysis of factors important to sustainability, such as rates of heat production and pollution formation/persistence, may require more elaborate models; (4) sophisticated chemical kinetic software to describe chemical rate phenomena in complex systems, like practical combustors and urban atmospheres, are available. Simulation modules will continue to improve because of continuing research on numerics, computation of chemical rate constants, and mechanisms of relevant chemical processes.

3.5 The Physical Transport of Heat

Heat transfer is vertically integrative and multi-faceted. In this section, we will examine a few foundational concepts at an introductory level. Much of this discussion is based on material from Deen (1998), Whitaker (1977), Holman (1976), Hottel and Sarofim (1967), Feynman et al. (1963), McAdams (1954), and Hottel (1954).

3.5.1 Foundations for quantitative analysis

Our study of power cycles (Section 3.2) provided several examples in which heat is transferred between a system and its surroundings or vice versa because of a temperature difference between two locations. We call this temperature difference a *temperature potential*. We also know from everyday experience that heat travels from a high temperature body to a colder one. An exception can occur if we perform work on a system (e.g., when a heat pump [Section 3.2.4, Figure 3-13] forces heat to flow uphill or from a lower to a higher temperature). To analyze the performance of heat engines and other energy-conversion devices requires quantitative characterization of heat-transmission rates. The discipline of heat transfer provides this understanding by experimental, theoretical, and numerical studies of transient (time-dependent) and steady state (time-independent) rates of heat flow between regions of space at different temperatures.

Rigorous study of heat transfer confirms that heat transfer rates depend on the temperature difference or driving force between the "heat sending" and "heat receiving" materials. A difference in temperature[7] causes heat transfer by three mechanisms: conduction, convection, and thermal radiation. However, for thermal radiation, the rates of heat transmission are decidedly nonlinear in the temperature difference, ΔT (Section 3.5.4). This has important ramifications for sustainable energy because solar energy

7 Heat can also be transmitted by differences in concentration, electrical potential, and magnetic potential. Some of these mechanisms are used to advantage in advanced energy conversion devices, such as thermoelectric refrigerators. Further details are found in Cadoff and Miller (1960), MacDonald (1962), Bird et al. (1960), de Groot and Mazur (1962), and Hirschfelder et al. (1964).

utilization and greenhouse warming are phenomena that depend strongly on radiative heat transfer. Heat transfer rates are also affected by various physical and chemical properties of the heat sources and receivers.

Conduction is energy transfer by the movement and interactions (collisions) of atoms and molecules over molecular distances (e.g., 1–10 Å in solids and liquids, and in gases at 1 atm pressure). *Convection* is energy transfer by *bulk motion* of a fluid. By bulk motion, we mean movement of a "large" volume of fluid over a distance that is much larger than the size of molecules composing the fluid and the distance a molecule travels between two successive collisions. A *large* volume of fluid is a volume big enough that its constituent molecules undergo many collisions during the time that the fluid is in motion. Note that the individual molecules in a convecting fluid move about in random directions. However, in convection, heat is transmitted by bulk displacement of macro-sized fluid parcels. By definition, convection always occurs by bulk motion of a liquid or a gas. However, solids, liquids, or cavities of gas are sometimes entrained in the moving fluid. In many practical situations, including synfuel manufacture (Chapter 7) and air pollution (Chapter 4), it is necessary to account for how multiple phases within a moving fluid affect heat transfer.

To solve practical heat transfer problems, it is almost always necessary to know several *bulk* (*macroscopic*) *properties* of the material(s) involved. Bulk properties include temperature, density, and heat flow rate, as well as attributes that measure the capability of matter to transmit, assimilate, and divest heat. Examples are thermal conductivity, thermal diffusivity, specific heat, heat transfer coefficient, emittance, and absorptivity (see below). Most introductory heat transfer analyses assume that all bulk properties are continuous functions of time and position (i.e., that they obey the *continuum approximation*). This assumption makes otherwise intractable heat transfer calculations manageable. Here's why. On a molecular scale, matter is structured more like a collection of discrete point particles separated by empty space than like continuous strings, sheets, or blobs of contiguous corpuscles. If we tried to measure temperature or any property for a small volume of matter, different values would be found at different locations, depending on the number of molecules within this volume. The lower this number, which we call the *local molecular number density*, the wilder the fluctuations that would turn up in our measurements. To perform reliable heat transfer measurements or calculations, we would need to keep track of these wild gyrations. Further, we would need to monitor how the molecules of our material are affected by containment vessel walls, as well as how fast these molecules respond when suddenly perturbed.

Calculations at these levels of detail are impossible for virtually all practical situations. Fortunately, many heat transfer problems relevant to sustainable energy involve types and sufficient volumes of materials that the continuum approximation is a good assumption.[8] Important exceptions are heat transfer in shock waves and in gases at low density. The former arises in high performance engines for supersonic flight and in advanced technologies for rock breaking that may open up new vistas for geothermal energy (Chapter 11), CO_2 sequestration (Chapters 4, 7), and exploration for oil and gas (Chapter 7). The latter is important in the behavior of thermal insulation materials (Chapter 20) and in understanding the role of the upper atmosphere in the earth's thermal budget (Chapter 4).

3.5.2 Thermal conduction

Spatial non-uniformities of temperature cause a *heat flux* or a flow of heat per unit time and area normal to the flow direction (occasionally called the *heat flux density*). The magnitude of the temperature difference per unit distance in the material is called the *temperature gradient*, which distinguishes it from the difference in temperature alone or the *temperature potential.* Transport theory provides mathematical expressions (*constitutive equations*) that relate energy (heat), mass, and momentum fluxes to properties of the transport medium and to gradients in bulk properties that drive the fluxes. Causal gradients occur in temperature, species concentration (rigorously known as chemical potential), and velocity. Fourier's law provides the essential constitutive equation for conduction:

$$\dot{q}_x = \frac{\dot{Q}}{A} = -\lambda \frac{\partial T}{\partial x} \tag{3-45}$$

where $\dot{q}_x$ is the heat flux in the direction x (in units of energy per area per time), $\dot{Q}$ is the corresponding rate of heat flow (in units of energy per time) where A is the area of the conducting region normal to the direction of heat flow, and $\partial T/\partial x$ is the spatial temperature gradient in the x direction. λ is the thermal conductivity, a specific property

8 *Bulk* or *macroscopic* properties (Denn, 1980) express at "large" scale a statistical average of numerous molecular scale events. Those events are collisions of molecules composing the material (*constituent collisions*). Some bulk properties are readily detected with human senses or rudimentary instruments. Others can be measured only with sophisticated apparatus. Bulk properties of various fuels and materials are essential for engineering and economic calculations important to sustainable energy. To be valid, the *continuum approximation* requires that there be enough constituent collisions to make bulk property values immune to statistical fluctuations in the number of these collisions and to effects of constituent molecule collisions with the boundaries of the substance (for example, container walls). Deen (1998) notes that fluctuation and boundary effects can be neglected if the system is big enough, e.g., $\geq 0.1 \mu m$ and $0.003 \mu m$, respectively, for a liquid of MW 30 g/mole and 1 g/cm^3 density, and ≥ 1 mm for a gas of MW 30 d at 293 K and 1 atm. The continuum approximation also demands that the molecules composing a substance respond rapidly when disturbed. For most materials this condition is readily met. For example, the molecular response (relaxation) times are typically of order 10^{-13} and 10^{-10} s for liquids and gases at the conditions just stated. An exception is molten or dissolved polymers where relaxation times can be several seconds.

of the conducting medium and, in general, dependent on temperature.[9] The minus sign is needed because $\dot{q}_x$ is positive, but the temperature gradient is negative because heat flows from a high to a low temperature. A key equation for practical conduction calculations is obtained by performing an energy balance on a differential volume element of material that receives and expels heat across its boundaries according to Equation (3-45) and generates heat internally at a rate of $\dot{q}_i$ per unit volume. The resulting equation for time-dependent conduction in one dimension is:

$$\frac{\partial}{\partial x}\left(\lambda \frac{\partial T}{\partial x}\right) + \dot{q}_i = \rho C \frac{\partial T}{\partial t} \tag{3-46}$$

where t is time, ρ is the density of the material, and C its specific heat capacity. The three-dimensional analogue to Equation (3-46) for Cartesian coordinates (x, y, and z) when λ is independent of position is:

$$\frac{\partial^2 T}{\partial x^2} + \frac{\partial^2 T}{\partial y^2} + \frac{\partial^2 T}{\partial z^2} + \frac{\dot{q}_i}{\lambda} = \frac{1}{\alpha}\frac{\partial T}{\partial t} \tag{3-47}$$

where α is the thermal diffusivity $\lambda/\rho C$. As α increases, the medium can equilibrate temperature non-uniformities more rapidly. Equations (3-46) and (3-47) are partial differential equations first-order in time but second-order in spatial coordinates.

To be solved, one initial condition and two boundary conditions must be specified. Specialized texts on heat transfer (e.g., Levenspiel, 1984, Özisik, 1980, Whitaker, 1977, Holman, 1976, Carslaw and Jaeger, 1959, McAdams, 1954) provide closed form or graphical solutions for 1-, 2- and 3-dimensional versions of Equations (3-46) through (3-47) for solids of various geometries and a range of initial and boundary conditions of practical interest. More elaborate versions of Equations (3-46) and (3-47), featuring complex initial or boundary conditions or thermalphysical properties that depend on temperature, position, or time, are solved numerically or by advanced mathematical procedures (Deen, 1998, Bird et al., 1960, Carslaw and Jaeger, 1959). Practical applications of the solutions to heat transfer equations require knowledge of the thermophysical properties of the conducting medium (i.e., λ, ρ, C, α, and their variation with temperature). Pertinent data are found in the textbooks noted above and in engineering handbooks like Perry et al. (1997).

3.5.3 Convective heat transfer

Convection transfers heat and matter in many situations that are important to sustainable energy. Convective heat transfer to and within boiler tubes helps raise steam for electricity generation and to power ships. Convective flows in the oceans and the atmosphere are among the processes that disperse heat and pollutants. An understanding

---

9 Equation (3-45) assumes that spatial effects on conduction arise only from property variations in the x direction, and, accordingly, treats λ as a scalar. More generally, because of structural anisotropy in the conducting medium, the heat flux density, spatial temperature gradient, and thermal conductivity are different in different directions.

of these flows is important in determining the role of ocean and atmospheric circulation in heating and cooling the earth. This information also elucidates the contributions of anthropogenic and natural sources of greenhouse gases to global warming (Chapter 4).

Newton defined the rate of heat transfer by convection, Q_{cv}, from a solid surface of area A and temperature T_w to a fluid of temperature T by the relation:

$$Q_{cv} = h_m A[T_w - T] \tag{3-48}$$

where h_m can be viewed as a mean apparent overall coefficient of heat transfer from the surface to the fluid. Equation (3-48) is sometimes referred to as Newton's law of cooling (or heating, depending on the sign of $[T_w - T]$). Rigorously, it is not a law but a definition of the heat transfer coefficient h_m. In practice, h_m may reflect effects of conduction and radiative heat transfer as well as convection. Equation (3-48) is a dimensional equation, so h_m depends on the units employed for temperature, and the heat flux $Q_{cv} \cdot h_m$ is affected by the thermophysical properties of the fluid, the geometry of the heat transfer system, and the temperature difference, $[T_w - T] = \Delta T$. In addition, h_m is influenced by the flow field. For example, h_m increases substantially when the flow changes from *laminar* (thin, parallel fluid lamellae slide smoothly past each other in the direction of flow) to *turbulent* (parcels of fluid move with fluctuating velocity in directions essentially random with respect to the average flow orientation). Further, a change in phase of the fluid at the surface (because of boiling or condensation) can cause a huge increase in h_m.

The fluid mechanical and thermophysical details of convective heat transfer are complicated. By substantially aggregating these complexities in h_m, Equation (3-48) is a reliable tool for many, but not all, introductory calculations in convective heat transfer. (Correlations for estimating numerical values of h_m for various geometries and fluid flows are found in the references cited at the beginning of Section 3.5. These works also provide alternative approaches when Equation (3-48) is too simplistic a model.)

3.5.4 Radiative heat transmission

The movement of *electromagnetic* (*em*) *radiation* (*waves*) transfers energy. This mechanism of energy transfer is vitally important in sustainable energy. In furnaces, it plays a major role in transmitting the chemical energy of fuels to the working fluid. It is also the means by which the sun supplies solar energy to the earth. Further, the greenhouse effect is heat transfer back to the earth by thermal radiation from CO_2 and other greenhouse gases in the atmosphere. Depending on its properties, em radiation can propagate through solids, liquids, gases, and even through nothing. Let's explain. Electromagnetic radiation consists of two disturbances (*fields*), one electrical, one magnetic, that move together at the speed of light, c. The intensity of each field continuously oscillates transverse to the direction of wave propagation. The number of these oscillations per second is called the *frequency* of the em radiation. When a parcel of matter or a volume of space is exposed to em radiation, it is said to be in an *electromagnetic field*. This means that a stationary electrical charge within the matter will feel oscillatory pulling and pushing forces from the electrical field. When that electrical charge moves, it will also feel continuously oscillating pulling and pushing forces from the magnetic field and from the electrical field.

But where does an electromagnetic disturbance actually come from and how does it move heat in practical situations encountered in sustainable energy? A constructive simplification is to view the origin of em radiation as regular back and forth gyrations of electrical charges at a location we will call a "source." A radio antenna is one example; the sun and space itself are others. During these to and fro motions, the electrical charges periodically accelerate and decelerate. These regular cycles of charge acceleration and deceleration enable em radiation to spread out over distances that can extend to cosmic scales.

The number of complete cycles in 1 second is the frequency of the em radiation, υ. The distance traveled through a particular medium, m, to complete 1 cycle, is the wavelength in that medium, λ_m. Frequency and wavelength are related to the speed of light in a vacuum, c, by:

$$c = n\,\lambda_m\,\upsilon \tag{3-49}$$

where n is the refractive index of the medium. You may be more familiar with the expression:

$$c = \lambda\,\upsilon \tag{3-49a}$$

which holds for a vacuum where $n = 1.0$. The faster the charges move back and forth in our source, the higher the frequency and the shorter the wavelength of the radiation. Electromagnetic wavelengths can vary from 10^{-13} m for cosmic rays (em radiation from outer space) to 10^{6} m for very long radio waves. Quantum theory teaches that em radiation behaves both as a flow of waves and as a stream of particles (photons) each with a fixed amount (a quantum) of energy E_p given by:

$$E_p = h\upsilon \tag{3-50}$$

where h is Planck's constant. Photons are released or absorbed when molecules, atoms, or subatomic particles relinquish or assimilate energy. However, it is not known exactly how photons form during these processes of microscopic energy exchange.[10]

The particulate nature of em radiation allows us to think of the mechanism of radiative heat transfer as a flux of photons from a radiator to a receiver. A radiating body "broadcasts" em radiation (i.e., it expels a stream of photons). A body heated by radiation absorbs some of those photons and converts their energy into heat. If the temperature of the radiating body (the photon emitter) exceeds that of the absorbing body (the photon receiver), there will be a net transfer of energy from the radiating body.

---

10 During a PBS television interview, Feynman spoke of his father's surprise when he told him that neither he (Feynman, a recent physics grad from MIT) nor anyone else, knew exactly where photons come from.

Historically, most studies of thermal radiation heat transfer have concentrated on em radiation of wavelengths between 1×10^{-4} and 5×10^{-7} m, or the infrared (IR) to visible to ultraviolet (UV) portions of the em spectrum. These wavelengths account for virtually all the energy transferred to the earth from the sun and for many radiative heat transfer situations important to humans, from industrial furnaces to home appliances for space heating. However, all em radiation, regardless of wavelength, carries energy and can be a means of radiative energy transfer. If you're unconvinced, just think of the ubiquitous microwave oven. These appliances radiatively heat food using em waves of $> 10^{-3}$ m wavelength (i.e., considerably larger than the longest IR waves). Em waves propagate even in a perfect vacuum and transfer heat from a high temperature source to a low temperature receiver, even if the two are separated by virtual nothingness. This is comforting news for earthlings, because most of the region of space between us and the sun is essentially a perfect vacuum.

So far we have learned that thermal radiation is energy transmitted at the speed of light by em waves or photons. The Stefan-Boltzmann law relates the rate of radiative heat emission per unit area of a body, sometimes called the total emissive power or radiant flux-density, $\dot{q}_r^{1-1}$,[11] to the temperature T of that body:

$$\dot{q}_r = \varepsilon \sigma T^4 \tag{3-51}$$

where σ is the dimensional Stefan-Boltzmann constant with units of [energy]/{[area][time][degrees absolute]^4}, e.g., 1.713×10^{-9} (Btu)/(ft^2)(hr)(°R)^4, or 5.67×10^{-8} W/(m^2)(K)^4, and ε is the emittance (or for a flat surface the emissivity), the ratio of the total heat radiation of a surface to that of an ideal or perfect surface, at the same temperature. This ideal radiating surface absorbs all incident thermal radiation, and the quality and intensity of the radiation it emits are completely determined by its temperature alone (i.e., it emits thermal radiation according to Equation (3-51), with $\varepsilon = 1.0$). It is called a black body or black surface. In general, the emissivity of a surface depends upon its temperature, its degree of roughness, and the particular wavelength of the em radiation. Note that in stark contrast to conduction (Equation (3-45)) and convection (Equation (3-48)), the radiative heat flux is directly proportional to the fourth power of the temperature. This strongly amplifies the effect of temperature on the radiative energy exchange. It can be shown that the total net rate of radiative heat transmission $\dot{q}_{r,bb}^{1-2}$ from a black body of area A_1 at a temperature T_1 to a second black body at a temperature T_2 that completely surrounds it, i.e., a black enclosure, is given by

11 Note that the symbol $\dot{q}_r$, Equation (3-51), denotes radiant heat flux density, i.e., energy per area per time, whereas dotted symbols, e.g., $\dot{q}_{r,bb}^{1-2}$, Equation (3-52), signify a radiant energy flow rate or current, i.e., energy.

$$\dot{q}_{r,bb}^{1-2} = A_1 \sigma \, [T_1^4 - T_2^4] \tag{3-52}$$

Equation (3-52) assumes that the black body cannot radiate back on itself, i.e., it lacks geometric irregularities that intercept and send back photons emitted from its surface. In practice, surfaces are imperfect so they absorb only a fraction, $\alpha_\lambda < 1$, of their incident radiant energy at each wavelength λ. Further, they emit less thermal radiation than that called for by Equation (3-51) with $\varepsilon = 1$, i.e., their emissivity is less than unity. Equation (3-52) can be modified to deal with these realities. If α_λ does not change with wavelength, the surface is called gray and $\alpha_\lambda = \varepsilon$. For a gray body at T_1 exchanging radiant energy from an area A_1, with a black body enclosure at T_2, the radiative heat flow rate (flux) is:

$$\dot{q}_{r,b}^{1-2} = \varepsilon \, A_1 \sigma [T_1^4 - T_2^4] \tag{3-53}$$

The gray body assumption gets better as T_1 and T_2 get closer, because it can be proved that $\alpha_\lambda \rightarrow \varepsilon$, as $T_1 \rightarrow T_2$.

When the above simplifications fail, complexities abound. To illustrate, let's see what happens if a radiating body can see part of itself so that it reabsorbs some of its own thermal radiation. For a black body radiating into a black enclosure, this is handled by prefixing Equation (3-52) with a geometrical view factor, F_{12}, expressing the fraction of the radiant energy emitted by a body at T_1, which is incident upon a receiving surface at T_2, i.e.:

$$\dot{q}_{r,bb,F}^{1-2} = A_1 F_{12} \sigma [T_1^4 - T_2^4] \tag{3-54}$$

It is tempting to assume that a similar patch job on Equation (3-53) would nicely handle a gray body that can see part of itself while it radiates into a black enclosure. This is conceptually correct, but there are some surprises in the details. Subject to certain limitations (Whitaker, 1977), the "correction factor," F'_{12}, is:

$$F'_{12} = \frac{\varepsilon_g F_{12}}{\left[\varepsilon_g - F_{12}(1 - \varepsilon_g)\right]} \tag{3-56}$$

where the emissivity of the gray body is denoted ε_g to draw attention to the fact that, in general, it is less than 1. The corresponding expression for the radiative heat flux is:

$$\dot{q}_{r,gb,F}^{1-2} = A_1 F'_{12} \sigma [T_1^4 - T_2^4] = \frac{A_1 \varepsilon_g F_{12} \sigma}{[1 - F'_{12}(1 - \varepsilon_g)]} \tag{3-57}$$

A further complication occurs if non-transparent substances (i.e., materials that absorb and emit thermal radiation) intersect thermal radiation, as occurs when greenhouse gases like CO_2 absorb and return IR radiation emitted by the earth (Chapter 4). (Several works, including Goody (1964), Hottel (1954), Hottel and Sarofim (1967, 1997), Siegel and Howell (1972), Sparrow and Cess (1966), and Whitaker (1977) show how to calculate radiative heat fluxes in the earth's atmosphere, industrial furnaces, and other real world systems beset by the above and other complexities.)

3.5.5 Heat transfer by tandem mechanisms

In heat transfer, we consider what happens if the rate of energy exchange between two regions is governed by two or more sequential processes. This is common to many practical energy and environmental systems. For example, we often need to know how rapidly heat flows between fluids at different temperatures separated by a barrier that can transmit heat by conduction, convection, or thermal radiation, or by two or more of these mechanisms in parallel. Each fluid may be in motion, or each may be initially stationary. Practical examples include cooling the cylinders of an IC engine (Section 3.2.4) and climate conditioning a building by transferring heat between living/working spaces and pipes that carry hot or cold water (Chapter 20). If the rate of heat flow does not change with time, we can analyze such problems as we would the flow of electrical current caused by a difference in electrical potential (voltage) across two or more electrical resistances connected in series or parallel depending on the situation. For heat transfer in series arrangements, the "potential" difference is the overall temperature difference, and the resistance is the product of a geometrical factor with the reciprocal of the thermal conductivity or the heat transfer coefficient.

Let heat be convected unidirectionally from a fluid 1 at a temperature T_{f1} to one face of a wall of thickness x_w and constant thermal conductivity λ_w, presenting area A_1 normal to the heat flow. Further, let heat be conducted through the wall and then convected from a second face of the wall of area A_2 by a fluid 2 at a temperature T_{f2}. The steady-state rate of heat flow from fluid 1 to fluid 2 is then:

$$\dot{q}_{f1-w-f2} = \frac{[T_{f1} - T_{f2}]}{\left[\frac{A}{h_1 A_1} + \frac{A x_w}{\lambda_w A_w} + \frac{A}{h_2 A_2}\right]} \tag{3-58}$$

where h_1 and h_2 are, respectively, the convective heat transfer coefficients for fluid 1 at the first face of the wall and for fluid 2 at the second face of the wall, A_w is a suitable surface area of the wall normal to the heat flux (typically its inside, outside, or mid-plane surface area), and A is any convenient surface area for the physical system at hand (e.g., the inside or the outside surface area of the wall, again normal to the heat flux). Notice that by defining a single equivalent overall composite heat transfer coefficient, h_{eq}, we could write Equation (3-58) in exactly the same form as Equation (3-48):

$$\dot{q}_{f1-w-f2} = h_{eq} A [T_{f1} - T_{f2}] \tag{3-59}$$

where[12]

$$\frac{1}{h_{eq}} = \frac{1}{h_1}\left(\frac{A}{A_1}\right) + \frac{x_w}{\lambda_w}\left(\frac{A}{A_w}\right) + \frac{1}{h_2}\left(\frac{A}{A_2}\right) \tag{3-60}$$

12 More generally Equation (3-60) should be written as: $\frac{1}{h_{eq}} = \frac{1}{h_1}\left(\frac{dA}{dA_1}\right) + \frac{x_w}{\lambda_w}\left(\frac{dA}{dA_w}\right) + \frac{1}{h_2}\left(\frac{dA}{dA_2}\right)$ because the area available for heat transfer may change along the pathway of heat movement, e.g., for radial heat flow in a cylinder or sphere.

Note that h_{eq} is an operational quantity synthesized to make our heat transfer problem fit into the mathematical form of Equation (3-48). The numerical value of h_{eq} will depend on our choice for A and on how many and what type of heat transfer resistances we have in series (i.e., conduction or convection), and upon the geometry of our heat transfer region. If we think of h_{eq}^{-1} as an equivalent heat transfer resistance, then Equation (3-60) is analogous to adding electrical resistances connected in series to obtain the total resistance of the circuit, (i.e., $Rt = h_{eq}^{-1}$) where the corresponding individual resistances for the different heat transfer mechanisms are $\frac{1}{h_i}\left(\frac{dA}{dA_i}\right)$ for convective heat transfer in fluid i, and $\frac{L}{\lambda}\left(\frac{dA}{dA_w}\right)$ for conductive heat transfer through a medium of thickness L and thermal conductivity λ, where the A_j terms represent the areas defined above.

Interestingly, for the particular case of a black body at T_1 radiating into a black enclosure at T_2, when $T_1 - T_2 < 110°C$, we can even define a corresponding radiative heat transfer resistance from an area A_r as $\frac{1}{h_r}\left(\frac{A}{A_r}\right)$, where:

$$h_r = 4F_{12}\sigma T_{av}^3 \tag{3-61}$$

Here, T_{av} is the arithmetic mean temperature of the radiator and receiver temperatures, T_1 and T_2:

$$T_{av} = (T_1 + T_2)/2 \tag{3-62}$$

Thus, for the stated constraints, we can reliably estimate the rate of heat transfer from thermal radiation using the linear difference in temperature between the radiating and receiving bodies incorporated in an equation analogous to Equation (3-48):

$$\dot{q}_{r,bb,F}^{1-2} = h_r A_r (T_1 - T_2) \quad \text{for} \quad (T_1 - T_2) < 110°C \tag{3-63}$$

3.6 Use and Abuse of Time Scales

Throughout this text, we will see that knowledge of time scales (characteristic times or time constants) for physical, chemical, biological, geological, and sociopolitical changes elucidates natural and anthropogenic forces that impact sustainable energy. By comparing characteristic times for pairs of phenomena, we can sometimes learn if one clearly dominates the other and, if so, under what circumstances.[13] For example, in heating a large piece of solid fuel, like a chunk of biomass, the chemical rate of decomposition may be so fast that

13 Ratios of characteristic times define *dimensionless* quantities called *groups, moduli,* or *numbers.* Most are named for distinguished engineers and scientists like Archimedes, Fourier, Galileo, Hooke, Mach, Newton, Nusselt, (Lord) Rayleigh, and Sherwood. Five different groups have been named for Damkohler!

heat transfer determines the rate of fuel consumption. For a finely ground fuel piece, the opposite may be true. By comparing the characteristic times for heat transfer and chemical reaction, we can map out conditions in which each rate is clearly dominant and an intermediate region where both phenomena are important.

Several geopolitical time scales impact sustainability. Important time scales include the time for the world population to double and the turnover times for elected officials. Also important are times for significant economic shifts or major wars—a few months (the Persian Gulf War) to a few tens of years (the Hundred Years War). At present, many experts are working hard to pin down time scales for changes in global climate arising from human activities (Chapter 4) and for other environmental impacts associated with the production and use of energy (Table 1.8). Estimation of probable time scales for change should be an inherent element of sustainability analysis.

Characteristic times are an informative starting point for evaluating the impacts of current or proposed sustainability measures. More detailed study of sustainability tactics and strategies must follow. This is because time scales are for discrete-rate phenomena. For many real-world systems, the underlying physical, biological, or sociopolitical mechanisms that determine the rates of associated macro phenomena may be unknown or too poorly understood to elucidate relevant time scales. Even if mechanistic underpinnings are well understood, it may not be known how to aggregate their impacts to deduce the rates of composite phenomena. One of the weak links may be oversimplification of mathematics so that less obvious, but important, processes are misrepresented or obfuscated entirely. Examples abound in all fields from chemical reactor engineering to modeling population changes.

As a simple illustration, consider our discussion of radiative heat transfer. If we used Equation (3-63) to derive a time constant for a linearized formulation of radiative heat flux, we would be in safe territory if we were studying a black body radiating to a black body and a temperature difference of up to about 110°C. However, we would quickly be in error if we applied Equation (3-63) to wider temperature ranges. In particular, we would badly underestimate how strongly our time constant changed with temperature. Yet another complication is that in large complex systems, multiple phenomena may be strongly correlated by their interactions where each is affecting all the others in a major way, so that time scales for discrete processes lose meaning.

There is no perfect recipe for sustainability analysis. One approach is to identify important processes of change associated with the technology or policy measure under scrutiny. Then, where possible, estimate time scales for these processes. Use the results to help bring the problem into focus by understanding how quickly or slowly crucial processes occur. In some cases, it may be possible to order processes of change from clearly rapid to woefully lethargic. Ultimately, to develop more solid foundations for eventual decision-making, recognize and enforce the imperative to carry out more detailed analysis of existing or proposed technology or of policy measures.

3.7 Energy Resources and Energy Conversion—Fertile Common Ground

The foci of Chapters 2 and 3, estimation of non-renewable energy reserves and technical analysis of heat engine performance, are seemingly unrelated. Further reflection reveals overlapping themes. One is that sustainable energy protocols must wrestle with constraints and uncertainty. For example, nobody really knows how much fossil energy remains to be extracted from the earth. Another is the inherent and practical bounds on the performance of real-world energy-conversion machines. Neither engineers nor policy analysts nor politicians can remove First and Second Law limitations on efficiency. But innovations in materials, heat transfer, and cycle design are closing the gap between current actual performance (Figure 3.12) and the ideal Carnot limit for closed-cycle heat engines. Likewise, good engineering, public policy, economics, and geology will almost certainly identify new fossil fuel deposits, generate better information on existing deposits, and release previously unrecoverable mineral stocks of fuels for human exploitation.

A second important conclusion is that technological and policy actions to achieve sustainable energy can and should move forward even though our current knowledge and hardware are imperfect. A third finding is the imperative to accelerate progress toward a more sustainable energy future—by research, development, pilot and field testing, public engagement, and commercialization.

References

Bird, R.B., W.E. Stewart and E.N. Lightfoot. 1960. *Transport Phenomena*. New York: J. Wiley & Sons.

Cadoff, I.B. and E. Miller, eds. 1960. *Thermoelectric Materials and Devices*. New York: Reinhold.

Carslaw, H.S. and J.C. Jaeger. 1959. *Conduction of Heat in Solids*, second edition. New York: Oxford University Press.

Clausius, R.J.H. 1865. *Über verschiedene für die Anwendung bequeme Formen der Hauptgleichungen der mechanischen Wärmetheorie*, Pogg. Ann. 125. Cited by Gibbs (1876).

Crank, J. 1975. *The Mathematics of Diffusion*, second edition. New York: Oxford University Press.

Cussler, E.L. 1997. *Diffusion, Mass Transfer in Fluid Systems*, second edition. New York: Cambridge University Press.

Deen, W.M. 1998. *Analysis of Transport Phenomena*. New York: Oxford University Press.

de Groot, S.R. and P. Mazur. 1962. *Non-Equilibrium Thermodynamics*. New York: Interscience.

Denn, M.M. 1980. *Process Fluid Mechanics*. Upper Saddle River, NJ: Prentice-Hall.

Feynman, R.P., R.B. Leighton, and M. Sands. 1963. *The Feynman Lectures on Physics*, volumes I, II and III. Reading, MA: Addison-Wesley.

Fogler, H.S. 1992. *Elements of Chemical Reaction Engineering*, second edition. Englewood Cliffs, NJ: Prentice-Hall.

Frank-Kamenetskii, D.A. 1969. *Diffusion and Heat Transfer in Chemical Kinetics*, second edition. J.P. Appleton, translation ed. New York: Plenum.

Fristrom, R.M. 1995. *Flame Structure and Processes*. New York: Oxford.

Gibbs, J.W. 1876, 1878. "On the Equilibrium of Heterogenous Substances." Trans. Connecticut Academy III: 108–248 (1876), 343–524 (1878).

Glassman, I. 1996. *Combustion*, third edition. New York: Academic.

Goody, R.M. 1964. *Atmospheric Radiation*. London: Oxford.

Heywood, J.B. 1988. *Internal Combustion Engine Fundamentals*. New York: McGraw-Hill.

Heywood, J.B. 2000. Private communication. Cambridge, MA: MIT, Department of Mechanical Engineering.

Hirschfelder, J.O., C.F. Curtiss, and R.B. Bird. 1964. *Molecular Theory of Gases and Liquids*. Corrected printing with notes added. New York: John Wiley & Sons.

Holman, J.P. 1976. *Heat Transfer*. New York: McGraw Hill.

Hottel, H.C. 1954. "Radiant-Heat Transmission." In *Heat Transmission*, ed. W.H. McAdams. New York: McGraw-Hill.

Hottel, H.C., and A.F. Sarofim. 1967. Radiative Transfer. New York: McGraw-Hill.

Hottel, H.C., and A.F. Sarofim. 1997. "Heat Transfer by Radiation." *Chemical Engineers Handbook*, eds. Perry, Green, and Maloney. New York: McGraw Hill.

Levenspiel, O. 1996. *Understanding Engineering Thermo*. Upper Saddle River, NJ: Prentice-Hall.

Levenspiel, O. 1984. *Engineering Flow and Heat Exchange*. New York: Plenum.

Levenspiel, O. 1972. *Chemical Reaction Engineering*, second edition. New York: John Wiley & Sons.

Lewis, B., and G. von Elbe. 1987. *Combustion, Flames, and Explosions of Gases*, third edition. New York: Academic.

MacDonald, D.K.C. 1962. *Thermoelectricity: An Introduction to the Principles.* New York: John Wiley & Sons.

McAdams, W.H. 1954. *Heat Transmission*, third edition. New York: McGraw-Hill.

Özisik, M.N. 1980. *Heat Conduction.* New York: John Wiley & Sons.

Perry, R.H., D.W. Green, and J.O. Maloney, eds. 1997. *Perry's Handbook for Chemical Engineers*, seventh edition. New York: McGraw-Hill.

Reid, R.C., J.M. Prausnitz, and B.E. Poling. 1987. *Properties of Gases and Liquids*, fourth edition. New York: McGraw-Hill.

Rosner, D.E. 1986. *Transport Processes in Chemically Reacting Flow Systems.* Boston, MA: Butterworths.

Sandler, S.I. 1999. *Chemical and Engineering Thermodynamics*, third edition. New York: John Wiley & Sons.

Siegel, R., and J.R. Howell. 1972. *Thermal Radiation Heat Transfer.* New York: McGraw-Hill.

Smith, J.M., and H.C. Van Ness. 1975. *Introduction to Chemical Engineering Thermodynamics*, third edition. New York: McGraw-Hill.

Sparrow, E.M., and R.D. Cess. 1966. *Radiation Heat Transfer.* Belmont, CA: Brooks/Cole Publishing.

Steinfeld, J.I., J.S. Francisco, and W.L. Chase. 1999. *Chemical Kinetics and Dynamics*, second edition. Upper Saddle River, NJ: Prentice Hall.

Tester, J.W., and M. Modell. 1997. *Thermodynamics and Its Applications*, third edition. Upper Saddle River, NJ: Prentice Hall.

Whitaker, S. 1977. *Fundamental Principles of Heat Transfer.* New York: Pergamon.

Whitaker, S. 1984. *Introduction to Fluid Mechanics.* Upper Saddle River, NJ: Prentice Hall. (Reprinted with corrections by Krieger, Malabar, Florida, 1984.)

Wilson, D.G. 1984. *The Design of High-Efficiency Turbomachinery and Gas Turbines.* Cambridge, MA: MIT Press.

Problems

3.1 In many regions of the world, electrically powered heat pumps are being considered to provide year-round space conditioning (heating and cooling). To improve the efficiency of heat pump operation in cold climates, a geothermally assisted system is under development. In this case, the earth is used as an energy "source" in the winter and as a "sink" in the summer. Plastic coils containing an environmentally friendly antifreeze, like an aqueous solution of potassium acetate, are placed a few meters below the surface where the earth's temperatures are not influenced by diurnal or seasonal variations.

An electrically driven compressor and an insulated expansion valve are linked to two heat exchangers in a closed-loop cycle. During the winter, the heat exchanger connected to the house operates as a condenser/desuperheater for the circulating refrigerant. During the summer, with the flow of working fluid reversed, it acts as an evaporator. The heat exchanger connected to the ground source/sink loop operates in a reverse manner; that is, as an evaporator in the winter and as a condenser/desuperheater in the summer.

a. Sketch flow diagrams showing the equipment arrangements and working fluid flow directions for summer and winter operation.

b. For the following conditions typical of a four-bedroom house located in New England, what is the minimum electric power requirement in the winter and summer?
- average winter heat load = 5,000 J/s and house temperature = 20°C
- average summer cooling load = 4,000 J/s and house temperature = 25°C
- average year-round earth temperature at 3 m depth = 10°C

3.2 For a specific chemical reaction that is first order, its half-life has been measured to be about 10 hours at 25°C. What would its rate constant be at that temperature? Let's assume that the activation energy for this reaction is typical for oxidation reactions carried out under combustion conditions, that is, about 60 kcal/gmole. Estimate its half-life and rate constant at a combustion temperature of 1,000°C.

3.3 Would it be possible to use a normal domestic refrigerator as a heat source for a small hot tub or whirlpool? If so, what would you estimate to be the maximum heating rate and temperature that a 300 gal whirlpool could reach using a refrigerator that is sized to use a 1 hp electric compressor designed with an evaporator temperature of 10°F and condensing temperature of 125°F?

3.4 Estimate the rate of heat loss due to radiation from a covered pot of water at 95°C. How does this compare with the 60 W lost due only to convection and conduction losses? What amount of energy input would be needed to maintain the water at its boiling point for 30 minutes? The polished stainless steel pot is cylindrical, 20 cm in diameter and 14 cm high with a tight-fitting flat cover. The air temperature in the kitchen is about 25 °C. State any assumptions you make in making your estimates.

3.5 A novel heat engine is being considered for producing work from the thermal energy contained in the wastewater discharge from a large food processing plant. On an average day, 35°C wastewater leaves the plant at and is discharged into the ocean at 1,000 gallons per minute . If the average annual ocean temperature is 10°C at that location, what is the maximum power that could be produced by this heat engine? What would be the thermal cycle efficiency of your maximum power engine? You can assume that the heat capacity of water at constant pressure is constant at 4,200 J/kg K.

3.6 Heat losses through windows in buildings are substantial. What percent reduction in heat loss would be mitigated by replacing a window containing a single pane of glass with (a) double pane low E insulating glass or (b) a 3-inch thick sheet of expanded polystyrene sheet? The quoted R values for these items are:

- single pane of glass—0.90 ft^2 hr °F/Btu
- double pane of low E insulating glass—2.3 ft^2 hr °F/Btu
- 1-inch thick sheet of polystyrene sheet—4.0 ft^2 hr °F/Btu

Local, Regional, and Global Environmental Effects of Energy

4

4.1 How Energy Systems Interact with the Environment

4.1.1 Known and potential environmental threats

Every step in the production, modification, and use of energy interacts with the environment and, consequently, modifies the earth and its inhabitants (Figure 4.1). The resulting changes may be small or large, short-lived or long-lasting, beneficial or dangerous, by whatever environmental impact metric is most appropriate (e.g., lbs of pollutant, loss of open space [Chapters 1, 6]). It is typical to categorize energy-environmental considerations in terms of various tangible products of energy operations that flow into and potentially harm the environment.

Environmentally consequential interactions flow into the energy system as well as away from it. Thus, as shown in Figure 4.1, the production of energy or the performance of energy-intensive operations may consume natural resources with the goal of providing useful goods or services to humans. Examples include consumption of fuel to provide useful thermal or mechanical energy and the processing of spent products to recycle metals, paper, monomers, and other values. Further, to define and evaluate tradeoffs for sustainability analysis, understand that outflows from (products of) energy systems can benefit the environment (by enabling recycling, preventing pollution, cleaning past environmental damage, etc.; Section 4.6).

Figure 4.1 illustrates the most important interactions of energy systems with the environment. Bear in mind that this figure says nothing about the duration of these interactions, how far they reach, or how long it may take for their harmful consequences (e.g., human health effects, irreparable damage to ecosystems) to become apparent. Consider the environmental impacts depicted in Figure 4.1.

One impact is the unavoidable release of waste heat to the surroundings during any thermal conversion process (Chapter 3). Adverse effects (thermal pollution) result when byproduct heat causes ambient temperatures to exceed safe levels for humans, plants, animals, and even materials. These pathologies are important outside but also indoors (NRC, 1981). To protect surrounding ecosystems from severe thermal pollution, most large-scale heat engines, such as electric power plants, utilize effluent cooling devices that enable the surroundings to assimilate large amounts of byproduct heat with only modest temperature rises. However, inadequately controlled emanations of waste heat from buildings, automotive engines, industrial processing, and smaller-scale operations like landscape appliances and outdoor cooking, can contribute. Thermal pollution has become increasingly important with the so-far unabated trend toward concentrating more and more people in large urban areas, giving rise to *megacities* that function as urban *heat islands*.

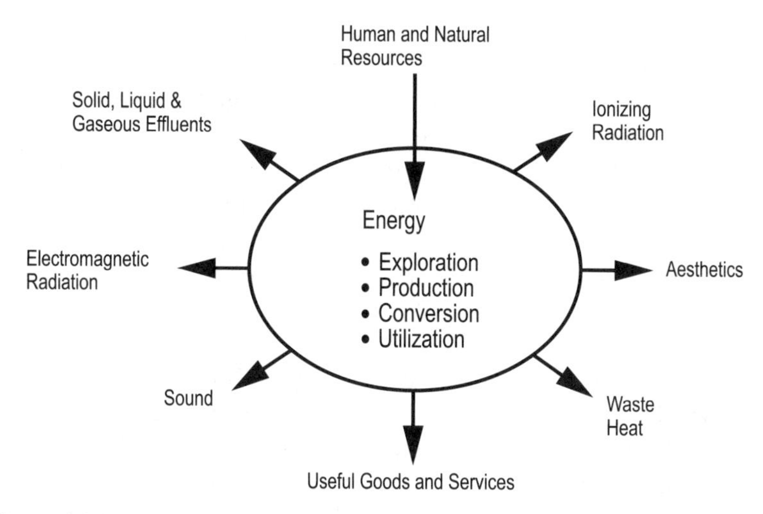

Figure 4.1. Important interactions of energy systems with the environment.

Energy systems also interact with the environment by release of solid, liquid, and gaseous phase byproducts that are known to be hazardous to human health or to threaten ecosystems. Material byproducts of energy systems are called, among other things, pollutants, (hazardous) emissions, effluents, and wastes. With the exception of carbon dioxide,[1] these substances arise from imperfections in the operation of the energy conversion/utilization process, from trace contaminants in fuels or oxidants, or from leaks in or corrosion of process vessel materials of construction (Section 4.1.2). Section 4.2 presents several examples of material pollutants from energy.

Figure 4.1 reveals another category of energy-generated environmental pollution, namely ionizing radiation, or radioactivity. Radioactive emissions may consist of energetic (i.e., short wavelength) electromagnetic waves such as gamma rays and X-rays (Section 3.5.4) and/or atomic scale particles (neutral, e.g., neutrons; negatively charged, e.g., electrons; or positively charged, e.g., alpha particles—helium atoms stripped of two electrons) released with high velocities as products of various natural or induced transformations of atomic nuclei (Chapter 8). Radioactive wastes come under close scrutiny in evaluating environmental impacts of nuclear electric power. Some decay so slowly (i.e., have such long half-lives) that they must be safely stored for up to 10,000 years. Moreover, some emissions from burning coal contain radioactive elements

1 CO_2 is an exception because, although it is a greenhouse gas (Section 4.3), it is not an unexpected or trace product of combustion but rather the natural (thermodynamically preferred) final state of carbon (regardless of its source, whether elemental or chemically combined with other elements) from combustion in air or oxygen.

(Brooks, 1979). Radioactive materials are of environmental concern because some induce carcinogenic or precarcinogenic responses in human tissues, and, in all probability, cause genetically related diseases, such as mutagenicity.

Certain energy systems may pierce their surroundings with sound at frequencies and intensities harmful to the hearing of humans or animals. In more extreme cases, these systems are capable of causing more severe damage owing to the force of shock waves on body organs and tissues. Sound emissions are another mechanism of energy wastage (i.e., some potentially useful thermal or mechanical energy is dissipated as useless vibrations of hardware or working fluids). These vibrations are broadcast as sound waves.

Electromagnetic radiation is another environmental impact of energy systems that has attracted considerable attention in the last two decades. Any flow of electrical current (AC or DC) will emanate electromagnetic (em) radiation (Section 3.5.4) to its surroundings. Leakage of radiative energy can be reduced or eliminated with insulation, but for standard electrical transmission and distribution networks, insulation would be expensive. Concern has arisen because some epidemiological studies have concluded that there is a statistically defensible link between em radiation from electric power transmission and distribution wires and human leukemias. Other epidemiological and supporting scientific studies conclude that a connection has yet to be proven, noting the absence of plausible underlying biochemical mechanisms to explain the proposed adverse effects.

Aesthetic damage is another possible impact of the interaction of energy systems with the environment. Energy-related infrastructure (e.g., electric generating plants, electricity transmission and distribution lines, mines, refineries, and utilization hardware like automobiles) occupies space, causing congestion and loss of open regions for other activities. It is easy to measure occupied area and volume, but a comprehensive evaluation of the apparent and prospective ecological damage from the consequent loss of space is less straightforward. For example, preservation of open space for human recreation is especially important in large urban areas—it would be lunacy to propose construction of a large energy facility for Central Park in New York City. Air pollutants from energy production and utilization (e.g., acid rain and its precursors) would cause various forms of aesthetic damage, such as decrepitation of building facades, injury to works of fine art, loss of visibility, and degradation of plant species.

4.1.2 Origin of harmful agents

Here we qualitatively examine the origins of energy-related environmental insults (Figure 4.1). (Sections 4.2 and 4.3 provide a more mechanistic perspective.) Aside from CO_2 emissions and certain aesthetic outcomes, all adverse impacts ultimately derive from imperfections in physical and chemical systems. The Second Law of Thermodynamics sets the lower limit on these imperfections (i.e., the maximum levels of desired performance that can be attained). Perfect performance is never possible (Chapter 3), and some level of undesired interactions of energy systems with their environment is inevitable. One of the major goals of modern energy and environmental science and technology is to enable the design and operation of energy systems that reduce adverse interactions to the point that they pose no serious risk to humans or other ecosystem residents. However, in practice, most system imperfections are worse than the minimum levels allowed by thermodynamics. Causes include inadequate or excessive rates of energy transmission and degradation of matter (e.g., fuels) and materials of construction (Chapter 3).

Chapters 7 through 20 provide details on how imperfections in specific technologies for the supply (Chapters 7 through 15), storage and transmission (Chapter 16), and utilization (Chapters 17 through 20) of energy give rise to unwanted environmental impacts. It is instructive to consider a few examples in the context of Figure 4.1. Emanations of em radiation, sound, and heat are all mechanisms by which the system dissipates energy to its surroundings. This energy originated as high-quality energy (e.g., in the form of solar radiation, wind, chemical energy in a fuel) but, owing to inherent and anthropogenic shortcomings in the energy system, was degraded to the forms indicated. Electromagnetic radiation can be thought of as one means by which equipment for the transmission and distribution of electricity also fulfills its obligations to the Second Law of Thermodynamics. Wires and transformers are energy conversion devices. Instead of converting heat into mechanical work, they expend some electrical energy to change the electrical potential in stepping up or down to various line voltages. By the Second Law of Thermodynamics, at least some of the electrical energy expended to change the electrical potential energy must be lost. Electromagnetic radiation is one of the loss mechanisms; heat (evidenced by a rise in temperature of the equipment) is another. Both mechanisms can be thought of as macroscopic manifestations of the electromagnetic forces that push and pull on the moving corpuscles of charged matter (largely electrons) that enable the flow of electricity. Owing to these forces, there is the electromagnetic equivalent of friction between the charge carriers and the equipment through which they move. Fortunately, energy losses during electricity transmission and distribution (5–15%), while not inconsequential, are appreciably less than the 40–50% losses in heat to work machines (Figure 3.12).

Sound emissions arise from mechanical vibrations, which in turn may be caused by agitation of solid, liquid, or gaseous substances in the energy system. For example, improperly balanced or aligned mechanical components (e.g., shafts, rotors, turbines, bearings, gears) in rotating machinery drain away useful rotational or translational mechanical energy, dissipating some of the energy as oscillatory motion of mechanical parts of the machine. Imperfections in the burning of fuels may cause some of the fuels' chemical energy to be converted to pressure waves that induce mechanical vibrations (sound) in the combustor hardware, rather than contribute to useful work (e.g., rotation of a turbine or translation of a piston). (See the brief discussion of knock in spark ignition engines, in Section 3.2.4, Footnote 6.)

Some waste heat is mandatory owing to limitations of the Second Law of Thermodynamics (Chapter 3). Losses substantially worse than those dictated by thermodynamics may arise from imperfections in thermal insulation and mechanical components and from incomplete combustion of a fuel. This last effect may arise because some of the fuel escapes burning, or because fuel is incompletely converted to the lowest energy state products (generally CO_2 and H_2O), and the maximum amount of chemical energy obtainable is not fully accessed during the combustion process. Insufficient temperature or time for burning, as well as inadequate mixing of the fuel with its oxidant (usually air or O_2), are root causes of combustion imperfections. (Chapter 7 elaborates on combustion imperfections.)

Perhaps the most commonly known adverse environmental impacts of energy are pollutant emissions and solid wastes. In 2000, combustion of fossil fuels supplied about 85% of the world's energy. Combustion of biofuels—mainly wood, crop and animal residues, and municipal solid waste—further adds to the contributions of combustion to global energy production. We define unwanted combustion effluents as all material products of combustion other than H_2O (i.e., we include CO_2 in recognition of its role as a greenhouse gas). Sources of these products are incomplete combustion of the fuel and the presence of certain pollutants and pollutant precursors in the fuel (e.g., inorganic matter [metals, mineral matter, ash], sulfur and nitrogen, and sometimes chlorine and phosphorus). During combustion, contaminants in fossil fuels and in biomass fuels (Tables 10.3, 10.4) may undergo partial or complete oxidation. However, their combustion products are not H_2O and CO_2 , but depending on combustion conditions, become various inorganic compounds: metal oxides, chlorides, sulfates, as well as oxides of sulfur (SO_x) and nitrogen (NO_x) and potentially HCl and Cl_2. If not captured or chemically transformed, these species can contribute to acid rain and various adverse human health effects. Even for fuels containing only carbon and hydrogen, serious pollutants may be formed during combustion. Examples include carbon monoxide, CO, owing to incomplete conversion (combustion) of the fuel carbon to CO_2; carbonaceous fine particulate matter or soot that may contain chemically or physically bonded ingredients, including metals, inorganic oxides, and a host of complex organic compounds, including biologically available polycyclic aromatic hydrocarbons spanning wide ranges of molecular weight and volatility. Interestingly, fuels containing only C and H, or even either element alone, including molecular hydrogen (H_2), can form NO emissions if the fuel is combusted in air (i.e., in an atmosphere of O_2 that contains appreciable amounts of molecular nitrogen [N_2]). Chemical reactions that give rise to this so-called *thermal NO* during combustion are (Glassman, 1996):

$$O\cdot + N_2 \rightarrow NO + N\cdot \tag{4-1}$$

$$N\cdot + O_2 \rightarrow NO + O\cdot \tag{4-2}$$

Our discussion has examined Figure 4.1 largely from the perspective of an energy system operating to serve its intended users, with recognition that normal or "routine" operations may entail off-specification episodes that further exacerbate adverse environmental effects or give rise to new problems, including industrial accidents and additional or different material or radiative effluents. Modern systems take steps to minimize or eliminate off-spec performance and counteract their consequences through the use of advanced hardware and software technologies that provide real-time assessments of process conditions and forewarn of dangerous upsets. Additional and different interactions of energy systems with the environment will occur during construction and initial testing of the facility (shakedown) and after its useful service time has expired. In the latter case, typical operations with environmental impacts are decommissioning, scrapping, and/or recycling of the system components, as well as temporary or permanent disposal or recycling of residual wastes of the process. Protection of air, surface and sub-surface waters, and land from hazardous materials are major pollution management issues during the formative and sunset periods of the system.

Few if any energy systems operate continuously during their useful service period. Thus, on a yearly basis, large base load electric generating plants may operate 80–90% of the time, whereas personal automobiles may only be used 5–10% of the time. During downtimes, energy systems undergo scheduled and emergency maintenance, which involves interactions of the energy system with its environment. For an individual auto, the interaction may seem trivial (e.g., disposal of spent lube oil, an air filter, and some grease-covered cleaning rags two to three times per year, and of a lead-acid battery perhaps every three years). Yet scaling these inputs by the roughly 150 million vehicles on the road in the US (Chapter 18) shows that impacts add up quickly. Further, for a large nuclear plant, routine maintenance includes replacement of fuel rods and the temporary storage of the spent rods at the plant site (Chapter 8).

4.1.3 Length and time scales for environmental impacts

Figure 4.1 tells us nothing about how long it takes for a given effect of energy to affect its environment, nor about how far from the energy system the effects will be felt. Both issues are vitally important. Some energy-derived pollutants will be transported over various distances, from small to global, before impacting human or other target sites susceptible to their damaging impacts. Many adverse effects, including those harmful to human health, will not become evident for periods from the time of first exposure, to a few hours, (e.g., in the case of acute respiratory insults), to over 100 years (for inheritable birth defects). Figure 4.2 provides length and time scales for several examples of environmentally harmful impacts of energy. By length scales, we mean the distance over which an adverse impact can be felt, owing to transport of a pollutant or its by-products in the environment. By time scales, we mean the time between first exposure to a threatening agent and the time at which adverse effects of that agent can be detected *at a macroscopic level.* We need this disclaimer because a potentially harmful agent may inaugurate molecular level transformations at risk target sites "immediately" (i.e., within milliseconds or less) upon exposure. However, it may take years to decades before it is known whether these insults have serious adverse consequences (e.g., disease, death, degradation of buildings or artifacts, etc.). Advances in toxicology and related biomedical sciences are now allowing cellular, genetic, and molecular level abnormalities in humans to be detected and interpreted as early warning signs of serious disease. Many of these technologies are in their infancy but may eventually allow humans to identify health risks and prevent diseases long before they become serious and to determine what exogenous environmental agents have contributed to the onset or persistence of the disease or its precondition (Section 4.4).

Figure 4.2 shows that environmental effects associated with energy are encountered at different length and time scales that, for each dimension, range in excess of seven orders of magnitude. Note, however, that various categories of human health and other adverse impacts are divided into more tractable sub-domains of the length-time space. This observation, plus the mammoth total span of distances and times, suggests that efforts to mathematically simulate and forecast (Section 4.3) energy-related environmental impacts will be most fruitful by disaggregating impacts of interest according to more manageable regions of their impact space, and that a single modeling platform would have little hope of reliably capturing all the effects shown in this figure.

Figure 4.2 contains a few examples of energy-related environmental effects. An examination of these examples reinforces the diversity and complexity of energy-related environmental consequences. At one end of the time space are acute respiratory impacts caused by SO_x, NO_x, photochemical smog, and other pulmonary irritants emitted by stationary and mobile combustion sources. Such sources may come extremely close to a vulnerable site, as when pedestrians or bicyclists encounter automobile exhaust. A lower length scale of 1 m (0.001 km) is appropriate. On the other hand, these irritants may be transported 100 km or more before being washed out of the ambient air by rain or diluted by clean winds. Hence, we have an upper length scale of 100 km. Time scales for respiratory irritation may be virtually instantaneous in extreme cases—one inhaled volume of highly polluted air is enough to cause discomfort, even to a healthy person—hence our lower bound of 10^{-7} years. Likewise, it may take a few days for adverse effects to become manifest, giving an approximate upper bound of 0.01 years.

At the other extreme of Figure 4.2 is global climate change. Here, the upper length scale is easy to estimate (i.e., one-half the circumference of the earth or roughly 20,000 km). The lower boundary is more problematic, given that some potentially harmful outcomes of climate change could include loss of arable farmland or an acceleration of certain diseases. However, keeping with our association of this length scale with the distance traveled by the energy-derived agent(s) (here fugitive or stack/tailpipe emissions of CO_2 and CH_4), a length scale of roughly one-half the width of the North American continent (i.e., 2,000 km) is reasonable.

For corresponding time scales, we have chosen a lower bound of 25 years, but we must specify our time zero. From what event do we measure the 25 years, and to what climate change event does this time scale refer? A greenhouse gas entering the atmosphere immediately begins contributing to the greenhouse effect (Section 4.3.1) by back-scattering to earth some heat that otherwise (i.e., in the absence of this greenhouse gas) would escape the earth's surface. Rather than this process, we choose the probable minimum time required to detect a human influence on climate that is discernible above natural background. We take as our zero time 1990. Average atmospheric concentrations of CO_2 for this year are taken as the benchmark for future greenhouse gas emissions targets in the 1997 Kyoto protocol. Based upon global circulation models (GCMs, Section 4.3.2), many scholars believe that it will be at least 2015 before it will be possible to discern above-natural background, a signal of human-induced global warming.

The upper time bound is even more problematic. If rates of warming fall in the lowest range of credible estimates, it may be 2050 or later before the unequivocal signal above background is seen. Further, it is not yet known how various plausible levels of global warming would impact sea level, desertification, regional climate, and a host of other

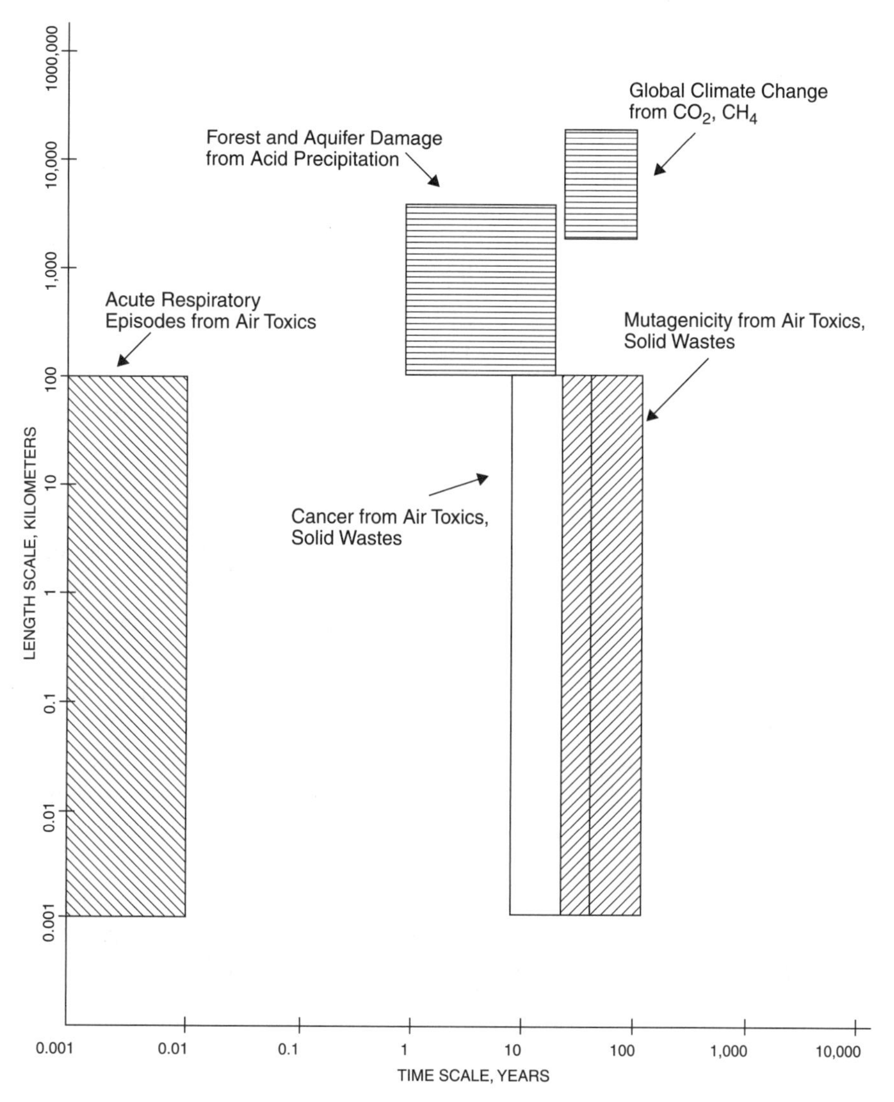

Figure 4.2. Approximate length scales and gestation times for selected environmental impacts of energy production and utilization.

attributes of the earth's ecosystem. Even if the more pessimistic scenarios for temperature increase pan out, it is uncertain how long it will take for adverse impacts to become manifest. In Figure 4.2, we assume the relevant time scale could be 125 years, but the true value could be appreciably longer or shorter. Time and further research will allow us to make better estimates.

Also instructive in Figure 4.2 are the blocks for cancer and mutagenicity. The role of exogenous environmental agents in these two pathologies is still a subject of intense research and, with exceptions such as cigarette smoke, well infused with controversy. For cancer, we took a lower time limit for gestation of 10 years recognizing the tragedy of childhood carcinomas without unequivocal proof of any linkage to energy-related environmental agents. Our upper bound of 50 years reflects the increased incidence of cancers with old age plus our uncertainty regarding gestation periods and the complicity of individual or multiple environmental agents. For mutagenicity, by definition the minimum time for an adverse impact must be one generation, which, in Figure 4.2, we took as 25 years. Likewise, a gestation of five generations is possible, giving rise to our upper bound of 125 years. For both pathologies, we take identical length scales to those given in Figure 4.2 for acute respiratory episodes induced by air toxics. This recognizes that some of the same pollutants are involved in respiratory and genetic diseases (in general by different mechanisms) and that putative carcinogens and mutagens in air toxics may be transported over similar length scales as respiratory irritants.

Acid rain caused by conversion of sulfur oxides and nitrogen oxides emitted by various combustion sources may degrade aquifers, forests, and other plant life over periods of 1–20 years from the time of exposure. The lower length scale is somewhat arbitrary in that, arguably, acid precipitation harmful to trees and water supplies could occur within 1 km of a source. Recognizing that forests and above ground aquifers are often 1–10 km or more in length or breadth and that those subject to acid rain damage are often 80 km or more from major anthropogenic sources of acid rain precursors, we take 100 km as our low bound. The upper bound of 3,500 km (roughly 2,000 miles) reflects the appreciable current evidence that forests and aquifers have suffered acid rain damage from anthropogenic pollutants generated as much as 2,000 miles away (e.g., damage to forests and lakes in New York, New England, Quebec, and Eastern Canada has been strongly linked to emissions of sulfur oxides from coal burning electric power plants in the midwestern US).

Let us not lose sight of an important message from Figure 4.2: neither distance nor time necessarily guarantees immunity from harmful environmental effects of a poorly operated energy system. Sites exposed to air pollutants, solid wastes, and other environmentally important emanations (Figure 4.1) face the insidious threats of subliminal vitiation and pollutant transport and transformation. The latter means that pollutants can jeopardize sites far from their sources, and the magnitude of the threat may be exacerbated. This is because pollutants or their precursors can undergo environmentally important modifications during their transport through ambient media, after emission from a source and before contact with a target site. These transformations (Seinfeld and Pandis, 1998) may detoxify hazardous emissions or strengthen their toxic potency. Subliminal vitiation refers to the degradation of human health, other living systems, or quality of an object by processes unrecognized by the victim or their care-givers (e.g., because of the early stage of the pathology where nanoscale damage is proceeding vibrantly yet undetectably with normal monitoring). The insidious consequence is that, without warning of an eventual problem, suitable countermeasures are delayed.

4.2 Adverse Environmental Effects Over Local and Regional Length Scales

4.2.1 Ambient air pollution

Many nations regulate the amount of pollutants allowed in ambient air. In the US, the EPA (2000) has established criteria for the cleanliness of ambient air. These are maximum allowable time-averaged concentrations, such as the National Ambient Air Quality Standards (NAAQS) for certain pollutants in outdoor air, measured over prescribed times. NAAQS currently exist for six *criteria pollutants*: carbon monoxide (CO), lead (Pb), nitrogen dioxide (NO_2), ozone (O_3), sulfur dioxide (SO_2), particulate matter ($PM_{2.5}$, PM_{10}, where the subscript refers to the mass mean diameter of the particles, in microns, i.e., units of 10^{-6} m). NAAQS are further divided into primary and secondary standards. The primary NAAQS are designed to protect human health, while the secondary standards are promulgated to protect against *welfare effects* (e.g., degradation of visibility, damage to buildings, crops, vegetation, and ecosystems). Averaging times can be long term: one year—designed to protect people and other target sites from adverse impacts caused by both short- and long-term exposure to air pollutants; and short term, or $\leq$ 24 hrs—aimed in particular at protecting humans from adverse health effects induced by abbreviated exposure to elevated levels of harmful agents. Table 4.1 presents the NAAQS and time averaging for the six US criteria pollutants. Note that secondary standards have been established for all criteria pollutants except CO, and that the primary and secondary standards are the same for each of the other five criteria pollutants except SO_2.

Table 4.1. National Ambient Air Quality Standards for the US EPA Criteria Pollutants as of December 1999

Pollutant	Primary Standard (Health Related)		Secondary Standard (Welfare Related)	
	Type of Average	Standard Level Concentration[c]	Type of Average	Standard Level Concentration[a]
CO	8-hour[b]	9 ppm ($10\ mg/m^3$)	No Secondary Standard	
	1-hour[b]	35 ppm ($40\ mg/m^3$)	No Secondary Standard	
Pb	Maximum Quarterly Average	$1.5\ \mu g/m^3$	Same as Primary Standard	
NO_2	Annual Arithmetic Mean	0.053 ppm ($100\ \mu g/m^3$)	Same as Primary Standard	
O_3	Maximum Daily 1-hour Average[c]	0.12 ppm ($235\ \mu g/m^3$)	Same as Primary Standard	
	4th Maximum Daily[d] 8-hour Average	0.08 ppm ($157\ \mu g/m^3$)	Same as Primary Standard	
PM_{10}	Annual Arithmetic Mean	$50\ \mu g/m^3$	Same as Primary Standard	
	24-hour[b]	$150\ \mu g/m^3$	Same as Primary Standard	
$PM_{2.5}$	Annual Arithmetic Mean[e]	$15\ \mu g/m^3$	Same as Primary Standard	
	24-hour[f]	$65\ \mu g/m^3$	Same as Primary Standard	
SO_2	Annual Arithmetic Mean	($80\ \mu g/m^3$)	3-hour[b]	0.50 ppm ($1{,}300\ \mu g/m^3$)
	24-hour[b]	0.14 ppm ($365\ \mu g/m^3$)		

[a] Parenthetical value is an approximately equivalent concentration. (See 40 CFR (Code of Federal Regulations) Part 50).

[b] Not to be exceeded more than once per year.

[c] The standard is attained when the expected number of days per calendar year with maximum hourly average concentrations above 0.12 ppm is equal to or less than 1, as determined according to Appendix H of the Ozone NAAQS.

[d] 3-year average of the annual 4th-highest daily maximum 8-hour average concentration.

[e] Spatially averaged over designated monitors.

[f] The form is the 98th percentile, that is no more than 2% of the daily observations can exceed the 24-hr standard.

Source: EPA (2000).

Figure 4.3 shows that combustion, including transportation and other energy-related sources, such as industrial processing, is the major source of all six criteria pollutants. (The figure does not show a pie-chart for tropospheric [i.e., near ground level] ozone, because this pollutant is a product of tropospheric reactions of volatile organic compounds (VOCs) and nitrogen oxides (NO_x) in the presence of sunlight.) Sources of NO_x include fixation of atmospheric nitrogen during high temperature combustion of various fuels in air (e.g., in motor vehicle internal combustion engines and boilers for electric power generation), as well as oxidation of nitrogen in fuel itself, and natural sources such as lightning and biological processes in soil. Common VOCs, shown in Table 4.2, come from emissions from motor vehicles, petroleum refineries, manufacturing, consumer and commercial products, and natural sources such as trees (Table 4.3). VOCs, NO_x,and O_3 are important contributors to regional air pollution because they can all be transported hundreds of miles from their primary sources. Figure 4.4 depicts pathways for the formation of fly ash and other primarily inorganic fine particulate during pulverized coal combustion.

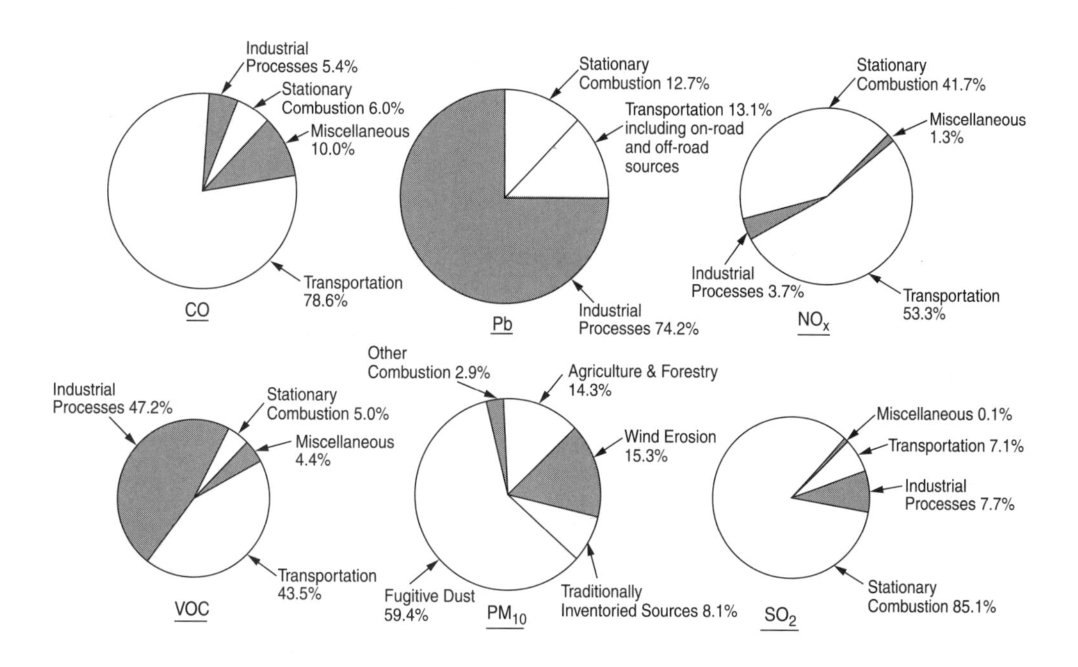

Figure 4.3. US EPA data for source apportionment of 1998 ambient air emissions of VOCs and five criteria pollutants. Source: EPA (2000).

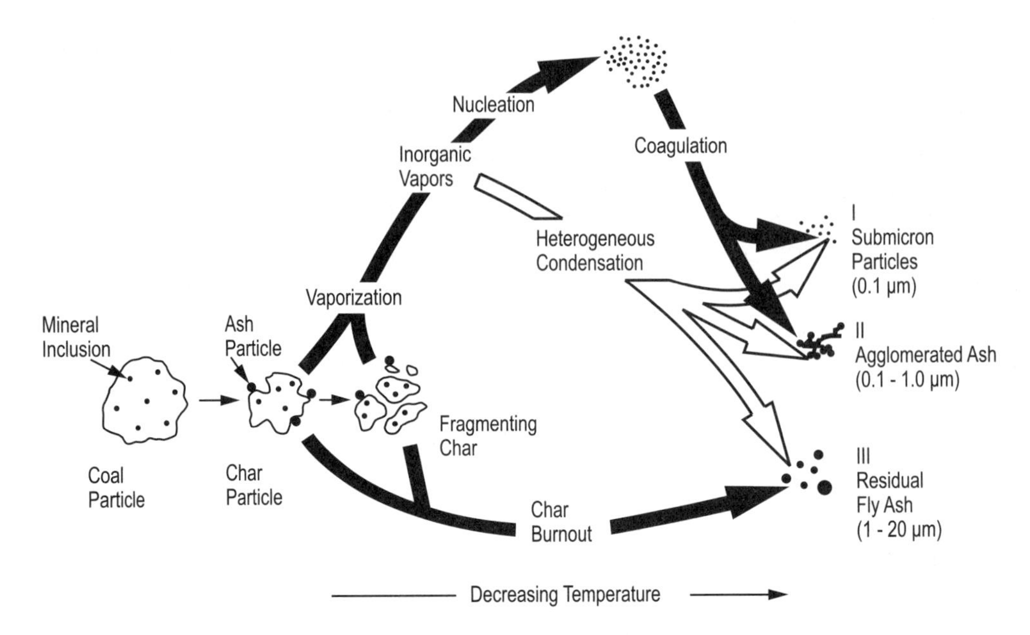

Figure 4.4. Processes responsible for the production of fly ash and other predominantly inorganic fine particulates during pulverized coal combustion. Source: Helbe et al. (1988). Reprinted with permission of The Combustion Institute.

Table 4.2. Examples of Volatile Organic Compounds (VOCs) in US Ambient Air

Propane	Isobutane	Isoprene
Isopentane	Benzene	Methylcyclopentane
Toluene	Acetylene	3-Methylhexane
Ethane	n-Hexane	Ethylbenzene
n-Butane	2,2,4-Trimethylpentane	m-Ethyltoluene
n-Pentane	Propylene	Formaldehyde
Ethylene	3-Methylpentane	Acetaldehyde
o, m, and p-Xylenes	1,2,4-Trimethylbenzene	Monoterpenes

Source: EPA (2000).

Table 4.3. Examples of Biogenic Sources of VOCs in the US

Region	VOC	Source
Southwestern United States	Isoprene	Oak (mostly), citrus, eucalyptus
	Monoterpenes	Pine, citrus, eucalyptus
Northeastern United States	Isoprene	Oak (mostly), spruce
	Monoterpenes	Maple, hickory, pine, spruce, fir, cottonwood

Source: EPA (2000).

Energy is among the sources of another category of air pollutants, denoted as *hazardous air pollutants* (HAPs) or *air toxics* or *toxic air pollutants* (EPA, 2000). HAPs are substances known or suspected to damage ecosystems or cause serious human health effects, such as cancer or neurological, cardiovascular, respiratory, liver, kidney, immune system, or developmental effects (Dockery et al., 1993). Under Section 112 of the US Clean Air Act, the EPA in the year 2000 identified 188 substances as HAPs. HAPs include heavy metals, such as arsenic, lead, cadmium, chromium, mercury, and nickel, and various organic compounds, including benzene, benzo[*a*]pyrene, formaldehyde, and various chlorinated solvents (Table 4.4). Among the energy-related sources of HAPs identified by the EPA are electric power generation, automobiles, and large industrial facilities. All of these are indispensable to human progress, so it is essential to devise means to mitigate their adverse effects on air quality and other indices of environmental sustainability. For example, it is important to know the mechanisms by which HAPs and other air pollutants are generated and released by various sources, the processes by which pollutants survive or transform in ambient airsheds, and how health impacts and other environmental effects depend on ambient concentrations of pollutants. Potential human health impacts of HAPs are estimated from studies of their toxicology in model biological systems, such as single cells and non-primate animals, and from the determination of the frequency of occurrence of disease and shortened life spans among defined populations, such as workers in a particular industry (*epidemiology*). Consequently, the level of certainty, as well as the relevance to humans of various studies of the health risk of HAPs, can vary widely (Section 4.4). Consistent with this, the EPA classifies more than half of 188 HAPs as "possible," "probable," or "known" carcinogens. In the last category are substances whose carcinogenicity is strongly supported by studies directly relevant to humans (e.g., benzene and arsenic are, respectively, linked to leukemia and lung cancer by epidemiological studies of workplace air and smelter workers). The EPA publishes yearly trends in US air quality, including extensive databases on annual emissions of criteria pollutants and HAPs by region and source.

Table 4.4. Examples of Various Substances in the US EPA List of Hazardous Air Pollutants (HAPs)

Trace Metals[1]	Various Organic Compounds
Nickel (fine)	1,2-Dibromoethane
Cadmium (PM_{10})	Ethylene dichloride
Arsenic (fine)	Styrene
Lead (fine)	Trichloroethylene
Manganese (fine)	Vinyl chloride
Lead (coarse)	1,2-Dichloropropane
Nickel (pm10)	Tetrachloroethylene
Nickel (tsp)	1,3-Butadiene
Manganese (PM_{10})	1,1,2,2-Tetrachloroethane
Chromium (coarse)	trans-1,3-Dichloropropene
Chromium (PM_{10})	Chloroform
Beryllium (tsp)	Carbon tetrachloride
Chromium (tsp)	Acrolein
Nickel (coarse)	Toluene
Chromium (fine)	Benzene
Lead (PM_{10})	Formaldehyde
Manganese (coarse)	Acetaldehyde
Chromium VI	Acrylonitrile
Arsenic (PM_{10})	Benzo(a)pyrene (total PM_{10} & vapor)
Arsenic (coarse)	Dibenz(a,h)antracene (total PM_{10} & vapor)
Beryllium (PM_{10})	Indeno(1,2,3-cd)pyrene (total PM_{10} & vapor)
Mercury (coarse)	Benzo(k)fluoranthene (total PM_{10} & vapor)
Mercury (fine)	
Manganese (tsp)	
Mercury (PM_{10})	
Mercury (tsp)	
Arsenic (tsp)	
Lead (tsp)	
Cadmium (tsp)	

[1]Fine and coarse refer to particle size ranges, PM_{10} = particles 10 micrometers or less in diameter, and tsp = total suspended particulate.
Source: EPA (2000).

Table 4.5 lists human health effects associated with five of the six criteria pollutants. Various studies implicate particulate matter, especially the fine particulates (nominally monitored as $PM_{2.5}$) in adverse human health effects. For example, in the early 1990s, results of a 15-year study of 8,111 adults in six US urban areas (the so-called six cities study of Portage, WI; Topeka, KS; Watertown, MA; Harriman, TN; St. Louis, MO; and Steubenville, OH) found a correlation between area-specific human mortality, corrected for various interfering co-factors, and ambient air concentrations of fine particles. However, significant controversy remains as to what attributes of the fine particles (e.g., chemical compostion, physical morphology, separately or in concert) are actually responsible for the adverse human health effects (Abelson, 1999). Combustion for the supply of energy, as in coal-fired power plants, oil-fired boilers, and diesel for engines, is a major source of these ultrafine particulates (Helbe et al., 1988).

Table 4.5. Examples of Human Health Effects Associated with Selected Ambient Air Pollutants

Pollutant	Effects
CO	Reduction in the ability of the circulatory system to transport O_2 Impairment of performance on tasks requiring vigilance Aggravation of cardiovascular disease
NO_2	Increased susceptibility to respiratory pathogens
O_3	Decrement in pulmonary function Coughing, chest discomfort Increased asthma attacks
Lead	Neurocognitive and neuromotor impairment Reduced hemoglobin synethesis and hematologic alterations
Peroxyacyl nitrates, aldehydes	Eye irritation
SO_2/particulate matter	Increased prevalence of chronic respiratory disease Increased risk of acute respiratory disease

Source: Boubel et al. (1994), see also Godish (1997), Table 5.2, p. 168.

Pollutant regulations typically limit the quantity of a pollutant that can be emitted per unit of operating intensity or output of the source under scrutiny, such as mass of pollutant per unit of heat expended (lb/Btu) or per time of operation (lb/hr) or mass per total volume of exhaust gas (g/standard m^3). Historically, US emissions standards have applied equally to all sources regardless of the quality of the air at the source location. This approach can make compliance and monitoring routine, but it may not produce sufficient emissions reductions in areas of poor air quality or it may cause excess pollution abatement expenditures in areas where air quality is already good. More recent regulatory practices in the area of acid deposition and air quality losses linked to automotive emissions respond to these potential deficiencies. Thus, automobile fuel formulations and tailpipe emissions are more stringently regulated in California and in other states where local air quality deterioration appreciably exceeds national averages.

Often, emissions standards take account of various indices of "practicality" (e.g., economic cost, technical viability, and political acceptability) (Godish, 1997). In Europe, this approach is designated "best practicable means." In the US, it refers to the use of "reasonably available control technology" (RACT). Under these regulations, emissions standards can become more stringent as technology improves or abatement costs decline. Sometimes, as in the US case of pollutants posing serious risk to human health (e.g., mercury, beryllium, benzene, certain radioactive isotopes, asbestos), higher level standards aimed at achieving the largest emissions reductions capable with available technology are called for. In the US, these are known as "best available control technology" (BACT) and the even-more-stringent "lowest achievable emission rebase" (LAER) standard of the 1990 US Clean Air Act (CAA) Amendments.

The 1990 Amendments to the US Clean Air Act codified a different approach to the reduction of acid deposition ("acid rain") caused by emissions of SO_x and NO_x from large coal-fired electric power plants and other sources. The objective (under Title IV) was to reduce SO_x and NO_x emissions by 10 and 2 million tons/year in two phases beginning in 1995 and 2000. To promote more economical and flexible approaches, Title IV replaced the "command and control" approach of the 1977 Clean Air Act Amendments, which mandated specific percentage reductions in SO_2 emissions from coal-fired power plants with a system of SO_x emissions allowances for fossil-fuel fired electric generating stations. Allowances equal to 1 ton of emitted SO_x are allocated each year for each unit in each emissions source. When a unit exceeds its allowance in a given year, it pays a penalty of $2,000/ton SO_x (adjusted by the consumer price index) *and* in a subsequent period specified by the EPA (e.g., the next year or a shorter time frame), it must offset the excess emission. The allowances were enacted to allow market forces to facilitate the realization of environmental goals. To this end, Title IV allows excess emissions to be offset through the purchase of allowances from other units that have realized their emissions goals, including trading of a restricted number of allowances in spot and advance auctions. Unused allowances can be banked for future use.

Industry and environmental activists both agree that this market mechanism approach to acid deposition reduction has been successful, with Phase 1 emissions reduction targets realized, and Phase 2 goals also expected to be met. The allowance approach does not discourage emitters from taking action to actually reduce their SO_x emissions. Title IV requires each source assigned emissions allowances to develop a plan and timetable to reduce its emissions, encouraging flexible approaches including use of renewable and clean alternative energy, pollution prevention, and energy conservation.

US clean air regulations address two other important issues linked to energy as a pollution source: maintaining good air quality in pristine areas such as National Parks, National Wilderness Areas, and National Monuments, and the impacts of new facilities for manufacturing or energy generation. In the 1970s, a split decision of the US Supreme Court resulted in promulgation of air quality standards for prevention of significant deterioration (PSD) in pristine air sheds. Table 4.6 displays, as amounts by which the concentrations of three criteria pollutants (PM_{10}, SO_x, and NO_x) can increase, levels of air quality deterioration allowed under the CAA for three categories of pristine areas. Class I includes International Parks, National Wilderness Areas, and National Monuments of more than 5,000 acres, and National Parks of more than 6,000 acres (Godish, 1997). As shown in Table 4.6, successively greater amounts of air quality deterioration are allowed in rural (Class II) and urban (Class III) areas, respectively. To prevent new pollution problems, the US Clean Air Act includes New Source Performance Standards (NSPS). These apply to all new sources and to sources that the EPA defines as significantly modified. The NSPS limit emissions of criteria pollutants and other pollutants by use of the best reduction system taking cost into account. Table 4.7 gives examples of sources and pollutants subject to NSPS. Many of these are facilities for energy production, petroleum refining, or energy-intensive manufacturing.

Table 4.6. Examples of US PSD Standards for Protection of Pristine Air Sheds

Pollutant	Standard	Class	Allowable increment ($\mu g/m^3$)
PM_{10}	Annual mean	I	4
		II	17
		III	34
	24-hour maximum	I	8
		II	30
		III	60
SO_2	Annual mean	I	2
		II	20
		III	40
	24-hour maximum	I	5
		II	91
		III	182
	3-hour maximum	I	25
		II	512
		III	700
NO_2	Annual mean	I	2.5
		II	25
		III	50

Source: Godish (1997).

Table 4.7. Examples of Atmospheric Emissions Sources and Pollutants Regulated Under US NSPS

Sources	Pollutants
Fossil fuel–fired steam generators	Particulate matter, opacity, SO_2, NO_x
Coal preparation plants	Particulate matter, opacity
Incinerators	Particulate matter, opacity
Primary lead, zinc, and copper smelters	Particulate matter, SO_2, opacity
Secondary lead smelters	Particulate matter, opacity
Secondary bronze and brass smelters	Particulate matter, opacity
Primary aluminum reduction plants	Fluorides
Iron and steel plants	Particulate matter, opacity
Ferroalloy production facilities	Particulate matter, opacity, CO
Steel plants—electric arc furnaces	Particulate matter, opacity
Sulfuric acid plants	SO_2, acid mist, opacity
Nitric acid plants	NO_x
Portland cement plants	Particulate matter, opacity
Asphalt concrete plants	Particulate matter, opacity
Sewage treatment plants	Particulate matter, opacity
Petroleum refineries	Particulate matter, SO_x, CO, reduced sulfur
Kraft pulp mills	Reduced sulfur, particulate matter, opacity
Grain elevators	Particulate matter, opacity
Lime manufacturing plants	Particulate matter
Phosphate fertilizer industry	Fluorides

Source: Godish (1997).

4.2.2 Adulteration of soil, water, and indoor air

Environmentally harmful products of energy production and utilization (Figure 4.1) may adversely impact not only ambient outdoor air, but also indoor air quality and the cleanliness of all the earth's ecosystems, including soil and water. (Important information sources include EPA, 1995; Leslie and Lunau, 1994; Sposito, 1994; and Schwarzenback et al., 1993.) A comprehensive examination of the environmental footprint of an existing or proposed energy technology must consider potential impacts on all vulnerable media. Federal and regional authorities typically require such assessments, sometimes referred to as *environmental impact statements*, as part of the process for seeking approval of a new energy project or a significant modification of an existing facility. Examples of typical problems that must be addressed are safe management and disposal of solids, such as radioactive waste from nuclear electric generating facilities (Chapter 8) and ash from coal-fired power plants, and avoidance of thermal pollution of adjacent rivers, streams, and bays by power plant cooling water. Potential sources of indoor air pollution include degassing of solvents and other volatiles from a host of anthropogenic products, such as carpets and paints, and emission of pollutants from gas-fired appliances, such as stoves and heaters (Wadden and Scheff, 1983, Boubel et al., 1994).

4.2.3 Transport and transformation of air, ground, and water contamination

Energy production and utilization facilities are sources of air, ground, and water pollution, directly through adverse emanations to these media and indirectly through release of pollutant precursors. The latter, through chemical or physical transformations within the air, ground, or water, are converted to undesireable agents. The medium itself, including the air, water, soil, sediment, or other terrain, may function as the host for the transformation process(es) or participate actively in the transformations. Physical transport of pollutants and their precursors into the region (e.g., waterway, airshed) of interest must be understood together with various chemical reactions that may occur within one phase, such as air or water, or between phases, such as air and droplets of polluted water. Detailed exegesis of the underlying mechanisms and practical consequences lies far outside the present scope. For information on pollutant transport and transformation in outdoor ambient air, see Seinfeld and Pandis (1998).

The concentrations, lifetimes, and propensity of energy-derived substances to generate pollutants in outdoor ambient airsheds depend upon tropospheric chemical reactions, many of which are induced by sunlight (Seinfeld and Pandis, 1998). The production of tropospheric ozone, an EPA criteria pollutant, accentuates an important message for sustainability, namely that thorough mechanistic understanding of pollutant formation and survival in all media is essential to devising meaningful ecosystem protection schemes, be they technological remedies, policy initiatives, or some blend of the two. Globally, we think of tropospheric ozone as resulting from sunlight-induced chemical reactions between one energy-derived pollutant, nitrogen dioxide (NO_2) (known to be released during energy production by combustion of fuels in air or of nitrogen-containing fuels with or without air), and other species that may be derived from energy utilization. These substances include hydrocarbons such as methane that are released by leakage from natural gas pipelines, from incomplete combustion of natural gas, and by fermentation of organic matter (Chapter 10). The underlying ozone formation chemistry is complex. This is illuminated by referring to Figure 1.17, which schematically plots contours of constant concentration of tropospheric ozone as affected by the tropospheric concentrations of NO_x (ordinate) and total hydrocarbons (abscissa). Note that, depending upon the concentration of hydrocarbons, a reduction in NO_x concentration might also reduce ozone, or it could have no effect or even *increase* the ozone inventory.

4.3 Global Climate Change: Environmental Consequences over Planetary-Length Scales

4.3.1 Introduction

Figure 4.5 shows that, after being relatively stationary for the previous six centuries, atmospheric concentrations of carbon dioxide gas (CO_2) have been increasing from their value of about 275 ppm around 1750. Moreover, this figure shows that CO_2 concentrations increased much more dramatically since about 1880—the dawn of fossil fuel-powered global industrialization. For comparison, Figure 4.6 shows the average global surface temperature variations for the earth for the period 1855–1999. Note an overall increase from 1900 to 1999, appreciable year-to-year variability, and a protracted period, ca. 1940–1960, of no overall temperature increase, and, arguably, a slight cooling trend.

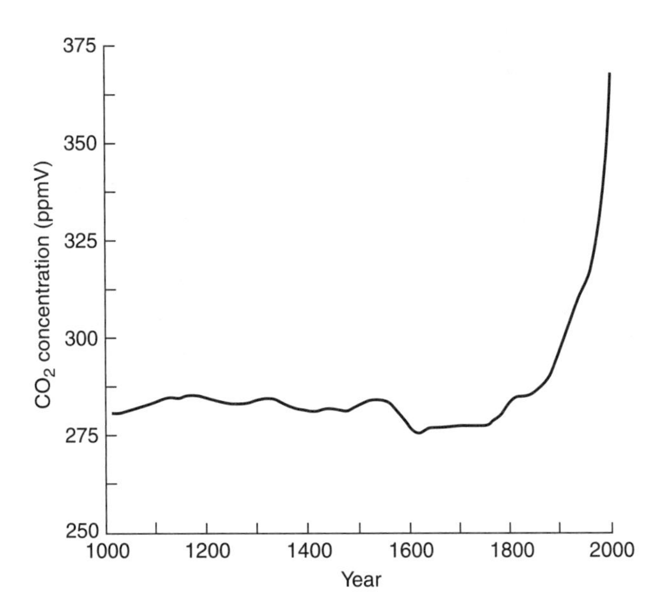

Figure 4.5. Trend of CO_2 concentrations in the atmosphere 1000–2000 A.D. (Data from Carbon Dioxide Information Anaylsis Center, 2000. Source: Fay and Golomb (2002).

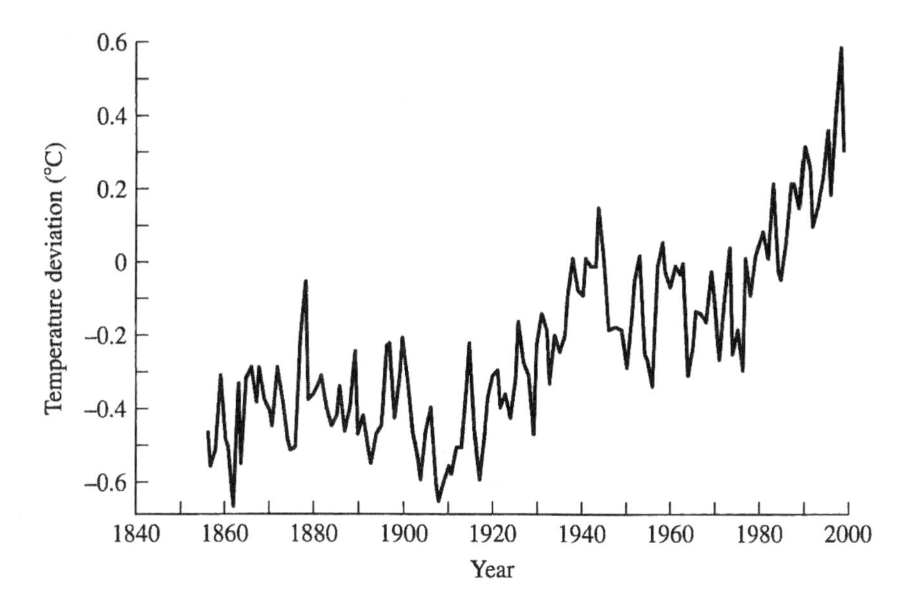

Figure 4.6. Observed average global surface temperature deviation 1855–1999. (Data from Carbon Dioxide Information Analysis Center, 2000. Source: Fay and Golomb (2002).

Many of us know qualitatively that atmospheric CO_2 helps warm the earth and that fossil fuel utilization emits appreciable quantities of CO_2 to the atmosphere. In the judgment of many experts, there is serious danger that too much warming could give rise to a number of adverse impacts for humans and threaten our survival on earth. Over the last 20 years, many scientific, technological, and policy experts have invested appreciable time and talent in seeking a clearer understanding of how and to what extent energy consumption, especially fossil fuels, contributes to global climate change and in proposing multinational steps to reverse the accumulation of greenhouse gases in the troposphere. During this same period, owing to increased attention from the print and electronic media and from politicians, the general populace has been exposed to the notion that unbridled growth in fossil fuel use will have dangerous consequences, variously termed "global warming," "the greenhouse effect," and "climate change." However, it seems probable that non-specialists have, at best, only a rudimentary awareness of how certain gases in the atmosphere cause greenhouse warming of the earth or of how energy utilization can increase the inventories of these gases and exacerbate warming.

In this section, we will examine how the greenhouse effect works and how energy use can upset this effect in unwanted ways. We will also look at potential adverse consequences of greenhouse warming of the earth and at what humans now and in the future might do to counteract or prevent greenhouse warming. We will also correct misconceptions about the consequences of greenhouse warming.

4.3.2 Basic science of the greenhouse effect

The wavelength or frequency of electromagnetic radiation (emr) can vary widely. Qualitatively, the scientific essence of the greenhouse effect is that the earth's atmosphere is largely transparent to some frequencies of emr but substantially reflects others. Frequencies well transmitted include much of the visible and ultraviolet radiation of the sun, although much of the infrared (heat) radiation from the sun is reflected (Figure 4.7). Why then is the earth's surface warmed by the sun? Some of the sun's UV and visible emr that reaches the earth's surface is converted by chemical and biological processes into heat energy. Further, some of that heat energy is re-emitted by the earth's surface in the form of infrared (IR) radiation. If the earth's atmosphere were completely transparent to IR radiation, all of that heat would be returned to space, and the average temperature of the earth's surface would be substantially lower. However, gases in the earth's atmosphere, such as water vapor and CO_2 among many others, are not transparent to all IR frequencies, so that some of the IR from the earth's surface is absorbed by these gases and then reradiated toward the earth's surface as IR radiation. The net effect is to provide additional warming to the earth's surface and to the lower atmosphere (Figure 4.7). Gases that perform this function of IR adsorption and reradiation are called *greenhouse gases* (*GHGs*). (The underlying chemistry and physics of how they garner and rerelease IR energy is rather complex. Readers may consult Seinfeld and Pandis (1998), and IPCC (1996).)

Human impacts on global climate can arise if human actions disrupt atmospheric concentrations of GHGs to the point that fluxes of IR radiation to the earth's surface are substantially reduced (*global cooling*, e.g., owing to lower GHG concentrations) or increased (*global warming*, e.g., because of higher GHG concentrations). A recent report (IPCC, 2000) states that "the preponderance of evidence" points to "a human influence on global climate." Specifically, this report identifies atmospheric emissions of greenhouse gases by human activities, in particular CO_2 released from combustion of fossil fuels, as contributing to increases in global mean temperature (i.e., to global warming). The report leaves open three crucial questions:

- By how much does global mean temperature increase for a given increase in atmospheric concentrations of CO_2?
- What fraction of the increase in global mean temperature is caused by human activities vs. natural processes?

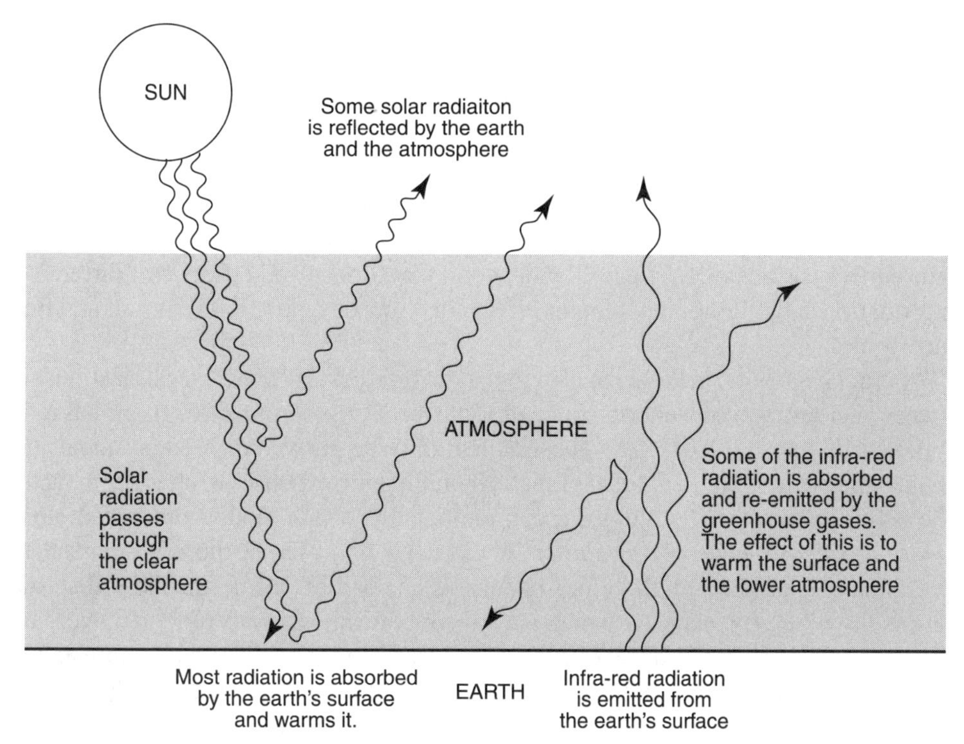

Figure 4.7. Simplified illustration of the mechanism of the "greenhouse" effect. Source: IPCC (1990). Reprinted with permission of IPCC.

- What will be the consequences of a given extent of global warming for humans and the earth's environment?

It is difficult to demonstrate that human activity is modifying global climate. Many atmospheric substances, not just CO_2, can increase or decrease the flux of energy to the earth's surface, and there is substantial uncertainty in their contributions. For example, Figure 4.8 shows estimates from the Intergovernmental Panel on Climate Change of the globally and annually averaged anthropogenic radiative forcing from changes in atmospheric concentrations of greenhouse gases and various aerosols (particulates) from before the Industrial Revolution to 1992, as well as natural changes in solar output from 1850 to 1992 (IPCC, 1996). The bar height denotes a mid-range estimate, and the error bars are based primarily on the spread in published values. Confidence levels reflect the judgment of the IPCC members who prepared the figure. The next-to-last bar denotes effects of indirect changes in contributions of clouds, an effect that is so uncertain that a mean value is not given. The figure omits contributions from aerosols emitted by volcanoes, which can be substantial. Note also that natural processes such as forest fires can change the impact of aerosols from anthropogenic biomass burning. A further complication is the role of atmospheric mixing. The greenhouse gases (first bar) are sufficiently long-lived that they can be well-mixed throughout the earth's atmosphere. In contrast, ozone and aerosols have shorter lifetimes and exert their impacts more regionally. Thus, the cooling effects of aerosols cannot be viewed as a trans-global offset to the warming of greenhouse gases. Further, land and ocean temperatures have exhibited different patterns of change over the last 120 years (Figure 4.9). Moreover, the global climate system itself is extremely complicated (Figure 4.10), involving solar radiation and its interactions with atmospheric gases and aerosols and with various parts of the earth's surface (e.g., oceans, land, snow) that absorb and reflect emr differently, the formation and migration of clouds, effects of winds, evaporation from salt and fresh water bodies.

The crucial question in the science of global warming is: are there causal relationships between increasing atmospheric concentrations of CO_2 and increases in global mean temperature (Stone, 2000)? This question can only be answered through models that simulate the behavior of the global climate, including its responses to changes in factors believed or known to affect global mean temperature. Such models are called *global climate models* or *general circulation models* (*GCMs*). These models simulate the three-dimensional state of the atmosphere, ocean, or biosphere, and how that state changes in time. By state, we mean wind velocities, humidity, cloud cover, and temperature at various heights above the earth's surface, as well as temperature and net flux to the atmosphere of water and energy at the earth's surface (Figure 4.11).

GCMs "work" by solving the mathematical equations that express the laws of conservation of material, momentum, and energy for the earth's atmosphere and its boundaries. However, the mathematical representation of these laws is extremely complex. The necessary equations are nonlinear and their solutions are chaotic (i.e., hypersensitive to the conditions chosen to describe the initial state of the atmosphere). These equations can only be solved approximately, even with the most powerful computers (e.g., about 10 million descriptive parameters are needed and of order 10 hours of CPU time on a supercomputer to simulate the evolution of the atmosphere for just one year; see Figure 4.11). Many important ocean-, land-, and atmosphere-based effects cannot be modeled from first principles and must be inputted to the GCMs parametrically (see Prinn et al., 1999).

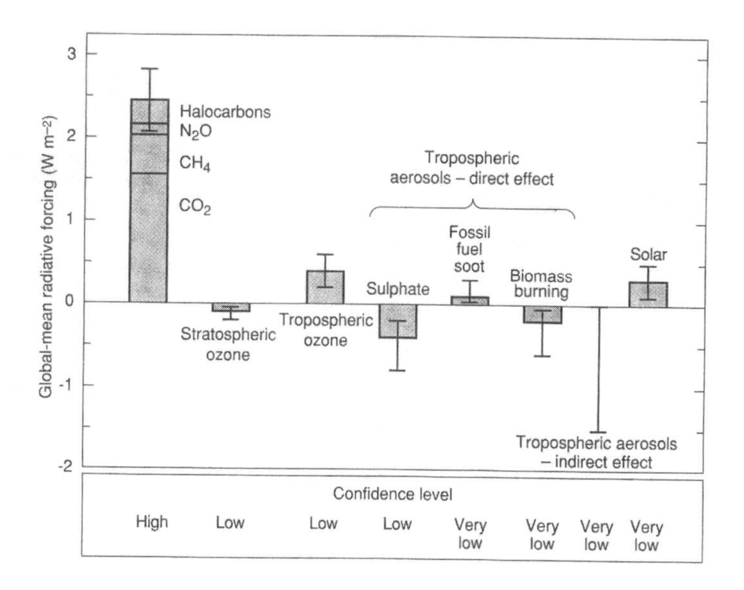

Figure 4.8. Estimates of the globally and annually averaged anthropogenic radiative forcing (in Wm^{-2}) due to changes in concentrations of greenhouse gases and aerosols from pre-industrial times to the present day and to natural changes in solar output from 1850 to the present day. The height of the rectangular bar indicates a mid-range estimate of the forcing while the error bars show an estimate of the uncertainty range, based largely on the spread of published values; our subjective confidence that the actual forcing lies within this error bar is indicated by the confidence level. The contributions of individual gases to the direct greenhouse forcing is indicated on the first bar. The indirect greenhouse forcings associated with the depletion of stratospheric ozone and the increased concentration of tropospheric ozone are shown in the second and third bar, respectively. The direct contributions of individual tropospheric aerosol components are grouped into the next set of three bars. The indirect aerosol effect, arising from the induced change in cloud properties, is shown next; our quantitative understanding of this process is limited at present and hence no bar representing a mid-range estimate is shown. The final bar shows the estimate of the changes in radiative forcing due to variations in solar output. The forcing associated with stratospheric aerosols resulting from volcanic eruptions is not shown, as it is variable over this time period. Note that there are substantial differences in the geographical distribution of the forcing due to the well-mixed greenhouse gases (CO_2, N_2O, CH_4, and the halocarbons) and that due to ozone and aerosols, which could lead to significant differences in their respective global and regional climate responses (see Chapter 6). For this reason, the negative radiative forcing due to aerosols should not necessarily be regarded as an offset against the greenhouse gas forcing. Source: IPCC (1996). Reprinted with permission of IPCC.

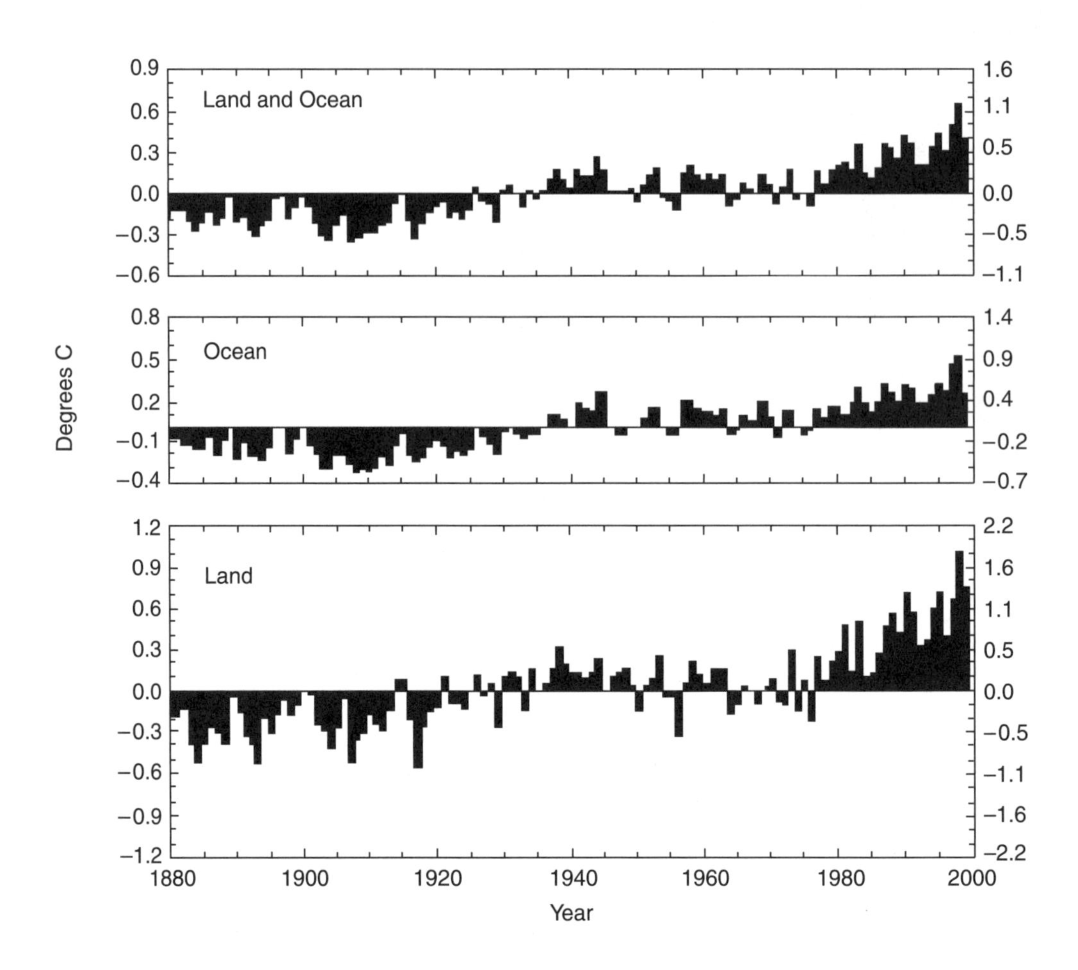

Figure 4.9. Differences in the 1880–2000 time series for the annual global surface mean temperature for land and ocean masses on the earth. Source: NOAA (2003).

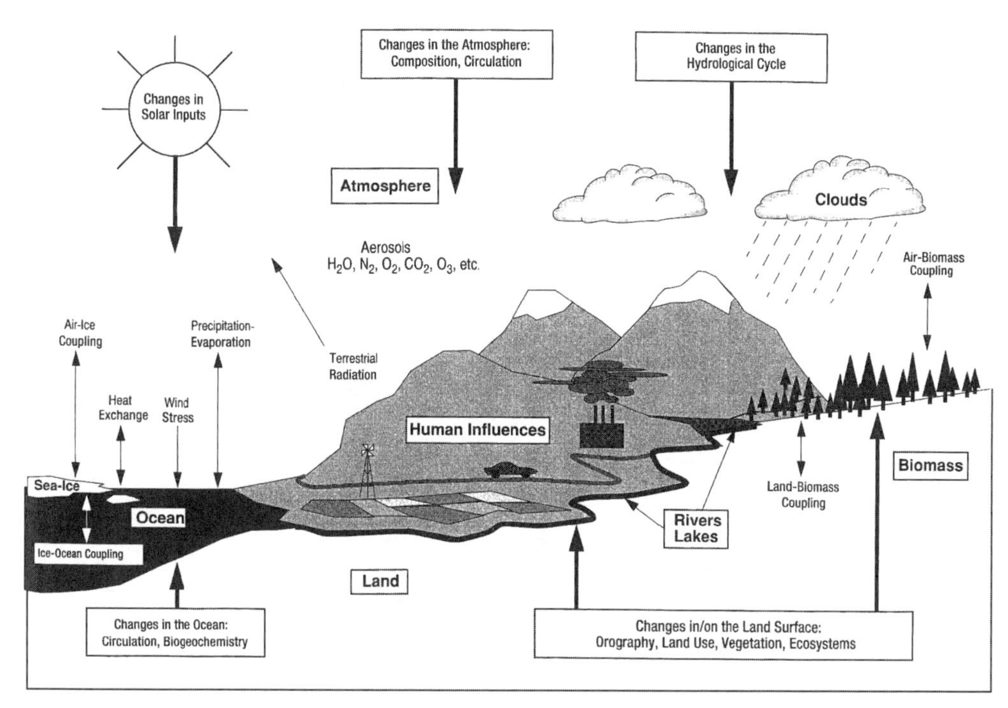

Figure 4.10. Schematic illustration of the complexity of the global climate system. Components of the system are depicted in bold; system processes and interactions are shown by thin arows; bold arrows denote some of the major system impact forces that may change over time scales relevant to global climate forcing. Source: IPCC (1996). Reprinted with permission of IPCC.

Figure 4.12 compares observed changes in global temperature with climate model predictions allowing for increases only in CO_2, and for increases in CO_2 plus direct effects of sulfate aerosols. Owing to natural effects, the year-to-year variability in the observed temperature changes can amount to 0.5°C. Thus, an anthropogenic effect on climate must exceed 0.5°C to be significant, and differences of this magnitude between model predictions, as occurs in Figure 4.12 for the sulfate and non-sulfate cases, cannot be taken as significant. Although even the best GCMs still have major uncertainties and omissions and cannot yet provide forecasts for specific regions, they are still the only means to detect and quantify human impacts on global climate.

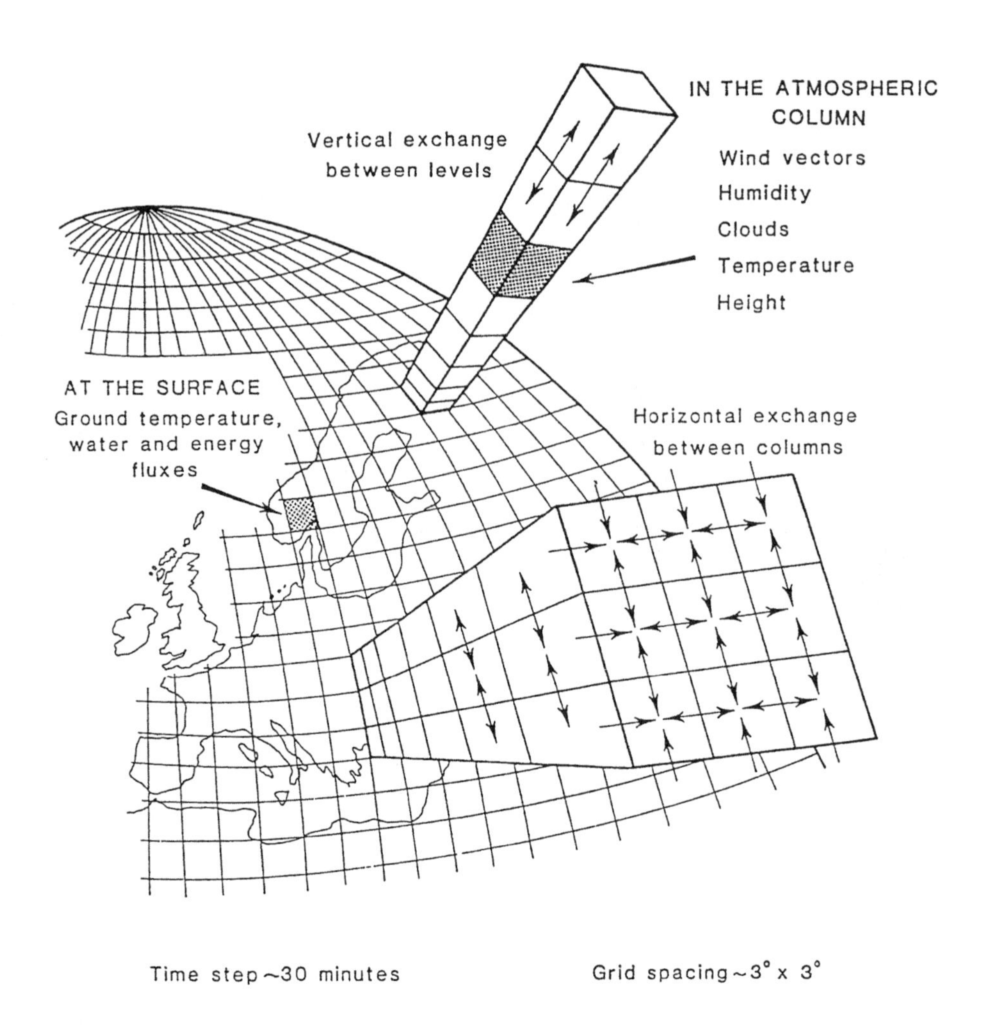

Figure 4.11. Spatial grids, time steps, and state characterization variables (clouds, temperature, etc.) used in global climate (circulation) models. Source: Stone (2000).

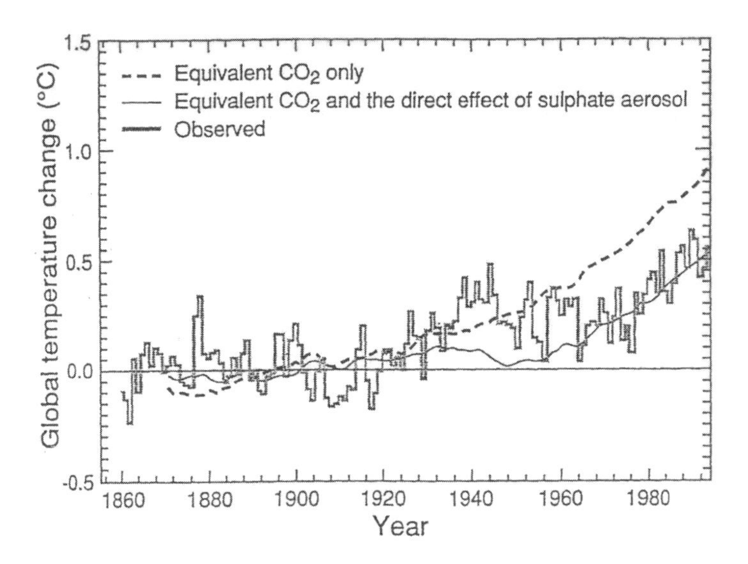

Figure 4.12. Simulated global annual mean warming from 1860–1990, allowing for increases in equivalent CO_2 only (dashed curve) and allowing for increases in equivalent CO_2 and the direct effects of sulphates (thin line) (Mitchell et al., 1995). The observed changes are from Parker et al. (1994). The anomalies are calculated relative to 1880–1920. Source: IPCC (1996). Reprinted with permission of IPCC.

4.3.3 Energy and the greenhouse effect

Since the early 1980s, the possibility of significant human impact on global climate linked to the supply and use of energy has attracted substantial attention from industry, the scientific community, policy makers, and the media. The extent to which the general public has become engaged with this issue is open to interpretation. Further, there has been considerable debate over the magnitude of human-induced climate change, what to do about it, and when. As of 2000, there appeared to be no universal consensus on how to proceed, although since 1990 there has been appreciable movement toward broader acceptance that the problem is real and merits timely action. It is essential to understand that (a) most who advocate prompt and substantial actions to limit greenhouse gases are serious analysts, genuinely concerned with preventing a potentially colossal environmental catastrophe within one to five generations; and that (b) most who propose more temperate measures are also serious thinkers, genuinely concerned with preventing a potentially colossal global economic catastrophe within one to five generations. Those of perspective (a) acknowledge that there are still significant uncertainties in many underlying scientific issues, while those identified with (b) do not question the underlying science of the greenhouse effect itself. They embrace greater caution because some proposed countermeasures would entail, over time, major increases in energy prices for all economic sectors, including individual consumers, and posit that the resulting economic dislocations would be unsustainable. Proponents of prompt action counter that unchecked climate change would have adverse economic consequences far more costly than mitigation measures. And thus goes the debate.

To avoid debilitative polarization, it is constructive to seek authoritative answers to three key questions:

1. How much climate modification is caused by a given change in a specific agent linked to human activity, e.g., what is the average global temperature rise for a given increase in stratospheric concentration of a greenhouse gas such as CO_2?
2. What are the relative contributions of human and natural factors in climate modification?
3. For a known level of climate modification, e.g., a 3°C increase in average global temperature, what will be the consequences for human activity (e.g., amount of sea level rise, loss of or gain in regional agricultural activity)?

Turning to question 1, some general circulation models (GCMs) predict that a doubling in current atmospheric concentrations of CO_2 will result in an increase in global mean surface temperature of 2.0 ± 0.6°C. This sounds definitive, but bear in mind that the global climate is a complicated outcome of synergistic interactions among numerous geophysical and geochemical processes operating on or within the earth's surface, oceans, and atmosphere. For example, temperature rise is expected to be about double the mean value in polar regions, and less near the equator. Because of these complexities, together with insufficient data over relevant time periods, GCMs must utilize approximate representations of many geophysical phenomena, omit others entirely, and be satisfied with a rather coarse resolution of length scales. Consequently, GCMs have limited capability to represent phenomena that entail short length scales, an especially vexing example being clouds, which may give rise to warming or cooling depending upon a plethora of complex factors. Indeed, it seems unlikely that the relationships between cloud lifetime, precipitation, and aerosols will be reliably represented in GCMs any time soon. Thus, the predicted temperature increases for prescribed CO_2 increases must be viewed with some caution. Nevertheless, there is no doubt as to the sign of the effect—that an increase in atmospheric CO_2 results in a temperature rise. Effects of atmospheric aerosols are somewhat more complicated to predict (Seinfeld, 2000) but, overall, an increase in sulfate particulates results in cooling owing to reflection of incoming solar radiation back into space, while an increase in carbonaceous (soot) particles can produce warming because soot functions analogously to a greenhouse gas, reflecting heat (infrared radiation) emitted by the earth's surface back to the earth.

The most recent report of the Intergovernmental Panel on Climate Change (IPCC, 2000) sheds some light on questions 1 and 2. This blue ribbon body of international climate science and policy experts concluded that there is a preponderance of evidence for human influence on global climate. Compared to their 1995 report, this document states that the case is strong for human involvement in global warming and revises upward the1995 predictions of the estimated range of expected global warming by the year 2010 from 1.8–6.3°F to 2.7–10.8°F. At present, the most widely acknowledged source of human influence on global climate is production of two greenhouse gases during the supply of energy from fossil fuels (i.e., generation of CO_2 from fossil fuel combustion and leakage of methane (CH_4) from pipelines). Some methane is also generated as a result of agricultural practices. Atmospheric CO_2 is a major greenhouse gas. When a fossil fuel is burned, essentially all of its carbon is emitted to the troposphere as CO_2. Further, combustion of fossil fuels at present supplies about 85% of the world's energy. Figure 4.13 shows that, since 1880, annual global atmospheric emissions of CO_2 from fossil fuel burning have increased about 20-fold. To produce a given amount of energy by complete combustion, different fossil fuels, owing to their different relative content of carbon and hydrogen, emit different amounts of CO_2. Clearly, a link between fossil-derived energy and the human impact on global climate must be taken seriously.

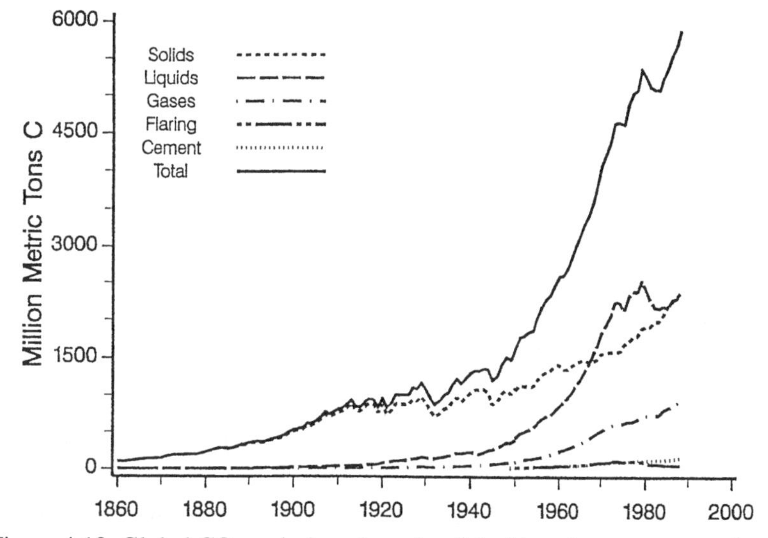

Figure 4.13. Global CO_2 emissions from fossil fuel burning, cement production, and gas flaring, 1860–1988. Source: Stone (2000).

From 1900 to 2000, the global mean near surface temperature of the earth experienced two extensive periods of warming, ca. 1910 to 1940 and 1976 to 2000 (and continuing), and one protracted period of overall cooling, ca. 1940 to 1975 (Figure 4.6). Stott et al. (2000) used a coupled ocean-atmosphere general circulation model to simulate variations in global mean temperature and in the temperature of large land masses.

The earth's climate is driven by transfer of energy to the planet from the sun (i.e., *solar irradiation*) (Zwiess and Weaver, 2000). Various natural and human processes (*forcings*) can modify this flow of solar energy (Figures 4.7, 4.10) and give rise to climate change. The initial flux of energy entering the earth's atmosphere depends on the intensity of the sun's radiation (irradiance), which varies in solar cycles over time scales comparable to those for appreciable climate change (i.e., of order 10 years). To reach the earth's surface, solar energy must first pass through the troposphere where sulfate aerosols (tiny submicron solid or liquid particulates, e.g., smoke, fog) spewed by volcanic eruptions may scatter a portion back into space (an example of an *external* but *natural forcing*) (Figures 4.7, 4.10). As it continues its journey toward earth, some of the surviving solar energy may be scattered back into space by clouds and aerosols, including sulfates produced by combustion of sulfur in fossil fuels (an *external anthropogenic forcing*). Solar energy that finally reaches the earth may be assimilated by various processes, including evaporation of water and raising the temperature of earthbound rocks, soil, and water. The earth reflects some of this radiation back toward space as thermal (i.e., *infrared* or *heat*) radiation. Greenhouse gases absorb some of this radiation and re-reflect (re-radiate) it back to earth providing an additional source of heat to the earth's surface.

Table 4.7 lists several greenhouse gases and their *global warming potential*, that is, their propensity, relative to CO_2, to cause greenhouse warming of the earth, as estimated by a climate model. When some natural or human process increases the concentration of a greenhouse gas in the troposphere, more energy can be reflected back to the earth's surface causing warming of the earth—a *positive radiative forcing*. Note that natural and human processes can contribute to increased tropospheric inventories of CO_2, water vapor, and methane. Chlorofluorocarbons (CFCs) are human-made and can only be of anthropogenic origin. Interestingly, methane, as well as methyl chloride, methyl bromide and methyl iodide, are all emitted from cultivated rice paddies. Mechanisms have yet to be established unequivocally, but evidence suggests methyl chloride may be formed without intervention of the rice plant itself, whereas soil, the plants, or both may be involved in generation of the other three (Redeker et al., 2000).

The Stott et al. study (2000) is important for several reasons. First, they did not fit their model to data, but rather drew on independent estimates of various natural and human forcings as inputs to their model and then used these to make independent predictions of climate change indices. Second, they found that both natural and human forcings were needed to explain the complete history of warming and cooling over the past 100 years. In particular, they found that human factors alone could not account for the first warming period of 1910 to 1945 or for the cooling of 1945 to 1976, but the combination of natural and human forcings, largely reflecting contributions of the former, accounted nicely for both. Third, they found that natural forcings could not account for the warming from 1976 to 2000 and actually predict an overall cooling from the high temperatures of 1945. However, anthropogenic forcings produce good agreement with this warming, with addition of both human and natural forcings suggesting an even better match. In summary for the entire century, they found that external forcings, both natural and human, needed to be accounted for to reliably account for multidecadal-scale global mean temperature variations, and that so doing accounted for 80% of these variations.

Table 4.7. Greenhouse gases and their global warming potential, referenced to the updated decay response for the Bern carbon cycle model and future CO_2 atmospheric concentations held constant at current levels.

Species[a]	Chemical Formula	Lifetime (years)	Global Warming Potential (Time Horizon)		
			30 years	100 years	500 years
CO_2	CO_2	variable[b]	1	1	1
Methane[c]	CH_4	12±3	56	21	6.5
Nitrous oxide	N_2O	120	280	310	170
HFC-23	CHF_3	264	9,100	11,700	9,800
HFC-32	CH_2F_2	5.6	2,100	650	200
HFC-41	CH_3F	3.7	490	150	45
HFC-43-10mee	$C_5H_2F_{10}$	17.1	3,000	1,300	400
HFC-125	C_2HF_5	32.6	4,600	2,800	920
HFC-134	$C_2H_2F_4$	10.6	2,900	1,000	310
HFC-134a	CH_2FCF_3	14.6	3,400	1,300	420
HFC-152a	$C_2H_4F_2$	1.5	460	140	42
HFC-143	$C_2H_3F_3$	3.8	1,000	300	94
HFC-143a	$C_2H_3F_3$	48.3	5,000	3,800	1,400
HFC-227ea	C_3HF_7	36.5	4,300	2,900	950
HFC-236fa	$C_3H_2F_6$	209	5,100	6,300	4,700
HFC-245ca	$C_3H_3F_5$	6.6	1,800	560	170
Sulphur hexafluoride	SF_6	3,200	16,300	23,900	34,900
Perfluoromethane	CF_4	50,000	4,400	6,500	10,000
Perfluoroethane	C_2F_6	10,000	6,200	9,200	14,000
Perfluoropropane	C_3F_8	2,600	4,800	7,000	10,100
Perfluorobutane	C_4F_{10}	2,600	4,800	7,000	10,100
Perfluorocyclobutane	$c\text{-}C_4F_8$	3,200	6,000	8,700	12,700
Perfluoropentane	C_5F_{12}	4,100	5,100	7,500	11,000
Perfluorohexane	C_6F_{14}	3,200	5,000	7,400	10,700
Ozone-depleting substances[d]	e.g., CFC and HCFCS				

[a]Water vapor has been omitted because of its shorter average residence time in the atmosphere (i.e., about 7 days).

[b]Derived from the Bern carbon cycle model.

[c]The GWP for methane includes indirect effects of tropospheric ozone production and stratospheric water vapour production.

[d]The Global Warming Potentials for ozone-depleting substances (including all CFCs, HCFCs and halons, whose direct GWPs have been given in previous reports) are a sum of a direct (positive) component and an indirect (negative) component which depends strongly upon the effectiveness of each substance for ozone destruction. Generally, the halons are likely to have negative net GWPs, while those of the CFCs are likely to be positive over both 20- and 100-year time horizons.

Source: IPCC (1996). Reprinted with permission of IPCC.

In a commentary on the importance and novelty of this study, Zwiess and Weaver (2000) note the importance of further studies to clarify still uncertain issues, such as feedbacks from clouds, the oceans, snow, and other land surfaces; better historical data on solar cycles before 1980; better information on tropospheric concentrations of sulfate aerosols and on their climate forcing potency; and on pre-1980 annual variations in stratospheric concentrations of volcanic aerosols. Despite these needs, the Stott et al. study strongly supports the proposition that human activities have played a substantial role in the climate warming of the past 2.5 decades, with about 80% of this variability arising from combinations of natural and human external processes that point to a clear and significant dominance by human impacts from 1976 to the present.

Returning to question 2, it is relatively simple to reconstruct human contributions to atmospheric emissions of sulfate aerosols and CO_2 from records of fossil fuel use and sulfur contents of those fuels. Further, the climate-forcing potency and atmospheric lifetime of CO_2 is well understood. However, tropospheric concentrations of sulfate aerosols and their climate forcing potential is less well-known. Also, there are numerous natural sources (e.g., natural fires in forests and grass lands) and sinks for carbon (e.g., soils and oceans) whose contributions to tropospheric inventories of this greenhouse gas are not well-differentiated from human contributions. Further, emissions of fine carbonaceous particulate matter (soot) from natural wildfires can result in tropospheric soot inventories that reflect infrared radiation back to the earth and function as greenhouse gases.

4.3.4 Greenhouse consequences: Consensus, unknowns, misconceptions

Climate is defined as the average of 10 or 20 years worth of weather (Jacoby et al. 1999). *Climate change* is *not* the year-to-year variability in weather or storm tracks. Climate change *is* global warming or cooling. A major cause is "radiative forcing," or an imbalance between the energy the Earth receives from the sun (largely as visible light) and energy the Earth transmits back into space as infrared radiation (loosely heat or light invisible to the human eye). "Greenhouse warming" of the earth occurs when gases and clouds in the atmosphere absorb some IR radiation and then send it back to the earth. Thus, atmospheric greenhouse gases function like the glass on the roof of an agricultural greenhouse, which, being largely transparent to visible light, admits solar radiation to the plants but, being substantially opaque to IR (heat) radiation, retains heat within the greenhouse. Snow and sand on the earth's surface, and clouds and aerosols in the atmosphere, combat warming by reflecting sunlight of various wavelengths directly back into space. Water vapor is the most important greenhouse gas, but its average residence time in the atmosphere is only about seven days. CO_2 may stay for 100 years. Atmospheric concentrations of CO_2 and other greenhouse gases have risen appreciably in the last 100 years, with human activity known to be a major source.

There is considerable uncertainty as to how climate responds to changes in radiative forcing, in part because crucial geophysical processes that can exacerbate or attenuate warming are poorly understood. For example, daytime clouds cool the earth by reflecting solar radiation away from the earth. At night, blanketing cloud cover or high humidity can warm the earth by reflecting IR back to the earth's surface. Oceans have a huge mass and high heat capacity and take a lot of energy to heat up or cool down. Oceans buffer climate change, but we know little about how rapidly they can store or release heat from their inner depths.

Just as human activities feed forward into climate intervention, so climate modification feeds back in terms of plausible and probable potential impacts on humans and their way of life. As shown in Figure 4.14, these interactions and drivers constitute a complex ensemble of causes, effects, and transformations. This figure depicts an Integrated Global System Model (IGSM) for assessment of climate policy issues. This model was developed by researchers in the MIT Joint Program on the Science and Policy of Global Climate Change and the Ecosytems Center at the Marine Biological Laboratory of the Woods Hole Oceanographic Institute to elucidate crucial issues linking climate science to public and industrial policy (Prinn et al., 1999). The model is unique in that it combines sub-models of economic growth and associated atmospheric emissions, natural fluxes between the earth and the atmosphere, atmospheric chemistry, climate, and natural ecosystems. The model addresses most of the processes of nature and human intervention that impact climate change. Note the many feedbacks between subsystem modules.

By considering the attributes of this IGSM, we can appreciate the complexities of climate change analysis and of relating those changes to public policy. Climate policy analysis must predict variables such as rainfall, ecosystem productivity, and sea-level change that can be connected to economic, social, and environmental consequences of climate change. Projections of atmospheric emissions of greenhouse gases and aerosol precursors must be related to economic, technological, and political factors, including mitigation strategies and international agreements, while accounting for uncertainties in climate science and in forecasts of greenhouse gas emissions, which must be based on vexing uncertainties in long-term forecasts of human population, economic growth, and technological progress. Further, climate models must be computationally efficient for policy-related predictive calculations. This necessitates judgment on what processes can be omitted or simplified and what ones must be treated in substantial detail. Brute force networking the most powerful subsystem modules comes up short because no existing computer could handle the number of computations required for informative policy analysis.

Human and natural emissions of radiatively and chemically important gases (Figure 4.14) combine to propel the coupled atmospheric chemistry and climate model that accounts for chemical and circulation processes in the atmosphere and the ocean. Predictions from this model drive a terrestrial ecosystems model that predicts *inter alia* CO_2 fluxes to the land and changes in vegetation and soil composition, all of which feed back to the models for climate, atmospheric chemistry, and emissions (Figure 4.14).

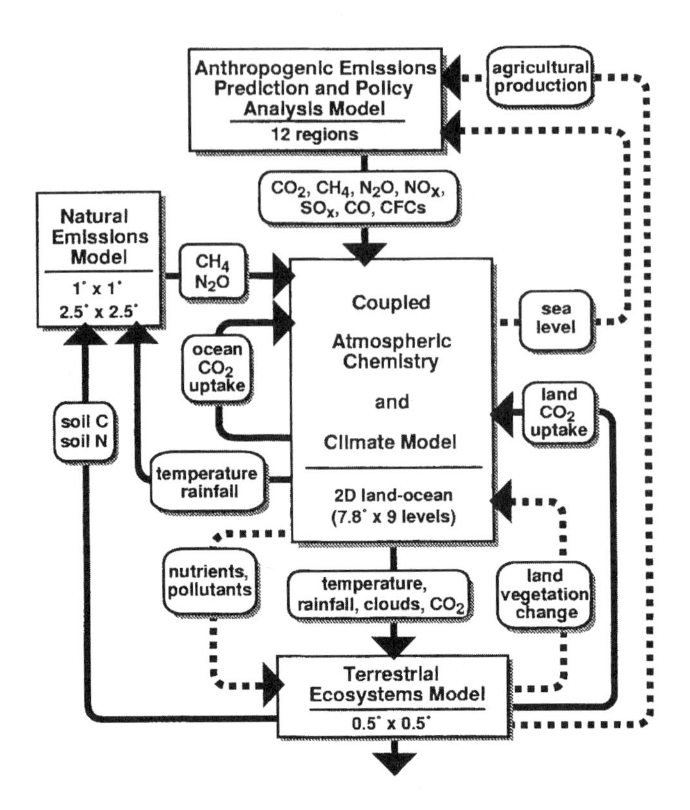

Figure 4.14. Schematic illustrating the framework and components of the MIT integrated global system model for climate policy assessment. Feedbacks between the component models which are currently included or under development for future inclusion are shown as solid and dashed lines, respectively. Source: Prinn et al. (1999).

Figure 4.15 shows predictions of their model for: (a) the change in global average surface temperature from its 1990 value and (b) sea level rise owing only to ocean thermal expansion. Model predictions were made for seven scenarios of economic development (assuming no restrictions on greenhouse gas emissions) and fundamental climate processes (Figure 4.16). Note that, by 2100, warming as little as 1°C or as much as 5°C is predicted. All of the seven scenarios are reasonable, and, at present, there is insufficient knowledge to know which of these, or perhaps other trajectories, best forecast the future. What *can* be said is that about two-thirds of the variability reflects uncertainty about the natural chemical and physical phenomena that determine climate, and the balance, uncertainty regarding atmospheric emissions of greenhouse and counter-greenhouse substances.

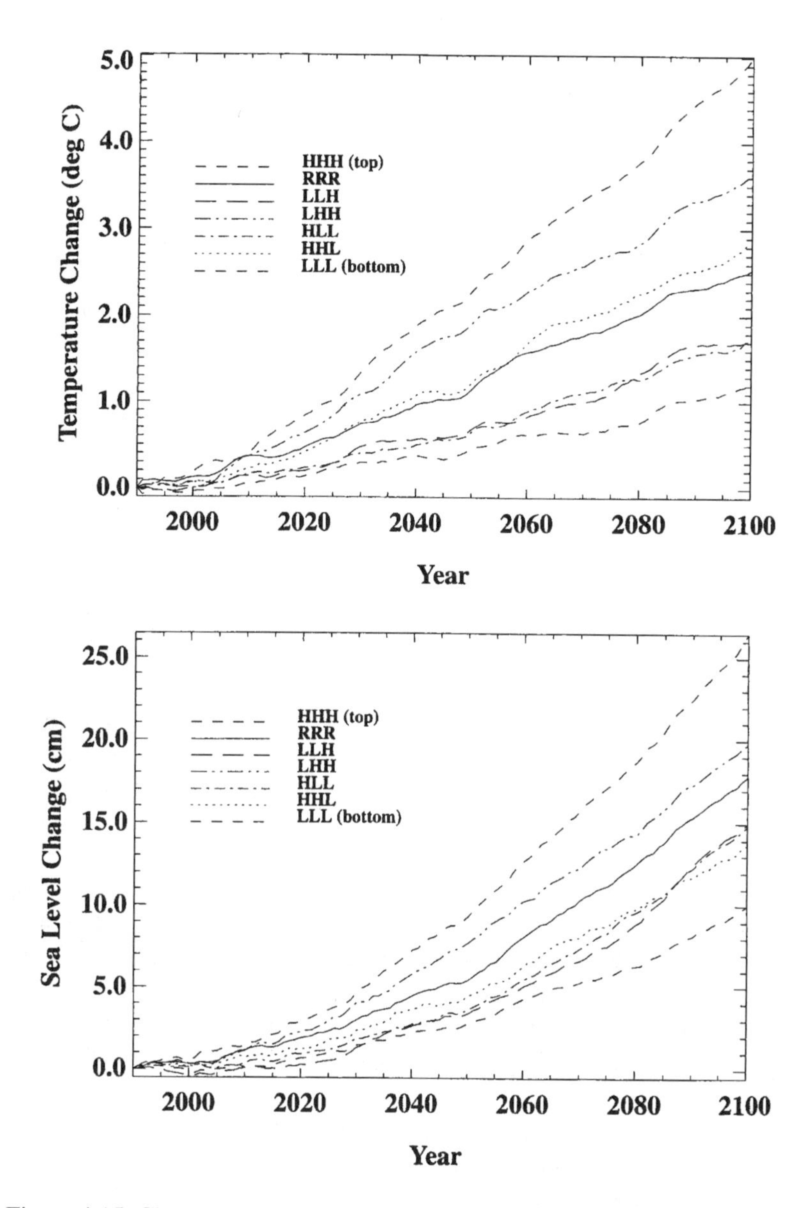

Figure 4.15. Changes in global-average surface air temperatures in °C (upper display), and sea-level, cm of rise owing to ocean thermal expansion only (lower display), predicted by the MIT Integrated Global System Model for the seven sensitivity scenarios (HHH to LLL) described in Figure 4.16. Source: Prinn et al. (1999).

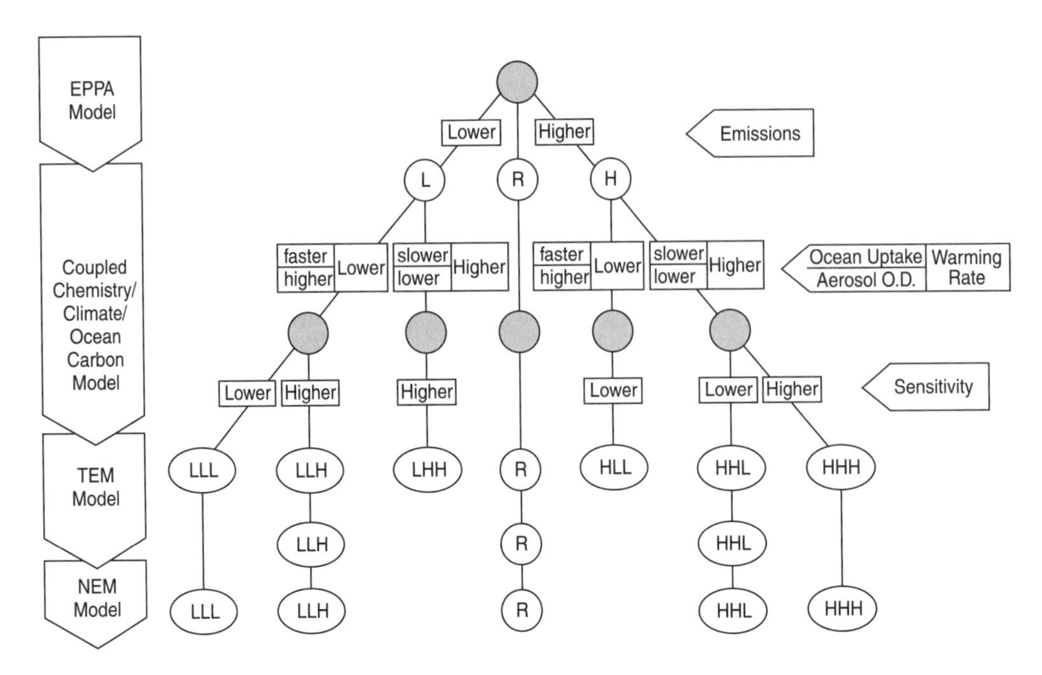

Figure 4.16. Sensitivity scenarios for the IGSM model predictions in Figure 4.15. Open circles denote points in sequence where output is available, with the letters in the circle denoting the identifying symbol for the output. Source: Prinn et al. (1999).

Even less is known about the *probable* consequences of various warming scenarios. For example, warming may give rise to increases or decreases in storm damage. There is some agreement among climate and policy experts that a warming of about 1°C over the next 100 years would cause minimal harm and might even benefit some regions of the earth, and also that the high-end prediction of Figure 4.15 would pose appreciable risks to ocean circulation, polar glaciers, unmanaged ecosystems, and to agriculture and other human activity. Sound policy analysis must take careful account of this wide range of possible outcomes.

The wide range of plausible temperature increases depicted in Figure 4.15 can create massive frustration for decision makers seeking scientific guidance in defining responsible climate change policy measures. If the warming over the next 100 years turns out to be the relatively modest 0.5°C, little harm is anticipated, and comparatively inexpensive corrective actions would be appropriate. On the other hand, the 5.8°C warming given as the upper estimate in the 2000 IPCC study is expected to cause appreciable harm and to merit serious preventative measures. But these would be expensive and possibly necessitate major reductions in the energy intensity of many lifestyles in the US and other developed nations. However, while any of the warming scenarios in Figure 4.15 are plausible, they are not equally probable. Thus, for meaningful policy analysis, it is essential to have quantitative understanding of the uncertainty in warming forecasts. Ultimately, even the most in-depth analysis cannot reduce uncertainties, given the inherent complexities of the global system. (See Jones et al., 2001, Moss and Schneider, 2000, Webster et al., 2001, Reilly et al., 2001, and Wigley and Raper, 2001.)

Two recent studies, Webster et al., (2001) and Wigley and Raper (2001) provide quantitative probabilistic assessments (see also Reilly et al., 2001). The latter study notes that the Third Assessment Report (TAR) from the IPCC (2001) projects global-mean warming, assuming no climate change–limiting policies, over the 1990 to 2100 period, of 1.4 to 5.8°C. Both the higher and the lower ends of these predictions are higher than those of the second IPCC assessment of 0.8 to 3.5°C (IPCC, 1996). The larger predictions are attributed primarily to revised emissions schedules, especially for SO_2, use of climate feedbacks in carbon cycle modeling, better treatment of radiative forcing and of effects of methane and tropospheric ozone, modified assumed rates of slowdown in the thermohaline circulation in the oceans, and better atmosphere-ocean general circulation models (AOGCMs) (Wigley and Raper, 2001). Accounting for uncertainties in emissions, climate sensitivity to those emissions, the carbon cycle itself, oceanic mixing, and aerosol forcing, their analysis found that the upper and lower TAR predictions have low probabilities and that the 90% probability interval is 1.7 to 4.9°C.

The latter value is in agreement with the probabilistic uncertainty assessment by Webster et al. (2001), projecting (Figure 4.17) a median global mean surface temperature increase (assuming no mitigation actions) of 2.5°C and a 95% confidence interval of 0.9 to 4.8°C. Their analysis used an integrated earth systems model (Sokolov and Stone, 1998) and took account of structural uncertainties in existing AOGCMs, cascading uncertainties in natural and human emissions of all gases and aerosols of climate forcing significance, in critical atmospheric, oceanic, and geochemical interactions, and in the carbon-cycle feedbacks from terrestrial ecosystems and the ocean. Their estimates of climate model uncertainties were constrained by climate system uncertainties from 1906 to 1995, and emissions uncertainties were deduced from errors in recent measurements of emissions, as well as expert judgments on variables impacting critical economic forecasts. This work also suggests that the upper limit of the TAR has a probability of far less than 1 in 100.

Nonetheless, environmental policymaking is often concerned with low probability but high damage incidents. Uncertainty analyses can illuminate the likelihood of extreme scenarios. For example, by combining all the 95% likely inputs for high warming outcomes, Reilly et al., (2001) projected that an extreme warming scenario (i.e., a global mean temperature increase of 7.5°C by 2100) has only one chance in 2.5 million. Naturally, processes currently unknown could change the conclusions of these analyses. Current analyses have made great progress but are still hampered by lack of key data, gaps in scientific and engineering understanding of key earth processes, and computers that are still too slow. Nevertheless, an important lesson from the work of Wigley and Raper (2001) and Reilly et al., (2001) is that assessments of complex environmental phenomena will be of greater value to policymaking if they provide formal analyses of the uncertainty in crucial projections and explicitly delineate the analytical methods employed.

4.3.5 Technological and policy response strategies: Evolutionary and revolutionary

Although other GHGs are important, anthropogenic emissions of carbon dioxide at present command the greatest attention as targets for a combination of technological or policy countermeasures or **both**.

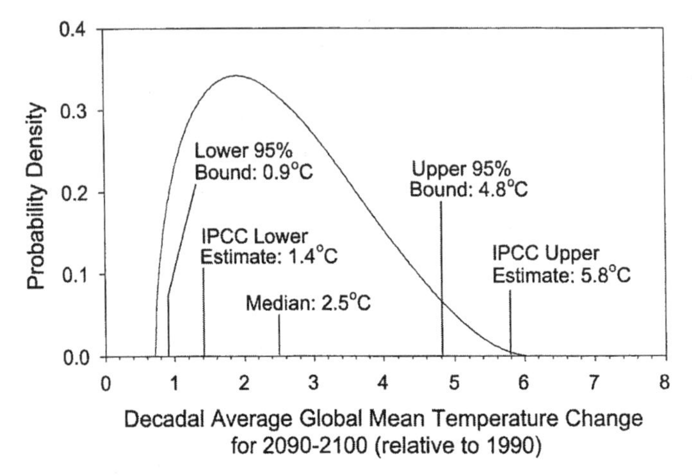

Figure 4.17. Probability density function for the change in global mean surface temperature from 1990-2100, estimated as a best-fit of a beta distribution to 100 simulations using Latin Hypercube sampling from input distributions. The IPCC upper estimate is well beyond the 95% confidence limit. Based on this distribution, there is a 17% chance that the temperature change in 2100 would be less than the IPCC lower estimate. Source: Webster et al. (2001).

Technological Approaches. Figure 4.18 summarizes the major technological means for preventing adverse climate impacts from anthropogenic CO_2. Under mitigation, geo-engineering refers to human interventions to counteract prospective warming (e.g., via changing the earth's albedo). Several strategies will reduce CO_2 emissions. People can improve the overall efficiency of their energy use (i.e., reduce the energy intensity term in Equation (1-23)) by extracting more useful heat and work in the supply and conversion of energy (*supply side management*) and by reducing their consumption of energy and energy-intensive services (*demand side management, DSM*). Another strategy is to reduce the carbon intensity of our energy system (i.e., the (CO_2/E) term in Equation (1-23)) by using fossil fuels with a lower C/H elemental ratio, such as natural gas (largely methane, CH_4) instead of coal (C/H typically ≥ 1.0) and by switching to non-fossil energy sources, such as nuclear and various renewables. A third reduction option is to prevent CO_2 emissions from entering the atmosphere through capture and then storage or recycle. "Removal" in Figure 4.18 refers to extraction of CO_2 from the atmosphere by manipulating natural phenomena, such as growth of terrestrial plants (biological sequestration), or possibly fertilization of the oceans with exogenous iron compounds to accelerate their uptake of CO_2. (A systematic examination of the technological, economic, and socio-political merits of these options is beyond the scope of this textbook. See Herzog et al., (2001) for an introduction to critical issues.)

Table 4.8 compares total global and US electric power plant emissions of CO_2 with order of magnitude estimates of the worldwide capacity of various CO_2 sequestration reservoirs. Long-term use of the sequestration option, even assuming substantial use of depleted oil and gas wells, would necessitate storage in aquifers and probably the oceans. Better estimates of the probable economic costs of various capture and sequestration options are becoming available as the result of analytical studies over the past 10 years and a number of on-going field studies of substantial annual capacity. The latter include a Statoil facility for capture and sequestration (by reinjection to a saline aquifer) of CO_2 emitted from a 140 MWe electric utility fired with natural gas derived from the Sleipner field in the North Sea. This plant, which is the first commercial-scale installation built in response to climate change issues, has been operating since September of 1996. ExxonMobil and Pertamina have announced that CO_2 capture and sequestration will be used to manage associated CO_2 brought up during extraction of natural gas from the Natuna field in the South China Sea. This will eventually be a gargantuan operation with plans for supplying 38,000 MWe of natural gas-fired electric generating capacity. Moreover, since 1978, about 10 CO_2 capture plants in the 100–1000 ton/day capacity range (which would correspond to 5–50 MWe of natural gas-fired electric generation) have been built to supply CO_2 for industrial uses. Herzog and Drake (1996) estimate that use of various technologies for the capture and disposal of CO_2 would result in 60% to as much as 230% increases in the cost of electricity (COE) from coal, and 40%–70% increases for generation by natural gas-fired gas turbine combined cycles (Table 4.9).

Table 4.9 expresses capture costs into an equivalent cost per tonne of avoided CO_2. These numbers give perspective on what levels of carbon taxes would be needed to motivate CO_2 capture. Recalling that \$1/(tonne CO_2) is equivalent to \$3.67/(tonne carbon), the capture costs for various coal-fired generation technologies range from about \$66–\$135/tonne carbon (or \$18–\$37/tonne CO_2). Lignites and coals typically contain 60%–90% carbon (Chapter 7), so the equivalent costs per tonne of coal would be 0.6 to 0.9 times the above values (i.e., they would range from about \$40 to \$122/tonne of coal). To set these numbers in perspective, typical US coal costs to electric utilities are roughly \$25–\$40/tonne, and, without CO_2 capture, coal costs account for roughly one-quarter to one-third the busbar cost of electricity from pulverized coal plants (Table 7.5). Thus, CO_2 capture costs alone are the equivalent of doubling to quadrupling the fuel cost to a coal-fired plant. There is considerable hope that advances in technology will reduce the costs of flue-gas capture of CO_2. However, carbon capture and sequestration face the serious challenges of winning public acceptance of the disposal of gigaton quantities each year in the earth and oceans and assuring that storage reservoir integrity will remain uncompromised without damage to the ambient ecosystem.

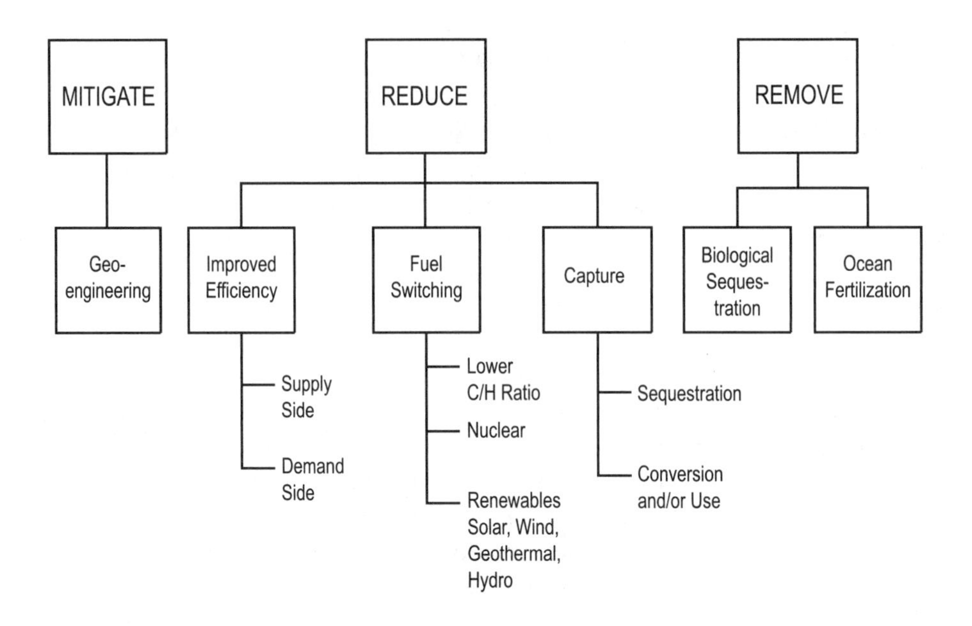

Figure 4.18. Options for mitigation of the release of fossil fuel-derived carbon dioxide to the earth's atmosphere. Source: Herzog (1998).

Table 4.8. Estimates of the CO_2 Capacity of Various Sequestration Options

Sequestration Option	Worldwide Capacity (Order of Magnitude)
Ocean	100,000 Gt
Aquifers	10,000 Gt
Depleted Oil and Gas	1,000 Gt
Active Oil	1 Gt/yr
Utilization	0.1 Gt/yr
Coal Seams	?

Notes:

Total CO_2 emissions worldwide = 22 Gt/yr
CO_2 emssions from US power plants = 1.7 Gt/yr

Source: Herzog (1998).

Table 4.9. Projected costs of CO_2 capture and disposal[a]

Power Plant Type	Capture Technology	CO_2 Avoided (kg kWhe^{-1})	Cost of Capture[b] (\$/tonne CO_2 captured)	Cost of Capture[b] (¢ kWhe^{-1})	Cost of Disposal[c,d] (¢ kWhe^{-1})	Increase in Cost of Electricity[e] (%)
Gas turbine combined cycle (GTCC)	MEA[f]	0.36	33	1.4	0.6–2.2	40–70
Pulverized coal (PC)	MEA	0.79	37	4.8	1.9–6.5	130–230
Pulverized coal (PC)	CO_2 recycle	0.93	35	4.9	2.0–6.7	140–230
Integrated coal gasification combined cycle (IGCC)	Shift/Selexol	0.76	18	1.7	1.5–4.9	60–130
Coal gasification/fuel cells (FC)	Shift/membrane	0.49	21	1.1	0.8–2.7	40–80

[a]From Herzog et al. (1993).

[b]Includes costs for compression (to over 100 bars) and dehydration.

[c]Includes transportation costs.

[d]Based on nominal range of \$15–50 per tonne of CO_2 disposed.

[e]Base electricity cost of 5¢ kWhe^{-1} assumed for all cases.

[f]Monoethanolamine; a chemical solvent for CO_2.

Policy Approaches. Policy measures include government-mandated, company-directed or voluntary actions to reduce one or more of the terms in Equation (1-23) (i.e., energy intensity per GDP, carbon intensity of the energy system, or even, in some jurisdictions, population). These measures may be punitive (e.g., federal, regional, and perhaps eventually global restrictions on emissions of GHGs enforced by civil penalties such as fines, seizure of assets or forced suspension of operations, or even by imprisonment). Other actions may take the form of mandating minimum levels of energy for appliances (green labeling) or automobiles (CAFE standards). Commonly, policy interventions are financial, for example, *carbon taxes* on GHGs, taxes on total energy consumption (*Btu taxes*), or on consumption beyond certain levels, such as in taxes on personal vehicles with fuel efficiencies well below average—a kind of "luxury tax" on energy. Other measures include tax credits or other subsidies from government or energy companies to purchase more energy efficient appliances or non-fossil energy sources.

Market-based methods have become popular in achieving environmental goals in emissions management (Ellerman et al., 2000). This may reflect the highly successful US national cap and trade program for sulfur dioxide emissions permits, established under Title IV of the 1990 Clean Air Act, as a means to combat acid rain by reducing SO_2 emissions from electric utilities by 2008 to 10 million tons below 1980 levels (Ellerman, 2000, Ellerman et al., 2000, Stavins, 1998, and Schmalensee et al., 1998).

The economics of tradeable permits are straightforward. They equalize marginal costs among candidate options for emissions abatement and enable least cost environmental compliance. Under this scheme, developed and developing countries both benefit, owing to the interdependence of the global economy, by reducing everyone's abatement tariffs and by providing developing countries with revenue-intensive export markets as well as advanced compliance technologies, probably at discounted costs (see Ellerman, 2000, Babiker et al., 1999, Ellerman and Decaux, 1998, Ellerman et al., 1998, Weyant, 1999).

Note that *tradeable permits* and *emissions trading* are not synonomous. Emission trading has two fundamental forms: (1) credit-based trading, in which there is no cap on total emissions and which entails trading of emission rights defined as differences from some pre-existing regulatory standard (e.g., an emission rate limit); and (2) allowance-based trading, also called "cap-and-trade," which has a fixed limit (cap) on aggregate emissions *and* tradeable emission rights, which are usually called allowances. Tradeable permits are essentially allowance-based emissions trading, and owing to their ability to realize a prescribed quantitative limit on GHG emissions, as well as the daunting information requirements for price or regulatory interventions to achieve the same, are strongly preferred as the policy instrument for the climate change problem (Ellerman, 2000). Moreover, credit-based emissions trading does not have a good performance record (Tietenberg et al., 1999).

The management of GHG emissions by policy or technical instruments provides an excellent illustration of one of the recurring themes of this textbook, namely the importance of identifying and interpreting length and time scales in energy and environmental systems. Once emitted, GHGs rapidly (in a few days or less) become mixed throughout the earth's atmosphere, so that, in contrast to other emissions to the air, water, or ground, the location of a GHG emissions source is irrelevant. Thus a ton of CO_2 released or avoided in Calcutta will have the same effect as a ton emitted or abated in Boston, São Paulo, Beijing, or Tokyo. GHG residence times in the atmosphere are a few to several decades to perhaps millennia. Thus, abating a ton of GHG this year or next, or even several years hence, matters little in terms of environmental impacts, provided the ton is abated in a time that is still short compared to GHG atmospheric residence times. Stockpiling and deferred trading ("banking and borrowing") tradeable emissions permits does not matter from an environmental impacts perspective (see Ellerman, 2000).

The cap-and-trade success story for SO_2 emissions has motivated studies to determine the viability of trading in CO_2 emissions permits to combat GHG emissions (Ellerman, 1999). For GHGs, a global trading system is preferred because there are major GHG sources *and* abatement mechanisms worldwide, and because costs of reductions in CO_2 vary appreciably by country, providing certain countries a means, via trading, to more economically achieve emissions targets by substituting lower cost *ex patria* reductions for more expensive domestic ones. Barriers to an international trading system are the lack of one good working example of a national system for CO_2 and deciding how to allocate emissions rights and the revenue realized from trading in these rights. According to Denny Ellerman (1999): "what has been heretofore free will be made scarce; and scarcity presents any society with a problem of how to allocate the use of the scarce thing and the associated rent." As to the former, lower-cost carbon abatement "reserves" outside the US will command market value only when demand for these reserves is generated by means of a working trading system in at least one nation. As to rent allocation, the problem is confounded by political and philosophical divergences, including abundant plausible claims for compensating non-government stakeholders and a school of thought that sees the government as the only legitimate "landlord."

By the year 2001, the fate of the Kyoto Protocol, as finalized in December 1997, appeared tenuous at best. However, emissions trading, although never clearly labeled, and arguably camouflaged in the treaty language, was there for the discerning eye to see. Ellerman (2000) illuminates the subtleties and their implications for GHG emissions management. Moreover, after the Kyoto meeting, several European countries seriously pursued GHG emissions trading as part of their strategies to comply with their emissions reduction targets. By 2000, the UK and Denmark were in the process of establishing an emissions trading system, and Norway had already devised a system that would begin once the Protocol came into force. Sweden, the Netherlands, France, and Germany also put forward actual proposals or announced their intentions to include emissions trading. By late 2000, the private sector was exploring trading in GHG credits (e.g., intra-company transfers at BP). These purely voluntary actions provided useful "pilot" data on various mechanisms of crediting, such as emissions reductions and source augmentation, as well as a green image for the participating companies.

For implementation of a broadly participatory system of international GHG emissions trading, patience will be a virtue. Ellerman (2000) discusses some of the challenges, including the vexing issue of the ability of the parties to reach agreement on what trading rules and legally binding country-specific emissions limits would equitably achieve environmental goals at acceptable costs. Moreover, there is serious doubt as to whether the Kyoto Protocol, in anything like its 1997 form, will enter into force. Those convinced of the environmental and economic advantages of GHG emissions trading should not predicate opening of trading systems on catalysis by some high-visibility international agreement to limit these emissions, although the future may well see such a policy instrument better designed to win ratification by the US and other skeptical nations. Ellerman (2000) posits that the likely course of events will be for individual countries to begin setting up domestic trading systems. These may stimulate trading between two countries with comparably valued emissions allowances followed by broader participation with multiple countries buying in eventually, somewhat in analogy to gradual adoption of a uniform currency. The main point is to get started so that one party or a small group can develop standards and document the credibility of the GHG emissions trading mechanism so that others will sign on.

The US Program for Trading in SO_2 Emissions Allowances. This first large-scale cap-and-trade system for environmental management performed so well in achieving cost effective reductions in SO_2 emissions that it has exceeded the expectations of economists *and* led a growing cadre of environmentalists to advocate cap-and-trade vehicles as more environmentally beneficial policy instruments than traditional command-and-control approaches. Moreover, experience with this program has taught a number of valuable lessons that can illuminate opportunities and challenges for applying this policy instrument to other environmental problems, most notably, reducing emissions of GHGs (see below). This highly successful program, however, cannot be casually adapted to other environmental problems, because each environmental problem has unique scientific, technological, and geopolitical attributes, including institutional and legal frameworks that may vary between countries. Nevertheless, this program teaches six valuable lessons, summarized in Table 4.10. (For a detailed examination of these lessons, see the appendix at the end of this chapter.)

Table 4.10. Lessons Learned from the US SO_2 Emissions Allowance Trading Program

1. Emissions trading does not compromise the effectiveness of environmental protection.
2. Mutually reinforcing simplicity, accountability, and flexibility are a natural consequence of emissions trading.
3. Markets for trading in emissions allowances will emerge.
4. An understanding of the politics of allocating emissions allowances illuminates program strengths and operability.
5. The consequences of allowing emitters to voluntarily join the program are complicated.
6. When devised and implemented to account for the idiosyncracies of a given environmental problem, the cap-and-trade mechanism "works," i.e., it provides lower cost attainment of environmental goals.

Adapted from Ellerman et al. (2000).

4.4 Attribution of Environmental Damage to Energy Utilization

Governmental bodies regulate SO_2 and other emissions from the production and use of energy because these substances harm the environment. But for environmental regulations to be most effective, policy makers and those who implement and enforce those regulations must have reliable answers to several questions:

- How do we know what substances generated by human activity are harmful to the environment?
- How harmful are they?
- Are their bad effects immediate or will they take some time to become known?

- How do we know that harmful substances originate with the production and use of energy?
- Do substances that we know are harmful change chemically or physically while they reside in the air, water, or ground, or even within the human body, and, if so, do they become more or less harmful?
- Are small amounts of these substances in the air, water, or soil less risky to people and their environment, or must these substances be totally eliminated?

We might be tempted to focus immediately on the last question and conclude that there is no point in taking any chances. Let's just eliminate all substances that we believe have any possibility whatsoever of harming people or their environment. That way, we can be 100% safe and sure. Unfortunately, the real world of energy and the environment is not that simple. There are tradeoffs in curtailing or eliminating the use of certain energy sources, chemicals, and other products of human endeavor. Some tradeoffs may be tolerable, desirable, or essential, whatever their consequences. Others may not be so straightforward. For example, consider how much you were inconvenienced the last time you lost electricity or your car broke down. How much are you willing to pay to be more certain that your sources of energy are clean? But don't frame the question in absolute monetary units (e.g., an extra dollar a gallon for gasoline). Instead, frame it in terms of a percentage of your income. That way, you can empathize with lower income consumers who may already need to pay 10% of their take home wages in energy costs. The real challenge is to learn how to reliably answer each of the above questions so that public and private resources can be informatively allocated to deal with environmental problems that truly pose serious harm to people.

4.4.1 Diagnosing receptor jeopardy and injury

Effective management of any environmental hazard requires reliable knowledge of the hazard's identity, origin, and danger to various targets (e.g., visibility, human health). To illustrate how this knowledge is obtained, we examine two decidedly different paradigms for detecting and assessing adverse human health effects of energy-derived products. Examples of such "products" are air emissions from electricity production or travel in an automobile. Figure 4.21 depicts typical processes that may occur in the pollutant "lifecycle," (i.e., synthesis, transport, transformation, impact on a susceptible target, inactivation). As discussed in Chapters 3, 7, and 10, pollutants are synthesized by imperfections in combustion or other energy generation processes. Pollutant transport may occur within the primary source, such as from an automobile engine to the tailpipe and then in the ambient environment—the air, water, and ground. Chemical and physical transformations of the pollutant may occur during pollutant transport in any location. The result may be a less hazardous pollutant or even its elimination, as occurs in post-combustion after-treatment devices such as catalytic systems in automobile tailpipes.

However, a more hazardous pollutant may arise. Such was the case for certain polycyclic aromatic hydrocarbons (PAH) captured on filters used to sample diesel soot. Owing to nitration of the PAH, strongly mutagenic nitro-PAH were formed. Transformation may also give the pollutant a new "home." Vapor phase pollutants may become absorbed onto fine solid particulates that journey through the environment by different pathways from a pollutant vapor. Target impact refers, in the case of humans, to physical contact of the pollutant with the human body (the skin, the lungs, etc.). At any stage, including virtually immediately after synthesis, the pollutant may be "inactivated" (e.g., owing to chemical modification or sequestration in a site where it can no longer contact humans). An example of the former is oxidation of an organic pollutant to carbon dioxide and water. An example of sequestration is vitrification, or binding of a toxic metal inside a glassy shell from which release by natural processes would take decades to centuries. We identify these steps because a thorough understanding of adverse environmental impacts and how to avoid them depends upon illumination of the importance of each of these steps in the system under consideration.

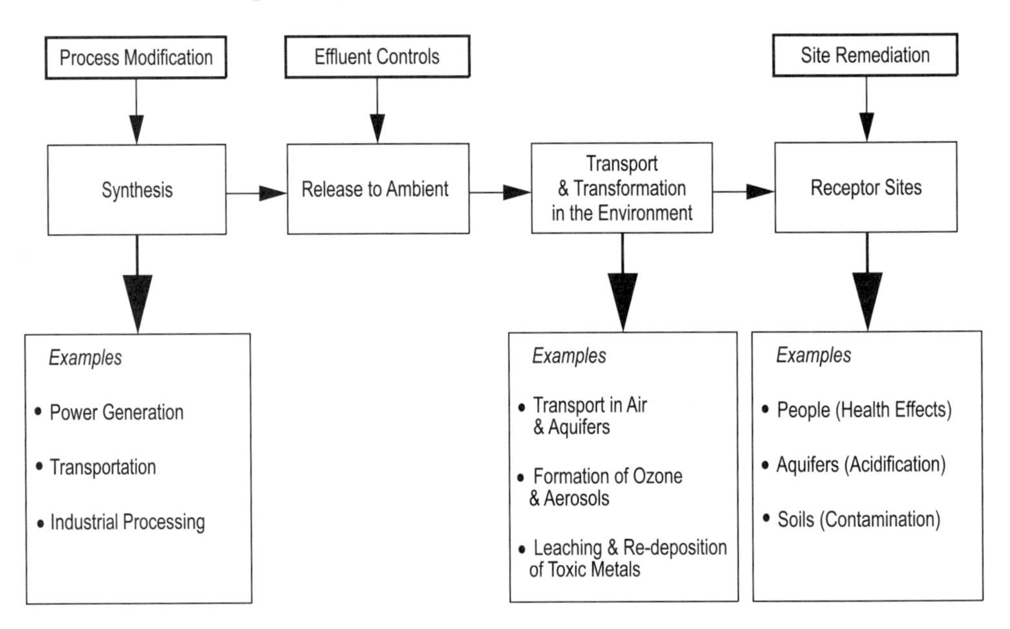

Figure 4.19. Processes and human intervention options during the lifecycle of environmental pollutants.

Figure 4.19 suggests two paradigms for gauging adverse human heath effects of energy emissions.[2] The first, more traditional paradigm of environmental toxicology focuses on a source that is suspected to emit hazardous substances. Steps are then taken to determine whether this suspicion is well founded. Typically, this entails the collection of a sample of the emissions from the exhaust of the source and asking the following questions:

- Is the sample toxic?
- If so, how toxic is it?
- If the sample is a mixture of substances, which is often the case in combustion effluents, what components are responsible for the toxicity of the sample?
- How much toxicity does the source emit per unit of useful energy or work provided?

A typical methodology to answer these questions is to test the sample for toxicological activity using model assay systems such as bacterial cells *in vitro*, human cells *in vitro*, or rodents such as mice. If one or more assay indicates toxicological activity, the sample is separated into components, which are individually tested for toxicological activity to track down the source(s) of the toxicity. The separation may be relatively coarse (e.g., producing 3–5 fractions based upon their different affinities for packed bed of alumina or other solid particles during elution under gravity with solvents of different polarity). Individual fractions may be further subdivided and bioassayed, the ultimate goal being to identify the specific chemical compound or compounds responsible for the toxicological activity of the entire mixture. Ideally, when the specific toxicants are identified and their weight fraction in the mixture illuminated, *and* if the bioactivity of those components per unit mass is unaffected by the presence of other substances, it should be possible to account for the toxicity of the entire mixture by a linearly weighted superposition of the toxicity of its bioactive components, i.e.,

$$H_M = \sum w_i H_i \tag{4-3}$$

Where H is toxicological activity per unit mass of substance, w is the weight fraction of that substance in the mixture, and subscripts M and i, respectively, refer to the entire mixture and component i in that mixture. There are cases where Equation (4-3) does a reasonable job of explaining the toxicological activity of a whole mixture, measured by some model bioassay system, in terms of the corresponding bioactivity of certain of its components. However, in general, Equation (4-3) is invalid because mixture components will interact, causing their specific bioactivity (H_i) to change. Nevertheless, in using this paradigm, it is wise to begin with the assumption that Equation (4-3) may be valid and to design an experimental testing program to provide the necessary data to

2 We thank Professor William G. Thilly (pers. comm.) of MIT for introducing us to these paradigms.

test whether this is so. If it does hold, it provides an additional measure of assurance that all hazardous substances in the test specimen have been accounted for. This is important because one motivation for studies of this type is to identify the offending components in the mixture to give guidance on how to devise engineering control strategies to eliminate the emission of these components. Examples of such control strategies are combustor redesign, fuel reformulation, and use of combustion after treatment.

H specifies toxicity per unit mass of emitted substance. For a reliable toxicological assessment on which to potentially base regulations or other public policy, there is need to index hazardous emissions to some measure of the practical benefit of the emissions source (e.g., toxicological activity per passenger mile of a vehicle or kWhr of electricity generated). If two sources consuming exactly the same amount of fuel and emitting the same toxicants in the same mass per unit of fuel consumed have different electricity generation efficiencies, clearly the higher efficiency source has less adverse impact per unit of useful "product" (i.e., electricity generated). There may still be a need to reduce emissions, but it is essential to take account of the environmental quality and the emission performance of the source to reconcile environmental and economic goals. Equation (4-4) illustrates the information typically needed for performing this calculation for combustion emissions from an electric power station:

$$H = (m_f \ m_e H_M)/(m_f Q[\eta/3{,}413]) \qquad (4\text{-}4)$$

where H is toxicity per kWh generated, m_f and m_e are, respectively, the mass of fuel fired per hour and the mass of emissions of specific toxicity H_M emitted per unit mass of fuel fired, Q is the heating value of the fuel in Btu/unit mass (usually given as lb), and η is the fractional efficiency of the generating plant for converting heat to electricity. Note that some electricity plants express efficiency as a "heat rate," which is the number of Btus of thermal energy (i.e., from the fuel) that the plant requires to generate 1 kWh of electricity. If the plant were perfect (i.e., had 100% efficiency), the heat rate would be 3,413 Btus. In practice, heat rates are typically two to three times this value. Note that we can also write $\eta = [(3{,}413)/(\text{heat rate in (Btu/kWh)})]$.

The above paradigm and variations of it have enabled great strides in understanding how human activity can contribute to the generation of environmentally hazardous substances. However, this methodology has imperfections that some view as especially serious when the ultimate goal is protecting humans from environmental pollutants. The use of model bioassay systems is somewhat suspect if the objective is to forecast health effects in humans, because no model system, even living systems of higher primates like monkeys or apes, can truly mimic human physiology or human biochemistry, and thus human susceptibility to, and defenses against, exogenous toxicants.

Modern toxicology has developed an alternative paradigm to overcome these limitations. Its essence is to focus on people who may have been exposed to hazardous environmental agents and then identify the origin(s), human or otherwise, of those toxicants. To better understand this methodology, we must recognize that "human toxicology" (i.e., the befoulment of humans by environmental agents) involves several or all of the following phenomena:

- Exposure of the person to a harmful agent or a precursor thereof
- Metabolism of the agent by the human biochemistry
- Detoxification by the body's defense mechanisms
- Conversion to a more toxic agent
- Interference with normal body functions including defense mechanisms
- Interaction with other exogenous agents to create or reduce harm

Genotoxic agents may chemically bond to human DNA. This may give rise to genetic changes that, if unchecked, may cause some form of human disease. The importance of these and other disruptions of normal body function induced by exogenous agents will vary from person to person, and by age and other factors, so the susceptibility of a given individual to adverse environmental agents is complicated.

The new toxicology paradigm seeks answers to the following questions:

1. What are the exogenous environmental agents that enter the human body and that, themselves or as products of their metabolism in the human body, undergo potentially harmful chemical or physical interactions with the molecules that compose the human body? For example, in the case of suspect genotoxins, which agents chemically bind to human genetic material?
2. Which of the chemically or physically interacting agents in question 1 cause significant change to the body's normal molecular functions? For example, in the case of suspect genotoxins that bind to human DNA, which if any cause significant genetic change?
3. Which of the agents in question 2 are causing changes sufficiently serious to eventuate an adverse health consequence? For example, which genotoxic agents, if any, contribute to genetic change so serious that it will lead to human disease such as cancer or inheritable birth defects?

Modern toxicology focuses on illuminating the molecular mechanisms for toxic effects in humans by exogenous agents. The investigative tools are the techniques of modern biological chemistry and genomics. The ultimate goal is to develop technologies that, in a harmless way, can determine if a person has been exposed to harmful chemicals, what those chemicals are, and whether, if left unchecked, those chemicals could cause disease or even premature death to that person. One approach is to draw a small sample of blood from a person and, through an ensemble of DNA measurements, deduce the information needed to answer questions 1 to 3, at least for some diseases.

Epidemiology is another tool for toxicological assessment. In this branch of medicine, researchers elucidate environmental factors in human disease and mortality. The approach is careful study of the incidence and history, including age of onset and of death, of particular diseases, in sets of people (cohorts) known to have experienced the environmental situation under scrutiny. The data from the medical case histories are compared to medical and other data for a second control cohort, ideally similar in all ways (age distribution, lifestyle, standard of living, ethnic background, etc.) to the people in the subject cohort, expect the control group lacked exposure to the suspect environment. Great care must be exercised to eliminate other factors that could also cause the disease under investigation.

An example of an epidemiological study is research at the Harvard School of Public Health years ago to determine the role of soot emissions from railroad diesel engines in causing lung cancer in railroad locomotive maintenance workers. Care had to be exercised to separate any effects of cigarette smoking, which is a known cause of lung cancer. This study benefited from the availability of good records on the workers' health.

From the perspective of sustainability and sustainable energy, all three of the above methodologies are important because they provide information on where to focus environmental resources to eliminate the most hazardous substances and avoid unproductive expenditures on agents posing little or no serious threat to humans. Likewise, illumination of the true "bad actors" can guide energy technologists in the design of new processes, and the correction of existing ones, to reduce emissions of hazardous pollutants.

4.4.2 Source identification

Pollutants in the environment do not come "labeled" with their composition and origin. Yet knowledge of the origin of pollutants is essential for formulating engineering and regulatory strategies for hazard mitigation, for identifying pollution sources, and for preventing unfair attribution of pollution to innocent parties. Physical characterization and chemical analysis of pollutants sampled from ambient environments, air, water, and soil can shed light on where pollutants originated, if care is directed to correcting for possible modification of the pollutants during transport through the environment and during sampling. (Readers may wish to consult Seinfeld and Pandis (1998), Godish (1997), Boubel et al. (1994), and Friedlander (1979), for more detailed information.) It is possible, however, to elucidate a pollutant source from analysis of an ambient pollutant. One approach is the use of so-called marker compounds unique to a particular source or operating procedure. For example, in a given region, a particular trace element may be emitted to the air only from a particular industrial operation. Specificity is crucial. It has been shown that a wide variety of sources from barbecuing meat to petroleum refining contribute to ambient air concentrations of particulates in the greater Los Angeles Basin. Sampling for total particulates without further clarification would do little to identify or eliminate a particular suspect emissions source. On the other hand, the presence of specific organic compounds absorbed onto the particulates can give a clue as to their possible origin.

One approach is to solve the equations that govern the transport and transformation of pollutants to compute the contribution of a given source to the concentration of a pollutant at its impact location (receptor) (Seinfeld and Pandis, 1998). While this is sometimes feasible, these models can be limited by inadequate data (e.g., on weather and spatially resolved time-dependent emissions inventories). Using "receptor models" it is possible to determine related pollutant concentrations at a receptor site to the responsible emissions sources, without taking detailed account of pollutant transport and transformation. Seinfeld and Pandis (1998) describe the following receptor modeling methods including crucial underlying assumptions:

- Chemical mass balance (CMB), which allocates pollutants to their various major sources (source apportionment) through chemical and physical analyses of particles or gases measured at the sources and at receptor sites
- Principal component analysis (PCA) for generally qualitative "fingerprinting" of pollutant sources by inversion of data for many emissions samples at a given site over multiple sampling periods
- The empirical orthogonal function method (EOF) for identification of emissions source locations and intensities, by inversion of data on many emissions samples from multiple sites taken over the same period

4.4.3 Risk and uncertainty

Uncertainty is a fact of life in virtually every approach relating environmental damage to specific human activities. Two major sources of uncertainty are inadequate data for parameterizing or testing models, or for emissions attribution, an inadequate understanding of the engineering, toxicology, and the underlying chemistry, physics, biology, and mathematics of pollution generation, transport, transformation, and receptor impacts. However, there has been remarkable progress in reducing and, in some cases, eliminating previously obstructive information voids (e.g., Seinfeld and Pandis, 1998). From a sustainability and policy perspective, it is essential to recognize and account for uncertainty in formulating and evaluating technological and non-technological approaches to environmental protection. (In Section 4.3.4, we discuss recent progress on accounting for uncertainty in forecasts of global warming owing to atmospheric emissions of anthropogenic greenhouse gases.) Risk is a concept for gauging the probability that a given activity will give rise to some form of damage or harm to a prescribed receptor. A thorough study of risk entails risk identification, risk assessment, and risk management. (A recent book by Kammen and Hassenzahl (1999) provides a good introduction including methods for risk analysis and illustrative cases from the energy and other sectors.)

4.5 Methods of Environmental Protection

4.5.1 Energy and the environment as an ensemble of coupled complex systems

In this context, we define a system as a collection of processes that in general may interact synergistically or adversely. The processes in the collection may, for example, include chemical or physical changes in matter, economic disruptions in an institution, country, or the world, and sociological transformations of an organization. We define a closed system as one for which clearly understood and typically narrow physical boundaries can be readily defined. Examples are the human body, the automobile, and a nuclear electric power plant. Note that all three of these systems are highly complex, physically and chemically, and in the case of the human body, biologically and sociologically, as well. Thus, a closed system can still be complex. An open system is one for which easily circumscribed physical boundaries are elusive and for which interactions of the system components extend over large length and time scales. One example of an open system is the assembly of processes, physical, sociological, and political, associated with the permitting, siting, construction, and operation of a nuclear power plant. Here, the system must account not only for the complex scientific and engineering processes that enable humans to produce electricity by safely harnessing the heat generated by nuclear fission, but also an array of complicated processes that determine how people and their social and political institutions are affected by and, in turn, affect the creation and operation of the nuclear plant.

Building on this example of an open system and revisiting Figure 1.2, we can see that, more generally, energy and its effects on the environment are part of a giant open system that spans the entire earth. Sustainable energy entails understanding not only energy but those items of the open global energy-environmental system that energy directly or indirectly affects. These items are many (e.g., potable water, food, shelter from the elements, nutrition, health care, fertile land, fertile waters, sanitation, education, economic advancement, preservation and extension of economic opportunity, and indeed the entire ecosystem of the earth). No text, or series of texts, could hope to present all the information needed to understand this open system sufficiently well to devise all the technological and non-technological prescriptions that will be needed to be good stewards of the global environment. Instead, we can draw upon this recognition of the terrible complexity of the energy-environmental system, abysmally qualitative though it may be, to inspire us to seek sustainable pathways that pay heed to protecting all system components rather than improving one vital component at the expense of degrading another of equal or greater importance.

4.5.2 Earth-system ecology as a working paradigm

Considerable progress in understanding and dealing with the complexity of the open system we call earth-human interactions has been made through a comparatively recent discipline known as industrial ecology and an even more instructive extension known as earth system ecology. To begin, we note the following definitions taken from Thomas Graedel (1996) (with some modification):

> *Ecosystem:* the ensemble of all the interacting parts of the physical and biological worlds
>
> *Ecology*: the study of the abundances and interactions of organisms
>
> *Organism:* an entity internally organized to maintain vital activities

> *Industrial organism:* any social or physical entity devised for the purpose of manufacturing useful products or supplying a useful service (e.g., generation of electricity by combustion of a fuel)

Thomas Graedel and Braden Allenby (1995) define and illuminate industrial ecology as follows:

> Industrial ecology is the means by which humanity can deliberately and rationally approach and maintain a desirable carrying capacity, given continued economic, cultural, and technological evolution. The concept requires that an industrial system be viewed not in isolation from its surrounding systems, but in concert with them. It is a systems view in which one seeks to optimize the total materials cycle from virgin material, to finished material, to component, to product, to obsolete product, and to ultimate disposal. Factors to be optimized include resources, energy, and capital.

Allenby (1999), Graedel (1996), and Graedel and Allenby (1995), elaborate on foundational axioms, representative applications, and implementation methodologies of industrial ecology (IE). Among the key concepts they illuminate are:

- There are instructive parallels between industrial ecology and biological ecology (Graedel, 1996).
- For example, the concept of an "organism," (i.e., "an entity internally organized to maintain vital activities") can be applied not only to biological species, where the "entity" is a plant or an animal, but also to social organizations and industrial practices (e.g., where the entity is a manufacturing plant or a geothermal power station).
- Industrial ecology keeps account of all inputs and outputs of materials and energy to products and processes throughout their lifecycle.
- By analysis of the budgets of these inflows and outflows, IE seeks to reduce the amounts of materials and energy allocated to achieve targeted functionality in goods and services, through various means including product and process redesign, materials substitution, and endemic permeation of energy efficient materials recycling.
- Industrial ecology draws heavily on the lessons of technological history and on the continuing evolutionary and revolutionary advances in technology that, in concert with enlightened public policy and industrial planning, can enable society to both sustain and extend economic progress and social equity.
- The territory of IE is the entire planet and, arguably, the entire universe. IE focuses on all impacts to the earth that derive directly or indirectly from the actions of humans.

- Components (organisms) in an industrial ecosystem interact with other components, sometimes over length scales of thousands of km and time scales of decades. For example, CO_2 emitted today by automobiles in Los Angeles may affect agricultural practices in Kenya in 2070.
- On the earth, industrial systems and totally natural systems are everywhere intermingled to at least some extent (Graedel, 1996).

Similar to the behavior of living organisms in biological ecosystems, the practice of industrial ecology carefully squires materials and energy to minimize cumulative overall waste of both through utilitarian reuse by other members of the industrial ecosystem (Graedel and Allenby, 1995, and Graedel, 1996). Indeed, IE rejects the idea that any material, however far degraded, must be viewed as "waste" in the sense that it has to be discarded because it has no further use in any "vital activity" of at least one member of the industrial ecosytem. To the contrary, in analogy with biological ecosystems, industrial ecology perceives materials substance and even energy emergent from a given activity not as wastes but as residues that can be a valuable input to some other activity of a nearby industrial organism. A trivial example is the management of "waste" heat in chemical processing and electrical power generation. In Chapter 3, we discussed the combined Brayton-Rankine cycle where the heat rejected at high temperature by the Brayton turbine is not allowed to immediately enter the ambient, but rather is corralled and used to generate steam to furnish more electricity before heat from the steam Rankine cycle is rejected to the atmosphere at a substantially lower temperature. Likewise, many chemical plants are now designed or refitted to configure process components so that cumulative heat flows are optimized to minimize the overall waste of heat in the process.

Were it not for the last bullet above we might conclude our discussion with the connection of industrial ecology with sustainable energy. However, we cannot decouple human activity from natural ecosystems. The most populous cities with intense human footprints show evidence of biological organisms organized and maintaining vital activities, and the most pristine part of the planet is not immune from effects of human activity. Thus, formal study of biological ecosystems and of industrial ecosystems, not as differentiated ecologies, but rather as one intertwined composite ecology, is preferred. This line of study, earth system ecology (ESE), offers great promise for illuminating means to harmonize human-earth interactions to actualize sustainable energy and sustainability more broadly to economic growth in a technology-based global society. ESE can be viewed as a model paradigm for achieving the balance between social, economic, and environmental goals essential to sustainability. ESE articulates the precepts of sustainability in a scholarly taxonomy and prescribes technologically and economically rational means to translate these precepts into workable operations throughout the plethora of interactive and complex industrial and social organisms that constitute the large, open system we call the 21st century global industrial ecosytem. The paradigm is not perfect. In particular, it cannot eliminate subjective judgments and personal biases in weighting the importance of various choices and tradeoffs. However, ESE does organize and unify an important set of precepts and implementation strategies that have the potential to nucleate the broadly inclusionary geo-socio-political consensus needed to translate good technological and policy ideas into reality at local, regional, and global scales.

4.5.3 Public policy instruments

In general, public policy instruments to reduce adverse environmental impacts from energy systems take the form of laws and regulations promulgated at the federal, state, or local level, or of a treaty among nations. An example of the latter is the Montreal Protocol, under which numerous countries agreed to phase out use of CFCs in order to reverse depletion of stratospheric ozone caused by these chemicals. The Montreal Protocol did not call for an immediate elimination of CFCs, but allowed reasonable time for development of alternatives that were harmless to ozone as well as nontoxic and non-flammable. The Montreal Protocol holds instructive lessons for sustainability. First, it shows that multiple countries representing different levels of economic well being and a range of political perspectives can reach a binding agreement to counter a global environmental problem, even though the agreement calls for some internal sacrifice. Second, this accord shows the importance of the earth system ecology perspective, as well as solid scientific and engineering understanding, in selecting new technologies and assessing their expected impacts on the environment. Third, it shows that technological progress can be a major contributor to sustainable practices especially when linked to a policy instrument that takes rational account of the time scales for technological innovation.

A more recent attempt of a treaty to achieve a global environmental goal—the 1997 Kyoto protocol—was not (as of August, 2004) ratified by the US, although several European nations have stated their acceptance of this policy instrument. (It is beyond our current scope to systematically examine the pros and cons of Kyoto 1997. Interested readers may wish to consult a series of works by the MIT Joint Program on the Science and Policy of Global Change, e.g., Ellerman (2000)). In brief, proponents view this instrument as an essential step to instigating global cooperation to reduce GHG emissions, and believe developed countries with high per capita energy consumption should set an example by adopting the protocol's measures. Several European countries adopted Kyoto precepts. Opponents view Kyoto's exclusion of countries such as China and India, which are expected to be the planet's two most prolific CO_2 emitters by 2025, as being at cross-purposes with the goal of reducing GHG emissions. Even if enacted as written, opponents believe Kyoto would delay a doubling of atmospheric CO_2 concentrations by at most two decades, and that the mandated time scales for reducing GHG emissions are too short compared to the time scales for viable transformation of energy technologies and their associated institutions.

Kyoto also provides lessons for sustainability architects. First, global environmental consciousness is real and sufficiently powerful to bring together most of the countries of the world. Second, as distinct from the Montreal Protocol, when the outcome of the global accord would entail appreciable potential economic sacrifice, great attention must be given to providing political decisionmakers with suitable offsets (e.g., unequivocal evidence that substantial benefits will accrue in the near term; clear evidence that each country is being treated fairly).

As with treaties, the impact of environmental laws and regulations is also mixed. So-called command-and-control limitations on pollution (e.g., automobile tailpipe emissions) began in the US in the early 1970s. These regulations and the technological responses to them by the automotive and fuels industries have dramatically reduced automotive emissions over the last three decades. But there is serious question as to whether command-and-control approaches are as effective as cap-and-trade methods. In the well analyzed and best documented case where cap-and-trade methods have been applied (i.e., to the reduction of sulfur dioxide emissions from roughly 370 coal-fired US electric power plants), the evidence is overwhelming that substantial pollution abatement was achieved at dramatically lower costs than would have been the case by applying stringent command-and-control measures to each individual plant (Section 4.3.5). There is now appreciable interest in extending cap-and-trade measures to reduction of CO_2 emissions. Implementation would be more complicated than for the SO_2 case, but the economically and environmentally encouraging results for the latter strongly motivate use of this instrument to manage atmospheric GHG emissions.

4.5.4 Technological remedies

All modern methods for the supply and use of energy incorporate technical means for environmental protection. These means may be as simple as provision to reduce soot emissions from a home heating system through timely maintenance and tuning of an oil burner. They may be as complex as the catalytic after-treatment devices used on passenger automobiles to meet tailpipe emissions regulations. (Several examples are presented in Chapters 7, 8, 10–15, and 17–20.) Overall, regardless of the specific energy system under scrutiny, there is an important role for technological measures, including steady innovations, in fulfilling environmental performance goals. Recall that a core precept of industrial ecology and earth system ecology is that humans must capitalize on continuing advances in technology in order to have hope of sustaining economic and social progress while protecting the earth. It is vital that governmental and private sector policies sustain technological innovation through commitment of financial resources and through enactment of legislation that stimulates research and entrepreneurship. Such measures will result in devices and processes that provide environmental protection.

4.6 Environmental Benefits of Energy

4.6.1 Pollution prevention and environmental restoration

Occasionally overlooked in the analysis of adverse environmental consequences of energy is the fact that energy can be a dynamo of environmental cleanup, restoration, and protection. Although some technologies for destruction of wastes can be energy neutral or even net energy producers, others require a net input of energy to achieve desired levels of cleanup and decontamination. Likewise, exhaust cleaning devices require energy for their operation. For example, scrubbers for removal of SO_2 emissions from power plant stack gases typically reduce the fuel-to-electricity conversion efficiency by 1–2%. Manufacture of environmentally cleaner fuels, such as ultra low sulfur diesel fuel, require expenditures of energy to operate fuel processing equipment and generate the hydrogen needed for extraction of the sulfur (by converting it to hydrogen sulfide). Similarly, the conversion of natural gas (Chapter 7) and of renewable resources (e.g., various forms of biomass discussed in Chapter 10) to clean fuels comparable in performance to today's liquid transportation fuels derived from petroleum, require substantial inputs of energy, in some cases, about equal to the amount of energy in the fuel itself.

4.6.2 Social and economic foundations for environmental stewardship

Affordable and accessible energy is an engine for preserving economic well being and for extending economic opportunity. If core human social needs, such as potable water, food, shelter, health care, and education are unavailable or inadequate, people will have little motivation to embrace personal practices or multi-national accords aimed at securing environmental benefits. When these needs are being met and there are clear near-term prospects for improved lifestyles, individuals and their political leaders will have incentives to entertain sustainability measures internally and even multi-nationally.

4.7 Implications for Sustainable Energy

4.7.1 Environmental footprints as sustainability metrics

Throughout this book, we embrace the proposition that to be "sustainable" a product or process must not cause long-term, irreparable harm to the environment. We can obtain some insight into the relative "sustainability" of different energy technology and energy policy options by determining their pollutant emissions per unit of desired product or service. In such estimates, we can begin with linear correlations such as Equations (1-22 to 1-25), while recognizing that these are approximate models and that interaction effects may be important for scientific understanding and policy decisions. More refined analyses take account of the plethora of possible interactions, not only among pollutants, but among technologies, economic activity, environmental impacts, and public policy. One result can be models of the type shown in Figure 4.14 for global climate change analyses. Recalling Section 4.6, we recognize that the ability of a technology or policy to improve the environmental performance of some other industrial organism merits sustainability credits. For example, a cap-and-trade methodology for implementing the SO_2 reduction mandates of the 1990 US Clean Air Act Amendments is a policy measure deserving of substantial sustainability credits. Likewise, an affordable technology that reduces automobile tailpipe emissions earns sustainability kudos. To reliably evaluate the sustainability potential of competing energy technologies and policies, we must devise rational methods to equitably credit or debit each of their relevant environmental attributes.

4.7.2 The unusual challenge of global climate change

Although there are different perspectives, there is substantial agreement that advances in technology can affordably overcome the environmental problems associated with the supply and use of energy (such as automotive tailpipe emissions, acid rain, particulates, trace metal emissions, solid wastes, etc.) and can be managed at acceptable economic costs through advances in technology. The exception is atmospheric emissions of CO_2 from the combustion of fossil fuels to support various human activities. At present, there is considerable uncertainty as to whether the CO_2 emissions abatement that would probably be required to prevent the extent of global warming predicted under the high-end scenarios of the IPCC and other analyses (e.g., Figure 4.15), can be achieved without seriously curtailing fossil fuel burning. In particular, three challenging obstacles would need to be overcome. First, carbon capture is technologically feasible today but the costs, including CO_2 disposal (sequestration), have been estimated to increase electricity costs over the base price by 40–70% for natural gas-fired generation and 130–230% for pulverized coal-fired plants (Herzog and Drake, 1996). Second, it must be demonstrated that the various places the carbon would be stored (i.e., drawn down oil and gas wells, coal seams not economically mineable with today's technologies, deep salt water formations, and the deep ocean) are safe, able to hold the CO_2 for many decades or longer, and immune to environmental damage by the CO_2. Third, there must be public and regulatory acceptance of carbon capture and sequestration technologies.

Sustainability architects must continue to vigorously pursue solutions to each of these challenges so that humans will have an option for an orderly transition away from fossil fuels should this be needed to reduce atmospheric CO_2 emissions. Moreover, it is essential that alternative energy technologies that could replace fossil fuel burning be pursued to assure that this option will be available when needed. Policy makers must then decide on what expenditures each technology merits at a given time.

Problems

4.1 Using the data on fuel H/C ratios and gross (higher) heating values (HHV) presented in Table 7.1, estimate the mass of CO_2 emitted per 1000 SCF of methane, and per lb of gasoline, methanol, and No. 2 heating oil. Also express your answers in mass of CO_2 per million Btu of each fuel.

4.2 Do you think the global warming potential of a given greenhouse gas would be different with a different climate, i.e., global circulation climate model? Explain your reasoning.

4.3 One way to reduce greenhouse gas emissions is to prevent leakage from natural gas pipelines. If the leakage from a pipeline transmitting 250 million SCF of natural gas per day is reduced from 5% to 0.5%, use the GWP data of Table 4.7 to estimate the equivalent amount of CO_2 emissions that would need to be curtailed, assuming the period of interest is 20 years, then 100 years. Assume that for this calculation natural gas is adequately approximated by methane.

Appendix: Lessons from SO_2 Emissions Trading

Lesson 1 derives from the fact that, although there is flexibility as to whether individual sources reduce emissions or not, the aggregate effect of actions plus inactions by all sources must be emissions reductions that at a minimum meet the cap. Lesson 1 would be void if the intent was to prevent emissions from each source from exceeding a prescribed level. There could be justification for this if the time required for a pollutant to mix with its cousins from all other sources was long compared to the time for adverse localized impacts from high concentrations of the pollutant emitted from a given source. This is not the situation with SO_2.

Lesson 2 reflects the fact that all SO_2 emitted is strictly tracked in the form of permits (denominated in tons of SO_2 emitted) that must be given up. Every emitter in the program must have equipment for continuous monitoring of SO_2 emissions and must report the data to the US EPA. However, the regulator need only focus on assuring that every ton of SO_2 emitted is covered by a valid permit. There is no need to focus on where, when, or how emissions reductions were accomplished, and there is no element of subjective judgment calls, special variances for hardship cases, or complex litanies of rules and exceptions to codify all contingencies. The single, uniform tracking device, the permit, allows simple yet thorough uncompromising accountability that, in turn, provides emitters flexibility in fulfilling their environmental obligations.

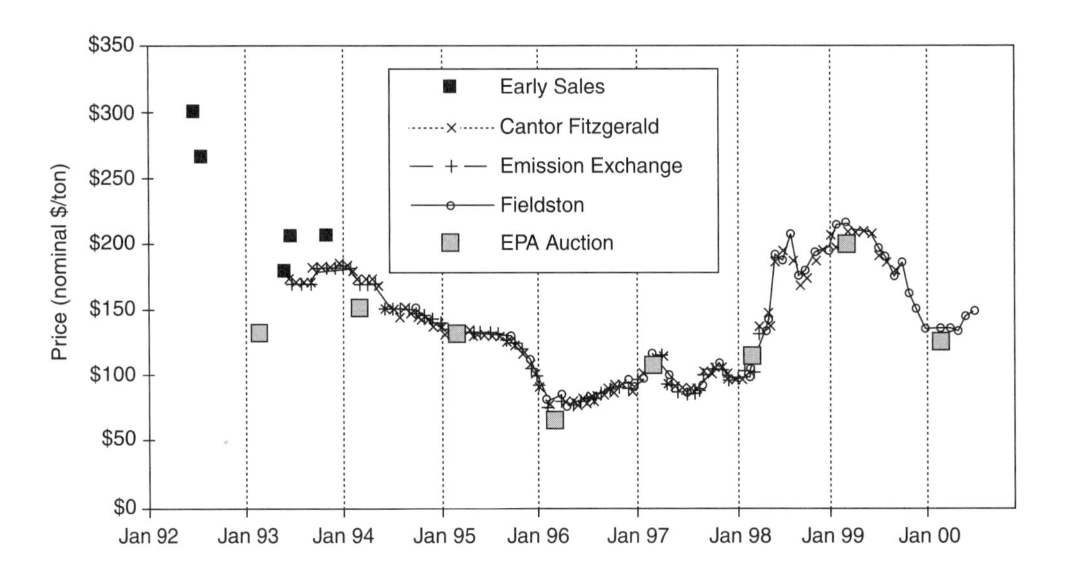

Figure 4.20. SO_2 Allowance Prices, 1993–2000. Source: Ellerman et al. (2000).

Lesson 3 is documented by Figure 4.20, which shows that after initial divergences, price indications on allowance permits converged by mid-1994, roughly six months before the program took effect, manifesting the "law of one price," providing a tell-tale signal of a workable market. Figure 4.20 also shows substantial long-run volatility in allowance market prices from 1995 onward, but no evidence of anything other than one price at a given point in time.

As to political teaching (Lesson 4), this program fractured a decade-long stalemate on acid rain legislation and provided a large enough tent to win buy-ins from a then new Republican administration in the White House, as well as legislators concerned with fairness issues. Industry appreciated grandfathering of allowances to those strongly impacted by the emissions reduction mandates and the decoupling of the allowances from particular sources (and the capitalization of those sources) by means of the trading mechanism, which also made the rights explicit and more secure for the next 30 years. This provided a finality that was a welcome change from the prior approach of government rulings on allowable conditions for emissions, which were often viewed by their recipients as arbitrary and prone to unpredictable changes.

The so-called "opt-in" provision (Lesson 5) allows an emissions source not mandated to be under the cap to voluntarily become part of the cap and receive emissions allowances. Theoretically, sources with more economical emissions reductions would wish to opt-in to reap financial gain from their cost advantage. However, this provision suffers from what can loosely be described as "subjunctive benchmarking." There are unavoidable time lags between when allowances are allocated and utilized, and, during these periods, the opting-in enterprise faces the vicissitudes of the continuous change in the economy. Consequently, it is impossible to prescribe a baseline for allocating allowances that accurately indexes what the opting-in source's emissions *would have been* if it stayed out from under the cap. Too harsh a baseline would discourage sources from joining, while one too lenient would inflate the true value of emissions permits as well as the emissions cap itself, in both cases diminishing environmental benefits. Stated another way, candidates for which economic change has created excess allowances will opt-in, even though they bring little low-cost abatement to the program. Those for whom the time lag has made their assigned allowances more severe will have less incentive to sign up, even though they could contribute relatively low-cost abatement. Ellerman et al. (2000) describes two plausible reactions to this problem (which he labels reactions to a "moral hazard"): (a) a purist view that the problem cannot be avoided and that the attendant risk of bloating the emissions cap is too serious for opt-ins to be allowed; and (b) a pragmatic position that opt-in rules can be devised to reduce the above risks and that the potential benefits of lowering costs and expanding the number of sources covered more than offset the undesirable impacts.

Purists might object to the meaning of "small," but the excess allowances that were allocated in Phase I of the US program (i.e., from 1995 to 1999) only inflated the total cap of 37 million tons by about 3%, and the sources that opted-in decreased emissions by about 1.2 million tons as a consequence of being in the program). If or when cap-and-trade instruments are applied to GHGs nationally and globally, opt-in provisions must be available to provide additional latitude to engage more countries and source types, and to cover other gases besides CO_2 that contribute significantly to radiative forcing in the atmosphere (Ellerman et al., 2000).

As to Lesson 6, Figure 4.21 documents just how well the US cap-and-trade program worked in achieving EPA-prescribed levels of SO_2 abatement on a year-by-year basis. The figure shows data on actual SO_2 emissions from 1985 to 1999, including Phase I (1995 to 1999) of the cap-and-trade program. Note that this was a transitional phase in which an SO_2 emissions cap of 2.5 lbs/million Btu, based on the 1985 to 1987 average heat rate, was placed on the largest and strongest emitting generating stations. Phase II, which began in 2000, assigned a cap of 1.2 lbs SO_2/million Btu to all generating units (again using the 1985 to 1987 average heat rate). Note the sharp decline in SO_2 emissions when the program began in 1995. The solid line in Figure 4.21 running from 1995 to 2010 denotes the emissions cap (i.e., the number of allowances allocated to the 374 affected generating stations over this period). The dashed line with closed circles forecasts emissions from these units in the absence of the Title IV emissions reduction program, while the dashed line with triangles projects what SO_2 emissions would have been in Phase I without Title IV, using actual electricity demand during that period.

The cap-and-trade program resulted in a cumulative SO_2 emissions reduction of about 20 million tons during Phase I, of which slightly more than 10 million tons were in excess of what was required to comply fully with the program. This "excess" compliance creates "banked" allowances that the title allows to be used to smooth the transition into the more demanding Phase II regulations by offsetting emissions that exceed annual allowances in the early years of Phase II. Earlier reductions in SO_2 are an additional benefit of this "over-compliance" during Phase I.

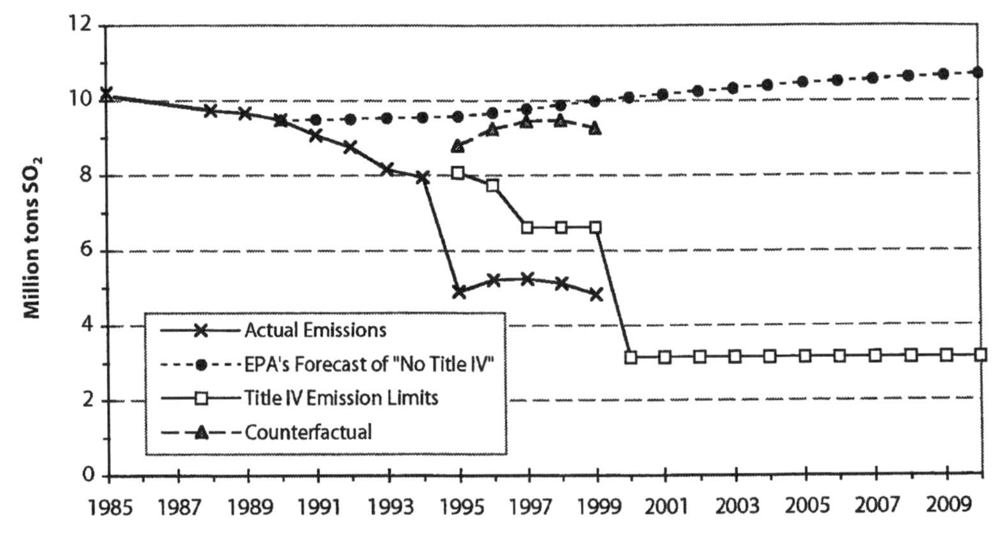

Figure 4.21. US SO_2 emissions trading: emissions, caps, and counterfactuals. Source: Ellerman et al. (2000).

References

Abelson, P.H. 1999. "Biological Warfare." *Science*. 286(5445): 1677.

Allen, M., S. Raper, and J. Mitchell. 2001. "Uncertainty in the IPCC's Third Assessment Report," Policy Forum: Climate Change. *Science*. 292: 430–433.

Allenby, B.R. 1999. *Industrial Ecology: Policy Framework and Implementation*. Chapter 1. Upper Saddle River, NJ: Prentice Hall.

Babiker, M., J.M. Reilly, and H.D. Jacoby. 1999. The Kyoto Protocol and Developing Countries. MIT Joint Program on the Science and Policy of Global Change, Report No. 56. Cambridge, MA: MIT.

Boubel, R.W., D.L. Fox, D.B. Turner, and A.C. Stern. 1994. *Fundamentals of Air Pollution*, third edition. New York: Academic Press.

Brooks, H., and J.M. Hollander. 1979. *United States Energy Alternatives to 2010 and Beyond: The CONAES Study*. Vol. 4, 1-70. Palo Alto: Annual Review of Energy.

Dockery, D.W., C.A. Pope, X. Xu, J.D. Spengler, J.H. Ware, M.E. Fay, B.G. Ferris, and F.E. Speizer. 1993. "An Association Between Air Pollution and Mortality in Six U.S. Cities (The Six Cities Study)." *New England Journal of Medicine*. (24)329: 1753–59.

Ellerman, A.D. 2000. "Tradeable Permits for Greenhouse Gas Emissions: a primer with Particular Reference to Europe." MIT Joint Program on the Science and Policy of Global Change, Report No. 69. November. Cambridge, MA: MIT.

Ellerman, A.D. 1999. *Obstacles to Global CO_2 Trading: A Familiar Problem.* Revised and published in *Climate Change Policy*, eds. C. Walker, M.A. Bloomfield, and M. Thorning. Washington DC: American Council for Capital Formation, Center for Policy Research.

Ellerman, A.D. and A. Decaux. 1998. Analysis of Post-Kyoto CO_2 Emissions Trading Using Marginal Abatement Curves. Joint Program on the Science and Policy of Global Change, Report No. 40. Cambridge, MA: MIT.

Ellerman, A.D., H.D. Jacoby, and A. Decaux. 1998. "The effects on developing countries of the Kyoto Protocol and CO_2 emissions trading." Joint Program on the Science and Policy of Global Change, Report No. 41. Cambridge, MA: MIT.

Ellerman, A.D., P.L. Joskow, R. Schmalensee, J.P. Montero, and E.M. Bailey. 2000. *Markets for Clean Air: The U.S. Acid Rain Program*. New York: Cambridge University Press.

EPA (U.S. Environmental Protection Agency). 2000. "Natural Air Quality and Emissions Trends Report, 1998." US EPA Report No. EPA 454/R-00-003. Washington, DC: EPA.

EPA (U.S. Environmental Protection Agency). 1995. *The Inside Story: A Guide to Indoor Air Pollution*. Document #402-K-93-007. Washington, DC: EPA.

Fay, J.A., and D.S. Golomb. 2002. *Energy and the Environment*. New York: Oxford University Press.

Freidlander, S.H. 1979. *Smoke, Dust and Haze: Fundamentals of Aerosol Behavior*. New York: Wiley-Interscience, John Wiley & Sons.

Glassman, I. 1996. *Combustion*, third edition. New York: Academic Press.

Godish, T. 1997. *Air Quality*, third edition. Boca Raton, FL: Lewis Publishers, CRC Press.

Graedel, T.E. 1996. "On the Concept of Industrial Ecology." *Annual Review of Energy and the Environment*. 21: 69–98. November.

Graedel, T.E. and B.R. Allenby. 1995. *Industrial Ecology*, chapter 1. Englewood Cliffs, NJ: Prentice Hall.

Helbe, J., M. Neville, and A.F. Sarofim. 1988. "Aggregate Formation From Vaporized Ash During Pulverized Coal Combustion." Proceedings, Twenty-First Symposium (International) on Combustion. Pittsburgh, PA: The Combustion Institute.

Herzog, H.J. 1998. "Understanding Sequestration as a Means of Carbon Management." Energy Laboratory Working Paper. Cambridge, MA: MIT.

Herzog, H.J., and E.M. Drake. 1996. "Carbon Dioxide Recovery and Disposal from Large Energy Systems." *Ann. Rev. Energy. Environ.* 21: 145–166.

Herzog, H.J., Drake, E., Tester, J., and Rosenthal, R. 1993. "A Research Needs Assessment for the Capture, Utilization, and Disposal of Carbon Dioxide from Fossil Fuel-Fired Power Plants." DOE/ER-30194. Washington, DC: US Dept. of Energy .

IPCC (Intergovernmental Panel on Climate Change). 2001. *Climate Change 2001: The Scientific Basis*. New York: Cambridge University Press.

IPCC (Intergovernmental Panel on Climate Change). 2000. *The Report of the Intergovernmental Panel on Climate Change*. New York: Cambridge University Press.

IPCC (Intergovernmental Panel on Climate Change). 1996. J.T. Houghton, L.G. Meira Filho, B.A. Callander, N. Harris, A. Katterberg, and K. Meshell. *Climate Change 1995: The Science of Climate Change*. New York: Cambridge University Press.

IPCC (Intergovernmental Panel on Climate Change). 1990. *The Report of the Intergovernmental Panel on Climate Change*. New York: Cambridge University Press.

Jacoby, H.D., R. Schmalensee, and I.S. Wing 1999. "Toward a Workable Architecture for Climate Change Negotiations." Joint Program on the Science and Policy of Global Change, Report No. 49. Cambridge, MA: MIT.

Jones, P.D., A.E.J. Ogilvie, T.D. Davies, and K.R. Briffa. 2001. *History and Climate: Memories of the Future?* Boston: Kluwer Academic Publishers.

Kammen, D.M. and D.M. Hassenzahl. 1999. *Should We Risk It? Exploring Environmental, Health, and Technological Problem Solving*. Princeton, NJ: Princeton University Press.

Leslie, G.B., and F.W. Lunau, (eds). 1994. *Indoor Air Pollution: Problems and Priorities*. New York: Cambridge University Press.

Mitchell, J.F.B., T.J. Johns, J.M. Gregory, and S.B.F. Tett. 1995. "Climate Response to Increasing Levels of Greenhouse Gases and Sulphate Aerosols." *Nature*. 376: 501–504.

NOAA (National Oceanic and Atmospheric Administration). 2003. See: www.noaa.gov

NRC (National Research Council). 1981. "Indoor Pollutants." *A Report of the Committee on Indoor Pollutants, Board on Toxicology and Environmental Health Hazards*, Assembly of Life Science. Washington, DC: National Academy Press.

Parker, D.E., P.E. Jones, C.K. Folland, and A.J. Bevan. 1994. "Interdecadal Changes in Surface Temperature since the Late Nineteenth Century." *J. Geophys. Res*. 99: 14373–14399.

Prinn, R., H. Jacoby, A. Sakolov, C. Wang, X. Xiao, Z. Yang, R. Eckhaus, P. Stone, D. Ellerman, J. Melillo, J. Fitzmaurice, D. Kicklighter, G. Holian, and Y. Liu. 1999. "Integrated Global Systems Model for Climate Policy Assessment: Feedbacks and Sensitivity Studies." *Climate Change*. 41: 469–546.

Redeker, K.R., N.-Y. Wang, J.C. Low, A. McMillan, S. Tyler and R.J. Cicerone. 2000. "Emissions of Methyl Halides and Methane from Rice Paddies." *Science*. 290: 966–969.

Reilly, J., P.H. Stone, C.E. Forest, M.D. Webster, H.D. Jacoby, and R.G. Prinn. 2001. "Uncertainty and Climate Change Assessments," Policy Forum: Climate Change. *Science*. 293: 430–433.

Schmalensee, R., P.L. Joskow, A.D. Ellerman, J.P. Montero, and E.M. Bailey. 1998. "An Interim Evaluation of Sulfur Dioxide Emissions Trading." *Journal of Economic Perspectives*. 12(3): 52–68.

Schwarzenback, R.P., P.M. Gschwend, and D.M. Imboden. 1993. *Environmental Organic Chemistry*. New York: Wiley-Interscience, John Wiley & Sons.

Seinfeld, J.H. 2000. "Clouds and Climate: Unravelling a Key Piece of Global Warming." *AIChE Journal*. 46 (2): 226–228.

Seinfeld, J.H., and S.N. Pandis. 1998. *Atmospheric Chemistry and Physics from Air Pollution to Climate Change*. New York: John Wiley & Sons.

Sokolov, A.P., and P.H. Stone. 1998. "A Flexible Climate Model for Use in Integrated Assessments. *Climate Dynamics* 14: 291–203.

Sposito, G. 1994. *Chemical Equilibria and Kinetics in Soils*. New York: Oxford University Press.

Stavins, R.N. 1998. "What can we learn from the Grand Policy Experiment? Lessons from SO_2 Allowance Trading." *Journal of Economic Perspectives*. 12(3): 53–68.

Stone, P.H. 2000. Lecture in MIT Sustainable Energy Course. Spring Semester.

Stott, P.A., S.F.B. Tett, G.S. Jones, M.R. Allen, J.F.B. Mitchell, and G.J. Jenkins. 2000. "External Control of 20th Century Temperature by Natural and Anthropogenic Forcings." *Science*. 290: 2133–2137.

Thilly, W.G. 1991. "What Actually Causes Cancer?" *Technology Review*. 49–54. May/June.

Tietenberg, T., M. Grubb, A. Michaelowa, B. Swift, and Z.-X. Zhang. 1999. "International Rules for Emissions Trading: Defining the Principles, Modalities, Rules, and Guidelines for Verification, Reporting, and Accountability." UNCTAD/GDS/GFSB/Misc. 6. Geneva.

Wadden, R.A., and P.A. Scheff. 1983. *Indoor Air Pollution: Characterization, Prediction, and Control*. New York: John Wiley & Sons.

Webster, M.D., C.E. Forest and J.M. Reilly. 2001. "Uncertainty Analysis of Global Climate Change Projections." MIT Joint Program on the Science and Policy of Global Change, Report No. 73. March. Cambridge, MA: MIT.

Weyant, J., ed. 1999. "The Costs of the Kyoto Protocol: A Multi-model Evaluation." Special Issue, *The Energy Journal*.

Wigley, T.M.L. and S.C.B. Raper. 2001. "Interpretion of High Projections for Global-Mean Warming." *Science*. 293: 451–454.

Zweiss, F.W. and A.J. Weaver. 2000. "The Causes of 20th Century Warming, Perspectives: Climate Change." *Science*. 290: 2081–2083.

Project Economic Evaluation 5

5.1 Introduction

Bankruptcy is the antithesis of sustainability. In common with all other business ventures, energy projects are undertaken in the expectation of profit, and alternatives are ranked in preference order according to this criterion. While formal analysis of project viability predates the Industrial Revolution, methods have evolved significantly in scope and sophistication ever since. Notably, over the past several decades, increased attention has been paid to quantification of "externalities," which are those costs incurred by society at large, or indeed by nature itself, as opposed to the conventional project costs and revenues actually appearing on the books of the firm engaged in the activity. Such costs are to be evaluated on a total lifecycle basis—from resource extraction through final disposal. These externalities can then be internalized, for example, by requiring abatement, imposing fees or taxes, or by assigning debits in the permitting process enforced by regulatory agencies. Insolvency, however, is a two-edged sword. In 1996, Kenetec, the largest US and world windpower manufacturer, declared bankruptcy, preceded in this act by LUZ, the largest US and world solar thermal electric power vendor, both due to the expiration of subsidies (or allowances for their lower externalities) (Berger, 1997). Thus, sustainable credentials may not always be a fungible virtue.

In this chapter, the rudiments of the economic assessment process will be presented at the simplest level capable of providing useful insight. No familiarity with microeconomics is presumed. However, considering the difficulty in (and poor past record of) predicting the future economic environment, one can plausibly argue that much more added complexity is unwarranted, if not occasionally delusional.

We also resist the temptation to digress into a detailed discussion of the fascinating past history of the theory and application of interest; the brief outline in Table 5.1 must suffice. (See Homer and Sylla, 1996, and *The Science of Money*, October 1998, for further background.) Despite persistent ambivalence, most now accept the concept of interest taking in domestic and international commerce.

The principal concern here is with the engineering economics of *project analysis,* where one has a fairly well-defined problem and a consensus on method of analysis such that different investigators would be expected to arrive at the same quantitative results. This is in contrast to the *program analysis* involved in energy policy planning, where the various stakeholders invariably start with different qualitative perspectives, which must be accommodated in a systematic manner, and a convergent system devised for reconciliation of the same.

Finally, "sustainability" is translated into econospeak:

> Sustainable energy systems are those that satisfy human needs cost-effectively, including quantified lifecycle impacts on the environment (externalities), for the foreseeable future.

For the time being, we will adopt the point of view expressed in the Portuguese proverb, "money is the measure of all things."

We begin the next section with a classical cost-of-product analysis using the generation of electricity as a concrete example, applying a conceptual framework entirely analogous to the thermodynamicist's approach of defining a system separated from its surroundings by a boundary crossed by various categories of transactions, as sketched in Figure 5.1. Note that the estimated costs of externalities can equal or exceed the conventional cost-of-product value. Also note that the externality costs are subject to considerable imprecision (see Section 5.7).

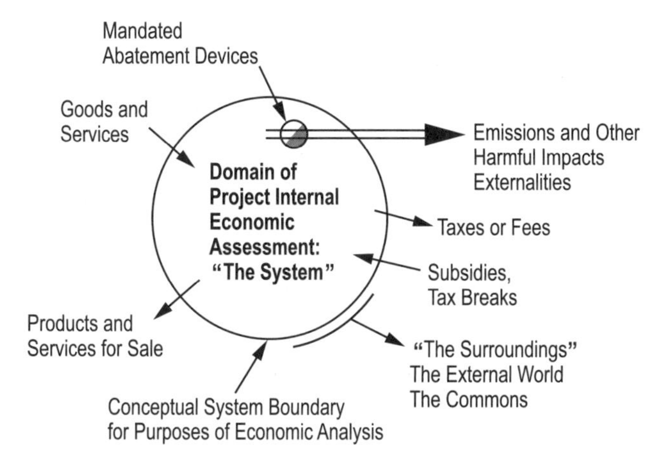

Figure 5.1. Schematic of our point of view for economic analyses. (See Chapter 4 for details on adverse environmental effects.)

Table 5.1. Historical Timeline Relevant to Project Economic Analysis

Year	Event
~3000 BCE	Sumerian records mention interest
~1800 BCE	Hammurabi's laws regulate interest rates in ancient Babylonia
1000 to 100 BCE	Old Testament/Torah compiled: Forbids charging interest on one's brethren
800 BCE	Coined money issued in Greece, (earlier in China)
594 BCE	Laws of Solon in Classical Greece do away with limits on interest rates
443 BCE	Roman "Twelve Tables" laws set maximum interest rate
350 BCE	Aristotle regards interest taking with repugnance
200 BCE to 220 CE	Han dynasty regulates interest rates in China
200-500	Mishnah/Talmud compiled; affirms prohibition of interest charges only to co-religionists
476	Fall of Rome ends organized commerce in western Europe
650+	Koran written down: forbids interest
800	Charlemagne forbids interest
1204	Sacking of Constantinople by Crusaders curtails Byzantine commercial network
1250+	Florence, Genoa, Venice, etc. re-establish a widely acceptable gold coinage money system
1312	Catholic bishops forbid interest taking under threat of excommunication
1510-1536	Martin Luther initially condemns, then accepts, interest taking
1545	Henry VIII legalizes interest
1582	Simon Steven publishes tables of interest
1602	Dutch East India Company is first true joint-stock enterprise, shares traded on the Amsterdam Exchange
1690	Massachusetts colony issues paper currency
1695	Bank of England established
1698	Royal Stock Exchange established in London
1716	First widespread (and soon disastrous) use of paper money in Europe (France, by John Law); used earlier in China
1776	Adam Smith publishes *Wealth of Nations*
1867	Marx publishes *Das Kapital*, which disparages interest on capital
1930	E.L. Grant publishes the first comprehensive textbook relevant to our concerns in this chapter: *Principles of Engineering Economy*
1946	World Bank established as an adjunct UN agency for loans to developing countries
1960s	Debate over appropriate discount rate for evaluating government projects
2000+	Debate continues over appropriate discount rate for future impact of externalities. Continuing problems in how interest is accommodated in strict Islamic societies.

5.2 Time Value of Money Mechanics

The goal here is quantification of the algorithm suggested by the English proverb, "time is money."

5.2.1 Basic aspects

Money deposited in a bank earns interest, and similarly, for large capital-intensive endeavors, money is invested in a business via purchase of bonds or stock to earn a rate of return.[1] Conversely, borrowers pay interest or dividends as their cost of money. When compounded continuously, the future worth, F, of a present amount, P, after t years at the rate i per year is just:

$$F = P\, e^{it} \tag{5-1}$$

or, equivalently, the present worth of a discrete cash flow in the future is:

$$P = F\, e^{-it} \tag{5-2}$$

Note that i here is assumed to remain constant over the period of concern; in real-life, it may well not and thereby constitute a source of complexity and uncertainty.

For example, one dollar of cost or revenue realized 40 years in the future has a present worth of only 1.83 cents today at a discount rate, i, of 10% per year. This explains why the far future has so little influence on conventional business planning.

Note that compounding continuously instead of at discrete intervals (for example, yearly) causes no loss in generality because of the following equivalence between continuous and discrete rates:

$$i_c = \ln\left[1 + \frac{i_d}{n}\right]^n \tag{5-3}$$

in which:

i_d = discrete compounding rate per year, %/yr/100

(hence i_d/n per period as per conventional terminology)

n = number of compounding periods per year

Thus if i_d is 10% per year compounded annually ($n = 1$), $i_c = 9.53$ %/yr

Note that compounding at i/n for n periods is *not* the same as i for one period, since compounding includes payment of interest on previous interest payments.

---

1 In the present chapter, in the interest of simplicity, we have succumbed to the temptation to not distinguish among "interest rate," "discount rate," "cost of money," and "rate of return." Nevertheless, different symbols (e.g., i, x, φ) are retained as a reminder that matters are not really so simple, and as inspiration for the reader to seek further enlightenment.

Another reason for preferring to work in terms of continuous compounding is that the present worth of any cash flow history is directly related to its Laplace transform (see Appendix B), and Laplace transforms are widely tabulated.

In more ambitious texts on engineering economics, it is customary to consider all four transaction combinations: discrete compounding and discrete cash flows; discrete compounding and continuous cash flows; continuous compounding and discrete cash flows; and finally, continuous compounding and continuous cash flows. Moreover, analytic expressions for the summed worth of common cash flow progressions (e.g., those representable by arithmetic or geometric series) are tabulated. For present purposes, this degree of complexity is unnecessary, but Table 5.2 summarizes how some of these prescriptions are arrived at.

Equation (5-2) allows us to consistently correct for the time value of money by expressing all costs in terms of their present worth and then computing an overall sum, P_T, following which a levelized (i.e., uniform) annual rate of expenditure $\overline{A}$ \$/yr can be calculated by equating present worth over a specified T-year time horizon.

$$\int_o^T \overline{A}\, e^{-it}\, dt = P_T \tag{5-4}$$

Thus:

$$\overline{A} = \left[\frac{i}{1 - e^{-iT}}\right] P_T \tag{5-5}$$

In the limit of large T:

$$\overline{A} \rightarrow iP_T \tag{5-6}$$

When P_T is an expenditure actually made at time zero (e.g., purchase of a machine), the rate $\overline{A}/P_T$ is often called the capital recovery factor or carrying charge rate. This is more familiar to us as the rate of uniform payment on a home mortgage or car loan.

Lifetime-levelized cost is a useful construct because it permits a single valued numerical comparison of alternatives having vastly different cash flow histories. For example, for generating electricity one can compare cheap machines burning expensive fuel (gas turbines) to expensive machines burning cheap fuel (nuclear fission reactors). These two cases highlight a classical tradeoff between up-front capital costs against long-term continuing expenses, a task faced in virtually all energy-related case studies. To further complicate matters, the most cost-effective overall *system* may contain a mix of alternatives (e.g., nuclear base load plus gas-fired turbine peaking units in the case of electric power generation).

Table 5.2. Present Worth of a Uniform Discrete Series of Cash Flows

The familiar geometric series encountered in algebra texts has a first term, A, at the *beginning* of the first interval and successive terms weighted by the ratio, R, to the power $(n - 1)$.

It has the sum:

$$S = A + AR + \ldots AR^{n-1} = A\left[\frac{1 - R^n}{1 - R}\right]$$

In engineering economics, we are often interested in the present worth of a series of uniform discrete cash flows starting at the *end* of the first interval.

Hence:

$$P = S - A$$

$$N = n - 1$$

$$R = \begin{cases} e^{-i} & \text{for continuous compounding} \\ (1 + x)^{-1} & \text{for periodic compounding at the end of interval} \end{cases}$$

Therefore:

$$P = A\left[\frac{e^{Ni} - 1}{(e^i - 1)\, e^{Ni}}\right]; \; P = A\left[\frac{(1 + x)^N - 1}{x(1 + x)^N}\right],$$

which also satisfies the equivalence $i = \ln(1 + x)$.

Note that the continuous compounding result is for *discrete* cash flows; for *continuous* cash flows of $\overline{A}$ \$/yr starting at time zero one has the substitutions:

$$A = \frac{\overline{A}}{i}(e^i - 1)$$

$$A = \overline{A}\left[\frac{x}{\ln(1 + x)}\right]$$

Details on derivations and applications of the above expressions are found in most engineering economics textbooks, such as Smith (1987).

5.2.2 Application to a typical cash flow scenario

Table 5.3 shows the result of applying present worth concepts and levelization to an appropriate example: the generation of central station electricity by a light water reactor (LWR–see Appendix A for derivations). The levelized unit cost of product, in cents per kilowatt hour, at the busbar (plant/transmission line interface) is obtained by equating levelized revenue to levelized expenditures for capital cost, operating and maintenance costs, and fuel costs. Note that the following additional embellishments are introduced, and see this chapter's Appendix A for relevant observations.

- The cost of money ("interest paid on borrowed funds") is given by a weighted sum of specified returns on bonds and anticipated returns on stocks. The carrying charge rate further considers that bond interest is tax deductible, which may or may not be the case everywhere and for all time.
- Future expenses are escalated at rate y per year.
- The plant capital cost at time zero is computed from an overnight cost (i.e., hypothetical instantaneous construction), corrected for escalation and interest paid on borrowed funds over a construction period starting C years before operation.

Three of the parameters listed in Table 5.3 deserve further comment:

The capacity factor, L, may vary widely among options. While all plants require maintenance outages and experience forced outages due to unexpected failures, nuclear units must typically shut down approximately one month for every 18–24 months of service for refueling. Renewable options are constrained by the diurnal and intermittent nature of sun and wind: wind turbine capacity factors are typically about 25%, and photovoltaic units as low as 15%. The capacity factor of typical auto engines is only on the order of 2%.

The rate of escalation, y, on fuel costs is in principle the sum of a component for monetary inflation plus an allowance for scarcity-related price increases. This latter term is usually taken as a positive quantity. However, over the past century, in constant dollars, the long-term average price of fuels and other mineral commodities has actually decreased (Chapter 2). In other words, economy-of-scale and learning-curve savings also apply to resource extraction. Thus, a third (and negative) term should really be added to allow for improvements in resource extraction and processing science and technology. This is almost never done in practice, which gives rise to false hope that renewable technologies need only outwait their fossil competition. Conversely, the fact that ingenuity has so far outpaced scarcity is no guarantee of future performance.

The useful life of the plant, T, is rather nebulous. Power stations are refurbished as time goes by. There are also annual capital expenditures, which we have neglected in the model of Table 5.3. For a typical US nuclear unit in the late 1990s, capital addition costs were on the order of 10% of the total production costs. Because of these renovations, operation significantly longer than the original design life is typical. Furthermore, the net rate of actual physical deterioration is only vaguely congruent with the depreciation schedule adopted for tax purposes in the determination of carrying charges.

The second and third terms in Table 5.3 constitute a "production cost," the sum of operating and maintenance costs plus fuel costs, which are important because they determine the rank of plants in the to-be-operated queue (capital-related costs are "sunk costs" and are in some sense irrelevant in the here and now). In 1998, US nuclear/coal/gas/ oil-fueled plants had production costs of 2.13/2.07/3.30/3.24 cents per kWh, respectively.

The prescription of Table 5.3 has been applied to the production of electric energy (kilowatt hours), but the general approach is easily transformed for other applications: for example, building a factory to produce automobiles and estimating the levelized cost per car, or the cost of facilities for extracting fossil fuels. Note, moreover, two contrasting points of view:

- One may estimate the likely cost of money (or the allowed rate in a highly regulated industry—as formerly in the US electric sector) and calculate the resulting cost of product directly from the tabulated equation.
- One may instead estimate the competitive free market allowable unit cost of product and back-calculate the achievable rate of return to see if it is attractive: the situation for most industries and increasingly the case in a deregulated electric power sector, both in the US and elsewhere.

Unlike our oversimplified example, the second approach will not, in general, permit a direct analytic solution.

Note that deregulation will increase perceived investment risk, hence increase an investor's expected rate of return and the cost of money to the utility. This will make high capital cost options less attractive than before. Because renewable options are generally capital intensive, it becomes doubly important to fully credit them with savings on externalities (a topic to be addressed in Section 5.7 and other chapters of this text).

5.2.3 Derivation of relations

The prescription set forth in Table 5.3 can be derived from no more than the preceding fundamentals plus some judicious approximations (see Appendix A). The procedure is a specific example of the more systematic general approach outlined in Figure 5.2, which, if religiously applied, should solve a wide range of problems. This introduction to engineering economics has been brief, covering as much in scope as is customary in a one-term full subject. The following worked-out example is offered as a palliative.

Table 5.3. Lifetime-Levelized Busbar Cost of Electrical Energy*

e_b cents per kilowatt hour (0.1 times mills per kilowatt hour) is the sum of:

Capital-Related Costs:

$$\frac{100\,\varphi}{8{,}766 \cdot L}\left(\frac{I}{K}\right)_{-c}\left[1+\frac{x+y}{2}\right]^{c}$$

Plus Operating and Maintenance (O&M) Costs:

$$+\frac{100}{8{,}766 \cdot L}\left(\frac{O}{K}\right)_{O}\left[1+\frac{yT}{2}\right]$$

Plus Fuel Costs:

$$+\begin{cases}\text{Nuclear} & \left[\dfrac{100}{24}\,\dfrac{F_o}{\eta B}\right]\left[1+\dfrac{yT}{2}\right] \\ \text{or} & \\ \text{Fossil} & \left[\dfrac{0.0034 f_o}{\eta}\right]\left[1+\dfrac{yT}{2}\right]\end{cases}$$

where:

			Typical LWR Value
L	=	plant capacity factor: actual energy output ÷ energy if always at 100% rated power	0.80
φ	=	annual fixed charge rate (i.e., effective "mortgage" rate)	0.15/yr
	=	$x/(1-\tau)$ where x is the discount rate, and τ is the tax fraction (0.4)	
x	=	$(1-\tau)b \cdot r_b + (1-b)r_s$, in which b is the fraction of capital raised selling bonds (debt fraction), and r_b is the annualized rate of return on bonds, while r_s is the return on stock (equity)	0.09/yr
$\left(\frac{I}{K}\right)_{-c}$	=	overnight specific capital cost of plant, as of the start of construction, dollars per kilowatt: cost if it could be constructed instantaneously c years before startup in dollars without inflation or escalation	\$1,400/kWe

Table 5.3. continued

y	=	annual rate of monetary inflation (or price escalation, if different)	0.04/yr
c	=	time required to construct plant, years	5 yrs
T	=	prescribed useful life of plant, years	30 yrs
$\left(\frac{O}{K}\right)_o$	=	specific operating and maintenance cost as of start of operation, dollars per kilowatt per year	\$95/kWe yr
η	=	plant thermodynamic efficiency, net kilowatts electricity produced per kilowatt of thermal energy consumed	0.33
F_o	=	net unit cost of nuclear fuel, first steady-state reload batch, dollars per kilogram of uranium; including financing and waste disposal charges, as of start of plant operation	\$2,000/kg
B	=	burnup of discharged nuclear fuel, megawatt days per metric ton	45,000
f_o	=	fossil fuel costs, at start of operation, cents per million British thermal units = (approximately) dollars per barrel times 16 for residual oil; dollars per ton times 4 for steam coal; cents per thousand standard cubic feet for natural gas; zero for solar or fusion energy	

Thus, for a light water reactor (LWR) nuclear power plant, using the representative values cited above:

		Cap	+	O&M	+	Fuel		
e_b	*=	4.1	+	2.2	+	0.9	=	7.2 cents/kWhre

*Note that these costs represent only the cost of generating the electricity (i.e., excluding transmission and distribution). These costs are lifetime-average (i.e., "levelized") costs for a new plant starting operations today.

1. Draw a cash-flow diagram showing all cost vectors as a function of time.

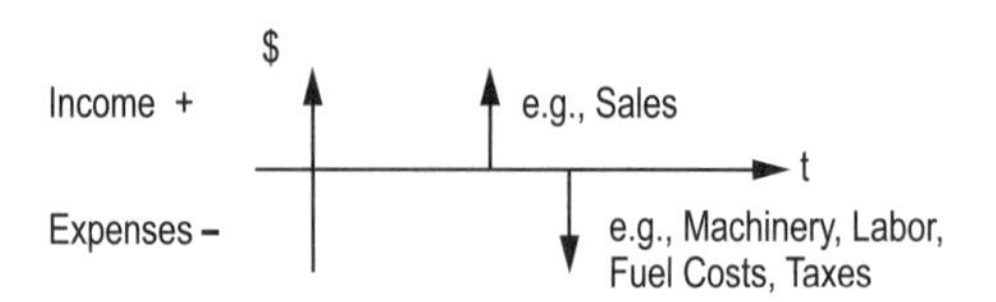

2. Bring all cash flows back to time zero using $P = F\,e^{-it}$ (Equation (5-2), separately summing revenues and expenses).

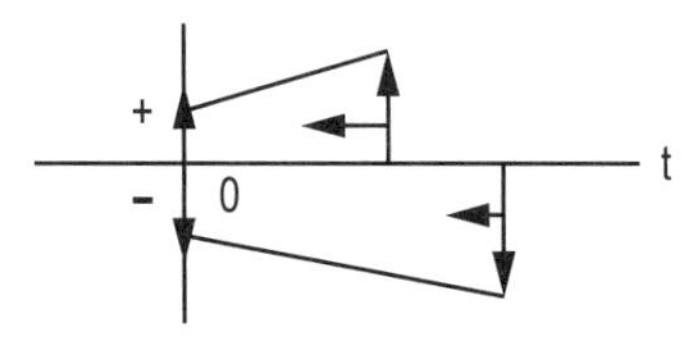

3. Using Equation (5-5), redistribute all j cash flows, $\sum_j P_j$, uniformly over the appropriate time horizon, T, with

$$\overline{A} = \frac{i \sum_j P_j}{(1 - e^{-iT})}$$

4. Equate levelized revenues and costs to calculate either
   - cost of product at a specified rate of return

   or

   - rate of return for a projected viable price for the product

Figure 5.2. How to calculate lifetime levelized cost and/or rate of return.

Sample problem

Question: In some jurisdictions, owners must contribute to a separate interest-earning account—a so-called sinking fund—to provide a future amount sufficient to decommission a nuclear power plant at the end of its useful life. Working in constant dollars, calculate the uniform rate of contributions in dollars per year at a real interest rate of 5% per year, which will total \$300 million in today's dollars 30 years from now.

Solution: We have the cash-flow time diagram:

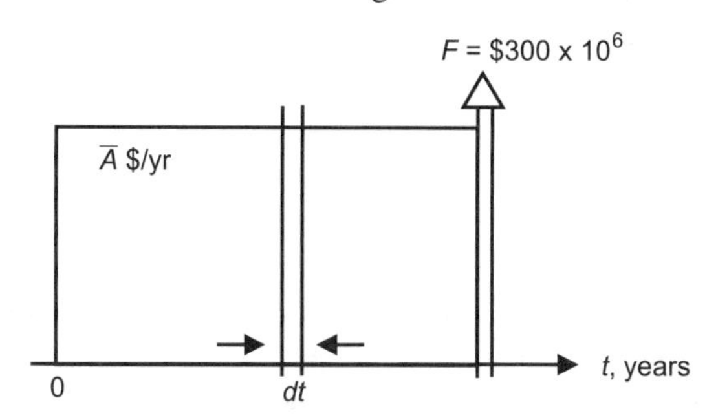

Following the algorithm recommended in Figure 5.2, we equate present worths (in millions of dollars):

$$\int_0^{30} \bar{A}\, e^{-0.05t}\, dt = 300\, e^{-0.05(30)}$$

Carrying out the integration gives:

$$\frac{\bar{A}}{0.05}(1 - e^{-0.05(30)}) = 300e^{-1.5}$$

or:

$$\bar{A} = \frac{(300)(0.05)}{(e^{1.5} - 1)} = 4.31 \text{ million dollars per year}$$

which is only a few percent of the annual carrying charge rate on a plant costing on the order of two billion dollars.

Question: Suppose instead that the utility was required to deposit a sufficient amount as a single up-front payment.

Solution: We merely need to compute the present worth of \$300 M at 5% yr for 30 years.

$$P = 300 \cdot e^{-0.05(30)}$$

$$= \$66.94 \text{ million (in today's, i.e., constant, dollars)}$$

Question: What would the actual fund accrue in $t = 30$ year dollars if the rate of inflation anticipated is 3% per year?

Solution: The market interest rate would then be 5 + 3 = 8%/yr and the future worth:

$$F = 66.94\ e^{0.08(30)}$$

$$= \$738 \text{ million}$$

Hence, with a large enough discount rate and a long enough plant lifetime, we could easily fulfill the critics' concerns that nuclear plant decommissioning costs will exceed their initial cost of construction.

5.2.4 Pitfalls, errors, and ambiguities

The bare-bones mathematics of engineering economics, just presented, is deceptively simple. However, there are many errors, both conceptual and specific, that can seriously compromise a too-casual analysis. The problem itself may be incorrectly posed, financial parameters incorrectly specified, system boundaries badly drawn, and ancillary cash flows omitted.

At the strategic level, it is important to note that the overall objective of an investor is to have all funds invested and to maximize net present worth, which may not necessarily favor an investment having the highest individual rate of return if it is small and precludes larger projects having a slightly lesser rate of return. Moreover, "doing nothing" properly defined, is always an option. One can always assume a universally available option for investment at a minimum alternative rate of return (e.g., treasury bonds). Retaining (perhaps refurbishing) a current asset may be properly considered by assuming its fictitious sale to a hypothetical outsider at its then-current market value (not book depreciation value) who then operates it as a competitor to the other new options being considered. One must be careful to ignore sunk costs, but not tax implications of premature retirement. Finally, when it comes to interpretation of output, optimality is *not* when total costs equal benefits, but when the next increment of cost is just offset by the incremental benefit. This situation often arises with regard to the purchase of safety equipment or other abatement hardware, which incur additional health and environmental costs in their own manufacture, thus leading to diminishing returns as the standard of achievement of abatement is tightened.

Financial parameters, such as the cost of money, must be selected appropriately. In particular, the analyst must be careful not to mix current and constant-dollar approaches (See Section 5.3)—market (so-called "nominal") discount rates and escalated costs versus deflated (so-called "real") discount rates and costs as of a reference year. Note that the market rate approach will yield a unit cost of product that is higher than today's market price and will appear somewhat unreal to the uninitiated. And, of course, the actual year-by-year cost of product is not constant, and only by chance equal to our levelized value. In addition, when comparing government-owned activities to private enterprise, note that the former usually have the considerable advantage of paying no taxes and of borrowing money at low rates (i.e., a low-risk premium). Furthermore, tax regulations can have a significant effect (e.g., by allowing accelerated depreciation compared to actual physical deterioration). Finally, the cost of money is not the same for all aspiring competitors. Interest charged can be considered as the sum of three components: a basic growth rate, an inflation increment, and an allowance for risk. The inflation component is an allowance for *expected* monetary inflation in the future, while the risk term accounts for the borrower's *perceived* likeliness of underperformance or default in the future. These latter components may be associated with considerable uncertainty, as well as bias.

Some care must be taken in constructing system boundaries in both space and time. Alternatives must be compared over the same time span. This may require sequential replication of some options, or sale at market value of others to make the future time horizons equal. Also, over a long time span neither the cost of money nor rates of inflation and cost escalation are likely to remain constant. Similarly, where beyond-the-busbar differences are involved (e.g., siting location), the added costs of transmission and distribution should be accounted for. In other words, all elements in the overall *system* in which the project is embedded must be evaluated.

Even when the system is appropriately isolated from its surroundings, it is easy to overlook certain cash flows. For example, byproduct credits should be recognized. This is hard to overlook for a cogeneration unit, which markets both electricity and hot water, but in other instances, more creative uses may be available, such as using otherwise waste heat for climate control in greenhouses and aquaculture. Even coal-fired plants can market fly ash as a concrete additive and scrubber sludge as feedstock for gypsum wallboard manufacture. Likewise, in real-life situations, the effect of subsidies and tax breaks on the project's cost-of-product must be credited. For example, windpower currently enjoys a 1.5¢/kWh_e federal production credit. From time to time, other favored technologies are awarded investment tax credits. However, subsidies *per se* are commonly attacked by free market advocates, whether they appear in policy-driven initiatives in developed countries or compassion-motivated programs in developing countries. Furthermore, there are always "cutoff" and "double-counting" questions. For example, is damage covered by paid-for insurance also reprised as an added externality? Are all options fairly burdened with future decommissioning costs? Some sectors, for example, nuclear, are often required to set aside funds for that purpose, while, for others, this expense is left for posterity. Finally, when evaluating energy generation costs, one must consider that different plant capacity factors imply unequal service. Replacement power and standby service costs should be considered.

One lesson to be drawn from this litany of disclaimers is that economic comparisons involving the compilation of results from independent analyses in the literature are fraught with considerable modeling uncertainties, not to mention outright errors of commission or omission. Side-by-side contemporaneous comparisons of alternatives by the same experienced analyst are preferable. (The Electric Power Research Institute (EPRI, 2000) has published a *Technical Assessment Guide*, which describes consensus good practice for such comparisons and is periodically updated.)

5.3 Current versus Constant-Dollar Comparisons

To overgeneralize, business people and engineers tend to work in then-current (i.e., market quoted) dollars and "nominal" interest rates, while economists prefer constant dollars and "real" interest rates (i.e., with the effect of monetary inflation removed). Constant dollars are preferable for international comparisons because the anticipated rates of monetary inflation differ from country to country (currently about 3%/yr in the US, but, for example, 200%/yr in Latin American in 1992–3). In the present chapter, the market (nominal) approach is applied except where explicitly noted.

While it is commonly put forth that these two approaches are compatible, in that they correctly rank alternatives in the same order, a more quantitative level of assurance is needed to ascertain whether the ratio of capital to on-going costs is distorted by one's accounting convention. A simple example will help illuminate this issue. Consider a generating facility having an initial capital cost $\$I_0$ and a lifetime of T years over which operation and maintenance (O&M) costs, initially at the rate $\$A_0$/yr, escalate exponentially with time at the rate of monetary inflation, y per yr, i.e.,

$$\overline{A}(t) = A_o e^{yt} \tag{5-7}$$

Discount rates are:

Constant dollars $x = x_0$ per year

Current dollars $x = x_0 + y$ per year

Furthermore, we neglect the effect of taxes and take x as the cost of borrowed money for our capital investment.

Then, the lifetime-levelized O&M cost is:

$$\overline{A}_L \int_o^T e^{-xt}\, dt = A_o \int_o^T e^{yt}\, e^{-xt}\, dt \tag{5-8}$$

From which:

$$\frac{\overline{A}_L}{A_o} = \left(\frac{x}{x-y}\right)\left(\frac{1-e^{-(x-y)T}}{1-e^{-xT}}\right) \tag{5-9}$$

Which is the same as Equation (A-2); it has the asymptotic limits, which we, on occasion, have employed in other instances:

For very long T (xT>>1)

$$\left(\frac{\overline{A}_L}{A_o}\right)_\infty \rightarrow \frac{x}{x-y} \tag{5-10}$$

For short-to-intermediate time horizons, series expansion gives to the first order (see Equation (A-3)):

$$\left(\frac{A_L}{A_o}\right)_T \rightarrow \left[1 + \frac{yT}{2} + \ldots\right] \tag{5-11}$$

For a capital expenditure at time zero having zero salvage value and in the absence of taxes, the annual carrying charge rate (capital recovery factor) is given by:

$$\varphi = \frac{x}{1 - e^{-xT}} \text{ per year} \tag{5-12}$$

With the limit at large T of:

$$\varphi_\infty \rightarrow x \tag{5-13}$$

Thus, from Equations (5-9) and (5-12), the ratio of levelized O&M to capital charges is given by:

$$R = \frac{\overline{A}_L}{\varphi I_o} = \frac{A_o\left(\frac{x}{x-y}\right)\left(\frac{1 - e^{-(x-y)T}}{1 - e^{-xT}}\right)}{I_o x\left(\frac{1}{1 - e^{-xT}}\right)} \tag{5-14}$$

Hence:

$$R = \left(\frac{A_o}{I_o}\right)\left[1 - \frac{e^{-(x-y)T}}{x-y}\right] \tag{5-15}$$

For constant-dollar accounting, one need only replace $(x - y)$ by x_0.

Thus, the relative magnitude of the term in brackets for the conventional and constant dollar approaches is:

$$\frac{R_x}{R_{xo}} = \frac{1 - e^{-x_o T}}{x_o} \Big/ \frac{1 - e^{-x_o T}}{x_o} = 1 \tag{5-16}$$

suggesting that the two approaches *are* consistent, at least at the level of this oversimplified example—a finding not immediately evident had we started with an incompatible set of asymptotic limit approximations to $\overline{A}_L$ and φ. But note that in neither model are levelized costs equal to today's cost of product—hence *relative* comparisons among options are essential.

In particular, constant-dollar analysis appears to understate both capital carrying charges and future fuel and O&M costs relative to everyday experience, while current-dollar analysis gives larger values for both capital charges and fuel and O&M charges. Thus, a current-dollar cost of electricity appears high and a constant-dollar cost appears low compared to what the observer sees on today's bills for electricity.

The US Office of Management and Budget (OMB) currently specifies a real discount rate of 7% per year, using constant dollars, for analyses of federal programs and policies. (See OMB, 1992.)

Figure 5.3 shows some selected results from a recent international survey by the Organisation for Economic Co-operation and Development (OECD) of projected electricity generation costs for 19 countries (OECD/IEA, 1998b). The ground rules called for the use of constant dollars and a corresponding real discount rate of 10% per year. The discounting mechanics employed are only slightly more sophisticated than summarized in Table 5.3. A variety of useful generalizations are suggested:

- The dominance of combined cycle gas turbine (CCGT) technology, qualified by the situation in Japan, where gas must be imported as liquefied natural gas (LNG) in cryogenic tanker ships
- A growing interest in renewable energy, as evidenced by the evaluation of windpower for Denmark and biomass for the US
- The persistence of coal as a competitor

The information in Figure 5.3 is an incomplete guide for option assessment. For one thing, the results represent, at best, most-probable projections, and lack any qualification due to differences in the associated degree of uncertainty. More importantly, only internalized costs are considered, whereas large externalities may apply, as in the use of coal. (Quantification of both shortcomings will be addressed in subsequent sections of this chapter.)

While on the subject of international comparisons, it should also be noted that the adjustment to a common US dollar basis using official currency exchange rates may be misleading. Many economists prefer correction for purchasing power parity (PPP)—the cost of an identical market basket of goods and services. PPP may over- or under-value the nominal expenditure by as much as plus or minus 50%.

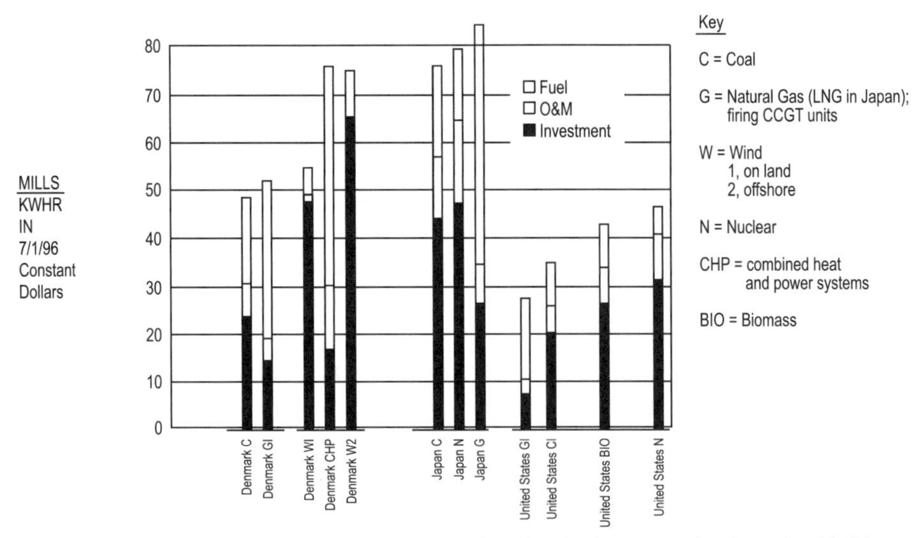

Figure 5.3. Levelized electricity generation (busbar) cost projections (at 10%/yr real discount rate). Adapted from: OECD/IEA (1998b).

5.4 Simple Payback

Although much denigrated by the sophisticated analyst, the "simple payback" approach is common because of its readily understandable nature. One merely has:

$$T_{PB}\text{, payback time, years} = \frac{\text{initial capital cost, \$}}{\text{incremental annual savings, \$/yr}} \quad (5\text{-}17)$$

Thus, spending an extra dollar now to save 20 cents per year thereafter corresponds to a five-year payback.

Note that neither discount rates nor rates of future cost escalation are involved. So long as simple and similar cash flow patterns apply to all alternatives, and the time horizon is short, the simple payback method is not terribly misleading. It has the merit of being immediately comprehensible to the general populace.

There is no hard-and-fast criterion of acceptability, but payback times of three or four years or less appear to be the order needed to inspire favorable action by the proverbial "man in the street." For the simple case of a capital increment at time zero followed by uniform annual savings *ad infinitum*, one has the rate of return:

$$i = \frac{100}{T_{PB}} \text{ \%/yr, or 20\%/yr in our example above} \quad (5\text{-}18)$$

5.5 Economy of Scale and Learning Curve

These two concepts are firmly established semi-empirical precepts of engineering economics and are central to the costing of most energy supply and end-use technologies (and a wide variety of other industrial systems).

Economy of scale refers to the general proposition that "bigger is cheaper" per unit output. In quantitative terms:

$$\left(\frac{C_i}{K_i}\right) = \left(\frac{C_o}{K_o}\right)\left(\frac{K_i}{K_o}\right)^{n-1}; \text{ or } \frac{C_i}{C_o} = \left(\frac{K_i}{K_o}\right)^n \quad (5\text{-}19)$$

where:

C_i, C_o = cost of size i and reference (o) units, respectively

K_i, K_o = size or rating of subject units

n = scale exponent, typically ~ 2/3

Thus, if a 50 MWe power station costs 2,000 \$/kWe, a 1,000 MWe unit would be predicted to cost:

$$\left(\frac{C_{1000}}{K_{1000}}\right) = \left(2000 \frac{\$}{kWe}\right)\left(\frac{1000}{50}\right)^{\left(\frac{2}{3}-1\right)} = \$737/\text{kWe},$$

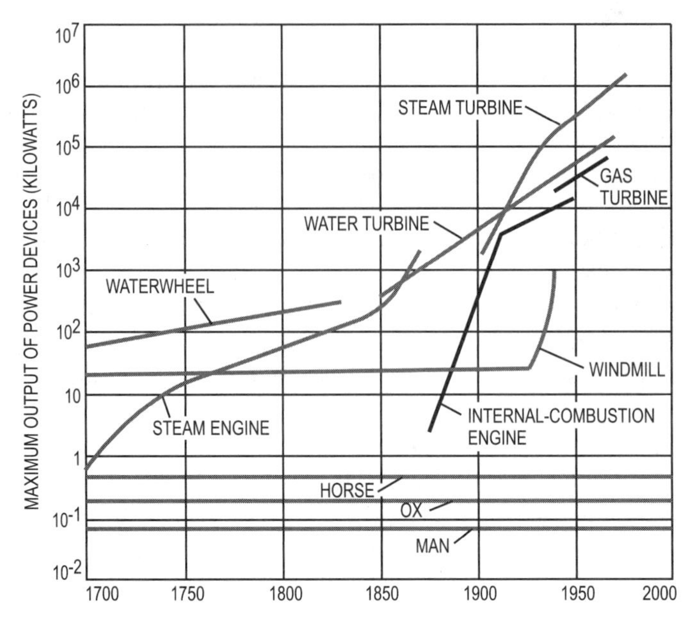

Figure 5.4. Output of power devices, 1700–2000. Source: Starr (1991). Reprinted with permission of *Scientific American*.

a substantial savings. This explains the steady increase in unit rating for steam, gas, and water turbines to meet increased demand as time progresses, as shown in Figure 5.4.

A simple example can be used to motivate n~2/3. Consider a spherical tank with a cost proportional to surface area; then the surface-to-volume (i.e., cost-to-capacity) ratio is:

$$\frac{C}{K} \approx \frac{S}{V} = \frac{4\pi R^2}{\frac{4}{3}\pi R^3} = \frac{3}{R} \approx \frac{1}{V^{1/3}}$$

The learning curve effect applies to the savings achieved by sequential mass production of a large number of *identical* units (same size or rating) i.e., "more is cheaper." The OECD/IEA report (2000) is a timely energy-oriented exposition on this topic. Automobile engines are a good example.

Experience shows that one can often characterize the cost of the Nth unit as follows:

$$C_N = C_1 \cdot N^{-\alpha} \tag{5-20}$$

with:

$$\alpha = -\left(\frac{\ln \frac{f}{100}}{\ln 2}\right)$$

where:

C_1, C_N = cost of first and Nth units, respectively

$f/100$ = progress ratio: second (or 2^Nth) unit is f percent as expensive as first (or 2^{N-1}th) unit; a typical value is ~85%

Thus, if the first first-of-a-kind production-run photovoltaic panel costs \$200/m^2, the last unit of a production run of 10^4 units would be predicted to cost:

$$C_{10^4} = (200)\ 10^{-4\alpha}$$

and:

$$\alpha = -\left(\frac{\ln 0.85}{\ln 2}\right) = 0.234$$

so that:

$$C_{10^4} = \$23/\text{m}^2$$

again a large decrease. Figure 5.5 illustrates this phenomenon for some technologies of current interest.

In addition to the unit cost of the *Nth* unit in a mass-production run or in a sequence of increasing size, the average unit cost of all N items, first through last, is also of interest.

For a learning curve sequence, the average cost is, to a good approximation:

$$\overline{C}_{1,N} = C_1 \left\{ \frac{N^{\alpha}}{1+\alpha}\left[1 - \frac{1}{N^{1+\alpha}}\right] + \frac{1}{2N}\left[1 + N^{\alpha}\right] \right\} \tag{5-21a}$$

with a maximum error of -1.5% for $\alpha = 1/2$ ($f = 71\%$) and $N = 2$; the error decreases as N increases and for larger f.

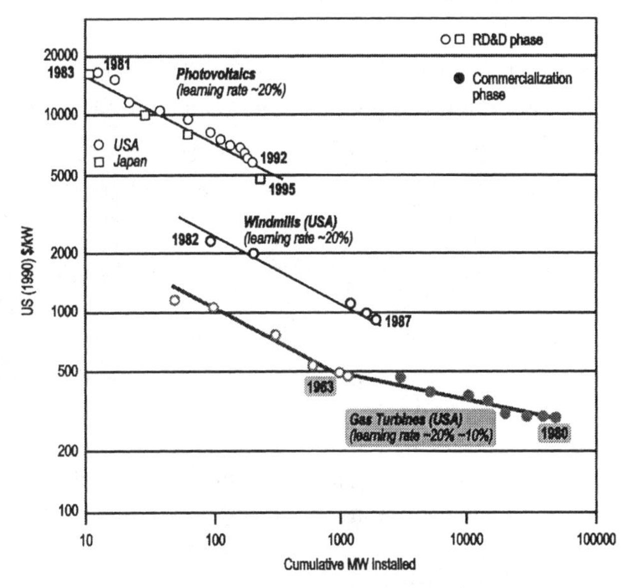

Figure 5.5. Technology learning curves: Cost improvements per unit installed capacity, in (1990) US\$ per kW, versus cumulative installed capacity, in MW, for photovoltaics, wind and gas turbines. Source: Nakicenovic et al. (1998). Reprinted with permission of Cambridge University Press.

For a sequence of N units each a factor of S larger than its immediate precursor:

$$\overline{\left(\frac{I}{K}\right)} = \frac{\Sigma I}{\Sigma K} = \left(\frac{I}{K}\right)_1 \left(\frac{S^{nN} - 1}{S^n - 1}\right)\left(\frac{S - 1}{S^N - 1}\right) \tag{5-21b}$$

where, as before, n is the scale exponent.

These relations provide estimates of the total expenditure for subsidized installations before a competitive product is produced. Note that for typical values of n and f, it generally does not pay to reduce size to increase number to satisfy a fixed total capacity.

Economy of size and learning curve effects, while dignified by a mathematical formulation, are by no means "laws of nature" of the same credibility as the Carnot efficiency equation (see Chapter 3). Hence, one must be aware of several caveats regarding the use of such projections:

- At some point, size increases may require switching to new materials—for example, to accommodate higher stresses, in which case the economy-of-scale relation has to be re-normalized.
- Larger size may lead to lower reliability (i.e., capacity factor) and therefore net unit cost of product may increase (i.e., there may well be dis-economies of scale).
- Learning curves apply to replication of the *same* design, by the *same* work force, in the *same* setting (e.g., factory), all of which are likely to change in the long run.
- Important factors such as materials resource depletion or technological innovation are not taken into account in an explicit manner.
- Shared costs of many units on a single site are also important (e.g., multi-unit stations save considerably on administrative infrastructure costs).
- Often both effects operate in series or parallel combination. For example, in nuclear fission and fusion reactor development, construction of a sequence of pilot/demonstration units of increasing size is usually envisioned, followed by replication and deployment of a fleet of identical large commercial plants.

The effects just described also highlight the essential issue of development cost: who pays for the early uncompetitive versions of a future technology? (See Chapter 23.)

5.6 Allowing for Uncertainty

5.6.1 Overview

Engineering economic analysis for assessment of future options is more akin to long-range climate forecasting than the precise "bean counting" involved in documentation of past transactions—the purview of accounting. It is not enough to merely compare only the expected values of lifetime-levelized costs. Some measure of uncertainty is essential. The standard deviation is the most common overall index.

Four approaches are common:

1. Poll the experts and report their consensus plus and minus absolute (or relative percentage) spread on values for estimates of the subject genre.
2. Propagate and aggregate uncertainties analytically for the governing equations under simplifying approximations, such as random independence of variables which are characterized by a normal (Gaussian) probability density function.
3. Carry out a Monte Carlo analysis. Repeat the calculation many times using randomly selected input variables from probability distributions characterizing their likely uncertainty, and compute the results for the output parameter. From this, compute the standard deviation.
4. Use a decision tree approach, which amounts to a crude Monte Carlo calculation in which variable probabilities are confined to only a few outcomes.

A brief sketch of the latter three approaches follows, beginning with the analytic.

5.6.2 Analytic uncertainty propagation

For independent variables distributed according to the familiar normal or Gaussian probability density function (the familiar bell-shaped curve), the uncertainty is readily and completely characterized as a standard deviation σ, from the mean, $\bar{e}$. A randomly selected large set of data will fall within $\pm 1\,\sigma$ of the mean 68% of the time, and within $\pm 2\sigma$, 95% of the time. Furthermore, the cumulative variance σ^2 in a dependent parameter e can be estimated from that for several uncorrelated independent variables x_i as follows:

$$\sigma_e^2 = \sum_i \left(\frac{\partial e}{\partial x_i}\right)^2 \sigma_{x_i}^2 \tag{5-22}$$

A simple example may help.

As presented in Table 5.3, the busbar cost of electricity is made up of three components, representing capital, operating, and fuel costs:

$$e = e_I + e_{O\&M} + e_F \tag{5-23}$$

where for fossil fuels:

$$e_f = \frac{0.34 f}{\eta} \tag{5-24}$$

where f is, for example, the cents per 1,000 standard cubic feet (SCF) paid for natural gas. We recognize that fuel cost is the principal uncertainty in estimation of the busbar cost of electricity from (for example) a CCGT unit.

Applying Equation (5-22) in a conveniently normalized form one has:

$$\left(\frac{\sigma e}{e}\right)^2 = \left(\frac{e_f}{e}\right)^2 \left(\frac{\sigma_f}{f}\right)^2 \tag{5-25}$$

Thus if the expected cost of natural gas is 60% of the busbar cost, while the one sigma uncertainty in the lifetime-levelized future cost of natural gas is judged to be ± 25%, this translates into a $\pm\sigma$ uncertainty in busbar cost of ± 15%, and if the latter is 4¢/kWhr it should be reported as 4 ± 0.6¢/kWhr.

Extension of this methodology to other terms in the busbar cost prescription is straightforward, but admittedly tedious.

5.6.3 The Monte Carlo method

Given a normalized probability density function (PDF), $P(z)$, one can integrate from $-\infty$ to z to obtain the cumulative distribution function (CDF), $P \leq z$, which gives the probability of a variable being z or less; or if one prefers, integrate z to $+\infty$ to obtain its complementary cumulative distribution function (CCDF), $P \geq z$ (refer to Figure 5.6). Then in Step 2, choose a random number between 0 and 1 and use the CDF(z) to select a value for z; repeat this for all independent variables to create an input data set. In Step 3, use this set in the governing analytic relation to compute a value of the dependent parameter (for example, the busbar cost of electricity). Repeat Steps 2 and 3 a large number of times (e.g., 1,000).

Then, the expected value $\overline{e}$ and its σ can be calculated from the accumulated output set as follows:

$$\overline{e} = \frac{\sum_{j=1}^{N} e_j}{N} \tag{5-26}$$

and for large N:

$$\sigma^2 = \frac{\sum_{j=1}^{N} (e_j - \overline{e})^2}{N(N-1)} \tag{5-27}$$

The widespread availability of powerful personal computers has made this formerly daunting task tractable.

Step 1: Concoct a PDF and integrate under it to develop a CDF distribution.

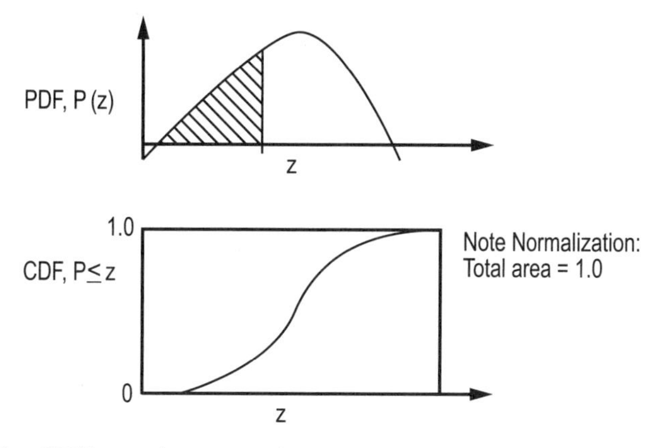

Step 2: Use the CDF to select a random entry for z by generating a random number to use as $P \leq z$.

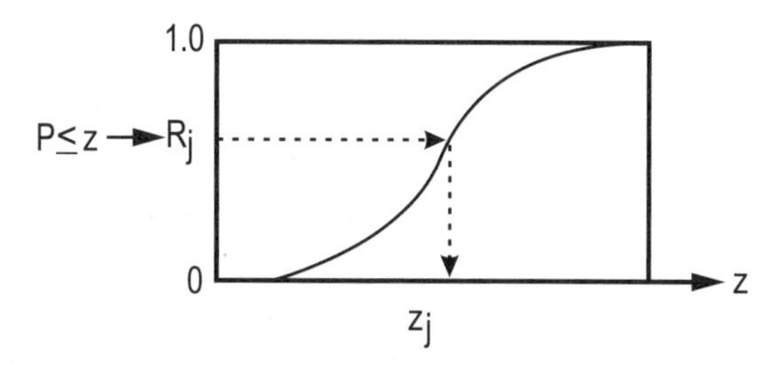

Step 3: Generate a set of dependent variable values

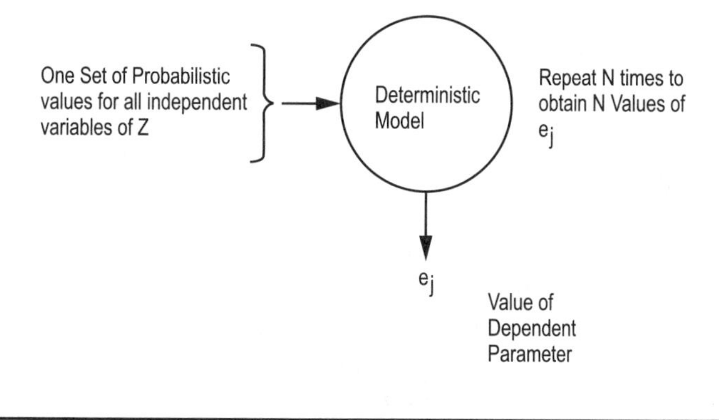

Figure 5.6. Outline of Monte Carlo approach (continued on next page).

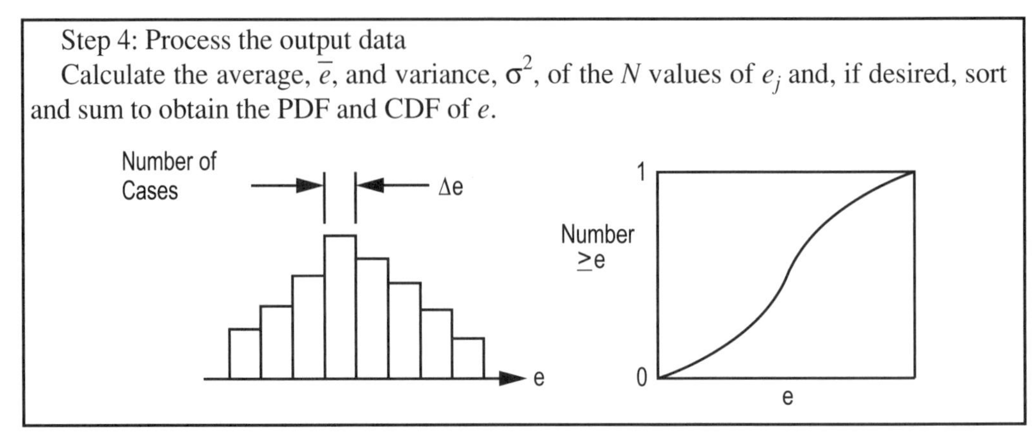

Figure 5.6 continued

5.6.4 Decision tree method

There is a choice of but two things, loss or gain.—Hindustani proverb

This approach is useful in that it displays in plain sight all of the assumptions made to develop the results. As shown in Figure 5.7, the decision tree method can be thought of as a degenerate form of the Monte Carlo method, in which only a few branches of outcomes are permitted in the selection of variables. Figure 5.7 also shows a simple tree in which annual revenues and their duration are the independent variables.

As will become evident in the section that follows, estimation of the cost of externalities is fraught with great uncertainty, a circumstance which makes it even more imperative to properly qualify any quantitative assessments.

5.7 Accounting for Externalities

Our next concern is with putting a dollar value on all off-book consequences of a commercial enterprise, the so-called "externalities." The boundary between costs internal to a project and those that are external is movable because mandatory abatement measures can be legislated, requiring add-ons to plant design. Examples include cooling towers for virtually all Rankine cycle electric-generation stations and stack-gas scrubbers for coal-fired units (see Chapters 4 and 8). These measures have well-defined costs. On the other hand, effluents discharged to the environment outside of plant boundaries impose real but hard-to-quantify economic penalties on the general public and the environment. For example, there are the sulfur and nitrogen oxides released by coal-fired stations, which create the notorious acid rain phenomenon. How one develops cost estimates for such effects goes far beyond the scope of this chapter, involving as it does atmospheric and aquatic transport, chemical and physical transformation of pollutants for all pathways leading to human exposure (and that of other biota), and the resulting epidemiological consequences (see Chapter 4).

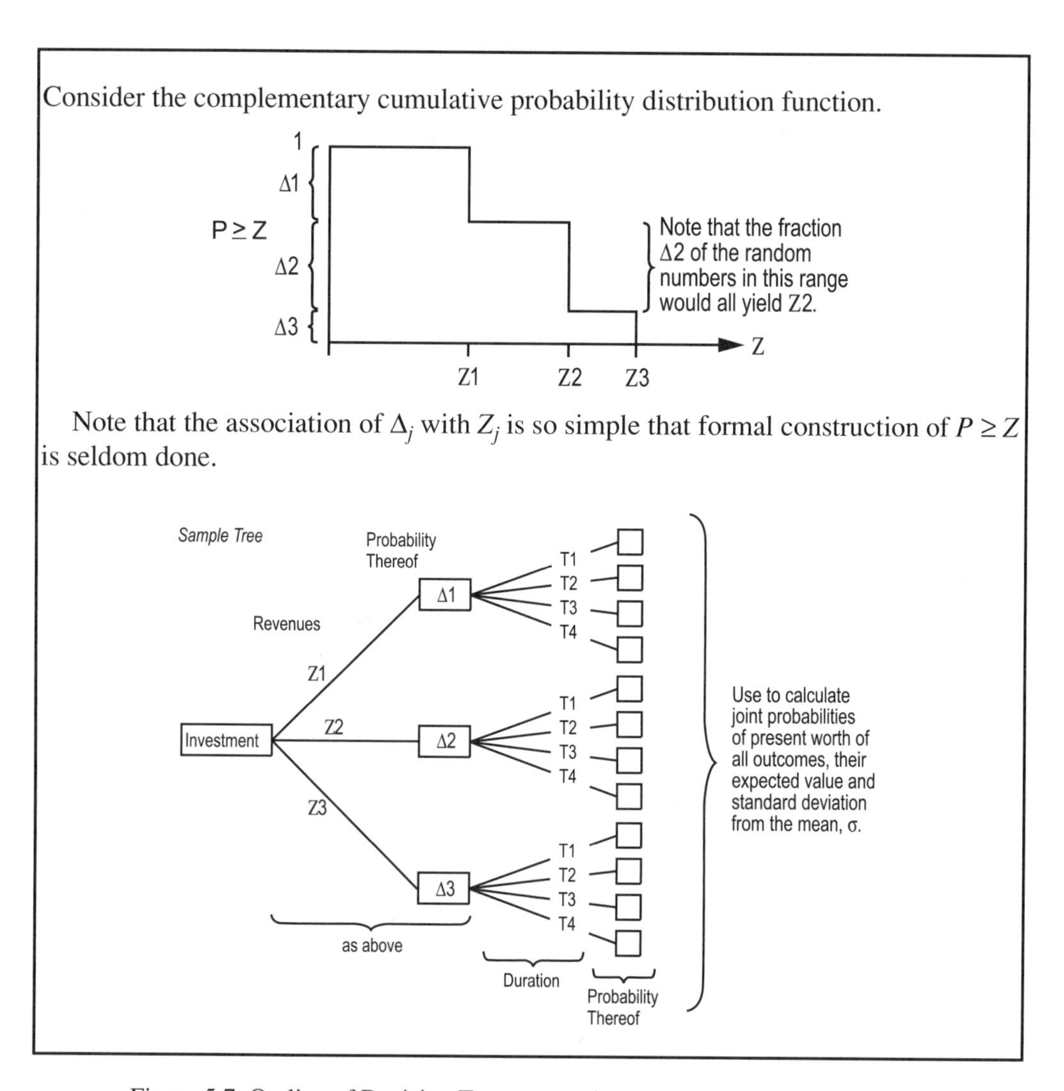

Figure 5.7. Outline of Decision Tree approach.

Detrimental effects on infrastructure are also of concern. Figure 5.8 shows the wide scope covered by a comprehensive lifecycle assessment. The generic procedure for externality accounting outlined in Figure 5.9 can be implemented if input/output coefficients are available for all important contributions. The magnitude of this task is daunting, as is the obvious large degree of uncertainty associated with most impacts due to the lack of scientific data, and the potential for introduction of socio-political biases. A further complication is that detrimental effects are often borne by those not benefiting from the primary activity. Moreover, impacts and benefits can take place at different times and in different locations.

Table 5.5 summarizes a set of externality cost coefficients used by electric utility regulatory commissions in several US states as input to assessments of this type. Note the wide range among the estimates. Table 5.6 displays representative externality assessment outputs, as reported by several investigators, in mills/kWhre for coal and nuclear power systems.

Figures 5.10 and 5.11 illustrate the wide variability among reported studies of environmental costs with and without the impact of global warming. In both cases renewables are superior, but only if credit is given for avoided greenhouse gas (GHG) emissions are the margins enough to tip the balance in favor of their preferential deployment. This is easily the case and can be appreciated from the following sample calculation:

Consider a coal plant having the following characteristics:

Thermodynamic efficiency	0.36 MWe/MWth
Coal heating value	0.35 MWd (thermal)/MT coal
Coal carbon content	0.7 MT carbon/MT coal
Lifetime-levelized cost at the busbar of electricity produced	6.5¢/kWhre

The added cost of product due to a carbon tax of $100/metric ton of carbon is then:

$$e_c = (100\$/\text{MTC})(100¢/\$)(0.7\ \text{MTC/MT coal}) \times (\text{MT coal}/0.35\ \text{MWdth})(\text{MWth}/0.36\ \text{MWe}) \times (1\ \text{d}/24\ \text{hr})(1\ \text{MWe}/1000\ \text{kWe})$$

= 2.3¢/kWhre, a 35% increase, which is significant and would virtually eliminate the construction of new coal plants and hasten the shutdown of existing units.

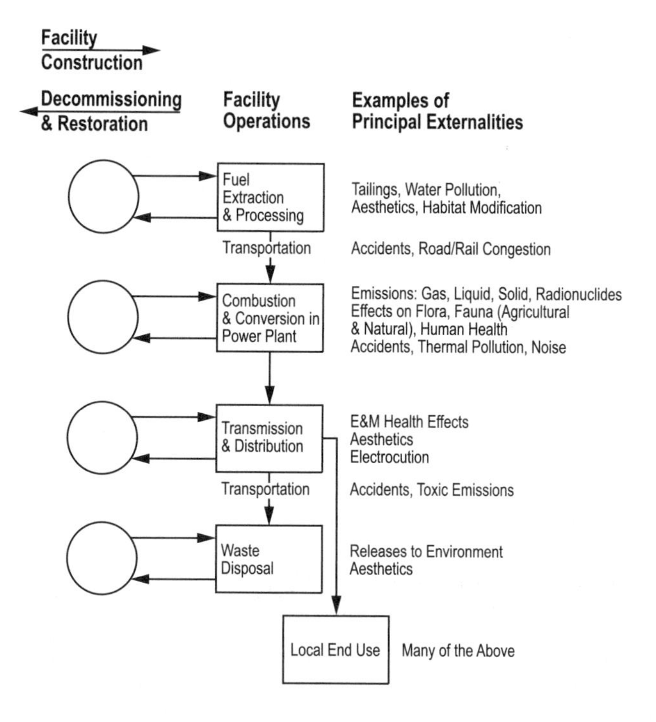

Figure 5.8. A Generic lifecycle roadmap for externalities assessment.

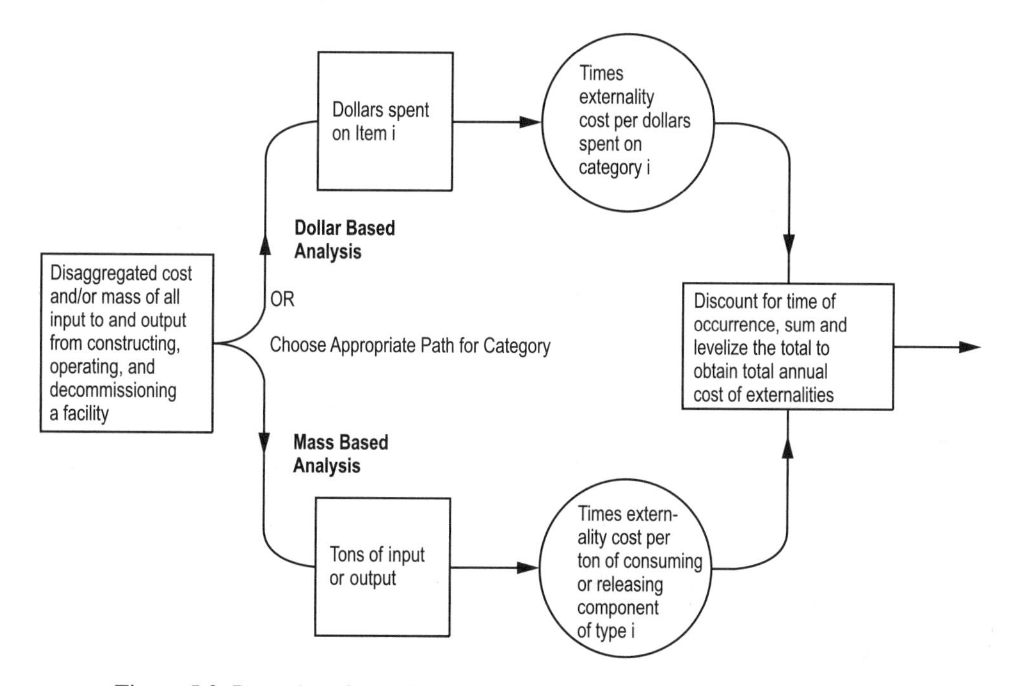

Figure 5.9. Procedure for assigning costs to externalities.

Table 5.5. Representative Externality Assessment Inputs

	State			
Emissions	**NY PSC**	**MA DPU**	**NV PSC**	**CA PU**
SO_2	900	1,650	1,720	4,500
NO_x	1,960	7,170	7,500	7,800
VOCs	—	5,840	1,300	3,600
CO	—	950	1,000	—
Particulates	350	4,400	4,600	2,600
CO_2	2.2	24	24	29
CH_4	—	240	240	—
N_2O	—	4,400	4,600	—

Units are dollars per metric ton

Source: Economic Research Associates (1992).

Table 5.6. Representative Externality Assessment Output

Summary of Impacts and Damage Costs for Coal Fuel Chain:

Study	**Impacts (deaths per TWh)**		**Damage Costs (in milli-Euro per kWh)**					
	Public Fatalities	**Occupational Fatalities**	**Public Health**	**Occupational Health**	**Environ-ment**	**Global Warming**	**Major Accident**	**Study Total**
Ottinger et al. 1991								22–55
Pearce et al. 1992			0.05		0.005	0.04		0.14
Pearce et al. 1995								0.11
Friedrich & Voss 1993			0.01–0.07		0.013–0.015			0.02–0.09
Ball 1994		0.04–0.14						
ORNL/RFF 1994			0.01–0.64	0.08	0–0.1	nq		0.7–1.4
Rowe et al. 1996			3–5		0.1	nq		3–5
ExternE 1995		0.13–0.23	4–13	1–2	0.2–0.8	10–18*		16–34
Rabl et al. 1996			5–14	nq	0.02	15		20–29
ExternE 1999			10–50		0.5–2	10–50		20–100

Notes: Numbers have been rounded.
nq = not quantified
*at 0% discount rate
1 milli Euro ≈ 1 US milli Dollar = 0.1 US cent
Source: Wilson et al. (1999).

Table 5.6. Representative Externality Assessment Output (cont.)

Summary of Impacts and Damage Costs for Nuclear Fuel Chain:

Study	Impacts (deaths per TWh)		Damage Costs (in milli-Euro per kWh)					
	Public Fatalities	**Occupational Fatalities**	**Public Health**	**Occupational Health**	**Environ-ment**	**Global Warming**	**Major Accident**	**Study Total**
Ottinger et al. 1991			4.9				18.5	23
Pearce et al. 1992			0.003–0.009			0.0012	0.002–0.006	0.007–0.017
Pearce et al. 1995						0.0012	0.006–0.044	0.006–0.044
Friedrich & Voss 1993			0.001–0.005		0–0.002		0.0005–0.004	0.002–0.01
Ball 1994	0.01–1.23	0.02–0.09						
ORNL/RFF 1994			0.012	0.08–0.09				0.09–0.1
Rowe et al. 1996								0.09
ExternE 1995	0.65	0.04	2.4	0.15				2.6
Dreicer et al. 1995	0.62	0.02	2.4	0.14			0.0005–0.023	2.5

Notes: Numbers have been rounded.

nq = not quantified

*at 0% discount rate

1 milli Euro ≈ 1 US milli Dollar = 0.1 US cent (depends on exchange rate)

The literature on externalities is diffusely published and the field lacks a definitive textbook or handbook. Two sources stand out as starting points for a contemporary literature search:

1. Publications under the auspices of the ExternE studies; in 1991 the Commission of the European Communities initiated this project to systematically quantify the external costs of electricity production (Commission of the European Communities, 1995). For a synopsis see NEA/OECD (2001).
2. The journal *Energy Policy* routinely publishes articles on the environmental externalities of energy use, which includes synopses of many of the individual reports developed under the ExternE project.

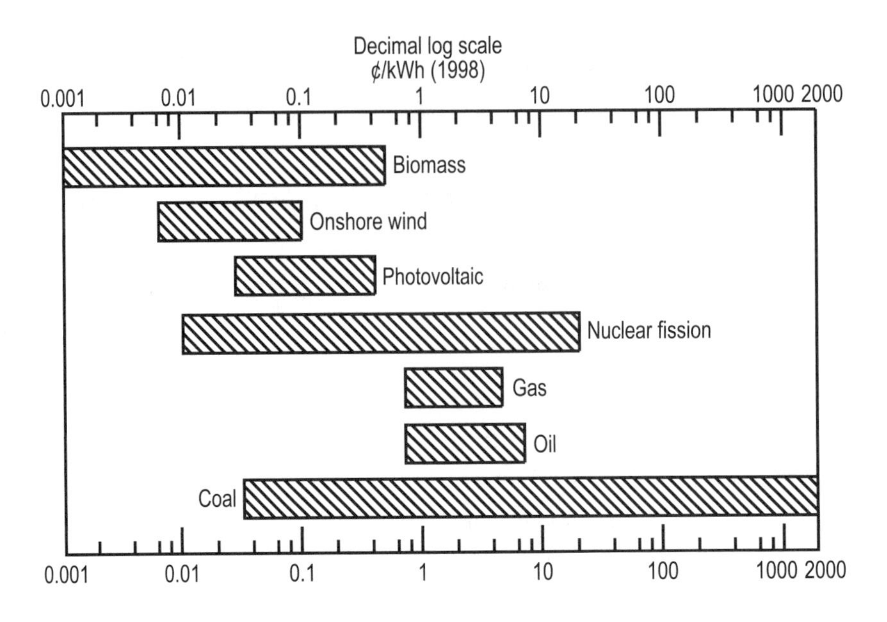

Figure 5.10. Environmentally related damage costs for selected electricity supply technologies (excluding global warming). Sources: Stirling (1992) and Toke (1995). Reprinted with permission of Pluto Press.

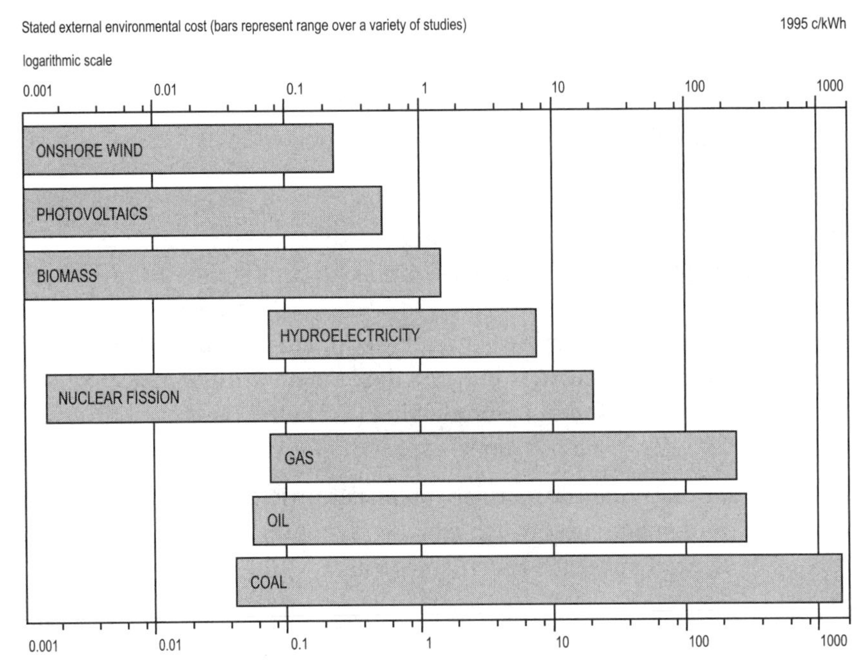

Figure 5.11. Range of externality study estimates (including global warming). Source: Stirling (1997). Reprinted with permission of Elsevier.

Apart from wide scope, there are several conceptual difficulties in the application of the procedure suggested in Figure 5.9. For example, what is the appropriate discount rate for effects that may persist for centuries, if not eons, into the future? Should different discount rates be applied to different categories of externalities, and to different time frames? In particular, how does one assign a monetary value to high value items such as human life, and how to (or even whether to) discount far-future deaths? A related point of potential confusion is that the "deaths" cited in most estimates of the effect of chronic (as opposed to acute) exposure are really a bookkeeping convenience, since they represent "statistical deaths"—the exposure to a whole population that would, if concentrated, kill an individual. Some investigators instead use life foreshortening as a metric, with equivalences such as 1 death = 6,000 person-days lost (PDL), and 1 injury = 50 PDL.

In addition to the intergenerational aspects of externality cost accounting, there are regional factors. For example, the California Energy Commission assigns different penalties to in-state and out-of-state externalities. Similarly, the cost assigned to a road accident death is 2.6 million dollars in the US compared to 20,000 dollars in Portugal.

There is well-documented public aversion to low-probability/high-consequence events, and to involuntary risks of the same imputed financial costs as commonplace and voluntary risks. Because of this and the other issues, it is common practice to apply additional weighting factors and a coarser set of ranking categories to move to a non-monetary scale reflecting stakeholder preferences. An elementary example is the use by economists of "utils" in lieu of dollars. (Chapter 6 discusses many of these broader issues.)

5.8 Energy Accounting

It is common sense and a requirement in the US (under the Federal Nonnuclear Energy Research and Development Act of 1974) to evaluate energy projects in another currency: energy itself. The objective here is to do a lifecycle comparison between the energy input to a facility for all phases: construction, operation, fueling, and decommissioning, and its net useful output. Applicability of the concept goes beyond generating plants alone; one can also gain useful insight from the study of conservation measures such as the production, installation, and benefits of insulation.

Reconsider Figure 5.9, but replace the encircled cost coefficients with energy coefficients: for example, the kWhre expended per dollar or unit mass of a commodity. The end result is the total energy consumed, which can then be compared to the projected energy output. For worthy projects the ratio of output to input should be at least four or more. In their early years of introduction, silicon solar cells were found to have a ratio of less than one. Even today, some ethanol from biomass (e.g., corn) schemes can fall below breakeven, in which case, other benefits such as reduced emission of certain harmful combustion products or subsidization of agriculture may be invoked as offsetting considerations. Spreng (1988) documents some investigations of this nature; Table 5.7 summarizes an energy analysis for producing ethanol from fermentation of corn, and Table 5.8 displays an analysis for nuclear power plants.

Taking the ratio of initial input to net output rate also yields an energy payback period, in analogy to the economic payback period discussed earlier. One major caveat in such analyses is the need to distinguish between unconverted thermal energy and work-ready energy, such as shaft work or electricity (see Chapter 3). Often, a conversion efficiency of heat to work of 1/3 is assumed (i.e., typical of a representative Rankine power cycle), and used to convert all data to a common basis. Even so, incongruities are apparent. Consider, for example, hydroelectric or windpower generators, which are only indirectly linked to solar thermal input.

Table 5.7. Energy Balance for Ethanol from Corn

	Inputs for 1 Gallon Ethanol:
Corn Production	55,300 Btu
Fermentation/Distillation	74,300 Btu
Total	129,600 Btu
Output in Combustion:	76,000 Btu

Notes:

(1) Analysis applies to typical US Midwest conditions and practices; current technology throughout.

(2) Production costs include fertilizer, operation of farm machinery.

(3) For comparison, gasoline has ~ 120,000 Btu/gal.

(4) In Brazil, using sugarcane and low-tech agriculture, the output/input ratio can be as high as 3 (Giampietro et al., 1997).

(5) For comparison, gasoline's output to input (for oil production, transportation, and refining) is about 7.

(6) In March 2000, the US Environmental Protection Agency (EPA) announced it would ban the gasoline additive MTBE and replace it with ethanol.

Sources: Pimental (1998) and Giampietro et al. (1997).

Table 5.8. Approximate Energy Balance for 1,000 MWe Nuclear Power Plant

Energy to Construct (and decommission)	4×10^9 kWh$_{th}$
Energy for Fuel (30 yrs)	7×10^9 kWh$_{th}$
Energy for Operations (30 yrs)	4×10^9 kWh$_{th}$
Total	15×10^9 kWh$_{th}$
Electric Energy Output (30 yrs)	210×10^9 kWh$_e$
Energy Ratio	$\frac{\text{electric energy out}}{\text{thermal energy in}} = 14 \frac{kWh_e}{kWh_{th}}$

Basis:

(1) 1,000 MWe PWR operated for 30 yrs at 80% capacity factor.

(2) Fuel enriched by gas centrifuge, 0.3 w/o tails

(3) Conversion between thermal and electrical energy: 3 kWhrth = 1 kWhre, in fuel cycle steps

(4) "Fuel" includes startup core, all front- and back-end steps.

Note that kWh(th) = kilowatt hours of thermal energy, kWh(e) = kilowatt hours of electrical energy

Similar considerations also motivate the concept of "exergy" or "availability" analysis, in which one disaggregates the thermal energy involved in a process as a function of its associated temperature and then weights it according to inherent worth as a producer of work, namely the Carnot Cycle Heat Engine Efficiency. Work itself has a relative weight of 1.0, but other manifestations require a more sophisticated analysis. Analogous to First Law efficiency (work output ÷ thermal energy input), a Second Law efficiency (exergy out ÷ exergy in) can be defined and evaluated. It is commonly low, 10% or less for many familiar devices, systems and processes. (See Ahern, 1980, Moran, 1982, and Kotas, 1995, for further discussion.)

5.9 Modeling Beyond the Project Level

The principal goal of this chapter has been to provide a minimal toolkit for choosing among alternatives on the project level. But this is only the first stage of the overall assessment process in planning for sustainability. As shown in Figure 5.12, project economic analysis is just a basic building block in the hierarchy of models of increasing scope. For example, new electric power stations must fit in an optimal way into the overall utility grid, which in turn must be harmonized with regional infrastructure, and strategic planning by both entrepreneurs and all levels of government. Table 5.9 is a selective list of systems models in each category and suggested references for further exploration. The low- and intermediate-level models are often programmed to ferret out optimized configurations given a set of criteria and constraints. At the highest level, where global climate change is the focus, the mode of usage is typically "scenario analysis," in which the future consequences of a range of plans spanning the plausible option space are computed. (For readers who are not motivated to dig into the specialists' literature, some flavor of its purview can be gained from more-popularized versions such as the book *Beyond the Limits* (Meadows et al., 1992), and the educational computer game SimEarth (1993).)

This is not the place to enter into a detailed critique of model capabilities and shortcomings. However, it should be appreciated that none are truly up to the task of predicting future innovation. To cite one example of a technology that would have revolutionary impacts on the energy sector, consider inexpensive room-temperature superconductors. At present, there is no accepted theory to tell us whether they are either potentially feasible or forbidden by the laws of nature. Since the volume of scientific and technological publications has doubled every 7–10 years dating back to the middle of the 20th century, one can only hope that pleasant and profitable surprises will continue to emerge.

Big models have little models
inside their guts to guide them;
little models have lesser models
and so on, *ad infinitum*

(with apologies to Jonathan Swift)

Table 5.9. Examples of Energy-Economics-Environmental Models

Category	Examples	Review Publications	This Textbook
Project Assessment	Table 5.3	Engineering economics textbooks, e.g., Smith (1987)	Chapter 5
Utility Planning and Energy Dispatch	EGEAS (EPRI)		Chapter 17
Regional Energy Planning	ENPEP (IAEA)	Weyant (1999) reviews the results of some thirteen models reported by US and international energy economists	Chapters 17, 21
Comprehensive World Models	HERMES-MIDAS	Weyant (1996) and OECD (1998)	Chapters 6, 21
Global Impact	40 scenarios in IPCC (2000)	Publications of the US Global Change Research Program, and IPCC reports	Chapters 4, 21

Sources: OECD/IEA (1998a), EPRI (1989), Hamilton et al. (2000), and Intergovernmental Panel on Climate Change (2000).

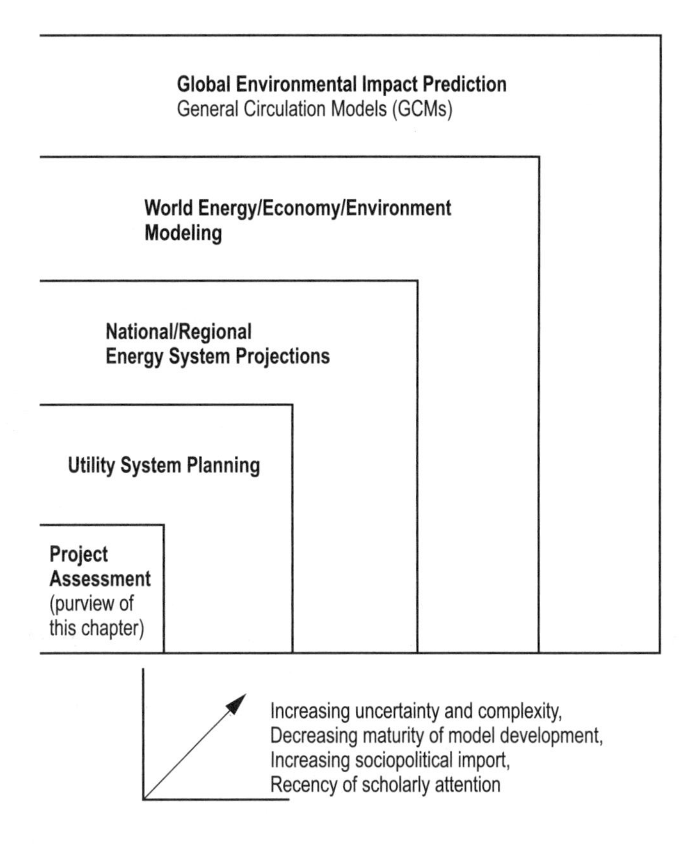

Figure 5.12. Hierarchy of energy models.

5.10 Chapter Summary

The main themes of this chapter are conceptually simple but devilishly complicated when it comes to the details of their practical application. Points to emphasize are:

- Money has a time value because of interest earned or paid, which requires that all cash flows be discounted from their point of occurrence in time; Figure 5.2 summarizes the basic conceptual framework. Once this is taken into account, options can be compared on a consistent basis. A simple but adequate way to do this using continuous compounding has been presented, leading to the concept of, and a prescription for, the single-valued levelized cost over the life of a project (Table 5.3). Market-based accounting, which includes monetary inflation, and constant-dollar/real interest rate accounting, which excludes inflation, are contrasted.

- Costs not reflected in the market price, known as externalities, must be considered to account for detrimental effects on the common environment and public health, an important step in ranking technology *and* public policy options according to their true total lifecycle cost. The elements of an input-output analysis for this purpose are sketched.
- Economy-of-scale and learning curve effects are noted: bigger is cheaper, and so is more of the same.
- The universal need for an explicitly quantitative recognition of uncertainty is stressed.
- Conceptual difficulties with a purely monetary measuring system motivate moving beyond the methods of this chapter in the formulation of actual decision-making processes that reflect stakeholder concerns and offer greater likelihood of decision process convergence.

We are left with the question of whether project economic analysis truly assures "sustainability." Lowest overall lifetime cost, including externalities, implies a least-harm (but not necessarily no-harm) option. Selecting the preferred option may eventually lead to an accumulation of hard-to-reverse (i.e., intolerably expensive) undesirable consequences—particularly if many others make similar choices in isolation. Also, because of the long life (about half a century) of infrastructure (e.g., power plants, buildings, transportation systems), trends reinforced by decisions made today are hard to redirect, let alone reverse. The best one can do is to periodically re-do assessments as more options are developed or perfected and better evidence on externalities becomes available. Although extrapolation from present status may be inadequate, the hope is that sustainability can be achieved asymptotically via a feedback process. In any event, it is essential that an overall framework be in place for decision-making in the face of great uncertainty.

Appendix A

Derivation of Relations for Levelized Cost

First, the annualized capital cost term readily follows from Equation (5-6), with i replaced by φ. Other ingredients require a bit more development.

Consider a uniform annual rate of expenditure, $\overline{A}$ \$/yr, but escalated at y/yr over a period T, discounted at x/yr.

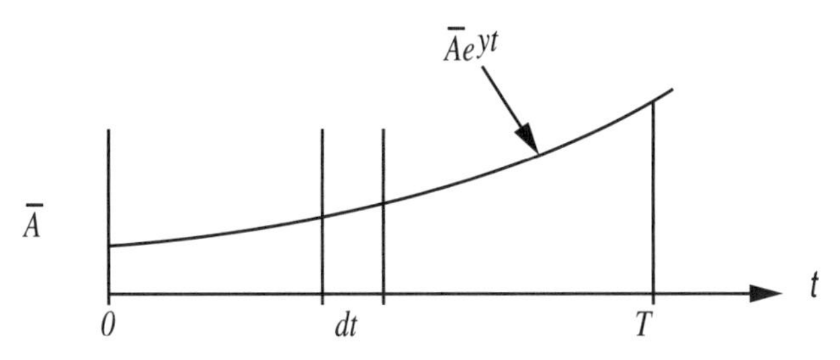

Levelizing to find the equivalent annual rate $\overline{A}_L$:

$$\int_o^T \overline{A}_L\, e^{-xt}\, dt = \int_o^T \overline{A}\, e^{yt}\, e^{-xt}\, dt \tag{A-1}$$

Then:

$$\frac{\overline{A}_L}{\overline{A}} = \frac{\int_o^T e^{-(x-y)t} dt}{\int_o^T e^{-xt}\, dt} = \left(\frac{x}{x-y}\right)\frac{1-e^{-(x-y)T}}{1-e^{-xT}} \tag{A-2}$$

Expand the exponentials as a Taylor series and retain terms through second order, which yields to first order:

$$\frac{\overline{A}_L}{\overline{A}} = \frac{1-\frac{(x-y)}{2}T+\ldots}{1-\frac{x}{2}T+\ldots} = 1+\frac{y}{2}T+\ldots \tag{A-3}$$

which is the multiplier used on today's O&M and fuel costs in Table 5.3.

Next, let's determine the capitalized cost of construction. The same diagram holds, where now $\overline{A}\,T$ = overnight cost of construction as of the start of construction (not plant startup, as preferred by some). We want the future worth, F, as of the date of completion (startup) namely:

$$F = e^{xT}\int_o^T \overline{A}\, e^{yt}\, e^{-xt}\, dt \tag{A-4}$$

Integration yields:

$$F = (\overline{A}\,T)\, e^{xT}\left[\frac{1-e^{yT}\, e^{-xT}}{(x-y)T}\right] \tag{A-5}$$

and series expansion gives to first order:

$$F = (\bar{A}\,T)\left[1 - \frac{(x-y)}{2}T\right][1 + x\,T]$$

or:

$$F = (\bar{A}\,T)\left[1 + \frac{x+y}{2}T\right] \approx (\bar{A}\,T)\left[1 + \frac{x+y}{2}\right]^{T} \tag{A-6}$$

which is the desired relation.

For the examples in Table 5.3, we can compare the "exact" exponential relations (Equations (A-2) and (A-5)) to their linearized versions (Equations (A-3) and (A-6)):

(1) Let $T = 30$ yrs, $y = 0.04$/yr, and $x = 0.09$/yr

Then exact $= 1.50$

Whereas $\left(1 + \frac{yT}{2}\right) = 1.6$; 6.7% high.

(2) Let $T = c = 5$ yrs, $y = 0.04$/yr, $x = 0.09$/yr

Then exact $= 1.39$

Whereas $\left[1 + \frac{x+y}{2}\right]^{c} = 1.37$; 1.4% low.

For present purposes, the above accuracy will be adequate, particularly in view of the uncertainty involved (see Section 5.6).

Appendix B

The Exponential Function e^x

This function is unavoidable in our area of interest because it appears in expressions for, among other things, continuously compounded interest, and exponential (geometric) growth or decay. The fundamental constant e is the base of the natural logarithm system. Fundamental properties include:

$$e = 2.71828...$$

$$\ln e = 1.0$$

$$\ln x = (2.30258) \log x$$

where $\log x$ is the base 10 system.

The exponential is important because it arises in the solution of many differential equations, the most common of which governs a rate of growth which is a constant fraction (or percentage) of the amount present:

$$dy/dt = ky$$

or:

$$\frac{dy/y}{dt} \equiv \frac{d \ln y}{dt} = k$$

with the solution:

$$y = e^{kt} + C,$$

where C is a constant to be determined by boundary conditions. Figure 5.13 illustrates the simplest case where $C = 0$.

The exponential function also appears as the limit of geometric compounding as the compounding interval shrinks to zero (continuous compounding):

$$e^x = \lim_{n \to \infty} \left[1 + \frac{x}{n}\right]^n$$

Important relations include:

The integral: $\displaystyle\int_{t_1}^{t_2} e^{xt}\, dt = \frac{1}{x}\left[e^{xt_2} - e^{xt_1}\right]$

The differential: $\displaystyle\frac{d}{dt} e^{xt} = x\, e^{xt}$

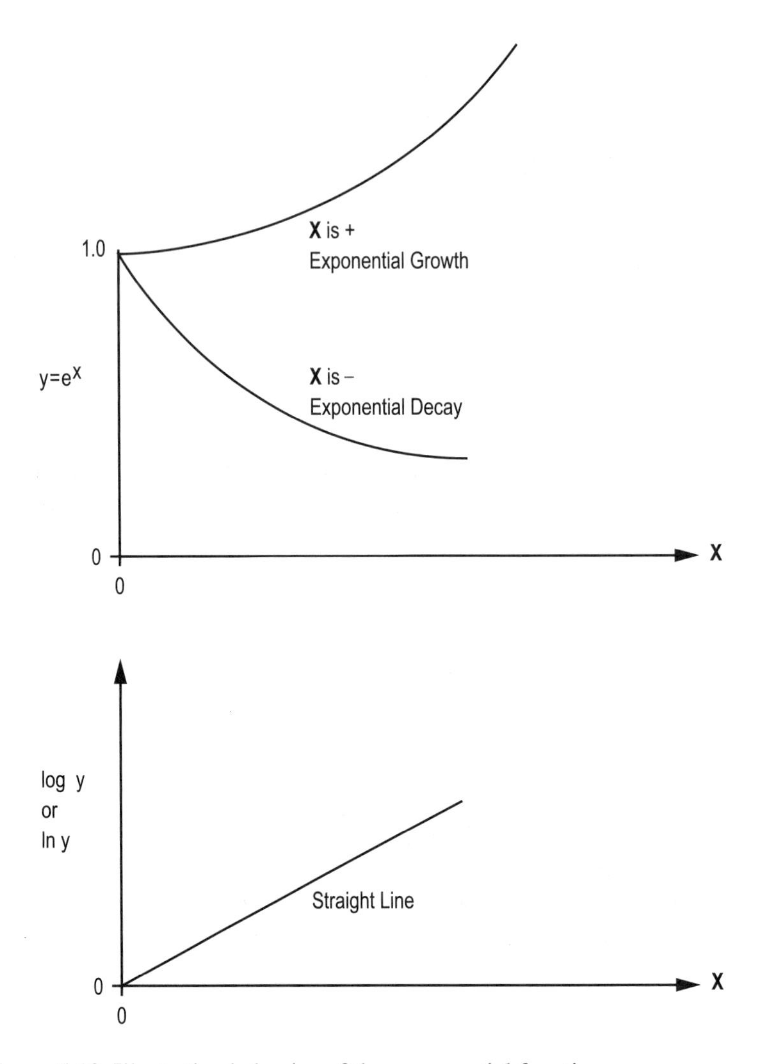

Figure 5.13. Illustrative behavior of the exponential function.

Pragmatically:

Just push the e^x button on your scientific calculator

or

Use the EXP(X) functions in most computer languages

Series expansions:

$$e^{xt} = 1 + xt + 1/2(xt)^2 + 1/6(xt)^3 + \ldots$$

$1 + xt$ is a common approximation:

i.e., linear interest

$$e^{xt} = \frac{1}{e^{-xt}} = \frac{1}{1 - xt + \frac{1}{2}(xt)^2}$$

Miscellaneous relations:

$$x^a = e^{a \ln x}$$

$$[e^x]^t = e^{xt}$$

Equivalence to geometric series:

$$\text{let } (1 + Z)^n = e^{xn}$$

$$\text{hence } x = \ln(1 + Z) = Z - 1/2\,Z^2 + 1/3\,Z^3 - \ldots$$

$$(\text{for} - 1 < Z \leq 1)$$

Alternative formulations:

let $e^{xt} = e^{t/T}$, where T is the period

or $1/T$ the mean lifetime

Doubling time or half-life:

$$\text{let } T = T_d = \frac{\ln 2}{x} = \frac{0.693}{x}$$

which gives rise to the rule of thumb:

time to double $\cong 70 \div x$, %/yr.
If one is dealing with a negative exponential, $e^{-t/T}$,
then $T_d = T_{1/2}$, the half-life.

The more mathematically experienced reader will recognize that discounting a cash flow history, *f(t)*, to obtain its present worth is equivalent to taking its Laplace Transform (Shafii-Mousavi, 1991), i.e.:

$$P = F(s) = £\{f(t)\} = \int_o^\infty e^{-st} f(t)dt$$

(Note that the conventional Laplace symbol *F(s)* is *not* the same as our future worth.)

For integrals between finite limits one can apply the relation:

$$\int_a^b e^{-st} f(t)dt = \left[e^{-as} - e^{-bs}\right] £\{f(t)\}$$

Laplace transforms are tabulated in many math handbooks and also in more specialized compilations (e.g., Oberhettinger and Bodii, 1973). In addition, tables of integrals list many entries for combinations of exponential and other functions (e.g., Gradshteyn and Ryzhik, 1980).

Example

Find the present worth of a periodic (e.g., seasonal) cash flow given by:

$f(t) = \hat{A} \sin^2 \pi t$, \$/yr for t in years, discounted at $i = 10\%$/yr continuously compounded. We have:

$$P = \hat{A} \int_0^{\infty} e^{-it} \sin^2 \pi\, t\, dt,$$

and with $\Theta = \pi t$:

$$P = \frac{\hat{A}}{\pi} \int_0^{\infty} e^{-\left(\frac{i}{\pi}\right)\Theta} \sin^2 \Theta\, d\,\Theta$$

Then, directly from a transform table (recognizing that i/π is the transform parameter s), one obtains:

$$P = \left(\frac{\hat{A}}{\pi}\right) \cdot \frac{2}{\left(\frac{i}{\pi}\right)\left[4 + \left(\frac{i}{\pi}\right)^2\right]} \cong 5.0\, \hat{A}, \$$$

Other interesting properties:

From the expression for the integral, one can easily show that:

> The amount accrued in the next doubling time is the same as all accumulated up to now since time immemorial—an apt illustration of the daunting prospect of exponential growth, say, in population, or consumption of energy and other natural resources.

(For an exposition on the subject, see Maor, 1994.)

References

Ahern, J.E. 1980. *The Exergy Method of Energy Systems Analysis*. New York: John Wiley & Sons.

Bailey, R., ed. 2000. *Earth Report 2000*. New York: McGraw-Hill.

Berger, J.J. 1997. *Charging Ahead: The Business of Renewable Energy and What it Means for America*. New York: Henry Holt & Co.

Commission of the European Communities. 1995. "Externalities of Energy, External Project, Volume 1 - Summary." *DGXII, Science, Research and Development, JOULE, Report No. EUR-16520 EN.*

Economic Research Associates. 1992. "Units Converted from Pounds to Metric Tons." In R.W. Larson, F. Vignota, and R. West, eds. 1992.

EPRI EL-2561. 1989. *Electric Generation Expansion Analysis System*, MIT Energy Lab., EGEAS Version 5 Capabilities Manual RPI-1529-2.

EPRI. 2000. *Technical Assessment Guide–Central Stations*. Report 1000578.

Frantz, C.C., and A.B. Cambel. 1981. "Net Energy Analysis of Space Power Satellites." *Energy*. 6: 485.

Giampietro, M., S. Ugliati, and D. Pimental. 1997. "Feasibility of Large-Scale Biofuel Production." *Bioscience*. 47(9): 587–600.

Gradshteyn, I.S., and I.M. Ryzhik. 1980. *Tables of Integrals, Series and Products*, Revised edition. New York: Academic Press.

Hamilton, B., G. Conzelmann,. Duy Thanh Bui, 2000. "Building Capacities for Comparative Energy Assessment: Expanding the Power Base." *IAEA Bulletin*, 42(2).

Homer, S., and R. Sylla. 1996. *A History of Interest Rates*, Third Edition Revised. New Brunswick, NJ: Rutgers University Press.

Intergovernmental Panel on Climate Change. 2000. *Special Report on Emission Scenarios*. Cambridge: Cambridge University Press.

Kotas, T.J. 1995. *The Exergy Method of Thermal Plant Analysis*. Boston, MA: Butterworths.

Kreith, F., and R.E. West. 1997. *CRC Handbook of Energy Efficiency*. Boca Raton: CRC Press.

Larson, R.W., F. Vignola, and R. West, eds. 1992. *Economics of Solar Energy Technologies*. Boulder: American Solar Energy Society.

Mack, R.S., H.S. Leff, S. Bell, S. Nestor, and H. McDuffie. 1980. *Industrial Energy Use Data Book*. Oak Ridge Associated Universities. New York: Garland STPM Press.

Maor, E. 1994. *e: The Story of a Number*. Princeton, NJ: Princeton University Press.

Meadows, D.H., D.L. Meadows, and J. Randers. 1992. *Beyond the Limits*. Post Mills, VT: Chelsea Green Publishing Company.

Meadows, D.H., D.L. Meadows, J. Randers, and W.W. Behrens. 1972. *The Limits to Growth*. New York: Universe Books.

Mishan, E.J. 1988. *Cost-Benefit Analysis*, Fourth Edition. New York: Routledge.

Moran, M.J. 1982. *Availability Analysis: A Guide to Efficient Energy Use*. Upper Saddle River, NJ: Prentice Hall.

Nakicenovic, N., A. Grubler, and A. McDonald, eds. 1998. *Global Energy Perspectives*. New York: Cambridge University Press.

NEA/OECD. 2001. "Externalities and Energy Policy: The Life Cycle Analysis Approach," Workshop proceedings, Paris, Nov. www.nea.fr.

Oberhettinger, F., and L. Bodii. 1973. *Tables of Laplace Transforms*. New York: Springer Verlag.

OECD/IEA. 1998a. *Mapping the Energy Future: Energy Modelling and Climate Change Policy*.

OECD/IEA. 1998b. *Projected Costs of Generating Electricity: Update 1998*.

OEDC/IEA. 2000. *Experience Curves of Energy Technology Policy*.

Office of Management and Budget. 1992. *Guidelines and Discount Rates for Benefit-Cost Analysis of Federal Programs*. Circular No. A-94 Revised. Washington, D.C.

Ottinger, R.L. 1993. "Incorporating Externalities—The Wave of the Future." In *Proceedings of the Expert Workshop on Life-Cycle Analysis of Energy Systems*, Paris, May 1992, OECD/IEA.

Ottinger, R.L., D. Wooley, N. Robinson, D. Hodas, S. Buchanan, P. Chernick, A. Krupnick, and U. Fritsche. 1991. *Environmental Costs of Electricity*. New York: Oceana Publications.

Phung, D.L. 1984. "Economics of Nuclear Power: Past Record, Present Trends and Future Prospects." In *Proceedings of the 46th American Power Conference*, Chicago.

Pimental, D. 1998. "Energy and Dollar Costs of Ethanol Production with Corn." *Hubbert Center Newsletter* #98/2. Golden: Colorado School of Mines.

Price, C. 1993. *Time, Discounting and Value*. Malden, MA: Blackwell.

Federal Nonnuclear Energy Research and Development Act of 1974. Public Law 93-577.

"The Science of Money." 1998. *Discover*, Special Issue. 19(10).

Shafii-Mousavi, M. 1991. *The Essentials of Laplace Transforms*. Piscataway, NJ: Research and Education Association.

SimEarth. 1993. Maxis, Inc. Downloadable free from sites like http://freeoldies.com.

Smith, G.W. 1987. *Engineering Economy: Analysis of Capital Expenditures*, Fourth Edition. Ames: Iowa State University Press.

Spreng, D.T. 1988. *Net Energy Analysis and the Energy Requirements of Energy Systems*. New York: Prager Publishing Co.

Starr, C. 1991. *Energy and Power*, a *Scientific American* Book. New York: W.H. Freeman and Co.

Stirling, A. 1992. "Regulating the Electricity Supply Industry by Valuing Environmental Effects." *Futures*, pp. 1024-47.

Stirling, A. 1997. "Limits to the Value of External Costs." *Energy Policy*, 24(5).

Toke, D. 1995. *The Low Cost Planet*. Boulder, CO: Pluto Press.

Weyant, J.P., ed. 1999. "The Costs of the Kyoto Protocol." 1999. *The Energy Journal*, IAEE. (Special Issue). (20): 1–390.

Wilson, R., M. Holland, A. Rabi, and M. Dreicer. 1999. "Comparative Risk Assessment of Energy Options: The Meaning of Results." *IAEA Bulletin*, 41(1).

Problems

5.1 a. A biomass plantation operator can invest $1,000 today to plant a new crop of trees, which, it is estimated, can be sold at a net gain (sales – harvesting costs) of $10,000 in 40 years.

Determine whether this is profitable if a rate of return of 7%/yr is desired.

b. The trees can be harvested early at reduced yield: the net profit is then 250T dollars per $1,000 initial investment, for T in years.

To the nearest whole year, at what T is the rate of return largest?

c. While the trees are growing, the farmer is paid $200/yr for carbon sequestration; but at the end of 40 years, the trees are, contrary to initial expectations, burned, releasing all of their carbon as CO_2.

At a discount rate of 5%/yr, calculate and comment on the time-zero present value to the farmer of carbon sequestration if he must return the $8,000 at the 40-year mark.

d. Briefly discuss any other internal costs and externalities which this simple analysis may have overlooked.

5.2 a. Determine whether it is advantageous to use compact fluorescent bulbs instead of conventional incandescent lights by comparing levelized annual expenditures, $/yr, over an appropriate time horizon, for bulbs having equivalent light output and the following characteristics:

	Compact Fluorescent	Incandescent
Power, W	15	60
Useful life, hrs	9,000	1,000
Today's cost, each, $	18	1

Other data:

Use rate: 1,000 hrs/yr

Cost of Electricity: 8¢/kwh (today, time-zero)

Interest (discount) rate: 7%/yr (continuous)

Inflation (escalation) rate: 4%/yr (continuous)

b. Economists find that consumers often behave as if their required rate of return is very high. Repeat the above comparison for a discount rate of 40%/yr.

c. List other major consumer sector examples of items where inexpensive first cost dominates purchasing decisions even though future continuing expenditures are actually high enough to justify behaving otherwise.

d. Briefly discuss any aspects not included in the above analysis which should be taken into account in a complete lifecycle analysis.

5.3 a. Recalculate the nuclear power plant generation cost in the example of Table 5.3, but for government ownership: zero taxes and 100% bond financing at the rate $r_b = 6\%/\text{yr}$.

b. For the commercially financed unit of Table 5.3, calculate the cost if the plant took 10 years to construct, as was the fate of some US projects in the 1970s.

c. Also redo the calculation for a plant capacity factor of 55%, which again was the early experience in some instances.

5.4 Assuming that both capital and O&M costs have the size scale exponent of 0.7 and a progress function of 85% for wind turbine powered electric generation:

a. Estimate the percentage decrease in the cost of electricity from such units for an increase in size from 200 kW to 2 MW.

b. Do likewise comparing the first and 1,000th unit of the same rating.

c. Discuss whether scale-up or mass production are more effective for cost reduction.

5.5 Show how the prescription of Table 5.3 for busbar cost, cents/kWhe, can be applied to energy storage devices. In particular clearly define "fuel" cost *f*, efficiency η, and capacity factor *L*.

5.6 a. Sketch the probability density function $P(x)$ and the cumulative PDF, $P \leq x$ for a variable which has an equal probability of falling in the interval $x_1 \leq x \leq x_2$. Normalize such that $P \leq x$ spans the interval [0,1].

b. Calculate the mean value of x (i.e., the center of gravity of the PDF)

c. Calculate the variance, σ^2 (i.e., the moment of inertia about the center of gravity).

d. Report $x \pm \sigma$, if x is a plant capacity factor ranging between 0.6 and 0.9.

5.7 Consider a NO_x removal system for a coal-fired power station that has a levelized-lifetime cost of

$$C = C_0 X^2, \text{ \$/ton}$$

where X = mol fraction of NO_x removed.

The corresponding benefit is linear in fraction removed.

$$B = B_0 X, \text{ \$/ton}$$

Calculate the economically preferred fraction removal if C_0 = \$12,000/ton and B_0 = \$8,000/ton.

5.8 In typical construction projects, the rate of expenditure starts and ends at a low level and peaks near the midpoint.

Consider a project having a total construction period of c years and an overnight cost as a function of time, t, in year zero dollars:

$$A_0 = \hat{A} \sin \frac{\pi t}{c} \text{ \$/yr}$$

where $\hat{A}$ is the peak rate.

Assume costs escalate at the rate y per year.

a. Find the future worth of the escalated total expenditure at time c at a discount rate x per year. Divide by the total overnight cost to obtain an expression analogous to the approximation in Table 5.3, namely

$$\left[1 + \left(\frac{x+y}{2}\right)\right]^c$$

b. For the numerical values of the parameters given in Table 5.3, compare your exact result to that given by the approximation.

5.9 Frantz and Cambel (1981) describe a net energy analysis for a space power satellite system in which solar cells in orbit generate electricity which is then microwave-beamed to earth. The net energy ratio (electrical output to primary thermal input) is predicted to be on the order of 2 to 5. Yet as we know, such stations are far from being economically viable. Discuss plausible reasons why energy and economic analysis might give such diametrically opposed results in this instance.

5.10 a. Using the approach outlined in Table 5.3, estimate the busbar cost of electricity generated by a CCGT power plant. It has the following characteristics:

Overnight Capital Cost: \$500/kWe

Construction time, $c = 3$ years

Thermodynamic efficiency, $\eta = 50\%$

Initial O&M cost $(O/K)_0 = \$27$/kWe yr

Initial Fuel Cost, $f_0 = \$2.60$/KSCF

and all other parameters are as listed in Table 5.3.

b. At what capacity factor, L, for the CCGT unit would its busbar cost of electricity equal that of the LWR of Table 5.3?

c. At what rate, y %/yr, would natural gas costs have to escalate for CCGT and nuclear busbar costs to be the same?

d. If a carbon tax of \$100/metric ton C were to be imposed, by how much would the busbar cost of the CCGT be increased? Natural gas contains 0.05 kg C per kWh_t of energy released in combustion.

e. Estimate the busbar cost if (as all too frequently done) one erroneously fails to consider future cost escalation (i.e., $y = 0$) in the analysis. How would this affect an option comparison study of, for example, CCGT units versus wind turbines?

f. Redo part (a) if (as is the case) the best available current CCGT units have a thermodynamic efficiency, η, of 60%.

5.11 An oil company contemplates investing 100 million dollars per year (in constant dollars) for five years in exploratory work to confirm the existence of new exploitable reserves. It will then face a potential delay until market conditions justify exploitation, which will require investing 200 million dollars for three years in production facilities, following which delivery of oil can start with a projected net revenue of 100 million dollars per year, essentially *ad infinitum*.

a. At a real discount rate of 5%/yr what is the longest delay tolerable between the start of exploitation and the start of oil delivery: i.e., how far in advance is it worth proving out reserves?

b. If the real discount rate were 10%/yr, what is the maximum time delay?

5.12 The original scenario for breeder reactor development in several different countries is roughly described as follows:

- Build a series of demonstration units of increasing rating: e.g., 50, 250, 1,250 MWe.
- Build a fleet of identical full size units, here 1,250 MWe; after (say) five have been completed, the fifth and subsequent units would be industrially cost competitive.

a. If the 50 MWe unit has a unit cost of $7,000/kWe, what is the unit cost of the last (fifth) full-size unit if the scale-size exponent is 0.7 and 80% learning is anticipated?

b. What is the total invested in pre-commercial units, and hence what is the total required subsidization (considering only capital costs): i.e., the dollars at or above that of a competitive unit?

5.13 Automobile engines are much cheaper than nuclear power plants, e.g., ~$20/kW for a 100 kW engine (engine only) versus nuclear at ~$1,400/kW (complete station) (as in Table 5.3).

Why doesn't your local utility just set up arrays of car engines in a large building to generate electricity?

a. Compare the useful life of a well-maintained engine, e.g., 220,000 miles at 50 mph, to the useful life of a central station nuke at 30 years.

b. Compare the annualized (i.e., per year) capital costs, engines versus nuke, at a carrying charge rate of 15% per year for the nuclear power plant (i.e., same as an annual "mortgage" rate).

Assume the auto engines are fully expensed when purchased and discarded for negligible salvage value at the end of useful life.

c. Compare fuel costs levelized over 30 years using the parameters in Table 5.3 and the following input for the gasoline engines:

Thermodynamic efficiency, $\eta = 25\%$

Cost of gasoline: $1/gal

Gallons per oil barrel = 42

Heat of combustion of gasoline is 90% of that for oil on a per-barrel basis

5.14 Inspired by the 1973 oil crisis, the US established a Strategic Petroleum Reserve (SPR), which now amounts to approximately 500 million barrels stored in underground reservoirs in Gulf Coast locations. This is roughly 90 days' worth of our foreign oil imports.

Discuss how you would carry out an economic cost-benefit analysis of the worth of the SPR, addressing key features such as the discount rate and time horizon to be used, etc.

Energy Systems and Sustainability Metrics 6

We have met the enemy and he is us!
—from "Pogo" by Walt Kelly

6.1 Introduction and Historical Notes

Energy is integrated into every aspect of human activity. In fact, the energy sector has played a vital role in the industrial and economic development of all the major Organisation for Economic Cooperation and Development (OECD) nations from at least the start of the Industrial Revolution. Along with growing industrialization have come major improvements in the standard of living, urbanization, and human health, but also appreciable acceleration in population growth. There were 0.6 billion people in 1700, double this number by 1850, and double again by 1950. World population today (2004) exceeds 6 billion and is *still increasing* by about 80 million per year.

Initially, industrial activities had minimal impact on the global ecosystem. Negative impacts were first discerned at the local level. Remedial action was taken either voluntarily or in the public interest through legislation. Today concerns about impacts extend to the global ecosystem. The major global concerns now include reduction in habitat, biodiversity losses, soil loss, damage to the ozone layer, and global climate change. Emerging global concerns relate to poverty and human health, societal disturbances driven by economic collapses or major inequities, availability of water of acceptable quality, and global quality of life. The present concerns about global climate change are most closely linked to the energy sector because of our increasing appetites for energy and for energy-intensive goods and services, as well as the present worldwide abundance and accessibility of affordable fossil resources to satisfy these appetites.

Early development was mostly driven by economics—where a resource was expensive, attention was paid to improving utilization efficiency or finding suitable substitutes (e.g., coal for wood). Concern about adverse impacts, such as industrial accidents, occupational safety, and environmental pollution, was rather limited. When problems occurred, they were addressed by incident-specific or site-specific remedial action. Industrial activity gradually improved in overall safety and environmental performance through trial and error—"learning by accident." When an impact became of repeated concern, designers would incorporate avoidance into their practice, either voluntarily or in response to social and political pressures.

This chapter will focus on those aspects of sustainability that pertain to the environmentally responsible use of energy. However, environmental stewardship applies to a diverse collection of interlinked human interactions with the global ecosystem, including food production, creation of habitat, movement of goods and people, and decontamination of wastes. All of these activities consume energy, often in appreciable quantities, but they also expend or degrade other natural resources, such as water, biomass, mineral ores, clean air, open space, and recreational areas. Thus, if we are to move in the direction of a sustainable world, we must learn how to apply the precepts and practicalities of sustainable development to all of these issues on a global scale.

The following sections present some of the basic ideas and concerns about sustainability, describe some of the analytical tools useful in assessing and comparing the sustainability of alternative energy technologies or policies, discuss indicators of sustainability, and give examples of some sustainable development principles. Chapter 21 will follow up on this introduction and examine more complex systems and describe some of the systems and decision analytical methods used to identify sounder pathways to a sustainable future.

What do we mean by "sustainable development"? As discussed in Chapter 1, Malthus (1798) was concerned about exponential population growth and the ability of his society to increase food production only linearly with time. He noted that population growth ceases when resources to survive are depleted. By not anticipating technology advances in agriculture, Malthus was overly pessimistic about the timing of such a disaster. Hardin (1968) wrote about the "Tragedy of the Commons," where free or undervalued common resources rapidly become overused and exhausted. In the same year, Ehrlich (1968) identified overpopulation as the key problem of the day and argued that further expansion at then-anticipated rates would soon create demands for resources that exceed the finite "carrying capacity of the Earth." More recently, Forrester (Meadows et al., 1972) advanced the concept of "limits to growth"; they concluded that then-current policies generating population and growth would lead to an overshoot of earth's carrying capacity and a major decline in both population and the economy sometime in the next 100 years. They noted that it is possible to alter growth trends to reach a state of global equilibrium designed "so that the basic material needs of each person on earth are satisfied and each person has an equal opportunity to realize his individual human potential." Finally, they note that the sooner people start striving for a sustainable world, the greater will be their chance of success. Twenty years later, in *Beyond the Limits* (1992), Meadows and Meadows found human activity has already overshot key aspects of the earth's carrying capacity and reaffirmed their belief that a sustainable society could be created.[1] Ostrom et al. (1999) updated Hardin's analysis by identifying barriers and catalysts of sustainable progress. They emphasized the importance of international cooperation in managing pervasive resources like water and large marine ecosystems.

Will the same potential for disaster or overshoot be true for climate change as well? With increasing concerns about current and future greenhouse gas (GHG) emissions to the atmosphere and their likely linkage to significant, but only partially understood, changes in global and regional climate, the international community is looking for effective ways to reduce GHG emissions without significantly compromising other factors that are important to our quality of life. These factors include stable social and economic structures, adequate food and housing, health, education, a clean environment, access to manufactured goods at reasonable cost, and convenient transportation. These factors cannot be ignored in a focus on GHG abatement because energy choices and

---

1 See Figures 21.1 and 21.3 and the further discussion of these results in Chapter 21.

their emissions are linked to population growth, economic well-being, infrastructure, institutional realities, human health, political stability, biodiversity, and other issues of environmental preservation.

As society begins to think about *sustainability* in both space (*intra-generational equity*) and time (*inter-generational equity*), a wide variety of possibilities emerge that reflect differences in cultural, societal, and individual values. Section 1.1 defined *sustainable practices* as those that responsibly utilize, conserve, recycle, or substitute raw materials, products, wastes, and technologies to preserve and extend economic progress and protect the environment now and hereafter. This definition emphasizes the need to harmonize economic and environmental stewardship. The definition also emphasizes the plausibility of sustainable progress by specifying means to make sustainable practices practical. Here are several other equally good and more widely recognized definitions:

- Sustainable development means the ability of humanity to ensure that it meets the needs of the present without compromising the ability of future generations to meet their own needs. (Brundtland, 1987)
- Biophysical sustainability means maintaining or improving the integrity of the life-support system of earth. (Fuwa, 1995)
- Sustainability means preservation of productive capacity for the foreseeable future. (Solow, 1992)
- Sustainability includes a participatory process that creates and pursues a vision of community that makes prudent use of all its resources—natural, human, human-created, social, cultural, scientific, etc. (Viederman, 1997)

Daly (1990) has proposed three operational principles of sustainable development, as follows.

- *For a renewable resource*: the sustainable rate of use can be no greater than the rate of generation
- *For a nonrenewable resource*: the sustainable rate of use can be no greater than the rate at which a renewable resource, used sustainably, can be substituted for it
- *For a pollutant*: the sustainable rate of emission can be no greater than the rate at which the pollutant can be recycled, absorbed, or rendered harmless by the environment

There are many more definitions of sustainability and sustainable development that span the gamut from total commitment to environmental preservation to a philosophy that everything, now and in the future, is on its own for survival. Ultimately, each of us will find some balance of concern between our immediate and selfish wants and our concern about the common good. In this chapter, we will identify those things that we wish to sustain and those factors that seem to be hindering sustainable living.

Some things we might want to sustain	Some trends hindering sustainable living
Our standard of living	Widespread poverty
Our health	Population growth
Our food and water supply	Unnecessary or excessive consumption
The environment (climate, water quality and availability, diversity of species, natural and recreational spaces, etc.)	Social inequity (including lack of health care, education, and jobs for the poorest, widening gaps between the rich and poor)
Personal freedom	Political self interests and short-term focus
International stability	Irresponsible industrialization
A healthy economy	Loss of habitat and species
Opportunities to improve status (individually, as a community, or as a nation)	Inadequate institutional systems to manage change
Global communications and mobility	Mal-distribution of resources, depletion

When we focus on the world from the perspective of energy supply and use, we need to keep all these issues in mind. Some will be more applicable in certain circumstances than others. However, in many countries, the concern about future energy use that receives most attention from politicians and the media is the potential of present and future fossil fuel utilization to cause changes in global climate that could, in turn, give rise to adverse regional impacts including sea level rise, shifts in rainfall patterns, modification of agricultural and natural crops and species, and changes in ocean circulation and storm patterns. If these shifts materialize, there will be economic and geopolitical winners and losers.

6.2 Energy from a Systems Perspective

Present rates of GHG emissions are causing a steady rise in GHG concentrations in the atmosphere. Because the primary GHG of concern is CO_2, we will use CO_2 as a surrogate for purposes of discussion.

About 85% of world commercial energy use is derived from carbon-based fossil fuels. If we look to the future, we see population growth and industrialization increasing energy demand in key countries of Asia and South America. Figure 6.1 shows one of several similar published opinions (Edmonds, 1995) on how world energy use might increase over the next century if technology improves as it has in the past without any intervention to reduce energy demand or fossil-fuel dependence (business as usual—BAU).

Today, the US produces about 20% of the total global CO_2 emissions and has a per capita energy usage about twice that of Europe and Japan and an order of magnitude higher than countries in early stages of industrial development. Figure 6.1 projects an increase of about 50% in total US energy use over the next century, while total world use will grow about fourfold. Although such projections are totally speculative, they give some idea of how GHG emissions might increase if the current energy supply mix continues through the 21st century. Note that Figure 6.1 only includes commercially produced energy use and excludes the gathered biomass fuels that can constitute over 50% of energy use in some of the poorest developing countries. However, combustion of biomass fuel is usually considered to be GHG-neutral because the biomass grew by taking CO_2 from the atmosphere. Replacement vegetation will also remove CO_2.

Figure 6.2 shows the distribution of commercial energy use by energy source and region. The developing world does not show up as a significant consumer of commercial energy because the extreme poverty of many makes access to commercial energy impossible. The World Bank (2001) estimates that 1.2 billion people live below the extreme poverty level of $1 purchasing power parity (PPP) per day and that 2.8 billion are below the $2 PPP per day level. Although these numbers have been improving slightly, primarily due to improvements in China, for many the poverty problem persists or is worsening. In Africa, use of biomass for fuel has led to defoliation of vast areas and is a major factor in desertification.

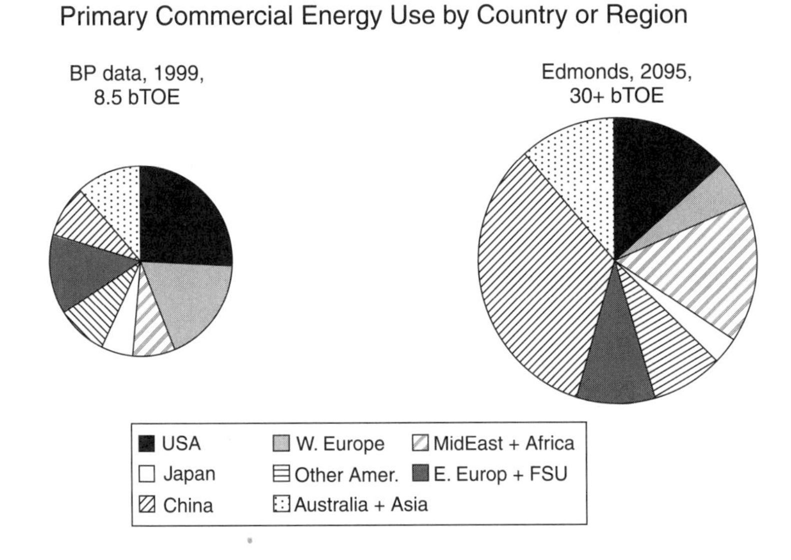

Figure 6.1. Future Shifts in World Commercial Energy Use—Business as usual. Source: Adapted from Edmonds (1995).

In extreme poverty, malnutrition is prevalent and health indicators show elevated infant-mortality rates and incidence of disease. Part of the solution to extreme poverty problems is related to energy. Studies have shown that improvements in standard of living of the poorest are linked to availability of commercial energy, which facilitates economic development—job creation and transportation for products to markets—along with education and provision of at least minimal health services.

The United Nations has defined a Human Development Index (HDI) (UNDP, 2001) which aggregates a variety of indicators relating to health, education, and economic status. There is a strong correlation between improvements in the HDI and electricity use as shown in Figure 6.3. The HDI improves rapidly with electricity use up to a level of 4,000 kWh per capita per year, where the rapid slope decreases and eventually plateaus. It is evident that additional energy will be required to make progress with the elimination of poverty. For example, if 3 billion people each consume an additional 4,000 kWh per year, world primary energy consumption would increase by about 100 EJ per year or 2.4 billion tonnes oil equivalent (TOE)[2] (40 EJ electric at a typical heat to work conversion efficiency of around 40%)—about a 25% increase in today's world energy consumption.

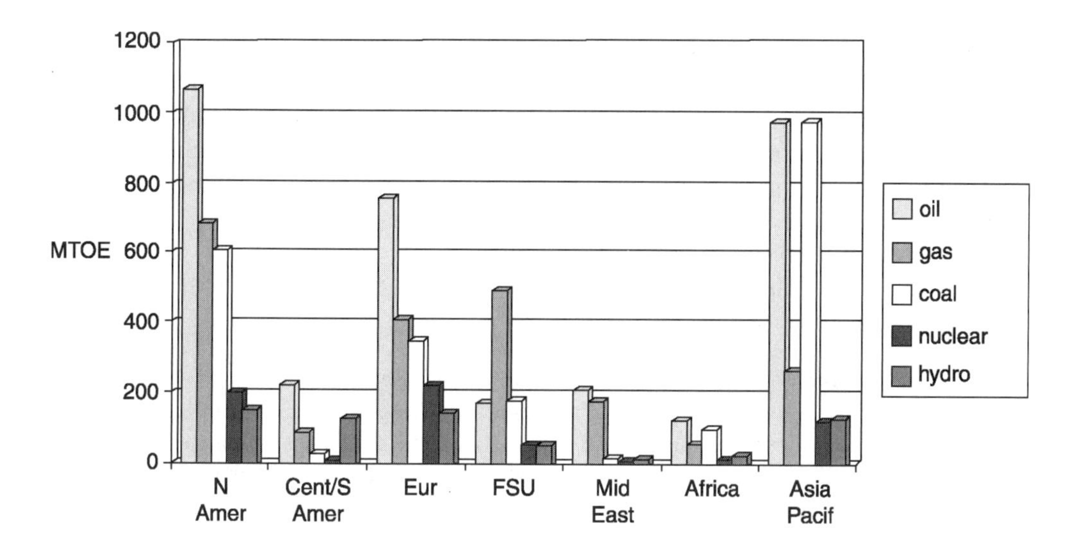

Figure 6.2. Commercial energy use, 2000, by fuel and region, MTOE (BP, 2001).

2 1 TOE equals 40 MMBtus, 7.35 bbls of oil, 11.8 MWh, 4.26×10^{10} joules

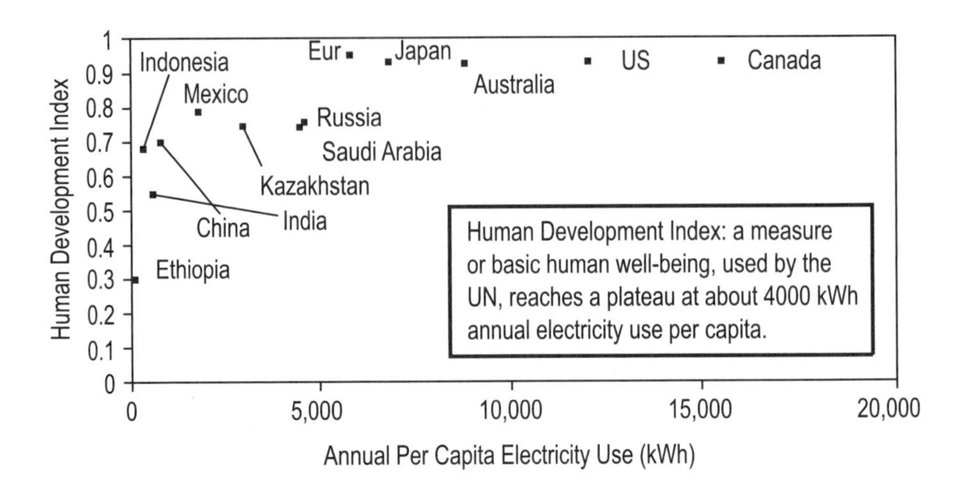

Figure 6.3. Electricity Use and HDI (UNDP, 2001).

Today, GHG concentrations in the atmosphere are steadily rising and are expected to grow at an increasing rate as more fossil energy is consumed. The first major effort to examine the potential extent and impacts of these changing concentrations on global climate was started in 1988 when the Intergovernmental Panel on Climate Change (IPCC) was formed. Charged with assessing available scientific information and formulating realistic response strategies for management of the climate change issue, the IPCC (1990) published its first climate change assessment report (*Working Group I*) and projected a range of scenarios resulting from increases in global temperature by 2100 from small increases up to about 5°C.

The Framework Convention on Climate Change (FCCC), which was ratified by the UN members in 1994, included a commitment to:

> stabilization of greenhouse gas concentrations in the atmosphere at a level that would prevent dangerous anthropogenic interference with the climate system. Such a level should be achieved within a time-frame sufficient to allow ecosystems to adapt naturally to climate change, to ensure that food production is not threatened and to enable economic development to proceed in a sustainable manner.

At present, the onset and rate of significant climate change resulting from human (anthropogenic) emissions of GHGs and the severity of potential effects of this climate change are still uncertain because the variability observed to date is still within the large "noise" of natural variability and the pace and carbon intensity of future economic development are uncertain (Chapter 4). We do not yet have conclusive evidence as to whether we are on a gradual or rapid path of climate variability—and we do not know much about how the consequences of climate change will be distributed geographically. Our emissions are rising, and CO_2 emitted today has a residence time in the atmosphere of about a century. The IPCC Third Assessment Report (2001) projects temperature rises of 1.4–5.8°C by 2100. If the next decade or so shows that our climate intervention is serious, we will not have the luxury of rapid corrective measures because the time scales for technological and institutional transformation in the energy sector are measured in years and decades, not weeks or even months (Section 1.6).

The ability to curtail GHG emissions is not distributed evenly over nations and time. The industrialized nations have been the major emitters in the past. The developing nations are projected to become major CO_2 emitters within two to three decades. Developing nations maintain that their development should not be hampered by GHG constraints. They argue that the developed world should take the lead in reducing emissions because their emissions predominate at present and because these countries are in a more favorable economic position to finance change. However, if it turns out that the earth is on a rapid path to climate changes, *all* nations need to be prepared to respond with effective and publicly acceptable technology and policy strategies. Otherwise, our limited resources will be expended on coping with the consequences of climate change.

GHG emissions are now affecting global climate. We are, in a way, conducting a global experiment without an emergency response plan. The Framework Convention on Climate Change (FCCC) recognizes the need to get started, but the world community has been unable to reach a consensus. Major points of contention are how, when, and by how much to change course, and who should play major and auxiliary roles in doing so at various times. Many believe time is running out.

Macroeconomic analysis has been widely used in the energy sector, although its aggregate nature hides many details of sectoral and developmental differences. One representation, popularized by Kaya (1991), which has been used to relate CO_2 emissions to the rates of change in key driving factors, is:

$$[CO_2] = \text{Population} \times \frac{\text{GDP}}{\text{population}} \times \frac{\text{Energy}}{\text{GDP}} \times \frac{CO_2}{\text{Energy}} - \text{sequestration}, \qquad (6\text{-}1)$$

where $[CO_2]$ is the net amount of carbon dioxide emitted to the atmosphere per year, sequestration represents the amount of emitted CO_2 prevented from entering the atmosphere owing to its capture and sequestration in a suitably stable sink (see below and Chapter 7), [GDP/population] represents the standard of living, [Energy/GDP] represents energy intensity of the economy, and [CO_2/energy] represents carbon intensity of the primary energy supply. Equation (6-1) is similar to Equation (1-24). (Refer to Section 1.7 for a discussion of simple extensions as well as physical and mathematical limitations of this equation.)

Table 6.1 shows the average annual percentage rate of change in these factors over a time period from 1980 to 1999. Sequestration, although not yet a major factor, is receiving increasing interest as we continue to study carbon storage in the biosphere and the potential for capturing and sequestering carbon from fossil-fuel use. Although the initial agreement under the FCCC involved a commitment to reduce emissions back to

1990 levels by 2000, it is clear that this is not happening. Those reductions that have occurred (Eastern Europe) have had more to do with economic collapse than with new policies, better technologies, or collaborative efforts to reduce emissions. It is ironic that the countries with the highest fraction of non-fossil sources in their present energy supply mix (Switzerland, Norway, France, etc.) are facing future increases in emissions if they phase out nuclear generating capacity. Moreover, major energy-consuming countries such as the US have until recently continued to devalue the role of two major low-carbon-intensity sources—hydropower and nuclear energy—in their energy supply mix (OSTP, 1997). For reasons discussed in Chapters 1 and 8, from about 1998 onward, nuclear energy is regaining substantial investor and political interest for central station electricity generation in the US.

All indications point to large increases in GHG emissions over the next century unless there are major decreases in one or more of the driving forces in Equation (6-1). Agricultural sector practices also strongly influence CO_2 and CH_4 emissions. Methane emissions are also important because methane is a more potent GHG than carbon dioxide by a factor of at least 20 over a 100 year period (see Table 4.8). This is usually outside of the energy focus unless decarbonization by switching to natural gas leads to a significant increase in CH_4 emissions through leakage. Modern gas installations generally have only small leakage losses, though a few percent loss of methane can offset any decarbonization advantage from switching to methane from coal.

Table 6.1. Kaya Equation Factors

	Average Annual Percent Change 1980–1999				
Region	**Population**	**Standard of Living**	**Energy Intensity**	**Carbon Intensity**	**Carbon Emissions**
Africa	2.54%	-0.58%	0.82%	-0.01%	2.77%
Australia	1.36%	1.98%	-0.37%	0.00%	2.98%
Brazil	1.61%	0.76%	1.83%	-0.80%	3.43%
China	1.37%	8.54%	-5.22%	-0.26%	4.00%
East Asia	1.78%	5.00%	0.92%	-0.70%	7.10%
Eastern Europe	0.44%	-1.91%	-0.14%	-0.61%	-2.21%
India	2.04%	3.54%	0.27%	0.03%	5.97%
Japan	0.41%	2.62%	-0.57%	-0.96%	1.47%
Middle East	2.98%	0.04%	2.45%	-1.14%	4.34%
OECD	0.68%	1.73%	-0.88%	-0.58%	0.94%
OECD Europe	0.53%	1.74%	-1.00%	-1.06%	0.18%
US	0.96%	2.15%	-1.64%	-0.21%	1.23%
World	1.60%	1.28%	-1.12%	-0.45%	1.30%

(See http://sequestration.mit.edu/pdf/understand_sequestration.pdf)

If we consider the possibilities for reducing CO_2 emissions resulting from fossil fuel usage, we find the general options presented earlier in Figure 4.18: reducing consumption, improving efficiency of use, decarbonization of the energy supply, and separating and storing GHG emissions from energy use or removal (e.g., by new biomass growth) directly from the atmosphere. Because of the magnitude of the response that might be required, we will probably need to use all the opportunities that exist. Technology is important in each of these options, and there are many different potential technology improvements and innovations distributed throughout the various energy-using sectors of the world economy. Emission-reduction opportunities come at different cost levels—some will actually create profit, some may have little cost impact, and others will entail higher energy costs.

Some developed countries may view their relatively lower cost of energy compared to other nations as a basic entitlement. Moreover, in the short run, an energy price advantage may become a disincentive for energy conservation by both industry and the general public. Countries with higher energy prices already tend to use energy more efficiently and conservatively, often by some sacrifice in living standards (e.g., the size and comfort features of passenger automobiles) but not necessarily in their satisfaction with life. However, these countries will have to move to still more efficient and/or less GHG-intensive technologies to reduce future GHG emissions if a crisis looms.

Energy efficiency improvements are usually the most cost-effective ways to achieve GHG and other emissions reductions—but inexpensive energy dissuades interest in making the needed incremental investments in efficient technologies. (More details appear in Chapters 16-20.) To facilitate energy-efficient development of the non-OECD world, it is likely that economic transfers will be needed, so that they may have access to clean and affordable energy to fuel economic development and improve their standard of living.

Renewable technologies are receiving much attention as a potential solution. Energy recovery from residual biomass reduces GHG emissions while helping with waste disposal. Major advances in wind and solar technologies are happening. But these sources are variable in both time and space and have low energy availability per unit area of ground surface. Solar technology also tends to be considerably more expensive at present, but it is attractive in remote areas where an electrical transmission infrastructure is not available. These technologies are well suited to providing power to rural areas and have niche uses in areas with high population density for supplementing base-load power systems. Renewable energy sources are evolving rapidly and will continue to increase in use, although they are not well matched to present urban and industrial energy-consumption use patterns. Solar systems mesh well with air-cooling needs in hot climates and with water heating. Emerging heat mining technologies show promise for increased use of geothermal energy. (See Chapters 9 through 15 for details.)

Considerable reductions in energy demand can also be achieved through the use of passive solar techniques coupled with building designs. Integration of energy systems and co-generation in urban buildings, use of ground source geothermal heat pumps in less dense population areas, smart windows, daylighting, insulation, shading, and control of the thermal mass of walls all provide opportunities for energy savings in cooling, heating, and lighting. Because of the diverse participants in the building sector, integrated design of buildings to achieve energy efficiency is not common. As developing countries industrialize, new devices (e.g., air conditioners), may be used in old buildings in inefficient ways. The economic drive for larger living space will probably increase energy consumption unless efficient design is made a priority and energy is priced appropriately. Improvements come slowly through changes in building codes and standards (see Chapter 20).

The transportation sector also faces major challenges because of the emissions from distributed fleets of vehicles. Air pollution problems can be addressed through emission regulations. Some benefit in CO_2 emission is possible from switching to less carbon-intensive fuels (e.g., from petroleum-based fuels to natural gas or hydrogen), but this would require a major infrastructure change and improvements in onboard generation and/or storage of H_2. A variety of alternate fuels are being examined. Fuel cells offer energy conversion efficiency improvements, but with the present state of technology, hydrogen fuel for fuel cells needs to be generated by "reforming" natural gas, methanol, or other fossil fuels. With this technology, the CO_2 is emitted from the reformer, rather than the fuel cell. There still may be some modest improvement in overall system efficiency, but no dramatic reduction in CO_2.

Pure electric vehicles are options that are especially attractive in urban areas with localized air pollution problems. However, in terms of CO_2 emissions, unless a non-fossil carbon power source is available, the overall system efficiency and emissions reductions improve only slightly. Both fuel cell and electric cars share an energy storage problem—hydrogen storage and electricity storage are still highly inefficient per unit weight of the storage medium. The biggest near-term opportunity is to improve fuel economy through use of advanced lightweight materials and smaller cars that still satisfy user demands and meet safety criteria. While consumers in the US like big cars and SUVs for reasons of perceived safety and prestige, future consumers may have to make some shift in the balance between personal wants and societal needs. As major developing countries, especially China and India, motorize, impacts from the global transportation sector will increase rapidly. Hopefully, the demand for affordable cars will result in a preference for smaller cars, although cost constraints may hinder the incorporation of more expensive advanced technology (see Chapter 18).

In high population density areas, the need will continue for bulk electric power supplies. Two possibilities for significant reductions in carbon emissions from bulk power generation are nuclear power (Chapter 8) and fossil fuel power with capture and sequestration of the carbon emissions (See Chapters 4 and 7). Both of these options have their own baggage, and whether the nuclear industry can solve the waste management, proliferation, and other safety concerns of the public will be keys to whether a nuclear option will continue. Likewise, reduced costs, assurance of storage reservoir integrity, and public acceptance hold the keys to the broad-based penetration of carbon capture and sequestration (CCS). To continue the use of fossil fuels for power generation, while combating CO_2 emissions through direct CCS, would require additional energy expenditures. While the purist might prefer to switch to a non-fossil source, the capture-and-sequestration option may be needed as a transitional strategy.

In the long term, it is likely that a creative, technologically sophisticated society will develop new bulk energy sources that are still dreams today—fusion power, heat mining from deep earth formations, solar collectors in synchronous earth orbit, artificial or direct photosynthesis of premium fuels and chemicals, or some other, not yet discovered, energy source. But national (OSTP, 1997) and international energy R&D investments are not sufficiently aggressive to produce economically viable alternatives to low-cost fossil fuels, at least in the near term. As in the past, the technological barriers we face today are likely to become tomorrow's opportunities. To capitalize on these opportunities, we need to continue and expand investments in R&D because the climate change problem is likely to increase.

6.3 Systems Analysis Approaches

At the start of the Industrial Revolution, technologists who were developing new products and processes focused primarily on making them serve their intended purpose. Little attention was paid to the side effects associated with production, use, and disposal. When serious accidents occurred or health problems developed, technologists made modifications or additions to their designs. To pass on the knowledge learned by experiences of this sort, industries, professional groups, and governments began to establish sets of codes and standards (representing recommended practices) or regulations. Today, each new industrial activity begins with a search for all the applicable codes, standards, regulations, and insurance/liability requirements, so that a design baseline can be established.

This approach usually assures some minimum level of human and environmental safety, but it does not represent the current best practices because of the time lag in establishing the standards and regulations and because of compromises made so that less-advanced members of an industry are not placed at a competitive disadvantage. Further, this mode of establishing "good practice" is usually responsive to problems on an individual basis—and the responses may be uneven with regard to the relative risk of the individual problems. Sometimes solutions to a particular problem may create other problems that will remain unidentified until later. For example, many years ago, ammonia and propane were commonly used as refrigerants. Both are flammable, and ammonia is toxic to humans. When non-flammable chlorofluorocarbon (CFC) refrigerants were developed, a major improvement in refrigerant safety was achieved. However, years later it became obvious that leakage of CFCs was damaging to the earth's upper ozone layer. The solution had generated another serious, unanticipated impact. Through international action (the Montreal Protocol, see Chapters 1 and 4), CFCs are now banned, and replacement refrigerants are being used.

We have been using fossil fuels for many years. To combat flammability hazards, the first half of the 20th century witnessed major governmental and industry progress in establishing regulations and standards for fuel properties and for fuel storage, transport, and utilization equipment. The era from around 1950 to the present has likewise seen a much more widespread recognition by governmental bodies, industry, and the general public that emissions from careless combustion of fossil and other fuels can have a wide range of adverse health and environmental impacts (Chapter 4). In response, emissions-control regulations encouraged the development of cleaner fuels, combustion processes, and after-treatment technologies to minimize harmful emissions. With time, these regulations have been getting more stringent, as the effects of the emissions have become better understood and as societal standards have risen in the developed world. Developing countries are following the developed countries' learning curve with the hope that they can "leap-frog" environmental protection by not repeating OECD mistakes. Cleaner technologies are sometimes more efficient than old technologies and can be used without significant cost penalty. Where added cost is involved, some outside investment may be needed to encourage developing countries to choose a cleaner path of development.

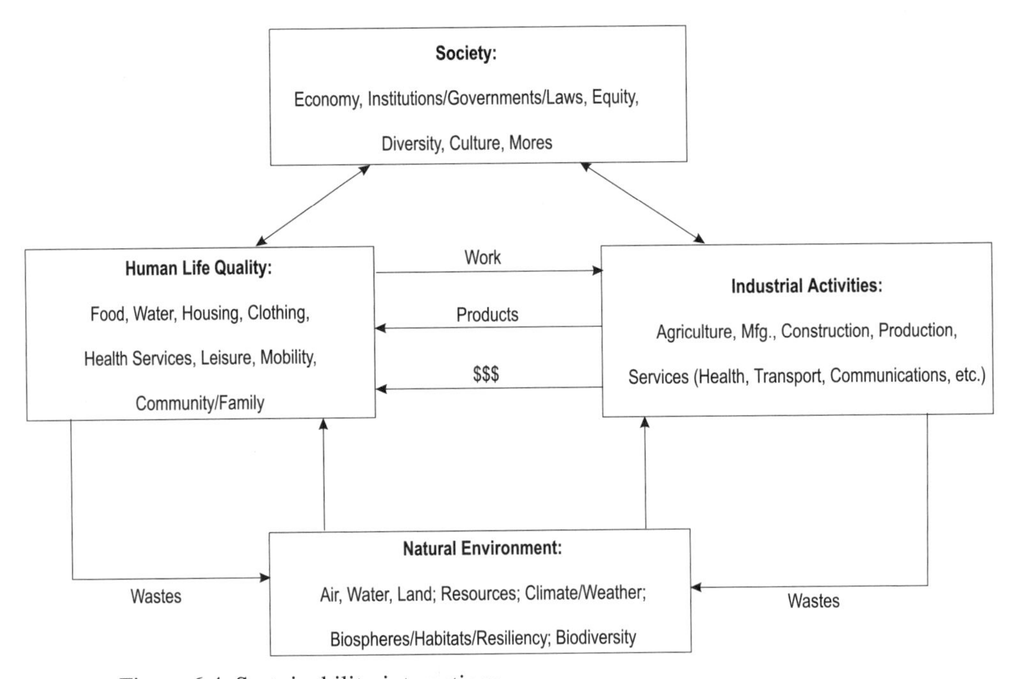

Figure 6.4. Sustainability interactions.

Finally, concerns are escalating over the connection between GHG emissions from fossil fuel combustion and changes in climate. Historically, these suspect problematic emissions have largely been produced by the industrialization of developed countries. In the future, countries currently categorized as developing will emit a larger proportion of anthropogenic GHGs (Figure 6.1).

Who will pay the costs for major reductions in GHGs and other adverse environmental impacts of energy is a question that will be explored further in Chapter 21. Figure 6.4 shows some of the complex system interactions that are important to recognize in moving toward sustainable development.

The systems view of even the GHG aspects of energy use is complex, and the interconnections among components of the system are not easily understood. The reasons include the number and complexity of these components (which are discussed in the example given in section 21.3.3) and the diversity of length and time scales that impact the dynamics of the GHG system. Progress occurs by describing these components with their own models and then devising means to link these sub-system modules. The eventual aims are to illuminate component behavior, not in isolation but under the collective influence of other system components, and to reliably simulate the behavior of the total system. The resulting integrated platform may (or may not) provide a basis for credible forecasting of the GHG system behavior in response to various inputs and constraints.

More generally, larger system models facilitate comparisons among future technology and policy options on a consistent basis and for a broad range of input variables under various regulatory and other operational constraints. Economists and business planners often use such models to assist in finding the most cost-effective pathways to manufacture products or provide services. However, it is difficult to assign monetary figures to environmental values, such as clean air, clean water, pristine land, quiet, and so forth, within a normal market framework. Most of these values are public goods that, until recently, have been largely overlooked in making predictions about consumer behavior. However, these issues are now commanding attention from resource economists and industrial and governmental sectors (NRC, 1999b).

6.3.1 Lifecycle analysis

Lifecycle analysis (LCA) provides a well-characterized methodology (Consoli et al., 1993, and Fava et al., 1994) that attempts to inventory all the impacts associated with each and every stage of a process or a product, from cradle to grave (i.e., from raw materials through materials processing, manufacture, distribution, use, repair and maintenance, and final disposition or recycling). LCA's strengths are rooted in the basic engineering principles of materials and energy balances and in the methodologies used for process analysis. LCA helps avoid the pitfall of a too narrow view of system attributes during environmental and economic assessments. For example, it is not atypical to need an evaluation of a specific technology such as a zero-emissions vehicle. An LCA approach recognizes that although the vehicle itself can well approach the ideal of an emission-free energy source, emissions are still produced during generation of the electricity to power that vehicle and in the manufacture of the energy production plant and the vehicle itself. By using LCA, the impacts of the entire system of technologies associated with one particular activity can be compared on a common basis with those of an alternative activity. Further, this technique allows environmental impacts to be evaluated quantitatively along with material flows and costs. (To see LCA in action, the reader may wish to study Weiss et al.'s (2000) assessment of some 15 different propulsion system–fuel combinations for mid-size passenger cars in the year 2020. Portions of this study are presented in Chapter 18.)

After the boundaries of the system have been defined, the next stage of an LCA is usually a complete inventory of all material and energy usage and all byproducts and wastes associated with each relevant stage of the lifecycle activity. Costs can be tracked in a similar manner. Next, the societal and environmental impacts of each of the stages are quantified in an appropriate manner—quantities of each waste stream emitted, health effects from toxic emissions, loss of land use, ecological damage, or noise, etc. This allows one to identify the most significant impacts and plan improvements or mitigation strategies. By comparing LCAs from alternative technologies, one can also make choices as to the "preferred" option(s), where it is recognized that preferences may differ according to preset parameters of the study, as well as the values and biases of the analyst.

LCAs can also be used to track inventories of hazardous materials and for many other purposes. For example, McDonald's was interested in whether it was better for the environment to use paper or Styrofoam coffee cups. An LCA comparison indicated that the manufacture of paper cups was estimated to consume 36 times as much electricity and to generate more than 500 times as much wastewater as the manufacture of polystyrene foam cups. The paper cups required a plastic coating and were unsuitable for paper recycling, whereas the Styrofoam cups could be recycled (Hocking, 1991).

As with any methodology, correct application is critical to the validity of results. Boundary conditions need to be well defined, and all significant attributes that might affect the comparison must be included. For example, in the McDonald's example above, the environmental and economic impacts of producing the paper raw material and the coating raw material, as well as the manufacturing and distribution of the cups, needed to be compared with those of producing the Styrofoam raw material and making and distributing the foam cups. It was important that the characteristics of both types of cups be fairly equivalent in terms of performance with respect to user and customer concerns—stackable, durable, not leaking, not being too hot to hold, etc. If recycling is possible, then costs and impacts of waste disposal could be reduced and a credit might be taken for reuse of the waste material.

Not only can LCAs be useful in choosing among competitive alternatives, but they can be used to identify the major sources of undesired impact during a product or process lifecycle. This way, designers can evaluate ways to reduce the impact and make a more desirable product or process. Often, reducing waste streams, choosing more benign process paths, and incorporating recycling options can lead to environmental and economic improvements.

Global resources that are directly used in human industrial activities can be described as "materials," where energy is considered a class of materials. As described in Chapter 2, some materials are depletable in the sense that time scales for their regeneration are huge compared to any conceivable societal utilization time. Other materials may be reused or recycled into other uses and some are truly renewable—for example, solar energy and its derivative forms of energy: wind, water, and biomass (Chapters 9–15).

In designing for sustainability, conservation of depletable materials is a parameter that is beginning to attract attention. This strategy of intentional product (or process) design for "dematerialization" contrasts with much of the historical pattern in which manufacturers had access to abundant supplies of raw materials or responded to perceived or real scarcity by materials substitution rather than materials conservation. Although dematerialization has not become a common design criterion, it is now gaining visibility. One example in the energy sector is the attention being paid to whether there will be sufficient platinum to support a widespread market penetration by fuel cell technology.

Another approach to lifecycle management of materials is to make the manufacturer responsible for taking back the materials in products at the end of their useful life. "Take back" regulations on bulk goods (e.g., major appliances, cars) encourage manufacturers to include the final disposition costs in their LCA, and they often promote more attention to recycling.

6.3.2 Simulation models

Another approach to modeling is to mathematically represent the system being studied and to simulate its operation over time or under varying conditions or scenarios. If the system of interest is fairly simple and has well-defined linear behaviors, the models can be straightforward. More sophisticated models are used in the design of new cars, for example, to understand the effects of body shape on the vehicle aerodynamics, as well as to refine the performance of various types of engines. However, as system complexity increases, including non-linearities, feedbacks, and uncertainties in the interactions among both individual variables and groups of variables, modeling becomes more challenging. Brute force attempts to represent the system may exceed the number-crunching capacities of present computers—even considering the rapid improvements in computer technology. The Global Circulation Models discussed in Chapter 4, which attempt to model the global atmosphere and its couplings to the ocean and biosphere, are limited by present-day computational ability, as well as by uncertainties associated with the couplings among the various variables and with the prediction of future behaviors. Further, since the global atmosphere is subject to inherent variability, it is difficult to separate long-term trends from the short-term noise. Using such models for scenario analysis, however, does provide useful insights if the limitations of the model are understood.

In the energy area, there are many types of models that are useful—from modeling of geological structures in the search for natural resources like petroleum and gas, to models of energy technology usage such as the National Energy Modeling System

(NEMS) used by the US Energy Information Administration (EIA) in preparing future estimates of US energy use (DOE, 2000). The model was developed for the US Department of Energy (DOE) to allow energy forecasting on a 20-year horizon. It is adjusted and improved based on annual energy reporting by major energy companies and organizations to the DOE. The model projects production, imports/exports, conversion, consumption, energy prices (based on assumptions about world energy markets, resource availability and costs, and other macroeconomic and financial factors), behavioral and technological choice criteria, technology cost and performance characteristics, and demographics. Thirteen major submodules, representing demand sectors (transportation, residential, commercial, and industrial), supply sectors (coal, renewables, and domestic oil and gas), natural gas transmission and distribution, international oil, and conversion (refineries and electric power) are linked to a macroeconomic model that relates energy prices to economic activity. In addition to its use in preparing the *EIA Annual Energy Outlook* for the US, it is used to evaluate policy options for influencing US energy choices and to forecast the effects of new technologies. Of course, it is subject to the usual uncertainties associated with complex models that are used to predict the future, but it does provide a consistent and quantitative basis for evaluating alternatives.

On the global energy prediction level, the IPCC (2001) has developed a number of major models for predicting the impacts of global energy use on the future. These models have been developed by teams of international experts and are used to aid the formulation of international policies under the FCCC. (More discussion and examples of some model results are presented in Chapter 21.)

6.3.3 Risk-based models

LCAs or other system models are often a starting point for risk-based assessments. Although all significant impacts are usually identified in such models, the procedures do not provide a quantitative basis for comparing the effects of one impact against another. Further, most systems do not normally consider the impacts of unusual accidents or upsets. Risk analysis techniques provide a quantitative method for describing, evaluating and managing risks, where risk is defined as any undesired impact. Risks can be continuous (e.g., car tailpipe emissions) or episodic (e.g., a car crash and fuel fire). In simple terms, risk involves two elements:

$$Risk = Frequency \times Severity\ or\ consequence$$

- *Frequency* can range from continuous exposure to daily or yearly problems to rare accident events that happen once in a million years or more. Its units are *events per unit time*.
- *Severity* is a measure of the undesired impact of the event and could be expressed in terms of costs, injuries, deaths, land contamination, etc. Thus, one common unit is *injuries per event*.

In any activity, a spectrum of risks is present ranging from continuous, low-level risks to rare disastrous events, such as flood losses or the release of a toxic vapor cloud from a chemical plant or a fuel storage facility. Whether we recognize it immediately or not, we all accept some level of risk, and our personal choices of "acceptable risk" may vary depending on how we view the desirability of the risky activity. Societal risk choices are evidenced in safety and environmental regulations and accepted practices. While "zero risk" sounds like a great goal, it is unachievable, and the costs of risk reduction tend to increase rapidly as risk levels are decreased to increasingly low levels. Risk models are useful for helping in the understanding of sources of risk, the relative importance of different types of risk, and in the effective management of risks.

The first step in a risk assessment involves careful consideration of possible ways an undesirable event might occur. This consideration should extend over the lifecycle of the process or product (e.g., hazards associated with misuse of the product, improper disposal, impacts of raw material extraction). Risk assessment techniques vary from qualitative "risk ranking" activities, based on judgment and experience, to much more sophisticated risk assessment of the type required for nuclear power plants (Rasmussen, 1975). Usually, the level of assessment is related to the potential magnitude of risk.

Boolean and other logic tools (e.g., fault tree analysis, event tree analysis, uncertainty analysis) are useful in describing and portraying the interconnections among more complex risks. A general overview of risk assessment and the various techniques that are useful in different applications is available in an American Institute of Chemical Engineers publication, entitled *Guidelines for Chemical Process Quantitative Risk Analysis* (CCPS, 1989). A compendium of safety and environmental risk management practices is provided in *Loss Prevention in the Process Industries: Hazard Identification, Assessment, and Control* (Lees, 1996). Kammen and Hassenzahl (1999) provide an introductory account of various risk assessment techniques and illustrate their applications with a diverse range of case studies on environmental issues that have gained public and media attention.

There are a number of challenges in risk-based management, although it offers, in theory, the possibility of allocating limited resources to mitigation of the most important residual risks. There have been advances in risk-based regulation in the recent past. For example, the US Environmental Protection Agency (EPA) has begun to use risk criteria in some of their guidelines. Now, according to the EPA, emissions of carcinogens from hazardous waste combustion facilities should not exceed a cancer risk level of 1×10^{-5} to a "maximally exposed individual" over a 70-year lifetime (EPA, 1994). This regulation looks at the nearest neighbor to the site, assumes continuous exposure over a lifetime, and can be applied to all carcinogens that might be emitted, including those low levels that might be emitted continuously, and episodic emissions of higher levels, for any foreseeable reason.

The Nuclear Regulatory Commission, which requires detailed risk analysis as part of the approval process for a nuclear power plant, uses acceptability criteria of less than 1×10^{-6}/yr chance of a core meltdown for an analysis based on "conservative" assumptions and 1×10^{-7}/yr for analysis based on "likely" assumptions. Note a 1×10^{-6}/yr accident has an expected recurrence interval of 1 million years. However, if there were 1,000 nuclear plants operating, one might expect an accident of this sort every 1,000 years. Thus, a person living close to a nuclear plant might have a 10^{-6}/yr risk from this type of accident, while society as a whole would have a 10^{-3}/yr risk. Note also that a 10^{-3}/yr risk does *not* mean that the subject accident is unlikely to occur until 1,000 years have elapsed. It does mean that in a given year the probability for an accident with this risk to occur is 10 times more likely than for an accident with a 10^{-4}/yr risk.

Safety risks can be measured in terms of deaths, injuries, lost days of work, and associated financial losses, but health risks are more difficult to estimate. Some chemical exposures may cause temporary discomfort, but no permanent effect. Others may have cumulative physiological effects that culminate in disease or even death. Thus, small exposures to an external environmental agent over a long time may eventually cause the human body to experience what may be thought of (very approximately in toxicological terms) as a critical "dose" (i.e., a time-integrated total assimilation of harmful agent(s) that may directly or indirectly contribute to some human disease). That disease may not become evident for years (Chapter 4).

Meaningful estimates of human risk from environmental agents (e.g., chemicals, ionizing radiation) require reliable data on human responses to known exposures to these agents. Historical records of disease and mortality for select populations whose environmental exposure can also be estimated are sometimes available. On occasion, these data are sufficiently comprehensive to be mined for meaningful conclusions on human health risks. This is the field of public health known as *epidemiology*. Reliable epidemiological data are generally preferred in forecasting potential human health effects of environment agents, because these data directly reveal human responses. However, epidemiological data are unavailable for many toxicants, including most suspect carcinogens. In such cases, the evidence for human risk is established from toxicological responses observed when single cells *in vitro*, or small animals *in vivo*, are exposed to suspect substances. Models are used to extrapolate cell and animal test results to humans, but uncertainties grow rapidly as data sources get sparser and more distant from human exposures.

Another trend in risk-based management grew out of the concept of Total Quality Management (TQM), which was developed by major industries and is now in widespread practice as International Standards Organization (ISO) Standard 9000. This standard revolutionized manufacturing by establishing methodologies that allowed individual companies to self-assess the quality of their manufacturing processes and products. Companies that wished to be certified as ISO 9000 compliant committed to voluntary certification. The voluntary certification included having a consistent corporate policy, internal assessment procedures, an internal monitoring and enforcement system, independent auditing, and a commitment to continual improvement.

The ISO approach has several features that enhance its effectiveness. The first is that companies have such varied practices that each has unique areas where improvements are needed. However, by setting performance criteria and requiring self-assessment, companies are encouraged to take a systematic view of their actual performance and identify areas that need the most improvement. Next, the assessment requires the involvement of a variety of employees who then become educated in the quality goals and the weaknesses of present practices. They become engaged in finding improvements and monitoring the results. Finally, periodic external auditing and continual improvement goals challenge workers to collaborate on making the highest quality products, efficiently and at competitive costs. Standards like ISO 9000 are incorporated through marketing advantages of certification, by national adoption, or through international agreements.

In 1996, ISO followed up with the ISO 14000 series on environmental management. These standards follow the same successful framework implemented in ISO 9000, but focus on: environmental management systems, environmental auditing and other related investigations, environmental performance evaluation, environmental labeling, lifecycle assessment, and environmental aspects in product standards. Companies that have signed on to ISO 14000 find it helpful. However, this standard has been somewhat less successful than ISO 9000 because it sometimes conflicts with existing environmental "command-and-control" regulations. One reason is that ISO 14000 would allow a company to be non-compliant with an established environmental regulation while a more serious environmental problem (regulated at the time or not) was being addressed. In the US, for example, the EPA is not at present sympathetic to this sort of violation. Thus, many US companies prefer to take the safe path of compliance to regulation, as opposed to the more challenging path of risk-based assessment and continual improvement. However, there are many advantages to risk-based environmental management, though it is much more challenging to audit.

6.4 Measures of Sustainability

The systems analysis methodologies described above require quantification of impacts. However, when one evaluates attributes of different types (e.g., cost, performance, safety, environmental impact), different measures are used. In order to judge whether a particular system is better than an alternative, one needs to have a way of putting these measures on a common basis. Some measures are simple (e.g., economic cost can be described in straightforward terms). But even for economics, as described in Chapter 5, there can be differences in the appropriate discount rate to apply to future costs and returns. When one moves to environmental impacts, it may be straightforward to quantify emissions in tons of sulfur or carbon or ppm of polycyclic aromatic hydrocarbons (PAHs) in an effluent stream, but the individual and societal health impacts of such emissions are much harder to estimate. And if we want to convert these impacts into monetary costs, the challenge is even greater. First, there will be great uncertainty in the impact assessment. Second, people will disagree on whether to be cautious by including all plausible risks or to minimize the importance of most risks. (In Chapter 21, there is more discussion of how to elicit risk preferences from groups of decisionmakers with disparate views.)

6.4.1 General indicators of sustainability

Indicators can be thought of as measurable quantities that can be related to some aspect of global health. For people's health, body temperature is one indicator. Statistically, we know a normal range of body temperatures that indicate a healthy state, other ranges of temperature that indicate the relative intensity of illness, and extreme temperatures that lead to irreversible brain damage or, ultimately, death. Other human health indicators include pulse rate, respiratory rates, and composition of components in various body fluids.

Because of the large number of systems that interact to form the web of human activity (Chapter 21), many different indicators will pertain to pieces of the whole picture. How to apply the right indicators to a particular assessment is a major challenge, so it is important to have some ways of prioritizing different levels and types of indicators. (There is much literature on measures or indicators of sustainability, and a comprehensive overview is provided in Murcott, 1996.)

The United Nations and the World Bank have defined major goals for sustainable progress in the 21st century and have specified three broad indicators that will measure progress toward main goals related to the environment, the economy, and society. Under each main goal are a myriad of sub-indicators. While the selection of these particular goals and indicators may seem somewhat arbitrary, they have evolved out of considerable international dialogue on the important future directions for our global society. We will now identify some of the sub-indicators and goals for these three categories.

Environmental.

Government and institutional commitment (countries with a national strategy for sustainable development)

Water resources (population with access to safe water, intensity of freshwater use, percentage of annual available resources used)

Biodiversity (nationally protected area as a percentage of total land)

Energy use (GDP per unit of energy use, total and per capita carbon dioxide emissions)

Economic.

Standard economic indicators include total and per capita GDP, gross national product (GNP), national debt, employment rates, investment rates, new housing starts, Dow Jones or other stock market averages, balance of trade, etc.

Social.

Social indicators include poverty, education, and health usually in the context of setting goals such as:

Reduce poverty by half (headcount of people with less than $1/day equivalent income, share of income accruing to poorest 20%, incidence of child malnutrition)

Provide universal primary education (net primary enrollment rate, progression to grade 5, literacy rate of 15- to 24-year olds)

Improve gender equality in education (gender differences in education and literacy)

Reduce infant and child mortality (infant mortality rate, under-5 mortality rate)

Reduce maternal mortality (maternal mortality ratio, births attended by health staff)

Expand access to reproductive health services (contraceptive prevalence rate, total fertility rate, HIV prevalence in pregnant 15- to 24-year olds)

On first inspection, it might appear that only the above indicators specifically focused on improved energy use, productivity, and reduced carbon dioxide emissions are concerned with energy. However, it is essential to realize that the role of energy in sustainable progress is far more pervasive. In particular, energy is a major enabling force for many of the above indicators. Electricity availability is key to facilitating education, to providing health-care services, and to providing an infrastructure for commercial and industrial development. Transportation services worldwide are heavily dependent on petroleum availability. And, as energy services become widely available in developing countries and living standards improve, birth rates are observed to drop. Affordable and accessible energy has historically been a major engine of economic progress (Chapters 1 and 2). In contrast, persistent poverty is a guaranteed roadblock to environmental stewardship.

The Worldwatch Institute (2000), in *State of the World 2000*, published a number of ongoing assessments using a variety of indicators. They identify a challenge for this new century of properly valuing important ecological and societal objectives as part of a future economic system. They raise another important issue—anticipating environmental "surprise." As human populations expand their influence on the global environment, some critical life-support systems may start to fail locally. Continued actions may create a global crisis. Thus, they stress the importance of looking for early warning signs of deterioration. Some signs may appear in the poorest parts of the world—others in areas of rapid development or where infrastructure is aging and neglected.

6.4.2 Categories of indicators

Stock indicators. Stock indicators are measures of total inventories of resources, such as national capital (defined by the UN System of National Accounts), total population, total undeveloped land, and fossil fuel reserves. These indicators provide points of reference for evaluating the impact of changes in inventories.

Flow indicators. These relate to changes in the stock indicators described above. For example, GDP is an indicator of the annual capital generated by domestic activities. The capital domestic stock will change annually depending on how much of the GDP is reinvested in the overall economy. Likewise, species become extinct and new species are generated by mutations. The net stock of diverse species will change according to the cumulative impact (the integral over time) of the rates of species generation and depletion.

Composite indicators. These include combinations of multiple individual indicators. For example,

- Pressure-State-Response Analysis (OECD, 1993). This approach was explored in the formulation of the UN's Agenda 21. It is a technique for analyzing progress with regard to sustainability goals. For example, for the concern with stratospheric ozone depletion, the pressure comes from present production of CFCs (which have been banned in countries signatory to the Montreal Protocol). The state indicator is the present concentration of CFCs in the atmosphere. The response is the creation of programs leading to the phase-out of these materials. A model can be formulated to indicate the future behavior of CFC concentrations in the atmosphere. A similar analysis can be done with regard to fish stocks, desertification, or water pollution. The main problem with this sort of analysis is that it does not capture the feedback between different analyses. For example, protection of a fishery may cause major local unemployment and other societal problems.
- Ecological Footprint (Rees et al, 1996). This approach seeks to estimate the impact of an activity on surrounding areas. Imagine a city like Tokyo that has a greater metropolitan area of about 125,000 square kilometers and a population of about 30 million. This gives an average population density of 240 people per square kilometer (or 2.4 people per hectare) with population densities much greater in the heart of the city. To support such a city, surrounding areas provide food, water, power, and

> materials for buildings and transportation systems. Recent studies by the Earth Council and at the University of Tokyo suggest that the "footprint" of the city is more than thirty times its physical area. Taking Japan as a whole, resource utilization requires about 6.25 hectare per capita, but Japan only has available land area of 1.9 hectare per capita. Because of the global diversity of commerce, the supporting activities may physically be distributed around the world.

The concept of an "ecological footprint" is useful in thinking about how industrial activities affect their surroundings, but it is difficult to quantify the concept in a consistent way that will allow direct comparisons of different activities. Suppose we look at the ecological footprint of a coal-fired power plant. The impacts start back at the coal mines that are the source of the coal. Coal transportation uses energy and infrastructure and causes some air emissions that spread in the vicinity of the travel route. The coal plant is constructed of materials that require the production of steel and concrete and other materials. Electricity generated by the plant is sent to a power grid where it is distributed widely, and emissions from the plant are dispersed into the atmosphere where they may impact downwind areas through acidification or low-level impacts on human health. It is clear that this is an LCA list, and it is incomplete—but the impacts of the different pieces of the footprint are different in nature, and many are shared with other activities. If the plant uses low-sulfur coal, acid rain impacts may be minimal, but some criteria are needed to calculate what is part of a damage footprint and what is not. If one focuses on GHG emissions that enter the atmosphere, then the footprint may even be considered global.

The footprint concept is qualitatively useful for comparing different approaches to supporting "megacities." These are high population density metropolitan areas with total populations exceeding 10 million. Some 20 such cities exist worldwide, with most expected to experience appreciable increases in population for the foreseeable future. The footprint is a measure both of consumption and of efficiency in the use of resources. Large per capita footprints in developing countries can be reduced through improvements in infrastructure and services. Footprints of megacities can be reduced through better infrastructure management, but they may ultimately require adjustments to a less consumptive lifestyle. (The Redefining Progress Group, based in Oakland, California, has a Web site that offers a short questionnaire and a calculated estimate of the user's personal ecological footprint: http://www.earthday.net/footprint/info.asp.)

Multiattribute indicators. Attempts have been made to find a single indicator that will represent progress toward sustainability. One of these is to inventory total national capital (TNC), which is defined to include not only economic capital but also human capital, resource base, and environmental capital. Changes can be tracked in terms of net national product (NNP), which subtracts use of natural capital and environmental and social degradation from GNP. Again, this technique is difficult to apply consistently because of the problem of pricing changes in environmental and human capital.

Moreover, because the primary function of GNP in democratic societies is social progress, it is important for multi-attribute approaches to compare, over time scales cognizant of economic cycles, indices of social degradation and social advancement, under various levels of environmentally suspect GNP. This is crucial because the loss of social progress from reduced GNP may be far worse than the social degradation attributed to unreduced GNP.

Other indicators include "literate life expectancy." This can be thought of as a composite measure that combines health and education and inherently reflects the standard of living needed to support these activities. However, using a single average value for a national indicator eliminates the opportunity to examine the distributional aspects. For example, a country with half the population in abject poverty and half twice as well off as the average would appear the same as a country where the entire population had a similar standard of living at the average value. The HDI shown in Figure 6.3 is a composite indicator developed by the United Nations Development Programme (UNDP) to represent the developmental status of a country in terms of economic, health, educational, and other statistics.

Distributional indicators. These types of indicators explore the distribution around mean values that are used in many indicators. Among the World Bank indicators, some scrutinize the incidence of poverty, while another looks at the share of income accruing to the poorest 20%.

Specific indicators. These are indicators that apply within a more limited context. They may be representative of local conditions or particular activities and their impacts. For example, in regional planning, attributes that measure local pollution, congestion, economic conditions, and so forth may be useful indicators in choosing among alternatives. If technological alternatives are being compared, indicators will usually measure the cost of the product or service, expected service life, performance, and environmental and safety impacts.

6.5 Drivers of Societal Change

If society wishes to embark on a transition to more sustainable living, there are a number of mechanisms available:

- technological innovation
- substitution of alternatives
- policy and regulatory requirements
- changes in consumer preferences

In affluent industrialized countries, many people hope that technological improvements will allow the full transition without any impact on habits and lifestyles. While technology has led to many remarkable advances in quality of life over the past years, we need to think about sustainability and the real possibility that we may be overstressing the life support systems of the planet. We need to consider a broader range of changes that may affect our life styles, but not our overall quality of life.

Beyond investment in alternative technology development, there are other challenges that present barriers to adoption of new technologies. The two main barriers are market barriers and overcoming inertia—increasing costs and human resistance to change. No one wants to pay more for anything, although we do so routinely when prices increase in response to normal economic processes. We do not want to discard a functioning large infrastructure in a shorter period than its normal renewal cycle. Institutions do not want to be displaced by new organizations. Consumers will accept a new product that enhances their way of life, but they may grumble at paying for mandated changes, such as air bags in personal vehicles. Companies will invest in products that they believe will expand their markets and profits. "Green" products appeal to many consumers, but only if performance and service standards are met. Businesses thrive on growing markets. Consumers globally want *more*. In developing countries, *more* is needed to sustain health and improve the quality of other life essentials (i.e., potable water, sanitation, human nutrition). The developed world has made material and economic goods an indicator of personal success.

In sharp contrast to the overly consumptive lifestyle, Thring (1998) has formalized an approach to the sustainable use of energy based on the concept shown in Figure 6.5. Suppose consumption is food. Below the lower limit, a person is malnourished and ill. Above the upper limit, a person is obese with resultant adverse health effects. Suppose consumption is wealth. Below the lower limit, financial resources are inadequate to sustain minimal quality of life. Excess wealth may bring added material benefits, but it also entails management and responsibilities that take time away from other pursuits that might enhance overall quality of life. Between the lower and upper limits, people have enough of the consumable resource to provide a satisfactory quality of life.

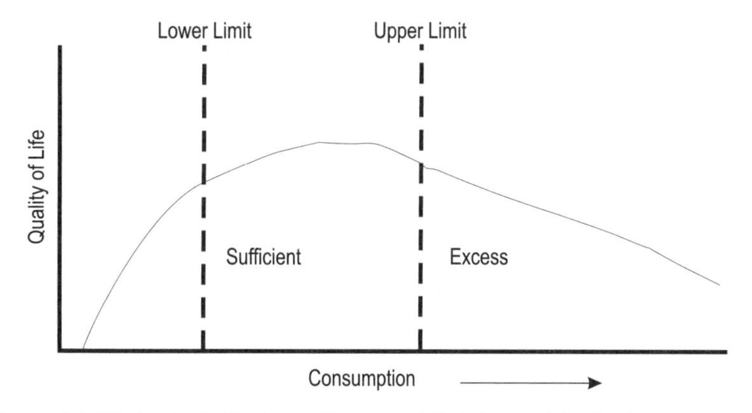

Figure 6.5. Thring's Sufficiency Concept (slightly modified). Source: Thring (1998).

Thring (1998) estimates that the lower limit for energy consumption per capita is about 0.3–0.5 TOE per capita per year; the upper limit is from 1.0–1.5 TOE per capita per year. Present energy consumption in the US is about 6 TOE per capita per year, about 3 in Europe, and well below 1 in most of the developing world. This illustrates the danger of looking at sustainability variables one at a time. China already has a per capita energy use that is at Thring's lower limit for sufficiency. But much of this energy is used inefficiently and without satisfactory pollution control. The distribution of energy across the population is also uneven, with many people still well below the lower limit for sufficiency. In the US, which is well above the Thring upper limit, energy is used for providing a higher level of environmental quality and for many other attributes that provide a high standard of living (transportation, industry and commerce, comfortable housing, etc.). Nevertheless, if humans worldwide are to curb GHG emissions substantially, some experts believe it will be necessary to transition to a state where those in the excess zone cut back net GHG emissions to allow the rest of the world room to move cleanly and efficiently to at least the lower range of energy sufficiency. We may also find ways to obtain the services we want in new ways that may substantially reduce the consumption of primary energy.

Is there any sort of a pathway from here to the sort of a future that Thring suggests and that weighs all the important attributes of sustainability? Thring believes that humankind must rediscover the human conscience to bring this about. If sustainability is viewed in a broader context involving issues beyond energy use—depletion of multiple natural resources, preservation of a healthy environment, economic equity, etc.—some see need for a major transformation in societal behavior and rules, with this sort of a change unlikely unless there is some major outside driving force. In the past, major changes resulted from catastrophes or the knowledge of impending catastrophes. People pulled together and helped each other in the aftermath of disasters. Once the crisis was resolved, the normal behavior returned and interactions with the group ceased, but the people usually remembered the social interactions with nostalgia. Others may view Thring's perspectives as closer to the axioms of "deep" or even "radical ecology" as defined by Allenby (1999) (see Section 1.6 and Table 1.9) than to "mainstream" sustainability. Still others may perceive a persuasive balance by drawing from all three.

The other type of driver involves governmental actions. International policies are pushing the early steps toward change, but the large existing infrastructures and competing economic interests and priorities make this mechanism slow—probably too slow to reduce GHG emissions in time if the world is on a rapid path of climate change. Even on a more gradual climate modification path, the world will eventually have to modify its energy use. In the long term, other issues of sustainability, such as resource depletion, water quality and availability, and biodiversity, will push society to change pathways. Is there a way to change human behavior in the direction of more sustainable consumption of energy and other depletable resources? Again, the essence and the challenge of sustainable development is learning and communicating publicly acceptable ways to accomplish this while avoiding a global economic collapse that might trigger a new Dark Age.

The political process appears to be taking on sustainability issues one at a time—climate change, biodiversity, development, trade, human rights, international stability, etc. And the trade off decisions will vary depending on the particular circumstances surrounding each individual decision—by geography, by sector, by affluence, and so forth. There is a need to forge institutional partnerships that can work together to set a general framework and then allow flexibility for individual decisions—by industries, governments, and consumers—that will move humankind toward more sustainable living.

6.6 Some General Principles of Sustainable Development

The International Institute for Sustainable Development has adopted the 10 "Bellagio Principles," based on work from a series of meetings held at the Rockefeller Foundation's Conference Center in Bellagio, Italy, in November 1996.

1. **Clarity**. Establishment of a vision of sustainable development and clear goals that provide a practical definition of that vision in terms that are meaningful for decisionmaking
2. **Holistic perspective**. Assessment of progress toward sustainable development should:
    - include review of the whole system as well as its parts
    - consider the well-being of social, ecological, and economic sub-systems, their state as well as the direction and rate of change of that state, of their component parts, and the interaction between parts
    - consider both positive and negative consequences of human activity, in a way that reflects the costs and benefits for human and ecological systems, in monetary and non-monetary terms
3. **Essential elements**. Assessment of progress toward sustainable development should:
    - consider equity and disparity within the current population and between present and future generations, dealing with such concerns as resource use, over-consumption and poverty, human rights, and access to services, as appropriate
    - consider the ecological conditions on which life depends
    - consider economic development and other, non-market activities that contribute to human/social well-being

4. **Adequate scope.** Assessment of progress toward sustainable development should:
   - adopt a time horizon long enough to capture both human and ecosystem time scales thus responding to needs of future generations as well as those current to short-term decision making
   - define the space of study large enough to include not only local but also long-distance impacts on people and ecosystems
   - build on historic and current conditions to anticipate future conditions—where we want to go, where we could go
5. **Practical focus.** Assessment of progress toward sustainable development should be based on:
   - an explicit set of categories or an organizing framework that links vision and goals to indicators and assessment criteria
   - a limited number of key issues for analysis
   - a limited number of indicators or indicator combinations to provide a clearer signal of progress
   - standardizing measurement wherever possible to permit comparisons on a consistent basis
   - comparing indicator values to targets, reference values, ranges, thresholds, or direction of trends, as appropriate
6. **Openness.** Assessment of progress toward sustainable development should:
   - make the methods and data that are used accessible to all
   - make explicit all judgments, assumptions and uncertainties in data interpretations
7. **Effective communication.** Assessment of progress toward sustainable development should:
   - be designed to address the needs of the audience and set of users
   - draw from indicators and other tools that are stimulating and serve to engage decision makers
   - aim, from the outset, for simplicity in structure and use of clear and plain language
8. **Broad participation.** Assessment of progress toward sustainable development should:
   - obtain broad representation of key grass-roots, professional, technical, and social groups, including youth, women, and indigenous people, to ensure recognition of diverse and changing values

- ensure the participation of decision makers to secure a firm link to adopted policies and resulting action

9. **Ongoing assessment**. Assessment of progress toward sustainable development should:
   - develop a capacity for repeated measurement to determine trends
   - be iterative, adaptive, and responsive to change and uncertainty because systems are complex and change frequently
   - adjust goals, frameworks, and indicators as new insights are gained
   - promote development of collective learning and feedback to decision making

10. **Institutional capacity**. Continuity of assessing progress should be reassured by:
    - clearly assigning responsibility and providing ongoing support in the decision making process
    - providing institutional capacity for data collection, maintenance, and documentation supporting development of local assessment capacity

These principles capture the essential pathways to sustainable development. However, at present, there is no international will to embark on such an ambitious endeavor. A recent US study, "Our Common Journey: A Transition Toward Sustainability" (NRC, 1999a), addresses these issues. That report suggests that our society already has the technologies and resources to make a successful transition to sustainability over the next several decades, but that the political will to choose this path is not present and would have to be developed to make this transformation a reality.

References

Allenby, B.R. 1999. *Industrial Ecology: Policy Framework and Implementation*. Upper Saddle River, NJ: Prentice Hall.

BP Statistical Review of Energy. 2001. London. See: www.bp.com/centres/energy/.

BP Statistical Review of Energy. 1995. London. See: www.bp.com/centres/energy/.

Brundtland, H.G. 1987. *In Our Common Future*, WCED. New York: Oxford University Press.

CCPS (Center for Chemical Process Safety of the American Institute of Chemical Engineers). 1989. *Guidelines for Chemical Process Quantitative Risk Analysis*. New York.

Consoli, F., D. Allen, I. Baustead, J. Fava, W. Franklin, A. Jensen, N. De Oude, et al., 1993. *Guidelines for Life-Cycle Assessment : A "Code of Practice."* Florida: Soc. of Env. Tox. And Chem.

Daly, H. 1990. "Towards Some Operational Principles of Sustainable Development." *Ecological Economics*, 2:1–6.

DOE (Department of Energy). 2000. *The National Energy Modeling System: An Overview 2000*. DOE/EIA-0581, Washington, DC.

Edmonds, J. 1995. *Energy Policy*, 23:4–5.

Ehrlich, P.R. 1968. *The Population Bomb*. New York: Ballantine.

EPA (US Environmental Protection Agency). 1994. *Exposure Assessment Guidance for RCRA Hazardous Waste Combustion Facilities*. EPA 530/R94/021. Washington, DC.

Fava, J.A., R. Denison, B. Jones, M. Curran, B. Vigon, S. Selke, and J. Barnum. 1994. *A Technical Framework for Life-Cycle Assessment*. Florida: Soc. of Env. Tox. And Chem.

Fuwa, K. 1995. In *Defining and Measuring Sustainability: The Biogeophysical Foundations*, eds., M. Munasinghe and W. Shearer. Washington, DC: World Bank.

Hardin, G. 1968. "The Tragedy of the Commons." *Science*, 162(5364):1243–8.

Hocking, M.B. 1991. "Paper Versus Polystyrene: A Complex Choice." *Science,* 251:504–505.

IPCC (Intergovernmental Panel on Climate Change). 2001. *Climate Change 2001: Synthesis Report* (Working Groups on: The Scientific Basis; Impacts, Adaptation & Vulnerability; and Mitigation). Cambridge, England: Cambridge University Press.

IPCC (Intergovernmental Panel on Climate Change). 1990. "First Assessment Report." *Third Assessment Report. Working Group I: Climate Change 1990: The Scientific Basis*. Geneva: IPCC Secretariat.

Kammen, D.M., and D.M. Hassenzahl. 1999. *Should We Risk It? Exploring Environmental, Health, and Technological Problem Solving*. Princeton, NJ: Princeton University Press.

Kaya, Y. 1991. Panelist in Plenary Session II, "Policy Strategies for Managing the Global Environment," *Energy and the Environment in the 21st Century*. ed., J.W. Tester. Cambridge, MA: The MIT Press.

Lees, F.P. 1996. *Loss Prevention in the Process Industries: Hazard Identification, Assessment, and Control*. London: Butterworths.

Malthus, T.R. 1798. *Essay on the Principle of Population.* Reprinted by Oxford University Press (1993).

Meadows, D., J. Randers, and D. Meadows. 1992. *Beyond the Limits.* London: Earthscan Publications.

Meadows, D., D. Meadows, J. Randers, and W. Behrens III. 1972. *The Limits to Growth.* New York: Universe Books.

Murcott, S. 1996. "Sustainable Development: Definitions, Principles, Criteria and Indicators." In *Sustainable Development: Criteria and Indicators.* Laxenburg, Austria: IIASA.

NRC. 1999a. "Our Common Journey: A Transition Toward Sustainability." *Panel on Integrated Environmental and Economic Accounting.* Washington, DC: National Academy Press.

NRC. 1999b. "Nature's Numbers: Expanding the National Economic Accounts to Include the Environment." *Panel on Integrated Environmental and Economic Accounting.* Washington, DC: National Academy Press.

OECD (Organisation for Economic Co-operation and Development). 1993. OECD core set of indicators for environmental performance reviews. *OECD Environment Monographs, No. 83.* Paris, France: OECD.

OSTP Report. 1997. *Federal Energy Research and Development for the Challenges of the 21st Century.* Washington, DC: PCAST (President's Committee of Advisors on Science and Technology).

Ostrom, E., J. Burger, C.B. Field, R.B. Norgaard, and D. Policansky. 1999. "Revisiting the Commons: Local Lessons, Global Challenges." *Science.* 284: 278–282.

Rasmussen, N. 1975. *Reactor Safety Study: An Assessment of Accident Risk in U.S. Commercial Nuclear Power Plants (WASH-1400).* Washington, DC: Nuclear Regulatory Commission.

Rees, W.E., and M. Wachemagel. 1996. "Our Ecological Footprint: Reducing Human Impact on the Earth." British Columbia: *New Society Publ.*

Solow, R. 1992. "An Almost Practical Step toward Sustainability." Invited MIT lecture on the 40th anniversary of *Resources for the Future.*

Thring, M.W. 1998. "A Permanently Sustainable Energy Strategy." *Energy World.*

UNDP (United Nations Development Programme). 2001. *Human Development Report, 2001.* New York: Oxford University Press.

U.S. Energy Information Agency. 1995. *Annual Energy Outlook.*

Viederman, S. 1997. "Key Issues Underlying Earth Summit II, Agenda 21, Globalization and Sustainable Development." Remarks at United Nations.

Weiss, M., J. Heywood, A. Schafer, E. Drake, and F. AuYeong. 2000. "On the Road in 2020." *MIT Energy Laboratory Report* No. 00-03.

World Bank. 2001. *World Development Report 2000/2001: Attacking Poverty.* New York: Oxford University Press.

World Commission on Environment and Development (WCED). 1987. *Our Common Future.* New York: Oxford University Press.

Worldwatch Institute. 2000. *State of the World 2000.* New York: Norton & Co.

Web Sites of Interest

http://www.iso.ch/
http://www.aiche.org/ccps/
http://www.worldbank.org/
http://www.ciesin.org/IC/wbank/sid-home.html
http://www.iisd.org
http://www.ecouncil.ac.cr/rio/focus/report/english/footprint/
http://www.earthday.net/footprint/info.asp

Problems

6.1 Webster defines "sustain" as: "to provide for the support of; to maintain, or cause to continue in existence or a certain state; to keep up or prolong..." Do you think energy use in the US today is "sustainable?" Explain why or why not.

6.2 Give three examples of unsustainable energy use practices and explain how they might be made sustainable in the future.

6.3 The developed world tends to blame unsustainability on population growth in developing countries. The developing countries, on the other hand, blame the developed countries for excessive energy consumption. The US per capita energy consumption is double that of most European countries and Japan. What are the reasons that US energy use is so high?

6.4 Energy use per capita in India today is about one tenth that of the US levels, or about the equivalent of 1,000 watts of electricity on a continuous basis. This includes all energy, including industrial and transportation uses. Roughly estimate your personal energy use and convert it to an equivalent continuous wattage of electricity use. (You may want to do summer and winter estimates!)

6.5 Sustainability has aspects of "intra-generational equity" and "inter-generational equity." Discuss the relative importance of these two values.

6.6 In economics, future costs are usually "discounted" at annual rates of 10–15% to differentiate between the present and future value of money (Chapter 5). What sorts of discount rates might be appropriate for various types of environmental damages?

6.7 Why should developed countries be concerned about social inequity in developing countries? List at least two reasons and give explanations.

6.8 What can you do right now to facilitate sustainable practices as an individual citizen of your country? As an individual citizen of the world? Do you think you will be able to do more or less as an individual five years from now? Why or why not? Try to be quantitative. Would it be feasible for you to consume 5% less energy, paper, water, and other resources than you do at present? Would a less-consumptive lifestyle be easier or harder five years hence? Estimate the maximum amount you could cut your consumption and not feel "pinched."

6.9 If you were leading a group of friends in launching a high-tech startup, would sustainable practices be a significant priority in your business plan and financing strategy? Why or why not? How would you succinctly explain the principles of sustainable development to a group of venture capitalists? How would you convince them that sustainability is a solid business value worthy of their investments?

6.10 How would you convince other busy decision makers of the merits of sustainability (e.g., your mayor or town council, an elected member of your state or federal government)?

6.11 Compare and contrast the Thring Sufficiency Concept (Section 6.5 and Figure 6.5) with the Bellagio Principles for Sustainable Development (Section 6.6). List specific applications of each approach to a particular energy issue of interest to you (e.g., urban transportation, expanding energy services in a remote region, recreation). For example, are playing outdoor sports like baseball, football, and soccer at night or in domed stadiums good sustainable practices?

6.12 Take the ecological footprint quiz at www.earthday.net/footprint/info.asp.

a. What is the footprint of your present lifestyle?

b. Recalculate for the future lifestyle you would like at the height of your career. Compare this with your present lifestyle.

c. Write two or three paragraphs evaluating this "footprint" analysis. Is it realistic? What does it omit?

Fossil Fuels and Fossil Energy 7

7.1 Introduction

7.1.1 Definition and types of fossil fuels

A *fossil fuel* is a substance that releases energy by a chemical reaction. This is a broad definition under which many substances can function as fuels. Most fuels release their energy by reacting with a separate substance (the *oxidant*), but there are cases (e.g., the explosive, TNT) where the oxidant is actually part of the fuel itself. Thus all fossil fuels are storehouses of chemical potential energy. Under the right circumstances (i.e., by chemical reaction with an appropriate oxidant) this chemical potential energy can be tapped for useful (e.g., home heating, propulsion) or destructive (e.g., fires) purposes. We typically think of oxidation as a chemical reaction that causes oxygen atoms to combine with some other substance, as in the oxidation of methane gas (the principal component of natural gas) by pure oxygen, to form carbon dioxide and water:

$$CH_4 + 2O_2 \rightarrow CO_2 + 2H_2O \tag{7-1}$$

However, many different substances, including the halogens and CO_2, can be oxidants, and some substances, such as boron, sulfur, and phosphorous (which are fuels when reacted with oxygen or fluorine) can themselves or as their oxides or halides be an oxidant (Fristrom, 1995). When fuel oxidation is rapid and primarily produces heat, the process is called *combustion*. We use this definition to distinguish combustion from electrochemical oxidation of a fuel to produce electricity and small amounts of heat. The equipment used for these two processes are a *combustor* (or *burner*) in the former and a *fuel cell* in the latter. To fathom the role of fuels in sustainable energy, we will study the two classes of fuels that are primarily responsible for supplying energy for human needs—fossil and biomass (Chapter 10)—and we will illuminate some of the basic features of combustion, including effects that give rise to pollutants.

Naturally occurring *fossil fuels* are solid, liquid, or gaseous substances that contain *organic* or covalently bonded carbon and are produced by chemical and physical transformations of plant and animal remains *over geological time periods*. The second attribute is crucial to differentiate fossil fuels from biomass fuels, which are produced and regenerated over time scales as short as 1–100 years (i.e., 1/1,000,000 or less of geological time scales). The major naturally occurring fossil fuels are (Glasstone, 1982):

- *Coal*: (Hendricks, 1945, and Corey et al., 1984) A compact stratified mass of metamorphosed plant that has, in part, undergone arrested decay to different extents of completeness. Coal originates from the arrested decay of the remains of trees, bushes, ferns, mosses, vines, and other forms of plant life that flourished in huge swamps and bogs many millions of years ago during prolonged periods of humid, tropical climate, and abundant rainfall. Peat is a coal precursor formed by bacterial and chemical action on the plant debris. Subsequent actions of heat, pressure, and other physical phenomena metamorphosed the peat to the various types (ranks, an index of geological age) of coal found today. Because of the various extents of transformation during

Basic Definitions

The *atomic weight* of any atom is the ratio of its mass to the mass of a prescribed standard, taken as a particular type (isotope) of the carbon atom, C, which is assigned a value of exactly 12 units. One atom of atomic hydrogen, H, weighs almost exactly one twelfth of this form of the carbon atom, so the atomic weight of H is very close to 1 (1.00797 ± 0.00001 to be more precise).

The *molecular weight* (MW) of any substance is the sum of the weights of all the atoms in the substance. Thus, the MW of CO_2 is the sum of the weights of one atom of carbon and two atoms of oxygen, i.e., (1 atom C × 12 units/(atom of C) + 2 atoms O × 16 units/(atom of O)) = 12 + 32 = 44. As defined here, these atomic and molecular weights have no units. They are merely ratios (i.e., one molecule of CO_2 has approximately 44 times more mass than 1 atom of hydrogen).

We define a *mole*, better referred to as a gram mole for reasons given below, as the amount of that substance that has a mass in grams equal to its molecular weight. A mole of CO_2 has a mass of 44 grams. Because one atom has such a small absolute mass, we conclude that a gram mole of a molecule must contain a large number of molecules. This is correct, and that number is an important physical constant known as Avogadro's number or N_o, which defines a mole as 6.023×10^{23} molecules of any substance. Thus, 6.023×10^{23} molecules of CO_2 have a mass of 44 grams.

The *pound mole* is useful in many practical engineering calculations. It is defined as the amount of a substance of mass equal to the molecular weight of the substance in pounds rather than in grams.

Ideal gases have a volume directly proportional to the absolute temperature of the gas and inversely proportional to the pressure of the gas. For many fuels-related calculations, it is accurate to treat all gases as ideal. The temperature, T, pressure, P, volume, V, relationship of any gas is described by a relationship known as an equation of state, which for an ideal gas takes the simple form:

$$PV = nRT$$

Where n is the number of gram moles or lb moles of the gas (calculated by dividing the corresponding weight of the amount of gas present in grams or pounds, w, by the molecular weight of the gas), and R is the universal gas constant in consistent units.

The standard volume of a gram mole of any ideal gas occupies a volume of 22.414 standard liters. A standard liter is defined as a liter of an ideal gas at a pressure of 1 atm and a temperature of 0°C (273.15 K). A pound mole of any ideal gas occupies a volume of 380 standard cubic feet (SCF). One SCF is defined as a cubic foot of an ideal gas at a pressure of 30 in of mercury and a temperature of 60°F (519.7°R). Note the use of 60°F as the reference temperature, rather than 32°F, the Fahrenheit equivalent of 0°C.

coal's geological history, coal is not a uniform substance, and no two coals are ever exactly identical. Coal is a rock-like combustible material, generally rich in organic matter, but exhibiting substantial variability in chemical composition, molecular structure, inorganic content (e.g., minerals), and physical constitution, depending upon geologic age and other factors. Coal is sometimes referred to as an "organic rock," but it is well to remember that all coal types contain some amount of inorganic material. Elemental sulfur and nitrogen are important components in coal as they influence emissions.

- *Natural gas*: A gas found, among other locations, in underground reservoirs of porous rocks, alone or physically mixed with petroleum (see below), and consisting primarily of a mixture of simple paraffinic hydrocarbon gases of which methane (CH_4) is the major constituent (70–90% by volume), with ethane, propane, butanes, and higher paraffins occurring in proportions that, in general, decline with increasing carbon number of the hydrocarbon. Natural gas may also contain H_2S and inerts, such as N_2 and CO_2.
- *Petroleum*: Literally "rock oil," petroleum (Elliott and Melchior, 1985) refers to deposits of oily material found in the upper strata of the earth's crust. Petroleum consists mainly (generally 90%) of a complex mixture of hydrocarbons spanning a wide range of molecular weight, and was formed by a complex series of chemical transformations of organic precursors deposited in previous geological periods. Carbon and hydrogen make up the bulk of petroleum, but it can also contain sulfur (trace to 8%), nitrogen (trace to 6%), and oxygen (trace to 2%), as well as nickel and vanadium.
- *Oil shale*: A fine-grained sedimentary rock consisting of compact mineral matter (e.g., carbonates, sand, and clay) associated with an organic material, kerogen, of complex and widely variable composition but consisting mainly of carbon and hydrogen and capable of producing petroleum-like liquids upon heating (typical yields being 10–30 gallons of liquid per ton of oil shale heated, depending primarily on the quality of the oil shale).
- *Tar sands*: A sand or sandstone impregnated with a highly viscous, crude, asphalt-like hydrocarbon substance known as *bitumen* (Towson, 1985). Bitumen is defined as a hydrocarbon substance soluble in carbon disulphide. Typically, tar sands are a mixture of sand, water, and bitumen, with the latter ranging from trace to 18%. Roughly 50% of the bitumen can be recovered from the sand by heating without serious degradation. A typical bitumen H/C ratio is 1.4–1.5, and bitumen can contain 4.5–5.0 % S and 0.4–0.5 % N.

Most of the fuels we use in everyday life are refined or manufactured from the naturally occurring fossil fuels discussed above. Table 7.1 lists some attributes for a number of fuel types. Thus, gasoline, automotive diesel fuel, jet fuel, and home heating oil, as well as numerous chemicals, are refined (Section 7.3.3) from petroleum. Coal is often washed or otherwise treated to effect physical separations of less-desired components before it is combusted to generate electricity. Even natural gas, a rather clean fuel in its own right, typically undergoes some processing (e.g., desulfurization and stripping of higher hydrocarbons and inerts to adjust its heating values) before use in the marketplace. Further information on the upgrading of naturally occurring fossil fuels for consumer applications is presented in Section 7.3. The ratio of hydrogen to carbon atoms in the fuel (H/C ratio) and the fuel's higher heating value (HHV) are important properties that affect practical utilization.

Table 7.1. Hydrogen-to-Carbon Ratio and Heating (Calorific) Value of Selected Fossil and Process-Derived Fuels

Fuel Type	**H/C, Atomic**[a]	**Gross Heating Value (kcal/g)**[b,c]
Gaseous		
High-Btu		
Methane	4.0	13.3
Natural gas	3.5–4.0	11.7–11.9
Intermediate-Btu		
Hydrogen	∞	33.9
Coke oven gas	4.9	9.6
Water gas (H_2 - CO)	2.0	4.3
Low-Btu		
Producer gas (N_2-diluted; from bituminous coal)	1.2	1.2
Liquid		
Methanol	4.0	5.3
Gasoline	2.0–2.2	11.2–11.4
No. 2 fuel oil	1.7–1.9	10.7–11.0
No. 6 fuel oil	1.3–1.6	10.0–10.5
Crude shale oil	1.6	10.3–10.4
Bitumen (Athabasca tar sands)	1.4–1.5	9.8–10
Solid		
Kerogen (Green River oil shale)	1.5	10
Lignite	0.8	3.9–5.4
Subbituminous coal	0.8	5.5
Bituminous coal	0.5–0.9	6.7–8.8
Anthracite	0.3	8.4
Low-temperature coke	0.4	8.2
High-temperature coke	0.06	8.0

Source: Howard (1981) and references cited therein.

[a]Mineral-matter-free C and total organic H (i.e., including H bound with O, S, and N).

[b]Coals—moist, mineral-matter-free; cokes—dry, ash-free; 1 kcal/g = 1,800 Btu/lb.

[c]Gross heating value is equivalent to the higher heating value of the fuel (HHV).

7.1.2 Historical and current contributions of fossil fuels to human progress

As of 2002, roughly 85% of the developed world's energy was supplied by fossil fuels—largely petroleum, natural gas, and coal—almost totally by means of combustion. Owing to a number of factors, the percentage contribution of fossil fuels will probably decline as time moves along, but it is not known how rapidly this will occur. Undoubtedly, the drawdown in fossil use will vary from country to country. Moreover, in the US, fossil fuels dominate all major utilization sectors, with nuclear energy being the closest competitor in the electric sector, accounting for well over 20% of the kWhs generated in recent years. In other countries, fossil fuel is less significant in electricity generation but still dominates total energy consumption in the developed world.

Fossil fuel has gained its dominant position through consumer convenience and the attractive costs of fossil fuel–derived energy products and services, such as electricity, transportation fuels, industrial process energy, and home heating. The maturity and size of the fossil industry, its existing infrastructure, and its attractive economics are major challenges to the penetration of alternative energy technologies. To appreciate the magnitudes involved, consider the following facts.

- In the US, premium gasoline, even with taxes, costs about 1/4 or less the cost of bottled water.
- During fueling of a motor vehicle with gasoline or diesel fuel, energy flows into the fuel tank at a rate of 5–10 MW. If this were electricity, it would be equivalent to running an electric generating station to supply the needs of 5,000–10,000 people.
- World crude oil production is roughly 70 million bbl/day or 3.8 billion tons/yr. World steel production is about 1/5 this value.
- World coal production is 5.2 billion tons/yr, about seven times that of steel.
- World primary energy production is 375 Quads (quadrillion Btu). If this were entirely petroleum, it would be the equivalent of 175 million bbl/day.
- The world population in 2000 was roughly 6 billion people and growing. Sadly, 2 billion of these people had no electricity and lived in poverty.
- In the Organisation for Economic Cooperation and Development (OECD) countries, roughly 30 million new passenger cars are sold each year.
- The daily market value of electricity production in the US is about $0.5 billion.
- The daily cash flow associated with global trade in crude oil is about $1 billion.
- The cost to replace the global infrastructure for liquid transportation fuels is estimated at $3–5 trillion.
- For comparison, the 1999 US gross domestic product (GDP) was $9.3 trillion.

Selected Fuel Properties Affecting Performance and Environmental Impacts

The *heating value* of a fuel is the maximum amount of energy obtainable by combusting unit weight or volume of fuel and then cooling all the products of combustion to room temperature. Chemical enthalpy is released by the cumulative effects of breaking and forming bonds during combustion. Moreover, the process of cooling contributes additional energy if combustion products condense. This energy equals the enthalpy of vaporization (also known as the latent heat of vaporization) of the condensed product. Water is the most common condensable combustion product, so cooling the products of combustion to room temperature releases an additional 971.2 Btu for each pound of water condensed (or equivalently 539.6 cal/g). Tabulated values of the heating value of a fuel may take credit for the additional heat released by condensation of water. For fossil fuels, the tables rarely account for the small amount of heat liberated by condensation of other combustion products such as metal vapors or for the conversion of gaseous sulfur and nitrogen oxides into acid aqueous solutions (i.e, to form sulfuric and nitric acid, respectively). If a tabulated heating value does include the energy released by condensation of the water of combustion, it is called the *gross* or *higher heating value* (HHV). Similarly, if the value does not include this latent heat it is called the *net* or *lower heating value* (LHV).

Another common term is the *adiabatic flame temperature*. This is a useful idealization that signifies the maximum temperature that would be obtained if the fuel in question were completely combusted with exactly the right amount of oxidant and with no loss of heat to the surroundings. In such an idealized case, the fuel is completely oxidized to the lowest energy state products obtainable from its constituent atoms.

- Time scales in the fossil fuels and related large throughput industrial sectors:
  - Conceptualize new ideas for technological innovation: hours to months.
  - Research and development to test the technical and economic promise of new ideas at scales up through pilot (1/1,000 to 1/100 of commercial scale) and demonstration (1/10 of commercial scale): 2 to 10 years.
  - Design, siting, permitting, construction, commissioning, and shakedown of commercial facility: 1 to 10 years.
- Time scales in the political sector:
  - Single term of chief executive (president, prime minister); 3 to 5 years or less.
  - Single term of legislative branches (Senate, House of Representatives, Parliament); 2 to 6 years.
  - Global scale jeopardy to international shipments of energy: 6 years, i.e., 1939 through 1945 during World War II.

- Time between successful negotiation of major international economic treaties: 50 years from the General Agreement on Trade and Tarriffs (GATT) to the World Trade Organization (WTO).

These numbers delineate the magnitude of materials flows and monetary transfers in the global fossil energy sector, the long times typically needed for technological transformation in this sector, and the gargantuan sums of money that would be needed to completely replace the supply infrastructure for just one of the four major fossil utilization sectors (e.g., transportation fuels) with some non-fossil alternative. The sums involved could total the entire GDP of the US. Such a transformation is arguably feasible, but would be challenging to accomplish more than once in a century. Likewise, the listed time scales for terms of elected officials in democracies and between treaties show that institutional transformation typically proceeds slowly.

7.1.3 Sustainability: Challenges and opportunities

Were it not for the climate-forcing potency of CO_2 emissions produced by fossil fuel combustion and the CH_4 emissions owing to leakage from pipelines and the like, fossil fuels might not face serious environmental challenges. This is because it seems probable that technological advances would conquer environmental impacts at costs acceptable to consumers. However, present understanding is inadequate to project whether technological measures would allow an economically viable fossil option to be preserved if serious restrictions on carbon emissions to the atmosphere are called for. The potential conflict between continuing to harvest the attractive benefits of fossil fuels and preventing adverse global climate impacts from greenhouse gases (GHG) emitted by the use of fossil fuels epitomizes the difficulty of translating the laudable ideals of sustainability into practicality. As discussed in Section 7.4.3, the quest to overcome this difficulty has spurred proposals, research, pilot testing, and field studies to mitigate GHG emissions or their adverse climate effects through geo-engineering, emissions reduction, and GHG removal from the atmosphere. However, each of these approaches has "sustainability" challenges.

7.2 The Fossil Fuel Resource Base

7.2.1 How long will fossil fuels last?

Nobody knows when we will "run out" of oil or other fossil fuels. In seeking a glimpse into the future, one always looks at the cost of incremental production from a given source. When that exceeds the then-prevailing costs of other wells or mines that produce fossil fuels, production at that source is suspended—maybe forever or until that source again becomes economical (e.g., owing to some technological advance). Indeed, the production of petroleum and other fossil fuels is a grand example of how the advance of technology has created pleasant surprises. The amounts of fossil fuels recoverable from the earth are determined by competition between depletion of the best resources and advances in technology that open up previously unavailable stocks. So far, technology has won this battle.

Table 2.3 shows that additions to petroleum reserves over the period 1944–1993 outpaced cumulative production. Table 7.2 shows estimates of the earth's resources of coal, petroleum, natural gas, tar sands, and oil shale, as well as peat and uranium as ^{235}U. Such data are useful in making approximate comparisons of how much various energy resources may contribute to future consumption, as well as rudimentary forecasts of the time scale to deplete a set fraction of these resources under a prescribed drawdown schedule. Sample problem 7.1 illustrates such a calculation. However, any specification of a fixed value of a fossil fuel resource automatically brings with it assumptions about how well technology can liberate the resource from the earth. Again, the record has always been in the direction of better and better technology liberating previously unrecoverable stocks. This will ultimately change because the time scale for replenishment of naturally occurring fossil fuels is so much longer than any imaginable societal utilization time.

7.2.2 "Unconventional" naturally occurring fossil fuels

Two other categories of naturally occurring fossil fuels are "ultra-heavy" (i.e., high specific gravity) oil and natural gas hydrates. Both are considered "unconventional" in that the form of their occurrence in the earth is substantially different from forms in which oil and natural gas are found. Both resources are of appreciable interest because the deposits are believed to be so large that, if they could be exploited, they would increase total global deposits of petroleum and natural gas several-fold.

Table 7.2. Estimates of Various Energy Resources on Earth ca. 1999 (See also Sample Problem 7.1)

Fuel	10^{15} kJ*
Coal	290,000
Petroleum	2,600
Natural Gas	5,400
Tar Sands	5,700
Oil Shale	11,000
Peat	3,000
Uranium (^{235}U)	2,600

*1 kJ = 0.948 Btu

Sample Problem 7.1a: Fossil fuel lifetimes at present consumption rates.

Take coal as an example. If the world suddenly switched all its energy consumption to coal, how long could the estimated resource last?

Resource for world $= 290{,}000 \times 10^{15}$ kJ $\times 0.948$ Btu/kJ

$= 275{,}000 \times 10^{15}$ Btu

At 1995 world total energy consumption rate $= \dfrac{275{,}000 \times 10^{15}}{325.4 \times 10^{15}} = 846$ yrs.

Sample Problem 7.1b: Fossil fuel lifetimes with exponential growth in consumption.

What effect would growth in energy consumption rate have? Q = energy, t = time, m = growth rate

$$\frac{dQ}{dt} = ce^{mt}$$

take $t = 0$ at 1995 for convenience in data use.

$$\left(\frac{dQ}{dT}\right)_{t=o} = ce^{m(0)} = c = \left(\frac{dQ}{dT}\right)_{1995}$$

$$\frac{dQ}{dT} = \left(\frac{dQ}{dT}\right)_{1995} e^{mt}$$

$Q_\infty \equiv$ the coal resource originally available (before any use)

$Q_\infty - Q_{1995}$ = resource yet to be consumed

$$\int_{Q_{1995}}^{Q_\infty} dQ = \left(\frac{dQ}{dt}\right)_{1995} \int_0^{\Delta t_{lifetime,\ growth}} e^{mt}$$

$$Q_\infty - Q_{1995} = \left(\frac{dQ}{dt}\right)_{1995} \frac{1}{m}\left[e^{m\Delta t_{lifetime,\ growth}} - 1\right]$$

$$m\left\{\frac{[Q_\infty - Q_{1995}]}{(dQ/dt)_{1995}}\right\} + 1 = e^{m\Delta t_{lifetime,\ growth}}$$

$$\Delta t_{lifetime,\ growth} = \frac{1}{m} \ln\left\{m\Delta t_{lifetime,\ no\ growth} + 1\right\}$$

Take $m = 2\%$/yr

$$\Delta t_{lifetime,\ growth} = \frac{1}{0.02\ yr^{-1}} \ln\left\{(0.02\ yr^{-1})(846\ yrs) + 1\right\}$$

= 144 years (would be 109 years for 3%/year).

Therefore, even small exponential growth can significantly reduce depletion times even for large resources.

Ultra-heavy oil (Glasstone, 1982) is a highly viscous and dense mixture of hydrocarbons. Its viscosity (resistance to flow) is typically so high that less than 10% of the oil can be recovered from the earth by conventional pumping. To lower the oil's viscosity, the oil is heated (e.g., by injecting steam from the surface of the well). The oil can also be heated *in situ* by injecting air to burn gas within the well or to burn some of the oil itself (*fire flooding*). Combustion of the oil is maintained by continuously injecting air so that a slow burning wave moves through the oil, heating it sufficiently to allow it to be pumped. Water may be introduced ahead of the burning front to yield hot water and steam to facilitate flow of the oil. In some locations, heavy oil is mined by penetrating oil deposits with holes that allow the oil to flow into collection sumps from which it is pumped to the surface. Heavy oil has a low H/C ratio and must be extensively processed to derive fuel products comparable in quality to present day liquid transportation fuels (Section 7.3.3).

Gas hydrates (Malone, 1990) are crystalline, ice-like solid cages in which frozen water molecules surround natural gas. For certain ranges of temperature and pressure, natural gas hydrates form naturally in sediments beneath the earth's oceans and permafrost. By some estimates, global deposits of natural gas hydrates are so vast that they may exceed all the earth's other deposits of fossil fuels combined. The challenge is learning how to safely and economically harvest the natural gas locked up in hydrates. Natural gas is an attractive fuel because of its high H/C elemental ratio. However, it is also a potent GHG, so fugitive emissions of natural gas to the atmosphere during recovery and processing need to be avoided. Moreover, in hydrates, there are many water molecules for each molecule of gas. Thus, it would be wasteful to "lift" the complete hydrate with all that water. Instead, we need to devise economical and efficient *in situ* processes to separate the natural gas from its water cage and to then either harvest the gas for use above the surface or to use the gas in place (e.g., to produce electricity or hydrogen for transmission to *ex situ* locations where they can be used).

7.2.3 Fossil resources and sustainability

The driving force for sustainability is *not* that the world is running out of fossil fuels and needs to seek alternative energy sources. To the contrary, the weight of evidence is that the combination of known deposits exploitable with current technologies plus the opening of additional deposits through advances in technology will continue to keep pace with demand growth for many decades. However, alternative energy sources do need to be developed, preferably in the context of a sustainability view of earth-human ecology:

- As insurance in case the higher levels of global warming predicted from a doubling of current atmospheric concentrations of CO_2 would prove to be correct
- As a means to smooth the transition to lower fossil dependency at whatever rate of transformation circumstances call for

- To diversify global sources of energy to prevent over-reliance on any one source and the resultant vulnerability to source unavailability, e.g., political, environmental, regulatory, or technological factors

7.3 Harvesting Energy and Energy Products from Fossil Fuels

7.3.1 Exploration, discovery, and extraction of fuels

Coal deposits are plentiful and globally distributed, so most coal companies focus on issues of extraction and bulk transportation. Major petroleum companies devote substantial resources to the so-called "upstream" portion of their industry (i.e., to finding additional exploitable deposits of oil and natural gas). Modern seismic and other technologies for illuminating underground and undersea geology are identifying previously well-camouflaged deposits of petroleum and gas for profitable development. Likewise, new generations of drilling and production hardware and software are liberating oil and gas from reservoirs that in the past were unassailable by human intervention. One example is deep-sea production. In the 1970s, offshore drilling was technically feasible only in water a few feet deep. In 2003, oil and gas were routinely produced from water depths of a mile or more.

7.3.2 Fuel storage and transportation

The storage and transport of fossil fuels pose considerable safety and environmental risks. (These topics are discussed in more detail in Chapter 16.) Safety problems, in particular fire and explosion risk, arise from ignition of spills of liquid fuels, such as gasoline or liquefied natural gas (LNG), or of pressurized releases of gases/aerosols from liquefied petroleum gas (LPG) or natural gas or even combustion of solid fuels. Careless handling of fuels during shipment, loading, or unloading may create flammable or even explosive mixtures of the fuel with air at particular locations, so that the presence of an ignition source can instigate a catastrophic fire or explosion. Unfortunately, effective ignition sources are ubiquitous, and the risks of human injury or death, as well as of property damage, can be substantial. Common ignition sources are tiny electrical discharges (sparks) formed by flipping a light switch or from relaxation of static electricity accumulated on a rubber conveyor belt or other electrically low-conducting material.

Fuel spills can also cause serious environmental damage. The tragic contamination of the Prince William Sound ecosystem by a major oil spill from the Exxon *Valdez* in 1989 and of the west coast of Spain by an oil tanker breakup in 2002 show the tragic consequences of human or technological failures in environmentally sensitive areas. Because every region of the planet is laden with its own unique ecological sensitivities and vulnerabilities, great caution must be exercised in planning and implementing the shipment of anthropogenic goods regardless of their location. The risk of human injury and environmental damage from fuel storage and transport can be reduced through modern technology and energetic vigilance by owners and operators to avoid human errors and prevent mechanical failures. However, the risk can never be made zero, unless the activity itself is eliminated. Once again, we witness a compelling example of the tradeoffs in the real world of sustainable energy—finding the proper balance in protecting humans and their environment through regulations, monetary expenditures, and monitoring of employees' on-the-job performance to assure minimal risk in the storage and shipping of oil, natural gas, and other fossil fuels.

Economics are another issue in fuel storage and transport. Many otherwise attractive deposits of natural gas go undeveloped because it is too expensive to ship the gas to consumers. Because shipping costs scale with volume, this has spurred R&D to increase the volumetric density of this so-called "remote" or "stranded" gas by economically converting it to useful liquid or solid fuels. (Recall that solids and liquids are about 1,000 times more dense than gases at atmospheric pressure.)

7.3.3 Fuel conversion

Fuel conversion, known as fuel processing, is the chemical or physical transformation of a naturally occurring or already modified fuel to improve the quality of the fuel. A fuel conversion process (Figure 7.1) may result in one or more upgraded fuel products that may be solid, liquid, or gaseous, and may generate chemicals or raw materials for chemical manufacture (chemical feedstocks). Each raw fuel feedstock has its own conversion requirements. Coal typically requires size reduction, washing, and removal of inert species. Natural gas may need removal of H_2S and CO_2, along with separation of gas liquids like LPGs and C_2 compounds. Petroleum refining and coal gasification are examples of more complex fuel conversion processes. Quality upgrades are typically driven by the need for a fuel that is more compatible with existing utilization equipment or to remove environmentally offensive components in the fuel. Examples are the manufacture of automotive gasoline by "cracking" (thermal decomposition) of petroleum components in a refinery and removal of sulfur and nitrogen from liquid fuels by reacting the fuel with hydrogen. An example of a non-chemical conversion of a fuel is the fractional distillation of a crude oil to separate it into several fractions of different volatility (boiling point) ranges.

The essence of fuel conversion is prescribed modification of fuel composition, chemically or physically, to augment fuel quality. In general, fuels of higher H/C ratio exhibit higher "quality" as measured by environmental impacts, ease of storage, handling, combustion (Section 7.3.4), and overall consumer convenience. Two basic mechanisms of fuel conversion are rejection of carbon and addition of hydrogen. Often both are practiced in a comprehensive fuel conversion plant. The detailed design and operating requirements of a fuel conversion plant are dictated by the properties of the raw material (fuel) being processed, by the properties desired in the product (refined) fuel, and by considerations of overall energy consumption, process economics, and process safety and environmental impacts.

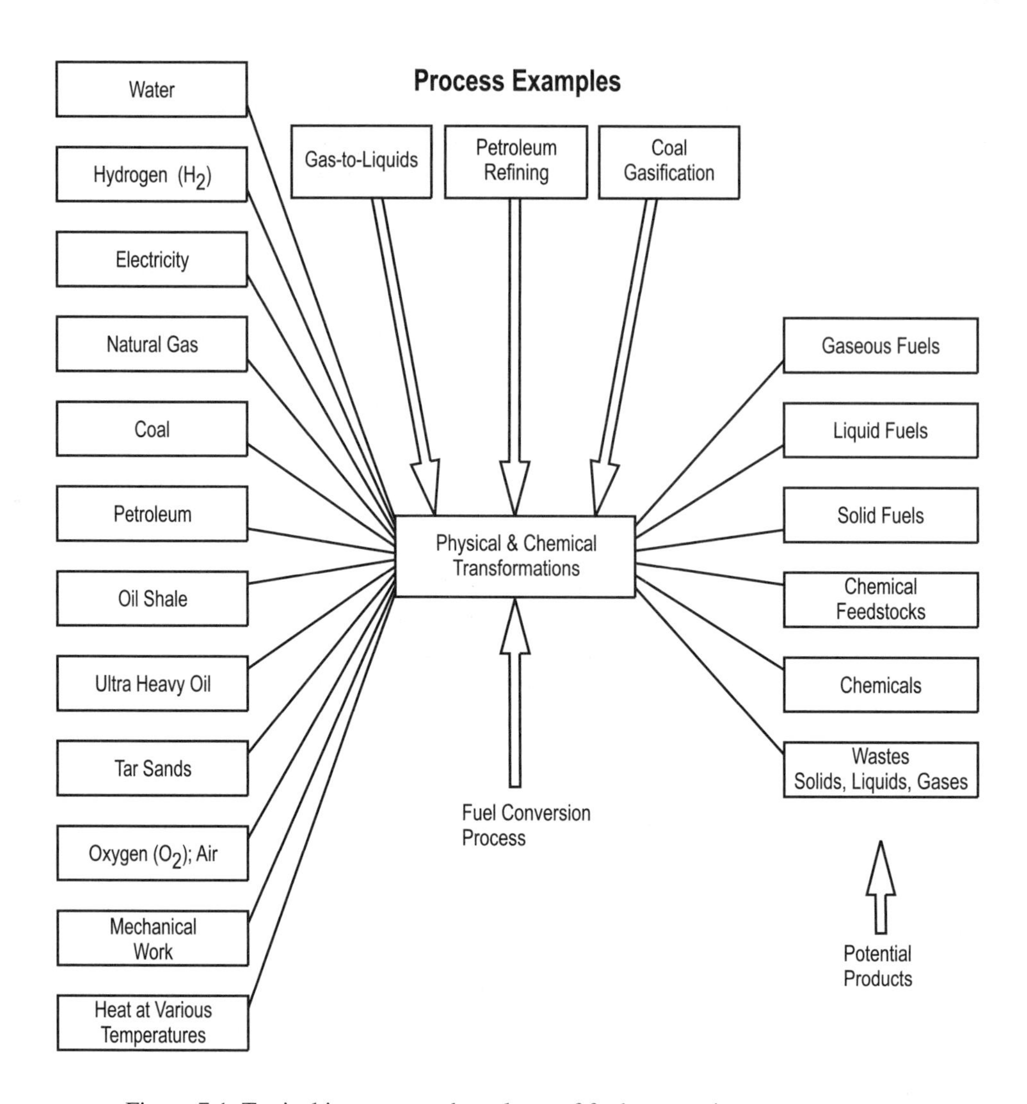

Figure 7.1. Typical inputs to and products of fuel conversion processes.

Technology for fuel conversion has been studied for roughly a century. The art is highly sophisticated and successful for petroleum refining. Major progress has been made in coal conversion, in thermal extraction of oil from shale and tar sands, and in the upgrading liquids derived from coal, shale, and tar sands. There is still abundant opportunity to develop improved fuel processing technologies for all fossil (and non-fossil, Chapter 10) substances.

There are also opportunities for applying biotechnology to fuel conversion (e.g., to removal of heteroatoms and improvement of volatility). However, thermal and catalytic (largely heterogeneous catalysis, i.e., fluid-solid processing) reactions dominate fuel conversion chemistry. Moreover, diverse physical separations are employed in fuel upgrading. Detailed information is found in Elliott, 1981, Lowry, 1963, 1945, Hottel and Howard, 1971, Gary and Handwerk, 2001, Satterfield, 1991, and Shreve and Brink, 1984. Recent progress can be found in publications from the US Department of Energy and the International Energy Agency. Katzer et al. (2000) summarize the historical trajectory of enabling technologies in producing liquid transportation fuels from petroleum and provide insights on expected technological changes including diversification of the raw materials base to include natural gas.

Petroleum refining is currently the dominant technology for fuel conversion. Modern refineries convert crude oil to an array of fuels and other valuable products. Refining technology is remarkable in that it transforms a low-quality, highly complex mixture of naturally occurring and often polluted substances (i.e., hydrocarbons and affiliated compounds) to a host of clean fuels, chemicals, and other products essential to modern living. In a modern refinery, all of this is accomplished at modest expenditures of energy, in an environmentally sound manner, while generating products so inexpensive that they are readily accessible to most consumers. The details of petroleum refining are discussed in Gary and Handwerk, 2001, Satterfield, 1991, Shreve and Brink, 1984, and Mark et al., 1985. Katzer et al. (2000) look at where refining technology may take us in the first half of the 21st century.

All refineries perform a series of physical and chemical operations to achieve three basic functions that we will call the Three "C"s: *compartmentalization*, *conversion*, and *cleaning*. These operations may occur sequentially or together, and they may be repeated at different stages of refining (i.e., they may be applied to the crude oil and to various products generated from the oil). *Compartmentalization* denotes separation into simpler components. It is exemplified by fractional distillation of crude oil but also includes any of a number of separations that occur at various stages of product manufacture and cleaning throughout the refinery (e.g., separation of H_2 gas from CO_2). *Conversion* refers to modification of the chemical composition of fuels to instill desired properties (e.g., catalytic cracking of higher molecular weight fractions to obtain hydrocarbons suitable as gasoline components). *Cleaning* encompasses the removal of pollutants from the fuel (for example, catalytic hydrodesulfurization, hydrodenitrogenation, and hydrodemetallation, in which the dirty fuel is reacted with H_2 in the presence of a catalyst to reduce sulfur, nitrogen, and metals, such as Ni and V).

Although most refineries exhibit these essential features, each refinery is unique. Refineries must adapt their operations to accommodate appreciable variations in the quality of the crude oil they process. Moreover, refineries must adjust the yield and the specific properties of their various products to meet regional and temporal variations in customer and regulatory requirements. For example, in the US, refineries are "tuned" to produce more gasoline in the summer to meet the demands of increased driving, and more middle-distillate range fuel in the winter months to meet increased demands for home heating oil. More and more specialization of product quality over shorter and shorter time frames is a trend for refineries of the future (Katzer et al., 2000). In addition,

US refineries must be operated to meet region-specific air-quality regulations. Consequently, a refinery in one part of the country with excess output cannot, without substantial adjustments, help allay shortages of fuel in a different region.

During the last 80 years, there has been substantial on-again, off-again interest in producing so called "substitute" or "synthetic" fuels from coal and other non-petroleum sources, such as oil shale and biomass (Chapter 10). There is substantial governmental and industrial interest in developing technologies to use the vast coal resources of the US as raw materials for the production of clean fuels comparable in environmental quality and consumer convenience to petroleum-derived stocks. The Clean Coal Technology Program of the US Department of Energy exemplifies current efforts (Section 7.8). The term *synthetic fuels* is somewhat unfortunate because, for commercial use, fossil fuels are rarely used "as received" (i.e., they have undergone some measure of cleaning or other processing), and therefore, most fossil fuels that we encounter are synthetic to some degree. A more precise nomenclature would be *non-petroleum-derived synthetic fuels*. As a compromise, we will use the words *substitute* or *alternative fuels*.

Historically, the development of technologies to manufacture alternative fuels has been motivated by concern that depletion of domestic supplies or political interference with access to foreign sources would create scarcity of petroleum or natural gas. More recently, environmental factors, such as carbon management and reduced sulfur emissions, have helped spur interest in the use of natural gas and of biomass (Chapter 10) as raw materials for alternative fuels manufacture. At various times, the objectives of alternative fuels initiatives have been to manufacture cleaner, more versatile fuels from coal or oil shale.

Here, we illustrate basic processes for converting coal to various alternative fuels. Most processes entail thermal treatment, at some stage causing the coal to undergo *pyrolysis* (i.e., thermal decomposition in the absence of oxygen), producing (Figure 7.2) a mixture of gaseous, liquid, and solid products in proportions that depend on the type of coal and on the treatment conditions. Each of these products could be used as fuels. In general, however, they require further processing to be compatible with today's utilization equipment and environmental requirements.

Pyrolysis itself can be a coal conversion process and many technologies have been developed to actualize pyrolysis commercially. The most common and successful process is the manufacture of metallurgical coke for steelmaking. There have been a plethora of process variations, including heating the coal in the presence of hydrogen (*hydropyrolysis*) and combining pyrolysis of the coal with downstream processing or combustion of (some of) the pyrolysis products.

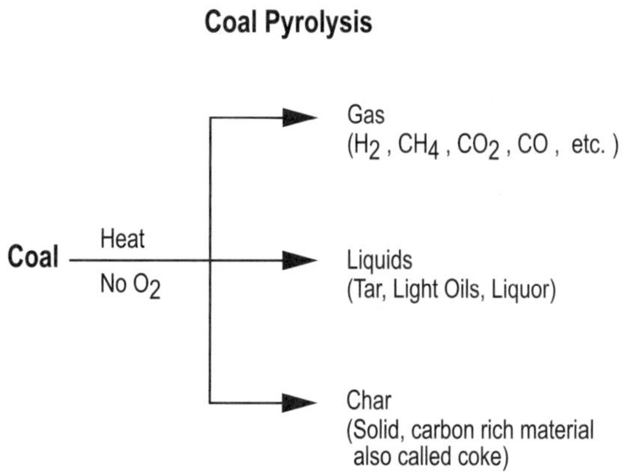

Coal Pyrolysis

Gas
(H_2, CH_4, CO_2, CO, etc.)

Coal — Heat / No O_2

Liquids
(Tar, Light Oils, Liquor)

Char
(Solid, carbon rich material also called coke)

Figure 7.2. Schematic of coal pyrolysis showing three major product categories.

Coal liquefaction involves converting coal, a solid at room temperature, to a liquid suitable, either directly or with further upgrading, as a substitute for petroleum-derived, consumer-grade liquid fuel. Coal is deficient in hydrogen, a typical empirical formula being $CH_{0.8}$ versus $CH_{1.8}$ for diesel fuel or No. 2 fuel oil used to heat single family homes. Thus, coal liquefaction entails the rejection of carbon, or the addition of hydrogen, or both. Pyrolysis is a carbon-rejection process (Figure 7.2) and produces a liquid (tar), so pyrolysis is a means of coal liquefaction. In general, coal pyrolysis tars require further upgrading before they are suitable as petroleum-substitute fuels. Another means of coal liquefaction entails reaction of the coal, perhaps slurried in coal-derived oil, with hydrogen:

$$CH_{0.8} + H_2 \rightarrow CH_{1.3\text{-}1.7} \text{ (Liquid Products)} + \text{Gas} + \text{Solids} \qquad (7\text{-}2)$$

Although only a schematic equation, Reaction (7-2) provides a lot of information about the requirements of coal liquefaction technology. Note that the key to a good liquefaction process is having a good source of low-cost hydrogen to upgrade the coal. This typically means having a good coal gasification process, because in most cases, unless there is access to low cost natural gas, the H_2 needs to be manufactured by gasification of the coal (see below). Note also that there are co-products including solids and gas. The gas may include H_2S, NH_3, or HCN formed by reaction of hydrogen with S and N from the coal. These gases would need to be separated from the liquid products and further processed to extract valuable products, such as elemental sulfur and ammonia for conversion to fertilizer or other substances. Somewhat paradoxically, solid byproducts are a major challenge in coal liquefaction technology. These solids consist of inorganic matter in the coal, which cannot be converted to hydrocarbon liquid, as well as solids derived from the organic portion of the coal. These solids need to be separated from the liquid product so that the latter can be conveniently pumped and used in modern combustion equipment. These solids are used, if possible, in road paving or similar applications, or they are disposed of as wastes. The gas can be separated and burned to provide some of the energy needed for the process.

Coal gasification is the production of gas from coal. Table 7.3 summarizes the major chemical reactions that occur in coal gasification, where it is recognized that coal is approximated by solid carbon (see footnote to this table). Reaction (7-3) is endothermic (i.e., it requires an input of energy to proceed from left to right as written). The energy needed to drive Reaction (7-3) can be generated by a suitable heat-liberating (exothermic) chemical reaction. Examples are burning some of the coal, partially to form CO (Reaction (7-6)) or more fully to form CO_2 (Reaction (7-9)). These reactions can occur in a separate section of the gasifier or in a separate reactor equipped with some means to transfer the heat back into the gasifier. Reaction (7-4), *the water gas shift* reaction, is important in fuels technology as it provides a simple means of generating hydrogen from carbon monoxide merely by reacting it with steam at temperatures between 500 and 1,000°C. Reaction (7-5) depicts direct production of methane by so-called hydrogasification (i.e., reacting the coal directly with hydrogen gas).

Table 7.3. The Basic Chemical Reactions for Simulation of Coal Gasification Chemistry[a]

$$C(s) + H_2O(g) \rightarrow CO(g) + H_2(g) \quad (7\text{-}3)$$

$$CO(g) + 2H_2O(g) \rightarrow CO_2(g) + H_2(g) \quad (7\text{-}4)$$

$$C(s) + 2H_2(g) \rightarrow CH_4(g) \quad (7\text{-}5)$$

$$C(s) + 1/2O_2(g) \rightarrow CO(g) \quad (7\text{-}6)$$

$$C(s) + H_2O(g) \rightarrow 1/2CH_4(g) + 1/2CO_2(g) \quad (7\text{-}7)$$

$$CO(g) + 3H_2(g) \rightarrow CH_4(g) + H_2O(g) \quad (7\text{-}8)$$

$$C(s) + O_2(g) \rightarrow CO_2(g) \quad (7\text{-}9)$$

[a]A reasonable empirical formula for the organic, i.e., non-mineral matter portion of coal is $CH_{0.8}S_xN_yO_z$, where x, y, and z vary but typically are each < 0.1. Here for simplicity coal mineral matter is neglected and the organic fraction of coal is represented as solid elemental carbon C(s), often taken as graphite in thermodynamic calculations.

Adapted from Hottel and Howard (1971).

In the US in the late 1960s and early 1970s, it appeared to many observers that substitute or synthetic natural gas (SNG) from coal would be needed because of depletion of domestic natural gas reserves. This precipitated substantial R&D by government and the private sector to develop technically and economically viable means to capitalize on basic gasification chemistry. Reaction (7-7) is the result of Reactions (7-3), (7-4), and (7-5), and depicts every coal gasification technologists "dream" reaction, namely the direct production of methane by reacting coal with steam in a single step. Unfortunately, this reaction is slow at temperatures where thermodynamics favor a decent yield of methane, ca. 500°C, so a suitable catalyst would be needed to accelerate the reaction rate without going to higher temperatures unfavorable to methane production. By "suitable," one means a catalyst that is inexpensive or readily separable and recoverable from the reacted (gasified) coal and without interference or catalyst poisoning by the coal mineral matter. In the 1970s, the Exxon Catalytic Gasification process accomplished this to a high degree of technical success, but the process was never commercialized at large scale owing to declining interest in producing SNG from coal.

Reaction (7-8) is another process chemistry for making methane from coal, this time indirectly, in that the reactants are gases, CO and H_2, that can be produced from coal (e.g., by reactions such as Reactions (7-3), (7-4), and (7-6)). Reaction (7-8) is referred to as *methanation*. As with Reaction (7-7), it is slow at temperatures sufficiently low to obtain high yields of methane, and it also requires a catalyst. Moreover, Reaction (7-8) releases tremendous amounts of heat, and, because Reaction (7.8) operates at low temperatures, this heat emerges at a low "availability" (i.e., it cannot be used to directly drive other reactions in coal gasification that require a high temperature such as Reaction (7-3)). Some of the heat can be recovered to preheat coal and steam fed to a gasification reactor. (Note that Reaction (7-8) is the same as Reaction (7-5) subtracted from Reaction (7-3).)

Figure 7.3 shows the wide range of gas heating values obtainable by coal gasification. Low heating value gas is of interest for electric power generation at locations proximate to the gasifier because its low quality precludes economical shipment over extended distances (e.g., > 100 miles). Intermediate heating value gas can be stored and shipped more economically because of its relatively higher energy density. Suitable applications are power generation and industrial process heat. Synthesis gas is a mixture of H_2 and CO in various proportions. Many valuable fuels, including high heating value gas (mainly methane, see Reaction (7-8) and Table 7.1) and chemicals, can be manufactured by reacting synthesis gas with the appropriate H_2/CO ratio over a suitable catalyst (Figure 7.4).

Sample Problem 7.2

Question: What is the mass of: (a) 0.5 g mole of carbon dioxide, and (b) 0.25 g mole of water vapor?

Solution: First find the molecular weights.

For CO_2:	1 atom C × atomic wt of C	1 × 12 =	12
	2 atoms O × atomic wt of O	2 × 16 =	32
		Total =	44
For H_2O:	1 atom O × atomic wt of O	1 × 16 =	16
	2 atoms H × atomic wt of H	2 × 1 =	2
		Total =	18

Part (a): 0.5 g mole = 0.5 × MW in g = [0.5 g mole CO_2] × [44 g CO_2/(g mole CO_2)] = 22 g
Part (b): 0.25 g mole = 0.25 × MW in g = [0.25 g mole H_2O] × [18 g H_2O/(g mole H_2O)] = 4.5 g

The last two decades have witnessed substantial interest in the development of technologies to convert natural gas to liquid fuels capable of displacing petroleum-derived transportation fuels, in particular diesel and gasoline. Drivers include the known and probable deposits of natural gas throughout the world. Some of this gas is in remote locations and would require densification for economical shipment to major markets. Other deposits are of so-called unconventional gas such as hydrates, but these could potentially swamp all known petroleum deposits. Thus, having technology to transform natural gas to storable and transportable liquid fuels able to displace petroleum-derived premium transportation fuels has appreciable market appeal. Moreover, there are indications that substitute diesel of higher environmental quality than petroleum-derived diesel, in particular near zero sulfur and nitrogen content, can be manufactured from natural gas. (Schmidt et al. (2001) elaborate on the opportunities.) Here, using methane as a proxy for natural gas, we describe one of the basic approaches to *gas-to-liquids* fuel conversion, namely steam reforming of the gas to make synthesis gas followed by Fischer-Tropsch (FT) catalytic synthesis of premium liquid fuels:

$$CH_4 + H_2O + \text{Heat} \rightarrow CO + 3H_2 \tag{7-10}$$

$$CO + yH_2 \xrightarrow[\text{Pressure} \sim 1 \text{ atm}]{\text{FT catalyst}} \text{gasoline, diesel, methanol, etc.} \tag{7-11}$$

The quantity y denotes that different H_2/CO ratios are selected in FT processing according to the desired product.

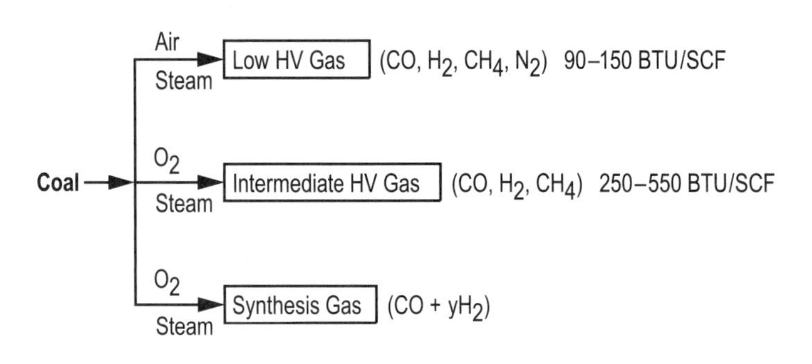

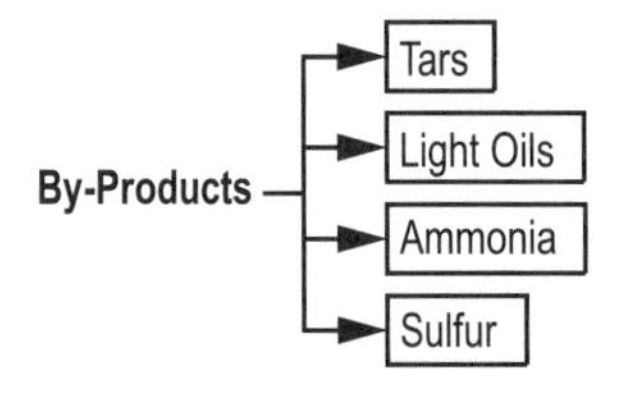

Figure 7.3. Major products and byproducts of coal gasification with steam using various process options.

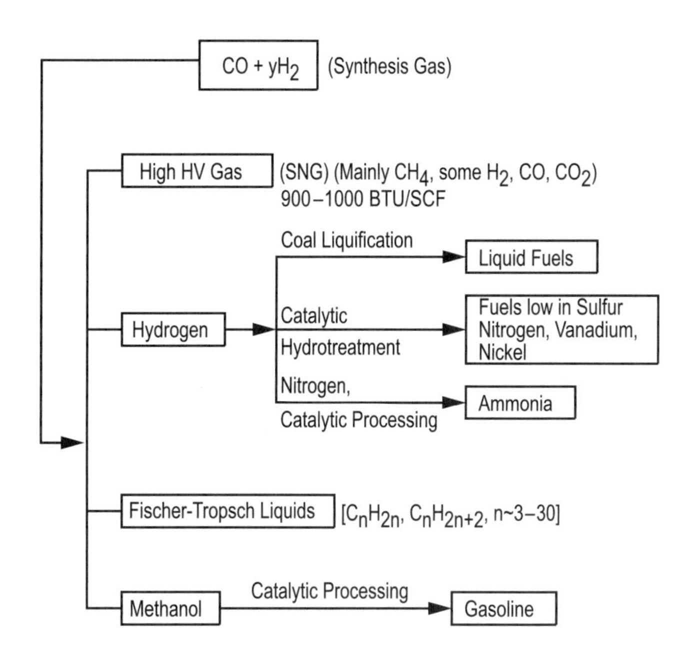

Figure 7.4. Examples of the wide variety of fuels and chemicals obtainable from synthesis gas.

Sample Problem 7.3

Question: Estimate the amount of carbon dioxide, in lbs and in SCF, emitted by completely burning one gallon of gasoline with the following chemical analysis: 85.7% carbon and 14.3% hydrogen. Assume the only products of combustion are water and carbon dioxide and that the carbon dioxide behaves as an ideal gas.

Solution: We first need to find a chemical formula for the gasoline, so that we can then write a chemical equation for combustion of this fuel. Even though a gallon of gasoline has a mass of roughly 7 lbs, to make the calculations easier in finding the chemical formula, assume that we have 100 lbs of this gasoline. Then, from the analysis we know that there are 85.7 lbs of carbon and 14.3 lbs of hydrogen atoms in this gasoline. To find the corresponding number of each carbon and hydrogen atom, we simply divide by their atomic weights:

Number of C atoms = (85.7 lb C)/(12 lb C/lb-atom C) = 7.14 lb atom C

Number of H atoms = (14.3 lb H)/(1 lb H/lb-atom H) = 14.3 lb atom H

The chemical formula for the gasoline is $C_{7.14}H_{14.3}$, or, normalizing to 1 atom of C, $C_1H_{14.3/7.14} = CH_2$.

Assuming only H_2O and CO_2 as combustion products, the chemical reaction for combustion is:

$$CH_2 \quad + \quad yO_2 \rightarrow CO_2 \quad + \quad H_2O \qquad (7.3\text{-}1)$$

Here y is an unknown *stoichiometric coefficient* equal to the number of moles of molecular oxygen needed so that the chemical reaction accurately accounts for all the atoms of each element involved in the combustion process. To compute y, we must "balance" this chemical reaction, which means that we must find the value of y so that the number of oxygen atoms are the same on both sides of the equation. This is already the case for the atoms of H and of C, but in more difficult problems it may be necessary to perform individual balances for several elements (i.e., to solve simultaneously for two or three or even more unknown stoichiometric coefficients). Equating the number of oxygen atoms on the left and right hand sides of the equation we find:

$2y$ = 2 (oxygen atoms from CO_2) + 1 (oxygen atom from H_2O) = 3, i.e., $y = 1.5$, and Equation (7.3-1) becomes:

$$CH_2 \quad + \quad 1.5O_2 \rightarrow CO_2 \quad + \quad H_2O \qquad (7.3\text{-}2)$$

This chemical equation tells us that complete combustion of 1 lb mole of this particular gasoline in pure oxygen emits 1 lb mole each of water and of CO_2. Note that if a ton mole of this gasoline were burned, the mass of water and CO_2 emitted would each be 1 ton mole. The present problem asks how much CO_2 is emitted by burning a gallon of gasoline, so we need to know how much mass is in 1 gal of gasoline, i.e. we need to know its mass density, ρ_m. A typical value is roughly 7 lb/gal. To use Equation (7.3-2) we also need to compute how many lb moles of gasoline are in a gallon. As in sample problem 7.2, first find the MW:

For CH_2:	1 atom C $\times$ atomic wt of C	$1 \times 12 =$	12
	2 atoms H $\times$ atomic wt of H	$2 \times 1 =$	2
		Total =	14

Then from Equation 7.3-2 we have:

(44 lbs CO_2)/(14 lbs gasoline) $\times$ (7 lbs gasoline)/(1 gal gasoline) = 22 lbs CO_2/gal of gasoline

To express this as standard cubic feet of CO_2 we then make the following conversion:

(22 lbs CO_2)/(gal of gasoline) $\times$ (1 lb mole CO_2)/(44 lbs CO_2)
$\times$ (380 SCF CO_2)/(1 lb mole CO_2) = 90 SCF of CO_2

7.3.4 Fuel combustion

Combustion is rapid release of heat by chemical reaction(s). The reactants are always a fuel (see above) and almost always a separate substance, an *oxidant*, which converts the chemical potential energy of the fuel to thermal energy. Combustion is enabled by feedback of some of its own output (e.g., energy, material, fluid motion, or various combinations). This is the case whether a small portion of fuel is burned over a short time (batch combustion), as occurs in the cylinder of an internal combustion engine (Chapter 3), or fuel is continuously burned over many hours or longer, as occurs in a coal or natural gas-fired electricity utility boiler. Thus, combustion is inherently a feedback process. This feedback creates and maintains an environment in which fuel and oxidant are continuously "mixed" (i.e., brought into chemical intimacy) for temperatures and times needed to sustain the desired fuel-oxidant reactions. The energy feedback can be by thermal radiation from the flame to the unignited combustible mixture, or by recirculated flow of hot combustion products entrained into the unignited turbulent fuel/air jet stream.

Other than rocket propulsion, most practical combustion reactions that produce energy for useful purposes involve combining the fuel with a source of molecular oxygen, typically air itself (a mixture of approximately 79 vol % N_2 and 21 vol % O_2) or oxygen enriched air (i.e., air in which the percentage of oxygen exceeds 21%, up to essentially pure oxygen). For several reasons, a basic comprehension of combustion is essential to understand the challenges and opportunities of sustainable energy:

- Combustion is the dominant means of transforming fossil (and biomass, Chapter 10) fuels to other useful forms of energy and is virtually certain to remain so, even with major technological and marketing gains by fuel cells (see Section 7.3.5).
- Combustion is the major source of anthropogenic emissions of carbon dioxide to the atmosphere.
- Uncontrolled combustion generates diverse pollutants with adverse ecological and human health impacts.
- Current scientific and engineering understanding of combustion is sufficient to allow design and operation of efficient, low environmental impact, stationary (Beér, 2000) and mobile (Heywood, 1988) combustors.
- Combustion technology will continue to improve at an evolutionary rather than revolutionary rate, but with sufficient dispatch to keep pace with demands for increasingly better environmental performance.

Reaction (7-12) depicts a generalized combustion reaction in a form that illuminates issues relevant to sustainable energy:

$$\text{Fuel} + \text{Oxidant} + \text{Diluent} \rightarrow \text{Desired Products} + \text{Energy} + \text{Diluent} + \text{Undesired Products} \quad (7\text{-}12)$$

Rapid release of the chemical potential energy of the fuel is essential to combustion (i.e., energy obtainable by breaking and/or forming *chemical bonds*). Continuous combustion requires that this release be self-perpetuating. In Reaction (7-12), the *oxidant* is a chemical reactant that converts the chemical potential energy of the fuel into thermal energy. In general, the oxidant is initially separate from the fuel (although some "fuels," such as the explosive TNT come with their own "built-in" oxidant, as part of the fuel molecule). The most common oxidant in energy practice is molecular oxygen (O_2) but a variety of solid, liquid, and gaseous substances can be oxidants. A *diluent* is a component of a combusting mixture that does not participate chemically in the main combustion reactions as either a fuel or an oxidant. The most common combustion diluent in practice is molecular nitrogen N_2, in air, but again, various solids, liquids, and gases may function as combustion diluents. Diluents can strongly influence the temperature, efficiency, and the environmental performance of combustion. They can serve as parasitic sinks for the energy liberated by combustion, causing energy to be lost as waste heat as diluent gases emerge from the combustor exhaust. Moreover, diluents can be a source of adverse emissions, perhaps the best example being nitric oxide, formed at high temperature by reactions of N_2 diluent with labile combustion intermediates, such as oxygen atoms, e.g.:

$$N_2 + O \rightarrow NO + N \qquad (7\text{-}13)$$

$$N + O_2 \rightarrow NO + O \qquad (7\text{-}14)$$

Reactions (7-13) and (7-14) are known as the Zeldovich "atom shuttle" reaction of "thermal NO" formation and allow the cycle to perpetuate as long as N_2 and O_2, i.e., air, is available.

Note that if combustion involved fuels containing only carbon and hydrogen in pure oxygen and worked perfectly, the only emissions product of concern would be CO_2. The idealized case of burning perfectly pure methane, Reaction (7-1) illustrates the point. However, Reaction (7-12) better captures the real world of combustion (i.e., that there are always some undesired products). These arise because of pollutants in the fuel and because combustion is an imperfect process. The amounts of unwanted co-products and their propensity to cause environmental malfeasance will depend on the composition and inlet ratio of the fuel and oxidant, the amounts and types of diluents, and the specific forms of combustion behavior.

To gain an understanding of these ideas, we examine steady combustion. For a given amount of fuel, there must be sufficient oxidant to fully combust all of the fuel, i.e., to chemically transform each of its reactive elements (generally carbon and hydrogen in practical fuels for energy production) to their highest oxidation state, i.e., CO_2 and H_2O, respectively. Next, combustion must release its energy fast enough to (a) heat up the fuel and oxidant (and any diluents) at least to a minimum temperature at which the enabling chemical reactions are sufficiently rapid to sustain the feedback(s) necessary to maintain combustion; and (b) compensate for heat and other forms of energy, such as the work of gas expansion, withdrawn and/or unavoidably lost from the combustion environment. Combustion energy is the product of chemical reaction(s) between the fuel and the oxidant. Thus, for any form of combustion, the requirements for a chemical reaction must be satisfied.

First, there must be physical contact between molecules of the fuel and the oxidant (i.e., there must be good molecular level mixing—*micromixing*—between the two). In some combustors, the fuel and oxidant are mixed ahead of time or before these reactants have been heated to the combustion temperature. In others, the combustion equipment mixes the fuel and oxidant. Often mixing is promoted by creating *turbulence* within and between flows of fuel and oxidant. Turbulence is fluid motion in which macroscopic, random (in both magnitude and direction) fluctuations in fluid properties, such as velocity and pressure, and potentially temperature and density, are superimposed upon the main stream motion, resulting in eddying motions of the fluid and attendant mixing over length scales that are large compared to molecular dimensions.

Second, the reactants must be able to react fast enough to achieve the desired heat release rates. Because rates of chemical reactions increase with temperature, this means that the fuel and oxidant must be heated up at least to a minimum temperature for fast reactions. Then the reactants must remain at this temperature for sufficient time for combustion to proceed essentially to 100% completion (i.e., the point in time at which most of the fuel has been productively consumed). We can summarize these requirements in terms of the Three "T"s of combustion: *temperature*, *time*, and *turbulence*, where the last is a proxy for mixing. The situation is depicted schematically in Figure 7.5, for combustion of a single particle or droplet of fuel. Heat, fed back from combustion of already burning fuel, raises the temperature of the particle or droplet. First, moisture is evaporated and then combustible components in the fuel vaporize owing to evaporation or chemical decomposition creating volatile products, or both. These components migrate away from the fuel particle (droplet) and at some point come into contact with oxidant vapor (O_2 gas) and react, forming a combustion zone. This "region of burning" may surround the entire fuel particle (droplet), creating a so-called circumambient flame. Sufficient time must be allowed for combustion to proceed. Depending on the geometry and operating conditions of the combustor, mixing of fuel vapors and oxidant vapors may occur only by non-turbulent processes (e.g., molecular diffusion) or turbulent mixing may also play an important role. Note that while combustion always requires chemical changes, the operation of most practical combustors also requires contributions from the physical transport of heat, material, and momentum, with momentum transfer being the mechanism of macroscale mixing of the combustion fluids.

Figure 7.6 depicts a typical temperature-time history for the surface of a fuel particle or droplet heated from room temperature to a temperature high enough to allow stable combustion of a steady feed of such particles. The heating rates in practical combustors can be enormous. For example, powdered coal (roughly 74 μm particle size) in a modern electric utility boiler heats up at a rate of 10,000–15,000°C/s. Pyrolysis of such small particles is complete in about 100 ms; burnout of the resulting char occurs in about 1 s.

In oil boilers and furnaces where the fuel is injected as an aerosol, the droplets do not burn as individual particles because they first vaporize in a fuel-rich flame zone. Then the generated vapor burns as a turbulent gaseous diffusion flame. This is shown in Figure 7.7, which illustrates a model for spray combustion. Some large droplets may survive the fuel-rich zone and burn with an envelope flame around them (as in Figure 7.5), but over 95% of the fuel burns as a diffusion flame.

The basic reaction kinetics for NO formation are slower than those for the combustion reactions, but become more important at higher temperatures. In fact, in the 1960s, the quality of a combustion process (good mixing and high combustion temperature) was monitored using NO as an indicator. The formation of NO in combustion processes can involve several different pathways:

- In fuel-lean flames, nitrogen oxides ("thermal NO") form by the attack of O atoms or OH^- radicals on molecular nitrogen in the air, as shown in Reactions (7-13) and (7-14). These reactions are highly temperature dependent; at 1,800K the reaction rate doubles for every 35K temperature rise.
- In fuel-rich regions of hydrocarbon-air flames, NO is formed by the attack of hydrocarbon radicals on molecular nitrogen, producing first HCN and then NO. This is a fast reaction that occurs at the flame front and produces what is termed "prompt NO." Prompt NO formation is not strongly temperature dependent.
- In combustion of fuels that contain organically bound nitrogen in their molecules (coals and oils), the pyrolysis and oxidation of these heterocyclic nitrogen compounds readily form NO in fuel-lean flame regions ("fuel NO"). The good news is that pyrolysis of these fuels for some extended residence time (100–300 ms) at high temperatures can convert these nitrogen species first from cyanides to amines, and then to N_2, which renders them innocuous for NO formation.

Examination of the chemical reaction paths and kinetics of nitrogen compound interconversions in flames has opened the opportunity for control of NO emissions by means of process modifications. Some of these modifications are:

- Reducing peak temperatures in fuel-lean combustion by heat extraction and/or by flue gas recirculation in boilers and furnaces. In gas turbines, NO formation can be greatly reduced by the use of a premixed ultra-lean feed stream.[1]

---

1 In gas turbines, the inlet temperature is limited by the structural integrity of the machine to about 1,570 K. At this temperature, an ultra-lean feed mixture will produce negligible NO over the short residence time (about 20 ms) in the gas turbine combustor. The problem is that this flame temperature is close to the lean limit of flammability. The combustion technologist's and turbine designer's arts are challenged to reconcile the conflicting requirements of low NO_x emissions and good flame stability over a wide range of operation.

- Staging the combustion air and/or the fuel injection to produce fuel-rich/fuel-lean flame zones to create conditions favorable for the conversion of fuel-bound nitrogen to molecular nitrogen.

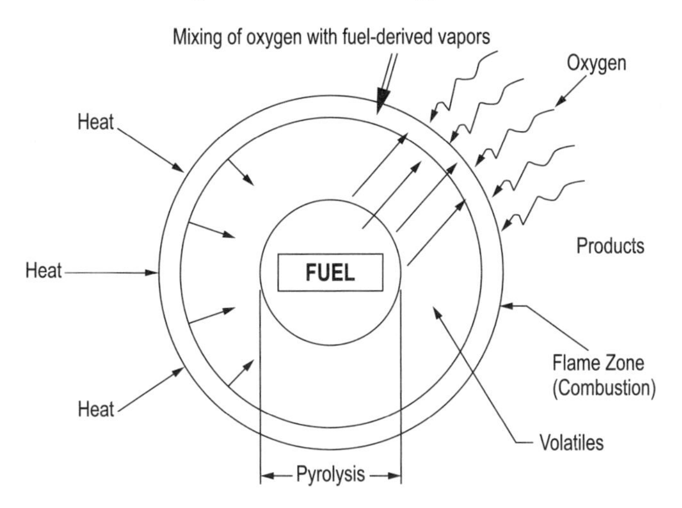

Figure 7.5. Schematic depicting heat transfer, mixing of fuel vapors with oxidant (oxygen), and two regions of chemical reactions (pyrolysis and the flame zone), in the combusion of an individual particle or droplet of fuel.

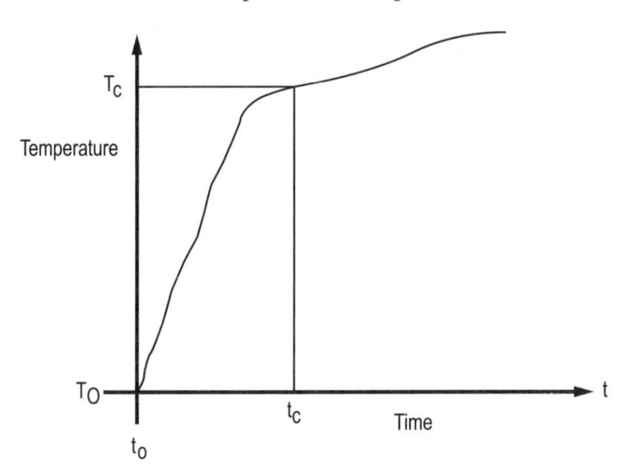

Figure 7.6. Schematic of a typical temperature-time history of the surface of a fuel particle or fuel droplet undergoing combustion (see Figure 7.5). T_c is the minimum temperature so that oxidation of the fuel vapors will be rapid and self sustaining; t_c is the time it takes to reach this temperature assuming the fuel was initially at a temperature T_o.

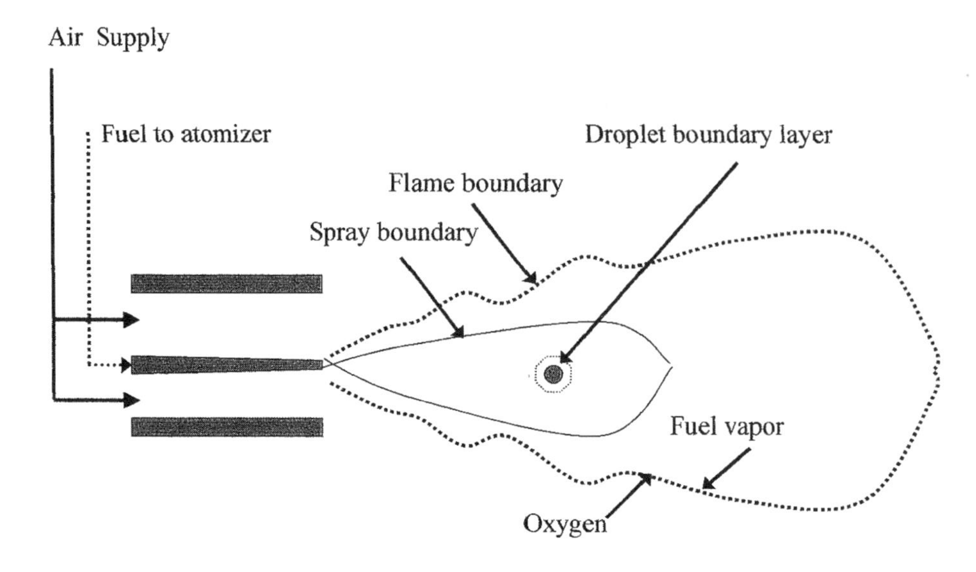

Figure 7.7. Typical behavior of atomized oil spray droplet flames in oil burners.

Figure 7.8 schematically depicts a gas turbine-steam turbine (GT-ST) combined cycle for combustion of natural gas and other clean fuels. This cycle is used extensively to generate electric power because it has relatively low capital cost and provides low-cost electricity even with high natural gas costs. (Quantitative details on these costs are presented in Section 7.6.) A GT-ST generation scheme integrates the Brayton and Rankine cycles (Chapter 3) and provides fuel-to-electricity generation efficiencies of about 60% with low NO_x emissions. Moreover, these cycles allow for incremental expansion of generation capacity in relatively small stages, avoiding the huge capital outlays of large nuclear and coal-fired plants. These cycles *do* require clean fuels, such as natural gas, petroleum middle distillates, or methanol.

The most typical example of a coal-fired power plant is sketched in Figure 7.9. As discussed in Chapter 3, this is a basic steam generator with a turbine that exhausts to a heat exchanger where spent (i.e., decompressed) steam is condensed and pumped up to pressure again. Typical operating efficiencies for such plants are about 35% based on the lower heating value (LHV) of the fuel.

A more modern type of coal power plant utilizes a pressurized fluid bed-combined cycle design as typified in Figure 7.10. Such plants are much more integrated for energy recovery than the basic plant and are capable of achieving 50% efficiencies, depending on the details of the design. As discussed in Chapter 3, there are many possibilities for energy integration and other efficiency improvements in power plants, and individual designs vary considerably. Technology improvements have gradually increased the efficiency of coal-fired power generation by up to 1% per year over the past decade.

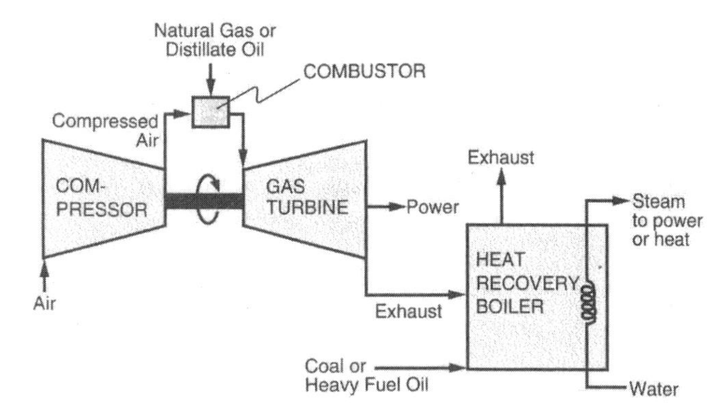

Figure 7.8. A schematic of a gas turbine-steam turbine combined cycle. Source: Beér (2000). Reprinted with author's permission.

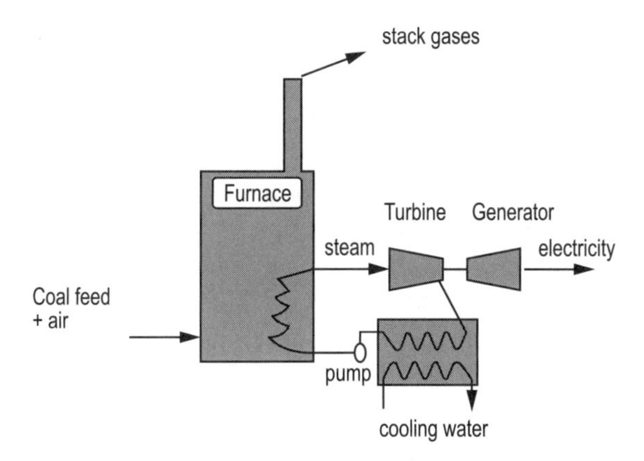

Figure 7.9. Typical coal plant.

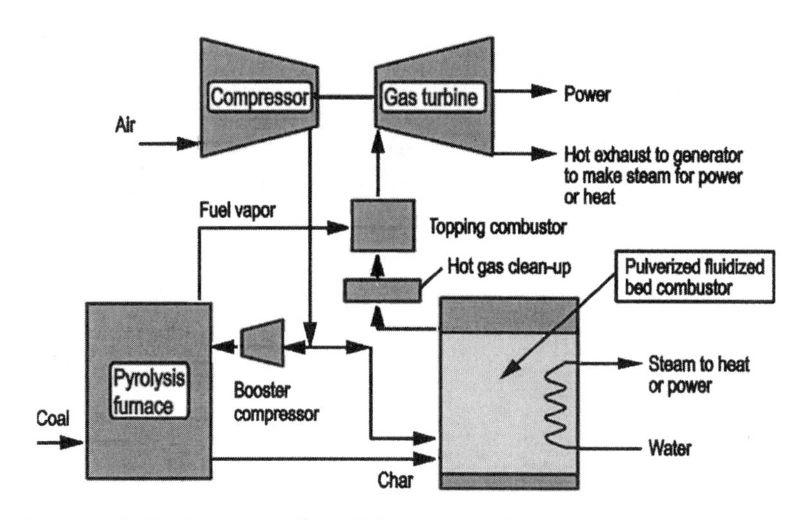

Figure 7.10. Typical pressurized fluid bed combined cycle.

7.3.5 Direct generation of electricity: Fuel cells

A fuel cell can be thought of as an "electrochemical combustor." Hydrogen or some other fuel is still oxidized, some heat is still released, and, as in any chemical reaction, electrons change hands (i.e., chemical bonds are broken). However, in a fuel cell, the fuel and the oxidant react separately in different regions that are connected to each other by two different conduits for charged particles. Each region contains a solid surface, an electrode, at which either the fuel is oxidized by giving up electrons or at which the oxidant is reduced by accepting electrons. The electrode for fuel oxidation is the anode, and the electrode for oxidant reduction is the cathode. In the fuel cell, the exchange of electrons among the reactants (the fuel and the oxidant) occurs through an electrical circuit outside the cell. The fuel cell converts chemical potential energy to usable electrical energy in the form of moving electrons. For electrons to journey through the external circuit, they must overcome any electrical barriers (impedance) to their transmission (i.e., they must do electrical work). Electrochemical reactions for use in fuel cells are purposely chosen so that the amount of electrical work attainable is sufficient to overcome the resistance to electron flow inherent in any circuit, but also to allow the electron flow to carry out useful electrical tasks (e.g., operating a motor or light bulb). The second conduit for charged particles is inside the cell and is called an "electrolyte." This can be an aqueous or other solution, a solid polymer, or an ion-conducting ceramic. Its job is to allow charged particles much more massive than electrons [e.g., positively charged hydrogen atoms, H^+, (if the electrolyte is an acidic solution) or negatively charged hydroxyl ions, OH^-, (if the electrolyte is a basic solution) or negatively charged O_2^- (if a solid oxide ceramic is used)] to pass between the two electrodes.

Figure 7.11 shows a schematic of a fuel cell in which hydrogen (H_2) is converted to electricity using oxygen (O_2) as the oxidant and an acidic electrolyte. The essential chemical and physical processes for this cell to operate are:

1. Oxidation of gaseous $H_{2(g)}$, the fuel, at a region of the anode in interfacial contact with the electrolyte:

$$H_{2(g)} \rightarrow 2H^+ + 2e^- \tag{7-15}$$

2. Physical transport of the hydrogen ion from the anode through the electrolyte to the cathode:

$$2H^+ \text{ (anode-electrolyte interface)} \rightarrow 2H^+ \text{ (electrolyte-cathode interface)} \tag{7-16}$$

3. Reduction of gaseous O_2, the oxidant, at a region of the cathode in interfacial contact with the electrolyte:

$$\tfrac{1}{2}O_{2(g)} + 2e^- + 2H^+ \rightarrow H_2O \tag{7-17}$$

4. Physical transport of electrons from the anode to the cathode through the external circuit:

$$2e^- \text{ (anode-electrolyte interface)} \rightarrow 2e^- \text{ (cathode-electrolyte interface)} \tag{7-18}$$

Table 7.4. Summary of Common Fuel Cell Technologies

Fuel Cell	Anode Reaction	Electrolyte	Transfer Ion[a]	Cathode Reaction	Operating Conditions: Temp, °C	Operating Conditions: Pressure, atm	H_2-to-Electricity Efficiency,%
Proton Exchange Membrane (PEM)[b]	$H_2 \rightarrow 2H^+ + 2e^-$	Solid Polymer	H^+	$2H^+ + 1/2O_2 + 2e^- \rightarrow H_2O$	80–100	1–8[c]	36–38
Phosphoric Acid (PAFC)	$H_2 + 2OH^- \rightarrow 2H_2O + 2e^-$	Phosphoric Acid	H+	$2H^+ + 1/2O_2 + 2e^- \rightarrow H_2O$	150–250	1–8[c]	40
Alkaline (AFC)	$H_2 \rightarrow 2H^+ + 2e^-$	Aqueous Base	OH-	$H_2O + 1/2O_2 + 2e^- \rightarrow 2OH^-$	80–250	1–10[c]	50–60+
Molten Carbonate (MCFC)	$H_2 + CO_3^{2-} \rightarrow H_2O + CO_2 + 2e^-$	Molten Salt (Metal Carbonate)	CO_3^{2-}	$CO_2 + 1/2O_2 + 2e^- \rightarrow CO_3^{2-}$	600–700	1–10	50–55
Solid Oxide (SOFC)	$H_2 + O^{2-} \rightarrow H_2O + 2e^-$	Solid Ceramic Oxide	O2-	$2H^+ + 1/2O_2 + 2e^- \rightarrow H_2O$	800–1,000	1	50–55

Source: Adapted from O'Sullivan (1999).

[a]Positively charged ions (cations) travel through the electrolyte from the anode to the cathode; negatively charged ions (anions) traverse the electrolyte in the opposite direction.

[b]Also called Polymer Electrolyte Fuel Cell

[c]Pressure must be sufficiently high to prevent boiling of water in PEM or aqueous electrolytes in PAFC and AFC.

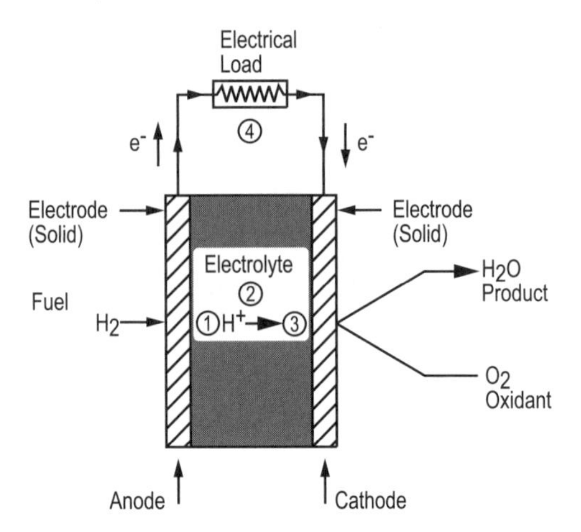

Figure 7.11. Schematic of a hydrogen-oxygen fuel cell with an acidic electrolyte and H^+ as the transfer ion. Personal communication, Sadoway; 1999.

Note that if we add up Reactions (7-15) to (7-18) we get the same overall reaction as if we had directly combusted hydrogen in oxygen:

$$H_2 + \tfrac{1}{2}O_2 \rightarrow H_2O \tag{7-19}$$

The difference is that, through use of the fuel cell, we are able to directly generate electricity. Fuel cells now command great interest as clean energy converters for use in producing electricity for consumers and as the energy source for electric vehicles. This interest is motivated by the potential for high fuel-to-electricity conversion efficiencies and by the fact that fuel cells fired with hydrogen emit only water. Table 7.4 summarizes several of the common types of fuel cells under development for stationary electric power and automobile propulsion applications, together with estimates of their fuel-to-electricity conversion efficiencies (assuming H_2 as the fuel). The ion flowing within the electrolyte may be H^+ or any of several negatively charged species (anions), e.g., OH^-, CO_3^{2-}, or O_2^-, (the anions flow from the cathode to the anode). Advanced electrolyte systems (i.e., proton-conducting inorganic oxides) may enable fuel-to-electricity efficiencies as high as 70% based on H_2 and its LHV. Fuel cells are also attractive because of their potential for low maintenance, high reliability, and low noise levels. Because they operate at lower temperature than typical combustion processes, they do not produce NO_x emissions from reactions involving N_2 diluent in air (see Reactions (7-13) and (7-14)).

Figure 7.12 shows that the output voltage of a fuel cell declines with the amount of current drawn from the cell. The decline at low current densities is caused by limitations on the rates of chemical reactions at the electrodes and at high current densities by the rates of ion transport through the electrolyte. The corresponding effect on power density (Figure 7.13) is a small plateau region where cell electrode area is used effectively at high current densities, but undergoes appreciable draw-down at lower and ultra-high current densities. The latter means that fuel cells are best operated at steady power densities and lose much of their electrode capability during load following. Because each fuel cell unit generates only about 1 volt, fuel cell systems are composed of stacks of individual fuel cells that are interconnected to produce the desired voltage and power densities.

Other major challenges with fuel cells include reducing their capital cost, assuring long-term durability (particularly in challenging environments, such as automobiles), and providing a fuel compatible with the cell. So far, this means providing H2 either directly from storage or by generating the H2 in situ by reforming some other fuel such as gasoline, diesel, or natural gas. This requires mating a fuel production technology to the fuel cell. Synchronization of the output of the fuel producer to the varying fuel input requirements necessary to load follow will be difficult, so some form of post-production storage will generally be required. There has been progress on direct firing of fuel cells with other fuels, such as methanol, and for higher temperature cells, such as solid oxide, with hydrocarbons. At high temperatures, reforming of hydrocarbon fuels can be incorporated into the cell system. Of course, the carbon from the fuel ends up as waste CO2. Fuel cell electrodes and electrolytes must be protected from poisoning (e.g., by sulfur in the case of platinum catalysis in the electrodes of low-temperature fuel cells, and by carbon monoxide or methanol in the case of polymer electrolytes). Figures 7.13 and 7.14 show that fuel cell power density varies appreciably with the current drawn and the cell voltage. Consequently, fuel cell performance changes considerably with the load demand on the cell and degrades precipitously at high loads.

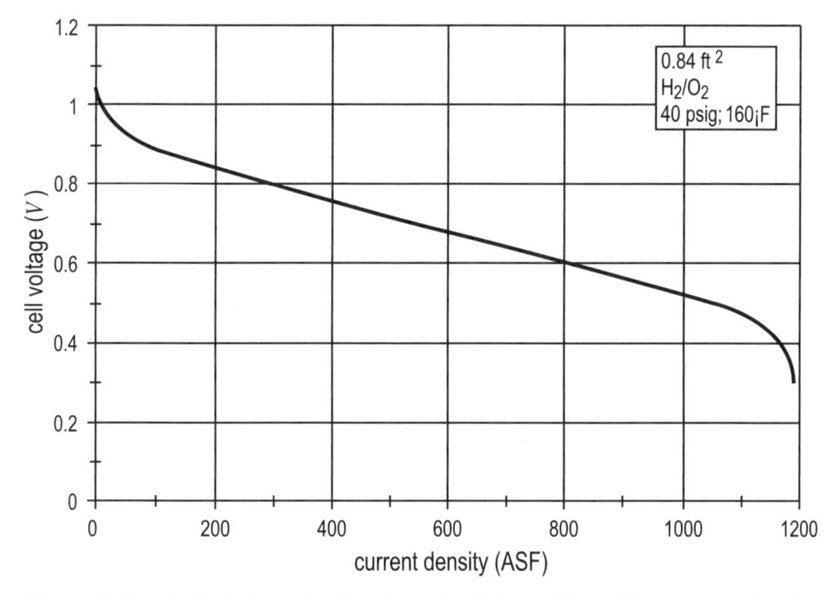

Figure 7.12. Typical plot of cell voltage (volts) as affected by current density amps/ft^2 (polarization curve) for a proton exchange membrane (PEM) fuel cell. Source: Babir, 1999. Reprinted with author's permission.

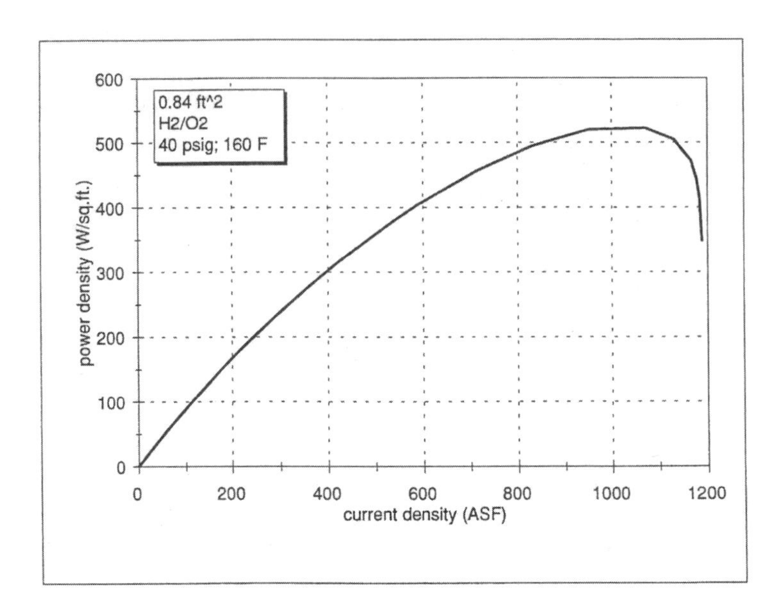

Figure 7.13. Power density (watts/ft^2) as affected by current density (amps/ft^2) curve for a typical PEM fuel cell. Source: Babir, 1999. Reprinted with author's permission.

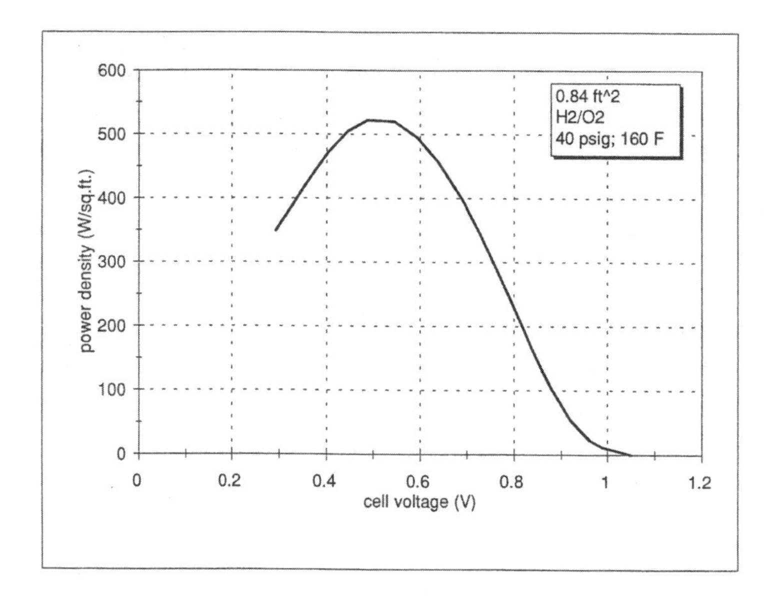

Figure 7.14. Power density (watts/ft^2) as affected by cell voltage (volts) for a typical PEM fuel cell. Source: Babir, 1999. Reprinted with author's permission.

Opinion varies on the future roles of fuel cells in stationary power generation and transportation propulsion. (See Chapters 16 through 18.) Some observers see significant roles in the electrical sector, especially in distributed power generation, and huge contributions to the automotive sector owing to the potential for low to zero emissions of various pollutants and of CO_2 when fueled by H_2 (assuming the H_2 is generated without carbon emissions). Some major automobile makers (e.g., Toyota and General Motors) are carrying out programs to develop fuel cell vehicles. Other analysts see a less certain future, pointing to appreciable competition from alternatives, including natural gas–fired turbines in the stationary power sector, and from hybrid diesel electric propulsion in the passenger vehicle sector.

7.3.6 Manufacture of chemicals and other products

As seen in Figures 7.1 through 7.4, naturally occurring fossil fuels, as well as by-products of their upgrading, are the input raw materials, or feedstocks, for the manufacture of a host of valuable non-fuel products, from ammonia to plastics to valuable chemicals. (Chemicals and other products manufactured from petroleum, natural gas, and coal are discussed at length in Shreve and Brink, 1984, Elliott, 1981, and Mark et al., 1985.)

7.4 Environmental Impacts

7.4.1 Pollutant sources and remedies: The fuel itself

During handling, processing, and combustion, adverse emissions may arise from the fuel itself. Specific sources are:

- Products of incomplete combustion (PICs), resulting from escape of fuel or fuel components unchanged or partially modified by the process.
- Fuel impurities chemically or physically liberated during combustion or other processing. Some impurities are present in only trace amounts (e.g., nickel and vanadium in heavy oil from certain regions of the world). Other impurities, such as the mineral matter, sulfur, and nitrogen in coal, occur in appreciable percentages (e.g., 1–10%).

When combustion systems, owing to poor design or unanticipated transients, result in conditions where these fuel components reach elevated temperatures without sufficient oxygen to convert them completely to CO_2 and water, they have a strong tendency to form soot or smoke, as well as polycyclic aromatic hydrocarbons, many of which are known genotoxins.

Fuel components that contribute substantially to the heating value of the fuel can also contribute to adverse pollutants. In an ideal combustor, all hydrocarbon components of the fuel, which we shall designate CH_x, will be completely converted to CO_2 and water, in the process making their maximum possible contribution to the energy obtainable from the fuel:

$$CH_x + (1 + x/4)O_2 \rightarrow CO_2 + (x/2)H_2O \quad (7\text{-}20)$$

In practice, a combustor may temporarily or indefinitely fall short of this ideal chemical reaction. The reasons can be varied. For example, there may be an inadequate supply of air or oxygen (e.g., owing to poor design), excess feeding of fuel, plugging of oxidant feed lines, poor mixing of fuel with O_2, and vitiation of the O_2 by recirculation of combustion products. The temperature of the fuel and oxidant may not be high enough for Reaction (7-20) to occur fast enough to consume all the fuel, or the fuel and oxidant may not spend enough time at the temperature needed for this reaction. The result is incomplete combustion including, potentially, escape of fuel totally unchanged, partially modified by physical processes (e.g., evaporation of more volatile components), or chemically transformed. The "combustion" chemistry may be represented by the following stoichiometry:

$$CH_x + [a + (b + 2c)/4]O_2 \rightarrow aCO_2 + bH_2O + cCO + dH_2 + eC_yH_z + C(1 - a - c - ey) \quad (7\text{-}21)$$

In addition to CO_2 and H_2O, Reaction (7-21) identifies several other products of "incomplete" combustion. These products include carbon monoxide, CO, molecular H_2, and, potentially, compounds of carbon and hydrogen with complex molecular structures, which we denote, collectively, as C_yH_z. Note that the products other than CO_2 and H_2O may represent a small percentage of the carbon and hydrogen in the original fuel (i.e., $[c + ey + f] << 1$, and $[2d + ez] << x$). Nevertheless, these parasitic products have important implications for the environmental performance of fuel-combustion systems and are important to sustainable energy.

Carbon monoxide is a serious pollutant owing to its ability to bind to hemoglobin in human blood. CO emissions from automobiles are severely restricted in North America and Europe. The substances C_yH_z may include polycyclic aromatic hydrocarbons (PAH), example structures of which are shown in Figure 7.15. Fine particulate carbon matter, commonly denoted soot or smoke, may also form. For simplicity, we have denoted soot in Reaction (7-21) by the symbol C, because, although it is typically rich in elemental carbon soot, it may also contain chemically bonded and physically absorbed components that contain other elements, including hydrogen. Soot and PAH are of particular interest from a sustainability perspective because soot, soot extracts, and many PAHs exhibit toxicological activity in model systems, and some PAHs such as benzo(a)pyrene are classified as known carcinogens. Moreover, soot particles have high emissivities for thermal radiation (Chapter 3). Thus, soot particles in flames increase radiative heat transmission, which is beneficial in boilers and furnaces but detrimental in gas turbine combustors, where they may lead to the overheating of combustion liners.

Given the "right" combustion conditions, any fuel containing carbon and hydrogen can form soot and PAH. Moreover, the propensity to form these pollutants is exacerbated when the fuel contains aromatic structures or components. These substances, which have empirical compositions of approximately CH, abound in coal and in higher boiling fractions of petroleum-derived liquids. The underlying chemical and physical mechanisms responsible for soot and PAH formation and growth in combustion are highly complex. Note that the pollutant products exhibited in Reaction (7-21) may represent only a fraction of the total amount of carbon and hydrogen in the fuel, but can degrade the energy efficiency and environmental performance of any given fuel-combustion combination.

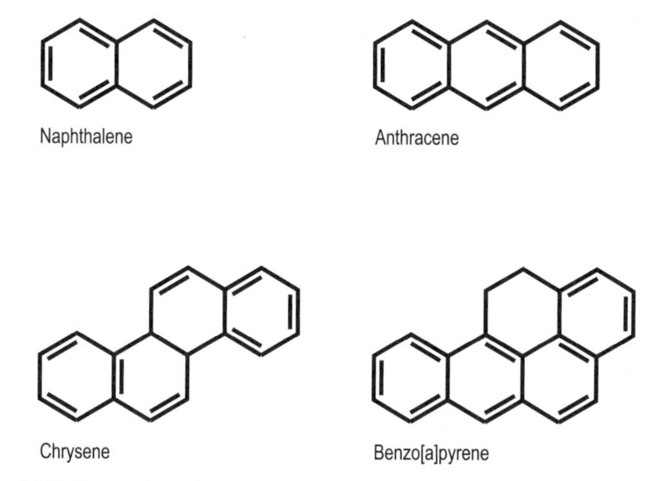

Figure 7.15. Examples of polycyclic aromatic hydrocarbons. See Bjørseth (1983).

Countermeasures to pollutants from the major components of fuels include combustion modification (see Section 7.4.2) and fuel reformulation. Generally, hydrocarbon fuels of higher H/C ratio are cleaner burning. More of the energy they provide comes from forming water through bonding of their hydrogen atoms with oxygen atoms supplied by the oxidant. Fuels of high H/C ratio supply less of their energy from their carbon atoms. This means there is less opportunity for that part of the combustion chemistry to go astray and form carbonaceous pollutants. For example, high H/C fuels have less inherent tendency to form soot and PAH compared to fuels of lower H/C ratio in the same combustion circumstances. The hydrocarbon fuel of highest H/C ratio is methane, CH_4, the principal component (typically 85%–90% by volume) of natural gas. Thus, there is great interest in the use of natural gas both as a stand-alone fuel and in co-firing with fuels of lower H/C ratio. The latter approach increases the overall H/C ratio of the fuel. Some scholars (e.g., Socolow and Williams, 2002) point out that production of molecular hydrogen (H_2) may be the preferred fuel reformulation strategy for a greenhouse gas-constrained energy economy, because carbon is then totally removed from the fuel, eliminating any possibility of adverse carbon-containing pollutants, as well as CO_2 emissions, if the carbon is stored outside of the atmosphere.

Fuel impurities contribute little to the heating value of the fuel, but they are another important source of pollutant emissions. Coal mineral matter can be volatilized from the coal during combustion and then condensed as fine particulate matter. If ingested into the lungs, these particles are toxic to human health. They can also degrade the efficiency of energy production by depositing on the surfaces of heat recovery tubes where they impede heat transfer and exacerbate corrosion.

7.4.2 Pollutant sources and remedies: Combustion pathologies

Combustion imperfections lead to pollutants and wasted energy. Sometimes the imperfections can be counteracted by redesign or the choice of a different fuel. Indeed, each of the essential ingredients of successful combustion (i.e., fuel and oxidant properties, stoichiometry, the Three "T"s) can be manipulated to improve the efficiency and environmental impacts of practical combustion equipment. On the other hand, uninformed adjustments of any of these can degrade combustion performance, resulting in incomplete utilization of the fuel and unacceptable releases of adverse emissions. Moreover, because of the complex interactions between chemical reactions and physical transport, simultaneous adjustments of two or more variables may produce negative results that are absent if each variable is changed separately or when all are changed in sequence, rather than all at once. Given sufficient time, all chemical reactions, including combustion, will attain *thermodynamic equilibrium* or, for the particular stoichiometry, temperature, and pressure of interest, the state in which all reactants and products are in their most chemically stable condition.

In actual combustors, realization of equilibrium for a given fuel/oxidant stoichiometry may be inhibited by slow chemical reactions or inadequate mixing of fuel and oxidant. Moreover, unwanted leakage (or inflows) of heat or material (air, fuel, etc.) may, at least for certain regions in the combustor, upset the prevailing temperature, pressure, and stoichiometry from that associated with the combustor as a whole. Any departure from design conditions can result in unburned fuel and creation of pollutants. A simple illustration of both is the situation in which, owing to inadequate mixing, there is not enough oxygen in contact with the fuel to completely convert it to CO_2 and H_2O. In consequence, the fuel is incompletely burned. Some escapes the combustor as unburned or partially converted fuel, so the combustor produces less energy than it could have if all the fuel had been completely combusted to CO_2 and H_2O. In addition, some of the products of partial (incomplete) combustion could well be pollutants such as soot or polycyclic aromatic hydrocarbons.

It is instructive to consider technological strategies to overcome these difficulties. Why can't we just return to the Three "T"s and employ higher temperatures, longer burning times, and more intense mixing to solve all our pollution and combustion inefficiency problems? The key to success is learning how to optimize these "control knobs," economic costs, and performance objectives. For example, use of higher temperatures can only go so far owing to limitations on the:

- Maximum temperature any given fuel can provide (adiabatic flame temperature)
- Operating costs (e.g., to separate O_2 in order to eliminate N_2 diluent and achieve a higher temperature)
- Materials compatibility with higher temperatures
- Intended use of the combustion-generated energy, e.g., to raise steam of a given "quality," i.e. temperature and pressure
- Propensity to form certain pollutants such as soot, PAH, and NO_x at high temperatures

The time available for combustion can be increased by use of a larger volume combustor, reducing combustor throughputs, or both. Volume increases are limited by capital costs, and volume increases and throughput reductions are both constrained by minimum requirements for power harvesting per unit volume of combustor. Mixing improvements entail higher capital and operating costs. Reduction of NO emissions through design and control of the combustor is discussed in Section 7.3.4. Atmospheric emissions of CO_2 are a more specialized environmental impact of fossil fuel combustion. Approaches for carbon management are described in Section 7.4.3.

7.4.3 Pollutant sources and remedies: Carbon management

Release of carbon dioxide and other carbon-bearing gases into the atmosphere by human activity is a human intervention in the environment. As discussed in Chapter 4, there is considerable uncertainty as to how serious this intervention is in terms of its consequences for global climate, as well as to the potential economic and environmental impacts of proposed curtailment strategies. Figure 7.16 summarizes a number of these measures. Before considering them in detail, we must remind ourselves of the amount of carbon involved. Each year, CO_2 emissions to the atmosphere due to human activity, almost all of it the combustion of fossil fuels, is 22 gigatons (i.e., 2.2×10^{10} tons or about 3.7 tons of CO_2 for every person on earth). A compact passenger sedan weighs about 1.5 tons, so, each year, the amount of CO_2 humans release to the atmosphere exceeds the weight of two Toyota Camrys or Ford Tauruses for every person in the world. Thus, any attempt to manage even 1% of this CO_2 would involve dealing with 74 lbs of material for every person on earth. Overall, there is wide variability around the average value of emissions, ranging from near zero for the poorest populations to many times the average for wealthy countries, especially the US.

One proposed approach to CO_2 management entails some sort of manipulation of the earth-atmosphere-sun relationship so as to counteract greenhouse warming. An example is to cool the earth artificially by adjusting its albedo to increase the fraction of incident thermal radiation reflected back into the atmosphere. Several approaches to reducing the amount of CO_2 emitted to the atmosphere (Figure 7.16) can be understood in terms of Equations (7-22) and (7-23), as introduced in Chapter 1:

$$S = (P) \times (GDP/P) \times (E/GDP) \times (CO_2/E) \tag{7-22}$$

$$S = (P) \times (GDP/P) \times (E/GDP) \times (CO_2/E) - (CO_2)_{sq} \tag{7-23}$$

CO_2 Mitigation Options

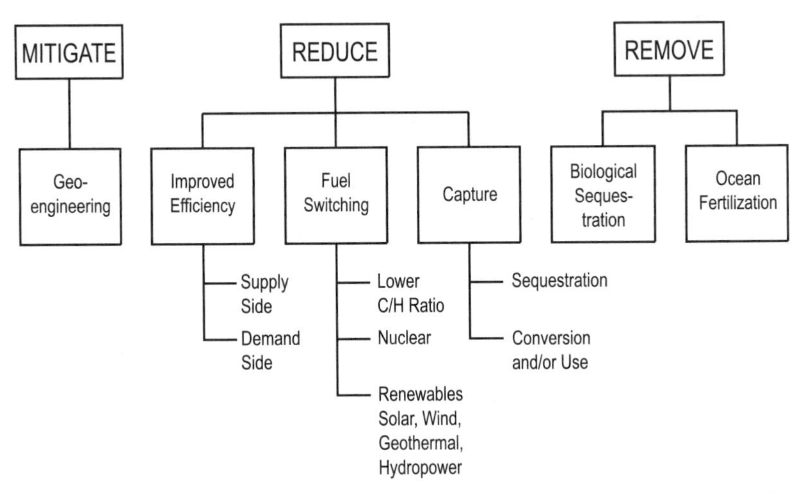

Figure 7.16. Options for mitigation of the release of fossil fuel-derived carbon dioxide to the earth's atmosphere. Source: Herzog et al., 2000.

Improved energy efficiency (Chapters 3, 7, and 16 through 20) brought about by less wasteful energy conversion equipment (supply-side management) or by reducing per capita reliance on energy (demand-side management) reduces CO_2 release by lowering the energy intensity, i.e., (E/GDP), term in these two reactions. Likewise, we can reduce the carbon intensity term, i.e., (CO_2/E), by so called *fuel switching* (Figure 7.16) or by placing greater reliance on fossil fuels of higher H/C ratio (e.g., natural gas instead of coal), and on energy sources that contain no fossil carbon, such as nuclear and renewables, including solar, wind, geothermal, and biomass. We can also capture CO_2 as it is emitted by human activity, such as from the stacks of electric power plants (where, in recent years, US emissions amounted to 1.7 billion tons/year). This is equivalent to increasing the sink term [-$(CO_2)_{sq}$] in Reaction (1-24). There are various chemical and physical processes that will separate CO_2 from flue gas but, at present, their broad-based application to flue gas capture would be expensive (Herzog et al., 1991).

Once captured, something must be done with the CO_2 (e.g., as implied in Figure 7.16). It can be converted to other products, employed in some useful purpose, or locked up (i.e., *sequestered* in a suitable repository). Successful storage of CO_2 requires the "lock-up" to be stable over time scales long compared to those for transitioning the global energy economy to non-fossil sources. They must be economical, cause negligible environmental damage over the storage lifetime, be acceptable to the public, and be of sufficient capacity to accommodate potentially huge quantities of CO_2. Table 7.5 displays estimates of the capacity of various earth-bound options for CO_2 sequestration. Recalling that current global releases are about 22 Gt/year, the table shows that, in principle, the oceans, aquifers, and depleted oil and gas wells could each provide multi-decade storage capacity. However, if only 10% of this capacity could be tapped and up to 50% of global CO_2 emissions required sequestration, depleted oil and gas wells would quickly become saturated, necessitating reliance on aquifers and the oceans. Environmentalists have expressed resistance to tapping aquifers and especially the oceans for carbon sequestration, so it remains to be seen whether these options could make a significant contribution as CO_2 storage reservoirs. Table 7.5 also shows that the yearly drawdown of oil wells worldwide, even if 100% utilization were possible, would not provide enough storage capacity for even 5% of the yearly CO_2 releases. In addition, at present, the market for CO_2 is minute in comparison with the amounts emitted (i.e., less than 0.5% if all current uses doubled).

Table 7.5. Estimates of the CO_2 Capacity of Various Sequestration Options

Sequestration Option	Worldwide Capacity (Order of Magnitude)
Ocean	100,000 Gt
Aquifers	10,000 Gt
Depleted Oil and Gas	1,000 Gt
Active Oil	1 Gt/yr
Utilization	0.1 Gt/yr
Coal Seams	?

Notes:
Total CO_2 emissions worldwide = 22 Gt/yr
CO_2 emissions from US power plants = 1.7 Gt/yr
Source: Herzog (2000).

Human intervention to remove CO_2 from the atmosphere can be used for carbon mitigation (Figure 7.16). One approach is *biological sequestration* in which the growth of trees or other plants fixes atmospheric CO_2. The mechanism is *photosynthesis* (Chapter 10), in which catalysts that occur naturally in the plant harness a portion of the energy of sunlight to force CO_2 and water to react chemically to form carbohydrates that become the building blocks of the plant. Photosynthesis uses solar energy relatively inefficiently (e.g., converting only about 1% of the flux incident on the earth's surface to plant matter). Biological sequestration requires huge amounts of land (or ocean) surface and copious amounts of water to significantly reduce atmospheric CO_2 concentrations (see Problems, Chapter 10).

Almost half of the earth's photosynthesis occurs by phytoplankton, tiny cells that in the first 100 meters of the ocean use sunlight and dissolved inorganic nutrients (compounds of nitrogen, phosphorus, silicon, and iron) to convert CO_2 to organic carbon. Most of this carbon becomes food for creatures in surface waters and is emitted back to the atmosphere as CO_2. However, some of this carbon ends up in the deep ocean (>3,800 m) where it may reside up to 1,000 years before it is then returned to the atmosphere by upwelling of deep ocean water. It has been proposed that atmospheric CO_2 can be reduced by artificially fertilizing the ocean surface waters where phytoplankton growth is nutrient limited, such as by iron in the equatorial Pacific and southern Pacific Ocean (Martin, et al. 1994). This approach is controversial. Small-scale experiments on artificial fertilization of ocean surface waters with iron have documented increased phytoplankton production. However, there was no evidence of increased transfer of carbon to the deep ocean, which would be essential for successful carbon sequestration.

Mann and Chisholm (2000) point out that artificial fertilization of the ocean in this way occurs over time scales much shorter than those used by the earth to adjust ocean inventories of iron, and that this form of human intervention entails risks of global ecosystem damage not worth the hoped for, and as yet unsubstantiated, benefits:

> The earth system consists of elements distributed between the land, air, and oceans by biological and geological processes over millions of years. Many environmental problems stem from our moving elements between these compartments at unprecedented rates that the system cannot accommodate (Chisholm, 2000).

Anthropogenic processes for chemical and biological conversion of CO_2 include manufacture of various industrial chemicals, growth of microalgae, production of carbonate minerals, and conversion to fuel. CO_2 is chemically far more stable than carbon and oxygen (Chapters 1, 3). Thus, conversion of CO_2 to a useful fuel entails high inputs of energy that must be supplied from some other, non-CO_2-emitting source. Moreover, current industrial demand for CO_2 is less than 0.5% of current emissions from US electric power plants. Geological sequestration options include enhanced recovery of petroleum from active oil wells and storage in depleted oil and gas wells, deep aquifers, coal beds, and mined caverns and salt domes. Research is needed to illuminate the long-term (100–1,000 years) integrity of the storage regions, their capacity, the mechanisms by which various depositories bind up CO_2, and the environmental and safety impacts of transporting and storing huge amounts of CO_2. Although vigorously challenged by some environmental groups, ocean sequestration offers by far the largest capacity for CO_2 sequestration (Table 7.5). The best near-term option for CO_2 injection appears to be dissolution at 1,000–1,500 m depths after release from pipelines or towed pipes. Research is needed to better understand the chemical and physical processes by which CO_2 interacts with seawater at depths relevant to sequestration, the efficiency of CO_2 incarceration as affected by ocean circulation, the effects on the ocean environment, and the infrastructure needed to operate effectively in the ocean.

Despite many challenges, government and industry continue to pursue options for CO_2 management. Roughly 10 capture plants (100–1,000 ton/day corresponding to 5–50 MWe) have been constructed since 1978 to provide CO_2 for industrial applications. In September 1996, Statoil began operation of the world's first commercial plant for capture and sequestration of CO_2 explicitly built to respond to concerns over climate change. This plant serves a 140 MWe natural gas-fired electric generating station located on a platform in the North Sea. Exxon and Pertamina have announced plans to include capture and sequestration of the CO_2 associated with the natural gas they will produce from the Natuna gas field in the South China Sea.

7.5 Geopolitical and Sociological Factors

7.5.1 Globalization of fossil energy sources

Over the last 40 years, international trading in fossil fuels has given rise to complex, often fragile, political relationships between exporting and importing nations. On the "receiving end" are many of the world's industrialized nations, some depending heavily or even predominantly on imported fossil fuels to power their economies. Imports make up about half of US petroleum consumption, and it has been argued that the economy of Japan would quickly collapse without the succor of Middle East oil. On the "sending end" are countries endowed with substantial deposits of oil, natural gas, or coal that can be profitably exploited in today's markets. Exporters are often (although not always: witness Canada, Norway, and the UK) less well-developed nations that depend heavily on fossil sales abroad to finance domestic economic progress. This ensemble of symbiotic interdependencies is further complicated by stark differences in governmental philosophies, economic aspirations, environmental objectives, and regional as well as global political agendas. Indeed, given the number of players (see Figure 2.5) and these disparities, it is remarkable that the export-import compacts have persisted for so long with so few serious disruptions. Three notable exceptions, all of relatively short duration, are the oil embargo of 1973 and the oil price spikes of 1979 and 1991, the last two owing in large measure to panic buying brought on by fears that the Iranian hostage seizure and the Gulf War (with Iraq) would precipitate sustained supply scarcities. Whatever the short-, near-, and longer-term prospects for fossil energy in the global economy, it is vital to fathom the known and potential sustainability consequences of the globalization of fossil energy trading. (Chapter 2 analyzes petroleum availability and prices.)

International accords on fossil carbon management are only workable if accepted by all major producers and consumers. The proposition that a curtailment in fossil demand is sufficient overlooks the attendant economic upheaval that would ensue for numerous exporters. This dislocation would almost certainly give rise to political instabilities that could disrupt regional and even global economic equilibrium for months and perhaps years. During a speech, President Nixon, referring to the imperative for social and economic equity in the US, said, "This won't be a good place for anyone to live unless it's a good place for everyone to live." In today's global village, this idea applies to all citizens of the planet. Thus, a crucial message for sustainability is that plans to transition the earth to a reduced- and eventually non-fossil economy must incorporate proactive measures to prevent contagious economic, social, and political instability in fossil exporting nations.

In Chapter 2, we learned that some exporters have been willing to band together and forgo sales to create artificial scarcities that bump up oil prices. However, economic forces invariably degrade these collegia in relatively short time scales (a year or less), as partners renege in favor of larger near-term profits and lower-cost, more reliable alternatives become available. Sustainability depends upon orderly transitioning to alternatives as those options become technically and economically available. For the future, it must be asked whether other factors will spur exporters into using fossil exports as a political weapon for *economic terrorism*. Radical groups driven solely by political objectives could gain control of major oil exporting nations and cut off exports or even demolish production infrastructure to propagate economic dislocation within and between importing nations. Serious advocates of sustainability must pay close attention to political developments throughout the planet and the socio-political factors that breed popular support for institutionalized radicalism.

7.5.2 Equitable access, Revenue scaffolds, "American Graffiti"

Sustainability requires people to reconcile the rates at which they consume goods and services with protecting the planet from the environmentally harmful human interventions that enable this consumption. But human attitudes toward consumption, social entitlements, political freedom, economic progress, conservation, and environmental protection vary dramatically. Fossil energy plays a prominent role in shaping those attitudes because today's global economy is based primarily (ca. 85%) on affordable fossil fuels. Many scholars are convinced that this dependency must be reduced in the near term (i.e., 10–20 years) to put the earth on a more sustainable path. However, the size of the global fossil footprint means that technological or policy measures to wean the planet from fossil fuels must win popular support. This in turn requires an appreciation of the current or potential role of affordable fossil energy in the day-to-day lives of people in industrialized nations and in poorer regions that aspire to some form of parity with their more industrialized kindred. Below, we discuss three examples.

Equity. With substantial justification, petroleum fuels are broadly perceived as the least expensive source of transportation fuels and, in many cases, of electricity, residential/commercial energy, and industrial energy as well.[2] Developing nations aspiring to the economic growth enabled in the developed countries by "cheap" fossil energy have little interest in adopting more expensive non-fossil alternatives without financing from the developed world. The issue is one of equitable treatment in which emerging economies desire to move rapidly to attain some of the socioeconomic benefits identified with their developed neighbors. Likewise, already developed countries eager to hang on to their fossil-enabled economic amenities are reluctant to pay more for energy and switch from fossil fuels. Advocates of higher energy taxes face an uphill road. In 1993, shortly after taking office, a new US federal administration, with a vice president strongly identified with environmentalism, proposed a so-called Btu tax on various forms of energy utilized in the US. Although the administration enjoyed a huge majority in the House of Representatives and a majority in the Senate, this energy tax initiative failed miserably because elected representatives of both political parties correctly fathomed that a majority of the US public was strongly opposed to higher energy prices.

Revenue Scaffolds. Owing to various taxes on gasoline, diesel and other fuels, fossil energy sales are a tremendous "cash cow" for governments. (Figure 7.17 shows the net earnings per barrel of oil in 1993 for Organization of Petroleum Exporting Countries (OPEC) and for various governments by fuel taxation.) Even in the US, where there is no *ad valorem* or national sales tax and where fuel taxes are low compared to Europe and Japan, tax revenues somewhat exceeded those of the fuel supplier. In several other countries, as well as the G7 and the EU11, government proceeds from fuel taxes dramatically outpace corresponding OPEC revenue (e.g., by almost two-fold in Canada, and approaching six-fold in Denmark). Thus, transitioning away from fossil fuels must include a plan to offset the loss of government revenue from fuel taxes (see Problem 19.5).

---

2 Exceptions are nuclear and hydro-powered central station electricity generation in certain regions of the world, and occasional, generally inconsequential (i.e., $< 5\%$ regional penetration), contributions from renewables. Dissenters posit that opaque government subsidies for infrastructure, as well as under-accountability for adverse environmental effects of fossil fuels, distort energy economics in fossil fuels' favor. This reasoning ignores the gargantuan income taxes and other fees (e.g., for oil leases) paid by fossil industries and the fact that the costs of energy and health care in the US are about comparable so that even if fossil fuels were responsible for every physiological and mental ailment suffered by Americans, they would be under-priced by, at most, a factor of two.

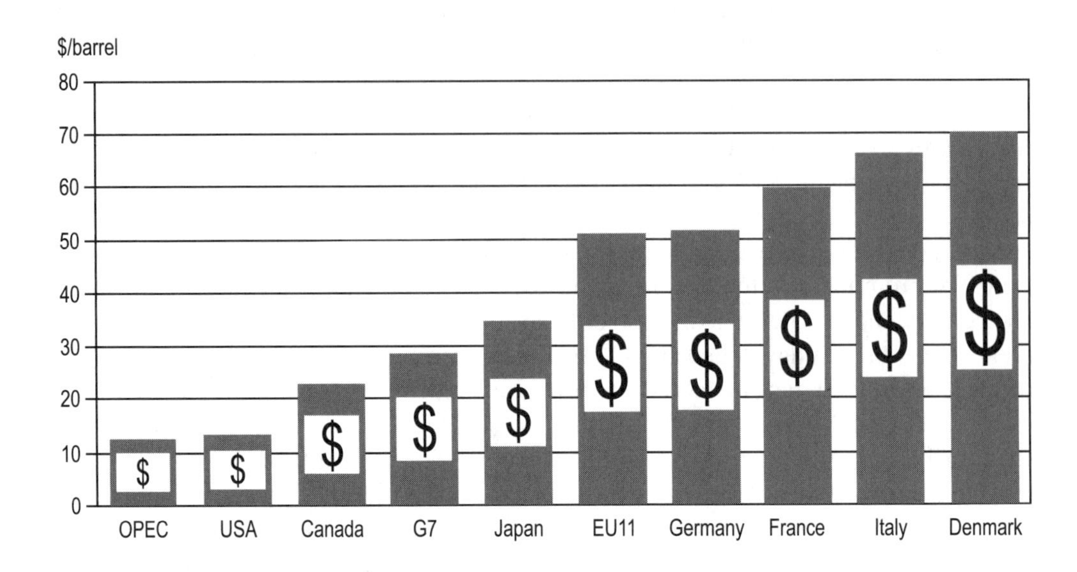

Figure 7.17. Net earning per barrel of crude oil in 1993 by OPEC and by various governments by means of fuel taxes. Source: Okogu, 1995. Reprinted with permission of OPEC.

"American Graffiti." Sustainable progress is stultified by finger-pointing over "unsustainable" practices by global stakeholders. On occasion, developing countries have been identified with an over-reliance on "dirty" technologies, while rich nations, especially the US, have been accused of unfair consumption rates, far exceeding those needed to maintain a decent standard of living. Much time can be spent weighing the merits of these claims. Alternatively, societies can focus on comprehending the forces that shape the current and expected behavior of energy stakeholders. That understanding can illuminate sustainable alternatives responsive to stakeholder needs and aspirations.

Modern technologies for environmental protection require not only financial capital but also good physical infrastructure and well-trained people to operate and maintain technically sophisticated equipment. Many developing regions are short on all of these. Use of inefficient or strongly polluting fossil or other energy technologies in a developing region probably indicates not a systemic indifference to environmental protection but the imperative to concentrate limited resources on immediate problems, such as nutrition, sanitation, health care, and other foundations of social well-being and economic progress.

Most of the US consumption of energy-intensive goods and services is enabled by fossil fuels. But what underlies these consumption patterns and must they be reformed to assure a sustainable future not only for the US but for the entire planet? Consider personal automobiles and light-duty trucks (see Chapter 18 for more details on the transportation sector in general), which account for almost 1/3 of US fossil fuel use. Some critics attribute much of this consumption to oversized "gas guzzling" cars or sport utility vehicles (SUVs) that waste huge amounts of energy to provide unnecessary luxury to the driving public. But SUVs didn't exist 20 years ago and demand for them was promoted through advertising. Such perspectives will do little to advance the cause of sustainable energy and could trigger backlashes that will impede progress. The auto sector is a major economic engine, and Americans have an almost 60-year history of enjoying increasingly consumer-friendly private vehicles at affordable purchase and operating costs. Roughly 16 million passenger vehicles including light trucks are manufactured in the US yearly—roughly 64,000 vehicles per day or one vehicle every 1.35 seconds (assuming 250 manufacturing days per year). Large automobiles, especially SUVs, have become popular with US consumers, as well as with manufacturers that reap an average profit of several thousand dollars per vehicle sold. One view is that sustainable progress will be far better served by understanding that the automobile is ingrained in US popular culture and devising high-efficiency, low-pollution fuels and vehicle technologies that preserve private SUVs and other large automobiles as affordable transportation for the middle class.[3] However, if future energy prices rise substantially, consumers may choose to downsize or modify their transportation choices.

7.6 Economics of Fossil Energy

Fossil fuels are available worldwide at attractive costs and affordable prices to end users. Fossil fuel costs are low (see Chapter 2 and below) in part because they benefit from over a century of steady technological improvements in our ability to discover and produce petroleum, natural gas, and coal and to transport, upgrade, distribute, and utilize these substances and their derivatives throughout the planet. These advances have created today's remarkable global fuel supply infrastructure, as well as automobile engines, electric utility boilers, industrial furnaces, and other combustion technology that have steadily improved in efficiency and environmental performance at manageable costs. Moreover, these advances will continue, further improving our ability to obtain and cleanly use fossil fuels. Part of the development of the present infrastructure has been subsidized through government investment, tax breaks, and other mechanisms. Some externalities such as pollution from local emissions have been managed through

3 The iconoclastic role of the automobile in US culture is aptly portrayed in a George Lucas motion picture that was a box office bonanza long before his *Star Wars* blockbusters. His *American Graffiti* depicts the exploits of several teenagers coming to grips with their post-high school futures on one Saturday evening (and Sunday morning) in a small town in California in 1962. The juxtaposition of diverse automobiles with the characters' lives transforms this movie from just another "teen flick" to an American masterpiece. Motor vehicles become central players, unifying some half-dozen sub-plots and instilling momentum, humor, and pathos. To appreciate why the private auto is likely to continue to play an important role in the US, study this movie carefully.

regulations which incorporate environmental costs into operations. More vexing, however, will be the issue of CO_2 emissions from continued use of fossil fuels (see Chapter 4 and Section 7.4.3).

Petroleum. It is sometimes stated that fossil fuels are "cheap." If we ask in what sense this is true we could answer, taking petroleum as a proxy for fossil fuels, substantially lower in cost than several other liquid consumer products, such as Coca Cola®, milk, orange juice, and bottled water (Figure 7.18). Again, keep in mind typical consumption rates. The average US consumption of petroleum (about 2.47 gal/day) is substantially greater than that of the other consumer products shown in Figure 7.18. But we can also argue that fossil fuels are, on average, becoming even cheaper. Recall Figure 2.6, which shows gasoline prices (before taxes) in real dollars (i.e., corrected for inflation), on average, declining over time from 1900 to 1996. Over the same period, crude oil prices were more or less constant over 3–5 year periods, with shorter run episodes of increasing but then decreasing prices. Can such trends continue? For petroleum, it is likely that prices will remain mercurial between $20 and $50/bbl for many years (Chapter 2).

Natural gas. For natural gas, there are two schools of thought on future US prices: (1) they will follow classic commodity cycles and decline to $2.00–2.50/MMBtu in two to three years as new supplies and transportation infrastructure become available; or (2) they will be governed by a different paradigm in which long-range prices appreciably higher than historical norms will be commonplace. Jensen (2001) illuminates a number of complexities and subtleties of North American natural gas markets, including the behavior of short-run prices and their relationships to oil prices. Moreover, he provides considerable evidence that gas drilling productivity has been declining with increasing drilling activity, which has important consequences for increasing US gas-fired electric generating capacity. To expand gas deliverability to meet the expected demand growth from the electricity sector (California alone is about 10 GWe short on generating capacity), it will be necessary to accelerate the creation of new gas supplies. This will necessitate higher gas prices to attract the multi-billion dollar investments needed to capitalize long-term supply projects.

Costs of electricity from fossil fuels. Further appreciation for fossil energy economics is gained by comparing the costs of electricity (COE) generated in large central station plants fired by natural gas and by coal with corresponding costs for nuclear fired power. (See also Chapter 17.) Tables 7.6 through 7.8 provide relevant data. Note first that typical coal costs are $1.00–1.50 per million Btu (i.e., about $24–36 per ton), and that representative natural gas costs fall in the range of roughly $2.00–5.00 per million Btu (i.e., to $12–30 per barrel oil on the basis of an equivalent energy content).

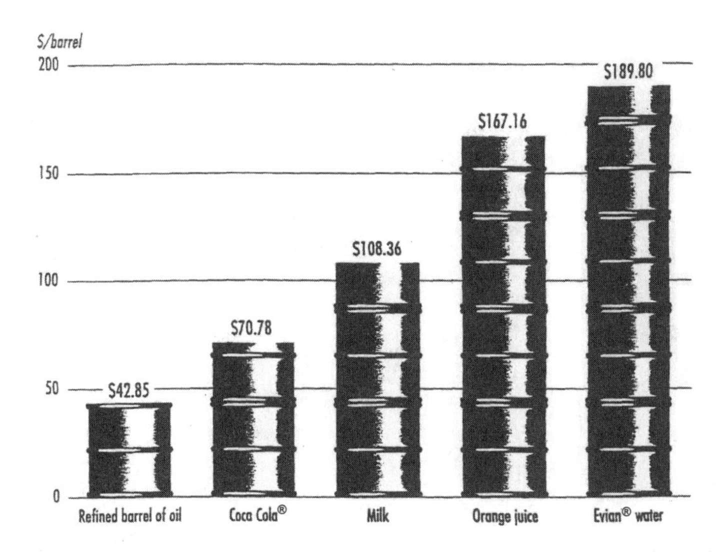

Figure 7.18. Prices paid by consumers for a barrel of refined oil (without taxes) and a barrel of various other consumable liquids. Source: Okogu (1995). Reprinted with permission of OPEC.

Table 7.6 shows that sub- and supercritical steam pulverized coal boilers have similar COEs for a given coal price, with the somewhat higher capital cost of supercritical units largely offset by their greater efficiency. Further, unless natural gas prices return to more traditional lows, i.e., below about $2.70/MMBtu, the COE for coal is lower than that for gas. This assumes that US coal prices will remain low and stable in the coming years, continuing this industry's multi-year trend of thin margins and intense competition. However, Ellerman (2001) draws attention to recent price surges for coal earmarked for US regions experiencing high gas prices. This may be evidence of a coupling between short-run coal and gas prices. (In Table 7.6, coal prices of $1.00, $1.25, and $1.50 per MMBtu, respectively, correspond to $24, $30, and $36 per ton coal, assuming a typical heating value for steam coal of 12,000Btu/lb.)

Table 7.6. Cost of Electricity Generated in the US from Coal and Natural Gas[d]

	Fuel Cost	Contributions to Cost of Electricity COE $/MWh			Total COE
Technology	$/million Btu	Fuel	O&M	Capital	$/MWh
Pulverized coal with Supercritical Steam Boiler[a]	1.00	8.50	3.50	19.62	31.62
	1.25	10.62	3.50	19.62	33.74
	1.50	12.75	3.50	19.62	35.87
Pulverized Coal with Subcritical Steam Boiler[b]	1.00	9.22	3.50	18.40	31.12
	1.25	11.52	3.50	18.40	33.42
	1.50	13.83	3.50	18.40	35.73
Natural Gas Combined Cycle[c]	2.00[e]	16.60	2.00	10.72	29.32
	3.40[e]	25.66	2.00	10.72	38.38
	4.60[e]	37.71	2.00	10.72	47.43

Source: Beér, 2001.

[a]Basis: Capital cost $960/kW; O&M (non-fuel); $3.50/Mh; average heat rage 8,500 Btu/kWh

[b]Basis: Capital cost $900/kW; O&M (non-fuel); $3.50/MWh; average heat rate 9,220 Btu/kWh

[c]Basis: Capital cost $600/kW; O&M (non-fuel); $2.00/MWh; average heat rate 6,800 Btu/kWh

[d]Other assumptions:
— Capacity factor 85%
— Book life 30 years for coal plants; 20 years for natural gas
— Levelized Capital Charge: Data in table calculated assuming 85% capacity factor and 15% interest. B&W information (supplied as bar charts, not shown) assumes "EPRI Tag." The Table 7.5 data and the bar charts are in good agreement.
— All heat rates are in HHV but, consistent with the B&W bar charts, fuel contribution to COE for natural gas is based on LHV.

[e]B&W quotes the following forecast of natural gas prices presented by Mr. Ken Lay of Enron at the Governor's Conference on Gas, Sept. 20, 2000:
— Yr 2000 $5.30/MBtu
— Yr 2001 $4.40/MBtu
— Yr 2002 $3.70/Mbtu

See text for discussion of concerns over future US natural gas prices.

Table 7.7 illustrates that in the US, a *new* nuclear plant, based on today's technology, has little hope of competing with coal or even rather highly priced gas (i.e., $4.60/MMBtu). However, several *existing* US nuclear plants have generation costs as low as $20/MWh because their capital costs have been paid off (Moore, 2000). This is consistent with the sum of the data for "Production Cost" and "Other Costs" (Columns 3 and 4, Table 7.7) for the top 50% of US nuclear plants averaged over 1997–1999. The prospect of attractive returns from these already-paid-down nuclear plants explains the resurgence of investor interest in acquiring existing US nuclear generation plants in the last two years (Moore, 2000).

Table 7.7. Costs of Electricity Generated in the US by Nuclear Fission Reactors

		Contribution to Cost of Electricity (COE) $/MWh					
Relative Plant Performance	Average Capacity Factor, %[a]	Production Cost[b]	Other Costs[a,c]	Capital Cost[d] @1,500$/kW	@2,000$/kW	Total COE $/MWh,@ 1,500$/kW[e]	2,000$/kW[f]
Bottom 25%[b]	74.6	28	5–8	39.43	46.58	70.93	82.58
2nd 25%[b]	82.3	18.4	5–8	31.67	42.22	58.07	68.62
3rd 25%[b]	87.9	15.8	5–8	29.65	39.53	53.45	63.33
Best 25%[b]	91.4	13.3	5–8	28.51	38.02	49.81	59.32

Source: Fertel et al. (2000).

[a]Three-Year Rolling Average 1997–1999.

[b]=Fuel + O&M + Waste Fee + Decommissioning

[c]Capital additions to existing plants, A&G, property tax, etc.

[d]Assumes capacity factor given in column 2. Levelized cost calculated assuming 30-year plant life and 15% interest

[e]Assumes higher value of 8$/MWh for "other" costs

[f]Disaggregated data for fuel and non-fuel O&M not provided by this source.

Table 7.7 also documents the dramatic impact of differences in nuclear plant performance on COE, owing to variations in capacity factor, as well as safety performance, fuel costs, and operation and maintenance (O&M) charges. Table 7.8 disaggregates fuel and non-fuel cost contributions (three-year average for all US nuclear plants) to the COE for 1987, 1991, and 1996. The overall improvement (additional data are provided by Kazimi and Todreas, 1999) is evident, but the relatively small contribution of these costs to total COE (i.e., <40% at most, Tables 7.7 and 7.8) reaffirms that reduced capital costs, without compromising safety or overall performance, are essential to a competitive future for electricity generation from new nuclear plants in the US. On-site storage and eventual sequestration in a national depository are technically viable strategies for management of wastes from existing and future nuclear plants, but these options are virtually certain to face continuing political challenges.

Table 7.8. Fuel (Uranium) and Non-Fuel O&M Costs for Electricity Generation in the US by Nuclear Fission Reactors[a]

	Contribution to COE, $/MWh[a]		Total, Fuel
		Non-Fuel	+ Non-Fuel
Year	**Fuel**	**O&M**	**$/MWh**
1987	10	15.1	25.1
1991	7.4	17.4	24.8
1996	5.7	13.8	19.5

Source: Kazimi and Todreas, 1999.
[a]Three-year average median cost for all plants.

Uncertainties in electricity costs. The data in Tables 7.6 through 7.8 are believed to be reliable. The capital costs for the fossil plants are probably accurate to within ± 5%. Nuclear capital costs are more variable. More stringent environmental regulations on pulverized coal plants would probably degrade plant efficiency by no more than 1% (i.e., fuel-to-electricity might drop from 40 to 39%). Plant location impacts the COE, because of considerable regional variations in fuel costs and fuel use regulations (see Table 7.6). These factors can be accounted for in cost projections. For example, increasing capacity factors clearly reduce capital charges per unit of output, but may jeopardize longevity and cumulative profitability by excessively delaying essential and preventative maintenance. Some natural gas plants facing "must run" contracts have learned how to achieve good returns on investment (ROIs) and operate at capacities as high as 90%. The profitability of these so-called merchant plants, which compete in a deregulated electricity market, depends on how much electricity they can sell and on how much gas they can obtain reliably and at what price. In contrast, a regulated electric utility agrees to provide electricity at high reliability in exchange for a guaranteed fixed, but generally lower, ROI. Somewhat less than 50% of the kilowatt hours generated in the US are deregulated at this time.

7.7 Some Principles for Evaluating Fossil and Other Energy Technology Options

Chapter 5 provides analytical tools to estimate the economic feasibility of energy projects. Here, we present additional criteria to apply in forecasting the potential for success of an energy project. These performance metrics apply to conventional and alternative energy projects. Many were rediscovered in the 1970s, when the public and private sector in the US and other countries, spurred on by fears of natural gas and petroleum scarcity, devoted financial and technological resources to produce substitute or synthetic fuels from coal, oil shale, and other non-traditional fossil raw materials, as well as from biomass.

We can complement the Three "T"s of combustion and the Three "C"s of petroleum refining with the Three "Es" of project evaluation: *environmental performance*, *efficiency*, and *economics*. In many countries today, no project has any hope of going forward without convincing governmental and private sector bodies (e.g., those developing or financing the project) that it can fully comply with strict environmental performance requirements—on its overall footprint, aesthetic impacts, air and water emissions, solid wastes, and occupational safety. Granted, environmental regulations are not universal and can vary substantially from jurisdiction to jurisdiction, especially between the developed and developing regions. Most proposed projects depend on successful performance of various technologies to fulfill their environmental requirements. Thus, technical evaluations of various environmental subsystems are essential to a thorough project evaluation. Efficiency means not only the overall energy efficiency of the project (Chapter 3) but how well the project minimizes waste of all substances (raw materials, cooling water, etc.), how easily its end-products can be recycled, and how easily, at the end of the project life, the project hardware can be decontaminated or recycled and the project site made suitable for other uses. Assessments of various technological sub-systems are critical to gauging overall "efficiency."

Several principles are useful in determining the technical viability of systems and their components:

- *Energy balances.* Identify all energy inputs including energy to operate equipment and the energy content of all raw materials. What fraction of the inputted energy is "recovered" in process outputs (First Law of thermodynamics, Chapters 1 and 3), and what is the availability of this energy output (Second Law of thermodynamics, Chapter 3)? In large fossil energy projects, capital costs scale roughly inversely with thermal efficiency. Projects to produce commodity fuels (e.g., substitute petroleum liquids) will face tremendous difficulty if their calculated thermal efficiencies are $< 50\%$. A lower overall efficiency is tolerable if the energy is generated at a high availability (e.g., as electricity or high temperature heat).
- *Material balances.* Have all needed inputs of raw materials, water, air, and the like been accounted for and can the fate of these materials in the process be accounted for? Does the mass of material exiting the process equal that entering the process? Acceptance criteria for material balances depend on the scale of development of the project. In fossil fuels, during the research, development, and demonstration project phase, material balance closures to within $\pm$ 25, 10, 5, and 1% would build confidence in work at laboratory, pilot, critical-subscale, and demonstration scale, respectively.
- *Technical feasibility essentials.* No project can succeed if it violates basic laws of science. Thus, projects that exhibit 100% efficiency are gibberish. This may seem obvious, but scientifically impossible notions can gain remarkable public visibility. Even when there is no violation of scientific principles, it is important to estimate the performance limits of a candidate process. Specifically,

thermodynamic calculations (Chapters 1 and 3) should always be performed to forecast upper bounds on thermal efficiency and yields and selectivities of desired products for the proposed operating conditions. If these results are encouraging, there is need to estimate the kinetics of rate limiting processes, chemical or physical, in the process. A process can be totally thermodynamically feasible, but so slow as to be of no practical use. In many cases, the necessary rate information can be assembled from the literature, but in others, considerable research involving experiments, as well as interpretive modeling, must be carried out to generate the necessary kinetic understanding.

Another important factor is whether materials are available to build the hardware needed to operate the process over long periods of time without safety risks, vessel failure, or product contamination owing to corrosion or erosion of process equipment. Successful performance by virtually all modern energy technologies depends on synergistic operation of an array of process components. Of necessity, the formative stages of research and development are typically concerned with demonstrating and improving the technical and economic viability of one or a few key technical components. However, it is essential to determine whether there is a high probability that the proposed technology will operate as an integrated system of all crucial components. This evaluation must occur before committing to major investments in large-scale demonstration, and more preferably, to the extent feasible at the scale of development, will be on-going throughout the process of technology scale-up. Examining the process from a lifecycle viewpoint (Chapter 6) is also useful to assure that feedstock will continue to be available within economic limits and that wastes and emissions will conform to emerging regulatory standards.

Related questions include: Is the overall process composed of an elaborate assembly of numerous technical sub-systems? Is there use of temperatures, pressures, or chemical reaction environments that push or exceed the envelope of experience at commercial scale? Is there heavy reliance on high-performance materials of construction? If the answer to any of these three questions is "yes," there is need for extra caution in approving further investments.

Scale at which performance has been satisfactorily proven. In the energy sector, innovative technologies undergo substantial *scale-up* before becoming acceptable for commercial service. This is because most consumer and industrial applications require reliable delivery of heat, mechanical work, electricity, and fuels at rates many times greater than those used in embryonic research to create and validate technological innovations (Table 7.6). For example, a fundamental experiment on the kinetics of coal thermal decomposition in a pulverized fuel boiler shows ability to process 10 mg of coal in 0.1 min (i.e., 0.013 lb coal/hr). By comparison, a large pulverized coal-fired electricity utility generating station might consist of three 340 MWe boilers (for a total capacity of 1,020 MWe, i.e., enough electricity to supply the needs of about 1 million people in a residential/light industrial commercial setting). Each of these boilers would consume about 218,000 lb/hr (109 ton/hr) of a good quality steam coal (14,000 Btu/lb HHV). Here, the difference in scale from research to full commercial service is about a factor of 16 million.

Many processes become less expensive per unit throughput as they are scaled up, but they also become more complex because components must be integrated and function harmoniously. For example, in the 1970s, unit capacity of pulverized coal boilers reached 1,300 MWe in a Tennessee Valley Authority power plant. Since then, no significant scale-up has occurred. The problem of scale-up to such large boiler sizes is that, under conditions of geometrically similar scaling, the surface-to-volume ratio of the combustion chamber is reduced in inverse proportion to increasing size. This makes it difficult to maintain a furnace gas exit temperature below the softening point of the fly ash without installing extra heat transfer surface in the combustion chamber (steam generating tube walls), changing the furnace geometry to a more slender form (taller and narrower), and/or reducing the volumetric heat rate in the combustion chamber. These design additions and changes, as well as the increased tendency for slagging and fouling that cut into boiler availability, have set practical limits on the technologically and economically feasible scale up.

In evaluating any energy project, it is imperative to determine at what scale the process performance has actually been demonstrated. Investors' confidence typically increases as the scale of proven performance goes up. Scaling factors between testing depend upon process complexity and on the extent to which its performance depends upon established versus new sub-systems. In evaluating technologies for producing synthetic fuels from coal, scale-up occurred in 3- to 10-fold increments beginning with bench scale tests at one to a few lbs of coal per hour. These scale-up factors are in part a consequence of the many technological challenges in processing coal and other solid raw materials: storage, size reduction, and feeding to pressure vessels of large quantities of erosive solids, some of which soften and become sticky upon heating; operation at elevated temperatures and often at high pressures under reactive gases such as oxygen, steam, and hydrogen; and difficulties in separating desired products in high yields without contamination. Scale-up nomenclature includes terms such as research (laboratory), bench, pre-pilot, pilot, critical subscale (proof of concept) demonstration, demonstration, and commercial.

Some technologies reach a maximum unit size for optimal performance. For example, coal plant boiler units are much larger in generating capacity than a typical gas power turbine, so multiple turbines are installed to produce more capacity. Solar panel and fuel cell installations also use multiple smaller units to meet larger demands. Wind turbines are examples of multiple-unit systems.

As scales increase, throughputs and the number of interacting systems to be demonstrated also increase. The overall objective is to increase the understanding of the process so as to provide reliable technological, environmental, and economic bases for advancing to the next stage or for implementing changes that will make that next stage viable. It is also wise to test, at various scales, non-technological proposals to achieve sustainability, such as regulations, company policies, and voluntary performance targets.

As a new technology moves from pilot scale to larger demonstration scales, the level of investment required increases substantially. Many promising technologies may die at this stage because of the large initial investments needed to commercialize a new technology. Returns on any investment of this type are subject to varying levels of risk. Vendors or large utilities, sometimes with financial support from the government, used to invest in new technology development. With deregulation, much of this ability has disappeared, and new ways are being found to promote energy technology development.

Other Questions of Diligence. It is vital to assess the reliability of technical and economic assumptions that are crucial to process success. Have alternatives to achieve the same objectives been thoroughly screened and how do these alternatives compare with the proposed technology? Does the proposed approach offer overwhelming advantages in one or more of environmental impacts, efficiency, and economics—or does it face appreciable competition? Will developers, financiers, and regulators fairly credit the advantages offered by the proposed technology? Have adverse contingencies been adequately identified and appropriate countermeasures incorporated in the process implementation plan? What other factors will challenge process performance or economics? For example, by 2005 Europe and the US will require substantially lower sulfur content in gasoline (30 ppm) and diesel (15 ppm) to help reduce automobile particulate emissions, adding an estimated 3 cents/gallon to the refiners' margins on these fuels. To be successful, does the process depend on yet-to-be-demonstrated advances in some sub-system technology or yet-to-be-enacted environmental or fiscal regulations? What is the probability that these changes will occur in time scales relevant to developing this process?

Technology readiness levels (TRLs). For roughly 20 years, NASA has used a series of well-defined R&D performance metrics known as technology readiness levels (TRLs; Table 7.8) to gauge the technical maturity of various products and processes important to space exploration and other NASA missions. In 1999, the US Army adopted TRLs (GAO, 1999 and Kumagai, 2002) to assist decision makers in assessing the state of development of military systems. TRLs facilitate communication during technology assessment and assist planners in discriminating among candidate technologies and in formulating technology development plans including the identification of where prescribed levels of R&D investment can have a greater impact in moving promising technologies closer to practical applications. TRLs are an important tool that can be applied in assessing the maturity and potential practical impacts of new energy technologies.

Table 7.9. Technology Readiness Levels and Their Definitions

Technology Readiness Level	Description
1. Basic principles observed and reported.	Lowest level of technology readiness. Scientific research begins to be translated into applied research and development. Examples might include paper studies of a technology's basic properties.
2. Technology concept and/or application formulated.	Invention begins. Once basic principles are observed, practical applications can be invented. The application is speculative and there is no proof or detailed analysis to support the assumption. Examples are still limited to paper studies.
3. Analytical and experimental critical function and/or characteristic proof of concept.	Active research and development is initiated. This includes analytical studies and laboratory studies to physically validate analytical predictions of separate elements of the technology. Examples include components that are not yet integrated or representative.
4. Component and/or breadboard validation in laboratory environment.	Basic technological components are integrated to establish that the pieces will work together. This is relatively "low fidelity" compared to the eventual system. Examples include integration of *ad hoc* hardware in a laboratory.
5. Component and/or breadboard validation in relevant environment.	Fidelity of breadboard technology increases significantly. The basic technological components are integrated with reasonably realistic supporting elements so that the technology can be tested in a simulated environment. Examples include "high-fidelity" laboratory integration of components.
6. System/subsystem model or prototype demonstration in a relevant environment.	Representative model or prototype system, which is well beyond the breadboard tested for TRL 5, is tested in a relevant environment. Represents a major step up in a technology's demonstrated readiness. Examples include testing a prototype in a high-fidelity laboratory environment or in simulated operational environment.
7. System prototype demonstration in an operational environment.	Prototype near or at planned operational system. Represents a major step up from TRL 6, requiring the demonstration of an actual system prototype in an operational environment, such as in an aircraft, vehicle, or space. Examples include testing the prototype in a test bed aircraft.
8. Actual system completed and "flight qualified" through test and demonstration.	Technology has been proven to work in its final form and under expected conditions. In almost all cases, this TRL represents the end of true system development. Examples include developmental test and evaluation of the system in its intended system to determine if it meets design specifications.
9. Actual system "flight proven" through successful mission operations.	Actual application of the technology in its final form and under mission conditions, such as those encountered in operational test and evaluation. In almost all cases, this is the end of the last "bug fixing" aspects of true system development. Examples include using the system under operational mission conditions.

Source: GAO (1999).

Time Scales. Has the total time to transition from basic concept to commercial scale been reliably quantified and addressed? (For many fossil fuel projects, this time can be of order 3 to 10 years.)

Supporting Infrastucture. Abrupt or even gradual replacement of one fuel with another will require facilities for the manufacture, storage, transport, and distribution of the fuel. The capital and O&M costs for these facilities must be offset in the customer's price of the new fuel.

Project Location. Remote or other unusual venues can substantially impact project costs and technical viability, owing to (Weiss et al., 2000):

- The need to construct infrastructure to support the project
- Higher costs for construction and operation of project and infrastructure facilities
- Local costs of fuel, electricity, raw materials (note that fuel costs, e.g., natural gas, may be appreciably discounted and benefit the project owing to their long distance from lucrative markets)
- Additional risks, e.g., from political instability, that must be countered by higher rates of return for investors

The resulting differences in project costs can be substantial. For example, Weiss et al. (2000) cite US Department of Energy (DOE) data that estimated the capital costs (1987$) of a 10,000 tonne/day methanol plant to vary from $588 million in Trinidad to $1,323 million at Alaska's North Slope.

Cost Elements for Petroleum-Derived Fuels. The pretax cost of liquid fuels produced by refining petroleum (gasoline, diesel, jet fuel, home heating oil, heavy fuel oil, etc.) consists of the refiner's cost to purchase crude oil (i.e., the raw material for fuel manufacture), the cost to refine the oil into the desired product (e.g., gasoline), and the cost to distribute the refined product to customers (Weiss et al., 2000). Crude oil prices are the most volatile and most uncertain of these three (Chapter 2), with spot prices varying from $10/bbl in December 1998 to $38/bbl in September 2000 to about $22/bbl in March 2002. Refining costs vary because each refinery is a unique collection of physical and chemical process operations that are harmonized in various ways to produce diverse marketable products whose specifications and throughputs vary by geographic region and season of the year. Weiss et al. (2000) point out that a convenient, although imperfect, proxy for refining costs is the *refiner's margin,* defined as the difference between the average price at which refiners sell a gallon of finished fuel at the refinery gate and the average price refiners pay for crude oil. From 1982 to 1999, refiner's margins for gasoline and diesel in the US went from 22 to 31 and 12 to 20 cents/gal, respectively. Distribution costs in the US were about 15 to 16 cents/gal for gasoline and diesel. They are expected to change little over the next 10 to 20 years (Weiss et al., 2000).

7.8 Emerging Technologies

For fossil fuels technology, R&D targets increased utilization efficiency and decreased environmental footprint. Numerous studies describe specific challenges and opportunities (see http://www.doe.gov). For example, Table 7.10 (NRC, 2000) summarizes the goals of the US DOE Vision 21 Program for use of petroleum, natural gas, and coal in the US. Many programs concentrate on reducing atmospheric emissions of NO_x, SO_x, and particulates and, increasingly, on developing means to lower emissions of CO_2 and other greenhouse gases to the atmosphere (see Chapter 4 and Section 7.4). Reduced adverse impacts on all media including water, open space, and lower solid wastes are related objectives. There are breakthrough opportunities, such as technology to convert natural gas directly to premium liquid fuels that are economically competitive with petroleum-derived liquids. Such technologies must be demonstrated at full scale at the large throughputs typical of modern fuel use patterns. Moreover, the historical pattern that technological progress in the fossil fuels industry has been evolutionary rather than revolutionary means that true paradigm-shifting breakthroughs, while possible, will probably be rare. The encouraging news is that this sector of the energy industry has an impressive record of remarkable progress in improved efficiency and environmental performance through steady evolutionary improvements in technology.

Most engineers are familiar with the idea that successful new technologies experience rapid improvements in the early stages of development. Significant improvements are also made as the technology is commercialized—in streamlining production, in system integration, and in achieving economies of scale. As the technology matures, the rate of improvement decreases to small improvements in efficiency or in recovery of energy from waste streams.

However, most major energy production systems consist of combinations of new and mature technologies, so introductions of a new control technology or new higher temperature materials can contribute to additional improvements in a mature technology. This is an added challenge to alternative technologies, which not only compete with existing technology but with continuing improvements in that technology.

7.9 Closure: Why Are Fossil Fuels Important to Sustainable Energy?

Fossil fuels are needed to sustain economic progress now and for the foreseeable future. Further, they must be used to enable the transition to non-fossil alternatives at a pace that will prevent economic disaster. Environmental concerns notwithstanding, the current economy; abundance; availability; proven performance; multi-trillion dollar production, transport, and utilization infrastructure; and widespread global use of fossil fuels teaches that fossil fuels will continue for some time to play a significant and, perhaps, dominant role in the supply of energy. No one knows what "for some time" means. We do know that fossil fuels have been indispensable in enabling social and economic progress in the developed world.

Table 7.10. Goals of the US DOE Vision 21 Program

Attributes	Goals
Efficiency-Electricity Generation	60% for coal-based systems (based on fuel HHV);[a] 75% for natural gas-based systems (LHV)[b] with no credit for cogenerated steam
Efficiency-Combined Electricty/Heat	Overall thermal efficiency above 85%; also meets above efficiency goals for electricity
Efficiency-Fuels Only Plant	When producing fuels, such as H_2 or liquid transportation fuels alone from coal, 75% fuels utilization efficiency (LHV)
Environmental	Near-zero emissions of sulfur and nitrogen oxides, particulate matter, trace elements, and organic compounds; 40–50% reduction in CO_2 emissions by efficiency improvement; 100% reduction with sequestration
Costs	Aggressive targets for capital and operating costs and RAM;[c] products of Vision 21 plants must be cost-competitive with market clearing prices when they are commercially deployed
Timing	Major benefits (e.g., improved gasifiers and combustors, gas separation membranes) begin by 2006 or earlier; designs for most Vision 21 subsystems and modules available by 2012; Vision 21 commercial plant designs available by 2015

[a]HHV = higher-heating value
[b]LHV = lower-heating value
[c]RAM = reliability, availability, maintenance
Source: NRC (2000) and adapted from DOE (1999).

Developing countries with fossil fuel reserves (almost all countries) will wish to use these sources for their own development or as a source of revenue for development. Under the international Agenda 21 framework (United Nations, 1992), developing countries define their own national clean development guidelines, under general guidelines that promote efficiency and environmental protection. Where cleaner technologies are significantly more expensive, international aid will be needed—but availability of such funding is limited. (See Chapter 21.) If alternative technologies can compete with fossil fuels in the absence of existing infrastructures, there is the possibility of "leap-frog" technology (e.g., cell phones). However, such opportunities seem limited for bulk energy production today.

To sustain progress largely without fossil fuels will require a carefully crafted transition period. This is needed to allow viable alternatives to be developed and tested at commercial scale and then to be deployed to the marketplace at a pace that will avoid untenable economic dislocations and the local, regional, and global political disruptions that would almost certainly accompany economic upheaval. The time scales for this transformation will be long enough that fossil fuels must play a significant buffering role throughout most of the transition. There can be no doubt that global use of fossil fuels will at some point decline. This may arise from concerns over climate or other untenable environmental impacts, or from irreversible scarcity—scarcity that is permanent, not temporary (Chapter 2). However, planners and decision makers cannot forecast when this will occur. Thus, we must forge ahead both with the development of non-fossil alternatives and with the improvement of fossil energy technologies to sustain the benefits fossil fuels currently provide and to assure that they can be used in an economically and environmentally sound manner.

References

Babir, F. 1999. "PEM Fuel Cells: Basic Principles and Equations." White Paper. West Palm Beach, FL: Energy Partners.

Beér, J.M. 2001. Personal communication. Data supplied by Byers Rogan of the Babcock & Wilcox Co. for a 2000–2001 study by the National Coal Council Advisory Committee for the US Secretary of Energy.

Beér, J.M. 2000. "Combustion technology developments in power generation in response to environmental challenges." *Progress in Energy and Combustion Science*. 26: 301–327.

Bjørseth, A., ed. 1983. *Handbook of Polycyclic Aromatic Hydrocarbons*. New York: Marcel Dekker.

Chisholm, S. 2000. "Stirring Times in the Southern Ocean." *Nature* 407: (685–687).

Corey, R.C., R. Barrett, R.C. Ames, H.F. Chambers, Jr., E.L. Clark, N.H. Coates, W.E. Fraize, Y.C. Fu, H.A. Grabowski, E. Mezey, and D.E. Stutz. 1984. "Energy Utilization, Conversion, and Resource Conservation," in *Perry's Chemical Engineers' Handbook*, sixth edition, R.H. Perry, D. Green, and J.G. Maloney, eds. New York: McGraw-Hill.

DOE (U.S. Department of Energy). 1999. *Vision 21 Program Plan: Clean Energy Plans for the 21st Century*. Morgantown, WV: US DOE Federal Energy Technology Center.

Edwards, J.B. 1974. *Combustion, the Formation and Emission of Trace Species*. Ann Arbor, MI: Ann Arbor Science Publishers.

Ellerman, A.D. 2001. "Energy and Environmental Policy Workshop." Comments during discussion period. Cambridge, MA: Center for Energy and Environmental Policy Research, MIT, May 4.

Elliott, J.J., and M.T. Melchior. 1985. "Petroleum Composition," in Kirk-Othmer Concise Encyclopedia of Chemical Technology, H.F. Mark, D.F. Othmer, C.G. Overbeyer, G.T. Seaborg, M. Grayson, D. Ecksoth, E. Graber, A. Klingsberg, and P.M. Siegel, eds. New York: Wiley-Interscience, John Wiley & Sons.

Elliott, M.A., ed. 1981. *Chemistry of Coal Utilization, Second Supplementary Volume*. New York: Wiley-Interscience, John Wiley & Sons.

Fertel, M.S., R.J. Myers, and V.E. Kait. 2000. "Outlook for Nuclear Energy In A Competitive Electricity Business." Washington, DC: Report by the Nuclear Energy Institute.

Fristrom, R. 1995. *Flame Structure and Processes*. New York: Oxford University Press.

GAO (General Accounting Office). 1999. "Best Practices—Better Management of Technology Development Can Improve Weapon System Outcomes." A Report of the United States General Accounting Office to the Chairman of the Ranking Minority Member, Subcommittee on Readiness and Management Support, Committee on Armed Services. US Senate. Report No. GAO/NSIAD-99–162.

Gary, J.H., and G.E. Handwerk. 2001. *Petroleum Refining: Technology and Economics*, fourth edition. New York: Marcel Dekker.

Glasstone, S. 1982. *Energy Deskbook*, DOE/IR/05114-1. Oak Ridge, TN: US DOE, Technical Information Center.

Hendricks, T.A. 1945. "The Origin of Coal." In *Chemistry of Coal Utilization*, Committee on Chemical Utilization of Coal, Division of Chemistry and Chemical Technology, National Research Council, H.H. Lowry, Chair. New York: John Wiley & Sons.

Herzog, H.J. and E.M. Drake. 1996. "Carbon Dioxide Recovery and Disposal From Large Energy Systems." *Ann. Rev. Energy. Environ.* 21: 145–166.

Herzog, H.J., B. Eliasson, and O. Kaarsted. 2000. "Capturing Greenhouse Gases." *Scientific American*, 282(2): 72–79.

Herzog, H.J., D. Golomb, and S. Zemba. 1991. "Feasibility, Modeling and Economics of Sequestering Power Plant CO_2 Emissions in the Deep Ocean." *Environmental Progress*, 10(1): 64–74.

Heywood, J.B. 1988. *Internal Combustion Engine Fundamentals*. New York: McGraw-Hill.

Hottel, H.C., and J.B. Howard. 1971. *New Energy Technology: Some Facts and Assessments*. Cambridge, MA: The MIT Press.

Howard, J.B. 1981. "Fundamentals of Coal Pyrolysis and Hydropyrolysis," in *Chemistry of Coal Utilization*, second edition, M.A. Elliott, ed. New York: John Wiley & Sons.

Jensen, J.T. 2001. "North American Natural Gas Markets—The Revenge of the Old Economy?" Oral presentation and written handout materials, "Energy and Environmental Policy Workshop." Cambridge, MA: Center for Energy and Environmental Policy Research, MIT, May 4.

Katzer, J.R., M.P. Ramage, and A.V. Sapre. 2000. "Petroleum Refining: Poised for Profound Changes." *Chemical Engineering Progress,* 96(7): 41–51.

Kazimi, M.S. and N.E. Todreas. 1999. "Nuclear Power Economic Performance: Challenges and Opportunities." *Annual Review of Energy and the Environment,* 24: 139–171.

Kumagai, J. 2002. "On Point for the Military." *IEEE Spectrum*, 4(3): 65–67.

Lowry, H.H., ed. 1963. *Chemistry of Coal Utilization, Supplementary Volume.* New York: John Wiley & Sons.

Lowry, H.H.. 1945. *Chemistry of Coal Utilization, volumes I and II.* New York: John Wiley & Sons.

Malone, R.D. 1990. "Gas Hydrates Technology Status Report." In Report No. DOE/METC-90-0270. Morgantown, WV: US DOE, Morgantown Energy Technology Center.

Mann, E. and S. Chisholm. 2000. "Iron Limits the Cell Division Rate of Prochlorococcus in the Eastern Equatorial Pacific." *Limnology and Oceanography,* 45(5): 1067–1076.

Mark, H.F., 1985. *Kirk-Othmer Concise Encyclopedia of Chemical Technology,* D.F. Othmer, C.G. Overbeyer, G.T. Seaborg, M. Grayson, D. Ecksoth, E. Graber, A. Klingsberg, and P.M. Siegel, eds. New York: Wiley-Interscience, John Wiley & Sons.

Martin, J. et al. 1994. "Testing the iron hypothesis in ecosystems of the equatorial Pacific Ocean." *Nature*, 371:123–9.

Moore, T. 2000. "License Renewal Revitalizes the Nuclear Industry." *EPRI Journal,* 25(3): 8–17.

NRC (National Research Council). 2000. "Vision 21: Fossil Fuel Options for the Future." Committee on R&D Opportunities for Advanced Fossil-Fueled Energy Complexes, Board on Energy and Environmental Systems, Commission on Engineering and Technical Systems, National Research Council. Washington, DC: National Academy Press.

NRC (National Research Council). 1980. "Refining Synthetic Liquids from Coal and Shale," Final Report of the Panel on R&D Needs in Refining Coal and Shale Liquids, Energy Engineering Board, Assembly of Engineering, National Research Council. Washington, DC: National Academy Press.

Okogu, B.E. 1995. *OPEC Bulletin*. May.

O'Sullivan, J.B. 1999. "Hydrogen Technical Advisory Panel Report and Fuel Cell Development Status." National Research Council Committee Meeting. May 11.

Satterfield, C.N. 1991. *Heterogeneous Catalysis in Industrial Practice*, second edition. New York: McGraw-Hill.

Schmidt, D., V. Wong, W. Green, M. Weiss, and J. Heywood. 2001. "Review and Assessment of Fuel Effects and Research Needs in Clean Diesel Technology." *Proc. ASME Internal Combustion Engine Division*, Spring Technical Conference, Philadelphia.

Shreve, R.N., and J.A. Brink, Jr. 1984. *Chemical Process Industries*, fifth edition. New York: McGraw-Hill.

Socolow, R., and R. Williams. 2002. Research in progress. Princeton, NJ: Princeton University.

Towson, D. 1985. "Tar Sands." In *Kirk-Othmer Concise Encyclopedia of Chemical Technology*, H.F. Mark, D.F. Othmer, C.G. Overbeyer, G.T. Seaborg, M. Grayson, D. Ecksoth, E. Graber, A. Klingsberg, and P.M. Siegel, eds. New York: Wiley-Interscience, John Wiley & Sons.

United Nations. 1992. *Agenda 21*. A/Conf. 151/26, Vols. I–III. Geneva: UN Division for Sustainable Development.

Weiss, M.A., J. Heywood, A. Schafer, E. Drake, and F. AuYeng. 2000. "*On the Road in 2020, A Life-Cycle Analysis of New Automobile Technologies*." Report No. 00-03. Cambridge, MA: MIT Energy Laboratory.

Problems

7.1 Assuming capture and recovery efficiencies of 98% and 97% respectively, show that 1,000 ton/day of CO_2 would correspond to capturing most of the carbon emissions from a 50 MWe plant fired with coal. State all your assumptions including fuel H/C ratio, and fuel-to-electricity conversion efficiency. How would your answer change if the electrical output of the plant remained the same but you switched from coal to natural gas as the fuel?

7.2 Estimate the rate at which energy flows from a gasoline pump to a motor vehicle during a "fill-up." Express your answer in kW. If this rate of energy delivery were electric power, roughly how many people would it serve assuming a typical US residential community? Assume: the HHV of 1 gallon of gasoline is 145,000 Btu; 5 minutes for filling up a 16 gallon auto tank; and a 1 GWe electrical plant serves a residential community of 1 million people.

7.3 Our discussion of electricity costs in Section 7.6 stated that coal costs of $1.00–1.50 per million Btu are equivalent to about $24–36 per ton, and that natural gas costs of $2.00–5.00 per million Btu correspond on an equivalent energy basis to $12–30 per barrel oil. Use the data on fuel properties in Table 7.1 to confirm these equivalencies.

7.4 One measure of the scale of a process is its throughputs of fuel or other input raw materials, e.g., lbs of coal per hour. Compare the coal throughputs for a small scale experiment that uses 10 mg of coal in 6 seconds, and a single boiler in a large commercial scale central station electric generating plant. Assume: the commercial facility has a total output of 1,020 MWe supplied by three boilers of equal capacity; the boiler has a heat rate of (8,980 Btu_t/hr)/(kWe) based on fuel HHV, and fires coal with a gross heating value (i.e., HHV) of 14,000 Btu_t/lb.

7.5 Your company is currently burning natural gas in air to provide high temperature heat for an important process. You discover that to reduce adverse emissions of certain raw materials used in the process, you need to increase the process operating temperature. Discuss qualitatively how you could accomplish this. What if you had to *decrease* the temperature? What other factors would need to be considered if you chose to replace the fuel entirely, say with "intermediate Btu gas" manufactured from biomass? (Hint: see Chapter 10.)

Nuclear Power 8

8.1 Nuclear History

The world is composed of atomic nuclei and their electron clouds. Since the "big bang," many of the naturally occurring nuclei have been unstable, undergoing transformations to more stable forms, and emitting particles and energy in the process. The time scales of these transitions are nuclei-dependent and can range from nanoseconds to billions of years. The naturally occurring emissions of these transitions have created a consistent radioactive environment in which all biological organisms have evolved. As time has passed, the overall level of radioactivity has consistently decreased, but it has always been with us and it always will be.

Despite this reality, nuclear reactions and radioactivity were unknown until 1898 when radioactivity was discovered by Henri Becquerel in Paris. This was followed two years later by the isolation of naturally radioactive radium-226. Until 1938, when the neutron-induced fission reaction was discovered by the team of Lise Meitner, Fritz Strassmann, and Otto Hahn, radioactivity had little practical importance—it was used mainly in luminous watch dials and as a supposed curative at spas. However, the years before 1938 were an interval of research to understand the nature of the ionizing radiations and particles released in nuclear reactions.

Nuclear fission is important in that it releases tremendous amounts of energy and can be made self-sustaining. Chemical reactions involve energy releases on the order of the differences in the energy levels of the quantized states of the atomic electron cloud, on the order of electron volts (eV). Similarly, the energy released by nuclear radioactive decay reactions is on the order of the differences of the nuclear quantized state energy levels, on the order of millions of electron volts (MeV). The fission reaction involves the release of almost 200 MeV in the neutron-induced fission of ^{235}U. Thus, when we compare the order of magnitude of the fission energy release to that in a chemical reaction, the ratio is on the order of 100 million. This basic physical fact is reflected crudely (taking differing fuel concentrations into account) in the annual fuel requirement of a 1,000 MWe electric power station: about three million tons of coal or about 36 tons of enriched uranium (including about 1.0 tons of ^{235}U) for a light-water (cooled) reactor. This fact is also fundamental to the ability of nuclear submarines to cruise for years without refueling and to the ability of a nuclear weapon, involving tens of kilograms of material, to devastate a city.

The most directly visible energetic nuclear reactions are those that power the sun, but their nature as nuclear fusion reactions was not understood until 1938 when the solar reaction cycle was deciphered by Hans Bethe. Fusion reactions typically release energy of the order of 10 MeV and can also be self-sustaining, as is illustrated by the sun. This is an example of the more general principle that all energy is nuclear. That is, any form of energy used on earth is ultimately seen to be a form of energy initially produced by a nuclear reaction (see Table 8.1).

Table 8.1. Fundamental Sources of Energy Used by Different Energy Technologies

Energy Source	Fundamental Nuclear Energy Source
Solar	Gravitationally confined solar fusion reactions transmitted via photons
Fossil Fuels	Gravitationally confined solar fusion reactions transmitted via photons and stored in biomass
Geothermal	Naturally occurring radioactive decays of materials within the earth and gravitational work
Tidal	Nuclear reactions following the Big Bang sustaining current gravitational work
Nuclear Fission	Neutron-induced fission reactions of heavy nuclei
Nuclear Fusion	Nuclear fusion reactions of light nuclei

When it was discovered that a neutron-induced fission reaction produced additional neutrons, then secondary reactions could be induced using the neutrons produced from the first (see Equation 8-1). This is because a single neutron is needed to stimulate a fission reaction, and ν neutrons are produced by it. The value ν will vary from one reaction to another, depending on the fuel nucleus and initial neutron energy, but $\nu = 2.5$ is a typical value.

$$n + {}^{235}\text{U} \rightarrow {}^{236}\text{U} \rightarrow 2 \text{ Fission Products} + \nu(\approx 2.5)n + 6\beta + 10\gamma + 10(\text{neutrinos}) + \text{E}(\cong 200 \text{ MeV}) \quad (8\text{-}1)$$

From Equation (8-1), we can see that as long as, on average, one of the neutrons produced from the fission reaction causes another fission, the neutron population and the fission rate in the reactor can be sustained. Should there be more than one on average, the population and reaction rate will increase. In investigations of this result, it was realized that the potential existed for a rapid release of enormous amounts of energy from a small amount of matter. These ideas were pursued by the World War II allies to their logical conclusions, resulting in the development of nuclear weapons in 1945. This was followed by the debut of nuclear naval propulsion with the startup, in 1953, of the first nuclear-powered submarine, USS *Nautilus*. Initially, the US held a monopoly on these technologies, but beginning in the 1950s, technological diffusion to other countries started. Today, many countries possess the potential to acquire practical military nuclear technologies, and most industrial countries utilize civilian nuclear technologies.

Civilian nuclear power technologies are focused on reactors that can be used for producing electric energy. The first nuclear power plant, at Calder Hall in the UK, started operation in 1956. Today, more than 440 nuclear generating units are active in most of the industrialized countries (Nuclear News, 2002), and after coal, nuclear power is the second most abundant source of electric energy worldwide (EIA, 2002a).

8.2 Physics

The atomic nucleus is composed of positively charged protons and electrically neutral neutrons (together termed "nucleons") and is surrounded by a swarm of negatively charged electrons (in number equal to that of the protons). The chemical nature, or elementary form, of the atom is determined by the number of its electrons, but the nucleus of a particular element can have differing numbers of neutrons with each combination of nucleon populations constituting an isotope of that element. Thus, uranium has 92 protons and electrons but is found in nature having totals of either 235 or 238 nucleons. The behavior of one isotope can be different from that of another. For example, the fission product ^{129}I has a half-life (time for half of an initially existing population of nuclei to decay) of 16 million years, while that of ^{131}I is only eight days (Parrington et al., 1996).

8.3 Nuclear Reactors

In a nuclear reaction, the nucleus may make a transition from one quantized state to another, emitting energy in the form of photons. Alternatively, the composition of the nucleus may also change, with emission of nucleons, electrons, and photons. At the simplest level, these emitted particles may be charged positively (α particles, e.g., ^{4}He nuclei), negatively (β particles, such as electrons), or neutrally (e.g., γ particles (photons) or neutrons). These particles are sufficiently energetic that, when they interact with other atoms, they are able to ionize them, thereby changing the properties and behavior of the latter. The charged radiation particles interact electromagnetically with neighboring atoms and have short (i.e., mm) ranges in solid matter. Because of this, they affect biological organisms primarily when ingested. In that way, they are most usefully viewed as toxins, often affecting only particular organs within the body. They can be blocked by light shielding such as a few mm of metal or even a layer of paper. The uncharged particles (i.e., neutrons and photons) interact with neighboring atoms more weakly and have longer ranges (i.e., m) in solid matter. They are termed "penetrating" radiations, affecting the entire body of an exposed organism. They typically require massive amounts of shielding (i.e., meters of lead, water or concrete) to provide adequate protection. Such changes are especially important in biological systems, where ionizing radiations can be used to kill cancerous cells, but they can also cause changes resulting in cancer induction among other effects.

Nuclear reactions can involve nuclear transitions in which ionizing radiations and energy are released. Harnessing them beneficially, or at least harmlessly, is the major difficulty. In nuclear applications, the environment of any device (e.g., reactor) often becomes radioactively contaminated and difficult to access should something go wrong. This creates an imperative for doing the engineering right the first time and understanding the full implications of any new application, as it can be both hazardous and expensive to learn via trial and error. These realities, coupled with the social controversies attached to some nuclear technologies, have made them expensive compared to non-nuclear alternatives. A resulting emphasis on high quality and reliability have emerged as means of countering the attendant economic penalties of this radiation environment.

The important uses of ionizing radiation and nuclear reactions include medical diagnosis and treatment and industrial and mineral resource deposit interrogation; but these are all overshadowed by the energetic applications. In the civilian sphere, use of the fission reaction to produce electricity has been the most important application. A program to do the same using the nuclear fusion reaction has remained in the research phase for many years. However, it offers the promise of a greatly expanded energy resource base should it prove to be practical.

Nuclear fission reactors include those that serve research as radiation sources (research reactors) and those that produce electricity at industrial scales (power reactors). Typically, a nuclear power plant will have an electric-generating capacity in the range 700–1,400 MWe, and a corresponding reactor thermal power about three times as large.

The neutron-induced nuclear fission reaction described in Equation (8-1) shows the mean outcomes in the fission of ^{236}U (produced by neutron capture by ^{235}U). In this reaction a neutron collides with a target nucleus, producing an excited, unstable, compound nucleus, which quickly decays to a more stable form by splitting apart and emitting many ionizing particles. These particles, and especially the fission fragments formed from the bulk of the compound nucleus' nucleons, carry 192 MeV of kinetic energy. As these particles interact with surrounding matter, they transfer this kinetic energy to the atoms with which they interact, thereby causing ionizations (removal of atomic electrons) and subsequent nuclear reactions. Eventually the transferred energy is apparent as an increase of the internal energy (i.e., heat) of the surrounding materials.

As Equation (8-1) shows, a single neutron is needed to stimulate the neutron-induced fission reaction, and about 2.5 neutrons are produced from it, on average. If one of these neutrons, on average, can be made to cause a subsequent fission, a self-sustaining fission chain reaction will be created. This is what is done in the creation and control of a fission reactor. A neutron born from the fission reaction has a mean energy of about 2 MeV. This "fast" neutron will transfer that energy to other nuclei via scattering collisions until it comes into energetic equilibrium with its surroundings (e.g., at an energy determined by the environmental temperature: at 300°C the "thermal" neutron energy is 1/40 eV). Neutron absorption and fission reaction probabilities typically increase as the neutron energy decreases, and it is preferable to design a reactor so that it utilizes low-energy neutrons to drive the fission reaction. The only naturally occurring material that can serve as the nuclear fuel in such a reactor is ^{235}U. However, high-energy neutrons can cause fission in many high-nuclear-mass materials (termed actinides), and "fast (energy) spectrum" reactors exploit this fact. The fraction of a neutron's kinetic energy that is transferred to a nucleus with which it collides grows as the mass of the target nucleus decreases. Materials that are efficient at slowing neutrons are termed "moderators" and are comprised of low-mass isotopes (e.g., ^{1}H, ^{2}H, C, Be).

When the first man-made reactor was built at the University of Chicago in 1942 by a team led by Enrico Fermi, achieving and controlling a self-sustaining chain reaction was a heroic goal, with many attendant uncertainties. When the rate of the chain reaction is steady, the reactor is said to be in a "critical" condition (not a dangerous one), reflecting the importance accorded by the early developers to success in reaching that condition. A reactor having a decreasing reaction rate is termed "subcritical," and one with an increasing rate is "supercritical." Neutrons are the entities that permit a reactor to function; the power level and the nuclear reaction rates are proportional to the neutron population within the reactor.

The most important nuclear reactions involving neutrons are scattering and absorption, the latter consisting of the alternative outcomes of capture (i.e., retention) of a neutron and neutron-induced fission. As scattering becomes more likely compared to capture, the fraction of neutrons that will escape from the reactor without causing a fission will grow. As capture becomes more likely compared to fission, the fraction of neutrons causing fissions will decrease. The fate of a particular neutron within a reactor is not known in advance; rather the odds of escape from the reactor will depend on the material composition and geometry of the reactor. When materials that can undergo fission are concentrated sufficiently, an assembly that can become critical or supercritical will be created. Otherwise, it will remain subcritical.

The design of a reactor begins by selecting from a menu of materials to perform essential functions. (These functions and some of the materials used most commonly are listed in Table 8.2.) If the neutron is sufficiently energetic, it is possible for it to cause a fission reaction with many heavy isotopes. Concepts for transmutation of high-level nuclear wastes into short-lived radioactive species utilize this fact, as many of the long-lived waste species have a high atomic number.

Of the 2.5 neutrons born in fission of excited ^{236}U (produced via neutron capture by ^{235}U), one is required in order to sustain the chain reaction. In steady state, the remainder will either be captured or escape from the reactor. Some materials that are not fissile are also needed in addition to reactor fuel in order to create a functioning reactor. The essential functions include neutron population control—resulting from absorption, fertile material conversion, neutron escape and neutron energy reduction (i.e., moderation), structural support, and fuel cooling.

In Table 8.2, three nuclear fuel isotopes are listed, but of these only ^{235}U occurs naturally (at least in macroscopic quantities). Most of naturally occurring uranium (99.3%) is ^{238}U, which does not undergo fission upon absorption of low-energy neutrons, such as are commonly used to sustain the chain reaction. The (small) remainder is ^{235}U.

Table 8.2. Essential Fission Power Reactor Functions and Commonly Used Materials

Function or Attribute	**Typical Materials**
Self-Sustaining Chain Reaction	Fuels: Uranium 233- (U) Uranium 235- (U) Plutonium 239- (Pu)
Conversion into Fissile Materials	Fertile Materials: Thorium 232- (Th) Uranium 238- (U)
Nuclear Fuel Material	U, Pu UO_2, PuO_2 UC, PuC
Reactor Internals and Reactor Vessel Structural Material	Steel Stainless Steel Zircalloy (Zr) Aluminum (Al) Graphite (C)
Neutron Moderator	Water (H_2O) Heavy Water (D_2O or ^2H_2O) Graphite (C) BeO_2
Reactor Coolant	Water (H_2O) Heavy Water (D_2O) Helium (He) Carbon Dioxide (CO_2) Organic fluids Lead (Pb) Lead-Bismuth (Pb-Bi) Sodium (Na) Sodium-Potassium (Na-K) Molten salts
Neutron Absorber (Control) Material	Boron (B) Cadmium (Cd) Gadolinium (Gd) Hafnium (Hf)

8.4 Burning and Breeding

A process of fundamental importance in all reactors is conversion of "fertile" materials into fissile isotopes, thereby creating new fissile fuels. The two most important such conversion reactions are:

$$n + {}^{232}Th \rightarrow {}^{233}Th \rightarrow {}^{233}Pa + \beta + \gamma \rightarrow {}^{233}U + \beta + \gamma \qquad (8\text{-}2)$$

$$n + {}^{238}U \rightarrow {}^{239}U \rightarrow {}^{239}Np + \beta + \gamma \rightarrow {}^{239}Pu + \beta + \gamma \qquad (8\text{-}3)$$

The new fissile isotopes created are ^{233}U (Equation (8-2)) and ^{239}Pu (Equation (8-3)). Of the neutrons produced from a single fission reaction, one is needed to sustain the chain reaction. Of the remaining neutrons, on average if at least one causes replacement of the fissile nucleus consumed in the original fission reaction via the reactions of Equations (8-2) or (8-3), the reactor is said to "breed" new fuel. If the original fission reaction leads to creation of less than one new fissile nucleus, the reactor is termed a "burner" or "converter." In breeder reactors, the amount of fissile fuel taken from the reactor during refueling can be greater than the amount loaded initially. (This is not an example of a "free lunch." The new fuel is obtained by converting fertile material nuclei.)

In order to be useful, the portion of these new materials not fissioned in the reactor must ultimately be separated chemically and incorporated into new fuel elements (termed "recycling"). These reactions proceed most rapidly in a "fast" neutron energy spectrum reactor because the number of neutrons produced by a single fission reaction is greatest, but they also occur in "thermal" reactors. They are important because they offer the potential to increase the world's effective supply of nuclear fuel greatly (by a factor of about 70 in reality, by a maximum theoretical factor of 99.3/0.7 = 142). If this were done on a large scale, the potential energy contributions of nuclear power technologies could be increased greatly. However, even without recycling, fissions of ^{239}Pu are an important energy source in "thermal" reactors.

However, ^{239}Pu (produced via the reaction of Equation (8-3) from ^{238}U) is also the most efficient nuclear weapons material. A large industrially based commerce in ^{239}Pu (and involving other Pu isotopes as well) would offer the potential for diversion of some of this material into nuclear weapons use. Ensuring that this would not happen appears to be a difficult task, which some societies may choose to avoid by foregoing the benefits of breeding plutonium. Deterring use of recycled plutonium is an anti-proliferation policy priority for the US, which sets it against countries such as Japan and France, which value reprocessing for national strategic reasons, including energy supply security. Regardless of how this conflict is resolved, in economic terms, for the next several decades, one may ignore the need for recycling, as adverse economics and abundant natural fuel will render it unattractive. However, in the long run, the ultimate social contributions of fission energy technologies may depend on the degree to which these conversion reactions are exploited.

8.5 Nuclear Power Economics

The economics of any energy technology are sensitive to local conditions and fuel prices such that a technology can be competitive under some circumstances and not others. With nuclear power light-water reactors (LWRs), the components of the cost of electric energy generation are typically about 57% capital, 30% operations and maintenance, and 13% fuel (see Table 5.3). Thus, the economics of nuclear power are most sensitive to factors affecting capital costs, including interest rates, construction duration, hardware modifications and replacements, and thermal efficiency. Existing nuclear power plants have been competitive with coal-fired units in most circumstances, but not with the more recently introduced gas-fired combined-cycle units. The economic prospects of nuclear power plants are likely to depend on the ability to reduce their specific capital costs enough to be in the neighborhood of the costs of the competitors. Means of attempting this include design modularization and standardization and construction automation and optimization. Whether these will be sufficient remains to be seen.

In some countries, even though the expected results are satisfactory, uncertainties in these capital cost factors have been sufficiently large as to deter investors. This has been especially the case in the US, where the range of specific capital costs of the existing nuclear power plants has varied by a factor of about 2.5 for what is essentially the same hardware (this volatility is illustrated dramatically by the Seabrook, NH, nuclear power station, which was initially estimated to cost $520/kW and ultimately came in at $2200/kW), and where more than half of the nuclear power plants initially ordered were not brought to completion (EIA, 2002b). Factors that have been especially important in the US context have been uncertainties in the nuclear safety regulations (which at times have resulted in expensive design changes), high interest rates, and determined nuclear power opponents. The last have taken great advantage of the opportunities presented by the US nuclear power regulatory system for a determined individual to introduce delay into the licensing of nuclear power plant projects. This is done by raising technological objections to the project, which can take substantial time to resolve and drive up project costs. Although technological objections are a strategy for some, they also reflect genuine concerns on the part of some citizens. Reflecting these factors, it is not surprising that nuclear power has done best in countries offering stable regulations, low-interest financing, and little opportunity for individuals to affect project licensing. The most conspicuous examples are those of Japan and France, each of which has a nuclear power economy about half the size of that of the US (Table 1.1) but substantially smaller populations and economies (Table 1.5).

In many countries, nuclear power plants are old enough that capital costs have been amortized, and the costs of production are only those of fuel, operations, and maintenance, which are favorable to the case of nuclear power.

8.6 The Three Mile Island 2 Nuclear Power Plant Accident

A milestone in the history of nuclear power is the accident at the Three Mile Island nuclear power plant unit 2 (TMI-2) (Knief, 1992) in Middletown, Pennsylvania. It occurred during the wee hours of the morning, involved human errors and several failures of minor hardware, and resulted in the melting of much of the reactor's fuel (the great majority of which remained within the reactor vessel). Nearly all of the gaseous and volatile radioactive material released from the core remained within the containment building, and no public injuries occurred. However, the public alarm that it caused resulted in a major blow to the nuclear power enterprise.

The March 28, 1979, core melt accident at the TMI-2 nuclear power plant began at 4:00 AM with operational shutdown of the main feedwater pump, which provided make-up water to the steam generator. Within a minute, this event led automatically to reactor shutdown and startup of the emergency system (i.e., back-up feedwater supply system). Flow of coolant from the emergency feedwater system could not reach the steam generators because a block valve was incorrectly closed (and its position incorrectly indicated). These events resulted in a decrease of the primary coolant system pressure to the point that, at about three minutes into the accident, the emergency core cooling system (ECCS) began providing coolant to the reactor.

Due to a combination of events and either incorrect or inaccessible instrumentation readouts, plant operators thought that the primary coolant system was filling with water and about to become overpressurized. In fact, due to a relief valve remaining open, it was losing water. During the initial hour of the accident, in order to prevent the feared overpressurization, the operators reduced flow from the ECCS into the primary coolant system. Subsequently, in order to protect the main coolant pumps from damage when their inlet pressures became too low, the operators shut them down. The result was that the reactor became starved for coolant, and during the next two hours about 80% of the fuel melted and relocated to the bottom of the reactor vessel. Ironically, had the plant operators simply permitted the automatic safety systems to operate, the core would have remained intact. At this point, most of the damage in the accident had been done, and the attendant media and social drama surrounding this event began.

The conventional wisdom of the time was that the reactor fuel in the bottom of the vessel would melt through the vessel wall and drop onto the containment building floor below. That did not happen, because the water remaining in the vessel was sufficient to cool the debris and vessel wall adequately. Subsequently, control of the plant was regained by its staff, and the reactor was cooled, initially by active, and, months later, by passive means. The fuel was finally removed from the vessel about 10 years after the accident.

The hours following core damage were characterized by high uncertainty and anxiety. The uncertainty arose from ignorance about the status of the plant and reactor (partially because needed instrumentation was either absent or destroyed) and about the physics of what would happen next. Anxiety occurred partially because unclear information was provided. Many people were concerned about what could be a dangerous situation.

Anxieties were then amplified as a result of competition for market share by the thousands of news organizations that reported the accident to the world, usually in dreadful terms. With hindsight, we know that needless alarm was created on day 2 of the accident over an incorrect radiation release measurement reading that led to widespread school closings (involving about 140,000 persons) and partial public evacuation from the surrounding countryside. This was followed by concern about a postulated, but actually impossible, hydrogen-fueled explosion within the reactor vessel. By day 4 of the accident, the president of the US, Jimmy Carter, visited the plant to assure that all was under control, and on day 11, the emergency alert ended. The subsequent cleanup of the power plant and the investigation into what had happened lasted about a decade and cost more than the plant had initially cost to build.

No member of the public was physically harmed by the accident, and the occupational effects were mainly those of radiation exposures associated with dealing with the accident's aftermath. However, the accident frightened and angered many people, shaking their confidence that those in charge of the nuclear enterprise could be trusted to do things right. For the worldwide nuclear power industry, and especially that of the US, the accident was an economic catastrophe, contributing to more than 100 canceled nuclear power plant orders and greatly increased operating costs (mainly due to a much more punitive and uncertain regulatory and political climate). It also showed how a compounding of minor malfunctions could lead to a serious accident. In response, the nuclear industry increased operational and maintenance vigilance and improved communications and emergency response training for employees. Three Mile Island was the most serious setback to the nuclear power enterprise until the Chernobyl accident in 1986, which killed and injured many people and forced widespread property abandonment.

Since the accident, many improvements have been made. These include modification of all control rooms to be easier to operate, modification to plant emergency operational algorithms from being knowledge-based to being symptom-based, and requiring each US nuclear power plant to have a plant-specific control room simulator, which is used for regular operator training and testing.

The recognition that all nuclear power plants are "hostages of each other" led to the creation of the Institute of Nuclear Power Operations (INPO)—an internal policeman of the nuclear power industry. Its functions include biannual evaluations of all member nuclear power plants, assistance to those who request it, training, and information evaluation and sharing. INPO has created minimal performance standards for member nuclear power plants and a means of information sharing and mutual learning among plants. Its enforcement capability is derived from INPO's ability to shut a plant down by having the plant's liability insurance canceled.

8.7 Reactor Safety

Fission energy production requires special care for protection against occupational and public dangers arising from the possibility of rapid energy production and human exposures to radioactive materials produced during reactor operations. Nuclear public safety requirements demand that the following essential safety functions be performed reliably:

1. Control of the reactor power, particularly reactor shutdown upon demand
2. Cooling of the reactor
3. Sequestration of radioactive materials from the biosphere

Ensuring that these functions are followed is the responsibility of the reactor operator, but they are also overseen by governmental regulatory authorities and industrial organizations, respectively, the Nuclear Regulatory Commission (NRC) and the Institute for Nuclear Power Operations (INPO) in the US. In the US, it is required that redundant systems be used for accomplishing these essential safety functions, under the principle termed "defense-in-depth." For reactor power control, at least two independent shutdown systems, as well as passive negative power-feedback mechanisms, are required. For the function of confining radioactive materials, a succession of barriers to material transport from the fuel matrix to the containment building (which is required by regulations) are used.

The safety function that has proven most difficult to perform is that of reactor cooling. Cooling is important because overheating the reactor could make the fission products produced within the reactor fuel mobile and able to enter the environment. Such an event would also ruin the reactor as an economic asset and could threaten the health of workers and the public. The problem is that the reactor is a perfectly reliable heat source in that once started, it cannot be fully shut off, but the system for cooling it is not. Energy is produced by a reactor mainly from fission reactions while operating in the power production mode, and it is also produced from post-fission radioactive decays after the reactor is shut down. No method exists to turn off the latter heat source, which, shortly after reactor shut down, is about 7% of the full reactor power. Thus, its failure probability (i.e., unreliability) is equal to zero. However, that of a cooling system is always positive. The engineering challenge is to make the cooling system unreliability values as close to zero as is feasible. In doing this, a large set of potential accidents (most of which have never occurred) are postulated, for which mitigation systems are required to be installed and made highly reliable.

The menu of materials of Table 8.2, arranged in suitable geometries, can be used in formulating a reactor concept. Many reactor concepts have been realized at least experimentally (particularly during the 1950s), and many more theoretically (see Table 8.3). Currently, only a small fraction of proposed ideas are in practical use as functioning nuclear power plants. The most important decisions made in designing a reactor are those of the coolant (which determines the reactor's thermochemical environment) and the fuel (which determines the associated fuel cycle structure). That decision plays a large role in determining the thermochemical regime, and thus, the practical problems of the nuclear power plant. Conceivably, many better reactor concepts await development. The types of power reactors in widespread use are listed in Table 8.4. In the following discussion, important aspects of some of the more widely used reactor concepts are presented.

Table 8.3. Examples of Previously Built Reactor Types

Moderator	Coolant	Fuel	Representative Projects
	H_2O	HEU/Th	Indian Point 1
H_2O	H_2O	HEU as Sulfate	HRE-1
	B, S, H_2O	LEU	BONUS
	N_2	HEU	ML-1
	B D_2O	LEU	Halden
D_2O	B H_2O	UNAT, LEU	SGHWR, Cirene, Gentilly
	CO_2	UNAT, LEU Metal	Bohunice, Lucens
	Organic	UNAT	WR-1
	D_2O	HEU as Sulfate	HRE-2
Graphite	Na	LEU	Hallam
	Na	HEU/Th, Metal	SRE
	He	HEU/Th, Carbide	THTR, Dragon
	Molten Salt	HEU, Fluorides	MSRE
Organic	Organic	LEU, Metal	Piqua
BeO	He	HEU	EBOR
ZrH_2	Na	MEU	KNK-1

B = Boiling Coolant; S = Superheated Vapor
UNAT = Natural Uranium (0.711% ^{235}U)
LEU = Low-Enriched Uranium (5% ^{235}U)
MEU = Medium-Enriched Uranium (5 « 20% ^{235}U)
HEU = Highly-Enriched Uranium (20 « 95% ^{235}U)

} as UO_2 except where noted

8.8 Light-Water Reactors (LWR)

There are about 441 power reactors operating worldwide. Of these, about 78% are cooled by light water (LWRs). The choice of water as a coolant dictates that the reactor system environment will be of high pressure (up to 150 bar), medium temperature (roughly 300°C), and corrosive. These factors have caused headaches for the plant operators, but they are tolerable ones. Light water (consisting of $^{1}H_2O$) absorbs neutrons more readily than does heavy water (D_2O). Consequently, an LWR cannot become critical using natural uranium. Enrichment (concentration) of the ^{235}U fraction in the fuel is required (to about 3–5%). This is done using gaseous UF_6 in an enrichment plant. Heavy-water cooled/moderated reactors can become critical using natural uranium fuel, as can graphite-moderated, gas-cooled reactors.

8.9 Pressurized-Water Reactor (PWR) Technologies

Of the LWRs, about 75% are pressurized-water reactors (PWRs) and the remainder are boiling-water reactors (BWRs). (The basic principles of the two are shown in Figure 8.1, which illustrates their steam production/Rankine cycle electricity production cycles.) With the PWR, water in the primary coolant circuit is kept at pressures so great (i.e., 255 bars) that it remains in the liquid state, even though heated to approximately 300°C. Steam is produced in a heat exchanger (steam generator) where heat extracted from the hot primary circuit water converts cooler, lower pressure (i.e., 70 bars) water into steam. This steam is used in a steam Rankine cycle system to produce electricity.

The reactor consists of an array of 40,000–50,000 fuel rods, each rod being about 1 cm in diameter and 4 m long. A rod uses a cylindrical zircalloy cladding to contain a stack of UO_2 pellets. The rods are arranged in fuel assemblies of about 270 rods each. Water fills the spaces between the rods, and all are held within a reactor vessel, to which the primary circuit piping is connected. Typically, two to four primary circuit loops will be arranged in parallel (Mascke, 1971).

The power is controlled using neutron-absorbing control rods that move within the core. A second independent shutdown system is required for every reactor. With the PWR and BWR, this is done by means of injecting water-soluble boron into the reactor and utilizing the large neutron absorption capabilities of ^{10}B.

The low temperatures of the coolant result in the thermal efficiency of an LWR plant being equal to about 33% rather than 41%, as is more common with fossil-fueled and other nuclear-powered electricity plants. This economic penalty is partially tolerable, because the fuel cost is a much smaller portion of the total energy cost in an LWR (typically 15%) than in a fossil-fueled plant (typically 40%). However, it increases the sensitivity of LWR economics to capital costs, which are the dominant components (typically 65%), in that a more efficient plant would have a lower specific (capital) cost and would require fewer generating units for a system of a given capacity.

Table 8.4. Summary of Types of Power Reactors Used Worldwide

Type	Coolant	Moderator	Maximum Coolant Temperature (°C)	Current Deployment	Current Population
Pressurized Water (PWR)	Light Water	Light Water	330	Most nuclear countries	259
Boiling Water (BWR)	Light Water	Light Water	288	Most nuclear countries	92
RBMK	Light Water	Graphite	270	Former Soviet Union	13
Pressurized Heavy Water (PHWR)	Heavy Water	Heavy Water	267	Argentina, Canada, China, India, Korea, Pakistan, Romania	43
Gas-Cooled (GCR)	Carbon Dioxide, Helium	Graphite	741	UK, Russia	32
Liquid Metal-Cooled (LMFBR)	Sodium, Lead, Lead-Bismuth	None	545	France, Japan, Russia, and India	2

Source: Nuclear News, 2002.

As with a political race, technology development can be viewed as a positive feedback system where the candidates who establish an early lead over the competition are better positioned to compete for the resources needed to increase that lead. This often happens (termed technology "lock-in"), however, without regard to which candidate is best. The history of the VHS video recorder technology winning over the Beta format and the early competition of Internet firms for market share in the absence of profits are cited as examples of this phenomenon. Similarly, the PWR technology was originally developed for ship propulsion by the US Navy (in total, more than 700 such power plants have been built worldwide) and was adapted for use in the Shippingport plant as the original US commercial nuclear power plant technology. From that point, it was improved and greatly scaled-up in an evolutionary fashion as the technology was deployed worldwide in the industrialized democracies. An independent parallel development occurred in the former Soviet Union, where about half of the nuclear power plants are PWRs. Critics of PWR technology have noted this military derivative history of PWR development, arguing that had it not had a "free ride" due to its initial military origins, it would never have been brought into widespread use. The suggested corollary of this argument is that future development funding should go to other technologies.

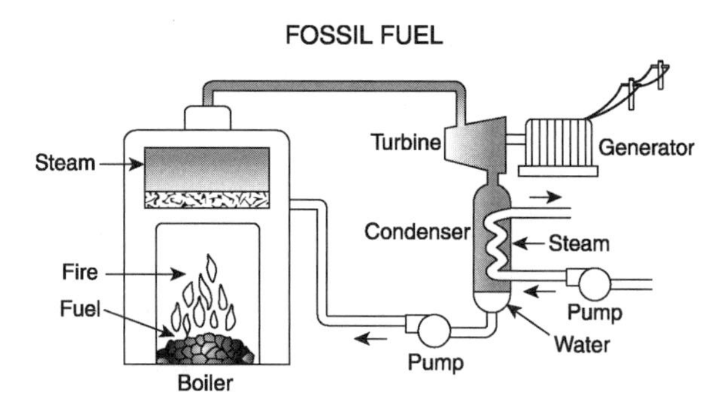

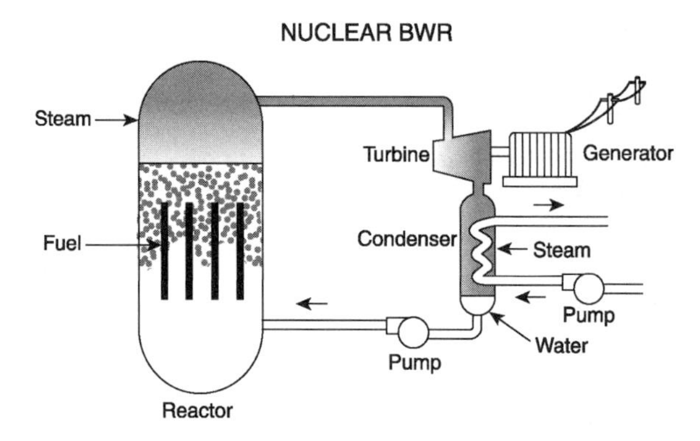

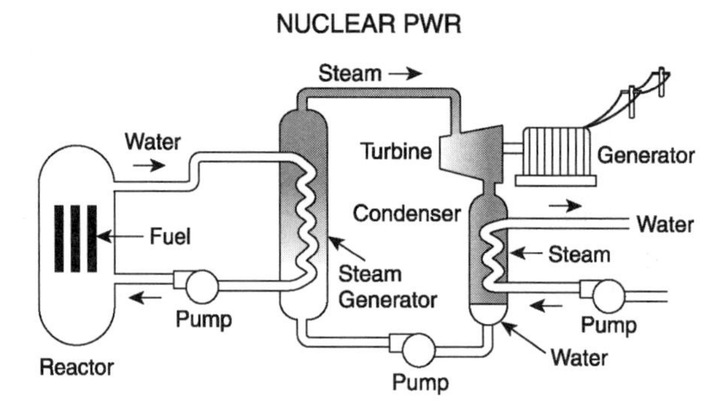

Figure 8.1. Comparison of pressurized-water reactor (PWR) and boiling-water reactor (BWR) to fossil-fueled steam rankine cycle electric power plant. Adapted from USAEC, 1973.

8.10 Boiling-Water Reactor (BWR) Technology

The BWR technology is a derivative of PWR. With BWR, water boils within the reactor core, and steam flows directly to the turbine (General Electric, 1972). In order to accomplish this, the coolant pressure is reduced to the approximate level of the PWR secondary coolant circuit, and the steam generator is eliminated, thereby converting the cooling system into a single circuit. This arrangement is similar to that of the direct steam Rankine cycle of the fossil-fueled electric power plant (see Figure 8.1). The boiling within the reactor core creates a strong interdependence between the distribution of neutrons and the spatial energy distribution in the water. The reactor power changes quickly in response to changes in coolant pressure that might be caused by events within the power conversion system. This fact requires the reactor control system to be more elaborate and quicker acting than with the PWR technology.

A second characteristic of BWR technology is that the power conversion system is typically more radioactively contaminated than that of the PWR, particularly during power-generating operations. This occurs because surface contamination, consisting of irradiated corrosion products and small amounts of fission products escaping from any defective fuel rod, becomes distributed throughout the power-conversion system, and the system is occupied by short-lived water-derived isotopes of nitrogen and oxygen carried from the reactor within the coolant. The latter decay to negligible activity levels shortly after reactor shutdown, while the former persist. As a result, the power-conversion system, including the turbine and condenser, must be shielded and maintenance involving them is more difficult than with the PWR.

Operators who have used the two technologies report that each has its strengths and problems, but both can be managed successfully. In some countries (e.g., Japan, Taiwan), the majority of nuclear power plants are BWRs. In other countries (e.g., US, France, Germany, Korea, Belgium), PWRs predominate.

8.11 RBMK Reactors

In the former Soviet Union, the RBMK reactor (large tube-type reactor) was developed as a type of BWR (see Figure 8.2). However, it utilizes graphite rather than water as the neutron moderator and water is the coolant. In that role, water is important more as a neutron absorber and is negligible as a moderator. Because of this, the reactor has positive power feedback, since boiling of the water results in a power increase that must be counteracted by means of control-rod motion (Knief, 1992). This characteristic played an essential role in the 1986 accident at the Chernobyl reactor in Ukraine.

With the RBMK, the volume of the reactor is occupied by a block of graphite through which vertical holes accommodate metal pressure tubes, each of which contains high-pressure water. Within each tube is a stack of fuel rod bundles. Water flows upward over the rods, boiling as it proceeds to the top of the reactor. Each tube has apparatus permitting fresh fuel bundles to be loaded into it and irradiated ones to be removed, thereby permitting refueling as the reactor operates. This is an operational advantage over pressurized vessel type reactors, such as the LWRs, which must be shut down and partially dismantled for refueling. The purposes of the RBMKs are production of electricity and of plutonium for use in nuclear weapons (a policy different from those of the industrialized democracies). The latter function is facilitated by online refueling and provided a reason for operators to tolerate the problems and dangers of using an active power control system. Approximately 13 RBMKs still operate, although international pressure to shut them down (or at least to improve their safety) has persisted since the Chernobyl accident. The operators of the RBMKs have displayed limited interest in actually closing these reactors, but have cooperated with the industrialized democracies to some extent.

The Chernobyl accident occurred as part of an operational experiment that violated both a competent understanding of reactor operations and the operational procedures of the power plant. In this experiment the reactor was placed in a liquid-filled configuration where the strength of its positive power feedback mechanism was unusually high and the marginal ability of the control rods to decrease the power was absent (most rods having been removed from the core). The result of deliberate initiation of boiling caused a rapid power increase that resulted in virtually instantaneous production of so much steam that the reactor was blown apart and the fuel fragmented.

Contrary to requirements outside the former Soviet Union, the reactor was built without a containment building. (It did have a limited emergency cooling and steam-quenching capability.) As a result, the steam explosion ejected much of the reactor's fuel onto the roof of its industrial shelter-style building. The fuel ignited the building's roof material, which resulted in the majority of the reported 31 fatalities from the accident when firefighters were exposed directly to the fuel's radiation.

The explosion rendered the reactor uncoolable, which resulted in overheating, release of about half of the reactor's fission product inventory to the biosphere, and melting of the remaining fuel and its relocation into the nether regions of the reactor building. The accident was first discovered outside the former Soviet Union when radioactive material was detected on the shoes of a worker *entering* the Forsmark nuclear power plant in Sweden, more than 1,000 km from Chernobyl. It had been transported there via a stratospheric plume of volatile fission products that eventually circled the globe. The Soviet response to the accident was improvised and only partially effective in protecting the public. Much of the land near the power plant was taken out of agricultural production and many people were exposed to low levels of radiation from the reactor. Persistent controversy and disagreement exists about the consequences of these exposures, with estimates of ultimate consequences ranging from small numbers to tens of thousands of fatalities. The US Department of Energy (DOE) estimated ultimate fatalities to be 4,100 among USSR evacuees (Anspaugh et al., 1988). The fear and controversy produced by this accident was such a serious blow to the prospects of nuclear power worldwide that the enterprise has not yet recovered from it.

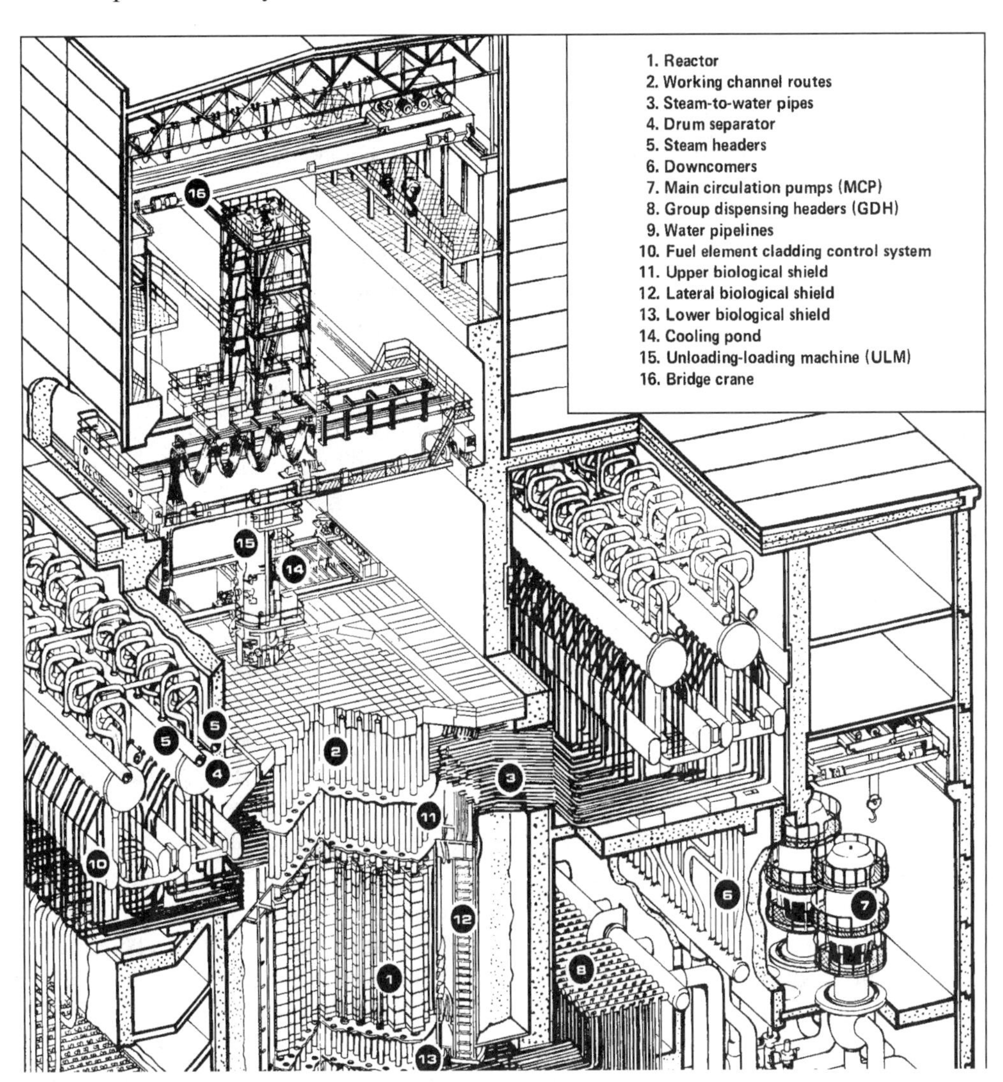

Figure 8.2. View of large pressure tube reactor (RBMK) similar to that at Chernobyl. Source: IAEA Bulletin, 1983. Reprinted with permission of IAEA.

Following the Chernobyl accident, the World Association of Nuclear Operators (WANO) was formed, echoing the earlier creation of INPO. WANO has been much less successful than INPO, however, due to a weaker financial base, greater worldwide diversity of social and legal environments, and the absence of effective mechanisms for punishing poor performance.

8.12 Heavy-Water Cooled Technologies

A different PWR reactor concept was developed in Canada as the CANDU reactor. Used in several countries, it is a horizontally oriented pressure tube reactor, using pressurized liquid heavy water as a coolant within the tubes, and cooled, low-pressure liquid heavy water as a moderator outside of them within a tank termed a "calandria" (AECL, 1978) (see Figure 8.3). The primary coolant flows to a steam generator, where steam is produced as with a PWR. The CANDU utilizes natural uranium as fuel, thereby avoiding the expense of uranium enrichment. However, this economy is counterbalanced by the additional expense of producing heavy water, which is water containing ^{2}H (deuterium, D) instead of naturally occurring ^{1}H (about one atom of hydrogen in 6,600 occurs naturally as D). The CANDU reactor utilizes online refueling as a way of dealing with the relatively rapid depletion of fissile nuclei associated with use of natural uranium fuel.

8.13 Gas-Cooled Reactor Technologies

The oldest large nuclear power economy is that of the UK, where many Magnox CO_2-cooled, graphite-moderated reactors were put into service during the 1950s and 1960s (Murray, 1961). These performed poorly (mainly due to problems with the magnesium-alloy fuel) and were succeeded by CO_2-cooled advanced high-temperature reactors (AGRs). Both types of reactors remain in use today, but neither is still being produced. More recently, the He-cooled high-temperature gas-cooled reactor concept (HTGR) has been developed in the US, and one of commercial size was built at Fort St. Vrain, Colorado. It was shut down prematurely due to gas-circulator bearing failures. That reactor's construction was preceded by smaller units in the US, UK, and Germany, and was followed by several orders for larger HTGRs, all of which were eventually canceled when the reactor manufacturer concluded that they were unprofitable.

In Germany, the Arbeitsgemeinschaft Versuchsreaktor (AVR), pebble-bed-type gas-cooled reactor operated successfully for 20 years, demonstrating the fuel's ability to withstand imposed loss of coolant and loss of coolant flow accidents. The key to the fuel's good performance is use of a microsphere design where a fuel kernel is surrounded by concentric spherical layers of SiC and pyrolytic graphite. The small dimensions of this system (a typical diameter is 300μm) render it capable of withstanding high internal pressures, resulting in robust structural integrity. The AVR was succeeded by the 300 MWe Thorium Hochtemperatur Reaktor (THTR), which was shut down prematurely due to internal structural support failures.

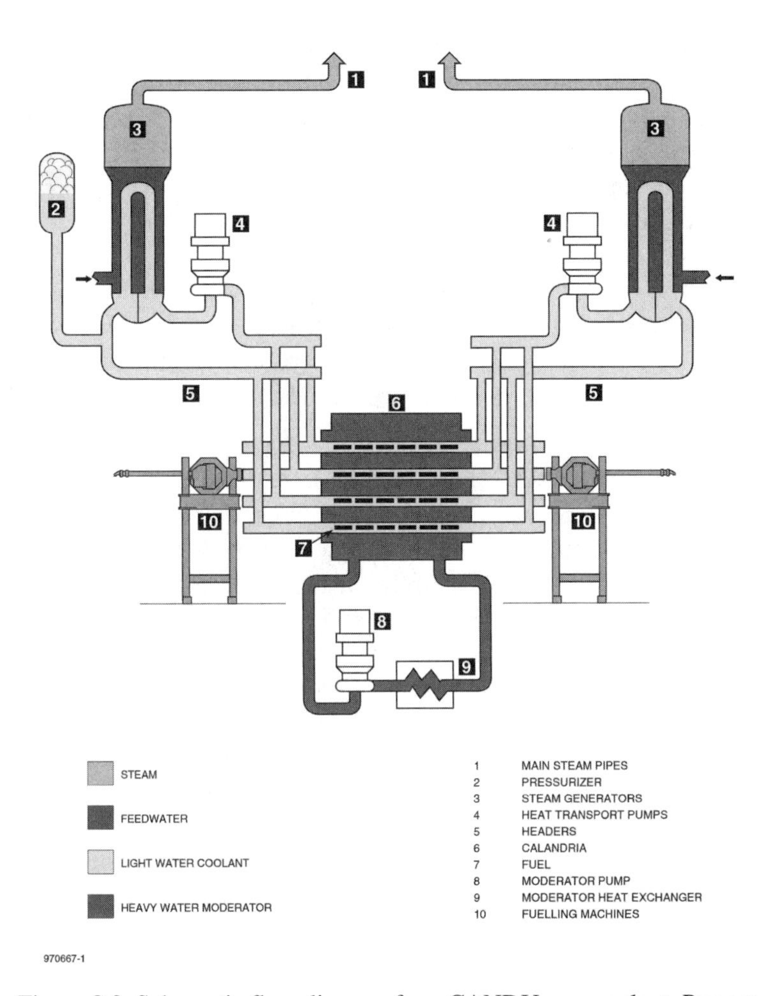

Figure 8.3. Schematic flow diagram for a CANDU power plant. Present CANDU systems are essentially pressurized heavy water reactors. Individual fuel channels pass through a calandria, which contains heavy water moderator with its own circulation system. Heavy water coolant, on the other hand, flows through the fuel channels and raises steam (from ordinary water) in the steam generator. Diagram courtesy of Atomic Energy of Canada, Ltd. (AECL). Source: Nero (1979). Reprinted with permission of Atomic Energy of Canada, Ltd.

Despite these setbacks, enthusiasm for this reactor scheme remains strong due to its impressive conceptual strengths. These include the use of a coolant that is inert in both a chemical and nuclear sense, the potential for high thermal efficiencies and for direct displacement of fossil fuel combustion, the replacement of the steam Rankine cycle by the less troublesome and, perhaps, more efficient gas turbine Brayton cycle for electricity production, and—most importantly—improved nuclear safety. Which of these goals can actually be realized remains to be seen.

Two gas-cooled concepts remain current (see Figure 8.4):

1. The prismatic reactor design, where the core consists of prismatic blocks of graphite (for moderation) held within a steel reactor vessel, using vertical channels to allow for flow of the helium coolant, and with the fuel microspheres containing kernels of UO_2 fuel held in graphite "compacts" (with shutdown being required for refueling); and

2. The "pebble bed" design (PBMR), where the core consists of a nest of graphite balls held within a steel reactor vessel with helium coolant flowing over the balls. Each ball contains many UO_2 kernel micropsheres, and, with online refueling, typically each ball flows from the top to the bottom of the pile several times over its lifetime (PBMR, 2003).

Each concept could utilize a gas-turbine power-conversion system (thereby permitting high thermal efficiency) and passive post-shutdown cooling. A particular virtue of this type of reactor is that its fuel has shown the capacity for surviving intact at much higher temperatures (i.e., > 1,600°C) than does the fuel of other reactors. Actually achieving such performance would require that nearly all of the millions of fuel particles to be used within a reactor be manufactured to exacting quality standards, but tests have demonstrated the capability of actual fuel particles to perform as described above. Seeing an opportunity in such results, reactor designers have attempted to utilize combinations of natural convection and radiative heat transfer to permit the reactor to survive a broad range of postulated accidents, such as those that LWRs are currently required to cope with, without the intervention of humans or active systems.

From analyses performed to date, it appears that this can be done if the reactor power is kept below about 600 MWt. Passive heat-transfer systems are reliable (i.e., gravity drives natural convection, while pumps drive current, active, reactor cooling systems) but less effective than those using active systems. The maximum heat flux that can be transferred passively is much less than can be accommodated actively, thereby leading to the need for the reactor power capacity to be kept small. Unless the diseconomies of small scale can be cleverly overcome (e.g., through such approaches as serial manufacturing of high-quality modular units in factories), it appears that such small reactors will be considerably more expensive than larger ones using active cooling systems.

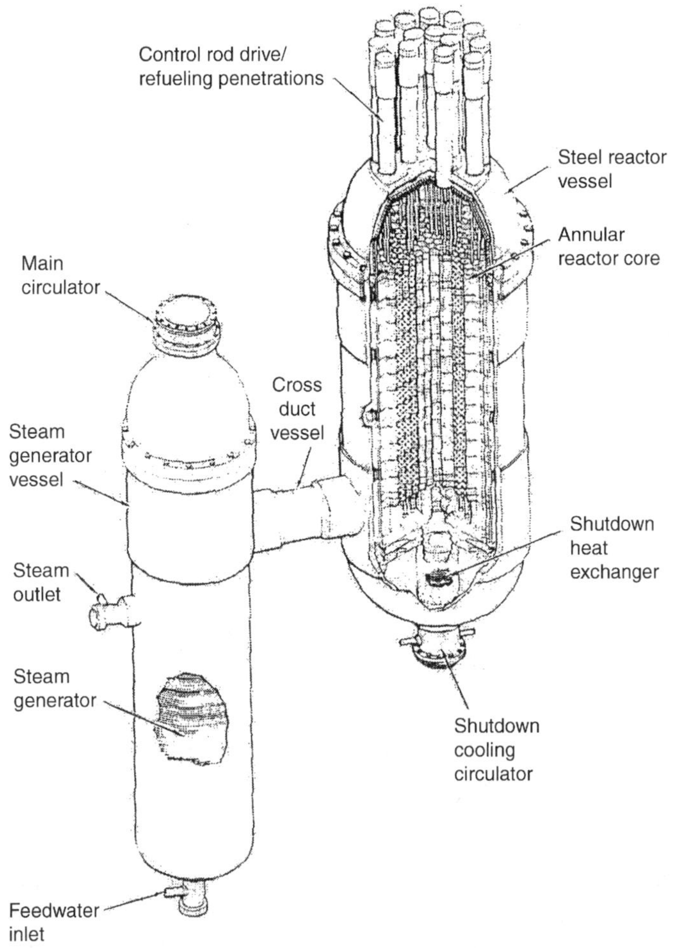

Figure 8.4. View of high temperature gas-cooled reactor (HTGR). Source: *Science* Vol. 244, July 1994. Reprinted with permission of the American Association for the Advancement of Science.

A controversial aspect of the development of gas-cooled concepts using passive safety features is that, for economic reasons, some have proposed relaxing the requirement for a sealed containment building, arguing that the fuel pellets can be made so reliable under all anticipated accident conditions that one is not needed (instead a "confinement" structure would be used). Thus, the manufacturing quality of the fuel microspheres becomes an essential aspect of the safety and economic attractiveness of this concept. The containment is an engineering response to uncertainty. The comparison of the Chernobyl and TMI-2 accidents demonstrates its value. However, the need for the containment cannot be demonstrated or quantified definitively.

The two outstanding questions regarding reactors using passive safety systems are:

1. Can they be made economically attractive?
2. Can some subtle, unanticipated failure mechanism or previously unobserved system performance mode arise that would defeat the performance of the passive safety systems?

In the absence of extensive experience, it is impossible to be confident of affirmative answers. The associated uncertainties, coupled with the need for substantial investment in conceptual development before an actual reactor can be built, are among the principal impediments to the realization of this reactor concept.

The gas-cooled reactor has been proposed principally as a thermal reactor concept, but it has also been proposed in a fast breeder reactor version (by using a new fuel form and omitting graphite from the core and reflector). Should a thermal reactor actually come to be realized, one of its potential merits would also be its capability to provide basic information needed to develop a fast breeder.

8.14 Liquid-Metal Reactor Technologies

The main reactor concept that has been developed for breeding new reactor fuel is the liquid metal (cooled) fast breeder reactor (LMFBR) (Murray, 1993). It utilizes the reaction of Equation (8-3) to convert ^{238}U into fissile ^{239}Pu. This is done most rapidly when the mean neutron energy is kept high. Thus, reactors lacking moderating materials are preferred for breeding. This requirement leads to the use of liquid metals or gases as reactor coolants. (Liquid metal coolants that have been used include those listed in Table 8.2.) The most exploited coolant to date is sodium, although recent interest in lead and the lead-bismuth eutectic has been high. The most ambitious breeder reactor is the 1,200 MWe Superphénix in France. It began operation in 1985 and operated successfully for several years as a demonstration experiment. However, sodium leaks due to piping corrosion have forced it out of service, and it is planned to be decommissioned shortly.

Thermal breeder reactors have also been built, utilizing the reaction of Equation (8-2), which results in the most rapid conversion of fertile material should a thermal neutron spectrum be used. However, the overall rate of new fuel production is greater in a fast reactor because fast neutron-induced fission produces more neutrons per neutron absorbed than does "thermal" fission. Using fast neutrons, the reaction of Equation (8-3) results in a greater neutron abundance, while that of Equation (8-2) is more abundant in thermal neutron-induced fission.

Sodium-cooled breeder reactors (see Figure 8.1) have been built experimentally in most countries that have advanced nuclear power programs. Today, these development programs remain active mainly in Japan, India, and Russia. Some countries, especially those lacking indigenous uranium resources, have pursued fast breeder-reactor programs for national strategic fuel-stockpile resource purposes (and, possibly, as a socially acceptable means of producing plutonium, which could be useful in a nuclear weapons program).

The economics of breeder reactors have turned out to be unfavorable because uranium is abundant and available at low cost, the capital costs of liquid-cooled reactors have proven to be higher than expected, and the costs of recovering uranium and plutonium in spent-fuel reprocessing have been unexpectedly high. In an earlier day, when rapid growth of the worldwide nuclear power economy and more meager uranium resources were anticipated, the urgency for developing industrial-scale LMFBRs was greater. For the next several decades the current situation is likely to continue, but societies will have to decide whether to pursue the benefits of breeding. The pressures to do this could grow as uranium resources become depleted, especially if nuclear power were to play a larger role in the energy economy, as might happen if social concerns about global warming become more acute. Factors that might avert such an outcome include discovery of new, feasible, and abundant sources of uranium and thorium and the advent of practical fusion energy technologies.

Sodium-cooled reactors are attractive for breeding, but are more difficult to use than the concepts previously discussed. With sodium-cooled reactors, the coolant is opaque, is flammable in air, reacts violently with water (a problem when a steam Rankine cycle is used for power conversion), becomes radioactive during use, and corrodes steel structural components. Operators of these systems report that these drawbacks can be accommodated in practice through means such as use of extensive shielding, robots, chemically inert cover gases, and numerically controlled refueling machines. Advantages of liquid metal systems are their low vapor pressures at high temperatures (resulting in coolant systems that operate at atmospheric pressure) and their excellent heat transfer properties, which make high-power-density cores feasible and aid in post-shutdown reactor cooling.

The most important safety limitation of liquid metal concepts developed to date is a much weaker negative power feedback than with most thermal reactors. This is a consequence of the high energy of the fast reactor neutrons. In thermal reactors, loss of the moderator typically renders the reactor subcritical, and the effects of fuel and moderator heating are to increase neutron absorption and escape. With fast reactors, these mechanisms are either irrelevant or much weaker and must be compensated for by such means as making the thermally induced distortion of the reactor's structural members much greater so as to increase neutron escape probability at the reactor heats up.

8.15 Actinide Burning

A potentially attractive use of fast reactors or accelerator-driven subcritical assemblies lies in "actinide burning" as a means of treating many of the longest-lived high-level nuclear waste (HLW) isotopic species. For the first thousand years after shutdown, the hazard of the reactor's fuel is determined primarily by the radioactivity of the fission products, which are contained within the fuel (see Figure 8.5). These decays produce heat and γ radiation, as well as β and α radiations. Thus, the fuel requires cooling and shielding, and presents an ingestion hazard. Among the important isotopes are ^{90}Sr ($t_{1/2}$ = 29 yr) and ^{137}Cs ($t_{1/2}$ = 30 yr). After about 1,000 years, most of the radioactive isotopes in spent reactor fuel will decay to such low concentrations that most of the fuel's toxicity is determined by the concentrations of the long-lived actinides and the fission products ^{129}I and ^{99}Tc ($t_{1/2}$ = 213,000 yr). The actinides remain toxic from α and β emissions, but the associated γ radiation field is considerably less hazardous than that created by the shorter-lived fission products, which will have decayed by that time. Neutron-induced fission of many of these isotopes and their conversion into shorter-lived fission products is feasible using fast neutrons. Reactors, accelerators and fusion-driven devices for this purpose have been proposed, but none has been developed.

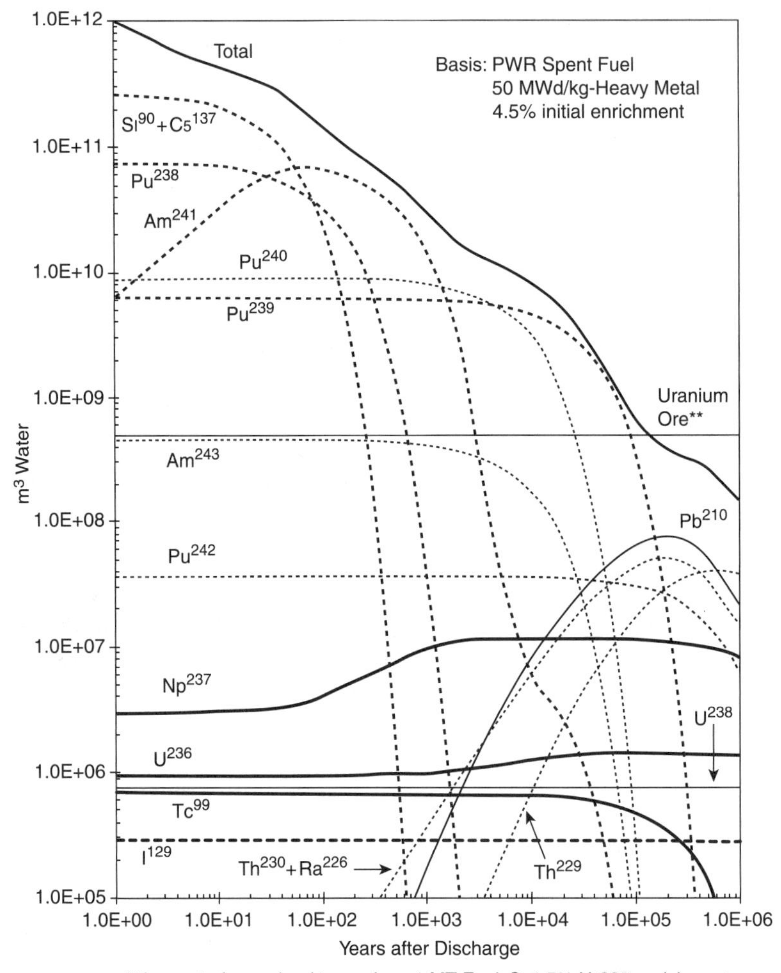

Figure 8.5. Radiotoxicity decay profile for spent PWR fuel (m^2 water/MT fuel) (Deutch and Moniz, 2003).

The effectiveness of actinide burning would depend upon its economics, the quality of the associated chemical separation processes, the degree to which "burning" eradicates the undesirable materials, and the methods available for sequestering the surviving undesirables from the biosphere. For burning to become practical and attractive, social attitudes concerning required treatments of hazardous materials, including radioactive ones, must change. To date, no one has been seriously injured, much less killed, by HLW, and analyses of future risks associated with terrestrial disposal of such wastes indicate little hazard but much uncertainty. To a large extent, proposals for other treatments of HLW are engineering responses to concerns about highly speculative perils in the far distant future. The degree to which the adequacy of terrestrial disposal of high level radioactive wastes will be considered unacceptably uncertain could have a strong effect on social requirements for alternative technological treatments. Such concerns could become much smaller with the passage of time, and terrestrial HLW disposal could become a routine, accepted treatment.

8.16 Advanced Reactors

A surge of work followed the 1979 core melt accident at the TMI-2 nuclear power plant. The goal of these efforts was to create safer reactor concepts (perhaps via physical tests and demonstrations) that would diminish public opposition to the use of nuclear power. Efforts to develop concepts using the coolants discussed earlier were mounted, although attention was first focused on LWR and gas-cooled concepts (see Section 8.13). Primary emphasis was on making the post-shutdown reactor cooling function more reliable, but strengthening the containment and passive negative power feedback were also featured in some programs. The approach taken was to rely as much as feasible on passive mechanisms (i.e., evaporation, natural convection, radiant heat transfer) to accomplish essential functions.

The developmental patterns split between concepts that attempted to emphasize safety improvements by the use of passive mechanisms and those that attempted improvements by both safety and economics, relying on a mix of active and passive mechanisms. The former group were typically low power ($\leq$ 500 MWe), such as the PIUS, modular HTGR, and PRISM fast reactor. The latter were typically of high power, such as the AP-600 PWR and the SBWR. As is mentioned above, the market has rewarded improved versions of evolutionary LWR technologies such as were available before the TMI accident. These included the ABWR (built in Japan), the N4 (PWR, built in France) and the KSP (PWR, built in South Korea).

8.17 Nuclear Power Fuel Resources

The important nuclear power fuels are uranium and thorium. Current consumption rates are approximately 100,000 and zero tonnes/years for natural uranium and thorium, respectively (EIA, 2002b). They are found in deposits in many countries, and have been little exploited, especially the latter, as the thorium conversion reaction (Equation (8-2)) is not utilized in any of the current nuclear power plant concepts. However, should uranium become scarce, thorium resources could become of practical importance.

Identified existing and likely additional uranium resources appear to be adequate for meeting current demands for about a century. Further, resource economists agree (see Chapter 2) that substantial additional resources are likely to be found should cumulative demand grow to much greater levels. As current consumption is caused almost exclusively by burner reactors, the possibility of plutonium recycling would potentially increase this effective resource base by more than an order of magnitude (but with attendant cost increases). Thus, if the scale of the worldwide nuclear power enterprise were to remain of the current magnitude, it would be highly likely that sufficient fuel resources are available to meet needs for many years.

This situation could be altered were nuclear power to be used on a much larger scale, as in displacing fossil fuel consumption in an attempt to mitigate global warming. If this were done, it would be required on a scale about two orders of magnitude greater than is current in order to be effective. This would result in serious compression of the time scales by which incentives for use of both plutonium and thorium breeders, and for use of speculative additional fuel resources, would become important. With the last option, abundant nuclear fuels are found in many mineral forms, but in more dilute forms. Exploiting them would imply greater costs and greater environmental disruption than with current nuclear mineral sources.

An extreme case is that of seawater, which is estimated to contain about 400 million tonnes of uranium at a concentration of 0.003 ppm (OECD, 1997). The costs and effects upon the aquatic ecosystem of its extraction could render it unattractive as a fuel source.

A greater problem is that of nuclear weapons proliferation, which would be aided by the existence of a breeder reactor economy. The great problem for a nation wishing to acquire a nuclear weapons capability is not that of mastering the associated technologies (after all, they are about 60 years old), but that of acquiring the materials needed for fabricating nuclear weapons. Both plutonium and thorium isotopes (termed special nuclear materials (SNM)) are valuable for this use, and could potentially be obtained from a nuclear power economy utilizing reprocessing of either material. This would be feasible if current reprocessing technologies, which create these materials in chemically isolated forms, were to be used.

The technological remedies to this problem appear to be controls upon diversion of SNMs in combination with new reprocessing technologies that create materials suitable for use in new fuel without creating separated SNM streams. This has been done with pyroprocessing technologies, and can presumably be improved. Additional research to create better technologies for both safeguards and reprocessing appear to be long-term needs of nuclear power technology.

8.18 Fuel Cycle

Any energy facility is supported with a fuel cycle, bringing energy to the facility—usually in the form of fuel—and removing wastes. With nuclear power, the steps in the light water reactor fuel cycle are those illustrated in Figure 8.6. This figure also summarizes the annual mass flows between the sequential fuel cycle facilities required to support a 1,000 MWe light water reactor.

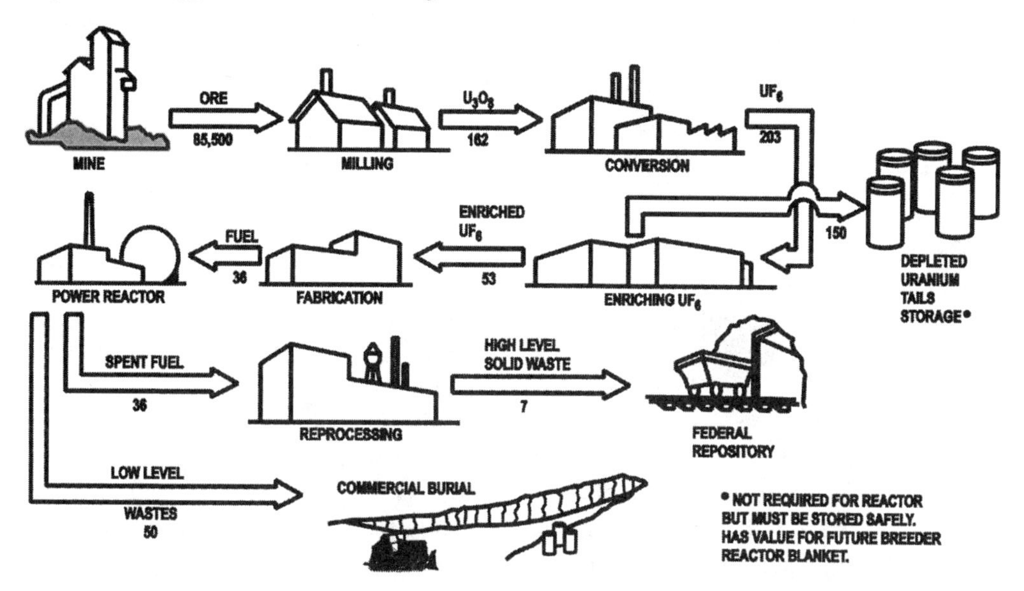

Figure 8.6. View of facilities used in LWR fuel cycle (including reprocessing). Typical annual mass flows needed to support a 1000 MWe LWR are shown between facilities (in tons). Source: USAEC, 1974.

Notes: The mill separates uranium from the ore (as U_3O_8), with the depleted ore being left at the mill site. At the fabrication plant the uranium is produced as UO_2 pellets held in zircalloy cladding and fuel assemblies. Spent fuel leaving the reactor consists of fuel assemblies containing less ^{235}U and ^{238}U than in the fresh fuel, as well as ^{239}Pu and fission products created during the reactor's operation. At the reprocessing plant the uranium and plutonium are removed and recycled, and the residues constitute the high level solid waste. The low level wastes leaving the reactor consist of materials that became contaminated during the reactor's operation.

At the "front end" of the cycle (the portion involving fuel before its delivery to a power plant), the steps are below:

- Mining of uranium ore at a mine
- Extraction of uranium from the ore at a mill
- Conversion of the uranium into the gaseous form UF_6 at a conversion plant
- Enrichment of the uranium to the desired amount at an enrichment plant

- Conversion of the enriched UF_6 into a solid fuel form (commonly UO_2) and its incorporation into fuel rods in a fuel fabrication plant

At the "back end" of the fuel cycle, the functions performed are the following:

- Storage of spent fuel in a storage facility (usually a water pool) after its removal from the reactor, until the radioactivity decreases enough that the fuel bundle is convenient to move
- In the "reprocessing" version of the fuel cycle, chemical separation and reuse in new fuel of the uranium and plutonium from the other species in the fuel bundle, the latter comprising various waste forms
- Disposal of the wastes (HLW), consisting of the spent fuel rod bundles should reprocessing not be employed as with the "once-through" version of the fuel cycle, or of the wastes from the reprocessing facility should it be used

8.18.1 Uranium mining

Uranium mining involves extracting uranium-bearing ore from the earth and moving it to a nearby uranium mill. Typically, the concentration of uranium in ore is low (e.g., 0.25% in high-quality ore found in the US). The highest-quality ore found in large quantities (in Australia and Canada) contains approximately 2–3% uranium (Deutsch and Moniz, 2003). Other important uranium-producing countries include Niger and Gabon (OECD, 1997). For economic reasons, most uranium mining is done at the surface. The mining operation involves the greatest annual mass flow of the LWR fuel cycle, as the great bulk of the material mined is not uranium. Reducing the expense of moving the ore requires the mine to be near the mill. The process of minimizing overall mining costs typically determines the size of a mine-mill complex. It usually has a maximum radius from the mill of about 10 km. Consequently, much uranium that has been identified has not been mined as it is not sufficiently clustered to justify construction of a mill.

8.18.2 Uranium milling

The uranium mill is a chemical leaching facility that separates the uranium from the ore. The resulting solid product form, U_3O_8, is then sent to a conversion (to UF_6) facility. The uranium-depleted ore is collected in settling ponds adjacent to the mill, where the liquids used in the separation process evaporate. The resulting solid residue (termed "mining tailings") must be stabilized so it will not be eroded by the actions of wind and water, and leached by the latter. Both ^{235}U and ^{238}U decay radioactively ($t_{1/2}$ = 700 million years and 4.5 billion years, respectively) via a sequence of α, β, and γ particle releases that ultimately produce stable isotopes of lead. The intermediate species in these decay chains are also radioactive. They, and the species involved in a similar decay chain beginning from ^{232}Th, contribute about half of the naturally occurring radiation exposure of an organism dwelling at sea level. These radioactive species are produced everywhere on earth, but in varying concentrations. The tailings are more radioactive than most of the earth's crust. This is because they contain greater initial concentrations of uranium in the ore. After the uranium is removed from the ore, the radioactive decay products remain in the tailings.

Most of these radioactive species are of no special concern, as they remain within the tailings and do not expose humans. However, among the species are ^{226}Ra (radium) and its daughter product ^{222}Rn (radon, $t_{1/2}$ = 3.8 d); both can become mobile in the biosphere—thereby providing the potential for radiation exposure of living organisms. Radium is soluble in water (detecting radium in groundwater is one of the ways of finding uranium deposits), and radon is a short-lived noble gas. Radium is kept out of the biosphere by eliminating pathways for water to enter or leave the tailings piles, and radon is kept within the pile by making its residence duration long compared to its half-life. Both goals can be accomplished by creating impermeable barriers surrounding the pile's exterior surfaces. This is most commonly done by use of a clay layer placed under and around the pile. To be effective, the clay must be kept moist. This is usually done by surrounding the clay with a layer of dirt, upon which vegetation is grown (when possible).

There were incidents during the early days of the nuclear enterprise when the need for these practices had not been appreciated, and some members of the public were exposed to low levels of radiation when tailings materials were used in construction projects. Seepage of naturally occurring radon into houses has also been a problem in some locales, which is now recognized and controlled.

8.18.3 Conversion

The commonly used enrichment processes require the uranium to be in gaseous form. A stable compound that has served well is UF_6, which is produced at the conversion plant.

8.18.4 Enrichment

The most commonly employed enrichment processes are the gaseous diffusion and centrifuge-based processes. With gaseous diffusion, a semi-permeable membrane is used to create a slight separation of ^{235}U from ^{238}U, taking advantage of the fact that in a gas at a given temperature, UF_6 molecules involving the former will strike the membrane slightly more often than will those involving the latter. Thus, the mixture on the downstream side of the membrane will become slightly enriched in ^{235}U. Use of many such membranes arranged in a series permits the ^{235}U concentration to be increased as desired. In practice, the highest enrichment produced is 93%, that of highly-enriched uranium (HEU). The gaseous diffusion process involves a large capital expense and high operating costs (mainly for the energy needed to power the various separation stages).

In order to achieve greater economies, the gas centrifuge process was developed. In this process cylinders containing UF_6 are spun at high speeds, creating a ^{235}U-rich region near the center and a ^{238}U-rich region near the periphery. As with the diffusion process, the enriched gas is extracted from within the cylinder and fed into a subsequent cylinder for further enrichment until the desired concentration is reached. The ^{238}U-rich gas is also extracted and fed back upstream to an enrichment stage operating at its inlet enrichment value.

From an enrichment facility, streams of both enriched and ^{235}U-depleted uranium are obtained (a consequence of conservation of isotopic species). The former are used in subsequent fuel fabrication. The latter (termed "enrichment tailings") are stored at the enrichment facility as UF_6. Ultimately, they may be solidified to render them less mobile in the environment, but that is not the current practice. These tailings are potentially valuable as a source of ^{238}U that could be used as a fertile material feedstock should a breeder reactor nuclear power economy come into being. However, this eventuality appears to be plausible only in the far (i.e., many decades) future. Other enrichment methods have also been used but have not achieved the level of industrial importance as those discussed above.

In a centrifuge, a spinning cylinder filled with UF_6 is slightly richer in the interior with ^{235}U. Successive extraction of gas from such locations in many centrifuges permits large enrichments to occur.

Enrichment above 20% in ^{235}U can permit use of uranium in a nuclear weapon. Consequently, countries doing this, especially should they not have a clear civilian need for enriched uranium, attract suspicion.

8.18.5 Fuel fabrication

In fuel fabrication, the enriched UF_6 is chemically converted into a fuel form such as UO_2 for LWR fuel (see Table 8.1), and enclosed in a cladding material (to keep the fission products that will be produced during the fuel's operation within the fuel). With most reactor concepts, the fuel is in the form of rods that are combined into fuel assemblies that can be loaded into the reactor as fresh fuel modules.

The fresh fuel assemblies can be handled without shielding (unless recycled fuel is employed). They are shipped by road and rail to the power plant. Typically, a single 18-month supply is provided.

8.18.6 Spent fuel

During power production operations in the reactor, neutrons bombard the fuel, causing fissions and activations. Ideally, this process proceeds until the concentration of fissile nuclei (mostly ^{235}U and ^{239}Pu, produced during operations via conversion of ^{238}U) becomes too low to maintain the reactor in a critical condition and/or accumulated radiation damage becomes too great. Following use in the reactor, the irradiated fuel bundle is removed and placed in a spent fuel storage pool. The pool provides a means of both cooling the bundle and shielding workers from its radiation. Cooling is required because the fission products created within the fuel continue to decay and produce heat (albeit at a steadily decreasing rate, see Figure 8.7).

At some power plants, the fuel assemblies have been stored long enough (>15 years) that the power levels of spent fuel rods have become low enough that it has been feasible to remove them from the storage pools and to place them in shielded containers (dry casks) that are cooled passively by the naturally convective flow of air. In countries that do not allow reprocessing and have not developed HLW repositories, fuel storage pools are becoming filled with fuel rods, and use of on-site dry cask storage is becoming more common. For the reactors of most countries, that is the extent of development of facilities for the back end of the fuel cycle HLW.

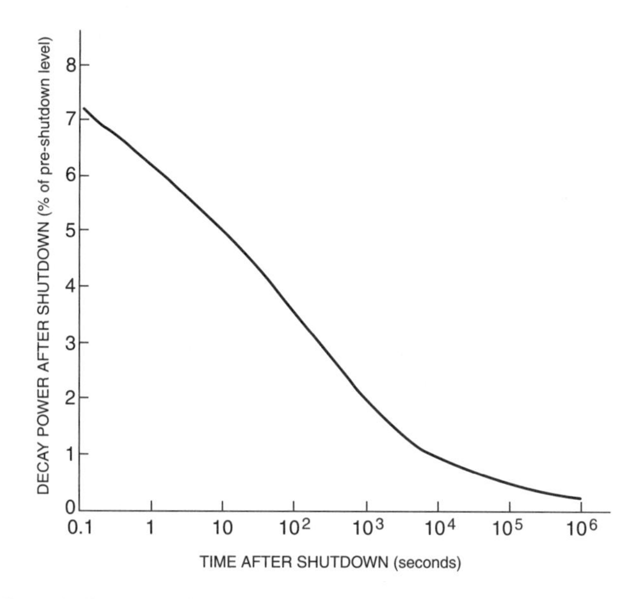

Figure 8.7. Radioactive decay power of spent LWR fuel as a function of time after removal from the reactor, normalized to steady operational reactor power. Source: Nero (1979). Reprinted with author's permission.

8.18.7 Reprocessing

The purpose of reprocessing is to recover uranium and plutonium from spent fuel for reuse. This is a commercially available service offered in the UK and France, and it is also performed in the nuclear weapons programs of any nuclear weapons state. Several processes have been developed to accomplish this function. The most widely used is the PUREX process, where fuel pellets are dissolved in nitric acid, and tributylphosphate is used in solvent extraction columns to separate the uranium and plutonium. Once the uranium and plutonium have been separated, the wastes left behind needing disposal are the fission products and neutron activation products of uranium and plutonium (actinides), and the structural materials of the fuel elements. Gaseous fission products are also liberated from the fuel.

Currently, reprocessing is widely recognized to be uneconomical and is likely to remain so into the foreseeable future—at least as long as uranium costs remain low and surplus Pu and HEU are available from worldwide nuclear weapons programs (e.g., those of the US and the former Soviet Union) for incorporation into fresh reactor fuel.

Reprocessing may also be performed in support of breeder reactor development programs or for national strategic reasons. However, for most nuclear countries since the 1980s, these have been insufficient reasons for maintaining reprocessing as part of their civilian nuclear power economies.

Once the Pu is liberated from the spent fuel, the potential for its diversion or theft arises. The need for safeguards against such eventualities is inherent in the reprocessing and downstream facilities. Safeguard methods include surveillance of operations, searches of personnel and shipments leaving the facility, and mass balance audits of the systems and inventories of the facilities. Since none of these approaches is completely reliable and accurate, operation of a reprocessing facility carries with it a state of uncertainty about whether diversion of Pu has occurred and to what extent.

Environmentally, reprocessing is a potentially important step in the fuel cycle, as it is the point where the greatest routine releases of radioactive materials to the environment occur under current PUREX practices. When fuel rods are cut up prior to insertion into a nitric acid bath for dissolution of the UO_2, the gaseous fission products (mainly ^{3}H (tritium, $t_{1/2} = 12.3$ yr), and ^{85}Kr ($t_{1/2} = 8$ yr)) and also ^{14}C ($t_{1/2} = 5700$ yr) are released to the atmosphere. They could be retained, but at increased costs. The noble gases are less important biologically, as they become dispersed in the atmosphere, and, by virtue of their chemically inert property, do not become incorporated into biological systems (although their daughter species may). Tritium is more problematical because it becomes incorporated into water molecules, which can be taken up by biological systems. This is especially important should the tritium become incorporated into reproductive organs, as its β emissions could thus disrupt the DNA of future generations.

8.18.8 High Level Wastes (HLW) disposal

No country has an effective, licensed HLW disposal system. Worldwide, HLW are being stored temporarily while societies try to agree on what to do with these wastes. The difficulty arises more from the reluctance of localities to accept the presence of disposal facilities than from the technical difficulties of assuring that they will perform acceptably (although some difficulties exist). This impasse is stable because HLW pose no near-term public health threat, and its persistence serves the interests of anti-nuclear pressure groups, which work hard to block any effective disposal scheme so as to perpetuate the belief that waste disposal (and thus, nuclear power) is impractical. Figure 8.5 shows the radioactivity decay profile for spent PWR fuel. While decay is slow, the health hazards are localized and minimized by shielding. However, great uncertainty is associated with assuring long term integrity of containment.

Nuclear wastes consist of any waste form produced at a nuclear facility that is radioactive above background levels. The activity allowed for an isotope is species dependent, with maximum US-allowed concentrations being specified by the NRC. The categories of wastes used in the US are the following (Uranium Institute, 1991):

- Low level wastes (LLW) are wastes that can be buried without dilution. There are three categories of such wastes (A through C—the most radioactive), where the quality of the waste packaging increases as the activity of the waste increases.

- Intermediate level, or greater than Class C wastes (ILW), are wastes requiring some dilution before they would be classified as LLW (e.g., irradiated reactor structural components). They must be buried in a monitored terrestrial waste repository.
- High level wastes (HLW) are wastes consisting of spent reactor fuel and reprocessing liquids. They must be buried in a monitored terrestrial waste repository.
- α-contaminated or transuranic wastes (TRU) are wastes contaminated only by α-emitting transuranic isotopes. They must be buried in a monitored terrestrial waste repository.

Within the US, LLW, ILW, and TRU are being disposed of in terrestrial facilities (either burial sites or repositories, with careful packaging of the wastes and monitoring of the waste facilities). High level wastes are being accumulated at the points of origin. For more than 25 years, a national effort has had the goal of opening a terrestrial repository for high level wastes. During recent years, this work has focused on demonstrating that the Yucca Mountain, Nevada, site is suitable for this purpose.

The experience in attempting to site an HLW repository at Yucca Mountain is a classic example of the dysfunction of the US system for resolving environmental conflicts. What is occurring is a power struggle between political interest groups supporting or opposing nuclear power. For the latter, no solution will be acceptable and the means of imposing their will is to render the project and thus nuclear power so expensive that it will be abandoned. For supporters, the project as proposed is acceptable, and the major question concerns whether they will be able to afford to resist their opponents. The tools in this struggle are scientific questions concerning the range and likelihoods of potential performance should the project be realized. Each side attempts to have the interpretation of uncertainties favorable to themselves accepted by the regulatory authorities. The process for determining what will be accepted consumes time and money, primarily from the project's proponents.

Ostensibly, the process is about scientific questions. In reality, these are pawns where the terms of the debate concern science while the substance of the debate concerns the future directions and structure of society but with this discussion conducted via use of surrogates. In that sense, the US system for environmental conflict resolution can be viewed more as a struggle than as a genuine scientific inquiry.

In the case of the Yucca Mountain project, Congress has voted to overrule the objections raised by the state of Nevada, and it has moved to obtain a license from the NRC. The latter could take more than a decade to conclude. Current spending is at a rate of about $600 million annually. Following NRC licensing, opponents to the project can still attempt to frustrate it via the regulations governing transportation between the nuclear power plant sites and Yucca Mountain and in the licensing of any facilities needed to package the fuel rods prior to burial (USDOE, 2001).

In the interim, spent fuel rods will continue to accumulate at the individual nuclear power plant sites. Most experts agree that this is much less desirable than some form of consolidated central storage, and that an alternative to the Yucca Mountain project, a long-term interim repository, makes greatest technological sense. Such a facility would function as a warehouse where resources and wastes could be concentrated to protect the public until an improved state of scientific knowledge would improve the confidence of permanent disposal. The peril of HLW is sufficiently low that even were this process to take a few centuries, it could be feasible and safe.

HLW presents an interesting problem. It is serious enough to alarm reasonable people, but not so serious that it needs to be solved. Thus, the opponents of nuclear power can both demand that permanent disposal be made operational before nuclear power can be used further, while opposing any attempts by its advocates to do so. In the meantime, the persistence of HLW as an unsolved problem serves as a powerful fund-raising vehicle for their cause.

In other countries, attempts to find socially acceptable schemes for HLW disposal have not done substantially better than in the US. In Sweden, greater progress has been made than elsewhere via the establishment of research facilities to study how to dispose of the wastes rather than establishment of permanent operational repositories.

The purpose of a terrestrial repository with the "once-through" fuel cycle is to store suitably packaged spent fuel rod bundles and other waste forms within the earth. The wastes can remain toxic for many millennia, although there is a substantial decrease of toxicity and penetrating radiation intensity after about 1,000 years of storage. The main postulated hazard of this scheme is that groundwater might intrude into the facility and, by corroding the waste packages sufficiently, could create a contaminated aquifer. Should organisms, including humans, ingest from this aquifer, they might, in hypothetical worst-case situations, become dangerously exposed. Analyses of the risks of such facilities are so uncertain as to be somewhat surreal, involving small individual exposures to perhaps many people over tens of millennia. The uncertainties of such estimates are large because of our ignorance about material behavior in the postulated environments of a repository and about future geological, climatological, and social behavior. Most analyses indicate that the repository risks are negligible and far into the future, but, for some people, that answer is not good enough. Societies have never previously attempted to be systematic in balancing such long-term risks and their attendant benefits, and they are now having difficulties in doing so in the instance of HLW.

With the "reprocessing" fuel cycle version, the HLW from a reprocessing facility would be packaged and stored. As this package lacks the uranium and plutonium of spent fuel, it is substantially less toxic than with the once-through fuel cycle, and it lacks the attractions to intruders of being a future source of plutonium. In either case, the storage capacity of a repository is determined by the total stored thermal power and geometry of the repository, as they determine the maximum temperature within each of the stored waste packages. For the first few centuries after removal of the spent fuel from a reactor, the power generated within a repository is derived mostly from the fission products, which are insensitive to which fuel cycle option is used. After a millennium, most of the activity and toxicity is derived from decays of the waste actinides.

The Yucca Mountain siting conflict has taken on an environmental guise, with scientific investigations used as pawns and environmental questions serving as surrogates for the actual points of conflict. Strategically, the goal of each side is to cause the other to exhaust its financial resources and quit the battle. Meanwhile, wealth is squandered that otherwise could be devoted to useful purposes. By 2000, about $6 billion had been spent "characterizing" the Yucca Mountain site (financed by a $0.001 per kWhr tax on nuclear electricity). Additional related expenses include those of the Yucca Mountain opponents and of nuclear power plants, as the latter store spent fuel in much greater amounts for longer durations than initially anticipated when the plants were built.

A similarly fruitless process has characterized attempts to develop an interim storage facility for spent fuel. It is argued that such a facility would provide a safer and more economical means of providing long-term storage of spent fuel than does the current *de facto* policy of improvised storage at the various nuclear power plant sites. Doing this would also reduce the urgency for opening a terrestrial disposal facility. To date, anti-nuclear pressure groups have blocked creation of such facilities.

Greatest concern is focused on HLW, as it would pose the greatest public health hazard should it become mobilized in the biosphere. Such mobilization has not happened outside of the former Soviet Union, but due to astonishing carelessness and indifference, several incidents that created widespread environmental contamination occurred at former Soviet Union weapons material production facilities.

In addition to terrestrial disposal, HLW disposal schemes receiving serious attention have involved separation and transmutation of the actinides, disposing of HLW into space as means of diminishing the waste-associated risks after 1,000 years of storage, and storage in polar ice sheets, deep sea sediments, and subducting tectonic plates. None of these schemes has progressed beyond the conceptual phase, and none appears to be economically more attractive than terrestrial disposal. However, interim HLW storage for a long time appears to be physically feasible. Should that be done, some of these disposal options could become more attractive over time.

8.19 Fusion Energy

8.19.1 Introduction

Fusion is a form of nuclear energy in which light elements, such as hydrogen and its isotopes, collide and combine with each other (i.e., fuse) to form heavier elements and, in the process, release large amounts of energy. Fusion is the dominant reaction that powers the sun. It is also the dominant reaction that takes place in thermonuclear weapons. The goal of the international fusion research program is to discover ways to safely and economically harness this powerful source of energy for the production of electricity and, perhaps, for the production of hydrogen.

At present, fusion is largely a research program, with commercialization not expected for about 30–50 years. Consequently, the contribution of fusion to the world energy supply over the next half century does not receive much attention. Even so, fusion offers potentially spectacular benefits in terms of fuel resources and environmental friendliness. For these reasons, it is a worthwhile topic for discussion.

To place fusion in a proper context, it is useful to compare fusion with the other well-known sources of electricity, including fossil fuels, fission, and renewables. What potential benefits does fusion have and what are its drawbacks? There are two major benefits. First, the ultimate fusion reactor will obtain its fuel (i.e., deuterium) easily and economically from the ocean. Because of the enormity of the ocean and the fact that deuterium occurs naturally as one part in 6,700 of the ocean's hydrogen, the fuel supply would be virtually unlimited, last for billions of years, and be easily accessible to all. Second, the fusion process is friendly to the environment. It produces no greenhouse gas, no long-lived radioactive waste products, and has a high power density compared to renewables, implying a small cost to the environment for the actual construction of the plant. The main waste products are the plant components that become activated during operation, but these have short half lives (<100 yrs). Thus, there is no need for long-term storage or disposal of radioactive materials. These potentially enormous benefits have fueled the dreams of fusion researchers for the last half century.

There are, however, drawbacks to fusion. The main problem is that fusion reactions are much more difficult to initiate than fission reactions. The difficulty involves both basic physics and basic technology issues. While researchers in the field currently believe that these problems are solvable, the solutions may lead to reactor cores that are substantially larger and more complex than fission reactor cores. The net result would be a larger initial capital cost, which would affect economic competitiveness.

8.19.2 Why is fusion more difficult than fission?

Basic nuclear physics shows that the use of low-energy neutrons, which is highly efficient in producing heavy-element fission, does not carry over to the light elements. There are two main points to be made. First, for light elements, one should focus on fusion rather than fission, as these are the reactions that produce rather than consume energy. Second, the reliance on neutrons to initiate fusion reactions must be eliminated, since these reactions consume rather than produce neutrons, and no external sources are readily available.

How then can light element fusion reactions be initiated? The basic idea is to replace the neutron with another light element, and thereby to generate a nuclear reaction by having two light elements bombard each other (e.g., two colliding deuterium nuclei). The advantage of this idea is that the lack of a chain reaction is overcome by simply providing a continuous supply of deuterium, which, unlike a neutron supply, is readily available and inexpensive.

The disadvantage is that for two deuterium atoms to undergo a nuclear reaction, their nuclei must be in close proximity to each other, typically within a nuclear diameter. At these close distances, the inter-particle Coulomb potential produces a strong repulsive force between the two positively charged nuclei, which diverts the particle orbits and greatly reduces the likelihood of a nuclear reaction. If the deuterium nuclei have sufficiently high energies, the repulsive Coulomb force can be overcome. This is the strategy behind all current fusion research. The required energies are typically in the range of tens of keV corresponding to temperatures of several hundred million degrees Kelvin, an almost unimaginably large value comparable to the temperature at the center of the sun. At these energies, the fuel is a fully ionized gas, called a plasma. The need for charged nuclei to overcome the Coulomb potential is a primary reason why fusion is so difficult.

Studies of the nuclear properties of light element fusion indicate that three such reactions may be advantageous for the production of nuclear energy. These involve deuterium, tritium, and helium-3, an isotope of helium. The relevant nuclear fusion reactions are given below.

- The D-D reaction (two branches with equal probability):

$$\mathrm{D} + \mathrm{D} \rightarrow {}^{3}\mathrm{He} + \mathrm{n} + 3.27\ \mathrm{MeV}$$

$$\mathrm{D} + \mathrm{D} \rightarrow \mathrm{T} + \mathrm{p} + 4.03\ \mathrm{MeV}$$

- The D-^{3}He reaction:

$$\mathrm{D} + {}^{3}\mathrm{He} \rightarrow \alpha + \mathrm{p} + 18.3\ \mathrm{MeV}$$

- The D-T reaction:

$$\mathrm{D} + \mathrm{T} \rightarrow \alpha + \mathrm{n} + 17.6\ \mathrm{MeV}$$

The D-D reaction is the most desirable in the sense of fuel resources since a virtually unlimited supply is readily available from the ocean. However, it is the most difficult of the three reactions to initiate (i.e., it has the smallest cross section). The D–^{3}He reaction is easier to initiate and produces mainly charged particles, which is desirable for high conversion efficiency to electricity. Its drawback is that the rare isotope ^{3}He is required as fuel, and there are no natural supplies on earth, nor easy ways to manufacture it.

The D-T reaction involves the fusion of deuterium with tritium. It is the easiest of all the fusion reactions to initiate, (although it is still much more difficult than $^{235}_{92}$U fission reactions—the cross section is smaller by a factor of about 100). In terms of energy desirability issues, D-T reactions produce large amounts of neutrons and require a supply of tritium in order to maintain continuous operation. Unfortunately, there is no natural tritium on earth. Furthermore, the tritium itself is radioactive with a half-life of 12.26 years. The D-T reaction, nevertheless, produces a significant amount of nuclear energy corresponding to 3.52 MeV per nucleon, which is macroscopically equivalent to 338×10^{6} MJ/kg. In spite of the problems associated with tritium and neutrons, the D-T reaction is the central focus of worldwide fusion research, a choice dominated by the fact that it is the easiest fusion reaction to initiate.

Having made this choice, how does one deal with the tritium and neutron problems? Many years of fission research have taught nuclear engineers how to handle the material activation and radiation damage resulting from high-energy neutrons. The same holds true for the radioactivity associated with tritium. The solutions are far from simple, but they are now well established.

The one outstanding problem is the tritium supply. The solution to this problem involves the breeding of tritium in the blanket surrounding the region of D-T fusion reactions. The chemical element that is most favorable for breeding is lithium. The corresponding nuclear reaction of primary interest is given by:

$$^{6}_{3}\mathrm{Li} + \mathrm{n} \rightarrow \alpha + \mathrm{T} + 4.8\ \mathrm{MeV}$$

If there were no loss of neutrons, then each T consumed in fusion would produce one new T by breeding with the fusion-produced neutron: the breeding ratio would be 1.00. In a practical reactor, however, there are always unavoidable neutron losses. Thus, some form of neutron multiplication is required, as well as a method of slowing down the high-energy fusion neutrons since the reaction above is most effective with slow, low-energy neutrons. For present purposes, one can assume that the issues are satisfactorily resolved. Consequently, breeding T from $^{6}_{3}\mathrm{Li}$ solves the problem of sustaining the tritium supply, assuming adequate supplies of lithium are available. The known reserves of lithium are sufficiently large to last many thousands of years, so fuel availability is not a problem. On the longer time scale, the goal would be to develop D-D fusion reactors.

8.19.3 Magnetic fusion energy

In magnetic fusion, power is produced in a steady-state manner by means of a donut-shaped reactor, where a D-T plasma with a molecular density of $10^{20}\ \mathrm{m}^{-3}$ (10^{-3} that of air) burns at a high temperature of 15 keV (10^{6} times that of air) in a volume on the order of 100 m^{3}. Steady-state is possible because the plasma is continuously self-heated by its own D-T-produced alpha particles. A magnetic field is absolutely essential. Its purpose is to provide steady-state confinement of the hot plasma, insulating it from the surrounding vessel wall.

A generic magnetic fusion reactor is illustrated in Figure 8.8. The reactor itself is in the shape of a torus (i.e., a donut). This mechanically complex shape is required because of the basic properties of magnetic fields. Outside the first wall is the blanket, where the energy conversion takes place. Surrounding the blanket is a shield to protect the magnets and the workers in the plant from neutron and gamma ray radiation. Finally, the coils producing the magnetic field are located outside the shield.

The magnets must, in general, be superconducting. This is because copper magnets dissipate substantial amounts of ohmic power during steady state operation, even though the electrical conductivity of copper is quite high. This power would be sufficiently large to seriously deteriorate the overall power balance of the reactor. Superconducting magnets dissipate literally zero power in steady state and require only a small amount of cooling power to keep the magnets superconducting. Thus, a generic fusion reactor consists of a toroidal plasma surrounded by a first wall, a blanket and shield, and superconducting magnets.

8.19.4 Inertial fusion energy

The second method to produce fusion power is "inertial fusion," an inherently pulsed form of power production. Here, a hollowed-out spherical pellet of solid D-T, weighing several milligrams with an initial radius of a few millimeters, is compressed to enormous pressures, densities, and temperatures by, for instance, a high-power laser or particle beam. This causes the fuel to ignite and release fusion energy. At peak compression, the pellet typically has a radius R of a few hundred microns, a mind-boggling pressure of hundreds of gigabars, a density of several hundred thousand kg/m^3, and a temperature exceeding 10 keV. Once ignited, the pellet freely expands at the thermal sound speed, c_s, disassembling in an inertial time, R/c_s. Hence the name "inertial fusion." The goal is to guarantee that a large fraction of D-T is burned before the pellet disassembles. Averaging the pulsed energy output from a continuing sequence of new pellets produces a source of steady-state power.

A generic inertial fusion reactor is illustrated in Figure 8.9. The reactor itself consists of a chamber, several meters in radius, in which pellets are continuously injected and burned at a rate of about five per second. Each pellet yields on the order of hundreds of MJ of fusion energy. The pellet itself initially consists of a hollow core of D-T gas surrounded by a high-density shell of cryogenically frozen D-T "ice." The driver ignites the hollow core of gas, which in turn ignites the shell—much like a spark plug ignites the fuel in an automobile engine.

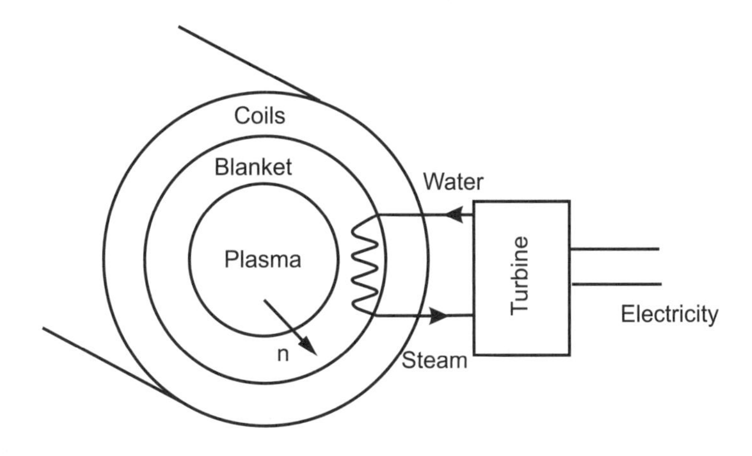

Figure 8.8. A generic toroidal magnetic fusion reactor.

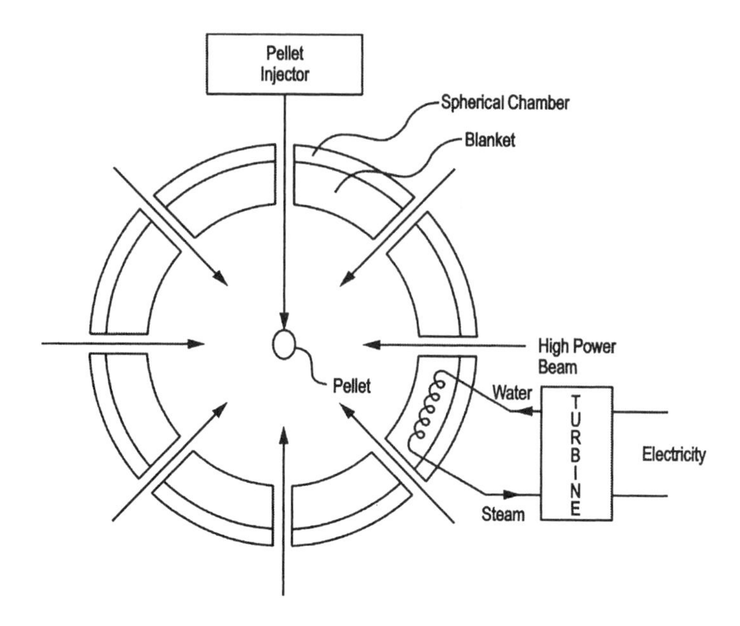

Figure 8.9. A generic inertial fusion reactor.

The fusion-produced neutrons are captured and converted to heat by a blanket containing lithium. This blanket can be either a solid or liquid and has a radius on the order of a meter. The driver, which is by far the most expensive component of the reactor, delivers a few MJ of energy to each pellet on a few nanoseconds time scale. Sophisticated optics are crucial in order that the driver energy be deposited uniformly over the surface of the pellet to avoid strong hydrodynamic instabilities. An inertial fusion reactor, thus, consists of continuously injected D-T pellets, surrounded by a blanket in a spherical chamber, driven by a high-energy, short-pulse source.

8.19.5 Prospects for the future

Both magnetic and inertial fusion researchers have made substantial progress in understanding the complex physics describing the fusion process. To date, large experiments have been built and operated, leading to a reasonable but still far from complete understanding of the multitude of physics phenomena that occur. Neither method, however, has yet to experimentally demonstrate a high-fusion burn system. This is the last major scientific challenge to overcome.

The international magnetic fusion community is in the process of addressing the burning plasma challenge by proposing construction of a large ignition experiment, the International Tokamak Experimental Reactor (ITER). This $4 billion project is expected to begin construction in 2006, with experiments starting about eight years later. A schematic diagram of the ITER is shown in Figure 8.10.

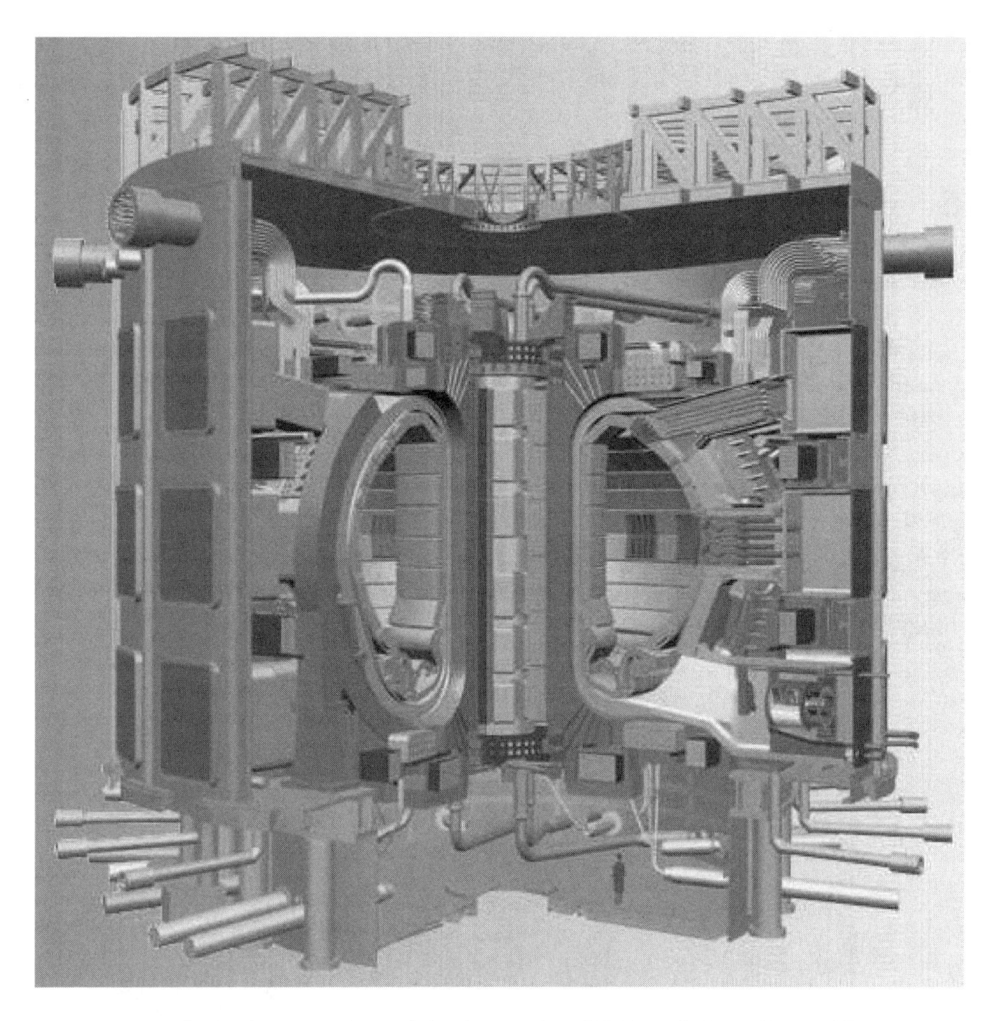

Figure 8.10. Cutaway view of the International Tokamak Experimental Reactor (ITER).

The US inertial fusion community is in the process of constructing a large, $3 billion laser-driven facility at Lawrence Livermore National Laboratory, called the National Ignition Facility (NIF). This facility is funded primarily by the US Department of Defense, but will also be used to ignite D-T pellets for energy applications. A schematic diagram is illustrated in Figure 8.11.

In summary, both the magnetic and inertial fusion communities have made substantial progress in understanding the basic physics of fusion plasmas. Attempts to actually demonstrate a D-T burn are under construction for inertial fusion, and, hopefully, will soon be started for magnetic fusion. The time scale for commercialization is still far into the future—on the order of 30–50 years. One can see from the scale of each ignition experiment that a fusion reactor of either type will require a large amount of highly sophisticated, high-tech equipment, which will likely lead to a high capital cost. Thus, economic competitiveness will be an issue. Still, the enormous benefits of fusion energy make fusion a worthwhile research investment.

8.20 Future Prospects for Nuclear Power

The future roles of nuclear power remain unclear. Nuclear technologies have been attractive and convenient when used with care. However, care will always be necessary concerning safety, the constraints of radioactivity, and the potential for nuclear weapons proliferation. Nuclear technologies have also shown themselves to be unforgiving physically and financially when used carelessly. They offer potential environmental benefits that may become more valued as the problems of other energy technologies become greater (e.g., CO_2 emissions). At the most basic level, nuclear power is judged according to its economic competitiveness. In stable business environments, nuclear power has been able to compete. Beyond that, some firms have valued nuclear power for helping to reduce regional air pollution. Some nations have valued it for increasing energy supply security, both by permitting compact fuel storage and by reducing imports of other fuels.

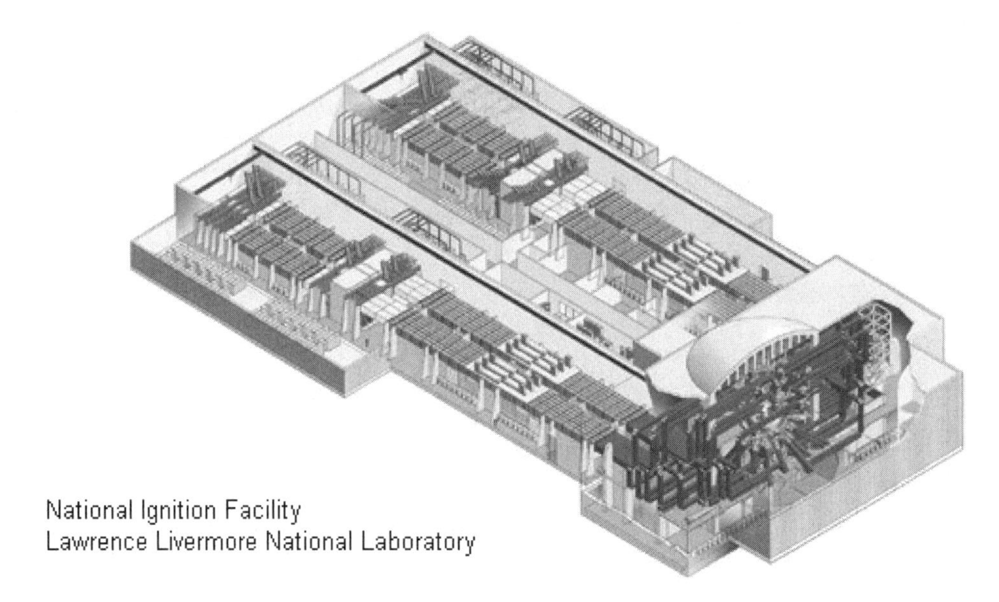

Figure 8.11. Cutaway view of the National Ignition Facility at Lawrence Livermore National Laboratory.

Societies may be unwilling to employ nuclear power on a larger scale unless its difficulties become more strongly balanced by benefits that go clearly beyond providing an electrical commodity. A problem-free operational record for the existing nuclear power plants is an important factor in gaining broader social acceptance. During the past two decades, nuclear power plants have operated safely, and the acceptability of nuclear power technologies has grown correspondingly.

The future use of nuclear weapons poses another important problem. Should nuclear wars occur—even if limited to a regional scale—social aversion to all things nuclear could become strong. We must hope that mankind will be sufficiently sensible to avoid such wars and to limit the number of states having nuclear military capabilities. Also, should global warming turn out to be the peril that some believe, keeping the nuclear power option could be important for the sustainability of the planet.

References

AECL. 1978. *CANDU Nuclear Power Station*. Mississauga: Atomic Energy of Canada Limited.

Anspaugh, L.R., R.J. Catlin, and M. Goldman. 1988. "The global impact of the Chernobyl reactor accident." *Science* 242:1513.

Deutch, J. and E. Moniz. 2003. *The Future of Nuclear Power*. MIT Report.

EIA (Energy Information Administration). 2002a. *Annual Energy Review 2001*. Washington DC: US Dept. of Energy.

EIA (Energy Information Administration). 2002b. *Uranium Industry Annual, 2001*. Energy Information Administration (EIA). Washington DC: US Dept. of Energy.

General Electric. 1972. *BWR/6 General Description of a Boiling Water Reactor*. Fairfield, CT: General Electric Company.

IAEA (International Atomic Energy Agency). 1983.Semenov, B.A. "Nuclear Power in the Soviet Union." *Bulletin* 25(2): 47–59. Viena: IAEA.

Knief, R.A. 1992. *Nuclear Engineering*. Washington: Taylor and Francis.

Mascke, G. 1971. *Systems Summary of a Westinghouse Pressurized Water Reactor Nuclear Power Plant*. Churchill, PA: Westinghouse Electric Corp.

Murray, R.L. 1993. *Nuclear Energy: An Introduction to the Concepts, Systems, and Applications of Nuclear Processes*. New York: Pergamon Press.

Murray, R.L. 1961. *Introduction to Nuclear Engineering*. Englewood Cliffs, NJ: Prentice Hall.

Nero, A.V. 1979. *A Guidebook to Nuclear Reactors*. Berkeley: University of California Press.

Nuclear News. 2002. "World List of Nuclear Power Plants." *Nuclear News*. 45: 39–53 (March 2002).

OECD (Organisation for Economic Co-operation and Development). 1997. *Uranium, Resources Production and Demand*. Report of OECD Nuclear Energy Agency and International Atomic Energy Agency. Vienna.

Parrington, J.R., H.D. Knox, S.L. Breneman, E.M. Baum and F. Feiner. 1996. *Nuclides and Isotopes*. Fairfield, CT: GE Nuclear Energy.

PBMR (Pebble Bed Modular Reactor Ltd.). 2003. *How the Pebble Bed Reactor Works*. See https://www.pbmr.co.za/

Uranium Institute. 1991. *The Management of Radioactive Wastes*. The Uranium Institute Report. Vienna: International Atomic Energy Agency.

USAEC (US Atomic Energy Commission). 1973. *The Safety of Nuclear Power Reactors and Related Facilities*. Report WASH-1250. Washington, DC: USAEC.

USAEC (US Atomic Energy Commission). 1974. *The Nuclear Industry*. Report WASH-1174-74. Washington, DC: USAEC.

USDOE (US Department of Energy) 2001. *Yucca Mountain Science and Engineering*. Report DOE/RW-0539. Washington, DC: USDOE.

Additional Resources

Golay, M.W. and N.E. Todreas. 1990. "Advanced Light Water Reactors." *Scientific American*, 262: 82–89.

Golay, M.W. 1993. "Advanced Fission Power Reactors." *Annual Rev. Nucl. Part. Sci.*, 43: 297–332.

Knief, R.A. 1992. *Nuclear Engineering*, Second Edition. New York: Hemisphere Publishing Corp.

Murray, R. 2000. *Nuclear Energy*, Fifth Edition. Boston: Butterworth Heinemann.

Nuclear Power. 1998. Vienna: International Atomic Energy Agency.

Renewable Energy in Context 9

9.1 Introduction and Historical Notes

In the next six chapters, we will examine six renewable energy types: biomass (Chapter 10), geothermal (Chapter 11), hydropower (Chapter 12), solar (Chapter 13), ocean-based energy (Chapter 14), and wind energy (Chapter 15). In each case, the nature of the resource and the technology associated with how energy is captured, transferred, and, if necessary, converted into other forms will be discussed. Environmental attributes and other characteristics connected to sustainability issues in general will also be presented, along with a prognosis on the current status of the technology in meeting its performance, economic, and deployment goals.

It is appropriate to repeat an earlier statement we made regarding the treatment of specific energy technologies. Given the rapidly evolving nature of renewable energy technologies in general, and the size limitations of this book, it is simply not possible to provide comprehensive, state-of-the-art reviews of each energy type. Our intent is to cover important features and attributes in order to provide a basis for evaluating specific technologies. Readers interested in current assessments are urged to consult the various resources (literature and Web-based) cited at the end of each chapter.

In a direct way, each renewable energy type is tied to one of three primary energy sources: solar radiation, gravitational forces, and heat generated by radioactive decay. Solar, thermal, and photovoltaic energy result from capturing a fraction of incident solar radiation. Wind, hydro, wave, ocean thermal, and biomass energy also require solar energy to produce energy. Gravitational forces between the moon and the earth directly cause the tidal flows needed for tidal power. These same forces also indirectly influence solar energy capture by controlling the earth's planetary motion around the sun. Fundamentally, geothermal energy results from radioactive decay of isotopes of certain elements contained in the earth's interior. However, other effects can be important, such as those associated with volcanic activity resulting from the motion and frictional forces of colliding tectonic plates. Such effects can last for millennia and directly influence the grade of geothermal resources at a particular location by increasing the local geothermal temperature gradient. (Figure 9.1 and Table 9.1 illustrate the interrelationships that exist between renewable-energy types and these primary energy sources.)

Renewable energy has a long history. Virtually all of the renewables have been used by humans for centuries to provide energy services. For example, wind power has been used for powering sailing ships for perhaps over 1,000 years, and the ancient Anasazi Indians made effective use of passive solar heating for their cave dwellings in the Southwest US. Geothermal hot water has been used for bathing, washing, and other direct heating purposes in many locations for all of recorded history. Hydropower first began to produce electricity at Niagara Falls in New York before the beginning of the 19th century, and before that, it provided mechanical energy for grinding grains and running machinery around the world. All of this activity predates our use of fossil energy and large-scale electricity generation.

Given its history, one might think that renewable energy would have evolved to be the major source of energy today. But only a small fraction (about 10%) of the world's primary energy is currently provided by renewable sources, with hydropower, biomass, and geothermal playing dominant roles in selected locations. Although renewable energy has real environmental benefits that could provide some insurance for stabilizing our planet's health, the short-term economics of renewable systems have not been universally competitive. Key issues include the low energy density, low capture efficiency, and lack of dispatchability of many renewable energy types. Such characteristics directly affect their costs relative to the competition—fossil and fissile fuels. Consequently, there remains much to be done to increase the performance of renewable energy systems—particularly in terms of the ability to capture, store, and convert it into more useful forms. In some cases, there can be significant environmental impacts, particularly those associated with land and water use.

Technology, as you will see in the forthcoming chapters, has and will play an essential role in increasing the effectiveness and competitiveness of renewable systems.

Table 9.1. Renewable Energy Types and Energy Transfer Processes (See also Goldenberg et al. (1987), Twidell and Weir (1986), and Johannson et al. (1993))

Primary Energy Source	Energy Transfer Process	Renewable Energy Type
Solar Radiation/ Photon flux	absorption on semi-conductor surface → conversion to electrons via photoelectric effect → conversion to electricity;	photovoltaic (PV)
	absorption on surfaces → conversion to thermal energy;	solar thermal
	differential absorption on the earth's land surfaces and oceans → conversion to kinetic energy;	wind, waves
	selective absorption of light energy to drive photosynthesis → conversion of CO_2 and H_2O to glucose → further metabolically driven conversions to carbohydrates, fats, and proteins;	biomass energy, bio fuels
	solar heat absorption → water evaporation → heat loss to the atmosphere → water condensation → rain → PE in water storage	hydro (solar)
Gravitational forces and planetary motion	PE and KE contained in stored water;	hydro (solar)
	dissipative forces induce periodic KE changes in ocean	tidal
Gravitational forces and friction	Stored and generated thermal energy in the earth's crust transported to surface by conduction and fluid convection, enhanced by tectonic plate motion;	geothermal
Nuclear energy	radioactive decay of isotopes deposits energy in the earth's interior (e.g., K, U, Th)	geothermal

PE = potential energy
KE = kinetic energy

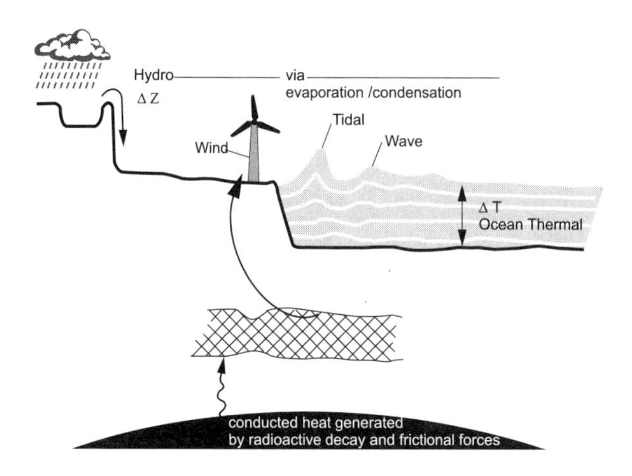

Figure 9.1 Illustration of renewable energy types and energy sources.

9.2 Resource Assessment

Classifying an energy form as "renewable" encompasses a range of assumptions regarding time scales. The implication is that renewable energy is just that—available continuously without depletion or degradation. For example, solar energy is available for some time period each day virtually everywhere on the surface of our planet. There is a natural 24-hour diurnal cycle, as well as seasonal variations because of changes in the relative angle of our rotating earth tilted on its axis as it makes its yearly orbit around the sun. Superimposed on these effects are the daily fluctuations that result because of weather—mostly cloud cover. Other renewable energy types such as biomass, hydropower and wind energy have analogous variations albeit over different time scales. Geothermal energy is different in that the earth conducts heat in a transient mode by virtue of the natural temperature gradient that exists between its inner core and the surface. The earth also stores heat continuously as generated by radioactive decay of long-lived isotopes, primarily those of potassium, thorium, and uranium. All of the renewable resources replenish themselves over times far less than those associated with fossil fuels. But on a short enough time scale, even renewable resources must be regarded as depletable. This feature will become critically important when we consider how and when renewable energy is captured and utilized. In many cases, the ability to dynamically store captured renewable energy is also crucial in determining whether a given renewable resource can provide a viable energy option. For example, energy storage is inherent to bioenergy with solar energy stored in the chemical bonds of the biomass itself and to hydropower in terms of the impounded water contained behind a dam.

Commonly, estimates of renewable energy resources are presented in support of a specific technology or in opposition to show that it is too small or too interruptible to make a difference. How can such different perspectives both be correct? It is useful to revisit the resource base concepts introduced in Chapter 2. In this context, Table 9.2 provides rough estimates of the average annual global energy flux and net stored energy for various renewable energy resources. For comparison, a rough estimate of the total fossil-energy resource base is included. With today's global energy consumption at more than 360 quads/year, one concludes that many renewable energy types can meet the demand for a long time into the future. Also, note that there is a massive resource base for fossil energy.

To arrive at a specific estimate for each energy type, several assumptions are needed. For example, the total amount of solar energy intercepted by the earth's land masses is taken to represent the resource base for solar thermal or solar photovoltaic energy, even though we recognize that a substantial fraction is needed to support life itself via photosynthetic and natural oxidative processes. In practice, only a small fraction of the total resource base can actually be recovered and counted as an energy reserve. This limitation applies to depletable resources, such as fossil, and renewable resources, such as solar or geothermal.

As with depletable resources such as oil and gas, the quality of a renewable resource can vary widely—from nation to nation, from region to region, or within a specific region from hectare to hectare. For example, consider the inherent difference in solar radiation intensity and duration between equatorial latitudes and northern latitudes or where the climate is frequently cloudy or mostly clear. Biomass energy quality is fundamentally connected to levels of solar radiation, water availability, and the presence of soil nutrients. Wind, wave, tidal, and geothermal energy are also widely variable from region to region depending on atmospheric and geologic conditions (e.g., wind speeds on mountain ridges tend to be higher than in deep valleys, and high-quality hydrothermal geothermal resources are frequently found in regions with relatively recent volcanic activity).

Table 9.2. Worldwide Resource Bases for Renewable Energy: Fluxes and Stored Energy*

Resource	Energy Type	**Total Flux TW (quads/yr)**	**Net Stored Energy TW-yr (quads)****
Solar	land flux	27,000	0
	ocean flux	53,000	0
	total flux	89,000 (2.7×10^6)	0
Wind	kinetic energy (land)	100 (3,000)	0
Waves	kinetic energy (ocean), gravity	200 (6,000)	0
Biomass	photosynthesis	30 (900)	750 (22,400)
Hydro	latent evaporative heat	40,000 (1.2×10^6)	80,000 (2.4×10^6)
Tides	gravitational forces between the moon and earth	> 3 (90)	3 (90)
Geothermal			
– hydrothermal – geopressured and – magma	convection	~2.5 (75)	1.710^5 (5.1×10^6)
– hot dry rock	conduction	≥ 30 (900)	3.3 10^6 (1×10^8)
All fossil fuels	chemical reaction enthalpies or heats ***	10 (300)	≥ 360,000 (1.1×10^7)

*estimates based on information provided by Armstead and Tester (1987), Grubb (1990), and Armstead (1983); US figures based in part on USGS estimates (e.g., see Muffler and Guffanti (1978), Sass (1993)).

**1 quad = 3.35×10^{-2} TW-yr = 10^{15} Btu

***does not include any energy stored from the solar greenhouse effect, e.g., in atmospheric CO_2, H_2O, or CH_4.

9.3 Environmental Impacts

Over a complete life cycle, there are no net emissions associated with the conversion of any primary renewable energy source to a specific energy form. For instance, no net accumulation of carbon dioxide in the atmosphere results from the complete cycle of producing energy by growing and burning biomass such as wood. Effectively, the CO_2 produced by combustion was previously removed from the atmosphere in the photosynthetic process that generated the wood. Nonetheless, the process of growing and harvesting biomass requires additional energy (e.g., in the production of ammonia for fertilizer or by the fuels consumed by equipment used in harvesting and transporting the biomass itself). Similar arguments exist for the embedded energy needed to manufacture solar photovoltaic or thermal panels, for drilling geothermal wells, or for building the structures needed to capture hydro, wind, ocean thermal, or wave power. The carbon-free energy produced from these renewable resources should be discounted by these embedded fossil energy requirements.

In some cases, the capture or conversion of a renewable resource may produce emissions that need to be controlled. For example, naturally occurring hydrogen sulfide (H_2S) found in many hydrothermal geothermal fluids or the particulate emissions that result from burning biomass fuels may require active abatement to reduce emissions. Further, even without the direct emission of toxic fumes or a net flux of CO_2, renewable energy systems are not without environmental impacts. Consider the siting issues associated with locating wind turbines, or the land and water use concerns of using agricultural land to produce bioenergy crops that could be used instead to grow food (e.g., ethanol production from corn in the midwestern US). Other natural resources and associated energy are consumed to produce the energy capture systems for renewables. For example, the cement and steel used to construct hydro dams, materials for PV collectors, wind turbine blades, and support structures have embedded environmental impacts. Other factors, such as noise, rejected waste heat, and induced land instabilities may lead to ecological impacts on animals and plants near a renewable energy system. Clearly, a full life cycle approach is needed to quantify the environmental benefits and impacts of any renewable system. Such quantification is necessary for understanding tradeoffs and making informed selections among all energy supply alternatives on a site-by-site basis.

9.4 Technology Development and Deployment

While many people are in favor of increasing our utilization of renewable energy, the rate of deployment in specific situations depends as much on energy policy as it does on the status of the technology relative to existing energy sources. There is considerable inertia both economically and politically in today's energy supply systems. When high-grade renewable opportunities exist, such inertia can often be overcome. Frequently, however, the high initial capital costs for a renewable technology are a barrier to its deployment. These higher costs should be offset by the environmental impacts and costs associated with other depletable energy options—particularly fossil. Currently, only a portion of these environmental costs have been internalized into the price paid for a given energy service. For example, the costs of reducing SO_x or NO_x emissions from fossil-fired electric power plants are part of the capital investment and operating costs, but environmental "damage" costs from the CO_2 emissions from those power plants are not.

In the future, limiting the CO_2 emissions from fossil-fuel-supplied energy in both developed and developing countries as a result of climate stabilization concerns would have a profound effect on the rates of development and deployment of renewable energy. Superimposed on this development would be policy incentives aimed at saving energy by increasing efficiency or those directed at reducing national dependencies on foreign oil and gas supplies. Conservation or demand-side management approaches are frequently coupled with deployment initiatives to promote the use of renewable energy. For example, measures for decreasing heat losses or gains from homes with improved window and wall insulation are often combined with designs that use both active and passive solar energy capture systems or employ geothermal heat pumps for heating and cooling. Perhaps with deregulated energy markets and "green" power incentives, renewable systems will get a boost.

9.5 The Importance of Storage

Renewable energy types such as solar, wind, and wave power are variable on a daily basis. The times when energy can be captured and converted efficiently do not always correspond to periods of high demand. Without the ability to dispatch energy on demand, power shortages or outages will occur. If the fraction of non-dispatchable renewable energy is large on a regional electric grid, similar shortages will occur when mismatches between supply and demand occur unless backup power is provided. Hybrid technologies that employ both renewable and fossil energy components provide an interesting alternative option. But without such options, an energy storage system is needed.

Renewable types such as biomass and hydro naturally have their own storage capacity— chemically in the biomass itself and in the stored potential energy of the contained water reservoir in the case of hydro. For geothermal, inherent storage capacity exists in the heat contained in the rock mass that comprises the active reservoir.

For non-dispatchable renewable systems to play a major role in supplying primary energy, robust energy storage systems are needed. For example, in small-scale applications at remote, off-grid sites, storage batteries are often used if electric power is being produced by a wind turbine or photovoltaic device. Having cost-effective, high-efficiency storage systems with low losses available for electricity and/or heat storage increases the attractiveness of these renewable energy types. As discussed in Chapter 16, there are many options for energy storage at various stages of development. Obtaining a good match between the type and size of the energy capture and a storage system depends on the nature of the resource at the site and the energy needs of the customer.

9.6 Connecting Renewables to Hydrogen

Given the increasing emphasis on the future role of hydrogen as an energy carrier for both stationary and mobile applications, it is appropriate to comment on the role of renewable energy in hydrogen-based energy systems. Renewable sources that are interruptible, such as solar and wind energy, could be used to produce hydrogen during off-peak demand periods with the hydrogen being stored on-site for later use or pumped into a national pipeline grid. Although one of the most attractive features of biomass is the conversion pathway that yields liquid transportation fuels, such as ethanol, biomass could be gasified to produce hydrogen as well.

Currently, most of the hydrogen we use is produced from chemically reformed methane:

$$CH_4 + 2H_2O \rightarrow CO_2 + 4H_2$$

Thus, hydrogen derived from methane as a fossil fuel represents a finite resource with an inherent carbon dioxide emission penalty. If a hydrogen economy is to prevail in the long-term, then, from a sustainability perspective, renewables eventually will have an important role in providing the primary carbon-free energy needed to make hydrogen either by electrolysis or by chemical splitting of water. Nuclear systems may also see a role in hydrogen production (see Chapter 8).

9.7 The Future for Renewable Energy

Enthusiasm for renewable energy is driven by three desirable characteristics:

- Renewable energy is abundant and available everywhere.
- It inherently does not deplete the earth's natural resources.
- It causes little, if any, environmental damage.

Although none of these attributes is rigorously true, renewable systems, if deployed correctly, come much closer to achieving these ideals than do their fossil and fissile counterparts.

Even with this enthusiasm, there has been reluctance to adopt renewable technologies because they appear to cost more or to not perform as reliably as fossil- or nuclear-powered systems. Are renewable technologies always more expensive and less reliable? The answer to this fundamental question depends critically on the grade of the renewable resource and the specific technology being employed to utilize it. In some instances, like a wind farm in North Dakota or Texas with production tax credit incentives, wind energy is competitive in 2003. In other situations, like solar hot water heating in New England with tax credits or other incentives, renewable solar energy is not the most economic choice. Nonetheless, both perceived and real risks often drive up investment and development costs to a point where renewable systems cannot compete.

In the past 25 years, investment in R&D for renewable energy has had a spotty and unbalanced record in the US and Europe. State and national governments have been inconsistent in supporting research for renewables, particularly in comparison to other speculative technologies such as breeder reactors and fusion energy. More recently in the US, ambitious deployment goals for specific renewable technologies (e.g., *xx* MWe on line by *yy* year) are frequently announced with insufficient funding and low commitments of staff to accomplish these goals. As a result, there is a dominant perception that renewables just won't work—they are too risky or too costly to make a difference. In contrast, the Europeans recently have taken a stronger stand on deployment of renewables such as wind and solar by providing a range of policy instruments that favor development. The Japanese have also made a much larger commitment to renewables with several long-range goals being emphasized and promoted.

Without robust international, national, and state policies in place that encourage the development of improved renewable technologies and internalize the environmental damages from fossil and fissile energy, the future for renewable energy is tenuous at best. Unfortunately, such barriers are inherent to most evolving technologies when alternative choices exist.

9.8 Additional Resources

There are many comprehensive reviews of renewable energy technologies. Interested readers should consider both published books and government reports (see the references section at the end of this chapter) and review articles in the series called *Annual Reviews of Energy and the Environment* as resources worth exploring. In addition, for the latest updates on technology developments, the electronic media, Web sites associated with the US National Renewable Energy Laboratory (www.nrel.gov), and associated links to various professional energy organizations (e.g., www.wcre.org) are valuable resources as well.

References

Armstead, H.C.H. 1983. *Geothermal Energy*, Second Edition. London: E. & F.N. Spon.

Armstead, H.C.H. and J.W. Tester. 1987. *Heat Mining*. London: E. & F.N. Spon.

Brower, M. 1992. *Cool Energy*. Cambridge, MA: MIT Press.

Conseil Mondial De L'energie (World Energy Council). 1993. *Energy for Tomorrow's World*. New York: St. Martin's Press.

Goldenberg, J., J. Johansson, A.K.N. Reddy, and R.H. Williams. 1987. *Energy for a Sustainable World*. Washington, DC: World Resources Institute.

Grubb, M.J. 1990. "The Cinderella options—A study of modernized renewable energy technologies," Part I-A Technical Assessment. *Energy Policy*, 18(6): 525–542.

Johannson, T.B., H. Kelly, A.K.N. Reddy, and R.H. Williams, eds. 1993. *Renewable Energy — Sources for fuels and electricity*. Washington, DC: Island Press.

Lovins, A., L. Lovins, F. Krause, and W. Bach. 1981. *Least-Cost Energy, Solving the CO_2 Problem*. Andover, MA: Brick House.

Muffler, L.J.P. and M. Guffanti, eds. 1978. *Assessment of Geothermal Resources in the United States*, Circular 790. Washington, DC: U.S. Geological Sciences.

National Academy of Sciences. 2000. "A Review of the U.S. Department of Energy's Renewable Energy Programs," Part A: Renewable Power Pathways. Washington, DC: National Research Council.

National Energy Policy Development Group. 2001. *National Energy Policy*. Washington, DC: US Government Printing Office.

PCAST (The President's Committee of Advisors on Science and Technology). 1997. *Federal Energy Research and Development for the Challenges of the Twenty-First Century*, Report of the Energy Research and Development Panel. Washington, DC: Office of Science and Technology Policy.

Sass, J.H., 1993. "Potential of hot-dry-rock geothermal energy in the eastern United States." Open file report: 93–77. Washington, DC: US Geological Survey.

Twidell, J. and T. Weir. 1986. *Renewable Energy Resources*. London: E. & F.N. Spon.

USDOE. 1990. *The Potential of Renewable Energy, An Interlaboratory White Paper*, SERI/TP-260-3674. Washington, DC: US Department of Energy.

USDOE. 1991. *National Energy Strategy, Powerful Ideas for America*, first edition. Washington, DC: US Department of Energy.

Biomass Energy 10

10.1 Characterizing the Biomass Resource

Defining biomass. In the study of sustainable energy, we define *biomass* as all living plant matter as well as organic wastes derived from plants, humans, marine life, and animals. Trees, grasses, animal dung, as well as sewage, garbage, wood construction residues, and other components of municipal solid waste are all examples of biomass.

In the past, biomass was the primary source of fuel for the world. As the Industrial Revolution progressed in Europe, forests were severely depleted and coal was gradually introduced as a replacement fuel. In the US, biomass was still a primary source of fuel in the 19th century, although, by 1885, its use was being outpaced by that of coal and by 1915, also by oil and gas (Hottel and Howard, 1971). Today, in many developing countries, biomass remains an important energy source for heating and cooking, and rudimentary combustion devices used for energy conversion are inefficient and polluting—especially if used in indoor spaces. Where population densities create demands that overwhelm the productivity of indigenous biomass, the land becomes depleted and, in extreme cases, desertification occurs (e.g., Africa). When the biomass production fails, the populations can no longer find food, much less biomass for energy. Obviously, the very poor have no income to purchase food or fuel, and have no access to transportation other than human power. Defoliation contributes to this destructive cycle, which locks about 20% of the world's population into extreme poverty.

As concerns about our increasing use of fossil fuels mount, many developed countries now are reexamining the potential for biomass energy to displace some use of fossil fuel. Biomass qualifies as a renewable resource because commercially meaningful quantities are regenerated in time scales that are comparable to or less than typical time scales for human use of the resource (see Section 10.1.2). Indeed, biomass is a natural engine for the conversion of solar energy to high-energy content products that can be stored, transported, and used conveniently. Plants grow by the process of *photosynthesis* in which sunlight transforms two naturally abundant raw materials, water and carbon dioxide, to carbohydrates and other complex organic compounds (see Section 10.9) of great natural and commercial value. For example, the global chemical pathway to form glucose ($C_6H_{12}O_6$) by photosynthesis can be represented by:

$$6CO_2 + 6H_2O \xrightarrow[\text{catalysts}]{\text{sunlight}} C_6H_{12}O_6 + 6O_2 \quad \Delta G^o = 480 \text{ kJ/mol} \tag{10-1}$$

Note that Equation (10-1) represents a process for pumping energy "up hill" because CO_2 and H_2O have zero heating value and reside at one extreme of the energy spectrum. The energy content comes from solar input, which plants convert to biomass energy with conversion efficiencies of about 1–2%. The catalysts, chlorophyll and other plant ingredients, facilitate the chemistry depicted in schematic form in Equation (10-1). The photosynthate is an intermediate in the formation of glucose/sucrose, cellulose polymers, fuel for plant respiration, or a variety of other molecules as depicted in Figure 10.1.

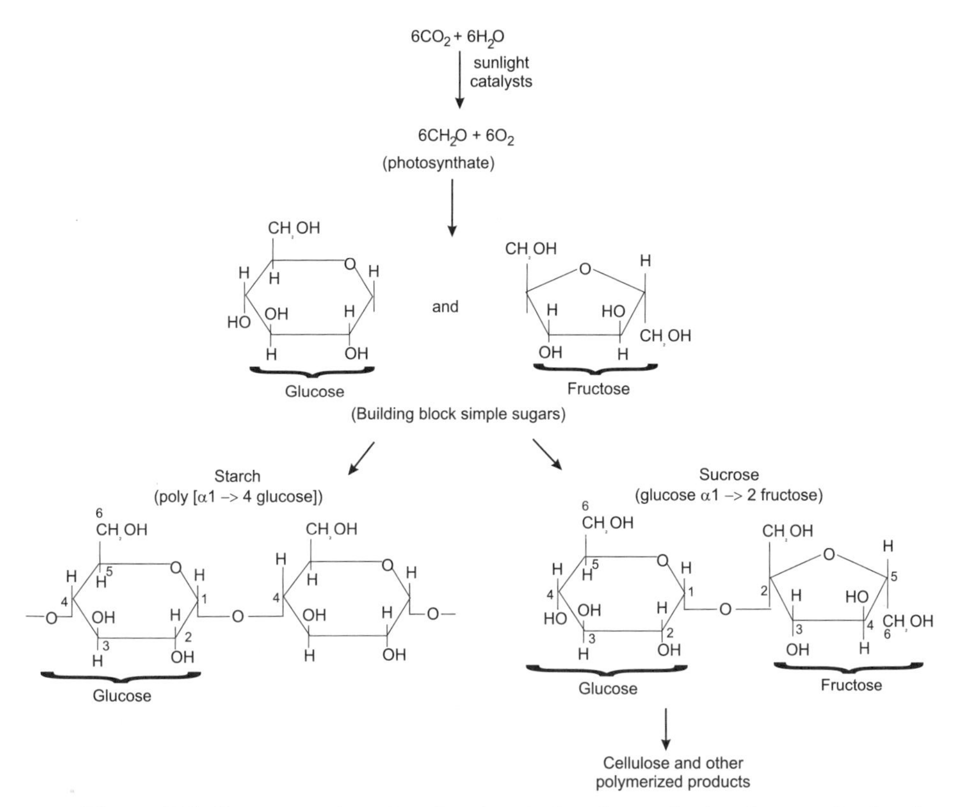

Figure 10.1. Representation of major chemical pathways in the photosynthesis of plants to convert carbon dioxide to carbohydrates and other biomass compounds.

Because biomass can capture and store solar energy as commercially useful and convenient products, it is no surprise that biomass is of great interest as a renewable fuel source. Biomass is an enabling technology for sustainability that is already economically attractive in selected applications. However, appreciable growth in its share of the energy market in developed countries is handicapped by its generally higher costs compared to fossil and nuclear-based alternatives. These higher costs reflect the relatively modest flux density of solar energy at the earth's surface combined with the low efficiency with which Equation (10-1) converts solar energy.

Renewability indices and biomass resources. Because biomass is renewable, any meaningful assessment of its resource base must consider how rapidly biomass is regenerated in quantities that are useful to humans. A semi-quantitative perspective on this notion is gained through an approach described by Professor Paul Weisz of the University of Pennsylvania (Weisz, 1978). His idea compares the time scale for regeneration of a commercially meaningful quantity of the resource under scrutiny, τ_r, to a time scale for some plausible period over which the resource can benefit humans—a measure of some characteristic societal utilization time, τ_u. Plausible scales are a generation, say 25 years, or some actuarially reasonable estimate of the lifetime of a person in the geographic region where the resource is likely to be used (e.g., 75 years in the US). Computation of the dimensionless ratio $[\tau_r/\tau_u]$ then gives a semi-quantitative measure of whether the resource in question is "strongly" renewable, $\tau_r/\tau_u << 1$, renewable, $\tau_r/\tau_u \approx 1$, or decidedly non-renewable, $\tau_r/\tau_u >> 1$. Table 10.1 presents calculated examples of τ_r/τ_u for various forms of biomass. Strictly speaking, fossil fuels are renewable, but the time scales for regeneration of petroleum, natural gas, and coal are so long ($\tau_r/\tau_u > 10^5$) that, from a practical perspective, we categorize them as non-renewable.

Table 10.1. Selected Dimensionless Time Constants Related to the Renewability of Biomass[a]

Activity	Time Constant, τ_r/τ_u
Regeneration of timber, northwest US[b]	1.43
Regeneration of timber, southeast US[b]	0.36
Production of 1 ton of biomass on 1 acre using "fast growing trees"	0.002
Production of 1 ton of solid refuse by 1 US resident	0.016

[a]Defined as the ratio of the time for resource regeneration, τ_r, to a representative societal utilization time, τ_u, here taken as 70 years. See Weisz (1978).
[b]Anon (1979).

A more quantitative approach for assessing the potential impacts of biomass as a sustainable resource compares biomass production rates or cumulative production over a much shorter period than τ_u, (e.g., one year) with human consumption rates or cumulative consumption over that period. Let:

$[dP_{i,q}/dt](t) =$ the rate of production of energy of a particular type i (e.g., heating value) from biomass of a particular type and quality q (e.g., trees), e.g., in units of kJ_{th}/s, i.e., kW_{th} where the subscript th denotes thermal as distinct from electrical energy (i.e., the energy content is assumed to be reliably quantifiable by the heating value of the biomass)

$[dC_{j,Q}/dt](t) =$ the rate of human consumption of energy of a type j (e.g., electricity) and quality Q (e.g., continuously available) that might be supplied by this biomass, e.g., in units of kW_e (electrical)

Because, in general, production and consumption rates vary over quite short periods of time, these two quantities depend on time as is indicated by the functional (t) notation. A biomass resource is load-matched to the market under consideration if, for example:

$$\int_{t_1}^{t_2} [dP_{i,q}/dt]dt + \int_{t_2}^{t_3} [dP_{i,q}/dt]dt = \int_{t_2}^{t_3} [dC_{j,Q}/dt]dt \qquad (10\text{-}2)$$

where t_1 is the time at which a renewal period begins (e.g., completion of renewal planting of a region dedicated to growing crops for energy purposes is measured), t_2 is a time long enough to build up an inventory of biomass sufficient to start up the energy supply operation and to smooth out modest short-term declines in biomass production rates, and t_3 is sufficiently long to provide a meaningful impact on societal needs (i.e., so that $(t_3 - t_2) >> (t_2 - t_1)$ and t_3 is comparable in magnitude to τ_u).

The following formulas illuminate factors that must be considered in gauging whether a given biomass resource is well matched to the demands of a particular commercial application:

$$[dP_{i,q}/dt](t) = Y_B A_g E_B f_q \tag{10-3}$$

where

Y_B	=	the rate of production of *dry* biomass per unit area of land (e.g., in tons/(acre-yr))
A_g	=	the total growing area available for biomass production (e.g., acres)
E_B	=	the unit energy content of the dry biomass (e.g., in Btu/ton)
f_q	=	the fraction of the biomass energy that is converted to the form of energy (e.g., electricity) desired in this application (e.g., 0.33 for dry biomass in some converters)

In general, any of these four terms may vary over time scales of order one year or less (e.g., crop yields Y_b may change from year to year). Similarly, we can write:

$$[dC_{j,Q}/dt](t) = Y_N A_N E_N f_j f_Q \tag{10-4}$$

where

Y_N	=	the population density in the area to be serviced by the biomass (people/acre)
A_N	=	the total area of the region to be serviced (acres)
E_N	=	the average per capita energy consumption rate in this service area [Btu/(person-yr)]
f_j	=	the fraction of the total average energy consumption that is made up of energy of the particular type j to be supplied by biomass (e.g., electricity) (typically $f_j \approx 0.33$)
f_Q	=	the fraction of energy of type j that is to be supplied by biomass (0 to 1.0)

In general, any of these five terms may vary during the course of a year. For example, owing to policy decisions, time-of-day pricing, availability, etc., the fraction of electricity to be supplied by biomass (f_Q) may undergo excursions for part or all of the year.

We can generalize the above discussion by defining a biomass renewability intensity index, I_B

$$I_B = \frac{\left[\int_{t_1}^{t_2} (dP_{i,q}/dt)dt + \int_{t_2}^{t_3} (dP_{i,q}/dt)dt\right]}{\left[\int_{t_2}^{t_3} (dC_{j,Q}/dt)dt\right]} \quad (10\text{-}5)$$

and noting that for: (a) $I_B \sim 1$ biomass production can just keep pace with consumer energy demand over some practically reasonable time; (b) $I_B < 1$ the subject biomass resource cannot keep pace with human needs; and (c) $I_B >> 1$, the resource can readily meet the identified demand. Note that new scientific and engineering understanding may lead to greater production from a particular biomass resource (e.g., higher productivity per unit of land Y_B, ability to use more and more land of marginal quality [e.g., deserts] A_g, increasing energy content E_B or energy quality f_q, [see Section 10.10] of the biomass). Similarly, decreasing availability of water or nutrients or financing may limit one or more of these same terms, and regulatory factors may open up or embargo certain lands or water resources for biomass development. These discussions are related to the concept of the ecological footprint (Rees, 2003) which was presented in Section 6.4.2.

10.2 Biomass Relevance to Energy Production

10.2.1 Utilization options

Figure 10.2 shows that by thermal or biological routes biomass can be converted to a wide range of useful forms of energy, including process heat, steam, motive power, liquid fuels, and electricity, as well as synthesis gas (syngas) and fuel gases of various heating value. Recall (Section 7.3.3) that syngas is a precursor to many other useful products such as methanol, substitute natural gas, ammonia (for fertilizer), and liquid transportation fuels. As a raw material, we can think of biomass as a nearly "universal feedstock" for producing energy and a plethora of energy intensive fuels and chemicals. Because of this versatility and the renewability of biomass, is it then the "preferred solution" for a sustainable planet? We shall see in this chapter that a variety of challenges must first be surmounted to push biomass beyond its current 5% contribution to energy needs in the developed world.

10.2.2 Advantages and disadvantages

Advantages. Major attractions of biomass as an energy raw material are its renewability and domestic availability. Biomass is often widely dispersed. This leverages biomass potential impact by enabling it to function as a distributed energy source (see also Chapters 16 and 17) rather than a raw material confined to a few limited sectors of a region or country. Moreover, as seen in Section 10.2.1, many forms of energy and energy-intensive products can be made from biomass (e.g., solid, liquid, and gaseous fuels, chemical feedstocks, lumber, finished wood goods, etc.). Biomass-derived products can be substituted for metallic or plastic goods that require huge quantities of energy for their manufacture. For example, a few years ago DaimlerChrysler was manufacturing undercarriage components for passenger buses from biomass.

A range of different processes can utilize biomass. Some have been adapted from existing technologies developed for processing or combusting coal. Biomass utilization for energy production offers another major benefit to sustainability, namely a pathway to manage municipal and agricultural wastes. An ability to collect and dispose of these wastes in an environmentally sound manner or on a regular basis commands significant revenue (*tipping fees*), which helps defray the otherwise high cost of biomass harvesting and may temper other process costs.

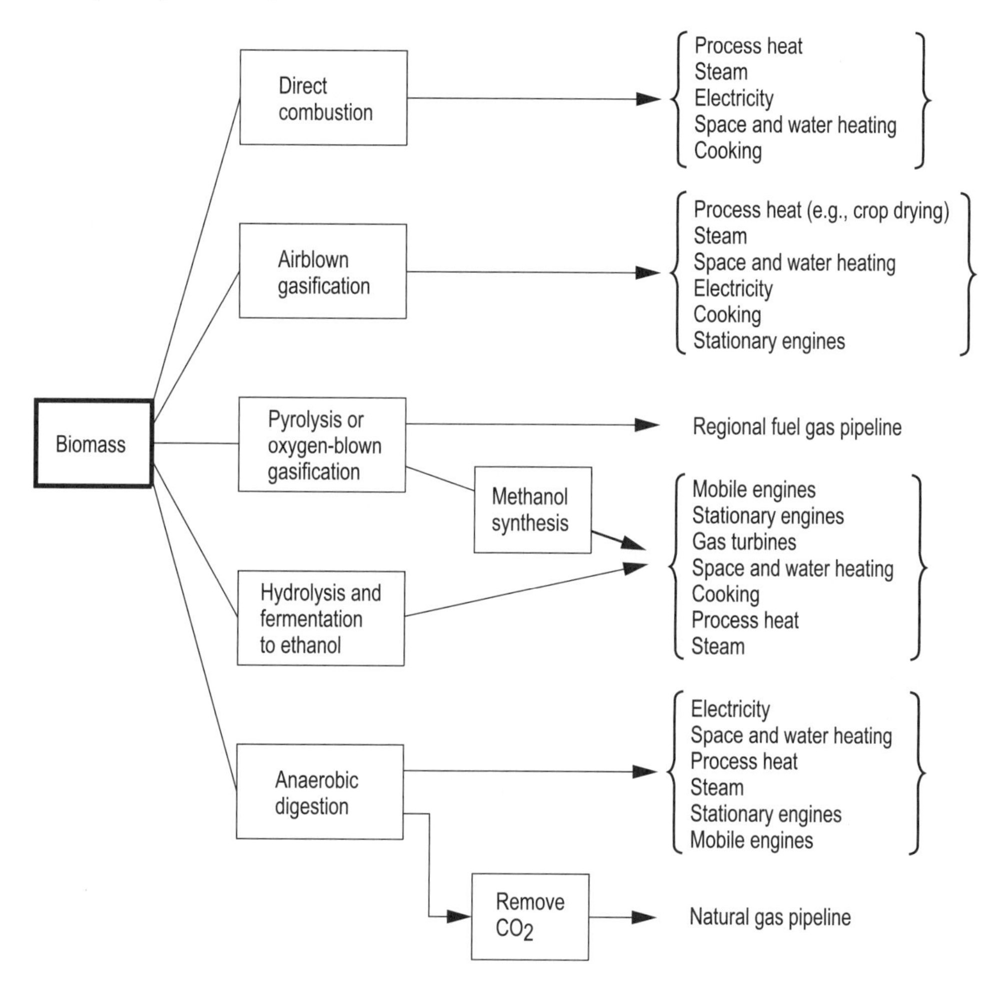

Figure 10.2. Processes for conversion of biomass to fuels, chemicals, and heat from various applications. Source: OTA (1980).

Particularly compelling from a sustainability perspective is the potential of biomass to combat atmospheric buildup of the greenhouse gas CO_2. Equation (10-1) shows that the growth of biomass by photosynthesis is a natural, solar-energy driven, CO_2 sink. We describe this as a *potential* means to counter CO_2 buildup because it must be demonstrated that energy expended to grow, collect, and dry the biomass (e.g., for fertilizer production, operation of harvesting machinery, and means of transport and storage) does not emit greater quantities of CO_2 (Chapter 7). One can imagine a CO_2-neutral biomass economy in which the growth rate of biomass exactly balances the release rate of CO_2 to the atmosphere. Some schemes are even promoting the idea of sequestering carbon by growing new biomass in the form of trees and keeping them standing in place. While this would partly offset the deforestation practices occuring around the world, one eventually has to face the finite "lifetime" of biomass.

Besides its renewability, biomass brings other advantages as a feedstock. With the exception of waste automobile tires, biomass has lower sulfur content (0.05–0.20 wt% S) than coal ($1\text{-}5^{+}\%$). This produces lower emissions of sulfur oxides during combustion. Per unit mass on a dry basis biomass is more reactive to combustion and reforming with steam than most common fossil fuels, including natural gas, most petroleum fractions, and most coal types. One practical consequence is that smaller process vessels can be used to provide comparable or greater outputs of fuels. This higher reactivity makes it more convenient to expand the output of biomass-to-electricity (Section 10.5) and other biomass-to-energy conversion processes in small increments (e.g., 50–100 MWe for electric power generation).

Biomass can also be mixed with other feedstocks, such as coal. For a fixed rate of energy or fuel production, co-firing can reduce the emissions of inorganic matter, sulfur oxides, and fossil-derived CO_2. Assuming adequate biomass resources, this can be a useful strategy for transitioning to a less fossil-carbon-intensive energy economy.

Disadvantages. Biomass has several disadvantages, some of which are a consequence of biomass characteristics that provide the above advantages. By all measures, the energy density of biomass is low compared to that of coal or liquid petroleum or petroleum-derived fuels. In other words, the yield of biomass per unit growing area of land or water is small (Section 10.4). The heat content of biomass, even on a dry basis (7,000–9,000 Btu/lb), is, at best, comparable with that of a low-rank coal or lignite, and substantially (50–100%) lower than that of anthracites, most bituminous coals, and petroleum. Most biomass as received has a high burden of physically adsorbed moisture, up to 50% of the total raw material (i.e., the mass of water is the same as the mass of dry biomass). Without substantial drying, the energy content of biomass per unit mass is even less than that of coal and petroleum. It is preferable to dry biomass "naturally" (i.e., by wind and solar heating) to avoid the expense of using fossil or biomass fuels to supply the energy needed to vaporize the moisture contained in biomass. Feeding high moisture content biomass to any thermal process is a surefire prescription for horrendous losses in thermal efficiency, explaining in part why some early biomass-to-electricity Rankine combustion cycles were clocked at fuel-to-electricity efficiencies below 20%.

Intensive cultivation of biomass may stress water resources, deplete soil nutrients, and displace open space by withdrawing land from natural use. Large-scale production of biomass for energy purposes could compete with use of land, water, energy, and fertilizer (natural or anthropogenic) for production of food or wood and grasses for construction of shelter. The wide dispersion of biomass combined with the low intensity of production per unit land areas are serious disadvantages when there is a need to supply huge amounts of energy to small land areas (e.g., in high population density megacities). Because of energy use associated with harvesting and transporting biomass, it is best suited to distributed, low-intensity use. If biomass has to be transported by farm equipment much over 100 km to a processing point or use facility, a substantial fraction of the energy content of biomass itself is consumed in the transportation process. Biomass production and use also has undesirable environmental impacts (Section 10.7).

As with other renewables, the major barrier to expanding the use of biomass for electricity and transportation fuel production is the higher costs of biomass fuels compared to fossil fuels.

10.2.3 More on resources

Because biomass is renewable, biomass resources are best expressed in terms of production rates (Section 10.1.2). Table 10.2 presents estimated potential production rates in 2050 for various regions of the earth, for biomass recoverable as crop, forestry, and animal residues, as well as for biomass intentionally grown to supply energy (*energy crops*). To "calibrate" these production estimates, recall that US total primary energy consumption in 2003 was roughly 100 EJ (or quads), of which roughly 3% was supplied by biomass. In the US, the pulp and paper industry is the major contributor, using lignin cellulosic wood wastes ("hog fuel") to supply captive energy for processing needs. Keep in mind, total world energy consumption that same year was about 360 EJ. Missing from Table 10.2 is any contribution to biomass resources from municipal wastes. In large urban areas, these resources could be substantial (see Problem 10.3). Growth of biomass to supply electricity has led to several agricultural field programs in different regions of the US (see Walsh et al., 2000):

- The Pacific Northwest: cross breeding eastern and black cotton wood trees to achieve a growth-rate target of 10 dry tons/(acre-yr)
- New York State: 2,000 acres of willow trees to produce biomass for co-firing with coal (10 wt% biomass/90 wt% coal)
- Minnesota: use of 5,000 acres of land subject to erosion, and thus poorly suited for typical Minnesota food crops such as wheat and sugar beets, to produce hybrid poplars for co-firing with coal
- Iowa: cultivation of 4,000 acres of switchgrass for co-firing with coal

These programs show significant interest on the part of electric utilities in expanding the role of biomass in their generation mix. Sufficiently dried biomass can be co-fired with coal or other fuels with little modification of an existing power plant if the biomass

mass fraction is kept low, typically about 0.10–0.15. Even small amounts of co-fired biomass from wastes or energy crops displace some fossil fuel; widespread use of co-firing could significantly increase biomass energy utilization without a major shift in our energy infrastructure, assuming nearby availability of the biomass. They represent important steps in the technical and economic assessment of the use of biomass intentionally grown to supply energy and in the application of biomass to a high energy demand commercial sector. Success would represent an important extension of biomass from its traditional large-scale energy applications in the forest products industry and in energy production as a product of municipal waste incineration.

Table 10.2. Estimated Potential Production Rates of Biomass for Energy in 2050, EJ/yr[a]

Region	Recoverable Residues					
	Crops	Forest	Animal Dung	Total Residuals	Biomass Energy Crops	Total Bioenergy Production
North America	1.7	3.8	0.4	5.9	34.8	40.7
Europe	1.3	2.0	0.5	3.8	11.4	15.2
Japan	0.1	0.2	—	0.3	0.9	1.2
Africa	0.7	1.2	0.7	2.6	52.9	55.5
China	1.9	0.9	0.6	3.4	16.3	19.7
World Total	12.5	13.7	5.1	31.2	266.9	298.1

[a]1EJ = 9.48×10^{14} Btu = 0.948 Q

Source: Larson (1993), quoting Hall et al. (1993).

It is difficult to estimate what the maximum available biomass might be for a region or country where all residuals are collected and utilized and all vacant land is planted with dedicated energy crops. In addition, there is wide variability in the amount of agricultural or forested land from country to country and region to region. For example, consider the vast differences that exist in the US between the Midwest and the desert Southwest. Or, compare the biomass capacity of a desert region country like Saudi Arabia versus the US.

Given the high potential for the US, let's consider what experts have estimated for the annual production of biomass residuals. Estimates of residual dry tonnage on an annual basis for forest and wood products, agricultural residues, municipal sludges, and food processing wastes are summarized in Table 10.3. Even with high uncertainty, the 400+ million dry tons per year total potential for biomass residuals is enormous. Furthermore, if low-intensity energy crops were planted on vacant US land, 20–30% more would be added to the estimate. There is also the possibility of growing biomass in the oceans and harvesting it for producing energy (see Table 14.2 and associated discussion).

Table 10.3. Estimated Residual Biomass for the US

Category	Estimated Annual Production (millions of dry tons/yr)	Source	Remarks
Forest Residues	109.9	1	Includes 65 mil. dt/yr from polewood, which normally would be left to grow for later harvest
Mill Residues	90.4	1	Includes residues currently used for other applications (fiber products, fuel, etc.)
Agricultural Residues	150.7	1	Includes corn and wheat stover only, assumes min. stover left to preserve soil quality (30–40% collected)
Urban Wood Wastes	36.8	1	Construction wood waste, yard trimmings, pallets
Municipal Sludges/ Biosolids	6.9	2	Solids remaining from treated domestic sewage
Food Residuals	21.9	3	Food residuals in municipal solid waste
Total	416.6		

Sources: (1) Walsh et al. (2000), (2) EPA (1998), and (3) EPA (2001).

From a scaling perspective, it is useful to consider how much energy biomass feedstocks represent if all this material could be collected. Given that biomass feedstocks have a distinct advantage over other renewable energy sources in that they can be chemically converted to liquid transportation fuels such as ethanol, methanol, and biodiesel, let's estimate how much imported oil could be displaced in the US by substituting biofuels. If one assumes that the harvesting, collection, transportation, and chemical conversion requirements for all residual feedstocks can be engineered in a high-energy and natural-resource-efficient manner, 400+ million tons of biomass per year could displace over 5 million bbl per day of imported petroleum. Of course, there are many questions and barriers. Environmental and ecological impacts, as well as economics, must all be carefully evaluated. In particular, water use; maintenance of soil nutrients and carbon levels; infrastructure requirements, including changes to farming practices, conversion plant location; and product distribution; and costs will all be important.

10.3 Chemical and Physical Properties Relevant to Energy Production

Fuel properties important to their use in clean and efficient production of high-quality energy forms are discussed in Chapter 7. These same requirements apply to biomass and fuels derived from biomass. Table 10.4 shows typical values for various fuel properties for biomass. Several properties are noteworthy. Biomass has a higher heating value on a dry basis of roughly 7,000–9,000 Btu/lb (i.e., roughly equivalent to that of a subbituminous coal or a lignite). Because most biomass is composed of lignin (H/C ratio 1.2), cellulose, and cellulosic compounds (hemi-celluloses H/C ratio 1.7), the H/C ratio of most biomass *must* lie between 1.2 and 2. Note in particular that "as received" (i.e., upon arrival from harvesting) biomass may contain large amounts of moisture—up to 75 wt% or three times the weight of the perfectly dry biomass. If not removed, moisture imposes dramatic penalties on the efficiency of biomass combustion processes because high-quality thermal energy must be expended to evaporate the moisture. The inorganic matter content can be significant even in wood—up to 2%. During combustion and gasification, these constituents can be vaporized from the biomass and later recondensed within the process, fouling heat exchange surfaces and degrading boiler performance. Some inorganics may be emitted as fine particulates that contribute to visibility loss and adverse health effects. Table 10.5 provides examples of trace inorganic element emissions from biomass combustion in modestly sized stoker furnaces for electricity production (9–17 MWe).

Note also that most biomass (other than waste tires) has low sulfur content (≤ 0.3 wt%) and, other than municipal wastes and animal residues, low nitrogen assays (≤ 0.3 wt%). These properties allow biomass to dilute the contributions of fuel sulfur and nitrogen to SO_x and NO_x emissions from fuels with much higher S and N inventories by blending those fuels with biomass (co-firing) before combustion. Unlike fossil fuels, biomass has high oxygen content (30–45 wt%), which dilutes the heating value per unit mass. The biomass carries around the weight of an extra element (the oxygen) which, unlike the carbon and hydrogen atoms of the biomass, does not combine with oxygen during combustion to release energy (by forming CO_2 and H_2O).

10.4 Biomass Production: Useful Scaling Parameters

Table 10.6 presents useful numbers for estimating the amount of biomass that can be produced by harvesting solar energy to grow crops. Such estimates must account for the flux density of solar energy to the earth's atmosphere, the fraction of that energy that actually reaches the growing location averaged to account for daily and season variations (the average solar incidence, which for the US is 13.6%, item D in Table 10.6), and the efficiency with which the plant's biochemical apparatus converts the solar energy into biomass (given as 1% in Table 10.6).

Table 10.4. Selected Properties of Biomass Relevant to Its Use as a Fuel

Property	Typical for Biomass
Energy Content (Heating Value)	7,000–9,000 Btu/lb (dry basis)
H/C Ratio (Atomic)	1.2 (Lignin), 1.7 (Cellulose)
Moisture	2–75%
"Volatiles"	65–90%
Mineral Matter (Ash)	0.2–2% (Wood) 25% (Mun. Solid Waste)
Sulfur	≤ 0.3%
Nitrogen	≤ 0.3% (Wood) 1.2% (MSW) 2.4% (Animal Waste)
Oxygen	30–45%

Table 10.5. Trace Element Emissions from Wood-Fired Boilers[a]

Trace Element	Amount (PPM by weight)[b]
Arsenic	90–230
Beryllium	< 20
Cadmium	10–190
Chromium	75–520
Copper	500–1,700
Lead	300–1,300
Manganese	2,000–13,000
Nickel	55–1,500
Zinc	6,200–26,000

Source: Bain and Overend (1992a).

[a]Data are for 9–17 MWe sized spreader stoker-fired boilers.

[b]PPM of particulate catch.

Table 10.6. Useful Numbers on Biomass Production and Utilization

A. Energy content	7,000–9,000 Btu/lb, on a dry basis, for most biomass
B. US solid refuse production	5 lb/person-day
C. Solar constant	1.94 cal/cm^2-min
D. US average solar incidence	13.6% of solar constant
E. Conversion efficiency of high yield "Energy Crops"	1% of solar incidence
F. (C) to (E) imply energy yields of	7 Tons/Acre-Yr. = 1.12×10^8 Btu/acre-yr
G. Water requirements for (F)	~200–300 lb. H_2O/lb biomass ~20 inches of rain per year

Sources: Energy Fact Book (1977), Fraser et al. (1976).

10.5 Thermal Conversion of Biomass

10.5.1 Biomass to electricity

Options for generating electricity from biomass include those listed below.

- Co-firing with coal to reduce sulfur emissions and smooth a transition to reduced fossil dependency
- *Repowering* (i.e., backfitting an existing generating station to switch to biomass fuel, or add a biomass-fired generator to an existing fossil-fueled unit)
- Direct combustion in a Rankine cycle to raise steam to operate a turbine generator
- Various configurations of combined cycles in which the biomass is first gasified and the gas then combusted to generate steam or, in a gas turbine, to provide motive power with the option for extracting additional electricity from the waste heat
- Thermal or hydrothermal conversion of the biomass to other fuels that, after substantial cleaning, are combusted to generate steam or motive power, or fed to a fuel cell for direct conversion to electricity

Table 10.7 provides data on the number and total electric generating capacity of various categories of biomass-to-electricity installations in the US (ca. 1998). The total installed capacity of about 10 GWe represents roughly 1.4% of the total US electric generating capacity of about 700 GWe. Much of the wood-fired capacity and all of the pulping liquor capacity represent captive uses in the forest products sector. These installations offset electricity that this industry would otherwise have to purchase from the national electricity grid.

Table 10.7. Biomass-To-Electricity Plants Installed in the US

Type of Biomass	Number of Installations	Total Capacity, MW
Wood	259	5,332
Pulping Liquor	6	443
Bagasse and Other Agricultural Residuals	39	669
Digester Gas	61	112
Landfill Gas	174	583
Tires	3	69
Total (Above + Other Sources)	678	10,006

Adapted from T. C. Schweizer (1998) "Renewable Power-Industry Status Overview," EPRI Report No. TR-111893, Table 5.2, p5–5, Palo Alto, CA.

These data document substantial growth from the roughly 200 MWe of installed capacity in 1978. However, other nations appear to be far more aggressive in their use of biomass as an energy source. Clean and efficient biomass energy production has become a priority for many developed and developing countries. Brazil, for example, supplied roughly 20% of its energy from biomass in 1996. Moreover, the mid-1990s saw appreciable biomass gasification technology efforts in, for example, Brazil (the World Bank's Global Environmental Facility Gasifier Project), and Scandinavia (e.g., the Varnarmo Bioflo project in Sweden). The more measured stance by US electric utilities reflects several factors, including the low efficiencies of traditional biomass-fired boilers (about 20–25%, probably due to high native moisture and inadequate pre-drying) versus up to 60% for natural gas-fired combined cycles. Some observers propose that electricity deregulation without environmental externalities imposed has been a disincentive to wider market penetration of biomass because of the generally higher cost of electricity compared to fossil and nuclear sources. Simply stated, deregulation leads wholesale vendors to seek the lowest possible electricity prices from their suppliers (i.e., the generator communities), which means that generators seek the lowest cost means of generating electricity. Deregulation has also spelled the end for many utility and government-based subsidy programs, as well as legislative mandates, all intended to stimulate wider adoption of biomass-to-electricity technologies. The lack of commercial-scale demonstrations of potentially enabling technologies, such as integrated gasification combined cycles including hot-gas cleanup and NO_x management, has been a further caution to the generally conservative and cost-conscious utility community— even among those responsible for the roughly 50% of US kWhrs that are not subject to deregulation.

In response to low conversion efficiencies, a major thrust of R&D activities spearheaded by the US Department of Energy has been to improve thermal technologies for conversion of biomass to electricity and to increase the fraction of US electricity currently supplied by biomass. Examples of thermal approaches are dedicated combustion of wood and of mixtures of wood and municipal solid waste, co-firing of biomass and coal, and gasification of the biomass followed by combustion of the resulting gas in a combined gas turbine and steam turbine power-generation cycle.

Figures 10.3 and 10.4, respectively, show schematic drawings of two biomass gasification technologies for production of electricity supported by the US DOE and the private sector: a steam/air or oxygen gasifier for conversion of sugar cane waste (bagasse) to electricity in Hawaii, and a coupled fluidized bed combustor (FBC)/ fluidized bed gasifier (FBG) to be interfaced to an existing 50 MWe generating station in Vermont. A major technical challenge to gasification for electricity is so-called hot-gas cleanup, or the removal of alkali metals and other trace inorganics from the gases generated in the gasifier and doing so without the need to cool down the gas prior to feeding it to the combustor. If gas cooldown were necessary, it would dramatically lower the overall efficiency of converting the biomass to electricity, in effect defeating the major purpose of the combined cycle. Note that the FBC/FBG cycle is designed to eliminate the need for an outboard hot gas cleanup system by incorporating the cleanup within the coupled fluidized beds. To succeed, all thermal cycles must handle deposits and emissions of alkali and other trace metals and of particulates and meet regulatory standards for NO_x emissions.

A further challenge is matching biomass availability to electricity demand on a sustainable basis. Harvesting costs are such that dedicated farms for supplying biomass must be within a 25-mile radius of the electric generating station (see Bain and Overend, 1992a). For example, with current growth rates and reasonable assumptions about land use and spacing between growth facilities, this would allow sustainable operation of a 150 MW_e generating plant near Sioux City, South Dakota, as depicted in Figure 10.5.

10.5.2 Biomass to fuels

Thermal and hydrothermal processes can also be used to convert various biomass feedstocks into gaseous and liquid fuels. Pyrolysis involves thermal treatment in the absence of oxygen to gasify the biomass to carbon monoxide and hydrogen (syngas). This mixture could then be chemically converted to liquid and gaseous fuels using suitable catalysts. Alternately, food processing wastes that have high levels of fats and oils can be easily hydrolyzed to produce low-Btu gas and a high-quality biodiesel-grade liquid fuel (Appel et al., 2003; Adams et al., 2004; Roberts et al., 2004).

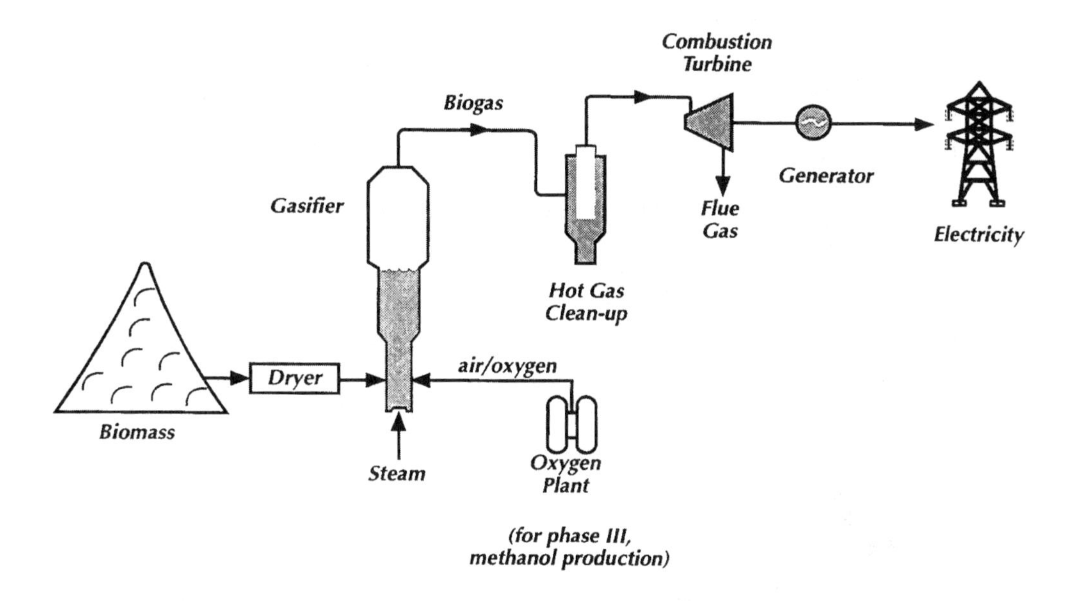

Figure 10.3. Schematic of Hawaii Gasifier Project for production of electric power from bagasse gasification. Source: US DOE Biomass Power Program.

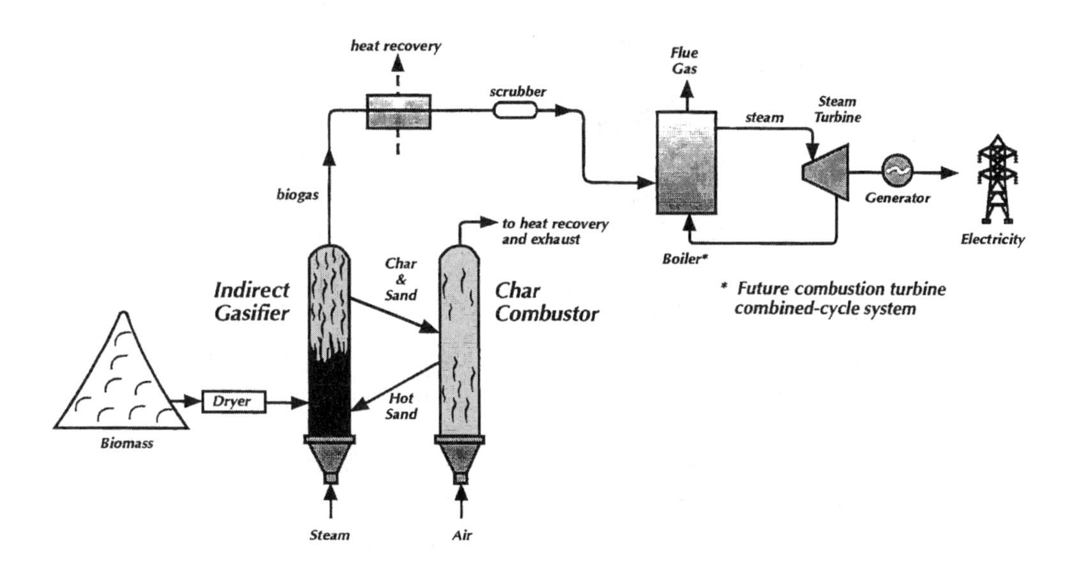

Figure 10.4. Schematic of Vermont Gasifier Project for production of electric power from biomass gasification. Source: US DOE Biomass Power Program.

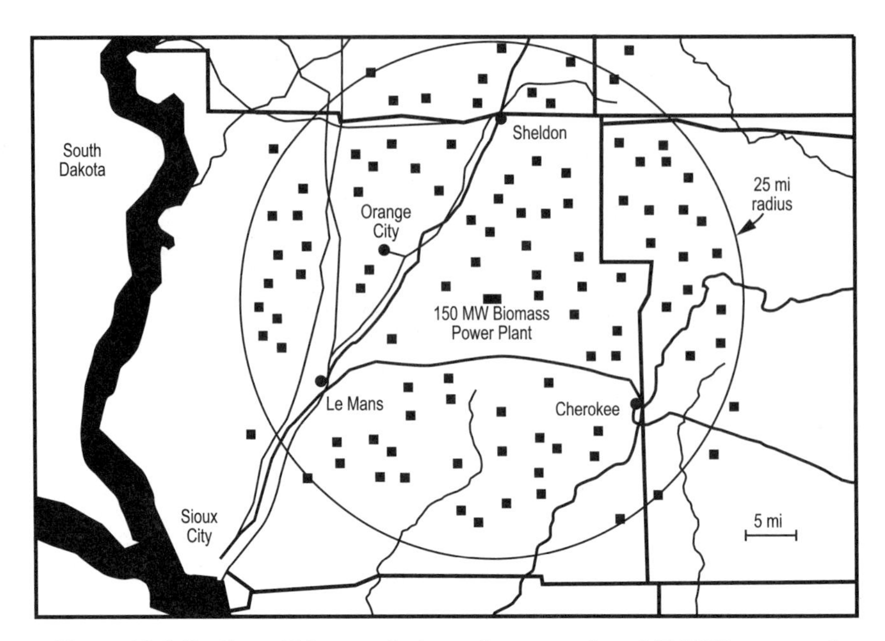

Figure 10.5. Dedicated biomass fuel supply system for a 150 MW_e power plant. Source: Bain et al. (1992a).

Despite the mixed record of these approaches, there are a number of factors that may catalyze appreciable expansion of the use of biomass for producing electricity and other energy forms over the next one to two decades.

- The need to manage huge quantities of solid waste in large urban areas
- "Green" pricing schemes adopted by some utilities to allow consumers, for a higher rate, to specify that their electricity, or a certain percentage of it, be generated from non-fossil, non-nuclear sources
- Voluntary or legislatively mandated programs to reduce atmospheric emissions of fossil fuel–derived CO_2
- Industrial or governmental policies that incentivize "self generation" (i.e., in-house production, rather than "sales generation" and extra-mural purchase of electricity and other energy forms). For example, the US pulp and paper industry has become almost 55% energy self-sufficient by using wood wastes, etc., to furnish energy.
- There is considerable potential to produce more biomass for energy purposes through recovery of residues and growth of energy crops (Table 10.3) and through harvesting municipal wastes (Problem 10.3).

- There is appreciable potential for further expansion of biomass to energy use in the forest products industry, e.g., repowering or replacing existing US pulp and paper electric power sources with biomass gasification combined cycle (BGCC) and black liquor gasification combined cycle (BLGCC) plants could generate a potential 8 GWe. (Black liquor, a waste product from making cellulose from wood, is a suspension of lignin substances plus residues of the inorganic chemicals used in extracting the cellulose from the lignin.) In addition, forest residues and residues from pulp and paper making in the US could fuel 30 GWe by BGCC.

10.6 Bioconversion

10.6.1 Introduction

Bioconversion or biochemical processing refers to the direct or adaptive use of the chemistry of living things to transform one substance to another. Fermentation is a bioconversion process known for centuries as a means to transform carbohydrates (sugars) to ethyl alcohol (ethanol or grain alcohol). It is the basis for production of a host of beverages, as well as ethanol for use as fuel. Bioconversion is appealing because it accepts feed materials (including fossil fuels) that vary appreciably in chemical composition and generates a range of useful products. Moreover, bioconversion enables the astonishing base of existing and rapidly increasing human understanding in biology and biochemistry to be applied to the manufacture of fuels and other energy-intensive products, such as chemicals. This provides scientists and engineers with new tools to devise processes that will run at milder conditions, synthesize chemically complex products from structurally simple starting materials, and mesh well with the chemical reactivity of biomass (see also Section 10.9). Two disadvantages of bioconversion in fuels manufacture are dramatically slower rates than thermal processes and the need to separate desired products from dilute mixtures. Slower rates translate into lower throughputs per unit time or the need for large process vessels, resulting in higher capital costs and/or higher plant footprint. Products recovery from dilute mixtures consumes energy and increases operating expenses.

10.6.2 Biogas

Biogas, approximately 50 vol% each of methane (CH_4) and carbon dioxide (CO_2), is produced by the decay of various forms of wet biomass in the absence of air (anaerobic digestion) (Larsen, 1993). Figure 10.6 depicts the overall process chemistry.

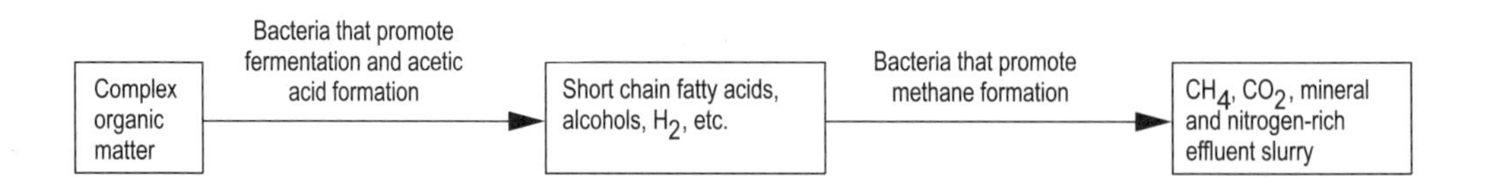

Figure 10.6. Schematic of the overall process chemistry for production of biogas by anaerobic digestion of wet biomass.

Even though CO_2 has zero heating value (for combustion in air or oxygen), biogas is still a high-quality fuel gas because pure methane has an HHV of about 1,000 Btu/SCF, giving typical biogas an HHV of 500 Btu/SCF. Biogas can be generated from animal and human wastes, sewage sludge, and crop residues but not lignin (the highly aromatic fraction of wood, Section 10.1). By-products of biomass production include a nitrogen-rich sludge that can be used as fertilizer and some pathogenic bacteria. Biogas is produced extensively in China and India. It is estimated that there are millions of small digesters ("biogas producers") in these two countries. Figures 10.7 and 10.8 show typical digesters widely used for biogas production from wet biomass in China and India, respectively. Biogas is also generated in more industrialized countries where perhaps 5,000 digesters are used to process animal wastes from stockyards and sewage from municipalities. Major goals in the operation of biogas producers are waste management and fertilizer generation. Another important issue is long-term operation and maintenance of small-scale digesters. It is relatively easy for them to fall into disrepair.

Digesters for biogas production operate at 35–55°C. Important process variables are the pH of the water-biomass-product mixture (broth) in the digester, the feed rate and elemental C/N ratio in the feed, the stirring intensity of the broth, and the time spent in the digester by the solids and the fluid. These two times may well be different. Larsen (1993) provides biogas production rates of 0.2 Nm^3 gas/(m^3 of digester volume-day) for village digesters in China and India (Figures 10.7, 10.8) and 4–8 Nm^3/m^3-day for larger scale digesters processing dilute industrial and municipal wastes. Biogas costs (Larsen, 1993) are given as 0.7–1.1, 5.8, and 11.6 \$/million Btu for industrial, village, and household size digesters, respectively.

In the US and many other countries, the largest biogas producers are landfills, which are now designed for the capture and use of the methane they generate from waste decomposition. In addition, higher efficiency anaerobic digesters are available for use in a variety of agricultural and industrial waste treatment processes (Roos and Moser, 1997).

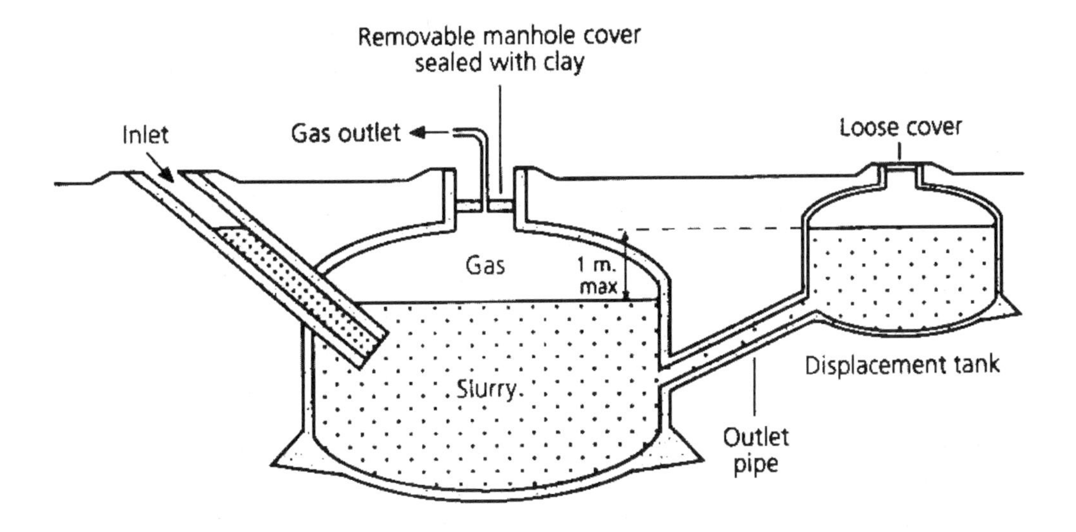

Figure 10.7. Fixed dome small scale anaerobic digester for biogas production from wet biomass (widely used in China). Source: Gunnerson and Stuckey (1986). See also Larson (1993). Reprinted with permission of *Annual Review of Energy and the Environment*.

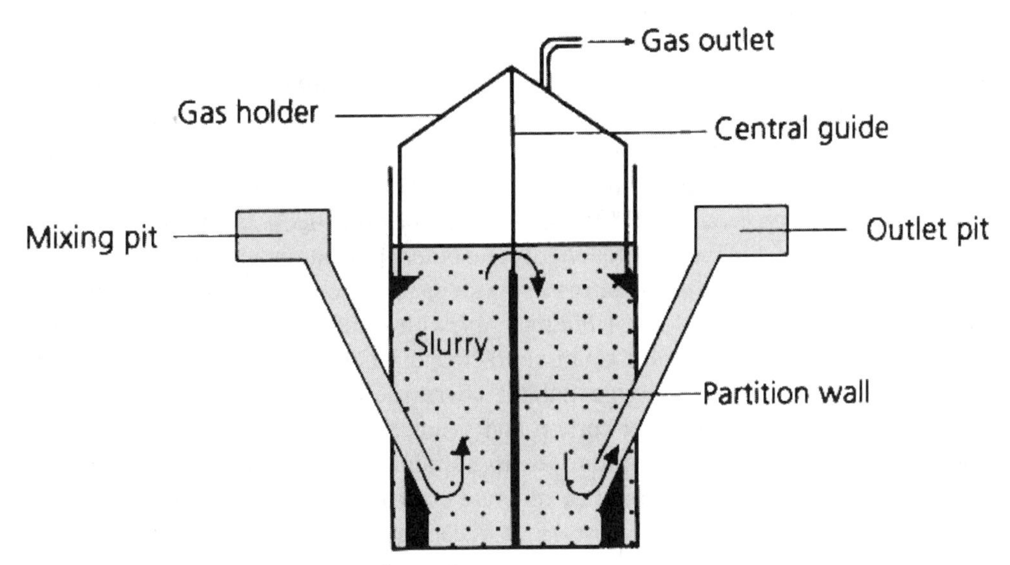

Figure 10.8. Floating cover small scale anaerobic digester for biogas production from wet biomass (widely used in India). Source: Gunnerson and Stuckey (1986). See also Larson (1993). Reprinted with permission of *Annual Review of Energy and the Environment*.

10.6.3 Fermentation ethanol from corn and cellulosic biomass

A schematic flowsheet depicting the major chemical and physical processes in producing ethanol by hydrolysis and fermentation of cellulosic biomass is presented in Figure 10.9. This technology has attracted considerable interest because it transforms biomass to ethyl alcohol, a high-quality liquid fuel for automotive transport and stationary gas turbines for electric power generation and a useful chemical intermediate. Fermentation of ethanol from biomass, particularly grains such as corn, has also generated appreciable controversy because of the poor energy balance (low first law efficiency), high production cost, and concerns over displacement of land and crops badly needed for food. Further, transport of ethanol by tankers (truck and train) for long distances also carries a significant energy penalty. Proponents of this approach point to the ability to convert a renewable resource to premium, energy-intensive products, the availability of considerable land that could be allocated to feedstock growth for energy production without interference with food production, and the potential to improve ethanol yields and process economics through advanced technology. (Table 10.8 provides useful data on US land availability and fermentation ethanol yields.)

10.7 Environmental Issues

Environmental issues in biomass energy are noted by Anath et al. (1981) and Braunstein et al. (1981). More recently, Larson (1993) identifies several environmental challenges to evaluate in the assessment and implementation of biomass energy options (Table 10.9). Environmental control becomes more challenging with smaller installations, such as residential wood stoves. Newer units are equipped with catalytic converters to reduce adverse emissions. Figure 10.10 depicts several US environmental regulations that affect electric power generation from biomass. Larsen (Table 10.10) identifies several adverse environmental impacts that could arise from expanded growth of biomass for energy production. On the positive side of the ledger, carefully managed growth and harvesting of biomass for energy and other applications can be used to restore and sustain forests and other sensitive ecosystems. One concern is the potential for ecological degradation and loss of biodiversity with some high-growth schemes that would use monoculture and short-rotation practices, leaving little land area or time for establishing sustainable and stable habitat (Braunstein et al., 1981). Importantly, biomass is a renewable source of carbon and a sink for atmospheric CO_2. Moreover, biomass can be converted to premium-quality liquid and gaseous fuels and chemicals, and biomass can be directly burned to generate electricity and high temperature heat. Thus, well-crafted plans for the growth, harvesting, and utilization of biomass are important in designing comprehensive strategies to reduce atmospheric buildup of greenhouse gases while preserving options for the supply and use of clean energy and energy-intensive consumer products.

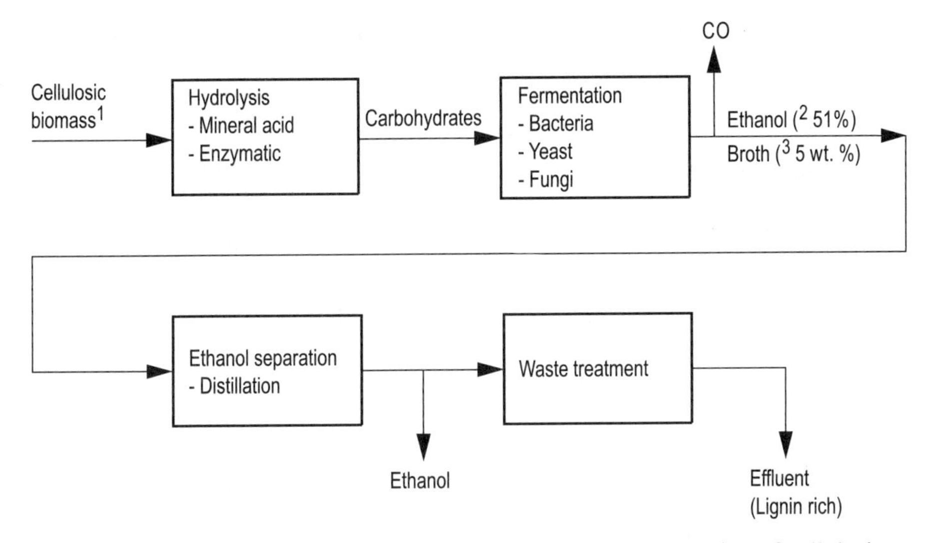

Figure 10.9. Schematic flowsheet for hydrolysis and fermentation of cellulosic materials to ethanol. Adapted from a discussion by Lynd (1996).

Notes:

1. Poor for softwoods conversion
2. Conversion of carbohydrate mass to ethanol
3. Typical ethanol concentration in broth

Table 10.8. Data on Land Availability and Product Yields for Production of Ethanol by Fermentation of Corn or Cellulosic Biomass

Federally idled US cropland	60 million acres[a]
Possibly available for energy crops	35 million acres[a]
Ethanol yield fermentation of cellulosics, advanced technology	107.7 gallons/ton[a]
At 8.4 ton/acre → 905 gallons/acre[a]	
Ethanol yield from corn fermentation (large plant)	275 gallons/acre[b]

Sources: [a]Lynd (1996) and [b]Pimentel (1991).

Table 10.9. Potential Adverse Environmental Impacts of Biomass Production and Utilization for Energy

1. Land, water, and nutrient consumption
2. Pollution from growing and harvesting
3. Effluents from thermal conversion processes
4. Combustion emissions
   - Centralized steam, electricity generation from refuse-based fuels:
     – Trace hydrocarbons (PAH), dioxins, furans
     – Metals
     – HCl
   - Wood stoves and fireplaces
     – PAH
     – Other complex organics
     – Particulates
5. CO_2 management
   - If fossil and biomass consumption offset by new biomass growth

Source: Larson (1993).

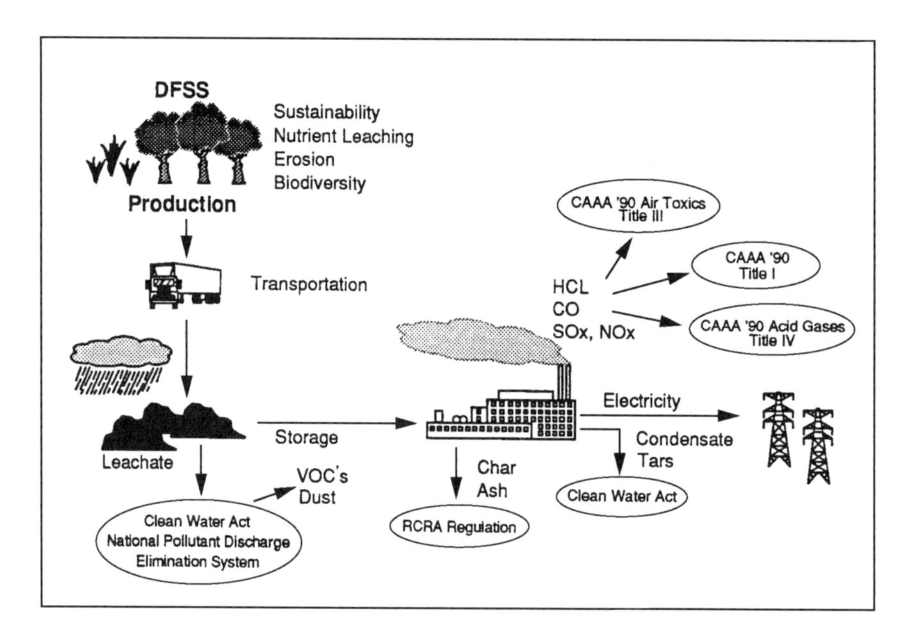

Figure 10.10. Examples of US environmental protection regulations relevant to the conversion of biomass to electric power. Source: Bain and Overend (1992a).

Table 10.10. Potential Environmental Problems from Expanded Growth of Biomass for Energy Production

Natural Crops

- Deforestation and loss of CO_2 sinks
- Loss of biodiversity
- Soil erosion
- Desertification
- Siltation of rivers
- Loss of agricultural productivity

Energy Crops

- Impact of growing only one plant species for multiple crop cycles
- Fertilizer contamination of ecosystems
- Changes in land use patterns
- Biodiversity modification or deterioration

Source: Larson (1993).

10.8 Economics

A major factor in biomass energy economics is the high cost of producing and harvesting the biomass. A 1978 study by the Institute of Gas Technology (IGT) reports estimates of feed costs as a percentage of final product cost for various biomass-to-energy processes ranging from 7% for manure to SNG, to 86% for kelp to SNG (Klass, 1978). When feed costs are tempered by collateral factors, the economics of energy from biomass improve substantially. One example is municipal or food processing wastes, where the cost of biomass is actually negative because waste collectors are paid to remove the biomass. A second case involves the forest products industry, where wood is harvested to manufacture higher-end products, such as pulp and paper or building products, and wastes from these manufacturing operations are a "free" energy feedstock. Use of such "waste" biomass to supply process energy needs has enabled some forest products manufacturing installations to become essentially energy self-sufficient. Likewise, when production and harvesting tariffs cannot be offset, the cost of energy from biomass can increase substantially. For example, on an equivalent performance basis (comparable miles per gallon), fermentation ethanol from biomass costs two to three times more than gasoline or diesel oil. Proponents maintain that progress in technology will lower this cost differential and that reduced dependence on fuel imports and lower emissions of fossil fuel-derived CO_2 will provide unaccounted-for economic benefits that will further offset the ethanol price premium. Trindade (2000) provides an excellent review of the bioethanol program in Brazil as well as other bioenergy systems.

10.9 Enabling Research and Development

R&D is needed to assure low environmental impacts from biomass production, harvesting, and utilization to improve the quality and supply of biomass raw materials and to reduce the cost of electricity and fuels from biomass. (Environmental challenges are discussed in Section 10.7.) Considerable attention is being directed to improving the efficiency of biomass-to-electricity production (Section 10.5) and of biomass conversion to fuels and chemicals (see NREL Web site). Two disadvantages of biomass as an energy crop are low production yields and substantial harvesting costs. In response, some current research (see ORNL Web site) is focusing on improving energy crop yields from the benchmark value of 7–9 dry tons/(acre-yr).

10.10 Disruptive Technology

Here we discuss a technological concept that, if successful, may overcome the problem of high harvesting costs while directly converting biomass to high-end fuels and energy-intensive products. The idea (Figure 10.11) is to create living plants that function as miniature manufacturing facilities for premium liquid fuels. These plants would use their internal biochemical apparatus, plus sunlight, to convert CO_2 and water to gasoline and diesel fuels that could be harvested automatically by tapping the plants at regular intervals. Thus, instead of harvesting the entire plant for direct combustion or conversion to clean fuels, only high-end liquids directly usable in current automotive engines would be collected. Ideally, the plants would be designed to grow without chemical fertilizers, to grow on poor-quality land, including deserts, and to readily accept saline or other brackish waters as their source of H_2O. A further variation would involve plants compatible with terraced terrain to facilitate liquid harvesting by gravity flow to centralized collection corridors and holding sumps.

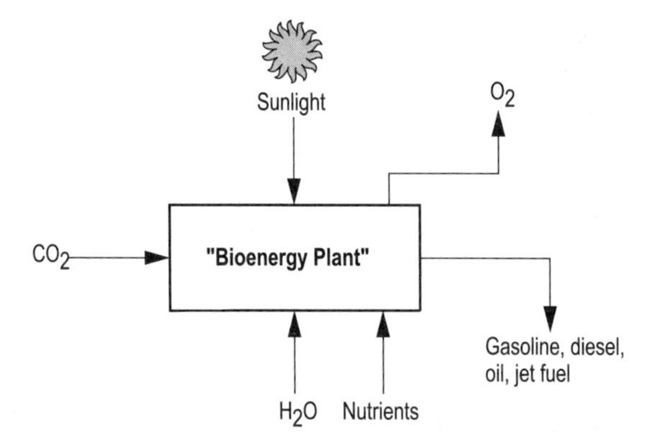

Figure 10.11. Example of a potentially disruptive biomass-to-energy technology: crops are naturally or genetically engineered to secrete premium liquid transportation fuels.

This goal makes sense for biomass but may not be achievable, especially since our automotive fuels are "complex mixtures" that include likely toxins for plants, but some version may evolve in the future. Various existing plants use their natural biochemistry to manufacture complex products. A compelling example for the proposed application is the rubber plant (*Hevea brasiliensis*), which "manufactures" natural rubber (*Hevea latex*), *cis*-1,4-Polyisoprene, a rather complex polymeric molecule consisting of approximately 5,000 isoprene monomers (Figure 10.12) (Shreve and Brink, 1997). The rubber is collected as a serum by tapping the plant. Calvin and co-workers (Calvin, 1978 and Nemethy et al., 1981) studied the growth of *Euphorbia lathyris* as a source of hydrocarbon-like liquid fuels. (Rubber plants are members of the *Euphorbiaceae* family.) Calvin (1978) reported that yields of an oxygen-containing hydrocarbon oil of about 10 bbl/acre-year could be obtained by cultivation followed by solvent extraction of *E. lathyris*. He noted in that 1978 paper that it would be possible to genetically modify plants to produce desired products. Indeed, the truly remarkable advances in biotechnology over the last 25 years, including novel drugs and genetically improved agro-seeds, suggest that there are tools ready to be applied to learning how to breed new plant species that provide these features without adverse environmental side effects. Success in this arena could lead to a future in which premium fuels, chemicals, and other energy-intensive consumer products would be produced in desired quantities from plant species tailored to grow in locations unsuited for food production. There are significant barriers, nonetheless. Initially, products are likely to be expensive and will be subject to the limitations of photosynthetic conversion with its inherently low efficiency for converting CO_2 and H_2O with sunlight to carbohydrates and other compounds. Tradeoffs involving land use will need to be resolved, as well.

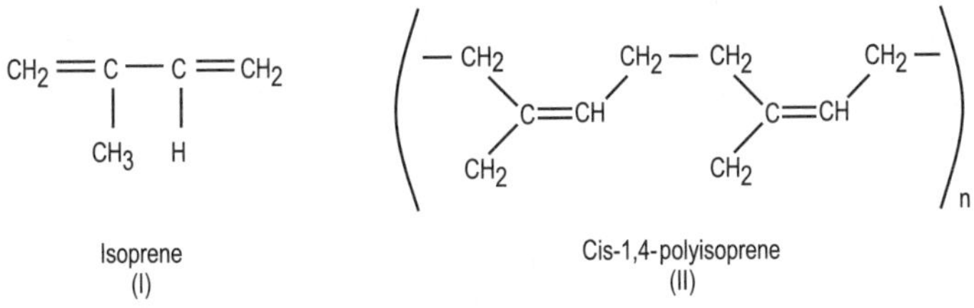

Isoprene (I)

Cis-1,4-polyisoprene (II)

Figure 10.12. Structural formulae for the monomer (fundamental building block) (I) and polymer (II) of natural rubber where $n \approx 2{,}500$. Note that the natural rubber plant first converts CO_2 and H_2O into (I) and then assembles many molecules of I in a particular orientation, to produce chain-like molecules containing roughly 2,500 of II linked together. Source: Shreve and Brink (1977). Reprinted with permission of The McGraw-Hill Companies.

10.11 Summary

Biomass currently contributes about 3% of total US energy consumption (i.e., roughly 65% the contribution of hydro and 40% that of nuclear). Since some biomass is used now commercially, it represents a slightly higher percentage of total primary energy use. The major uses are combustion of various biofuels, roughly 2 quads in the industrial sector and 0.5 quad in the residential and commercial sector. Biomass accounts for a considerably higher percentage of the total energy consumption of other countries. Biomass has several potential benefits in the electric power supply sector, including being dispatchable, accommodating gradual expansion in capacity, low SO_x emissions, co-firing with coal or other fossil fuels to smooth supply disruptions and facilitate gradual transitioning to reduced fossil dependency, and the potential to be CO_2 neutral. Other biomass-to-energy benefits include an opportunity for management of municipal, industrial, and agricultural wastes, a source of renewable (non-fossil) carbon energy, a means of providing low-carbon-emission traditional carbon products, and controlled growth management of ecosystems together with removal of CO_2 from the atmosphere.

In the long term, it may be possible to apply innovations in biotechnology to breed plants that directly convert sunlight to gasoline and other premium products. To achieve this goal will require methods, yet to be discovered, that utilize the modern tools of biotechnology, including genomics, metabolic engineering, and molecular-level understanding of biocatalysis. Environmental impacts of biomass growth, harvesting, and utilization deserve careful scrutiny. A major challenge for increasing the fraction of energy supplied from biomass is the economic disadvantage of biomass compared to nuclear and fossil sources, and its diffuse nature, which requires the use of large land areas.

References

Adams, T.N., and B.S. Appel. 2004. "Converting Turkey Offal into Bio-derived Hydrocarbon Oil with the CWT Thermal Process. Power-Gen Renewable Energy Conference, Las Vegas, NV.

Anath, K.P., M.A. Golembiewski, and H.M. Freeman. 1981. "Environmental Assessment of Waste-to-Energy Conversion Systems," in *Fuels from Biomass and Wastes*, eds. D.L. Klass and G.H. Emert. Chapter 27:505–515. Ann Arbor, MI: Ann Arbor Science Publishers.

Anon. 1979. "Biomass Potential in 2000 Put at 7 Quads." *Chem. and Eng. News,* 57(7):20.

Appel, B., and T. Adams. 2003. Technical bulletins and literature, West Hempstead, NY: Changing World Technologies, Inc.

Bain, R.L. and R.P. Overend. 1992a. "Electricity from Biomass: A Development Strategy." Solar, Thermal and Biomass Power Division, Office of Solar Energy Conversion, US DOE, Report No. CH10093-152 DE92010590.

Bain, R.L. and R.P. Overend. 1992. "Biomass Electric Technologies: Status and Future Development," in *Advances in Solar Energy: An Annual Review of Research and Development*, Vol. 7, ed. K.W. Boer. Boulder, CO: American Solar Energy Society.

Braunstein, H.M., F.C. Kornegay, R.D. Roop, and F.E. Sharples. 1981. "Environmental and Health Aspects of Biomass Energy Systems," in *Fuels from Biomass and Wastes*, eds. D.L. Klass and G.H. Emert. Chapter 26: 463–504. Ann Arbor, MI: Ann Arbor Science Publishers.

Calvin, M. 1978. "Green Factories." *Chem. and Eng. News*. 56:30–36.

Energy Fact Book. 1977. Prepared by Tetra Tech, Inc. Arlington, VA, under the direction of the Director, Navy Energy and Natural Resources Research and Development Office. XXI-1. (Available from NTIS, Report No. AD/A-038 802.).

EPA (U.S. Environmental Protection Agency). 2001. "Food Residuals Management Issue Paper: 2001." JTR Recycling Market Development Roundtable (www.epa.gov/htr/docs/food_res.pdf).

EPA (U.S. Environmental Protection Agency). 1998. "Biosolids Generation, Use and Disposal in the United States." *EPA530-R-99-009*.

Fraser, M.D., J.-F. Henry, and C.W. Vail. 1976. "Design, Operation and Economics of the Energy Plantation™." Symposium Papers, Clean Fuels and Biomass, Sewage, Urban Refuse and Agricultural Wastes, Orlando, FL.

Graboski, M. and R. Bain. 1979. "Properties of Biomass Relevant to Gasification," in *A Survey of Biomass Gasification, Volume II—Principles of Gasification*, ed. T.B. Reed. (Chapter 3) Report No. SERI-TR-33-239. Golden, CO: The Solar Energy Research Institute.

Gunnerson, C.G. and D.C. Stuckey. 1986. "Integrated Resource Recovery: Anaerobic Digestion," in *Technical Paper 49*. Washington, DC: World Bank.

Hall, D.O., F. Rosillo-Calle, R.H. Williams, and J. Woods. 1993. "Biomass Energy Supply Prospects," in *Renewable Energy: Sources for Fuels and Electricity*, eds. T.B. Johansson, H. Kelly, A.K.N. Reddy, R.H. Williams, and L. Burnham. Washington, DC: Island Press.

Holdren, J.P., G. Morris, and I. Mintzer. 1980. "Environmental Aspects of Renewable Energy Sources," in *Annual Review of Energy*, Volume 5, ed. J.M. Hollander. Palo Alto, CA: Annual Reviews, Inc.

Hottel, H.C. and J.B. Howard. 1971. *New Energy Technology—Some Facts and Assessments*. Cambridge, MA: The MIT Press.

Klass, D.L. 1978. "Energy from Biomass and Wastes: 1978 Update," in *Symposium Papers in Energy from Biomass and Wastes*, Chicago: Institute of Gas Technology.

Larson, E.D. 1993. "Technology for Electricity and Fuels from Biomass." *Annual Review of Energy and the Environment*, 21:403–465.

Lipinsky, E.S., D.A. Ball, and D. Anson. 1982. "Evaluation of Biomass Systems for Electricity Generation." Report No. AP-2265. Palo Alto, CA: Electric Power Research Institute.

Lynd, L.R. 1996. "Overview and Evaluation of Fuel Ethanol from Cellulosic Biomass: Technology, Economics, the Environment, and Policy." *Annual Review of Energy and the Environment*. 21:403–465.

Nemethy, E.K., J.W. Otvos, and M. Calvin. 1981. "Natural Production of High-energy Liquid Fuels from Plants," in *Fuels from Biomass and Wastes*, eds. D.L. Klass and G.H. Emert. Ann Arbor, MI: Ann Arbor Science Publishers.

OTA (Office of Technology Assessment). 1980. "Energy from Biological Processes." NTIS Report No. PB30-211477. Washington, DC.

Pimental, D. 1991. "Ethanol Fuels: Energy Security, Economics, and the Environment." *Journal of Agricultural and Environmental Ethics*, 4:1–13.

Reed, T.B. 1979. *A Survey of Biomass Gasification, Volume II—Principles of Gasification*, Report No. SERI/TR-33-239. Golden, CO: The Solar Energy Research Institute.

Rees, W.E. 2003. "Ecological footprints: a blot on the land." *Nature*, 421(6926):898.

Roberts, M., J. Williams, P. Halberstadt, and D. Sanders. 2004. "Animal Waste to Marketable Products." Natural Gas Technologies Conference, Phoenix, AZ.

Roos, K.F., and M.A. Moser, eds. 1997. *The AgSTAR Handbook*. EPA-430-B-97-615. Washington, DC: USEPA.

Schulman, B.L. and F.E. Biasca. 1989. "Liquid Transportation Fuels from Natural Gas, Heavy Oil, Coal, Oil Shale, and Tar Sands: Economics and Technology." A Report prepared by SFA Pacific, Inc., Mountain View, CA, for the National Research Council, Committee on Production Technologies for Liquid Transportation Fuels.

Shreve, R.N. and J.A. Brink, Jr. 1977. *Chemical Process Industries*, fourth edition. New York: McGraw-Hill.

Stenzel, R.A., B.T. Kown, M.C. Weelies, B.R. Gilbert, C.M. Harper, J.D. Ruby, Y.J. Kim, and R.T. Milligan. 1980. "Environmental and Economic Comparison of Advanced Processes for Conversion of Coal and Biomass Into Clean Energy." Report No. PB81-234239 by Bechtel National Inc. for the US EPA.

Thomas, R.J. 1977. "Wood: Structure and Chemical Composition," in *Wood Technology: Chemical Aspects*, ed. I.S. Goldstein. ACS Symposium Series 43(1). Washington, DC: American Chemical Society.

Trindade, S.C. 2000. "Beyond Petroleum and Towards a Biomass-based Sustainable Energy Future." *UNEP Industry and Environmental Journal*, 23(3), 31p revised.

US DOE Biomass Power Program. See www.eere.energy.gov/biomass

Walsh, M.E. et al. 2000. "Biomass Feedstock Availability in the United States: 1999 State Level Analysis." Bioenergy Information Network Web site. (bioenergy.ornl.gov/resourcedata/). [NB: Walsh et al. estimate a potential biomass feedstock of 510 million dry tons annually at a price of $50/dry ton or less.]

Weisz, P.B. 1978. A view of renewability based on a concept discussed by P.B. Weisz in *Energy and Society*.

Web Sites of Interest

www.nrel.gov
www.ornl.gov
www.usda.gov

Problems

10.1 In the renewable energy community, municipal solid waste is considered biomass. Explain qualitatively why we can write $A_g \approx A_N$ for municipal solid waste (see Equations (10-3) and (10-4)). Why are these quantities not equal exactly? Estimate whether a family saving all their municipal solid waste and burning it could harvest enough energy to heat their home. Use the following assumptions: family of four in a well-insulated New England home that burns 800 gallons per year of No. 2 home heating oil (140,000 Btu/gal); MSW production is 3 dry lb/person-day at a heating value of 8,000 Btu/lb (dry); furnace efficiencies: oil 90%; MSW 80%.

10.2 A stock of biomass (waste wood) is found to have a sulfur content of 0.1 wt% (dry basis) and a heating value on a dry basis of 8,000 Btu/lb. This fuel will be used to replace a Wyoming subbituminous coal which has a sulfur content of 1.0 wt% (dry basis) but a heating value (dry basis) of 10,000 Btu/lb. By how much will the emissions of sulfur dioxide, in lb per Btu, be lowered when the coal is replaced by the biomass?

10.3 Assume the population of the greater Tokyo area remains at its 2002 value of roughly 12 million people and that municipal solid waste is produced at a rate of 2 dry lb/person day, and that this material has a heating value of 7,500 Btu/lb. If 50% of this could be "harvested" for energy, by how much would it increase the total estimated 2050 biomass production rates for Japan, i.e., from recoverable residues and from biomass crops (Table 10.1)?

10.4 "Prove" line F in Table 10.6 using the data in lines A–E of that table. Discuss how the growth rate of biomass could be increased. What environmental and socio-political issues would need to be considered in translating your plan into practicality?

10.5 Chemical analysis of a sample of biogas from a digester gives the following results: CO_2: 44%; CH_4: 56% (all percentages are by volume). What is the HHV of this particular sample of biogas? Is this above or below average? It is later discovered that this analysis was in error and that the correct assay for this sample is CO_2: 44%; CH_4: 51%; H_2: 5%. What is the corrected HHV of the biogas? If the objective is to produce a gas with as much heating value per unit volume as possible, is hydrogen a good co-product? Assume the HHV of pure CO_2 and pure CH_4 are 0 and 1,000 Btu/SCF respectively. (Note: the presence of hydrogen is generally an indicator of abnormal digester operation.)

10.6 On a particular day, natural gas prices at Henry Hub are quoted as $2.70/SCF. How does this price compare with the cost of biogas from household, village, and industrial digesters? Express your results in $/million Btu and $/SCF. Assuming a barrel of oil contains 5.7 million Btu, further express this natural gas price and these three biogas costs as a barrel of oil equivalent (i.e., the amount of money needed to supply the energy content of a barrel of oil by each of these four gaseous fuels).

10.7 US residents consume roughly 10 million barrels per day of gasoline for transportation fuel. Using the ethanol yields projected to be attainable from advanced fermentation technology (Table 10.8), how many acres would be needed to be devoted to cellulosic biomass production to supply 10% of the miles traveled on gasoline using fermentation ethanol as the automotive fuel? Note that ethanol provides roughly 2/3 the miles per gallon attainable from gasoline. Compare your answer with the total area of France and of the state of California.

10.8 The US has a lot of federally owned land. Unfortunately, uncontrolled natural forest fires destroy large areas in the western US every summer. In 2002, about 1,000,000 acres of standing timber in national forests were consumed. Some consideration is being given to improved management practices that could produce electrical power from residual forest thinnings. Estimate the lost energy content of burned US forests during 2002. Assuming the US average electricity demand load is about 300,000 MWe, how much forested land would be needed to produce all the country's power? Is this a sustainable alternative? A few facts to consider:

- The total forested area on US federal lands in the lower 48 states is about 600 million acres with a standing stock density of about 100 dry metric tonnes of wood per acre

- Woody plants and trees capture solar energy via photosynthesis at an average rate of about 0.8 W/m^2 which corresponds to producing about 5–10 dry tons of biomass per acre annually with an average heating value of 8,000 Btu/dry lb. Note that 1 acre = 43,560 ft^2 = 0.405 hectare = 4,047 m^2
- A representative heat-to-work conversion efficiency of a biomass fired electric power plant is about 35%

10.9 Massachusetts is considering growing energy crops and burning them to produce electric power as part of a renewable energy deployment initiative that state lawmakers approved some time ago. Assuming the state's average electricity demand load is about 4,000 MW_e, how much forested land would be needed to produce all the state's power? Is this a feasible alternative? The total land area in Massachusetts is 8,284 square miles or 21,385 km^2 (see facts listed in Problem 10.8).

Geothermal Energy 11

11.1 Characterization of Geothermal Resource Types

11.1.1 Definition in general

In general terms, geothermal energy consists of the thermal energy stored in the earth's crust.[1] Practically speaking, the exact specification of a geothermal resource depends in part on the specific application or energy service that is provided. Thermal energy in the earth is distributed between the constituent host rock and the natural fluid that is contained in its fractures and pores at temperatures above ambient levels. These fluids are mostly water with varying amounts of dissolved salts. Typically, in their natural *in situ* state, they are present as a liquid or supercritical fluid phase but sometimes may consist of a saturated or superheated steam vapor phase. Most geothermal resources presently usable for electrical power generation result from the intrusion of magma (molten rock) from great depths (>30 km) into the earth's crust. These intrusions typically reach depths of 0 to 10 km.

Geothermal fluids of natural origin have been used for cooking and bathing since before the beginning of recorded history, but it was not until the early 19th century that geothermal energy was harnessed for industrial purposes. One of the first cases was the use of geothermal steam for heating evaporating ponds at the boric acid works near Larderello, Italy. In 1902, electricity was first produced using geothermal steam at Larderello. Since that time, other developments, such as the steam field at The Geysers, California, and the hot water systems at Wairakei, New Zealand; Cerro Prieto, Mexico; and Reykjavik, Iceland, and in Indonesia and the Philippines, have led to an installed world electrical generating capacity of more than 10,000 MWe and a direct use, nonelectric capacity of more than 100,000 MW_{th} (thermal megawatts of power) at the beginning of the 21st century.

The source and transport mechanisms of geothermal heat are unique to this energy source. Heat flows through the crust of the earth at an average rate of almost 5.9×10^{-2} W/m^2 [1.4×10^{-6} cal/(cm^2-s)] on the average. The intrusion of large masses of molten rock can increase this normal heat flow locally but, for most of the continental crust, the heat flow is due to upward convection and conduction of heat from the mantle

1 Author's note: Much of what is included in this chapter updates earlier review articles and books co-authored with other colleagues. In that context, we want to acknowledge their contributions, in particular: *Geothermal Energy* by H.C.H. Armstead (1983); *Heat Mining* by H.C.H. Armstead and J. W. Tester; "Geothermal Energy" (1982), a chapter in *Encyclopedia of Chemical Technology* by J.W. Tester and C.O. Grigsby (1997); a review article by J.E. Mock, J.W. Tester, and P.M. Wright "Geothermal Energy from the Earth: Its Potential Impact as Environmentally Sustainable Resource." *Ann. Rev. of Energy Environ.*, 22, 305–356, (1997).

and core of the earth and heat generated by the decay of radioactive elements in the crust, particularly isotopes of potassium, uranium, and thorium. Local and regional geologic and tectonic phenomena play a major role in determining the location (depth and position) and quality (fluid chemistry and temperature) of a particular resource. For example, regions of higher-than-normal heat flow are associated with tectonic plate boundaries and with areas of geologically recent volcanic events (younger than about one million years in the case of large magmatic intrusions of 10–100 km^3). This is why people frequently associate geothermal energy with specific places like Iceland, New Zealand, or Japan (plate boundaries) or with Yellowstone National Park or the Larderello field in Italy (recent volcanism), and they neglect to consider geothermal energy opportunities for other regions.

In all cases, certain conditions must be met before one has a viable geothermal resource. The first requirement is accessibility. This is usually achieved by drilling to depths of interest—frequently using conventional methods similar to those used to extract oil and gas from underground reservoirs. The second requirement is sufficient reservoir productivity, which depends on the type of geothermal system that is being exploited. Normally for indigenous resources, one needs to have sufficient quantities of hot, pressurized natural fluid contained in a confined aquifer with high rock permeability and porosity to insure long-term production at economically acceptable levels. In other situations, one only needs to have a sufficiently hot rock reservoir that can be artificially stimulated to produce a system for extracting energy at acceptable rates.

The term "hot" is a relative term as it depends on the specific application. The geothermal resource actually spans a continuum in at least three dimensions: temperature, depth, and permeability/porosity (see Figure 11.1). Low-grade systems involve temperatures from just above ambient to about 150°C, which may or may not contain natural fluids. The geothermal gradient $\nabla T \equiv \partial T/\partial Z$ quantitatively establishes the relationship between temperature (T) and depth (Z). Generally, low-grade systems have lower gradients and are located deeper in the crust. If fluids are not present, then *in situ* permeabilities and porosities are intrinsically low or the reservoir system is located above the natural water table. The converse is true if natural fluids are available for heat extraction. High-grade resources are at the other end of the continuum, characterized by hot fluids contained in high permeability and porosity host rock and at relatively shallow depths (typically less than 3 km). Geochemistry also plays a role. Reservoirs that spontaneously produce fluids of low salinity under artesian pressure and low concentrations of dissolved non-condensable gases are easier to exploit. If such systems are formed with high temperatures (>250°C) at shallow depths, then the resource grade is at its highest condition.

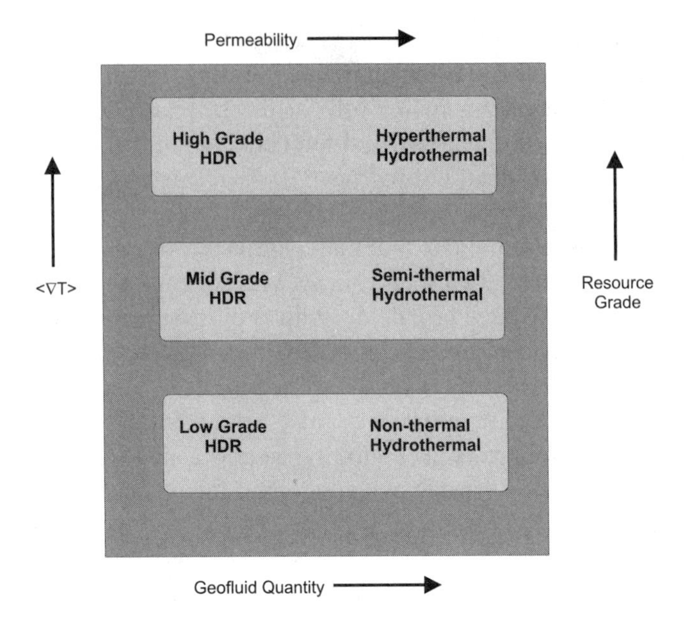

Figure 11.1. The geothermal continuum characterizes the resource in three dimensions: temperature gradient (∇T), geofluid quantity, and permeability/porosity.

Thermal energy is extracted from the reservoir either by coupled transport processes, such as convective heat transfer in porous and/or fractured regions of rock, and conduction through the rock itself. If the physical extent and heat content of a deposit are large enough and the deposit is near the surface, a heat extraction scheme must be designed based upon the hydrologic and geologic situation, including the *in situ* rock properties. After such a scheme is successfully implemented, hot water or steam is normally produced at the surface and converted into a marketable product (e.g., electricity, process heat, or space heat). Also, spent fluids are normally reinjected to avoid surface discharges that could have environmental impacts. Many aspects of geothermal heat extraction are similar to those found in the oil, gas, coal, and mining industries. Because of these similarities, equipment, techniques, and terminology have been borrowed or adapted for use in geothermal development, a fact that has accelerated the development of geothermal resources.

Commercial utilization of the resources requires that the process be economically competitive. Consequently, the commercial geothermal systems developed to date have been limited to a relatively few, accessible, high-grade deposits scattered throughout the world. Improvements in extraction technology to lower production costs or increases in the prices for conventional fuels would make lower-grade geothermal resources commercially feasible.

There are two basic types of geothermal systems: those that spontaneously produce hot fluid (steam or water) from a reservoir and those that do not. Systems that produce fluid spontaneously are obviously easier to exploit. However, these natural hydrothermal systems are not always located near a user. For example, of the 11 most promising new geothermal areas in the US, six are within about 400 km of Los Angeles, California. The other five are all more than 400 km from any city with a population greater than 300,000. Furthermore, nine of the 10 most populous cities in the US are more than 800 km from any known high-temperature, natural geothermal system. Economic utilization of these remote hydrothermal systems necessarily involves the generation and transmission of electricity rather than direct use of the geothermal fluid for space or process heating. A system of the second type, one lacking sufficient quantities of hot fluids, presents somewhat different challenges with respect to heat extraction, but because of the broader-based location of such resources, the remoteness problem is partially alleviated.

Geothermal resources are commonly divided into four categories: hydrothermal, geopressured, hot dry rock, and magma. The characteristics of each are discussed in the sections that follow.

11.1.2 Natural hydrothermal systems

Systems that spontaneously produce hot fluids are called hydrothermal or convection-dominated. Hydrothermal systems require a source of heat (usually a magmatic intrusion at depth), formations with enough permeability to allow fluid mobility, an adequate supply of indigenous fluid (water or steam), sufficient contact surface, time for the fluid to be heated, and a return path to the surface (see Figure 11.2). These systems are frequently located in or near zones of recent volcanism and tectonism (mountain building) that are associated with the boundaries of crustal plates (Muffler, 1975, Healy, 1975, White, 1973, and Tamrazyan, 1970). In addition to providing sources of heat, crustal plate margins are areas where tectonic forces have caused considerable fracturing of rock and high local permeability. Indeed, it is believed that the contributions of faults and joints to *in situ* permeability are far more important than from intergranular (matrix) permeability.

Water or steam in hydrothermal systems is usually of meteoric origin, typically located at depths of 1–4 km at temperatures up to 350°C (Mock et al., 1997, Kavlakoglu, 1970, and Craig, 1963). Water falls as rain or snow and percolates downward through sediments or fissures until it comes to a heat source. There, it is heated and buoyantly rises toward the surface where it usually appears as geysers, hot springs, fumaroles, or solfataras.

If the pressure on the fluid in the reservoir is insufficient to prevent boiling, a vapor phase forms in the upper portion of the reservoir. This vapor phase consists of steam (often superheated or dry) and noncondensable gases that separate from the liquid phase. Most of the dissolved minerals concentrate in the liquid phase, leaving the vapor relatively free of dissolved solids. Hydrothermal systems that produce superheated steam are called vapor-dominated. These systems are the easiest and most efficient to exploit for electricity generation because the steam can be transported to and expanded directly in low-pressure steam turbines with little risk of scaling or turbine blade damage.

High-grade, vapor-dominated systems occur rarely and several have been developed for base load electricity generation. The major ones are The Geysers field in California, the Larderello field in Italy, and the Matsukawa field in Japan.

Systems that are pressurized above the vapor pressure do not form a vapor cap, and production from these types of fields consists of hot water or a mixture of hot water and steam. Such liquid-dominated resources are common and widely distributed. Because dissolved solids remain in the liquid phase, liquid-dominated systems can pose serious scaling and corrosion problems in surface piping systems and in production and injection wells. If the fluids are not reinjected, they can pose a waste disposal problem. Usually, the fluid in liquid-dominated systems is flashed (or subjected to a pressure drop that allows a separate vapor phase to form) and separated so that the vapor phase can be piped directly to the turbine generator. The liquid may be flashed more than once (multistage flashing), but ultimately a significant fraction of the hot water and minerals remains to be disposed of by release to surface waters, reinjection, evaporation, or some alternative use, such as process or space heating.

High-quality liquid-dominated fields containing relatively low-salinity water under pressure at temperatures up to 350°C have been identified in many regions, including the western US, New Zealand, Iceland, Indonesia, the Philippines, Italy, Turkey, and several countries in eastern Africa. Liquid-dominated resources with fluid temperatures ranging from about 80 to 300°C are being used commercially throughout the world for generating electricity and for providing process and residential heat.

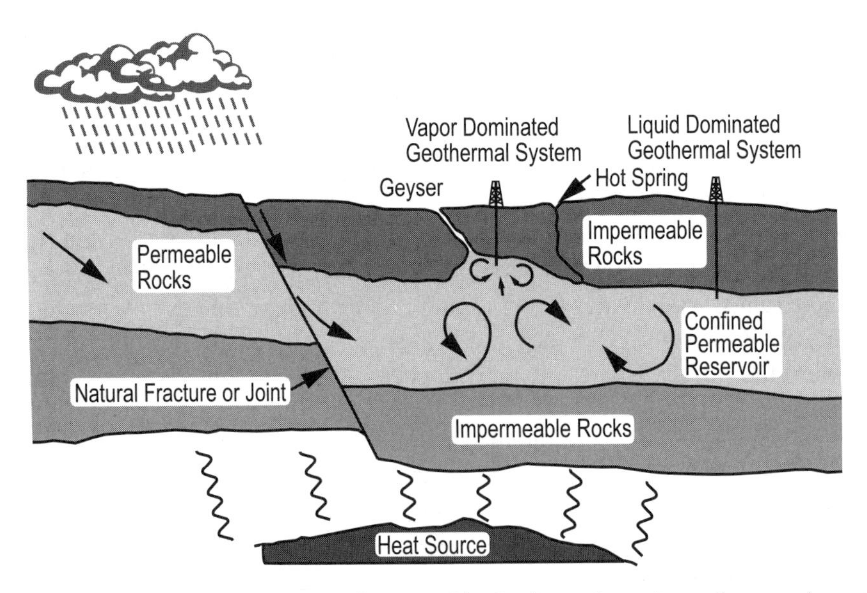

Figure 11.2. Typical features of a natural hydrothermal geothermal reservoir system. Adapted from Mock, Tester, and Wright (1997). Reprinted with permission of *Annual Review of Energy and the Environment*.

As shown in Figure 11.2, many aspects of a typical hydrothermal geothermal system are similar to that of a typical oil reservoir, and a number of techniques for measuring field properties, such as permeability, porosity, reservoir size, and fluid content in oil reservoirs, have been applied directly to hydrothermal systems with success. Geophysical prospecting involving heat flow, electrical resistivity, gravity mapping, and other methods provide exploration data for inferring the extent and depth of a thermal anomaly (Wright, 1995, Rowley, 1982). Geochemical techniques, including geothermometry, stable isotope geochemistry, analysis of dissolved gases, and ratios of major elements in hot springs are also used to locate and define geothermal fields.

11.1.3 Geopressured systems

Large sediment-filled reservoirs containing pore fluids under confining pressures much greater than the hydrostatic head are called geopressured basins. The sediments usually consist of sandstones and shales along with other sedimentary rock types that are more or less interbedded. Generally, the fluid contained in massive sandstone sediments is under near-hydrostatic pressure and is mobile, owing to relatively high permeability. The fluids contained in sandstone-shale and massive shale facies are generally overpressured by up to 60 MPa (600 bar) at temperatures of 150–180°C. These higher temperatures are caused, at least partially, by the low thermal conductivity of shale, which gives an apparent high-geothermal gradient. The heat flow is actually near normal. Pressures in the Texas and Louisiana geopressured reservoirs are expected to approach lithostatic pressure. The lithostatic pressure gradient in this Gulf Coast region is about 23 kPa/m (1 psi/ft).

In addition to hot water under pressure, geopressured reservoirs usually contain methane in solution. For the purposes of estimating the geopressured resource, the geofluids are usually assumed to be saturated with methane with (6.9–8.9) × 10^3 m^3 CH_4/kg water (about 40–50 SCF of CH_4/bbl water). Field tests of a geopressured well in southern Louisiana indicate this may be the case. The relatively high salinity (100,000 ppm total dissolved solids–TDS) and expected subsidence effects of geopressured resources may seriously restrict their utilization.

11.1.4 Hot dry rock

Throughout most of the world, one or more of the necessary components of a hydrothermal reservoir is missing. In particular, the reservoir rock is often hot enough (≥ 200ºC) but produces insufficient fluid for commercial heat extraction either because of low formation permeability or the absence of naturally contained fluids. Such formations form a part of the geothermal resource referred to as hot dry rock (HDR) or enhanced geothermal systems (EGS). From the early 1970s to about 2000, the term HDR was used to represent both conduction-dominated, low permeability and low porosity formations, and unproductive hydrothermal systems at the margins of or within known geothermal fields. More recently the US DOE has adopted the EGS classification to more rationally represent this resource.

In principle, HDR systems are available everywhere just by drilling to depths sufficiently deep to produce rock temperature useful for heat extraction—usually taken to be >150°C for producing electricity and >50–100°C for direct heat use. Therefore, for baseload electric power generation in low-grade, low-gradient regions (20–40°C/km), depths of 4–8 km are required, while for high-grade, high-gradient systems (60°C/km), 2–5 km are sufficient. HDR resources have the potential to provide a high quantity of primary energy with a resource base in excess of 10^6 quads, and, if developed economically, could provide an alternative to fossil and nuclear fuels.

Techniques for the extraction of heat from low permeability HDR have been under investigation in a number of laboratories worldwide (Sato et al., 1995, 1994, Parker, 1989, Armstead and Tester, 1987, Kapplemeyer et al., 1991, and Batchelor, 1984). For low permeability formations, the basic concept is simple: drill a well to sufficient depth to reach a useful temperature, create large heat transfer surface areas by hydraulically fracturing the rock, and intercept the fracture(s) with a second well. By circulating water from one well to the other through the fractured region, heat can be extracted from the rock (as depicted in Figure 11.3).

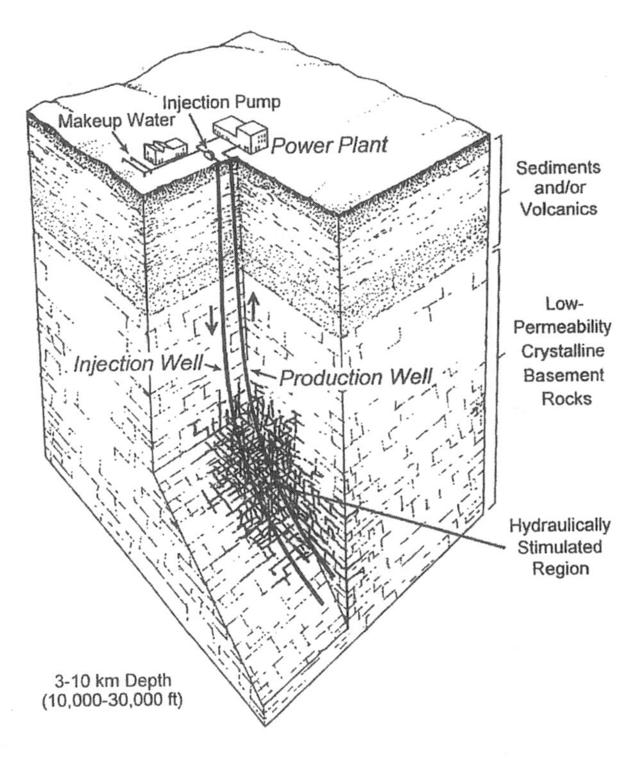

Figure 11.3. Hot dry rock reservoir concept for low-permeability formations. Source: Mock et al. (1997). Reprinted with permission of *Annual Review of Energy and the Environment*.

Although the concept is simple, many questions remain to be answered before it can be considered economically feasible. Significant progress has been made, however, in understanding reservoir characteristics, including fracture initiation and propagation, the specific details of fracture geometry (area, aperture distribution, connectivity, etc.), its thermal drawdown, water loss rates, flow impedance, fluid mixing, and fluid geochemistry in field tests of an HDR reservoir in low permeability crystalline rock (Armstead and Tester, 1987). In addition to hydraulic fracturing, permeability and surface area can be created by explosive fracturing, chemical leaching, and thermal stress cracking (Armstead and Tester, 1987). Studies of explosive fracturing were conducted in the UK (Batchelor, 1984, 1987) and in the USSR (Diadkin and Pariisky, 1975, Aladiev et al., 1975).

If rock of sufficient natural permeability exists in a confined geometry, techniques similar to waterflooding or steam-drive employed for oil recovery might be used to advantage (Tester and Smith, 1977, Bodvarsson and Hanson, 1977). Other techniques for heat extraction, such as downhole heat exchangers, downhole pumps, or alternative methods of operation, such as injection-recovery (huff-puff), have also been proposed.

Because many of these methods involve the creation of permeability in rock, HDR systems are not limited to particular tectonic settings. Until the heat-extraction technology is suitably developed and the economics are demonstrated on a large scale, the first commercial HDR systems will be most likely confined to regions of high geothermal gradient (> under 50°C/km).

11.1.5 Magma

The magma resource consists of rock that is partially or completely molten, encountered at accessible depths. Considerable development of drilling and heat extraction concepts and technology must be accomplished before heat can be extracted usefully from magma. Central to the problem of extracting heat is the ability to find a body of magma at drillable depths. Unfortunately, many suitable magma systems are located within the boundaries of protected regions such as Yellowstone National Park in Wyoming. Drilling into magma requires the development of equipment, lubricants, and cements that can operate at 700–1,000°C and depths to about 7 km. While this depth is substantial, it is not beyond the capability of commercial drilling today.

Magma's high temperatures, normally greater than 650°C, make this resource particularly attractive for efficient generation of electricity or for high-temperature industrial process heat application.

11.1.6 Ultra low-grade systems

Direct use of low-grade geothermal energy usually manifests itself as a thermal energy source or sink for heat pump applications or for agricultural uses such as greenhouse heating or fish farming. Geothermal heat pumps (GHPs) normally operate at shallow depths (2–4m) where the earth's temperature is relatively constant. The coefficient of performance (COP) of a typical GHP is about 4 or more, meaning that 4 units of thermal energy are transferred for every 1 unit of electrical work (see

Bloomquist, 1981, Keller, 1977; and see Section 11.3.3 for more details). Thus, the efficiency of electrical energy utilization is increased more than four-fold using a GHP in such combined heat and power applications.

11.1.7 Markets for geothermal energy

Although the natural hydrothermal, magma, and geopressured systems have reasonably large resource bases and will be exploited when technical and economic conditions are favorable, their distribution worldwide is controlled by prevailing natural geologic conditions. Figure 11.4 illustrates the relationship between resource base and reserve for a hypothetical 40°C/km geothermal system. In terms of total energy use, electrical power generation is now and will continue to be for the immediate future the predominant end-use for geothermal resources. Essentially all generated electricity is base load and goes to the transmission grid. To date, only high-grade hydrothermal systems have been developed for commercial electric power generation and district heating applications. Hydrothermal resources have been utilized most effectively when resource location and user demand coincide. Such is the case in developing countries like Indonesia and the Philippines, in parts of Central America, and in the western US.

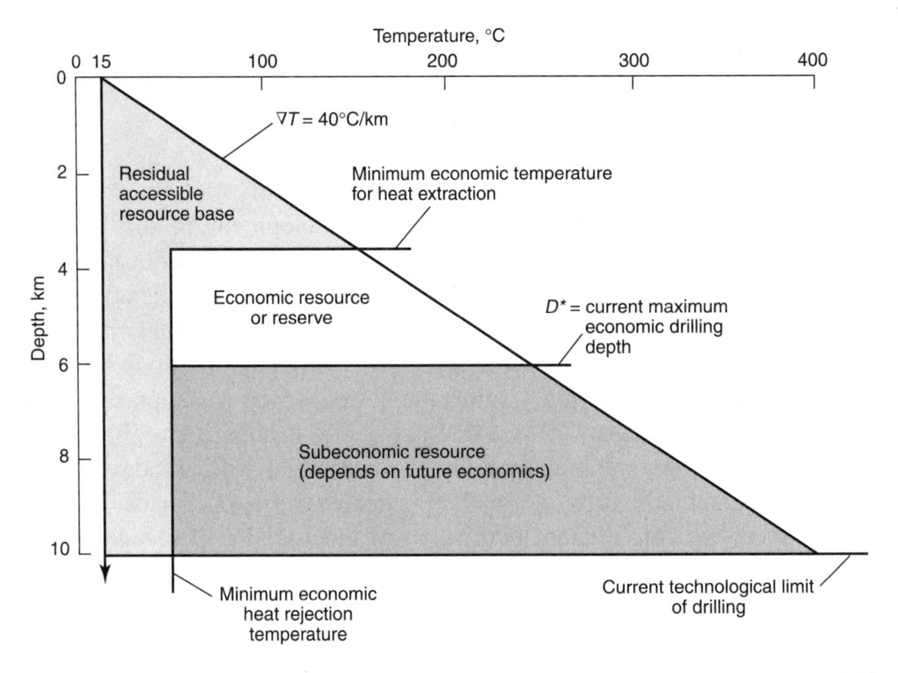

Figure 11.4. Idealized depth versus temperature profile for a hypothetical 40°C/km geothermal resource. The resource base and economic resource or reserve regions are shaded in different tones to illustrate the factors limiting the portion of the total resource base that can be economically produced with a specified set of technologies and economic factors. Adapted from Tester and Grigsby (1980). Reprinted with permission of John Wiley & Sons.

Although geopressured and magma resources have substantial potential, in order for geothermal energy to become a major player in supplying energy worldwide in the next decade, HDR systems must be utilized. Fortunately, HDR is by far the largest, most widely distributed resource, but appropriate heat mining technology has not become commercially available. Competitive universal geothermal energy requires the development of advanced heat extraction techniques for HDR to stimulate production and improved drilling methods to reduce development costs so that even low-gradient regions can be economically developed.

Given the nature of geothermal energy in the form of hot water or steam, direct heat applications have persisted long before electricity was first generated from geothermal steam in 1902 in the Lardarello field. Low-temperature geofluids, where they are available, have seen widespread use for heating, cooling, and bathing for centuries (Armstead, 1983). More recently, even the thermal capacity of the earth at shallow depths is being utilized in operating GHPs. In fact, a fairly aggressive deployment program in the US in the 1990s increased the adoption of GHPs in residential and commercial buildings from 40,000 units per year in 1994 to about 400,000 units per year in 2000 (Mock et al., 1997).

Given the remoteness of many hydrothermal fields, electricity is mostly generated at the field site and tied into an existing grid system at an acceptable cost. Geothermal power plants have smaller capacities (about 50–100 MW) than nuclear and fossil-fuel plants and consequently can be used as baseload units in much larger total-generating-capacity systems. Of the 600,000 MWe of generating capacity in the US in 2000, only 2,500 MWe was provided by geothermal sources. With cost-effective, reliable geothermal systems available, a substantial market exists for both new and replacement generating capacity during the 21st century.

The situation with the nonelectric market is different. Utilizing geothermal space and process heat systems may require some retrofitting by the consumer. For example, a fuel-oil or natural-gas furnace usually is designed with in-house storage or is attached to an existing distribution network. The use of geothermal hot water or steam would require a completely new fluid-distribution system within the community. Heat pumps are the exception, as the geothermal part is developed at each building site.

Reistad (1975) discusses energy usage as a function of utilization temperature and concludes that about 40% of our annual fossil energy consumption is severely degraded thermodynamically. Typically, fossil combustion temperatures of 1,000–1,500°C are used to produce space and process heat at temperatures below 250°C. The distribution of energy consumed versus use temperature is illustrated in Figure 11.5 for the US. A wide variety of applications are contained in the 32 EJ (30 quads) of energy used below 250°C. Industrial and other applications of geothermal heat are discussed by Armstead (1983) and Lindal (1973).

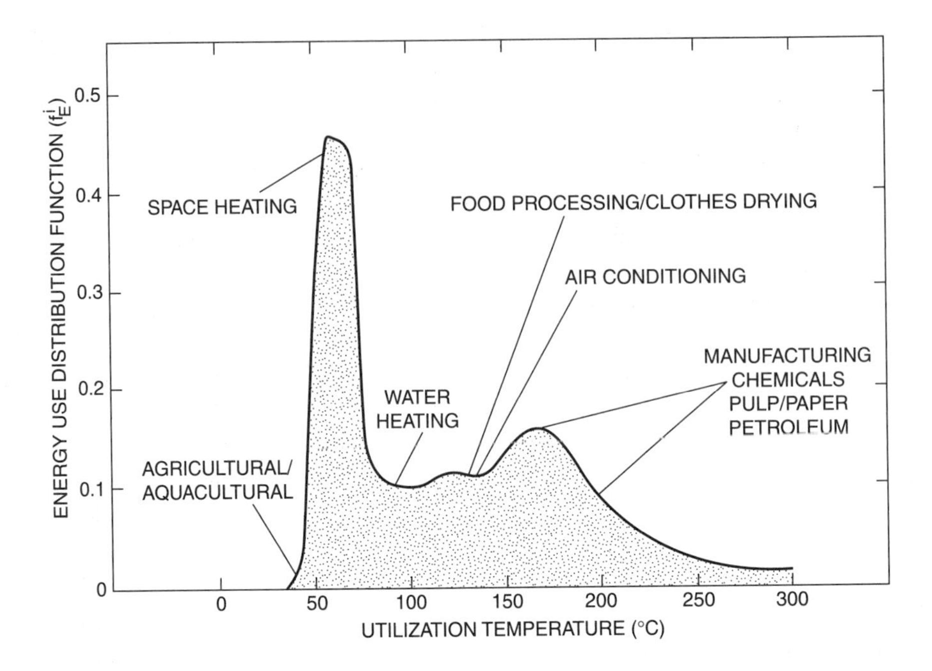

Figure 11.5. Fractional energy use distribution as a function of end-use temperature for non-electric applications below 300°C. The function f_E^i at T_i is simply the derivative of the cumulative energy use at that specific temperature T_i. Source: Tester (1982).

11.2 Geothermal Resource Size and Distribution

11.2.1 Overall framework and terminology

Geothermal resources are commonly considered as mineral resources like oil, gas, and coal. However, because they contain water and their production can affect ground water flows, geothermal resources may be classified as a particular type of water resource. In addition, there is considerable non-uniformity in the terminology used to describe geothermal resources. For example, the geothermal resource base has been defined several ways: (1) to include all stored heat above 15°C to 10 km depths (White and Williams, 1975); (2) the potentially useful heat in the earth's crust at temperatures greater than 80°C and depths less than 6 km (Armstead, 1983); or (3) the total heat contained in subsurface rocks and fluids to a depth of 3 km and at temperatures above 15°C (EPRI, 1978). The most widely accepted terminology for geothermal resources is that proposed by Muffler and Guffanti (1978), described below:

Resource Base. All of the thermal energy contained in the earth's crust, whether its existence is known or unknown and regardless of cost considerations.

Accessible Resource Base. All of the thermal energy between the earth's surface and a specific depth in the crust beneath a specified area and referenced to a mean annual temperature.

Resource. Thermal energy that could be extracted at costs competitive with other forms of energy at a foreseeable time, under reasonable assumptions of technological improvement and economic favorability.

Reserve. That part of the geothermal resource that is identified and also can be extracted at a cost competitive with other commercial energy sources at present.

It would be misleading to imply that this or any terminology is accepted throughout the geothermal industry. Indeed, these definitions are somewhat simplistic compared to the definitions applied to the same terms in the oil, gas, or mining industry (see Figure 11.4 for a slightly modified version). For example, in the US in 2000, an oil reserve can secure a bank loan, but a geothermal reserve can not. Currently, there are simply not enough cases of economic exploitation of geothermal energy for it to be considered a good financial risk. Nonetheless, at certain high-grade sites in the US, Europe, and Asia, investments in commercial developments have been significant.

11.2.2 Quality issues

Investment in the extraction of natural resources usually proceeds when the likelihood of return on invested capital is high. Consequently, the quality of the resource needs to be carefully characterized. For geothermal resources, this includes depth to the reservoir, consideration of rock type, fluid composition and production rates, and the energy content of the reservoir. Typically, mineral deposits or ore bodies are evaluated according to the grade and accessibility of the deposit, processing costs, and distance to the nearest market. Based on the evaluation, a deposit might be classified as a reserve or a resource. A similar body of terminology is under development in the geothermal energy field; thus far, geothermal terminology has borrowed heavily from the other mineral-related fields.

Gaining access to the fluids or hot rock is key to assessing the quality of any geothermal resource. Usually depth is the critical parameter. Drilling costs typically scale exponentially with depth. Rock hardness and stability are also critical as they determine the ease of drilling through the formation.

The chemistry of natural geothermal fluids can pose significant challenges to utilization. For instance, the salinity content of a liquid-dominated, hydrothermal reservoir fluid can vary from a few hundred ppm total solids (drinking water quality; 100 ppm = 0.01% by weight) to a concentrated brine (>30% by weight solids). The fact that it is hot and under pressure contributes to difficulties in processing. The presence of dissolved gases, such as H_2S and NH_3, usually require special abatement procedures to meet air emission standards.

11.2.3 Resource base and reserve estimates

First, we will consider the heat content of a mass of rock at a specific depth relative to its planned use as primary energy as a measure of the resource base. Then, we will discuss how much stored thermal energy can be removed by either extracting a portion of the hydrothermal fluids contained in a natural reservoir or by extracting energy from a suitably engineered artificial reservoir in HDR.

Accessible Geothermal Resource Base. Estimates of all or part of the accessible geothermal resource base are numerous, and each estimate is based on slightly different assumptions. If the definition of accessible resource base as given above is strictly followed, the total amount of heat Q contained in rock beneath the earth's surface to a depth is expressed by:

$$Q = \sum A_i \left[\int_o^z (\Phi \rho_f C_{p,r} + (1 - \Phi) \rho_r C_{p,r}) (\nabla T_i \cdot Z + T^* - T_{ref}) dZ \right] \quad (11\text{-}1)$$

where Z = depth, km; T = rock temperature, °C; A_i = surface area having a characteristic regional gradient of ∇T_i; T^* = ambient surface crustal temperature, about 15°C; Φ = porosity of rock; ρ_f = fluid density, kg/m^3; ρ_r = rock density, kg/m^3; $C_{p,f}$ = fluid heat capacity, J/(kg·K); $C_{p,r}$ = rock heat capacity, J/(kg·K); ∇T_i = geothermal gradient, °C/km; and T_{ref} = reference temperature, °C.

Integrating Equation (11-1) for each unit of area A_i assuming a constant geothermal gradient and taking $T^* = T_{ref}$ = 15°C, yields:

$$Q_i = A_i [\Phi \rho_f C_{p,f} + (1 - \Phi) \rho_r C_{p,r}] \nabla T_i \frac{Z^2}{2} \quad (11\text{-}2)$$

Below about 3 km, Φ approaches 0, and Equation (11-2) can be simplified to give:

$$Q_i = A_i (\rho_r C_{p,r}) \nabla T_i \frac{Z^2}{2} \quad (11\text{-}3)$$

Given the distributions of land areas overlying each range of geothermal gradient, the resource base to a given depth can be calculated. Likewise, given the heat contained per unit area to a given depth, one can extrapolate to another depth to find the accessible resource base by simply substituting the new depth in Equation (11-2). Estimates of the worldwide accessible resource base commonly are referenced to a specific mean annual surface temperature and a specific maximum depth. For example, the Electric Power Research Institute (EPRI, 1978) uses 15°C and a depth of 3 km assuming (1) a normal gradient of 25°C/km for all non-geothermal areas, (2) a gradient of 40°C/km for 90% of the area of a country that lies inside a geothermal belt and (3) a gradient of 90°C/km for the remaining 10% of the area within a geothermal belt. Typical values for rock (r) and fluid (f) properties used in these calculations are: ρ_r = 2,000 – 2,800 kg/m^3, $C_{p,r}$ = 800 – 1,200 J/kgK, λ_r = 2.5 – 3.0 W/mK; ρ_f = 800 – 1,000 kg/m^3; $C_{p,f}$ = 4,000 – 4,200 J/kgK.

Comparisons of accessible resource base estimates are given in Table 11.1 for the US from several sources all referenced to a mean annual temperature of 15°C and a maximum depth of 10 km. Two important conclusions can be drawn from these estimates. First, the amount of energy is enormous. Total US energy consumption is currently 91.4 EJ/yr (87 quads/yr) and is increasing at a rate of about 1 EJ/yr (≈1 quad/yr). If only 0.1% of the accessible resource base can be economically extracted, this resource would supply the complete energy needs of the US at its current rate of increase for almost 170 years. Second, probably 97% of the energy contained in the upper 10 km of the earth's surface is contained in the rock itself as a conduction-dominated resource and, therefore, requires an HDR extraction scheme.

Geothermal Reserves. Several factors limit the amount of energy that can be economically extracted from the earth. Some of these are now and always will be fixed; Figure 11.4 shows a depth-versus-temperature profile for a resource having an average gradient of 40°C/km. The resource base is depicted by the large triangle enclosed by lines extending from 15°C at 0 km to 400°C at 10 km depth. The economic resource or reserve is contained within the resource base by imposed minimum temperature and maximum drilling depth constraints. The first major limitation to the estimates of the amount of heat that can be economically extracted is the effective geothermal gradient. This largely determines the potential for a given area. If the gradient is small, the depth to the minimum acceptable initial rock temperature is large. A minimum initial rock temperature (T_{ref}) of 150°C is assumed to provide reasonable thermodynamic conversion efficiencies (see Section 11.3.3). A normal gradient (25°C/km) requires drilling to at least 5.4 km to reach 150°C. If the gradient is 40°C/km, as depicted in Figure 11.4, only 3.4 km deep holes are required. Another major limitation is the maximum economic drilling depth D^*. For an assumed maximum drilling depth of 6 km, only gradients above about 40°C/km represent substantial amounts of stored energy. The final limitation is imposed by heat rejection conditions for the geothermal fluid in the power plant (taken here to be 50°C). Imposing these limitations on an area of 1 km^2 with a gradient of 40°C/km, the total geothermal resource to 6 km (assuming average values of $C_{p,r}$ = 772 J/(kg·K) and ρ_r = 2,500 kg/m^3) is:

$$\left[\frac{Q_i}{A_i}\right] = \rho_r C_{p,r} \left[\frac{\nabla T_i Z^2}{2} + (T^* - T_{ref})Z\right]_{3.4km}^{6km} \tag{11-4}$$

$$\left[\frac{Q_i}{A_i}\right] = 0.81 \text{ EJ/km}^2 \text{ or } 0.77 \text{ quads/km}^2$$

The total energy available to the economic depth D^* = 6 km is 0.81 EJ/km^2 or 0.77 quads/km^2. The fraction of this amount that can be extracted economically will depend on the prices for other forms of energy and the degree of success in developing commercial-sized systems.

Estimating the economic resource or reserve for a particular type of geothermal resource shares much in common with estimating reserves for oil and gas and other mineral stocks. Technological improvements to reduce the cost of producing geothermal heat or electricity will increase reserves, while lower energy prices for fossil and fissile fuels will decrease reserves. At this stage of development, we can only say that the geothermal resource base is huge and warrants R&D expenditures to increase the fraction of the resource base that can be produced in competitive world energy markets.

Table 11.1. Geothermal Worldwide Resource Base Estimates—Total Thermal Energy Content in Place (Q)

	Total Q in 10^3 quads[1]	
Resource Type	**US**	**World**
Hydrothermal (vapor and liquid dominated)	9.6	130
Geopressured[4]	170	540
Magma[3]	500 – 1,000	5,000
Hot Dry Rock[2]	30,000	105,000
Moderate to high-grade (∇T > 40°C/km)	6,000	26,500
Low-grade ($\nabla T \leq 40$°C/km)	24,000	78,500

1. Q = 1 quad ≡ 10^{15} BTU ≈ 10^{18}J with
   2003 worldwide commercial energy demand = 400 quads and
   2003 US commercial energy demand = 100 quads
2. includes hydraulic and methane energy content
3. to depths of 10 km and initial rock temperatures >650°C
4. to depths of 10 km and initial rock temperatures >85°C

Sources: Mock et al. (1997), Armstead (1983), Armstead and Tester (1987), Duchane (1994), and Rowley (1982); US figures based in part on USGS estimates (Muffler and Guffanti (1978), Sass (1996) and Sass et al. (1993)).

11.3 Practical Operation and Equipment for Recovering Energy

11.3.1 Drilling and field development

The first step in developing a geothermal field evolves through evaluation of potential sites using all available lithologic, hydrologic, geophysical, and geochemical data. Locating and evaluating a reservoir can be difficult and expensive, particularly in regions that have not been drilled extensively or where surface manifestations of geothermal activity are not present. Once a site has been located, it must be purchased or leased for its geothermal rights before exploratory and production drilling can begin.

Drilling and completing wells for geothermal energy applications involve methods similar to those used in drilling for oil or gas, but are generally more difficult and expensive because formation temperatures are higher and the rock itself is typically harder to drill, with many fractures, lost circulation regions, or more abrasive mineral matter. In fact, drilling-related costs are usually the single largest cost component in any geothermal development. Drilling is inherently risky, including as much "art" as science (not to mention sophisticated engineering and an extraordinary infrastructure of specialized disciplines). Given this situation, one might expect the cost of drilling and completing a well to a specified depth would be highly variable. While this is certainly true to some degree, there are general trends.

Average costs for drilling tend to scale exponentially with depth whether they are conventional oil and gas wells or geothermal wells. Tester and Herzog (1991) and Armstead and Tester (1987) examined a substantial set of drilling cost data to provide guidelines for estimating costs. Figure 11.6 provides a summary of recent individual well cost data for geothermal, oil, and gas wells, along with several correlations. On the semi-logarithmic coordinates of Figure 11.6, a straight line results for average cost of a geothermal well versus depth such that an empirical exponential relationship for well cost as a function of depth results:

$$\Phi_{well} = \text{cost per well in } 10^6\$ = (2 \text{ to } 3)\,(0.082)\,\exp\,[7.51 \times 10^{-4}\, Z] \qquad (11\text{-}5)$$

Figure 11.6 demonstrates that, without exception, all hydrothermal and HDR well costs are higher than a typical oil or gas well drilled to the same depth. The line for the "oil and gas average" is based on costs for hundreds of wells compiled by the Joint Association Survey (JAS).

Completed well diameters for geothermal wells range from 8 to 12 inches (20–30 cm), which is somewhat larger than found for oil and gas wells. Larger diameters increase costs, as do the slower penetration rates often encountered in geothermal drilling (see Armstead and Tester, 1987). For depths between 1 and 5 km, geothermal wells are about 2–4 times as expensive as a comparable oil or gas well. At larger depths, we expect geothermal well costs to approach those for oil and gas wells because ultra-deep oil and gas drilling frequently encounters a similar set of problems as geothermal drilling, including hole collapse, hard rock, and higher temperatures and pressures. As we can see from the ultra-deep JAS data, the range of costs gets large for wells deeper than 6 km, with the most expensive wells about a factor of 10 more costly than the least expensive ones.

11.3.2 Reservoir fluid production

To justify the cost of developing a geothermal field, estimates of the total amount of extractable energy and the production rate must be made. Typically, computer models are used to simulate performance for a given set of reservoir properties. Two general categories of reservoir models, one involving matrix and one fracture-dominated flow, are usually encountered. The first type applies to high-matrix permeability hydrothermal formations and the second to fractured, low-permeability media common in HDR resevoirs. A simple example of each case with schematic thermal history curves is shown in Figure 11.7.

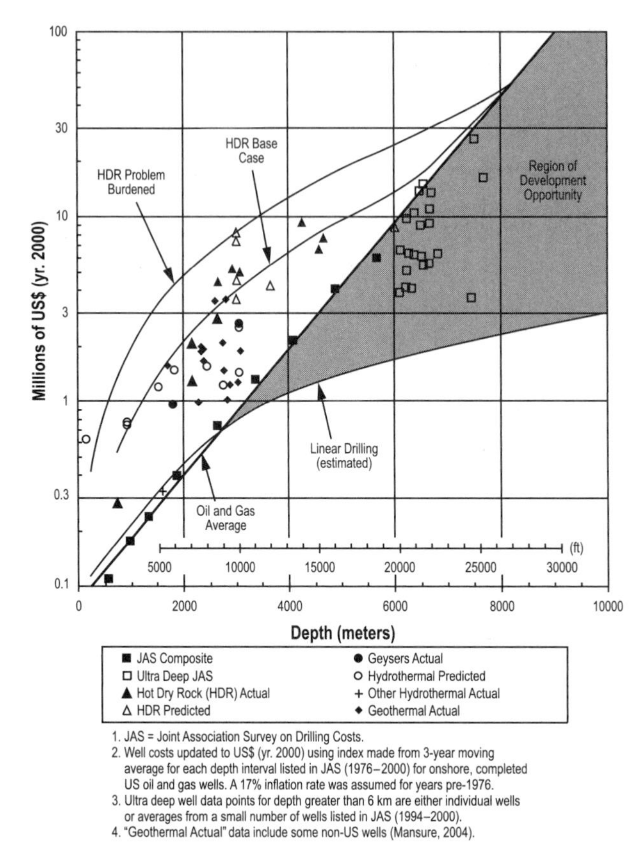

Figure 11.6. Drilling costs for completed wells showing the effect of advanced linear drilling technology. Updated by Augustine, personal communication (2004).

Porous and permeable reservoir. Extraction of heat from hydrothermal systems now in operation is straightforward. Because the reservoirs are pressurized, the fluid passes directly to the surface under artesian flow when the reservoir is penetrated. Productivity of the wells may occasionally be enhanced by stimulation at the wellbore or downhole pumping, but this is often unnecessary. Efficient production of a pressurized field is largely a matter of well spacing for a specified set of *in situ* properties. When the pressure of such a field drops to the point where it is insufficient to produce hot fluid, stimulation techniques are used such as injecting water to repressurize the system and literally force fluid to move through the porous rock to be heated as it flows toward the production well. Such water-drive flooding methods are commonly used to enhance oil recovery as well. Because reinjection of fluid is required in geothermal applications today, water-drive methods can be used throughout the life of the field. Also, in pressurized geothermal systems, most of the thermal energy is contained in the rock and not the pore fluids, so enhancement methods can greatly extend the lifetime of a field.

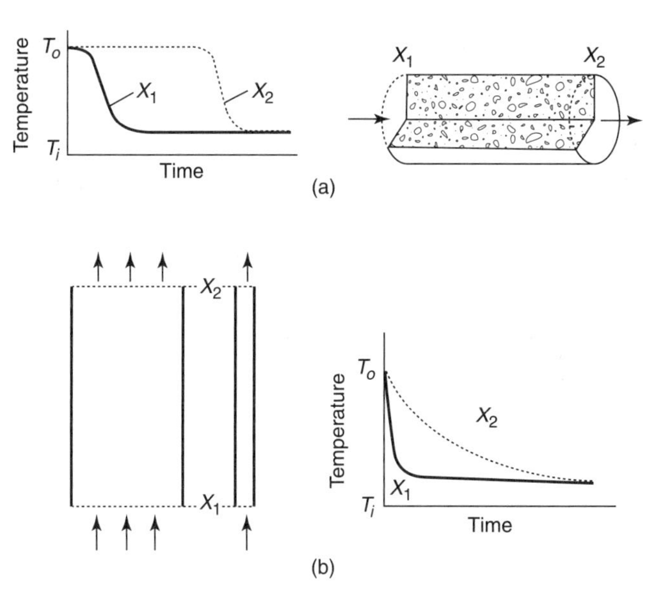

Figure 11.7. Limiting idealized cases used in modeling the thermal drawdown of geothermal reservoir (a) isotropic porous media and (b) flat fracture with a rectangular aperture. Adapted from Tester and Grigsby (1980). Reprinted with permission of John Wiley & Sons.

Discretely fractured, low-permeability reservoirs. Many hydrothermal systems and most HDR reservoirs consist of widely spaced open fractures between sections of low-permeability and low-porosity rock. The efficiency of electric power production over the 20- to 40-year lifetime of an HDR power plant depends primarily on the temperature and composition of the geothermal fluid produced. Knowledge of the anticipated thermal drawdown and of fluid chemistry is crucial in developing optimal plant designs and reservoir management strategies. The predicted lifetime and the rate at which energy can be extracted from a single fracture depend on several major factors: (1) thermal properties of the rock, (2) accessible surface area of the fracture, and (3) distribution of and impedance to fluid flow across the fracture surface. Because of the inherently low thermal conductivity of rock, conduction through the formation to the circulating fluid contained in the fracture controls the rate of heat extraction.

A simplified approach to estimating reservoir performance assumes that a certain fraction of the recoverable power, η, corresponding to uniform flow across the face of a circular fracture, could be extracted. Assuming only conduction of heat toward the fracture face and no thermal-stress cracking enhancement, the recoverable thermal power $P(t)$ can be expressed as (McFarland and Murphy, 1976):

$$P(t) = \eta\, \dot{m}_w\, C_{p,w}\, (T - T_{\min})\, \mathrm{erf}\left(\sqrt{\frac{(\lambda \rho C_p)_r}{t}} \cdot \frac{A}{\dot{m}_w C_{p,w}} \right) \tag{11-6}$$

where $A = \pi R^2$ = area of one face of the fracture, m^2 for a circular fracture or $A = L \times H$ for a rectangular fracture (see Figure 11.7); $C_{p,w}$ = heat capacity of water, about 4.2 kJ/(kg·K); $C_{p,r}$ = heat capacity of granite, about 1,000 J/kgK; $\dot{m}_w$ = water flow rate through the fracture, kg/s; t = time, s; T = mean rock temperature, °C; T_{min} = fluid reinjection temperature, °C; λ_r = thermal conductivity of granite, approximately 3 W/(m·K); and ρ_r = density of granite, about 2.7 g/cm^3. Note that "erf" indicates the error function of the quantity contained in the large parentheses.

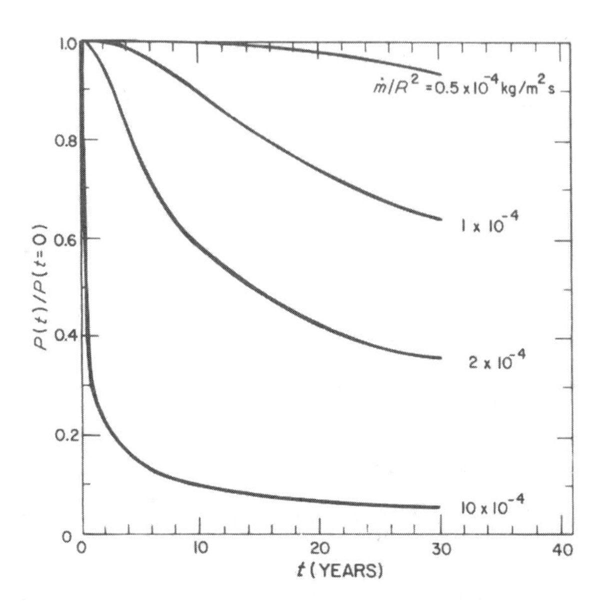

Figure 11.8. Parametric thermal power drawdown curves for a single idealized circular fracture of radius R. Adapted from Armstead and Tester (1987) and following Equation (11-6). Reprinted with permission of Thomson Publishing Services.

Equation (11-6) provides estimates of reservoir lifetime for an ideal fractured system with a specified area A and production rate. A few simple calculations reveal that large surface areas are required for low drawdown rates with wellbore flows in excess of 40 kg/s (see Figure 11.8). For a given set of rock properties, the parameter $\dot{m}_w/A$ will determine the thermal drawdown rate. In designing systems, one can utilize single or multiple fracture concepts with specified drawdown rates at given initial power extraction levels. Figure 11.9 shows the required wellbore separation distance as a function of initial thermal power or mass flow rate and the number of fractures. In this case, 50% drawdown is achieved after 20 years of continuous operation with an overall $\dot{m}_w/R^2$ of approximately 2×10^{-4} kg/(m^2·s). Because of the rock's low thermal conductivity, fractures separated horizontally by more than 50 m show negligible thermal interference over a 20-year period. Using simplified models to simulate reservoir thermal-hydraulic behavior, substantial enhancement of HDR reservoir performance might be possible by exploiting the effects of thermal stress cracking. However, the mechanisms associated with thermal stress cracking are extraordinarily complex and not well understood even under well-defined laboratory conditions (see Tester et al., 1989, Harlow and Pracht, 1972 for further discussion).

Economically acceptable rates of energy extraction and lifetimes can be achieved in low-permeability HDR reservoirs either by growing large single fractures (*R, L,* or *H* ≈ 1,000 m), by using smaller multiple fractures in parallel, by exploiting thermal stress cracking, or by remedial redrilling and refracturing to generate new surface area in nondepleted regions of hot rock.

11.3.3 Non-electric, direct-heat utilization

Low-temperature, hydrothermal resources in the form of hot water or steam have been used in commercial direct-heating applications ranging from soil warming (10°C) to district heating (60–90°C) to power drying (200°C). Figure 11.10 illustrates three modes of utilization: for heating with extracted geofluids indirectly involving a heat exchanger and for both groundwater and ground source heat pumps.

The oil price shocks of the 1970s catalyzed interest in non-electric applications of geothermal energy worldwide. In the US, for example, various federal and state tax credit programs were adopted to incentivize the direct use of geothermal resources. By 1994, geothermal energy provided over 1.4×10^{16} J (13 trillion Btus) annually for direct-heat applications in the US, mainly in California, Oregon, Idaho, Nevada, Utah, New Mexico, Wyoming, South Dakota, Texas, and New York.

Although the usable resource base of low- and moderate-temperature geothermal resources is large, development of direct-heat uses is proceeding slowly. There is essentially no direct-heat geothermal industry or infrastructure in the same sense that there is an electrical-power generation industry. Each direct-heat system is a separate design, and few consultants or contractors are trained and experienced in the direct use of geothermal resources. With low costs for natural gas and heating oil, this situation may not change soon (see Geo-Heat Center, 1988).

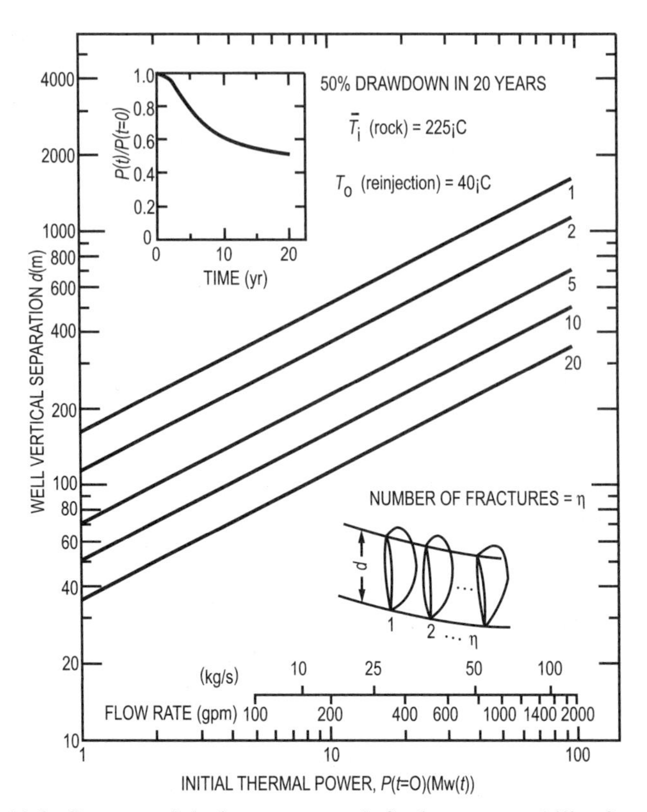

Figure 11.9. Conceptual design nomograph for low-permeability formations with a specified level of thermal drawdown (50% in 20 years of operation). Wellbore separation distances (*d*) that define single or multiple fracture sizes are shown as a function of initial thermal power level ($P(t = 0)$). Flow rates given in kg/s and in gal/min (gpm). Adapted from Armstead and Tester (1987). Reprinted with permission of Thomson Publishing Services.

Direct-heating applications. Industrial applications generally need moderate to high temperature geothermal resources. Industrial uses include: enhanced oil recovery (90°C), ore or heap leaching operations using selective chemical dissolution to extract precious metals (110°C), dehydration of vegetables (130°C), mushroom growing (60°C), pulp and paper processing (200°C), hay drying (60°C), timber drying (90°C), and diatomaceous earth drying (182°C) (Lienau, Lund, and Culver, 1995).

The first modern industrial use in the US was a vegetable dehydration plant at Fernley, Nevada. When the plant was built in 1978 to process onions, the geothermal fluid was used only to supply heat for the dehydration process. However, the advantages of using the fluid in the wet-preparation stage quickly became apparent. Because of its oxygen-free and essentially bacteria-free nature, the geothermal fluid maintains a bacterial count well below prescribed health standards. Advantages of using geothermal fluids include significantly increased production rates, elimination of the potential fire hazards of other fuels, and no degradation of the products through scorching.

Geothermal fluids are also finding increased use in aquaculture to raise catfish, trout, perch, bass, tilapia, sturgeon, shrimp, and tropical fish. The benefit of a controlled rearing temperature through the use of geothermal fluids can increase growth rates by 50–100%, significantly increasing the number of harvests per year.

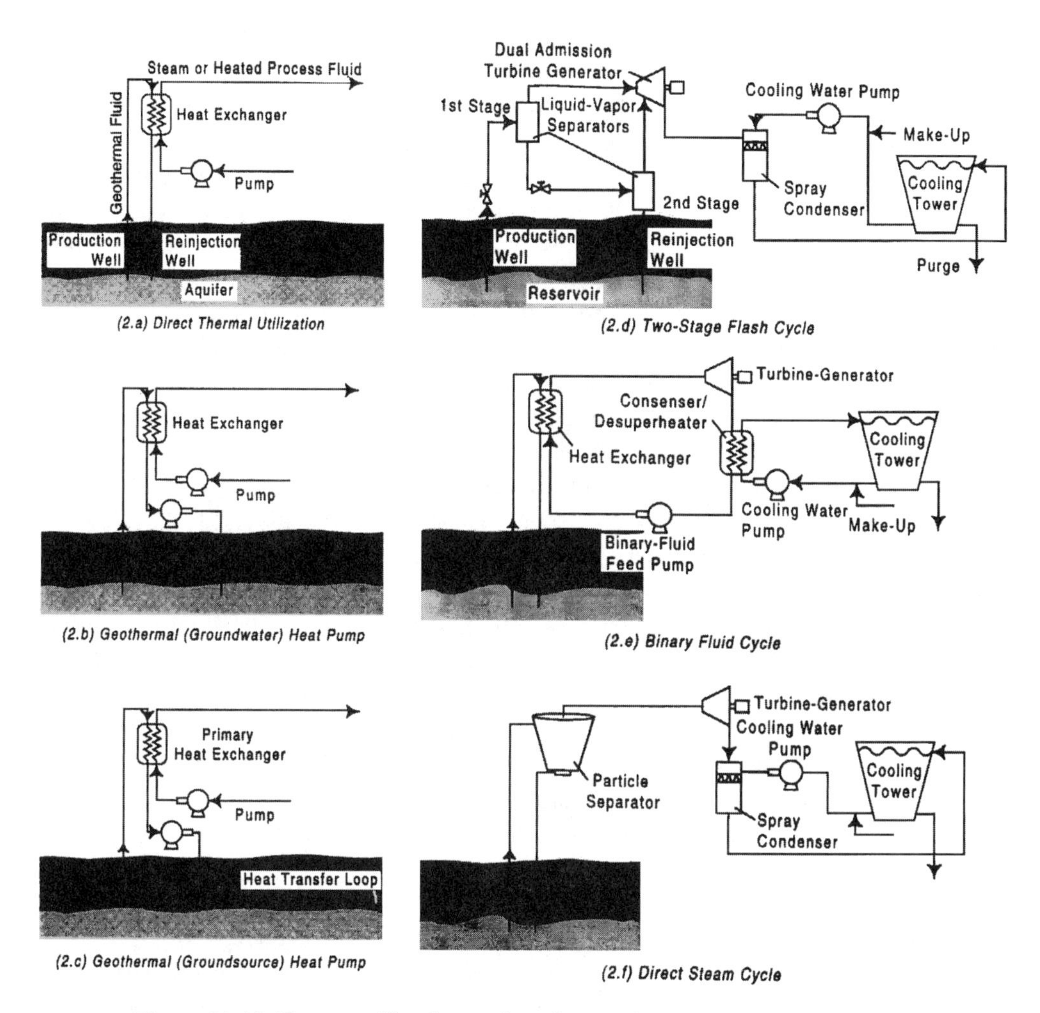

Figure 11.10. Energy utilization options for geothermal resources. Adapted from Mock et al. (1997). Reprinted with permission of *Annual Review of Energy and the Environment*.

Greenhouses are among the fastest growing applications. Many commercial crops (including flowers, house plants, vegetables, and tree seedlings) can be raised profitably, making geothermal resources economically attractive— especially in cold climates. Large geothermally heated greenhouses are currently operated in Italy, Iceland, Hungary, and the US (in New Mexico, Utah, and California). Currently, there are 23 geothermal district heating systems in the US, including the nation's oldest at Boise, Idaho, the 65,000 m^2 (700,000 ft^2) system on the Oregon Institute of Technology campus, the municipal detention facility system in Yakima, Washington, and the nation's largest geothermal district heating system in San Bernardino, California. To assist communities and developers in evaluating the costs and benefits of such systems, Washington State University and the Geo-Heat Center have prepared comprehensive computer programs for geothermal direct-use cost evaluation (Rafferty, 1994). The most recent compilation of low- and moderate-temperature geothermal resources in 10 western states contains information on 8,977 thermal wells and springs that are in the temperature range of 20–150°C. Data and maps are available for Arizona, California, Colorado, Idaho, Montana, Nevada, New Mexico, Oregon, Utah, and Washington. Resources with temperatures greater than 50°C and located within 8 km of a population center were identified for 271 cities and towns; 50 sites were judged high in priority for near-term comprehensive resource studies and development (Lienau, Ross, and Wright, 1995). Older data are available for the rest of the US (Reed, 1983). Like the nation's high-temperature geothermal resources, the resources suitable for direct-heat applications are generally underdeveloped.

Internationally, geothermal district heating has been popular where there is proximity of a good hydrothermal resource to a reasonable load center. Reykjavik, Iceland, Paris, France, and many locations in Russia, Japan, and Hungary (Edwards et al., 1982) provide good examples of such developments.

Geothermal heat pumps. The term "geothermal heat pump" (GHP) is generic for all heat pumps that utilize the earth's thermal capacity as an energy source (for heating) or energy sink (for cooling). At shallow depths greater than about 1–2 m (3–6 ft), the earth maintains a relatively constant temperature, warmer on average than the air above it in winter, cooler in summer, making possible typical GHP COPs of 4.0 or better. Such high efficiency can reduce energy consumption by 23– 44% over air source heat pumps and by 63–72% over electric resistance heating and standard air-conditioning equipment (L'Ecuyer et al., 1993).

The earth's thermal capacity can be utilized directly or indirectly, such as by using groundwater as an intermediary heat-transfer agent. The concept of GHPs is occasionally expanded to encompass the relatively few heat pumps that utilize lake or river water. Rafferty (1996) classifies GHPs in three groups: ground-coupled (direct thermal), groundwater, and hybrid (ground source, see Figure 11.9). In the first type, a closed loop of pipe is buried horizontally beneath the frost zone, or vertically 30–120 m (100– 400 feet) deep and filled with a water-based antifreeze solution that extracts heat in a closed circulating loop from buildings in summer, depositing it in the earth. In the winter, the system is reversed, extracting heat from the earth and carrying it into the building. Waste heat from a GHP during summer cooling periods can provide domestic hot water as an additional benefit.

The second type of GHP, first used in the US in the 1930s and, until recently, the most widely used approach, is the system in which groundwater is delivered to a heat exchanger installed in the heat pump loop then disposed on the surface or in an injection well (Pratsch, 1995). These systems have dropped in popularity as increased environmental regulations designed to prevent contamination of surface waters or aquifers have favored the adoption of the totally enclosed ground-coupled system.

The third type, the hybrid system, combines a ground-coupled system with a cooling tower, and is used primarily in commercial buildings. Due to the high cost in meeting peak cooling loads, hybrids incorporate a cooling tower, allowing the designer to size the ground loop for normal heating loads and to use the tower to help meet the much larger peak cooling loads occurring during the summer.

To help mitigate the perceived energy crisis in the early 1970s, the US government initiated activities to assist the commercialization of GHPs; for example, several large projects were supported including the 1982 installation of a 300 ton (Note: 1 ton = 3517 W = 200 Btu/min) heat pump in the county courthouse in Ephrata, Washington. Previously the annual bill for operating the oil-fired boiler was $14,000–22,000, depending on the severity of the winter. After GHP installation, costs fell to approximately $2,400 per year (Painter, 1984). Other demonstration projects included the 37,200 m^2 (400,000 ft^2) climate-controlled Oklahoma state capitol using 277 independently controlled GHPs, and New Jersey's Stockton State College, home of the largest school GHP system (1,600 ton).

The US lags behind other countries in taking advantage of geothermal heat-pump technology. To accelerate commercialization of GHPs, the government (mainly the US DOE and EPA) and public/private organizations (including Edison Electric Institute, National Rural Electric Cooperative Association, Electric Power Research Institute, Consortium for Energy Efficiency), and the electric power industry (with more then 70 electric utilities and over 50 trade allies) established in 1994 the Geothermal Heat Pump Consortium (GHPC), which initiated a $100 million, six-year program. Specific goals of the program were to reduce greenhouse gas emissions, improve energy efficiency, reduce heating/cooling costs, and increase annual GHP sales in the US from 40,000 units per year in 1994 to 400,000 units per year by the year 2000. Clearly there are significant opportunities for reducing the primary energy needs of buildings using GHPs as a part of an integrated design.

11.3.4 Electric power generation

Most geothermal, non-concentrating solar thermal, and ocean thermal resources share the common disadvantage of having inherently low resource temperatures. One technique for increasing their quality, as well as their transportability, is to convert their available energy into electric power. In practical terms, the efficiency of such conversion processes is limited by resource temperatures and prevailing ambient conditions for heat rejection. Efficiencies for converting geothermal resources at temperatures below 200°C to electricity are substantially less than those of fossil fuel–fired or nuclear-powered plants, typically 5–20% versus 35– 60% in terms of the ratio of work produced to heat supplied (Carnot, Second Law efficiency). New technologies have been improving the practical efficiency of power production toward ideal thermodynamic limits to provide an economical process. For example, for a number of these low-temperature, alternative energy sources, several hydrocarbons and their halogenated derivatives have been proposed as working fluids in Rankine cycles rather than the conventional choice of water/steam (Milora and Tester, 1976, Tester, 1982).

Figure 11.11 shows the typical range of cycle efficiencies expected for geothermal power plants. The Second Law of Thermodynamics imposes a real limitation on the production of electricity from low-temperature geothermal heat sources. The total maximum work that can be produced by cooling a condensed geofluid from its wellhead condition T_{gf} to the so-called dead state, or ambient condition T_o, is given by the thermodynamic availability B. The maximum work per unit mass of geofluid is then given by:

$$\Delta B = \left[\Delta H - T_o \Delta S\right]_{T_o}^{T_{gf}} = W_{max}/m_{gf} \tag{11-7}$$

where ΔH is the enthalpy difference and ΔS is the entropy difference between states for the geofluid on a per unit mass basis. This maximum work quantity can then be compared to the actual amount of work produced by any real power conversion process W_{net}.

Comparisons of this type are usually achieved by defining a cycle efficiency η_{cycle}, which represents the net useful work W_{net} obtained from the system divided by the amount of heat transferred from the geothermal fluid Q_H.

$$\eta_{cycle} = \frac{W_{net}}{Q_H} \tag{11-8}$$

As the cycle efficiency decreases, the amount of heat rejected to the environment increases. For an ultimate sink at 25°C with a geothermal fluid heat source at 50°C, cycle efficiencies would be less than 5% with 95% of the extracted geothermal energy rejected to the environment as heat. As a source temperature rises, the efficiency increases as shown in Figure 11.11.

An alternative approach to using cycle efficiency would be to compare directly the real work to the maximum possible work by defining a utilization efficiency, η_u, as,

$$\eta_u \equiv \frac{W_{net}}{W_{max}} = \frac{P}{\dot{m}_{gf}\,\Delta B} \tag{11-9}$$

where P = net power extracted, ΔB = availability per unit mass, and $\dot{m}_{gf}$ = geothermal fluid-flow rate. Then η_u is a direct measure of the effectiveness of resource utilization, because at a fixed T_{gf}, higher values of η_u correspond to lower well-flow rates for a given power output. In contrast, the cycle efficiency, η_{cycle}, is a measure of how efficiently the transferred geothermal heat is converted to work. The utilization efficiency concept is particularly useful in comparing different power conversion options for the same resource conditions. Ideally, η_u should be as high as possible, but there are limitations in the efficiency of work-producing machinery (turbines and pumps), as well as in the size of heat-transfer systems associated with generating power.

System options. Figure 11.10 illustrates several utilization schemes for electric power production and direct use. Typically, a Rankine cycle of some type is employed. In the multistage flash system shown, the pressure of the produced geofluid is reduced to generate saturated vapor and liquid phases. The vapor fraction is then expanded in a condensing steam turbine/generator to produce electric power. The liquid fraction is flashed again to a lower pressure with the generated vapor expanded in the turbine from a lower starting pressure. Although the process can be repeated for multiple stages, two-stage systems are usually the best choice economically.

For vapor-dominated resources, such as those at The Geysers field in California, a direct steam turbine condensing cycle is used. Particulate matter is removed from the steam before it enters a low-pressure turbine that employs conventional materials and designs.

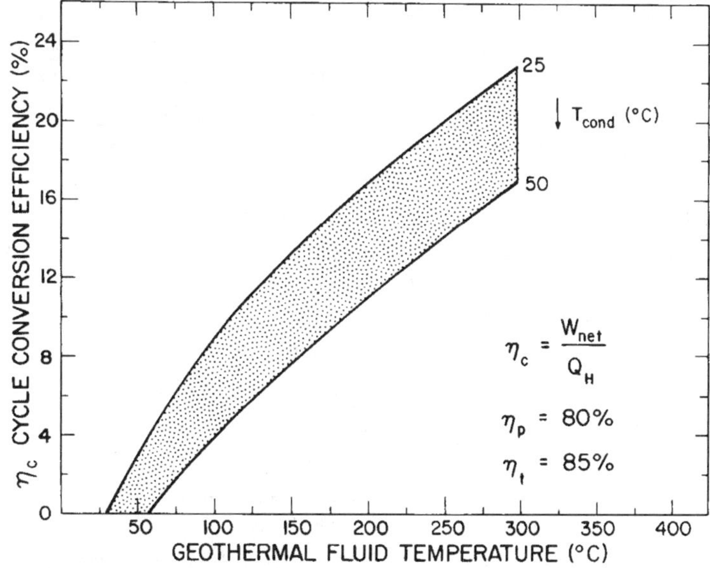

Figure 11.11. Typical range of anticipated cycle efficiencies for geothermal power plants. Adapted from Tester (1982).

When natural geofluids contain significant amounts of non-condensable gases, a direct steam or flashing cycle is not a good choice. Indirect binary-fluid or two-phase expander (Tester, 1982) cycles are more efficient for these applications. Binary-fluid cycles are closed-loop Rankine cycles that involve a primary heat exchange step in which a secondary working fluid is vaporized, expanded through a turbine/generator, and condensed (as shown in Figure 11.10). Binary-fluid systems tend to operate at somewhat higher work-to-heat rate efficiencies for geothermal fluid temperatures below 200°C. They are particularly well-suited to take advantage of lower ambient temperatures (25°C or less) for cooling, and are relatively insensitive to the presence of non-condensables. Furthermore, because these nonaqueous secondary working fluids have low-temperature vapor densities considerably greater than steam, much smaller and less expensive turbines would be required for the same power output. Flashing cycles are, of course, simpler in that they do not require a primary heat exchanger or secondary fluid feed pump, but flashing can cause difficult scaling problems.

If the geothermal fluid contains large amounts of dissolved material that may corrode or deposit on heat-exchange surfaces, a combined multistage flash and organic binary cycle may offer a reasonable solution. This system can be designed in a number of ways. For example, a two-stage flash unit could be used to produce steam for fluid vaporization in an organic Rankine cycle. Other designs may involve a dual-cycle system where the flashed steam fraction drives a turbogenerator, and a binary cycle operates using a remaining liquid fraction to vaporize the secondary working fluid. Organic binary systems are state-of-the-art technology requiring some developments, including the design of larger capacity hydrocarbon units, control and valving systems, turbine seals for hydrocarbon vapors, and improved resistance to corrosion, erosion, and scaling of certain components.

Working fluid selection. Accurate values of the thermodynamic properties of proposed working fluids are required to calculate electric power cycle performance. Heat capacity at constant pressure in the ideal gas state C_p^*, saturation vapor pressure P^{sat}, pressure-volume-temperature (*PVT*) behavior, enthalpy and entropy changes, and liquid density at saturation ρ_l^{sat} can be expressed with semiempirical equations. Many hydrocarbons, fluorocarbons, and other organic working fluids have been examined for use as a replacement for water in low temperature power conversion cycles. For example, R-600a (isobutane, i-C_4H_{10}), R-32 (CH_2F_2), R-717 (ammonia, NH_3), RC-318 (C_4F_8), R-114 ($C_2Cl_2F_4$), and R-115 (C_2ClF_5) were studied to provide a range of properties including molecular weight and critical temperature and pressure (Milora and Tester, 1976). All of these compounds have relatively high vapor densities at heat rejection temperatures as low as 20°C and would result in very compact turbines in comparison to steam turbines of similar capacity.

Factors other than desirable thermodynamic properties frequently determine practical working fluids. These include fluid thermal and chemical stability, flammability, toxicity, material compatibility (e.g., resistance to corrosion), and cost. The major disadvantage of hydrocarbons such as propane, pentane, and isobutane, is their flammability, which requires costly explosion-proof equipment and ventilating systems. The situation with fluorocarbons is more complex, but thermal instability is an important issue.

Engineering design criteria. Any real process for generating electricity or heat has inefficiencies or irreversible steps that result in a net work less than W_{max} given in Equation (11-7). Efficient use of the resource may be necessary for commercial feasibility if reservoir development costs are high. Efficient utilization requires that:

1. Most of the heat is extracted from the geothermal fluid before disposal or reinjection
2. Temperature differentials across heat-transfer surfaces are maintained at minimum practical levels
3. Turbines and feed pumps are carefully designed for optimum efficiency
4. Heat is rejected from the thermodynamic cycle at a temperature near the minimum ambient temperature T_o

Whether or not these conditions can be met depends on the choice of thermodynamic cycle and fluid working medium, the geothermal fluid temperature, and the temperature of the coolant (water or air) to which the power plant rejects waste heat. For example, if waste heat is to be rejected from the thermodynamic cycle at a constant temperature by a condensing vapor (as in condition 4 above), then the working medium's critical temperature must be greater than the temperature of the power-plant coolant. Fulfillment of conditions 1 and 2 suggests the use of supercritical Rankine cycles. Thus, the critical temperature of the working fluid should also be below the maximum geothermal fluid temperature.

11.3.5 Equipment

Heat exchange and prime-mover fluid production. As illustrated in Figure 11.10, a geothermal electric or heating system may require a variety of heat-transfer steps to produce a suitable prime-mover fluid. For example, for binary-fluid cycles, a primary heat exchanger operating as a combined preheater, boiler, and superheater at subcritical pressures or as a vaporizer at supercritical pressures will be required to remove heat from the geothermal fluid and generate the organic vapor for injection into the turbine. In a flashing cycle, prime-mover steam is produced by reducing the pressure of the fluid in one or more steps. In both cases, the design and component sizing of the heat exchanger or flashing unit require specification of the appropriate duty factor for each unit (J/h, or kg steam generated/h) and of temperature and pressure differences.

Because power conversion efficiencies are low, the amount of heat transferred may be 5–15 times the power produced. Large heat exchangers are required at significant cost. For example, a 100 MW geothermal plant with a 12% cycle efficiency requires about 60,000 m^2 (650,000 ft^2) of heat-exchange surface area. Thus multiple, parallel units of 1,900–5,500 m^2 (20,000– 60,000 ft^2) would be utilized.

For a primary heat exchanger in a binary cycle, the overall heat-exchange surface area, A, can be expressed as a function of the heat load, $\dot{Q}$, an appropriate mean $\overline{\Delta T} = < T_{hot} - T_{cold} >$, and an effective overall heat-transfer coefficient U:

$$A = \frac{\dot{Q}}{U\overline{\Delta T}} \tag{11-10}$$

Practical ΔT should be selected to keep the heat exchanger cost at a reasonable level. In some cases, the exchanger might reach a pinched condition where ΔT approaches zero at one or more locations. Then, a minimum ΔT would have to be specified for any potential pinch points, consistent with practical heat exchange area requirements.

To estimate the surface area of any exchanger, realistic values for heat transfer coefficients of both fluids, wall resistance, and fouling factors may be required. Because many different geothermal brine compositions and working fluids operating in sub- and supercritical modes are under consideration, no single optimum design can be proposed.

Because the chemical compositions and temperatures of geothermal fluids vary widely from site to site, it is difficult to estimate accurately the extent of a potential scaling or fouling problem for primary heat exchangers without field testing. For example, the 300,000 ppm total dissolved solids (TDS) and 370°C temperatures of Salton Sea geothermal brines in the Imperial Valley in California present a more challenging problem than the brines at Cerro Prieto with < 20,000 ppm TDS and 300°C temperatures. HDR systems introduce another factor in that the circulating water is not indigenous to the reservoir. Dissolution products build up in the circulating fluid and might present scaling problems particularly at higher reservoir temperatures (>250°C).

Only certain types of heat-exchanger designs are appropriate for geothermal applications. These include shell-and-tube exchangers, fluidized-bed exchangers, and direct-contact exchangers. Direct-contact and fluidized-bed systems may be the only usable candidates when scale deposition rates are high.

When steam produced by flashing is the prime-mover fluid and is used directly in a condensing turbine, no primary heat-exchange step is required. In its place, pressure reduction is used to produce a saturated vapor fraction. In a typical installation, the first stage of flashing is done at the wellhead. The two-phase mixture produced is piped to a separator system—frequently a cyclone unit—where the steam and liquid brine are separated. The steam fraction is transmitted to the turbine-generator unit located in a central area, and the liquid fraction is reinjected, discarded, or flashed to a lower pressure to produce additional steam for turbine expansion. This process can be continued as many times as are economically feasible. A two-stage flash system may produce 20–30% more power than a single-stage unit at the same total geothermal fluid-flow rate. But the improved performance of multistage systems must be balanced against the additional complexity and cost of utilizing the secondary steam (Eskesen, 1980).

Turbines. In selecting nonaqueous working fluids for power-cycle applications, turbine sizes should be small to reduce costs and offset the additional heat-exchange surface area required for binary-fluid cycles.

It is important to operate geothermal turbines at high efficiency whether they use steam or organic binary fluids. A similarity analysis of performance shows that turbine efficiency is essentially controlled by two dimensionless numbers involving four parameters: blade pitch diameter, rotational speed, stage enthalpy drop, and volumetric gas-flow rate (Baljé, 1962). Maximum turbine efficiency requires a specified relationship among these parameters. Therefore, turbine sizes, operating conditions, and, consequently, costs can be estimated. For fluid-screening purposes, a generalized figure of merit that relates directly to turbine size has been developed (Milora and Tester, 1976). Many organic fluids offer a significant reduction in turbine size. For example, a 100 MWe capacity plant designed for a 150°C liquid-dominated resource would require numerous, large turbine-exhaust ends with steam flashing. An ammonia binary cycle would require only a single exhaust end (Milora and Tester, 1976). Low pressure steam turbines for geothermal operation with 50 MW capacities or greater have been commercially produced for a number of years. Organic binary cycles using geothermal energy have been placed in commercial operation, albeit at a smaller scale than flashing systems (Mock et al., 1997).

Pumps. Pumping requirements for geothermal systems are of four major types:

1. Downhole pumps for fluid production
2. Fluid-reinjection pumps
3. Power-cycle feed pumps
4. Geothermal-fluid transport pumps

According to a published survey and assessment of existing geothermal pumping equipment (Nichols and Malgieri, 1978), surface-mounted, conventional multistage centrifugal designs can be used for commercial-size applications for 2–4 above. Reinjection pumps may present problems where highly corrosive and erosive geothermal brines are used. The most challenging problems are with downhole pumping, because of severe environmental and design constraints.

Several types of downhole pumping systems have been proposed. These include shaft-driven and submersible electric-motor-driven, multi-stage, vertical centrifugal designs that are commercially available and have been tested by a number of companies. There are complexities associated with downhole shaft pumps driven mechanically on the surface. For example, line-shaft bearings need to be effective for such high temperature service. Electrically driven submersibles also experience serious problems at operating temperatures above 180°C because of the effects of heat, moisture, and brine on the motor. The other general pump types under development include steam- or organic-fluid-driven, downhole turbine pumps (Sperry Research, 1977) and hydraulically driven systems. In these cases, the problems of shaft or electric-motor-driven units are replaced with a complex concentric wellbore-piping system.

Condensers and heat rejection. Because conversion efficiencies are low, a large fraction of the extracted geothermal heat is ultimately rejected to the environment. For example, a geothermal plant operating with a cycle efficiency of 10% rejects about three to six times as much heat as a similar capacity fossil- or nuclear-fired unit. Consequently, the substantial costs associated with condenser-desuperheaters and cooling towers in geothermal systems strongly influence economic feasibility. Seasonal and diurnal variation of ambient dry-bulb temperatures can also significantly affect cycle performance. For dry cooling systems, condensing temperatures change seasonally, causing changes in the power output of the plant. Because of the size of geothermal units (≤100 MWe), operating with a floating power output might be desirable.

The availability of water for cooling is also important. Fossil and nuclear plants are often located near abundant water supplies. Geothermal plants, of course, have to be located at the resource regardless of the water situation. Even so, if evaporative cooling is used, water consumption is high. For example, a 50 MW geothermal plant operating at 10% efficiency would consume cooling water at a rate of approximately 145 L/s (2,300 gal/min). In many existing plants, cooling water is provided by the geothermal fluid itself. For example, evaporative cooling at The Geysers consumes approximately 80% of the geothermal fluid, so that only 20% is reinjected (Holt and Ghormley, 1976). For liquid-dominated and geopressured reservoirs, total reinjection may be required to avoid subsidence or for other environmental or regulatory reasons. Furthermore, the brine may have to be treated before reinjection.

Design options for waste-heat rejection include direct-contact spray condensers, multipass water-cooled surface condensers, direct air-cooled condensers, spray ponds, and wet, wet/dry, and dry cooling towers with forced, induced, or natural-draft design. For many power-cycle operations, evaporative cooling has been commonly used and will continue to be employed in future designs. However, when total reinjection is required and water is scarce or expensive, totally dry and wet/dry cooling are the only alternatives. Although high capital costs have prevented widespread use of completely dry systems in the US, many such systems have been operating successfully in Europe. The major disadvantage of dry systems, besides higher cost, is that performance of the power cycle depends on dry-bulb temperature fluctuations. If floating power capacity is an acceptable operating mode for small base-load units, then variations in power output would not be a serious detriment to using totally dry systems.

An alternative to totally dry systems is a wet/dry cooling tower, which combines advantages of both systems. Low condensing temperatures can be maintained during peak periods with high dry-bulb temperatures by using supplemental evaporative cooling, and water can be conserved during periods of low dry-bulb temperature. Fogging and drift are also better controlled. Wet/dry towers cost more, however, and are somewhat more difficult to operate and maintain.

11.3.6 Power cycle performance

In principle, the utilization efficiency, η_u, can be somewhere between zero (P=0) and unity ($P = \dot{m}_{gf} W_{max}$). Recall from Equation (11-7) that W_{max} is a function only of the fluid's condition as produced at the wellhead temperature (T_{gf}) and the ambient or "dead state" temperature (T_o). In practice, its value is determined from economic considerations by balancing the cost of obtaining the heat (i.e., drilling and piping costs) against the cost of processing it (e.g., heat exchangers, turbines, pumps) to generate electricity in the power station. If η_u is small, then the resource is being utilized poorly, and a large investment in wells is required (cost per unit power goes to ∞ as η_u approaches 0). On the other hand, if we approach utilization of the full potential of the resource, then total well costs decrease but the required investment in highly efficient power conversion equipment is high (cost per unit power goes to ∞ as η_u approaches 1.0). The economic optimum occurs when η_u assumes some intermediate value (e.g., at The Geysers, $\eta_u \approx 0.55$ with $T_o = 26.7°C$).

In a typical analysis of cycle performance, a set of equations is developed to describe the work and heat-flow rates to and from the chief plant components. After selecting a working fluid and a geothermal-fluid inlet temperature, the principal independent design variables are the maximum cycle-operating pressure, P_{max}, at the turbine inlet, the condensing temperature, and the heat-exchanger approach temperature.

For any given working fluid, there is an optimum set of operating conditions yielding a maximum η_u for particular geothermal-fluid and heat-rejection temperatures. Computer optimizations for several different working fluids were conducted for geothermal fluid temperatures ranging from 100 to 300°C for a heat sink temperature of 16.7°C (Milora and Tester, 1976). At each point, cycle pressures were varied until an optimum was determined at that temperature. A characteristic maximum, of η_u, is observed at a resource temperature that is different for each fluid but is generally in the range of 55– 65% efficiency. Figure 11.12 compares maximum utilization efficiencies for different conversion methods including single- and dual-stage, flashing, binary, direct-steam injection, and dual flash-binary cycles combining data from a number of sources (Holt and Ghormley, 1976, Milora and Tester, 1976). The output of a multistage steam-flashing cycle primarily depends on:

1. Initial fluid temperature and composition
2. Flashing conditions of each stage (P_i^{sat}, T_i)
3. Turbine efficiencies
4. Condensing temperature (Eskesen, 1980)

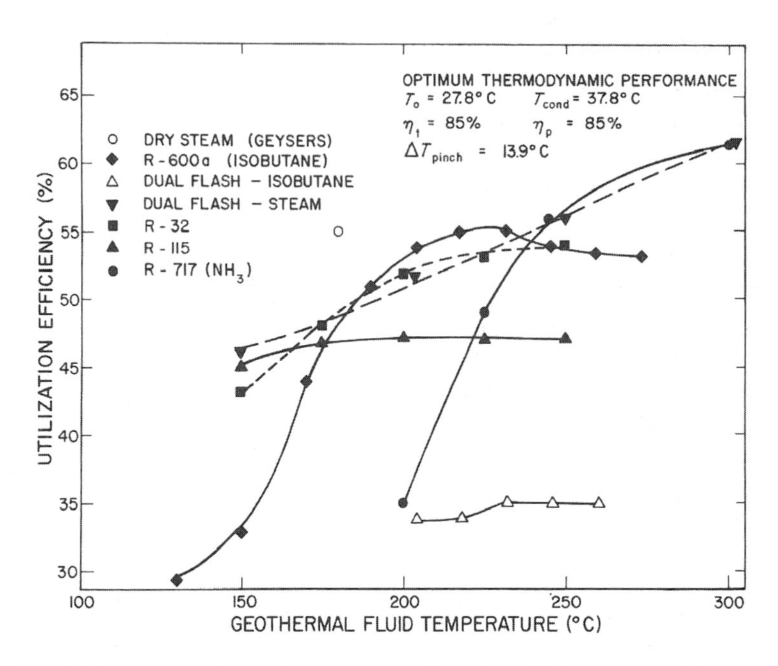

Figure 11.12. Comparison of maximum utilization efficiences for different conversion methods as a function of geothermal fluid temperature. Optimum thermodynamic ΔT_{pinch} = 13.9°C. Adapted from Armstead and Tester (1987) and Tester (1982).

Generally, the work produced by a given flash stage is proportional to the amount of vapor created by the isenthalpic throttling step (see Figure 11.10) and the enthalpy difference that the vapor experiences when isentropically expanded in the turbine to the condensing pressure. If the temperature of the flashed fluid is only slightly below its wellhead temperature, then the fraction of vapor produced is small but the isentropic enthalpy difference is large. The opposite is true if the fluid is flashed to just above the condensing temperature. Optimal performance in terms of power output occurs at some intermediate flashing temperature. The performance of a two-stage flashing system is shown in Figure 11.13, assuming an average turbine efficiency of 85%. Maximum output ($\eta_u = 0.50$) occurs when the first flashing stage is at 169°C (T_{s1}) and the second is at 107°C (T_{s2}).

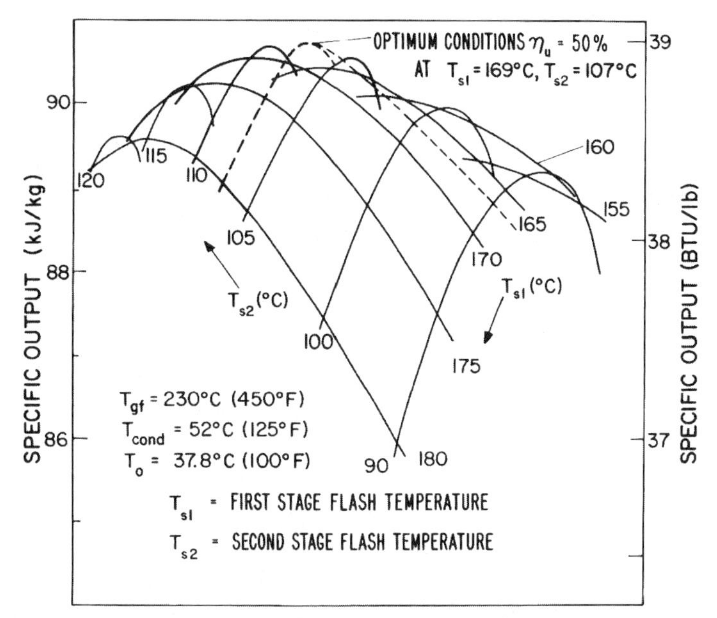

Figure 11.13. Effect of flashing conditions on the performance of a dual-flash/dual admission steam turbine cycle operating on a 230°C, CO_2-free geothermal fluid. Performance shown for the following conditions: turbine efficiency η_t = 85%; feed pump efficiency η_p = 85%; brine containing 2.5 wt.% NaCl T_{cond} = 52°C; T_o = 37.8°C; T_{s1} = first-stage flash temperature; T_{s2} = second-stage flash temperature. Adapted from Eskesen (1977, 1980).

11.4 Sustainability Attributes

11.4.1 Reservoir lifetime issues

Geothermal energy is considered by many to be the only renewable that does not depend directly or indirectly on sunlight. However, "it must be tapped slowly enough so as not to deplete the accessible reservoir of heat, and thus be truly renewable" (Brown et al., 1990). However, given the magnitude and distribution of the global geothermal resource, this restriction is neither practical nor necessary since the resource is large and improving technology will allow geothermal energy to contribute for centuries (see, for example, treatments by Wright, 1995, Armstead and Tester, 1987, Rowley, 1982).

Whether or not consumption of a resource can be said to be renewable depends on the time frame under consideration. Kozloff and Dower (1993) suggest that a perspective of 300 years or more of continuous production is adequate for an energy fuel to be considered as renewable, since technical advances during that time will have rendered today's perspective obsolete. For geothermal resources, the renewability concept has a special connotation. In all practical situations some degree of local depletion of geothermal energy within a specific reservoir occurs during production. But renewal of the thermal energy content will occur by natural processes of heat conduction from surrounding hot rock, by decay of indigenous radioisotopes, such as those of potassium, thorium, and uranium, and by convective heat transfer due to water recharge. With average rock thermal diffusivities of 10^{-6} m^2/s, heat conduction rates are slow enough to require a substantial time period to restore the thermal energy content of a locally depleted reservoir. Typically, within a period of time less than 10 times the production period, essentially complete recovery of original rock temperatures will occur. Of course, for hydrothermal systems natural water recharge rates may be extremely slow unless artificially supplemented.

We can also express the renewability of geothermal energy using the methodology introduced for biomass in Section 10.1. In terms of that nomenclature, the dimensionless time constant, τ_r/τ_u, representing the ratio of time needed for resource generation to a characteristic societal utilization time, could range from about 10 to 1 depending on the specific characteristics of the geothermal field. For example, there are wells at the Larderello field in Italy that have been in full production for more than 70 years, while others have provided fluid at useful rates for only a few years.

Wright (1995) argues that the traditional geothermal project analysis underestimates the sustainability of production from a geothermal resource. Such analyses are usually done to demonstrate the financial viability of the project using constant and relatively high discount rates, and make pessimistic estimates of resource depletion. Field production is not only simulated in feasibility studies but carried out in practice in a conservative way. The geothermal field is said to be depleted when it will no longer sustain some chosen level of generation. Even at this point, however, significant quantities of heat remain in the rock-fluid reservoir system.

11.4.2 Environmental impacts

The main environmental concerns facing a geothermal development include land and water use, noise, seismic and subsidence risks, gaseous emissions, liquid effluents, and solid waste resulting from the development, production, and decommissioning of the geothermal field. In terms of minimizing point-source pollution, many operating geothermal plants have nearly emissions- and waste-free operation.

An important feature of land and water use requirements for existing hydrothermal and future advanced systems is that the entire fuel cycle is located at the plant site. There are no strip mines, gas or oil pipelines, or waste repositories. Water consumption can be controlled by using total reinjection, non-evaporative cooling, and general pressure management in closed-loop recirculating cycles. Because the major elements of a geothermal system are underground, the surface footprint of a geothermal plant itself is relatively small. Typical requirements include structures to house the power-generating and heat-exchange equipment, land space for wellheads, and a pipe distribution system.

When geothermal energy is used to generate electricity, there is always waste heat rejected to the environment. Typically, the largest visible feature of a geothermal electric power plant is its battery of cooling towers. Thus, the impacts of the waste heat itself on the local ecology and the means used to reject it need to be considered. Fortunately, individual plant sizes are usually limited to 50–100 MWe, resulting in waste heat rates that usually have small local environmental consequences. There is potential to utilize some of the waste heat for direct-use, low-temperature applications such as greenhouse and soil heating and aquaculture/fish farming. These activities are usually carried out on a small scale so that additional capital investments are marginal.

During field development, drilling operations, and plant construction, noise and disruption of normal activities are of concern. After the plant begins operation, noise levels are usually controlled by silencers and other active noise abatement measures. At many geothermal sites, the surface land remains in service for residential, agricultural, recreational, and industrial uses.

Induced seismicity and subsidence due to geothermal fluid extraction with reservoir pressure changes are possible. However, the magnitude of these environmental concerns is often tempered by the high natural seismic activity levels that are commonplace in most active hydrothermal regions. Typically, high frequencies and densities of small microseismic events occur within and at the margins of these fields. These relieve *in situ* stresses in a regular fashion and may actually help mitigate massive fault movement or a major earthquake. Monitoring of seismic activity in geothermal areas is commonly employed to verify general behavior and to assess risks. Current data suggest that seismic risks in geothermal developments are low.

When large volumes of fluid are removed from underground reservoirs, the overburden confining stresses can cause compaction of the rock formation leading to observed subsidence of the land surface. In vapor-dominated fields, stable formations with *in situ* sub-hydrostatic pressures are common, and subsidence is minimal and rare. However, in liquid-dominated fields with super-hydrostatic fluid pressures, subsidence effects have been observed if replacement fluids are not injected to maintain reservoir pressures. Similar effects are observed in oil and gas fields where water injection is routinely used to mitigate subsidence. In a geopressured reservoir, subsidence can also be a problem, as the reservoir itself is supporting a major portion of the lithostatic stress. Here, fluid injection is needed to replace fluids that are extracted to maintain formation pressures and formation stability and production rates. With HDR systems, closed-loop arrangements with total reinjection of fluids are envisioned, thus both seismic and subsidence risks should be minimal.

Because all hydrothermal and geopressured systems contain steam and/or water phases with dissolved gases (CO_2, H_2S, NH_3, etc.) and minerals (silicates, carbonates, metal sulfides and sulfates, etc.) present in varying concentrations, depending on *in situ* conditions, there is a possibility of enhanced release rates over those naturally present. Nonetheless, technologies exist to separate and isolate most components in gaseous or liquid streams to control concentrations within regulated guidelines. For example, The Geysers field located in Lake County, California, has been able to operate well within California Clean Air Standards, which are currently the strictest in the US, by using appropriate H_2S abatement techniques. In addition, reinjection of spent brines or condensed vapor streams back into the formation is used to limit emissions and effluents and keep the reservoir pressurized.

11.4.3 Dispatchable heat and power delivery

One of the attractive features of geothermal energy is that it can be applied for both baseload and peaking needs for electricity and process heating. Furthermore, the reservoir itself has built-in storage capacity in the hot fluid and rock that comprise it.

For electrical applications, with the inherently low conversion efficiency of geothermal systems (10–20%), multiple production and reinjection wells are used in installations to supply one power plant, typically with a generating capacity of 10– 60 MWe. Therefore, if the power plant is designed properly, it would be possible to lower the output by throttling down the flow in one or more production wells. In addition, for HDR applications, a periodic injection-production cycle can be used with the reservoir providing dynamic storage.

11.4.4 Suitability for developing countries

Given the typical range of sizes of geothermal field developments from a few MWe to 100 MWe or more, their application in distributed generation systems is attractive. In addition, the conversion technology associated with the power plant is relatively conventional and fully dispatchable, making geothermal plants attractive for new installations in developing countries or remote regions. Indeed this has been the case in Indonesia, the Philippines, and Central America.

Admittedly, some infrastructure for exploration, drilling, reservoir stimulations, and power plant construction is needed to bring geothermal systems on line. Fortunately, many of the tasks involved can use existing labor forces with on-site training provided by the developer. Power plant components can also be shipped to sites easily, as they are typically modular.

11.4.5 Potential for CO_2 reduction and pollution prevention

For those geothermal systems that employ flashed steam cycles and/or do not use total reinjection of fluids, dissolved CO_2 can be released. The range of possible CO_2 emissions from plants utilizing hydrothermal resources in this manner is estimated to be 0.01–0.05 million metric tonnes (MMT) of carbon per quad of energy (10^{15} Btu or 10^{18}J). However, this is considerably less than for fossil-fueled alternatives (coal: 29 MMT/quad, oil: 21 MMT/quad; and natural gas: 15 MMT/quad) (Wright, 1995). Hydrothermal reservoirs also have very low emission levels of SO_x, no NO_x, and minimal particulates. For closed-loop HDR concepts, no CO_2, NO_x, SO_x, or particulates are emitted, assuming all captive plant power needs are provided by geothermal sources.

There are substantial opportunities for using geothermal energy to reduce greenhouse gas emissions and atmospheric pollution levels by displacing existing or planned fossil-fired electric power generation plants or fossil-fired boilers used for direct heating applications. These stationary systems currently make up about 80% of worldwide total energy demand. The other 20% is in the transportation sector, which also requires a transportable and storable fuel. There are a few options for using geothermal energy to produce transportation fuels as well (e.g., electrolysis of water to generate hydrogen which could be used directly or used to process biomass to produce liquid hydrocarbon fuels that are CO_2 neutral). Currently, these are not economically attractive, but concern about burning fossil fuels could change the balance in favor of such options.

11.5 Status of Geothermal Technology Today

11.5.1 Hydrothermal

International geothermal power industry. Providing energy services with robust infrastructures drives social and economic development. Although many developing countries have more than quadrupled their energy usage since 1960, they still face energy shortages that decrease the efficiency of their industrial infrastructure and lead to economic loss. Geothermal power is particularly attractive for developing countries because it can provide uninterruptible, indigenous baseload electric power and heat for both on-grid and remote, off-grid applications. Furthermore, many developing countries are located in areas of active geologic processes—areas that contain high-grade geothermal resources. The Geothermal Energy Association (1993) estimates that as much as 78,000 MWe of geothermal electrical power generation from hydrothermal resources are available for development from already known resource areas in some 50 developing countries. Realization of this amount of clean power generation would be of immeasurable value to the economies and environments of these countries. Table 11.2 and Figure 11.14 show the historical trends for geothermal energy utilization worldwide, with the exception of the period during World War II, when the Larderello field was out of service. Geothermal generating capacity has been exponentially increasing at about 8.5% per year since about 1920 as shown by the dashed line in Figure 11.14.

The most rapid development of geothermal energy for electrical power production is currently taking place in the Philippines and in Indonesia. The Philippines have become the world's second largest producer of geothermally generated electricity, growing from a capacity of 890 MWe in 1990, to 1,227 MWe in 1995, to about 2,000 MWe in 1998. Plans for the future include developing new fields, building new plants in collaboration with private industry, installing topping and bottoming cycles where feasible, and recovering of waste heat for industrial uses (Huttrer, 1995).

Geothermal capacity has grown rapidly in Indonesia, from 145 MWe in 1990, to 310 MWe in 1995, to almost 1,100 MWe installed by 2000. The government of Indonesia has promulgated a series of regulations that will stimulate the geothermal industry by permitting steam field development and power plant construction by private industry and by significantly reducing the tax burden on such projects. Further stimulus is the recent commitment of the World Bank to fund large (20–55 MWe), small (1–20 MWe), and "mini" (35–1,000 kWe) power projects at diverse sites in Indonesia (Huttrer, 1995).

In Japan, electrical power produced from geothermal resources continues to increase with the construction of new plants. At present, about 570 MWe of installed capacity have been built, and recent estimates of available (but still undeveloped) geothermal power indicate 5,820 MWe for 30 years for conventional steam generation, with an additional 14,720 megawatts for 30 years from binary generation (Ehara and Fujimitsu, 1995).

In many developing countries, electrical power generation and sales are enterprises undertaken exclusively by their governments. However, before international investors can or will undertake investment in a country, they must be assured of a reasonable economic return, which depends on the existence of an appropriate business and financial, legal, and regulatory environment in the country, along with reasonable assurances that the facilities created by private investment will not later be nationalized without fair compensation, and that retroactive laws, taxes, or other unforeseen setbacks will not be put in place. Only then can the quality of a business investment be judged against other opportunities available to the same investor.

A key question is who will develop the geothermal reservoir field? Some developing countries prefer to use state-owned companies to develop their natural resources (e.g., petroleum, geothermal energy) and then sell fuel (steam) to the builder and operator of a power plant. The national utility or a direct customer purchases the electricity, and the power-plant developer recovers his investment from the income stream of the electricity sales. In other deregulated markets, completely private ownership is more common.

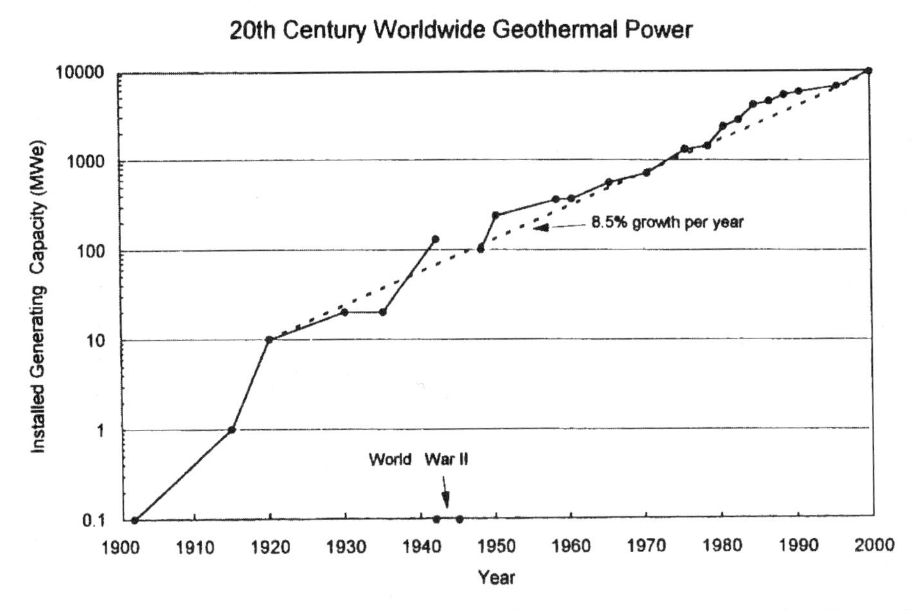

Figure 11.14. Worldwide geothermal electric power generating capacity. Updated from Mock et al. (1997). Reprinted with permission of *Annual Review of Energy and the Environment*.

Table 11.2. Geothermal Electric Generating and Nonelectric Capacities

	Installed (1979)		Installed (2000 est.)	
Country	Electric MWe	Nonelectric[a] MWt[b]	Electric MWe	Nonelectric[a] MWt[b]
United States	663	17	2,800	30
Italy	421	24	800	38
New Zealand	203	196	400	380
Japan	165	2,900	570	3,500
Mexico	150	—	150	680
Iceland	64	410	150	570
El Salvador	60	—	180	—
USSR	5	4,860	~5	NR
Philippines	4.2	4.9	2,000	—
Taiwan	1	0.6	1	5
Turkey	0.5	0.2	400	NR
Indonesia	—	—	1,100	—
Nicaragua	—	—	150	—
Costa Rica	—	—	100	—
Guatemala	—	—	100	—
Honduras	—	—	100	—
Panama	—	—	60	—
Argentina	—	—	20	—
Portugal	—	—	30	—
Spain	—	—	25	—
Kenya	—	—	30	—
Thailand	—	—	~10	—
Canada		—		10
England		—		~2
France		24		490
Hungary		1,050		NR
Czechoslovakia		93		NR
Yugoslavia		4.9		58

[a] Nonelectric does not include geothermal heat pump contribution

[b] MWt = thermal megawatts of power; capacities less than 0.1 MWt are not listed.

NR = no reported amount

Sources: DiPippo (1980, 1995), Huttrer (1995), and Mock et al. (1997).

US Geothermal Power Industry. The US geothermal industry is composed of more than 50 companies. The major US field developers are Caithness Corporation, CalEnergy Company, Inc.; Calpine Corporation; Constellation Energy, Inc.; ESI Energy, Inc.; Ormat International, Inc.; Oxbow Power Services, Inc.; and Geothermal Division, Unocal Corp. US utilities generating or purchasing geothermal power include Hawaiian Electric Company, Northern California Power Agency, PacifiCorp, Pacific Gas & Electric, Sierra Pacific Power, Southern California Edison, and Utah Municipal Power Agency. The US geothermal power industry currently has about 2,800 MW of generation capacity and produces about 17 billion kWh/year, in four states—California, Hawaii, Nevada, and Utah. Geothermal energy is the second largest grid-connected renewable electricity source, exceeded only by hydropower. In 1995, geothermal systems generated 200 times more electricity than solar energy and 5 times more than wind energy.

Nonetheless, hydrothermal geothermal resources must be managed properly or depletion rates can be large. The Geysers geothermal field, located about 120 km north of San Francisco, California, the most productive geothermal field in the world, provides a useful example. Power generation there has been decreasing (7–8% per year) due to declining reservoir pressure since 1987. The problem appears to be that the natural recharge rate of fluids is inadequate to keep pace with the reservoir production rate. In contrast, the thermal energy content of the reservoir rock within the field is far from depleted— only 5% of the thermal energy has been consumed in almost 40 years of production. Recent joint research work by the industry and the US DOE has indicated that injection of supplemental water into the reservoir will help arrest the pressure decline and enable the reservoir to produce power for many more decades (see Voge et al., 1994, Enedy et al., 1991). However, studies by Enedy and others indicate that injection will have to be done with care to avoid adverse interference effects leading to short-circuiting of cooler fluids into nearby production wells. Construction of a 46-km, 51-cm diameter waste water pipeline from the Lake County Sanitation District to the southeast part of The Geysers field was completed in 1998 (see Dellinger, 1996 for details). This pipeline brings 29.5 million liters (7.8 million gallons) per day of waste-water to the geothermal field to help maintain reservoir pressure and at the same time disposes of the waste water in an environmentally advantageous way. Reservoir engineers anticipate that the project will increase generation at The Geysers field by at least 70 MWe. With the successful start of this project, the city of Santa Rosa is planning a project to determine the feasibility of building a pipeline from their waste water treatment plants to bring additional water for recharging the central part of The Geysers geothermal field.

Potential for geothermal power in the US. Moderate- and high-temperature hydrothermal resources, in the range 105–350°C, are located at accessible depths mainly in the western third of the continental US as well as in Alaska and Hawaii. At present, there are about 2,800 MW of installed generating capacity in the US, enough to serve the domestic needs of about 3 million people. Thus, geothermal energy has moved beyond the experimental stage. Approximately 7% of California's electricity and about 25% of Hawaii's electricity comes from hydrothermal resources. There is substantial expansion and new development potential in these states, and there is potential for developing power plants in the near term in Alaska, New Mexico, Idaho, Washington, and Oregon. The rate of expansion of geothermal power in the US will depend on many factors, including availability of high-grade resources, economics, and policy incentives.

There are no recent, comprehensive, and well-documented assessments of the potentially recoverable hydrothermal resource in the US. Brook et al. (1979) gave the estimate of 95,000–150,000 MWe from known and undiscovered hydrothermal reservoirs having a projected 30-year lifetime, but most people now believe this number is too high. Wright (1995) gave a rough estimate of 4,800 MWe of electrical power available for future development, over the next 10 to 20 years, from high-grade hydrothermal systems in the contiguous 48 states. Hawaii could add 100–200 MWe and Alaska could add 50-100 MWe to the total in this time frame. Most of the disparity between these estimates is tied to the uncertainty of the magnitude of undiscovered hydrothermal resources, especially in the low to moderate temperature range. New and significantly improved technology will be needed to locate and evaluate these less accessible hydrothermal systems.

11.5.2 Advanced systems

Geopressured. Geopressured geothermal brines are hot, pressurized waters that contain dissolved methane and lie under the earth's crust at depths ranging from 12,000 to more than 20,000 feet (3,600– 6,000 m). Brine temperatures range from 50 to 260°C; pressures from 560 to 1,380 bar (7,500–20,000 psi); dissolved salts from 20,000 to 300,000 ppm; and gas content from 0.65 to 2.85 SCM (standard m^3 [23–100 standard ft^3]) per barrel of brine. Geopressured resources occur worldwide. In North America, such resources are found in Alaska, in the Rocky Mountain regions, in California, and along the coast of the Gulf of Mexico (Beeland, 1986, Negus-de Wys, 1992). The northern Gulf of Mexico sedimentary basin contains a deep sequence of relatively permeable sandstone layers interbedded with relatively impermeable shales. Both formations are relatively porous and contain large quantities of hot brine with dissolved methane under abnormally high pressures.

Geopressured resources possess a unique advantage; they embrace three forms of energy—thermal (hot brine), chemical (natural gas), and mechanical (hydraulic). It is possible to exploit each form singly or in combination to satisfy a variety of energy needs. In 1974, the federal government established a program to determine whether the geopressured aquifers along the Gulf Coast could be exploited technically and economically as a major source of domestic energy. The USGS has estimated that the northern Gulf of Mexico basin contains approximately 170,000 quads of energy (107,000 quads thermal at temperatures 100°C, 63,000 quads from combustion of the dissolved natural gas) (Muffler, 1979 and Wallace et al., 1979).

The geopressured R&D program supported by the US DOE successfully identified the geopressured fairways in the onshore Gulf Coast areas of Texas and Louisiana with an estimated 5,700 quads of methane in the sandstone reservoirs. In addition, the program established the feasibility of locating geopressured reservoirs using borehole petrophysical logs (or just well logs) from previously drilled petroleum wells and active seismic data, and demonstrated the feasibility of adapting US petroleum industry drilling and production technology to geopressured systems, proving it possible to extract the gas from the brine by simple and economical gravity separation techniques. The program also verified that geopressured pressure gradients range to more than twice the normal hydrostatic gradient, over temperatures ranging from 50 to 260°C. Further, the solubility of methane in geopressured brines was shown to decrease with increasing salinity. New technologies for assessing and using the resource were demonstrated, establishing that brine can be produced at rates up to 40,000 barrels per day from a single well (note that 1 barrel equals 42 gallons US) and reinjected into shallow aquifers without affecting the surface and near-surface water and without causing subsidence or associated seismic activity.

The research program also characterized two large sandstone aquifers, each estimated to contain in excess of several billion barrels of brine. Technology was developed for successfully controlling the formation of calcium carbonate scale in wellbores using phosphonate scale inhibitors. Electronic data bases were created using data from thousands of geophysical logs in Gulf Coast oil and gas wells and with technical bibliographic material at the University of Texas, Louisiana State University, and at the USGS facility at Bay St. Louis.

An important element of the geopressured program was the successful demonstration of a hybrid power system at Pleasant Bayou in Brazoria County, Texas, which supplied about 3,500 MWe-hr to the Houston Power & Light Company's grid over a one-year period (Campbell and Hattar, 1991). In a collaborative effort sponsored by the US DOE and EPRI, the Ben Holt Company designed and constructed a 1 MWe hybrid cycle system in which gas was burned in an engine to generate electricity directly. Exhaust heat from the engine was then combined with heat from the brine to generate additional electricity in a binary cycle. Heat from the gas-engine was available at high temperature, markedly improving the efficiency of the binary portion of the hybrid cycle resulting in 30% more power than stand-alone geothermal and fossil fuel-fired plants operating on the same energy resources.

Magma. Recovery of energy from hot magma bodies is a goal worth pursuing because it would open up possibilities for both high-efficiency electricity generation and high-temperature chemical or metallurgical processing that is not possible with any other geothermal resource. Magma bodies represent localized regions of immensely concentrated thermal energy consisting of molten or near molten material at temperatures in excess of 650°C. To be classified as a geothermal resource, they must be located no deeper than 10 km (30,000 ft) in the earth's crust at depths accessible with conventional drilling methods. In rare cases, magma bodies are found at the surface such as at or near the sites of active volcanoes. The US, the former USSR, and Japan have focused advanced research and development on engineering technologies for extracting heat. In particular, scientists at Sandia National Laboratories during the 1980s and 1990s experimented in the field with different methods at two sites in the US: at the Kilauea Iki lava lake in Hawaii, where actual heat extraction tests were conducted, and in California at the Long Valley Caldera, where exploratory drilling was carried out (Colp, 1982 and Gerlach, 1982). Other potential sites in the US include the Valles Caldera in north central New Mexico and 12 active volcanoes of the Cascade range in the Pacific Northwest.

The possible use of magma for carrying out high-temperature chemical reactions to produce transportable fuels, such as methane and hydrogen, has been proposed by Gerlach (1982). In one concept, particularly suited for magmas rich in iron oxides, one could carry out the thermochemical reaction of ferrous oxide (FeO) and water to produce ferric oxide (Fe_2O_3) and hydrogen. In another concept, biomass would be gasified to produce methane (CH_4) at moderate temperatures and then synthesis gas, a mixture of hydrogen and carbon monoxide (CO), by further reaction of CH_4 with CO_2 at higher temperatures.

Modeling efforts at Sandia, based on preliminary field testing in Hawaii, suggest that a single well drilled into a magma body at 1,000°C could generate 30 MWe. Another more recent estimate of power generation costs by Crewdson et al. (1991) for Long Valley predicts that commercially competitive busbar costs of 5.6¢/kWh may be possible. Although these are estimates for high-grade magma resources located near the surface, drilling and its associated costs may not be the limiting factor. Of more importance is engineering a heat-extraction system that will work for extended periods at high rates of thermal power production. Critical issues include maintaining high heat-transfer rates when solidification of magma occurs during extraction, finding economic materials that can withstand the temperatures and chemistry of molten rock (including drilling hardware, well casing and cements, and tubular goods involved with heat extraction), and finding ways to insure that the reservoir itself remains stable and safe during drilling and production. Meeting these challenges will require an active R&D effort (which was underway under US DOE sponsorship until recently). Unfortunately, current R&D levels for magma are now at a low level, and it is too early in the development of this resource to provide reasonable projections of costs for providing electricity or process heat.

Hot Dry Rock (HDR) or Enhanced Geothermal Systems (EGS). The greatest potential source of geothermal energy is contained in hot rock formations at technically accessible depths (currently 10 km [30,000 ft]) in the earth's crust that do not contain sufficient fluids and/or permeability and porosity to permit heat extraction at commercially viable rates. Figure 11.3 illustrates the heat-mining concept being pursued to stimulate production in low-permeability formations in competent rock by creating an open network of fractures that emulate many features of existing hydrothermal reservoirs. The primary technique for engineering these so-called HDR systems utilizes fluid pressure to open and propagate fractures from wells placed in a region of rock at temperatures ranging typically from 150 to 300°C. The main idea is to enhance natural permeability hydraulically by opening old and/or creating new fractures and connecting them to a set of injection and production wells. Energy would be extracted by circulating pressurized water in a closed loop from the surface plant down one well, through the fracture network where it is heated, and up the second well to return it to the plant. Engineering the reservoir to allow adequate flow and heat transfer performance while minimizing water losses at its periphery is a major challenge. Within the plant, thermal energy could be used directly for residential or process applications or to generate electricity in a power cycle similar to the ones employed for hydrothermal resources. With this closed-loop design, emissions and effluents from HDR systems are practically nonexistent. Of course, the impacts of waste-heat rejection, water and land use, and potential seismic risk are still present.

Because HDR systems do not require contained hot fluids or *in situ* permeability, the HDR resource base is much larger and more widely distributed—practically speaking it is ubiquitous, varying only in grade. Table 11.1 provides estimates of the HDR resource base for low- and high-grade systems. Assuming that the rock formation is amenable to stimulation, its grade, in economic terms, is largely specified by the average geothermal gradient, as this determines drilling depths to reach certain temperature levels. The average baseline gradient for the world is about 25°C/km—this establishes the low-grade HDR resource, one that would have to be exploited if geothermal energy is to be universally available. Hyperthermal areas with gradients in excess of 60°C/km characterize the high-grade end of the resource. For example, in the western US, Iceland, and parts of Japan generally higher heat flows and other desirable geologic conditions have led to large regions with gradients of 60–80°C/km and smaller areas with 100–200°C/km, such as in The Geysers Clear Lake part of northern California.

The resource estimates for HDR given in Table 11.1 are orders of magnitude larger than the sum total of all fossil and fissile resources. Although these estimates refer to the total usable thermal energy content of the geothermal resource that is accessible with current technology, even if only a small fraction can be economically extracted, the impact of HDR as a provider of sustainable, emissions-free energy would be far-reaching. In fact, it is this great potential of universal heat mining that has encouraged many to advocate pursuing HDR with national and international R&D efforts.

For about 25 years (1974–2000), the Los Alamos National Laboratory, with the sponsorship of the US DOE, led the research effort of developing HDR technology (Armstead and Tester, 1987, Tester et al., 1989, Duchane, 1994). Most of the field work was concentrated at the Fenton Hill site in north central New Mexico, which is in a high-gradient region on the western flank of an extinct volcano known as the Valles Caldera. For a period during the 1980s, the Federal Republic of Germany and the government of Japan supported a portion of the US effort at Fenton Hill under a collaborative agreement within the International Energy Agency (IEA). The Japanese, as reported by Sato and Ishibashi (1994) and Sato et al. (1995), are now developing HDR technology on their own, and the Germans have joined forces with other members of the European Economic Community (EEC) to pursue HDR with modeling and field work at several sites in Europe. Duchane (1994) and Kappelmeyer et al. (1991) described European R&D on HDR. From about 1974 until 1992, the British were carrying out extensive field tests at the Rosemanowes site in Cornwall under sponsorship of the UK Energy Technology Support Unit at Harwell (for details see Parker, 1989, Batchelor, 1984, 1987). Recently, they joined the EEC project in Europe and have suspended operations at Rosemanowes. There are other important R&D activities in other countries, most notably in Russia (see Duchane, 1991).

These field programs have made substantial progress in developing methods to drill, stimulate, and characterize the structure and performance of HDR reservoirs in a range of different geologic settings. Most of the field work has focused on creating HDR reservoirs in low-permeability formations that have little fluid content (i.e., low porosity) where the dominant features consist of sealed natural fractures or joints contained in otherwise competent rock of extremely low matrix-permeability (10^3–10^7 times less permeable than a typical oil reservoir). In these cases, hydraulic fracturing methods employing the injection of pressurized water with or without special rheological additives have been used to create an open labyrinth of connected fractures. In other efforts, where open joint systems containing significant amounts of water or steam exist naturally (so-called "hot wet rock"), field development work concentrates on reservoir characterization to insure proper well placement to optimize the distribution of fluid flowing in the fractured system.

US and Japanese testing have demonstrated that conventional drilling methods can be adapted for the harsh environments encountered in reaching zones of rock from 250 to 350°C, which are hot enough to be suitable for commercial power production in today's energy markets. Field testing has also verified that hydraulic pressurization methods can create permanently open networks of fractures in large volumes of rock. Experiments at Fenton Hill and Rosemanowes, for example, have extended open, connected fracture networks to kilometer dimensions, producing systems that are large enough for long-term commercial production. Techniques using chemical tracers, active and passive acoustic methods, and other geophysical logging techniques have been used to map out the geometric features of HDR reservoir systems to validate thermal hydraulic models of heat extraction performance.

While this early work has demonstrated the technical feasibility of the HDR concept, none of the field testing to date has operated a commercial system. Two major issues remain as constraints to global commercialization: (1) the demonstration of sufficient reservoir productivity with low-impedance fracture systems of sufficient size and thermal lifetime to maintain economic fluid production rates of 50–100 kg/s per well pair at wellhead temperatures above 150°C, and (2) the high cost of drilling wells deep into hard rock. In certain geologic situations, controlling water losses is also important, as it can have both negative economic and environmental impacts.

A key economic parameter is the cost of producing a unit of fluid at a specified temperature. This can be expressed in $ per kg/s of fluid produced or in $/kW of thermal power. High fluid-production temperatures, large sustained reservoir flow rates, and lower individual well drilling costs reduce the overall cost of HDR energy. On the other hand, higher flow rates through the reservoir lead to increased pumping power losses and accelerated rates of thermal drawdown, which lead to higher costs. The economic choice is to balance these effects.

This balance can be illustrated quantitatively by dividing the total installed cost of a geothermal development into major components associated with the field itself and its reservoir and with the power plant that converts the thermal energy into electricity. In mathematical terms, the total capital cost is

$$\Phi = \Phi_{geofluid} + \Phi_{powerplant} \tag{11-11}$$

where:

$\Phi_{geofluid}$	=	total field/reservoir development costs including those for geophysical exploration, well drilling, reservoir stimulation, and fluid transmission, collection, and distribution
$\Phi_{powerplant}$	=	installed capital costs of power conversion and electrical generation equipment and switch gear for connecting to the grid

The *geofluid* component can be represented roughly as:

$$\Phi_{geofluid} = 2n_w A^* (\Phi_{well}) \tag{11-12}$$

where n_w is the number of production wells (the factor of 2 accounts for an equal number of reinjection wells), Φ_{well} is the individual completed well cost in $/well and A^* is an empirical constant that includes costs for rig mobilization and demobilization, reservoir stimulation, and for the geofluid collection and distribution system to the power plant. Φ_{well}, in turn, is expressed as an exponential function of depth as depicted by a straight line in the shaded region of Figure 11.15,

$$\Phi_{well} = C \exp (D Z) \tag{11-13}$$

where Z is the reservoir/well depth in meters = $(T_{gf}(t=0) - T_{surface})/\nabla T$, $T_{gf}(t=0)$ is the initial reservoir fluid production temperature in °C or K, $T_{surface}$ is the average ambient surface temperature at Z=0 in °C or K, ∇T is the average geothermal temperature gradient in °C/km and C and D are constants fit to the data presented in Figure 11.16. The number of wells, n_w, is given by,

$$n_w = \frac{P}{\eta_u \dot{m}_w \Delta B} \tag{11-14}$$

where P is the power capacity of the plant in kW, η_u is the utilization efficiency (that is the fraction of thermodynamically-limited power, typically 0.50– 0.60), $\dot{m}_w$ is the geofluid mass flow rate per well in kg/s, and ΔB is the thermodynamic availability or maximum work producing potential of the geofluid in kJ/kg. For a liquid-phase geofluid,

$$\Delta B \approx \langle C_p \rangle \; [T_{gf}(t) - T_o - T_o \ln(T_{gf}(t)/T_o)] \tag{11-15}$$

where, $\langle C_p \rangle$ is the mean heat capacity of the geofluid in kJ/kg K, and T_o is the minimum thermal sink temperature in K. Plant capital costs typically decline as T_{gf} increases due to improved cycle efficiency and lower heat transfer area requirements. Empirically, we can represent this decline as a linear function,

$$\Phi_{powerplant} = \Phi^{o}_{powerplant} - ET_{gr} \tag{11-16}$$

applicable for $150°C \le T_{gf} \le 300°C$ where $\Phi^{o}_{powerplant}$ and E are empirical constants (Tester and Herzog, 1990, 1991, 1994).

Ideally, to minimize costs, we need to find a minimum for certain choices of $\dot{m}_w$ and T_{gf}. Unfortunately, for a given-sized reservoir, these variables are connected in a non-linear fashion and lead to certain economic tradeoffs. In real reservoirs, some decline in productivity is anticipated. This usually is expressed in terms of a reduction in T_{gf} over time. To optimally extract energy from a HDR reservoir, one must inject fluid at a sufficiently high rate ($\dot{m}_w$) to cause some thermal drawdown. But as T_{gf} decreases, so does ΔB from Equation (11-15), and consequently n_w (Equation (11-14)) and $\Phi_{geofluid}$ (Equation (11-12)) both tend to increase. To partly compensate for finite thermal drawdown, deeper, more costly wells (Equation (11-13)) can be drilled. Hence, to reach an optimum for a given HDR resource, defined generically by specifying both ∇T and reservoir size, a tradeoff exists between pumping more fluid (higher $\dot{m}_w$) and drilling deeper. With reasonable values for equipment and drilling costs and acceptable levels of reservoir performance, initial reservoir temperatures of 250–300°C are favored when $T > 50$°C/km while somewhat lower temperatures are better for lower ∇T.

Given these characteristics how does HDR, and more generally EGS, fit into the global energy picture? As we move into the new century, we should be exploring means of meeting our future energy needs without increased use of fossil fuels. Key criteria for any new energy technology are that it should be relatively simple to implement, safe to operate, and have reasonable costs, high availability, and low environmental impact over its life cycle. In addition, the "scale" of the technology should be compatible with anticipated demand, and the resource should be adequately distributed to meet both baseload electricity and other distributed primary-energy needs. Although HDR/EGS concepts seem to meet all these criteria as well as or better than other alternatives, they have not been pursued as vigorously as other technologies. HDR and geothermal in general have much smaller constituencies, and, as a result, their positive attributes and the current state of development of geothermal energy systems are not as widely known to the public.

Commercialization of any new energy system depends critically on how well it competes with existing energy supplies. As with any developing technology, competitiveness and risk have real and perceived elements. HDR is unique in that certain elements, such as the surface power plants, have relatively low cost and performance uncertainty as they consist of commercially available components (e.g., pumps, heat exchangers, turbines, etc.). On the other hand, the underground reservoir systems, consisting of the wellbore-fracture network, carries much higher risk, partly because the technology is not yet fully mature and partly because drilling and reservoir stimulation are perceived to be speculative because of natural parallels that are drawn to oil and gas production. One must remember that exploration and production uncertainties for an HDR reservoir are intrinsically lower than for a petroleum reservoir. In principle, all we require is sufficiently hot rock at accessible depths that is amenable to stimulation to create an open, connected fracture system.

The main technical obstacle for HDR/EGS is centered on formation and connection of the fractured network to the injection and production well system. A key requirement is low impedance access to sufficiently large rock areas and volumes with acceptable water losses. For example, for an average geothermal gradient resource of 60°C/km, production flowrates of about 40–75 kg/s with water losses less than 5% and thermal drawdown rates of about 2% per year or less from an initial temperature of 250–300°C are required to achieve breakeven electricity prices of 6–8¢/kWh (in 1996 $) (Armstead and Tester, 1987 and Tester and Herzog, 1990, 1991, 1994). These estimates include the amortized capital costs of drilling and stimulating the wells and building the power plant along with operating and maintenance costs of about 0.3–0.4¢/kWh.

Economic assessment studies conducted at Los Alamos, EPRI, and elsewhere have been dissected to reformulate a generalized economic model for HDR with revised cost components (Tester and Herzog, 1990, 1991). The main results of this comparative study are shown in Figures 11.6, 11.15, and 11.16 where drilling costs and breakeven heat and electricity prices are presented. In Figure 11.6, one sees an exponential dependence of individual well cost on depth in general, and that geothermal wells, on average, are 2–3 times more expensive than an oil or gas well to the same depth. In Figures 11.15 and 11.16, we see the strong dependence of projected breakeven price on average gradient, which reflects the grade of an HDR reservoir. As gradients increase, the drilling cost component decreases relative to the surface plant cost. The bandwidth shown in Figure 11.16 illustrates the effect of technological improvements on costs. In order to make HDR geothermal energy commercially competitive for low-grade resources with gradients of 20–30°C/km, revolutionary changes in the way in which wells are drilled are required to lower costs substantially. The linear drilling line illustrates a desirable limit for drilling costs (see also Figure 11.6).

Economic assessments for HDR/EGS assume reservoir productivities comparable to commercial hydrothermal systems. While field testing to date has not yet reached these productivity levels, a key problem is lowering flow impedance and not in creating reservoirs of sufficient size, which has already been achieved in tests at Fenton Hill and Rosemanowes. The hydraulic connectivity between the wells and the fractured reservoir needs to be improved to reduce pumping losses due to high flow impedances (Fenton Hill) or to lower water loss rates (Rosemanowes).

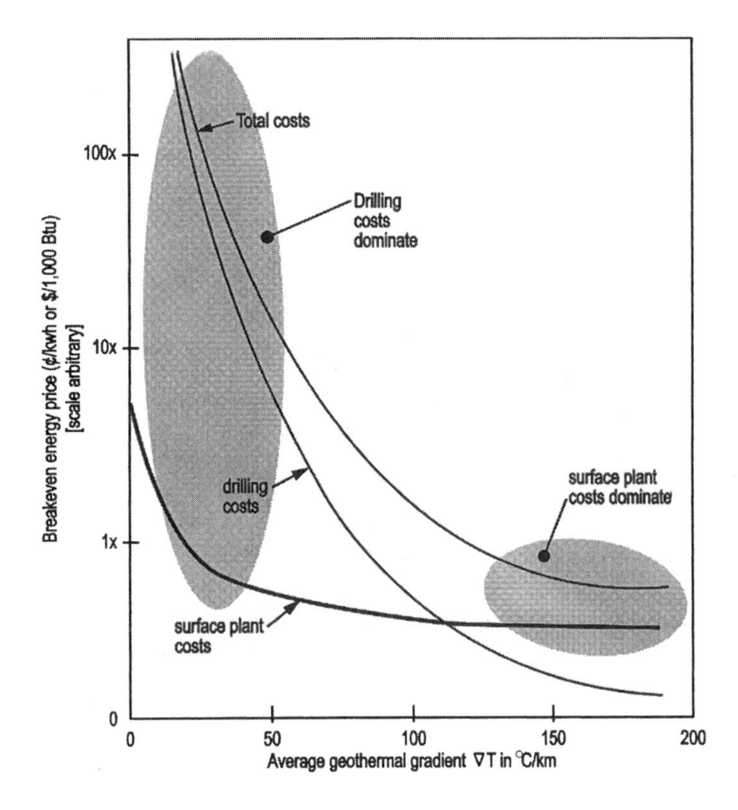

Figure 11.15. Effect of reservoir performance and resource grade on electricity or heat breakeven prices. Adapted from Armstead and Tester (1987).

Critics of HDR technology have argued that, with over $180 million invested during the past 24 years in the US alone, development toward a commercially viable system has been too slow. However, this criticism needs to be weighed against the R&D investments made in other energy technologies. For example, the total US investment in HDR in 24 years is less than half of what we are currently spending annually on fission, fusion, or solar energy technologies.

HDR technology development in the US, at least up to 2000, seemed to lie solely with the status of reservoir testing at the Fenton Hill experimental site in northern New Mexico. Unfortunately, with a decreasing federal budget for advanced energy research within the US DOE, funding for further testing at Fenton Hill or other sites needed to develop the potentially large HDR resource, in general, has become grossly inadequate. The pioneering field efforts at Fenton Hill, Rosemanowes, Soultz, and Hijiori have provided the basis for developing economically viable stimulation and extraction technologies. Regrettably, the Fenton Hill site was decommissioned in 1999. It remains to be determined whether a robust rejuvenation of R&D for advanced HDR/EGS technology will appear in the 21st century.

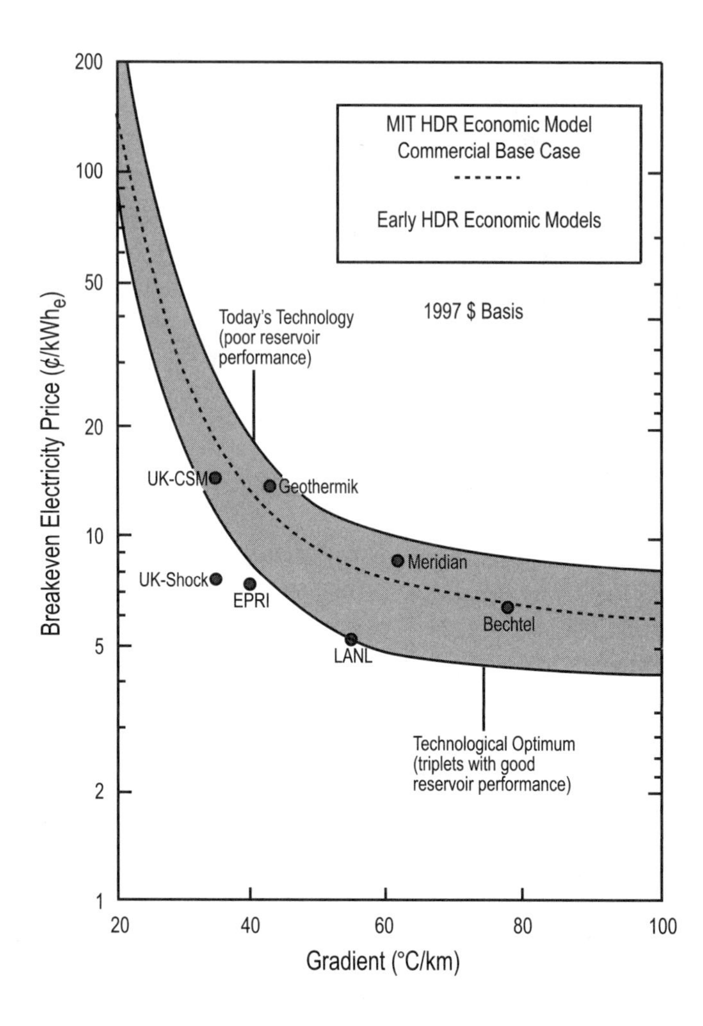

Figure 11.16. Estimated breakeven busbar electricity prices as a function of the geothermal gradient for HDR/EGS systems (updated from Tester and Herzog (1990)).

11.6 Competing in Today's Energy Markets

Geothermal energy is already commercially competitive in many locations worldwide where high-grade hydrothermal resources are found. In fact, by 2000, over 10,000 MWe of generating capacity from geothermal resources had been achieved. Even more significant is the fact that subsidy-free geothermal power is frequently the least-costly option for developing countries, such as Indonesia, the Philippines, and many countries in Central America. This is remarkable given the low cost of oil and gas on international markets and the capital-intensive nature of a geothermal development where all the "fuel" costs are embedded in the initial capital investment in drilling wells and connecting the fluid production and reinjection piping.

In the US, geothermal development has been slower for the last 10 years, as many regions do not see an immediate need for additional generating capacity. In addition, many states in the US are currently restructuring their electric utilities. Deregulation frequently focuses on the breakup of the vertically integrated utilities that constitute monopolies controlled primarily by state regulation. Generation, transmission, and distribution functions will be separated in an effort to increase competition in each area and lower electricity prices to consumers without disrupting the excellent electrical energy services Americans enjoy. As a result, many utilities will opt to sell their generating capacity to become transmission companies (transcos) or distribution companies (distcos). Generation companies (gencos) will comprise companies spun off from utilities, large independent power producers (IPPs), and the myriad small IPPs currently operating under the Public Utilities Regulatory Policy Act of 1978 (PURPA), which created markets for renewable energy in general. Gencos may sell their power into a pool from which distcos would make purchases on a long-term or a spot-market basis (or, more likely, a mixture of both), and transcos would wheel the power to distcos at an agreed transmission rate.

To preserve the benefits of clean, renewable energy supplies already in place, as well as to create a positive environment for further development, federal and state lawmakers are considering several options. For example, one option would be to require a specified amount of the overall electricity supply mix to be renewable energy, such as geothermal energy—this is the so-called *renewable-energy portfolio standard*, or RPS. If the RPS percentage were to rise with time, such an option would steadily increase the amount of renewable electricity in the supply, ensuring a growth environment for geothermal power. Another option, perhaps in addition to the RPS, would be to allow and encourage distcos to market a certain amount of their power as "green power," derived from clean sources, and offer such power to their customers at a marginally higher price. (Such issues are being discussed in the US Congress and state legislatures as this volume goes to press, and we can anticipate that several years will be required before all aspects of deregulation and restructuring of the electricity industry will be ironed out.) This is both good and bad for geothermal development: good in the sense that this trend decentralizes utility decision making and therefore provides a greater number of choices, and bad because geothermal electricity is capital intensive and more risky than options using low-cost fossil fuel. For example, a 10–50 MWe gas-fired, combined-cycle cogeneration plant with a long-term contract for purchasing inexpensive natural gas is hard for renewable energy to compete with. Tables 11.3 and 11.4 summarize estimates for baseload electricity and process heat for fossil, nuclear, and a range of hydrothermal and HDR resource grades.

For the advanced technologies—HDR, magma, and geopressured—the competition is even tougher, as risk factors will initially be higher for investors. Higher risks will be partially offset for HDR systems by the increased flexibility in locating the system near a load center and the fact that *in situ* fluids and high-formation permeability are not required. Economic projections for HDR or for any pre-commercial technology must be considered somewhat speculative given the assumptions made regarding reservoir performance. For HDR specifically, we have divided the range into high- and low-grade resources. Both cases are assumed to meet baseline reservoir productivity and other performance parameters—including mass flows per well pair of about 75 kg/s (to generate 10 MWe), effective heat transfer volumes and areas sufficient to support 10 years of production with 20% thermal drawdown or less, flow impedances (resistance to fluid flow) less than 0.1 GPa s/m^3 (an overall impedance of 1 GPa s/m^3 or less is a reasonable goal for a commercial HDR system), and water consumption rates less than 5% of the injected flow (Armstead and Tester, 1987). These baseline values are probably achievable with active R&D programs, as they more or less replicate the productivity levels already achieved for commercial liquid-dominated hydrothermal systems.

High-grade hydrothermal resources are competitive now for electric power generation, and high-grade HDR resources would be competitive in some fraction of today's energy markets. But the low-grade resources of either will require higher prices for fossil fuels or lower development costs before they can compete. Of course there are other forces—environmental, health-related, security-related—and general sustainable development concerns that could create a different economic situation.

Table 11.3. Estimated Busbar Electric Generating Costs for New Baseload Capacity (in 2003 $US)

Resource Type	Installed Plant Cost $/kW	Annualized Plant Cost[1] ¢/kWh	O&M Costs ¢/kWh	Annualized Well Drilling or Fuel Cost ¢/kWh	Total Breakeven Busbar Price ¢/kWh
Oil[2]	800	2.0	0.3	3.4–8.5 ($20–50/bbl)	5.7–10.8
Coal	1,200	3.0	0.3	0.6–4.0 ($15–100/ton)	3.9–7.3
Gas[2]	600	1.5	0.3	0.8–3.1 ($1–4/MBtu)	2.6– 4.9
Nuclear[3]	2,400	5.9	0.4	1.0	7.3
Hydrothermal					
high-grade	1,000 – 1,500	2.4–3.6	0.3	2–3	4.7– 6.9
low-grade	2,000 – 2,500	5.1–6.3	0.4	4–10	9.5–16.7
Hot dry rock					
high grade (≥60°C/km)	1,000–1,500	2.4–3.6	0.3	3–4	5.7–7.9
low-grade (ca .30°C/km)	2,000–2,500	5.1–6.3	0.4	20–40	25.5–46.7

Notes for Tables 11.3 and 11.4

Sources: Armstead and Tester (1987), DiPippo (1980, 1995), Tester et al. (1989, 1990, 1991, 1994), Herzog et al. (1997), Deutch, Moniz et al., (2003), and EIA (2003).

1. Annual cost based on 17% fixed charge rate, 80% load factor for 50 MW (e or t) plant.

2. Combined-cycle plant assumed for oil and gas.

3. Nuclear fuel estimate includes allowance for decommissioning and waste handling, treatment, and storage; fission plants only.

Table 11.4. Estimated Busbar Heat Supply Costs for New Baseload Capacity (in 2003 $US)

Resource Type	Annualized Plant Capital Cost $/MBtu[1,4]	Heat Supply[5] Fuel Cost $/MBtu	Annualized Heat Distribution Cost $/MBtu	Total Breakeven Price $/MBtu
Oil[2]	0.50	4.4– 11.0 ($20–50/bbl)	1.5	6.4–13.0
Coal	0.70	0.8–5.3 ($15–100/ton)	1.5	3.0–7.5
Gas[2]	0.40	1.2–4.7 ($1–4/MBtu)	1.5	3.1–6.2
Nuclear[3]	1.4	0.8	2.0	4.2
Hydrothermal[6]				
high-grade	—	2–3	2.6	4.6–5.6
low-grade	—	3–6	2.1	5.1–8.1
Hot dry rock[6]				
high-grade (≥60°C/km)	—	2.5–3.5	2.6	5.1–6.1
low-grade (ca 30°C/km)		4–10	2.0	6.0–12.0

See Table 11.3 for notes 1–3.

4. 1 MBtu = 10^6 Btu = 1.055×10^9 J.

5. Heat supplied at 150°C as steam or pressurized hot water, not cogenerated.

6. Fuel cost includes all heat exchange equipment and drilling costs.

11.7 Research and Development Advances Needed

The growth of geothermal energy development is not limited by resource availability. Rather, it is limited by inadequate technology. Only the highest-grade geothermal resources can be economically used today for the generation of electricity. Development of the vast majority of geothermal resources is not possible because, at most resource sites, power-generation costs are higher than those for competing fossil fuels, especially natural gas. With reasonable assumptions regarding discount rates, capital and drilling costs, and plant construction and field development periods, geothermal costs today range from 4¢ to 7¢ per kWh for new power development at high-grade geothermal resource sites, whereas new generation capacity from natural gas, where it is readily available, produces power at 2.6¢ to 6.2¢ per kWh. (Operation and maintenance costs at geothermal power plants whose capitalization has been paid may be as low as 0.3¢ per kWh.) Power-generation costs at the much more plentiful lower-grade geothermal resource sites are not able to compete today with natural-gas generation costs in the US. A core R&D program aimed at improving existing technology and developing new, advanced technology is critical to enable the geothermal industry to compete in the domestic and global energy marketplaces. Well funded R&D programs in Japan and Europe are aimed at advancing heat mining technology beyond that of the US.

Drilling. Drilling is one of the most expensive activities in geothermal development, but it is necessary to drill production wells to bring geothermal fluid to the surface and to drill injection wells to return the spent fluid to the reservoir. Because of the high temperatures and corrosive nature of geothermal fluids, geothermal drilling is much more difficult and expensive than conventional oil and gas drilling. In addition, geothermal wells are of larger diameter than oil and gas wells in order to support high flow rates, e.g., 25–320 kg/s, (200,000–2,500,000 lbs/hr or 15,000–190,000 BPD). A typical cost is $1–3 million (see Figure 11.6) for a geothermal well that will support 3–5 MWe of dry-steam or flashed-steam electrical capacity. Drilling costs account for one-third to one-half of the total costs for a geothermal project. Improvement in existing drilling techniques and development of new, advanced drilling techniques would significantly lower the cost of electricity generated from geothermal resources.

Exploration and Reservoir Technology. The major problem in exploration is how to remotely detect producing zones deep in the subsurface so that drill holes can be sited and steered to intersect these producing zones. No two geothermal reservoirs are alike, and their permeabilities vary widely over short distances. Present surface exploration techniques are not specific enough, and they lead developers to drill too many dry wells, driving up development costs. Further, inadequate knowledge of the physical and chemical properties of the subsurface makes it impossible to mine the heat in the most efficient way and ensure the sustainability of geothermal reservoirs. Increased knowledge of rock mechanics is needed to provide improved techniques for creating new fractures and better well stimulation methods to open existing fractures and lower flow impedance. Better geochemical and geophysical techniques as well as improved computer methods for modeling heat-extraction strategies from geothermal reservoirs are needed. It is possible that geothermal development costs could be lowered by 10–20% and reservoir lifetimes extended significantly with improvements in earth science techniques.

Energy Conversion. The efficiency in conversion of geothermal steam into electricity in the power plant directly affects the cost of power generation. During the past decade, the efficiency of dry- and flash-steam geothermal power plants has been improved by 25%. Power plants installed at The Geysers geothermal field in California during the 1960s required 9 kg (20 lbs) of steam to produce 1 kWh of electricity (i.e., 2.5 kg/s of dry steam per MWe, or 10–15 kg/s per 5 MWe well). The newest plants at The Geysers, installed in the 1980s, required only 6.6 kg (14.5 lbs) of steam to produce that same kilowatt-hour of electricity (i.e., 1.8 kg/s of dry steam per MWe or 9–11 kg/s per 5 MWe well). Geothermal power-plant efficiency can probably be improved at least 25% more over the next decade with a modest investment in R&D.

11.8 Potential for the Long Term

The global geothermal energy resource base is large and well-distributed. Geothermal power, generated from these resources, has grown steadily since the early 1920s at an average rate of 8.5% per year.

Geothermal power plants offer several advantages. They are simple, safe, modular (1–50 MWe), can be built rapidly (approximately one year for a 50 MWe plant), and are capable of providing base-load, load-following, or peaking capacity. Geothermal plants provide significant societal benefits, including indigenous energy for the long term with benign environmental attributes (negligible emissions of CO_2, SO_x, NO_x, and particulates, and modest land and water use). These features are compatible with sustainable development objectives at all levels, making geothermal energy an attractive option.

The growth in geothermal power capacity has been based exclusively on the use of high-temperature (T>150°C), high-grade, hydrothermal resources. If geothermal power is to become more universally available and have a significant impact on global energy supplies in the next century, then lower temperature hydrothermal resources and other advanced concepts, including HDR and enhanced geothermal systems, geopressured, and magma, must be vigorously pursued to make them economically competitive. This will require a robust advanced research program to reduce field development (especially drilling and stimulation) costs and increase energy conversion efficiencies.

Lower-temperature hydrothermal resources (T<150°C) provide an economical source of energy for geothermal heat pumps (GHPs) and for direct use in domestic, industrial, agricultural, aquacultural, and district heating applications. The installation of GHPs in the US has been growing rapidly over the past decade, at a growth rate of over 15% per year. GHPs enable users to obtain an inexpensive source of space heating and cooling, along with domestic hot water, while offering utilities the benefits of reduced peak demands for power and the deferred need for additional plant capacity. New concepts are also appearing that would increase the application of geothermal energy. For example, the idea of using a 2.5 km borehole as a deep GHP heat exchanger for heating and cooling applications in large building complexes is being pursued in Germany (Fell, 2003).

Research efforts continue in Europe, Japan, and the US to improve technology for heat mining in deep, hot formations of low permeability. If these efforts receive sufficient support and are successful, they could make geothermal energy universally available.

References

Aladiev, I.T., V.P. Peredery, E.M. Strigin, E.V. Saperov, and V.K. Fardzinov. 1975. "Heat and Mass Transfer Processes in Aquifer Systems with Artificially Increased Fracturing." *J. Eng. Power. Trans.* 3:1529.

Armstead, H.C.H. 1983. *Geothermal Energy*, Second edition. London: E. & F.N. Spon.

Armstead, H.C.H., and J.W. Tester. 1987. *Heat Mining*. London: E. & F.N. Spon.

Baljé, O.E. 1962. "A Study on Design Criteria and Matching of Turbo Machines" *J. Eng. Power. Trans.* ASME 84:83.

Batchelor, A.S. 1987. "Development of hot dry rock systems in the UK." *IEE Proceedings. A.*, 134(5):371–380.

Batchelor, A.S. 1984. "Hot dry rock geothermal exploitation in the United Kingdom." *Modern Geology,* 9:1–41.

Beeland, G.V. 1986. "A decade of geothermal development in the United States." *GRC Bulletin*, 15(8):27–33.

Bloomquist, R.G. 1981. *Proc. Low Temperature Utilization, Heat Pump Applications, District Heating*. WAONG-81-05. Olympia, WA: Washington State Energy Office.

Bodvarsson, G., and J.M. Hanson. 1977. "Forced Geoheat Extraction from Sheet-like Fluid Conductors." *Proceedings of the Second NATO-CCMS Information Meeting on dry hot rock geothermal energy*. Los Alamos Scientific Laboratory report, LA-7021:85.

Brook, et al. 1979. "Hydrothermal convection systems with reservoir temperatures ≥ 90°C," in Muffler, L.J.P. (ed.) *Assessment of geothermal resources of the United States*, US Geol. Surv. Circ. 790, Washington, DC:

Brown, L.R., C. Flavin, and S. Postel. 1990. *Picturing a sustainable society: State of the World*, a Worldwatch Institute Report on Progress Toward a Sustainable Society. New York/London: W.W. Norton & Company.

Campbell, R.G., and M.M. Hattar. 1991. "Design and Operation of a Geopressured-Geothermal Hybrid Cycle Power Plant." *Holt* Rep. 30008. Pasadena, CA: Ben Holt Co.

Colp, J. 1982. "Sandia National Laboratories report SAND 82-2377," Sandia National Laboratories, Albuquerque, NM.

Craig, H. 1963. "Isotopic Geochemistry of Water and Carbon in Geothermal Areas." A chapter in *Nuclear Geology on Geothermal Areas*, Consiglio Nazionalle della Richerche Laboratorio de Geologica Nucleare, 17–53. Pisa, Italy.

Crewdson, R.A., W.F. Martin, D.L. Taylor, and K. Bakhtar. 1991. *An evaluation of technology and economics of extracting energy from magma resources for electric power generation*, Bakersfield, CA: Mine Development and Engineering Co.

Dellinger, M. 1996. "Turning community wastes in sustainable geothermal energy—the S.E. Geysers effluent pipeline project." *Transactions*, Geothermal Resources Council 20:205–208.

Deutch, J. and E.J. Moniz, co-chairs. 2003. *The future of nuclear power: An Interdisciplinary MIT Study*, Massachusetts Institute of Technology, Cambridge, MA.

Diadkin, Y.D. and Y.M. Pariisky. 1975. "Theoretical and Experimental Grounds for Utilization of Dry Rock Geothermal Resources in the Mining Industry." *Proceedings of Second UN Symposium on the Development and Use of Geothermal Resources*. 3:1606. Lawrence Berkeley Laboratory, San Francisco, CA.

DiPippo, R. 1995. "Geothermal electric power production in the United States: a survey and update for 1990–1994." Proceedings of the World Geothermal Congress 1995. 18–31. Florence, Italy.

DiPippo, R. 1980. *Geothermal Energy as a Source of Electricity*. Washington, DC: US Government Printing Office.

Duchane, D. 1994. "Geothermal Energy." *Encyclopedia of Chemical Technology*. 12:512–539. New York: Wiley.

Duchane, D. 1991. "International Programs in Hot Dry Rock Technology Development." *Geothermal Resources Council Bulletin*. 135–142.

Edwards, L.M., G.V. Chilingar, H.H. Rieke III, and W.H. Fertl, eds. 1982. *Handbook of Geothermal Energy*. Houston, TX: Gulf Publishing.

Ehara, S. and Y. Fujimitsu. 1995. "The nature and occurrence of geothermal resources in Japan—a review." *GRC Transactions*, 19:311–314.

EIA (Energy Information Administration). 2003. Washington, DC: US Dept. of Energy.

Enedy, S.L., K. L. Enedy, and J. Maney. 1991. "Reservoir response to injection in the Southeast Geysers: Monograph on The Geysers Geothermal Field." *Geothermal Resources Council Special Report No. 17*:211–219.

EPRI (Electric Power Research Institute). 1978. "Geothermal Energy Prospects for the Next 50 Years." EPRI ER-611 SR.

Eskesen, J.H. 1980. "Flashed Steam-Steam Turbine Cycles." *Sourcebook on the Production of Electricity from Geothermal Energy*, ed. J. Kestin. Washington, DC: US Government Printing Office.

Eskesen, J.H. 1977. "Study of Practical Cycles for Geothermal Power Plants." *General Electric Company Report* COO-2619-1 UC-66, Contract No. EY-76-C-02-2619.

Fell, N. 2003. "Deep heat." *New Scientist*, 177:40–41.

Geo-Heat Center. 1988. "Geothermal direct use developments in the United States." *GHC Tech. Rep.* Klamath Falls, OR: Oregon Institute of Technology.

Geothermal Energy Association. 1993. "Geothermal Energy in Developing Countries—Opportunities for Export of US Goods and Services." Brochure originally developed by the National Geothermal Association.

Gerlach, T.M. 1982. "Analysis of magma-thermal conversion of biomass to gaseous fuel." *Sandia National Laboratories report* SAND 82-0031. Albuquerque, NM.

Gringarten, A.C., P.A. Witherspoon, and Y. Ohnishi. 1975. "Theory of heat extraction from fractured hot dry rock." *J. Geophys. Res.* 80(8), 1120.

Harlow, F. and W. Pracht. 1972. "A theorectical study of geothermal and energy extraction." *J. Geophys. Res.* 77:7038.

Healy, J. 1975. "Geothermal fields in zones of recent volcanism." *Proceedings of the Second UN Symposium on the Development and Use of Geothermal Resources*, Lawrence Berkeley Laboratory, San Francisco, CA, 1:415.

Herzog, H.J., J.W. Tester, and M.G. Frank. 1997. "Economic Analysis of Heat Mining." *Energy Sources*, 19:19–33.

Holt, B. and E.L. Ghormley. 1976. "Energy Conversion and Economics for Geothermal Power Generation at Heber, California, Valles, Caldera, New Mexico, and Raft River, Idaho—Case Studies." *Electric Power Research Institute Report*, EPRI-ER-301, Topical Report 2. Palo Alto, CA.

Huttrer, G.W. 1995. "The status of world geothermal power production 1990-1994." *Proceedings of the World Geothermal Congress 1995*, 18–31. Florence, Italy.

JAS. 1976–2000. *Joint Association Survey on Drilling Costs*, Washington, DC: American Petroleum Institute (API).

Kapplemeyer, O. et al. 1991. "European HDR Project at Soultz-sous-Forets General Presentation." *Geothermal Science and Technology*, 2 (4):263–289.

Kavlakoglu, S. 1970. "Origin of geothermal waters or natural steam." *Proceedings of the First UN Symposium on the Development and Utilization of Geothermal Resources*, Pisa, Italy: 1250.

Keller, J.G. 1977. "Heat Pumps: Primer for Use with Low Temperature Geothermal Resources." INEL Rep. GP-118. Idaho Falls, ID: Idaho National Engineering Laboratory.

Kozloff, K.L. and R.C. Dower. 1993. *A new power base—renewable energy policies for the nineties and beyond*. Washington, DC: World Resources Institute.

Kruger, P. and C. Otte. 1972. *Geothermal Energy Resources, Production, Stimulation*. Stanford, CA: Stanford University Press.

L'Ecuyer, M.L., C. Zoi, and J. Hoffman. 1993. "Space Conditioning: The Next Frontier." 430-R-93-004. Washington, DC: US EPA.

Lienau, P.J., H.P. Ross, and P.M. Wright. 1995. "Low temperature resource assessment." *Geothermal Resources Council Bulletin*: 19, 63–68.

Lienau, P.J. and H.P. Ross. 1996. *Final Report: Geo-Heat Center*. Report 35. Klamath Falls, OR: Oregon Institute of Technology.

Lienau, P.J., J.W. Lund, and G.G. Culver. 1995. "Geothermal direct case in the United States update: 1990–1994." *GHC Quarterly Bulletin,* 16 (2):1–6.

Lindal, B. 1973. "Industrial and Other Applications of Geothermal Energy." *Geothermal Energy: Review of Research and Development*, LC No. 72-97138: 135–148. Paris: UNESCO.

Mansure, A.J. 2004. personal communication, Sandia National Laboratories, Albuquerque, NM.

McFarland, R.D. and H.D. Murphy. 1976. "Extracting Energy from Hydraulically Fractured Geothermal Reservoirs." Proceedings of the Eleventh Intersociety Energy Conversion Engineering Conference, State Line, NV, Sept. 12–17. *AIChE:* 828.

Milora, S.L. and J.W. Tester. 1976. *Geothermal Energy as a Source of Electric Power*. Cambridge, MA: The MIT Press.

Mock, J.E., J.W. Tester, and P.M. Wright. 1997. "Geothermal Energy from the Earth: Its Potential Impact as an Environmentally Sustainable Resource." *Annual Review of Energy and the Environment*, 22:305–356.

Muffler, L.J.P., ed. 1979. *Assessment of Geothermal Resources in the United States*. US Geological Circular 790. Washington, DC: USGS.

Muffler, L.J.P. 1975. "Tectonic and Hydrologic Control of the Nature and Distribution of Geothermal Resources." Proceedings of the Second UN Symposium on the Development and Use of Geothermal Resources, Lawrence Berkeley Laboratory, San Francisco, CA. 499.

Muffler, L.J.P. and M. Guffanti, eds. 1978. *Assessment of Geothermal Resources in the United States*. US Geological Circular 790. Washington, DC: USGS.

Murphy, H.D. and R. Wunder. 1978. "Thermal Drawdown and Recovery of Singly and Multiply Fractured Geothermal Reservoirs." *Los Alamos Scientific Laboratory Report* LA-UR-78-739. Los Alamos, NM.

Negus-de Wys, J. 1992. "Geopressured-geothermal energy: the untapped resource." *INEL Rep*. BP633-0792. Idaho Falls, ID: Idaho National Engineering Laboratory.

Nichols, K.E. and A. Malgieri. 1978. "Technology Assessment of Geothermal Pumping Equipment." Final Report, US Department of Energy Contract EG-77-C-04-4162. Arvada, CO: Barber-Nichols Engineering.

Painter, J. 1984. "Warm-water well cuts power costs." *Public Power,* 34–35

Parker, R. 1989. "Hot Dry Rock Geothermal Energy Research at the Camborne School of Mines." *Geothermal Resources Council Bulletin,* 18 (9): 3–7.

Pratsch, L.W. 1995. "Geothermal: a household word by the year 2000." Presented at International Energy and Environmental Cong. Assoc. of Energy Engineers, Richmond, VA.

Rafferty, K. 1996. "A capital cost comparison of commercial ground-source heat pump systems." *Proceedings Geothermal Program Review XIV*, ed. A.J. Jelacic, 261–265. Washington, DC: DOE.

Rafferty, K. 1994. "Greenhouse heating equipment selection spreadsheet." *GHC Quarterly Bulletin,* 16(1):12–13.

Reed, M.J., ed. 1983. *Assessment of Low-Temperature Geothermal Resources of the United States, 1982*. US Geological Survey Circular 892.

Reistad, G.M. 1975. "Potential for Nonelectrical Applications of Geothermal Energy and their Place in the National Economy." *J. Eng. Power. Trans. ASME*, 84:2155–2164.

Rowley, J.C. 1982. "Worldwide geothermal resources." *Handbook of Geothermal Energy*, Edwards et al., eds. Chapter 2:44–176. Houston, TX: Gulf Publishing.

Sass, J.H. 1996. "Hot dry rock and the US Geological Survey: a question of priorities." *Geothermal Resources Council Bulletin,* 25(8):313–315.

Sass, J.H., ed. 1993. "Potential of hot-dry-rock geothermal energy in the eastern United States." USGS Open File Report 93–377. Washington, DC: US Geological Survey.

Sato, Y.K. and K. Ishibashi. 1994. "Technology developments in the Japanese HDR project at Hijiori." *Geothermal Resources Council Transactions*. 18.

Sato, Y.K., K. Ishibashi, and T. Yamaguchi. 1995. "Status of Japanese HDR project at Hijiori." *Proceedings of the World Geothermal Congress 1995,* 18–31. Florence, Italy.

Sperry Research. 1977. "Feasibility Demonstration of the Sperry Down-Well Pumping System." Final Report, Contract No. EY-76-C-02-238. US Department of Energy, SCRC-CR-77-48. Sudbury, MA: DOE.

Stuart, C.A. 1970. "Geopressures." *Proceedings of the Second UN Symposium on Abnormal Subsurface Pressures*. Louisiana State University, Baton Rouge, LA: 121.

Tamrazyan, G.P. 1970. "Continental Drift and Thermal Fields." *Proceedings of the First UN Symposium on the Development and Utilization of Geothermal Resources*, Pisa, Italy. Geothermics Special Issue: 1212.

Tester, J.W. 1982. "Energy conversion and economic issues for geothermal energy," in *Handbook of Geothermal Energy*, Edwards, L.M., G.V. Chilingar, H.H. Rieke III, and W.H. Fertl, eds. Houston, TX: Gulf Publishing.

Tester, J.W. and C.O. Grigsby. 1980. "Geothermal Energy," in *Kirk-Othmer Encyclopedia of Chemical Technology*, third edition, New York: Wiley. 11:746–790.

Tester, J.W. and H.J. Herzog. 1991. "The economics of heat mining: an analysis of design options and performance requirements for hot dry rock (HDR) geothermal power systems." *Energy Systems and Policy,* 25:33–63.

Tester, J.W. and H.J. Herzog. 1990. "Economic predictions for heat mining: A review and analysis of hot dry rock (HDR) geothermal energy technology." *MIT-EL-90-001*. Cambridge, MA: MIT Energy Lab.

Tester, J.W. and M.C. Smith. 1977. "Energy Extraction Characteristics of Hot Dry Rock Geothermal Systems." *Proceedings of the Twelfth Intersociety Energy Conversion Engineering Conference*, Washington, DC. American Nuclear Society, 1:816.

Tester, J.W., D.W. Brown, and R.M. Potter. 1989. "Hot dry rock geothermal energy—a new energy agenda for the 21st century." *LA-11514-MS*. Los Alamos, NM: Los Alamos National Laboratory.

Tester, J.W., H.D. Murphy, C.O. Grigsby, R.M. Potter and B.A. Robinson. 1989. "Fractured Geothermal Reservoir Growth Induced by Heat Extraction." *SPE J. Reservoir Engineering*, 3:97–104.

Tester, J.W., H.J. Herzog, Z. Chen, R. Potter, and M. Frank. 1994. "Prospects for universal geothermal energy from heat mining." *Science and Global Security,* 5:99–121.

Voge, E.B., B.J.S. Smith, S. Enedy, J.J. Beall, et al. 1994. "Initial findings of the Geysers Unit 18 cooperative injection project." *Transactions*, Geothermal Resources Council 18:353–357.

Wallace, R.H., T.F. Kraemer, R.E. Taylor, and J.B. Wesselman. 1979. "Assessment of geopressured-geothermal resources in the northern Gulf of Mexico basin," in *U.S. Geological Survey Circ: 790*, White, D.E. and D.L. Williams, eds.: 132.

White, D.E. 1973. "Characteristics of Geothermal Resources." *Geothermal Energy*, P. Kruger and C. Otte, eds. Stanford, CA: Stanford University Press.

White, D.E. and D.L. Williams, eds. 1975. Circ. 726.

Wright, P.M. 1995. "The sustainability of production from geothermal sources." *Proceedings of the World Geothermal Congress 1995*, 4:2825–2836. Florence, Italy: International Geothermal Association.

Web Sites of Interest

www.geothermal.org

www.eere.energy.gov

www.usgs.gov

www.nrel.gov

www.aeees.org

www.nedo.go.jp/english/index.html

Problems

11.1 Some argue that geothermal energy is a dilute, low-grade energy resource. Estimate the minimum mass of hot rock needed to produce 1,000 MWe-yr of electricity assuming that the rock mass is solid granite with a density of 2,500 kg/cubic meter and a heat capacity of 1,000 J/kg K at an initial temperature of 250°C. An ambient temperature of 25°C can be assumed.

11.2 A good geothermal well produces 75 kg/s of hot water at 200°C. For the same amount of delivered energy, what would the production rate be from an oil well? How does this compare with a typical US oil well producing about 100–1,000 bbl of oil per day? You can assume that the average heat capacity of the geothermal water is 4,200 J/kg K and that the minimum useful temperature is 50°C and that crude oil has a heating value of about 5.5 million Btu per bbl.

11.3 In conventional air-to-air heat pump systems, the atmosphere is used as a source of energy in the winter during heating season and as a sink for heat rejection in the summer during air-conditioning season. Geothermal ground source heat pumps use an underground reservoir as a thermal heat source and sink. These reservoirs are usually 3–10 meters deep, below the depth where seasonal fluctuations occur (i.e., below the frost line). At these depths the temperature is about 15°C. A non-freezing, non-corrosive fluid, like an aqueous solution of potassium acetate, is circulated through a coil of pipe buried in the ground to transfer thermal energy to and from the ground. How would you expect a geothermal heat pump system to perform in comparison to an air-to-air heat pump system operating under the following conditions in North Dakota?

a. Summer day when the outside temperature is 100°F (37.8°C)

b. Winter day when the outside temperature is -30°F (-34.4°C)

People in North Dakota like to keep their homes at a constant 70°F (21.1°C) year round. Comparisons should be made on an ideal basis for a fully reversible system and should be expressed in terms of units of heat (or cooling) transferred per unit of electrical work consumed, which is called the coefficient of performance (COP). What factors will limit the performance of practical heat pumps below their ideal limit?

11.4 What is the maximum amount of electric power that could be generated from a 250°C geothermal fluid flowing at 40 kg/s in Reykjavik, Iceland? The average ocean temperature off the coast of Iceland is about 5°C.

11.5 Just imagine a situation in the future where HDR/EGS technology has succeeded to a point where it could provide one-half of the US electrical capacity. In 2003 the US capacity was about 750,000 MWe. How many wells would have to be drilled and what would their average depth be? What would the total capital investment need to be to achieve such a level of penetration? Some argue that this change would drastically alter our electricity transmission and distribution (T&D) infrastructure, and this might lead to a more distributed system for supplying combined heat and power. Do you agree or disagree with that statement? Explain. Be sure to state all the assumptions you made in making your estimates.

11.6 Two-stage flashing systems are commonly employed for generating power from hydrothermal resources that contain only minor amounts of non-condensable gases (see Figure 11.10 and the discussion in the text). Here we see that a key design variable is the selection of the pressure or steam saturation temperature for each stage. Develop a simplified analysis to estimate the optimal flash temperatures for each stage. Compare your results with those shown in Figure 11.13 for a geofluid production temperature of 230°C with the same condensing (T_c) and dead state (T_o) conditions.

Hydropower 12

12.1 Overview of Hydropower

Hydropower is a renewable energy resource resulting from the stored energy in water that flows from a higher to a lower elevation under the influence of the earth's gravitational field. Ultimately, it is connected to solar energy and the natural hydrologic cycle of evaporating water from lakes and oceans re-deposited as rain or snow. Water flowing in rivers from upstream regions above sea level toward the oceans is continuously converting part of its potential energy into kinetic energy associated with the flow velocity. Such energy exchange creates opportunities for hydropower in the form of converting stored potential energy contained by a dam structure or extracting a portion of the total energy contained in flowing water itself. The former is frequently called impoundment hydropower, and the latter is called diversion or run-of-river hydropower. There is also a third type of hydropower, referred to as pumped storage. During periods when electricity demand is low, water is pumped using electricity from a lower reservoir up to a higher elevation where it is stored. When demand increases, the flow is reversed, and electricity is produced as water passes from the higher storage reservoir back to the lower reservoir.

In most installations today, hydro energy is converted to electricity by flowing water through turbines to produce rotating shaft work, which turns an electric generator. Technologies for carrying out this conversion are highly developed and efficient, with hydropower installations ranging in scale from a few kWe to over 10,000 MWe. For several centuries, hydropower had been used to produce mechanical power to perform a range of activities, including grain milling, textile processing, and other light industrial operations. In fact, a large fraction of the Industrial Revolution in the 18th century in Europe and the US was "fueled" by access to hydropower. For example, many industrial plants during that period were located on rivers to provide both power production and transportation of goods. In early times, water wheels were commonly used for mechanical power production (see Reynolds, 1983). Even today, in remote regions of less developed countries such as China and India, small-scale hydropower still plays an important role.

This chapter focuses on the conventional role that hydropower plays as a major supplier of electricity around the world. Hydropower now produces 20% of the world's electricity, with about 635,000 MWe of generating capacity in operation in more than 150 countries in 2002. According to the National Hydropower Association (2001), 103,800 MWe of hydroelectric capacity exists in the US, of which 78,200 MWe is conventional (about 10% of the US total) and 25,600 MWe is associated with pumped storage installations. In developed countries, hydropower can still be a major player. Switzerland, Canada, and Norway and regions like the Pacific Northwest of the US all rely heavily on hydropower. In developing countries, hydropower can be even more important, supplying on average about 1/3 of their electricity needs with less than 10% of the total hydropower potential exploited.

Hydropower installations have a reputation of being robust and durable, often operating successfully at specific sites for more than a century. Plants can be large. In fact, the 10 largest electric power stations in the world today are hydroelectric (see Table 12.1 for representative examples), and substantial hydropower potential remains undeveloped worldwide.

The key environmental issues involving hydropower center on impacts on fish migration, water quality, land inundation, and aquatic ecology. Although existing and new hydropower developments have come under attack in many countries during the last several decades, hydropower is important as a major, renewable, and potentially non-greenhouse-gas-emitting energy source and as an economic and political force in many countries. The multiuse aspects of hydropower for electricity, water supply, agricultural irrigation, flood control, and recreation are important from a sustainability perspective.

Table 12.1. Representative Mega-scale Hydropower Projects

Name	Location	Type	Capacity, MWe	Reservoir Size
Grand Coulee	Columbia River, Lake Roosevelt, Washington	Impoundment dam, 550 ft (168 m) high	6,480	9.4 million acre ft[a]
Niagara Falls	Niagara River, New York	Diversion, run of river	1,950	nil
Hoover Dam	Colorado River, Lake Mead, Nevada	Impoundment dam, 726 ft (223 m) high	1,500	28.3 million acre ft 146,000 acres
Norris Dam TVA	Tennessee River, Norris, Tennessee	Impoundment dam		
Glen Canyon	Colorado River, Lake Powell, Arizona	Impoundment dam, 710 ft (216 m) high	1,500	27.0 million acre ft
La Grande complex	St James Bay, Quebec, and Labrador, Canada	Impoundment, Multiple dams	10,000	9,600 km^2 or about 100 Quabbins[b]
Itaipu	Paraguay/Brazil	Impoundment dam, 150 m high	12,600	
Three Gorges	Yangtze River China	Impoundment dam	17,000	
Guri	Venezuela	Impoundment dam	10,300	
Krasnoyarsk	Russia	Impoundment dam	6,000	

[a]1 acre ft = 326,000 gal

[b]1 Quabbin = size of the major Massachusetts reservoir system (39 square miles surface area with 12,640 acre ft. = 4.12×10^9 gal of water contained)

12.2 Hydropower Resource Assessment

The unexploited potential for hydropower is large. For example, the World Energy Council (2001) estimates the gross theoretical potential to be about 40,000 TWh/year, of which 9,000 TWh/year is technically and economically feasible. This is more than three times the current hydropower production level of 2,600 TWh/yr. Without doubt, hydropower is strategically important worldwide. (Table 12.2 provides a regional breakdown of current capacity.)

Currently, further hydropower resource development is limited mainly by available capital and environmental concerns. For hydropower to grow significantly in the 21st century, significant national financial investments will have to be made in the face of more economically and environmentally attractive energy options. A key issue to keep in mind is that generating electricity is not the sole reason for building a dam. Broader water management concerns often dominate policy decisions. These may include flood control, agricultural irrigation needs, or even recreational considerations.

Canada and the US are currently the largest producers of hydropower, but this situation is likely to change in the decades ahead as China, Brazil, and other developing countries expand their capacity (see Table 12.3). Even though hydropower installations in developed OECD countries are mature, with higher-grade resources already exploited, there still is considerable growth potential. For example, consider the expansions that occurred from 1945 to 1990 under the auspices of the Bonneville Power Administration and the Tennessee Valley Authority. The gigantic Grand Coulee Dam in eastern Washington had its generating capacity increased by over 2,000 MWe in the 1980s to its current level of 6,480 MWe.

Table 12.2. Global Hydropower Capacity and Investment Estimates

North America 743,187 GWh/yr	Europe 647,000 GWh/yr
South America 470,992 GWh/yr	Asia 555,000 GWh/yr
Africa 59,283 GWh/yr	Australia 39,000 GWh/yr

1,560 North American Plants (5,000 Units)
13,000 International Plants (42,000 Units)
World total capacity = 654,000 MWe
World total output = 2,517,500 GWh/Yr
World total investment = approx. $2 trillion US total or annualized to $50 billion/yr

Sources: World Energy Council (2001), International Commission on Large Dams, ICOLD (2001); World Commission on Dams (2002).

With newer, more environmentally friendly technologies and proper policy incentives, expanding, re-powering, and upgrading turbine generators at existing dams could significantly increase capacity. For example, there is the potential to install 35,000 to 70,000 MWe of new capacity in the US alone using existing dam structures and reservoirs (US Hydropower Association, US Army Corps of Engineers, FERC, USDOE, 2001–2003). If lower-head, run-of-river technologies become feasible, the potential is much higher (see Section 12.3).

As we move from country to country and from region to region, one notes large differences in the relative importance of hydropower as a producer of primary electricity (see Table 12.4). For example, in Norway, Switzerland, Austria, and Brazil, 80–100% of electricity comes from hydropower, while in the US and China, it is about 10% or less. In South America, about 75% of the continent's electric power is supplied by hydropower dams. The major national players are Brazil, Argentina, Paraguay, Colombia, Peru, and Venezuela, with Chile and others not far behind. Large increases in capacity are currently under construction or planned for the near future in South America (e.g., for Brazil [53,000–94,000 MWe], Argentina [8,750–12,000 MWe], and Venezuela [12,200–27,600 MWe]).

Capacity increases in China are not as well documented except for large projects like the Three Gorges Dam complex that eventually will generate 17,000 MWe or more. Although only 1,855 dams are officially listed for China in the *World Register of Large Dams*, other sources have estimated that an additional 22,000 non-registered dams exist in China (*The Economist*, 2003). Assuming these estimates are correct, China has more than 45% of all the world's dams followed by the US (6,600), India (4,300), Japan (2,700), Spain, and Canada (ICOLD, 2001, and WCD, 2002). If a significant fraction of China's existing dams are equipped with turbines, then its hydropower capacity could increase markedly.

One of the strong attributes of hydropower is the dispatchability that results from the system's ability to store energy in the water contained behind a dam or by periodically pumping into a temporary storage reservoir. Given the general increase in electrification that is occurring worldwide, the demand for using hydropower reservoirs for both base-load and peaking applications is rising. In addition, pumped energy storage capacity is likely to grow as well. For example, in 1950, the US had less than 5,000 MWe of pumped storage capacity. Today, there is over 25,000 MWe. Worldwide, pumped storage capacity now exceeds 100,000 MWe with about 40% in Europe, 25% in Japan, and 25% in the US (extrapolated from Moreira and Poole, 1993). In the long term, other factors may also lead to increased interest in hydropower. The variable nature of other renewable energy sources, like wind and solar, make pairing with hydro energy storage an attractive option for integrated supply systems.

Table 12.3. Hydropower Capacity Estimates By Continent, Based on Large Dam Technology

Continent	Capacity in 2001		Maximum theoretical potential	Technically possible	Economically possible
	GWe	TWh/yr	TWh/yr	TWh/yr	TWh/yr
North America	154	743.2	6,150	2,700	>1,500
South America	99	471.0	7,400	3,000	>2,000
Africa	21	59.3	10,120	1,150	>200
Europe	210	646.9	5,000	2,500	>1,000
Asia	157	555.0	16,500	5,000	>2,500
Oceania *	13	42.4	1,000	300	>100
Total world	654	2,518	46,170	14,650	>7,300

Sources: World Energy Council (2001), International Commission on Large Dams, ICOLD (2001), World Commission on Dams (2002), and Moreira and Poole (1993).

* includes Australia and New Zealand

Table 12.4. Potential for Hydropower Development in Selected Countries Based on Technical Potential and Economic Potential in Today's Energy Markets

Country	Hydro as % of total electricity	Ratio of theoretical potential to actual	Ratio of economic potential to actual
Norway	100	5.77	1.8
Brazil	91.7	5.4	3.0
Switzerland	80	—	1.1
Canada	63	3.81	1.54
India	25	4.2	3.0
France	20	1.15	1.0
China	17	10.1	6.6
Indonesia	14	31.3	3.13
United States	10	1.82	1.3
World total	19	18.34	2.78

Sources: World Energy Conference, United Nations, MIT Energy Laboratory, Paul Scherrer Institute.

12.3 Basic Energy Conversion Principles

The primary energy sources for hydropower are solar and gravitational. The overall process is tied to the natural hydrologic cycle of evaporation and condensation in the earth's atmosphere, which redistributes water from lower elevations (sea level in the oceans) to higher elevations on land. This redistribution increases the potential energy of the water, which then flows in rivers back to the ocean under the influence of gravity. Given the intermittent nature of rain and snowfall, the amount of water stored or flowing at any time varies diurnally and seasonally. The change in potential energy that occurs as water makes its way back to the ocean provides an opportunity to extract a portion of that energy in the form of hydropower. In principle, hydro energy can be produced from any change in water elevation, but for practical purposes, changes due to tidal flows or ocean waves or currents are classified differently (see Chapter 14).

In today's hydropower applications, changes in both potential and kinetic energy of the flowing water are used to generate mechanical power to drive a generator to produce electric power. Before 1900, direct mechanical power applications were prevalent for a number of industries, including weaving, fiber spinning, and grain grinding.

There are three general types of hydropower systems: *impoundment,* which uses a natural or man-made dam for maintaining a water supply, *diversion* or run-of-river systems that intercept a portion of the natural flow of a river without employing an artificial dam, and *pumped storage*. In pumped storage applications, when the demand for electric power is low, water is pumped from a source to a storage reservoir located at a higher elevation. During peak load periods, the stored water is released, passing through a hydraulic turbine to generate power.

While all hydro resources differ in some detail, hydroelectric plants have many common components. For instance, water is brought to a hydro plant via a conduit, called a penstock, then enters an energy converter or turbine generator, and is discharged to the river in a tailrace. (Figure 12.1 schematically illustrates these components in a conventional impoundment or dam structure.) The range of power generation capacity for hydro systems is enormous, varying by over seven orders of magnitude from 1–100 kWe for micro hydro systems, from 0.1–30 MWe for small hydro systems, and to up to 12,000 MWe for large or mega-sized installations.

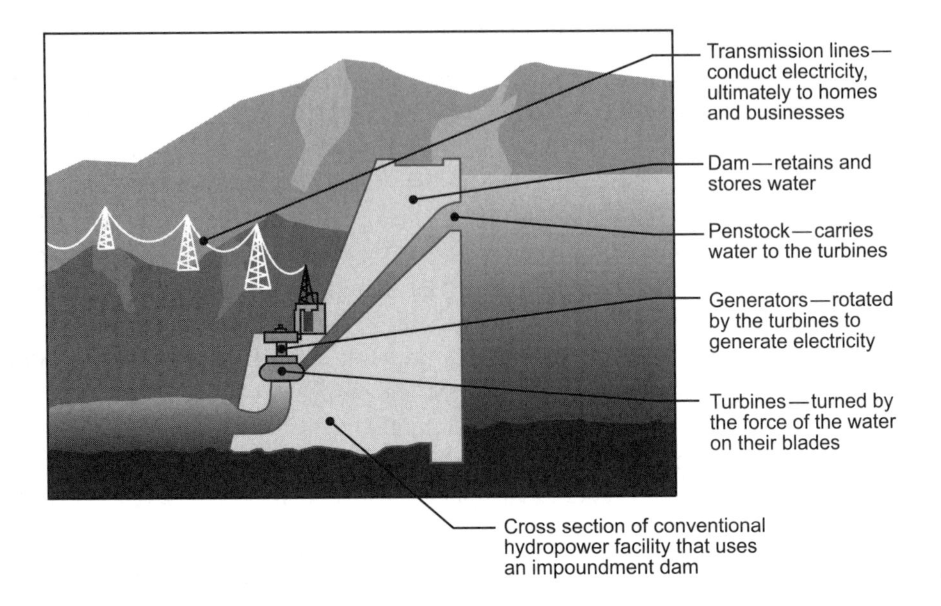

Figure 12.1. Schematic of a typical impoundment type hydropower installation. Source: INEEL (2003).

The main device used to capture hydro energy is the hydraulic turbine, which produces rotating shaft work that powers the electric generator. Although there are many types of hydraulic turbines, their basic approach is similar. They use a change in potential energy to increase fluid pressure and/or velocity (kinetic energy) and then deposit a portion of this hydraulic or kinetic energy on a turbine bucket to rotate a centrally located shaft. Thus, as fluid passes through the turbine, the change in its potential energy is continuously converted to mechanical power. The step to electric power is straightforward and achieved by connecting the rotating shaft from the hydraulic turbine to an electric generator. The hydro generator operates in a manner similar to those used in fossil-fired, gas turbine or steam, or even wind power applications. Hydro machines tend to be larger and slower in rotation speed than vapor or gas turboexpanders and may be oriented vertically or horizontally.

The overall power that can be extracted from any device will depend on the available potential and kinetic energy as reflected by the magnitude of the total (static plus dynamic) hydraulic head and conversion efficiency of the particular hydraulic turbine/ device– electric generator combination. Power output can be represented by a simple formula:

$$\begin{gathered} Power = (total\ hydraulic\ head) \times (volumetric\ flowrate) \times (efficiency) \\ Power = (\rho g Z + \tfrac{1}{2}\rho\Delta(v^2)) \times Q \times \varepsilon \end{gathered} \tag{12-1}$$

The first term in parentheses on the right-hand side contains the static head, ρgZ, and the dynamic head, $\frac{1}{2}\rho\Delta(v^2)$, contributions in units of kg/m s^2. Q is the volumetric flow rate in units of m^3/s, Z is the net height of the water head in m, ρ is density of water in kg/m^3, g is the acceleration of gravity 9.8 m/s^2, and $\Delta(v^2)$ is the difference in the square of the inlet and exiting fluid velocity across the energy converter. (Note that $v = Q/A$ where A is the cross-sectional area of the energy converter that is open to flow.) For hydro installations that are impoundment structures with the static head providing the energy, the dynamic head, given by the $\frac{1}{2}\rho\Delta(v^2)$ term, is effectively zero. For a low-head, run-of-river system, the dynamic head could be comparable to or greater than the static head, ρgZ.

The efficiency of the conversion process is represented by the term $\varepsilon < 1.0$, which captures the losses that occur due to friction and other dissipative effects. Using state-of-the-art technology, turbine-generator efficiencies can approach 0.9 for large flow machines. Older, poorly serviced plants or smaller (micro) installations typically have lower efficiencies in the range of 0.6– 0.8 or less.

By using a representative value of 1,000 kg/m^3 or 62.4 lb/ ft^3 for the density of water in an impoundment hydropower application where there is no dynamic velocity head effect, Equation (12-1) is simplified to:

$$Power = 9.81 \times 10^3\, ZQ\,\varepsilon \text{ in watts} = 9.81 \times 10^{-3}\, ZQ\varepsilon \text{ in MWe} \qquad (12\text{-}2)$$

where metric units are used for both Z and Q. We see immediately that the power generated is directly proportional to the head and the volumetric flow rate. For example, consider a hydropower dam with a static head of 100 m and a volumetric flow rate of 1000 m^3/s through the hydraulic turbine with an efficiency of 0.8. By using Equation (12-2), the power generated is 785 MWe. If the head was increased to 1,000 m, then a flow rate of only 100 m^3/s would be required to achieve the same power output.

To understand the scale of mega-sized hydro projects, let's take a look at the Grand Coulee hydropower complex, which is the largest in the US. According to Table 12.1, its output is 6,480 MWe and the dam height is 168 m. Assuming $\varepsilon = 0.8$ and $Z = 160$ m, then the required flow rate is 5,160 m^3/s or 182,000 ft^3/s.

12.4 Conversion Equipment and Civil Engineering Operations

12.4.1 Civil engineering aspects of dam construction and waterway management

The natural conditions that exist at each site, including surface topography, river flows, water quality, and annual rainfall and snowfall cycles, determine the particular design selected for a hydropower installation. When suitable hydraulic heads are not present, dams are constructed across rivers to store water and create the hydraulic head needed to drive the turbomachinery. Dams are typically designed to last for 50 to 100 years and, as such, are constructed of durable materials, such as reinforced concrete, earth, and crushed rock. They vary substantially in terms of height and storage volume, depending on the local topography. There are several design approaches that are used for concrete dams, including solid and hollow, gravity and arch geometries. On a life cycle basis, the CO_2 emissions associated with the production of concrete for dams should be considered (see Section 19.4). Many of the largest disasters associated with energy systems and their infrastructure have been a result of dam failure. As a result of these failures, improved construction methods and materials, and new technology for diagnostic testing, the reliability and integrity of dam structures has improved markedly during the past century to keep pace with public concerns. For example, roller-compacted concrete for improved performance, easier construction and lower costs, and polymeric materials for waterproof liners, are now widely used (Moreira and Poole, 1993).

In addition to the actual dam structure, there are a number of other major design considerations. For example, the turbine inlet manifold, or penstock, which usually includes screens to keep debris and fish from entering the turbine, and the discharge or tailrace system must be designed to maintain the hydraulic head and minimize the effects of sedimentation and silt buildup. Figure 12.2 illustrates a common arrangement of these features.

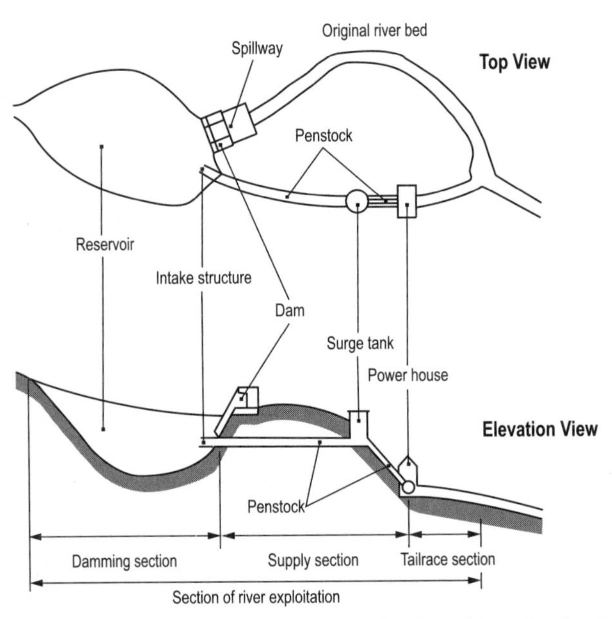

Figure 12.2. The characteristic components of a river-diversion hydroelectric plant. Adapted from Raabe (1985). From *Renewable Energy* by Laurie Burnham, et al., eds. Copyright © 1993 by Island Press. Reproduced by permission of Island Press, Washington, DC.

12.4.2 Turbines as energy convertors

Conventional hydraulic turbines are typically categorized as either impulse or reaction machines. Impulse or Pelton-type turbines convert the fluid's potential energy change (hydraulic head) into kinetic energy (fluid velocity) by expansion in a stationary nozzle to form a jet, which is then directed toward buckets attached to a rotating turbine wheel to create extractable rotating shaft work. Francis and Kaplan turbines are reaction machines that utilize both hydraulic pressure and kinetic energy to create rotating shaft work. As we can see from Figure 12.3, each turbine type has a specific operating range in terms of hydraulic head and power output. Although there is some overlap between types, choices are made depending on site characteristics.

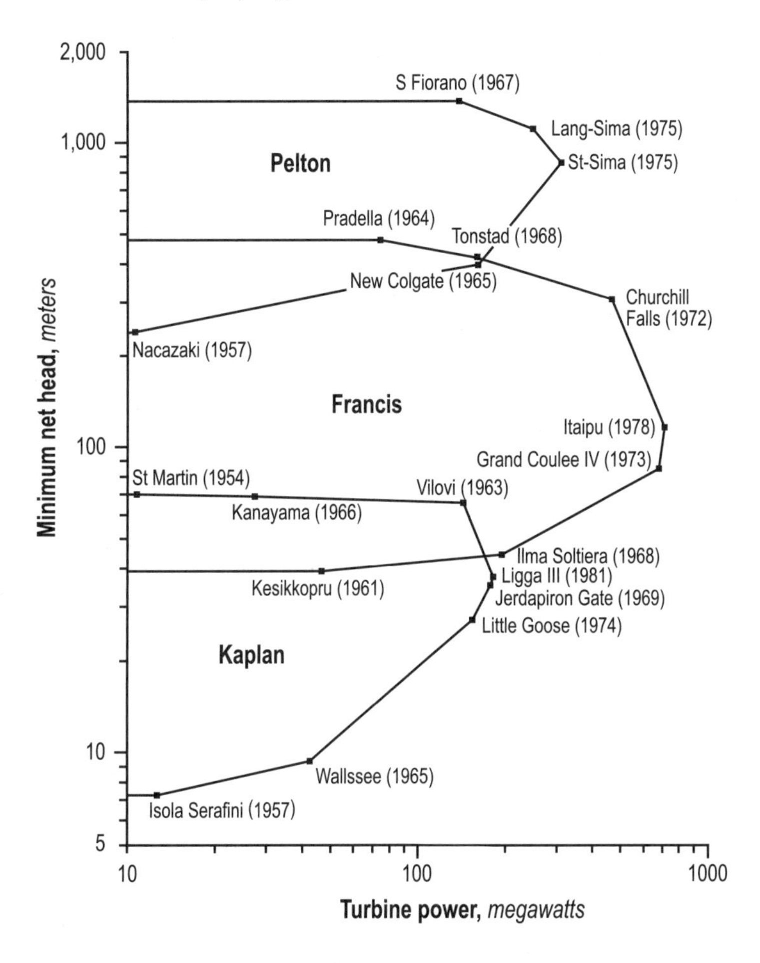

Figure 12.3. Practical operating regions for Pelton, Francis, and Kaplan hydro turbines. Adapted from Moreira and Poole (1993). From *Renewable Energy* by Laurie Burnham, et al., eds. Copyright © 1993 by Island Press. Reproduced by permission of Island Press, Washington, DC.

Pelton turbines may have single or multiple nozzles that accelerate flow to produce high velocity jets that impinge on a set of rotating turbine buckets to transfer their kinetic energy. Mechanical torque from the rotation of the wheel is transmitted directly to an electric generator. In contrast to a reaction turbine, the fluid contained in the impulse machine does not completely fill all available void space, and the turbine wheel operates at ambient pressure.

Although Pelton wheel devices were commonly deployed in the early days of hydropower, their lower efficiencies at low hydraulic heads in comparison to reaction machines made them less attractive. Nonetheless, Pelton turbines are ideally suited for high head resources (200 m) and are typically used for smaller power outputs (<50 MWe). Figure 12.3 shows that operating Pelton wheel plants can be as large as 300 MWe.

Reaction turbines convert a portion of the hydraulic head to fluid kinetic energy by expansion through a stationary nozzle impinging on the rotating turbine buckets, while the remainder is converted to kinetic energy within the runner structure of the rotating turbine blades themselves. When fluid enters the device, it fills all void passages. Operating at pressures above atmospheric, the dynamic forces created by the processes of fluid acceleration and impingement on a set of turbine blades lead to a net torque on the turbine shaft, which translates to rotational power.

The Francis turbine utilizes a set of fixed vanes that guide the fluid to the buckets that make up the turbine runner and are mounted on a central shaft (see Figure 12.4). Note that water enters the machine radially, perpendicular to the rotating shaft, and it exits the machine axially, parallel to the shaft. Francis turbines have a large optimal operating range, with heads from about 40–500 m and unit sizes approaching 1,000 MWe.

Kaplan turbines have many similar features to Francis turbines, but as seen in Figure 12.4, the turbine blade angle can be adjusted to improve performance under different flow conditions. Kaplan machines work well at low heads from less than 10 m to about 100 m with power outputs up to 200 MWe or more.

Hydro turbine development has had a long, rich history, spanning over a century. This has led to the evolution of designs that have been focused on rotary energy converter concepts that operate at high efficiency and are durable. For example, James Francis built his first water wheel turbine in 1849 with many of the basic features of today's designs, including a fully flooded chamber with adjustable blades. The Niagara Falls project, which began in the late 1880s, introduced hydropower for providing large amounts of base-load electricity using rotary turbine-generators with a high-head, diversion resource. This established a landmark precedent for future development.

As a consequence, hydro turbine technology has evolved to require hydraulic heads from 10 m to over 300 m, and has led to dam construction on rivers when nature does not provide such a resource as exists at Niagara Falls. The resulting impoundment of water, while providing useful seasonal storage, can lead to substantial inundation if the local topography is relatively flat, as well as a number of other environmental issues (see Section 12.5).

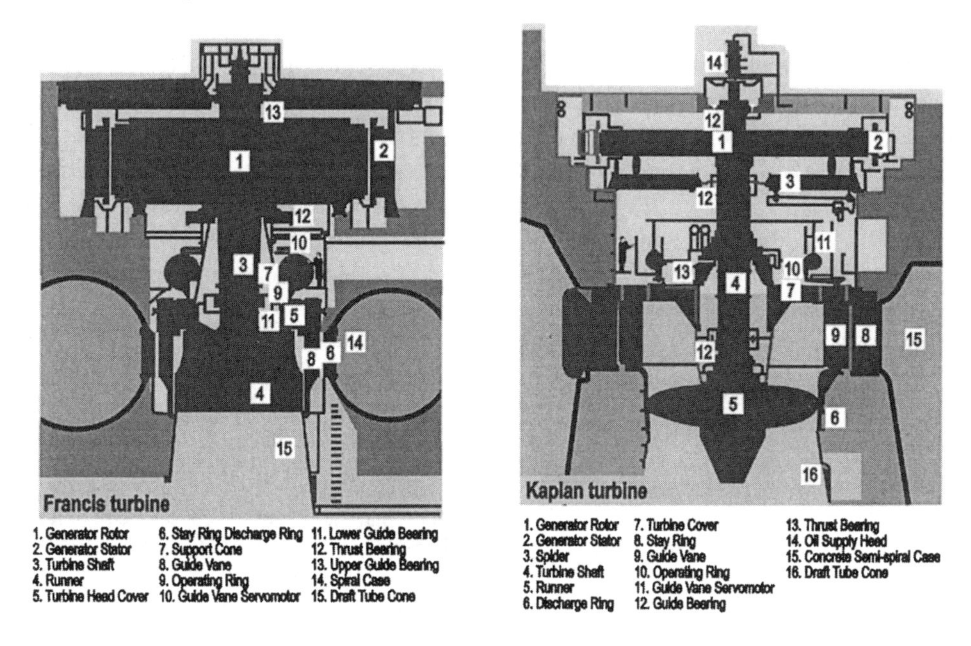

Figure 12.4. Francis and Kaplan hydroturbine schematics. Source: Franke et al. (1997).

Alternative approaches to conventional hydraulic turbines are being developed, where the kinetic energy of a flowing stream or river may be converted directly to mechanical shaft work. Such designs could lead to practical low-head, run-of-river technology that would eliminate the need to impound large volumes of water. (A few specific concepts are described later in Section 12.6.)

12.5 Sustainability Attributes

Hydropower and the multiuse aspects of hydropower developments, particularly those that involve mega-scale dam construction, illustrate both positive and negative attributes relating to sustainability. Table 12.5 outlines the major attributes of hydropower (see also Moreira and Poole, 1993, Brower, 1991, and *The Economist*, 1997a, 2003). The central issue is water management. From a sustainability perspective, it is important to keep in mind the magnitude of global water supplies and their current utilization, as well as their ecological importance for maintaining aquatic life and habitat. Figure 12.5 provides estimates of global water flows and their use. Hydropower is a renewable, and, in many settings, an emissions-free resource and should be considered for its role in displacing our dependence on depletable fossil fuels and in reducing the emission of greenhouse gases such as CO_2.

Many also argue that the construction project itself can improve the local economy of developing regions. This view appears in the political dogma of the massive Three Gorges project in China and other mega-scale projects in India or South America. Think back to what happened in the 1930s with the Hoover Dam project in Nevada. The US was in a deep depression, and projects like Hoover Dam offered relief by employing hundreds of construction workers and was viewed as a sustainable means of providing reliable, low-cost electric power for millions of people. Other positive elements of hydropower dams include their role in mitigating floods and in storing water for agricultural irrigation and human consumption. In addition, providing opportunities for recreational activities is often viewed as plus.

Table 12.5. Major Attributes of Hydropower

Positive	Negative
Emissions-free, with virtually no CO_2, NO_x, SO_x, hydrocarbons, or particulates	Frequently involves impoundment of large amounts of water with loss of habitat due to land inundation If not properly designed and constructed can lead to silt buildup with shortened lifetime and/or reduced productivity GHG emissions from concrete used in dam construction and from land inundation
Renewable resource with high conversion efficiency to electricity (>80%)	Variable output—dependent on annual rainfall and snowfall
Dispatchable with storage capability	Impacts on river flows and aquatic ecology, including fish migration and oxygen depletion
Usable for base load, peaking, and pumped storage applications	Social impacts of displacing indigenous people
Scalable from 10 kWe–10,000 MWe	Health impacts in developing countries
Low operating and maintenance costs	High initial capital costs
Long lifetime—>50 years typical	Long lead time in construction of mega-sized projects
Multiuse dams often provide flood control and stored water for agricultural irrigation and managed water supplies.	

Nonetheless, the negative side of the ledger is full, with public resistance to both proposed and existing hydropower installations. Issues center on the environmental, social, and health impacts (see, for example, Rosenberg et al., 1995). In addition, many mega-hydro projects need to be sited where the resource exists, and long distance electric transmission lines are needed. For example, the James Bay-Churchill Falls project in Canada has led to the construction of about 8,000 km of high voltage AC transmission lines.

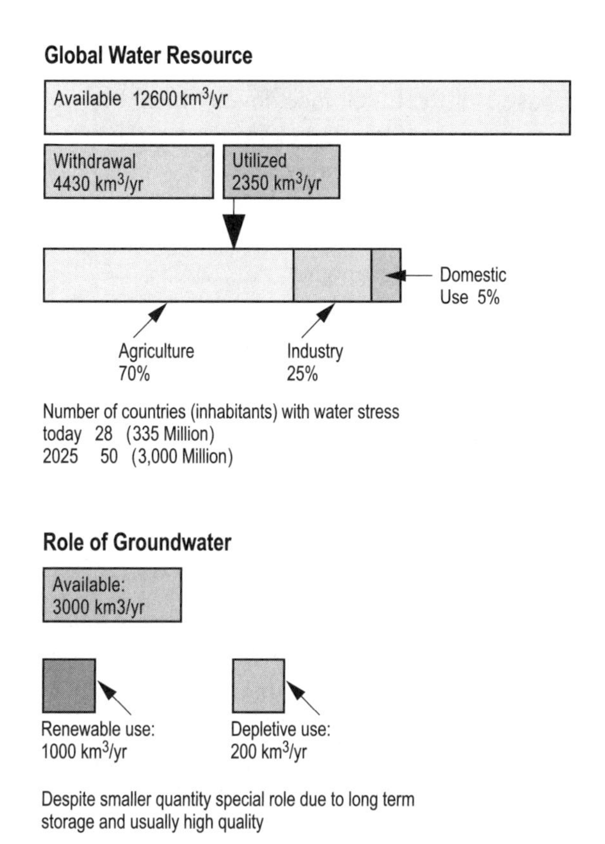

Figure 12.5. Global water supply and utilization. Adapted from AGS/ETHZ (1998).

In some instances, induced seismic risks are large and can lead to structural failure both of the dam itself and in surrounding communities. Effects associated with land inundation include displacement of indigenous people, loss of agricultural land, and in some cases, the production of greenhouse gases (both CO_2 and CH_4) from decaying organic matter submerged by the inundation (see, for example, Gardner and Perry, 1995, Lei, 1998, Grossman and Shulman,1990). Dams upset the natural flow of rivers and streams and have an impact on aquatic ecology and biodiversity, including altering normal flooding cycles and impeding fish migration (a major issue in the US). The scale of land inundation can also be massive. For example, consider the James Bay-Churchill Falls project where Phase 1 of the La Grande development has flooded over 9,600 km^2 of boreal forest (Rosenberg et al., 1995), largely a result of the minimal topological relief in that region. As a consequence, dams may reduce or remove habitats of indigenous plants and animals that depend on natural river flows, and they may compromise water quality in general. For example, mercury contamination of fish has been attributed to hydropower reservoir formation (Rosenberg et al., 1995).

While these effects are clearly present, quantifying their impacts and environmental costs is often difficult and mired in tradeoffs that are based more on qualitative values or perceived impacts than substantive facts. To frame the debate with all of the environmental, ecological, and social elements included, consider the opposing views of those who think creating a large water body behind a dam is of higher value than those who are proponents of sustaining our wild rivers.

Erosion and sedimentation can be a big issue. If not controlled properly in river systems that contain large quantities of suspended matter, rapid silting out can occur and greatly reduce the effective head and power output of a project. In addition, it disrupts the flow of nutrients contained in the sediment downstream. The oxygen content of waters contained behind dam structures is often lower than normal; such oxygen depletion may have a detrimental effect on native fish and other wildlife. The presence of large stagnant bodies of water can become a breeding ground for bacterial and viral infections, which, in developing countries, in particular, can pose significant health risks. For instance, the appearance of malaria, lymphatic filariasis, and schistosomiasis are frequently associated with hydro development projects where large quantities of water are impounded in warm climatic regions (Moreira and Poole, 1993).

The fish migration issue has been particularly polarized in the western US where salmon migration on major rivers, such as the Columbia, is threatened (PCAST, 1997; *The Economist*, 1997b, and Odeh, 1999). Even though the migration of adult salmon upstream to spawn can be achieved using well-designed fish ladders and similar concepts, the main problem is with the migration of the newborn young salmon fry downstream as they pass through the hydraulic turbines, where their mortality rate is high. A modest amount of research is underway in the US to understand what is causing the problem and to develop more "fish-friendly" turbine technologies (see Section 12.6.3 for more details). Without such technologies, the current approach has been to divert a portion of the flow around the turbine. While this diversion technique saves a few migrating salmon fry, it also de-rates the power output of the plant, thereby diminishing its value as a renewable energy resource.

Environmental concerns have led to flow de-rating and other systemic changes aimed at mitigation as regulated by the US Federal Energy Regulatory Commission (FERC) in their relicensing of existing hydro facilities. These concerns often lead to long delays in approving new hydropower facilities. On average, the US is experiencing an 8% decline in hydropower generation for relicensed facilities, mainly as a result of regulated stream flows. To maintain existing US hydropower capacity, let alone increase it in response to national objectives for more renewable energy, a better quantitative understanding of environmental, social, and ecological effects and tradeoffs, as well as the development of new, more environmentally friendly technologies for hydropower, are essential.

12.6 Status of Hydropower Technology Today

12.6.1 Economic issues

At least $2 trillion US have been spent on hydropower developments worldwide during the last century (WCD, 2002), and this has created over 635,000 MWe of capacity and contributed roughly 19% of the world's total generation of electricity (see Figure 12.6). In many instances, hydro provides the lowest cost option for generating electric power in a given area. Furthermore, its dispatchable characteristic makes hydropower an important component for meeting peak and seasonal loads in the generation mix of a particular company or utility.

Based on USDOE/EIA statistics, hydropower's annual revenues in the US are in excess of $25 billion for an estimated capital investment of $150 billion. As a renewable asset, its value in displacing carbon dioxide emissions and reducing dependence on depletable fossil resources is significant. Given the maturity of the technology, the costs for hydropower have a long history that correlate with the age of the facility and the environmental concerns that continue to grow.

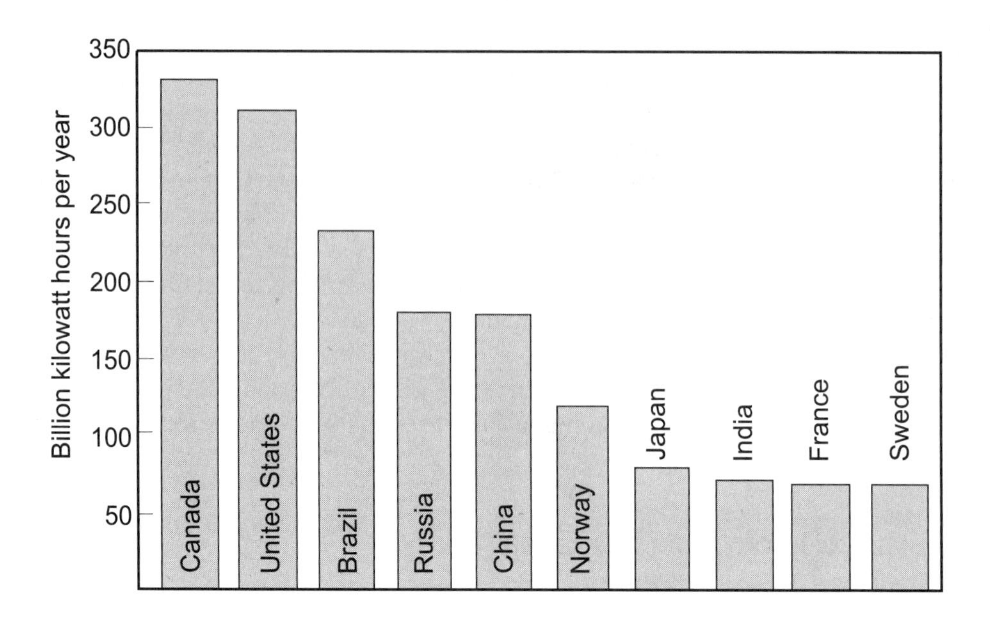

Figure 12.6. World's leading producers of hydropower. Source: EIA, Annual Energy Review 1999, July 2000, Table 11.15. Reprinted from INEEL (2003).

Hydropower costs vary considerably depending on the extent of civil engineering works required, which, in turn, depends on the natural terrain and climate of the region. Compare, for example, placing a hydro plant in the Alps of Europe with significant mountainous relief to the flatter, dryer regions of the US Midwest. As there can be large seasonal variations in available water (by factors of two to ten), the effects that needed reserve capacity has on reservoir and dam structure can be large. Furthermore, there can be significant economies of scale for the energy conversion equipment and effects of different capacity or availability factors.

Another aspect that makes hydropower different from other energy sources in economic terms is its low maintenance costs and long lifetimes, with 50 or more years typical and many systems approaching 100 years. For example, most of the mega-scale hydropower projects that were developed in the US between 1930 and 1970 during a period of rapid expansion under the auspices of the TVA, BPA, BLM, the US Army Corps of Engineers, and other government agencies, still produce power today. The busbar costs for these mature installations are in the 1–2 ¢/kWh range, which represents mostly operating and maintenance costs as the initial capital investment was paid off during the first 20 years of operation at most. For new installations, the situation is completely reversed as projects are capital intensive. The larger the hydro project is the more time it takes to plan and build, thus increasing the impact of invested capital before any revenues from power sales appear.

We estimate a range of about $1,000–2,500 per installed kWe of capacity for larger plants (>250 MWe), while smaller-sized plants would be proportionally more costly. For plants <20 MWe, costs might range from $4,000 to $6,500/kWe, while for those between 20 and 250 MWe, from $2,000 to $4,000/kWe. With these ranges of capital investments and the other risks of hydropower projects in developing countries, agencies like the World Bank and the Inter-American Development Bank provide financial sponsorship only after careful analysis of risks and benefits. Externalities are playing a larger role in determining the feasibility of sponsoring hydro projects, as environmental and social impacts are factored into the total project cost.

In the US, the DOE/EIA estimates capital costs for an average sized hydro installation of 31MWe from $1,700 to 2,300/kWe—annualized to about 1.8 ¢/kWh, operating and maintenance costs of about 0.8 ¢/kWh, for a total levelized generating cost of about 2.6 ¢/kWh assuming a 50 or more year plant lifetime and a 40–50% capacity factor.

12.6.2 Potential for growth

Although hydropower is currently the largest and most important producer of electricity from a renewable energy source, with over 600 GWe of capacity and 2,600 TWhr produced annually, its future role is less certain for the long term. While the potential for adding additional hydropower worldwide is substantial in terms of availability and reasonable capital investment (7,300 TWh/yr or more), environment-related concerns, particularly those associated with mega-scale projects that involve dams and their subsequent land inundation, pose substantial barriers to deployment and growth of hydropower as a renewable resource.

The possibility of expanding hydropower capacity by utilization of existing dams in countries that already have substantial hydro assets developed also has barriers. For example, a number of studies from credible sources like FERC and the USDOE estimate that the US has the potential to expand its hydro capacity by 30,000 to 73,000 MWe, using currently available technology and with reasonable financial investments on a $/kWe basis. When environmental issues are debated in the current licensing process, the FERC regulatory machinery usually imposes long delays or flow and power reductions, or rejects proposals outright. The result is that many informed groups, including advocates for hydropower like the National Hydropower Association, predict that no new US hydro capacity will be added unless policies are changed.

One way of addressing environmental concerns is by accelerating the level of scientific attention being directed at achieving quantitative understanding of the impacts and benefits of hydro and to developing new technologies that will mitigate these effects. There are a number of opportunities for achieving more sustainable hydropower systems, but one must keep in mind that the current level of R&D support for such undertakings is far too low to achieve much in the short term. (For example, see the PCAST recommendations for hydropower R&D for the US, 1997.)

12.6.3 Advanced technology needs

Advanced technology needs can be divided into two categories: *near-term improvements* for existing hydro installations to address fish migration and oxygen depletion issues and *long-term innovations* for utilizing low-head and run-of-river resources in an environmentally and economically sustainable manner.

Near-term improvements. Many people have the perception that because hydropower is a mature technology with substantial capital investments in place it cannot be markedly influenced by modern technology. While it is certainly true that hydropower has been around for awhile, the opportunity for retrofitting existing facilities, where fish-migration and oxygen-depletion impacts are significant, has led to a number of new technological approaches for reducing these problems. Regarding fish passage, the first steps have been focused on understanding exactly what was causing such high levels of mortality in young fish, followed by turbine redesigns to make them more "fish-friendly." The Alden Research Laboratory and Voith Hydro have been working on improved hydro technology for a number of years with private, EPRI, and public (USDOE) funding (see, for example, Odeh,1999 and Franke, et al.,1997). Their work has led to better understanding of what causes fish mortality in hydro turbines and has generated a number of innovations that would reduce the problem. Advanced modeling methods employing computational fluid mechanics (CFD) have identified locations and conditions inside existing turbines that are problematic to successful fish migration. The main injury mechanisms are driven by rapid pressure changes, impingement and abrasion on turbine blades, and damage induced by cavitation. One optimistic approach that uses both CFD modeling and experimentally validated methods with electronically tagged fish has resulted in proposed designs of the internal turbine blading that could be retrofitted in Francis and Kaplan units to reduce mortality. Another important aspect of these proposed fish-friendly retrofits is that the conversion efficiency would be preserved or even increased. A team of engineers at the Alden Research Lab and the Northern Research and Engineering Corporation (ARL-NREC) are also designing a completely new turbine design (as illustrated in Figure 12.7) that uses a centrifugal screw concept that would facilitate the migration of small fish and operate at efficiencies approaching 90%.

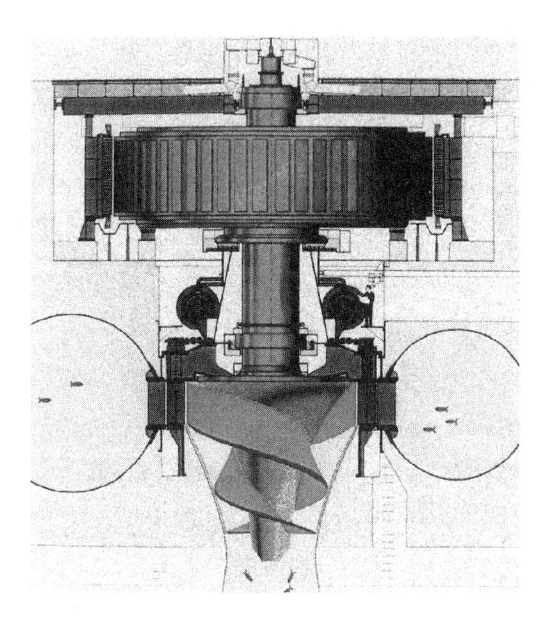

Figure 12.7. ARL-NREC fish-friendly hydro turbine design. Courtesy of Alden Research Laboratory, Worcester, MA (2002).

Oxygen depletion in the water discharged from hydro turbines also is a problem in many installations. Aerating weirs and turbine runners are being developed by Voith Hydro and others to increase oxygen content. Voith Hydro is also looking at possible retrofits to existing low-head, smaller hydro dams that would increase power output with little environmental damage.

Long-term innovations. If the so-called ultra-low-head (<1 m) or run-of-river energy converter concepts could be developed economically, then a large jump in the potential for hydropower would materialize that would match most, if not all, of the desirable sustainability attributes of energy systems. These concepts would allow for fish migration, maintain the natural flow and flooding cycles of rivers by eliminating or minimizing impoundment, and keep water quality at high levels.

Matrix turbines and specially designed ultra-low-head turbines are being considered by a number of groups. For example, Gorlov (1998, 1992) at Northeastern University has developed several low-cost alternatives using (1) slow rpm turbines made of composite plastics that operate efficiently with ultra-low heads (<1 m) and can capture both the potential and the kinetic energy of flowing water in rivers or tidal basins and (2) high rpm, air-driven Francis turbines that are powered by hydraulically activated chambers that compress air using river flows and low hydraulic heads (1–3 m).

Schneider and associates (1995–2001) have taken a different approach in which a river or tidal basin's hydroenergy is captured directly as kinetic energy ($1/2\ \rho v^2$) using a "hydroengine" that consists of a horizontal cascade of foils that are mechanically connected to the drive mechanism by looping around two axles resembling a venetian blind structure. The Schneider hydroengine utilizes natural river flows enhanced by low hydraulic heads (<3 m), while keeping fluid pressure changes and velocity and acceleration levels within safe ranges for fish passage. Voith Hydro (1999) is also working on ultra-low head machines employing matrix turbines and a redesigned power wheel concept. Although initial testing of these concepts has been encouraging, much work to validate performance including efficiency and durability and to ensure reasonable costs remains to be done before these advanced machines will be deployed commercially.

References

AGS/ETHZ 1998. Alliance for Global Sustainability Proceedings, ETH Zurich.

Brower, M. 1991. "Energy from rivers and oceans." *Cool Energy*, Chapter 6 (11-118). Cambridge, MA: The MIT Press.

The Economist. 1997a. "Asia - Stopping Yangtze's Flow." (August 2)

The Economist. 1997b. "Dam-busting—victory for the fishes." 345(8046): 28.

The Economist. 2003. "Damming evidence—The pros and cons of big earthworks." 368(8333): 9–11.

Franke, G.F., D.R. Webb, R.K. Fisher, D. Mathur, P.N. Hopping, P.A. March, M.R. Headrick, L.T. Iaczo, Y. Ventikos, and F. Sotiropoulis. 1997. "Development of environmentally advanced hydropower turbine system concepts." Voith Hydro, Inc. Report #2677-0141. Contract No. DE-AC07-96ID13382. Idaho Falls, ID: INEEL.

Gardner, G. and J. Perry. 1995. "Big-dam construction is on the rise." *Vital Sign, World Watch*, 8(5):36–37.

Gorlov, A. 1998. "Turbines with a twist" in *Macro-engineering and the earth: world projects for the year 2000 and beyond,* Kitzinger, U. and Frankel, E.G., eds. Chichester, UK: Horwood Publishing.

Gorlov, A. 1992. "A new opportunity for hydro: using air turbines for generating electricity." *Hydro Review*. 11:5.

Grossman, D. and S. Shulman. 1990. "The price of power—examining the hidden cost of New England's demand for energy." *Boston Globe Magazine*, August, 1990.

INEEL. 2003. *Hydropower Facts*. US DOE and www.hydropower.inel.gov

International Commission on Large Dams (ICOLD). 2001. Various reports.

Lei, X. 1998. "Going against the flow." *Science* 280:24–26.

Moreira, J.R. and A.D. Poole. 1993. "Hydropower and its constraints," in *Renewable Energy: Sources for fuels and electricity*. T.B. Johannson, H. Kelly, A.K.N. Reedy, and R.H. Williams, eds. Washington, DC: Island Press.

Odeh, M. 1999. "A summary of environmentally friendly turbine design concepts." Report, DOE/ID/13741. Idaho Falls, ID: USDOE.

PCAST (The President's Committee of Advisors on Science and Technology). 1997. "Chapter 6—Renewable Energy," in *Federal Energy Research and Development for the Challenges of the Twenty-First Century*, report of the Energy R&D Panel. Washington, DC: Office of Science and Technology Policy.

Raabe, I.J. 1985. *Hydropower—the design, use, and function of hydromechanical, hydraulic, and electrical equipment*. Dusseldorf, Germany: verlag.

Reynolds, T.S. 1983. *Stronger than a hundred men: A history of the vertical water wheel*. Baltimore, MD: Johns Hopkins University Press.

Rosenberg, D.M., R.A. Bodaly, and P.J. Usher. 1995. "Environmental and social impacts of large scale hydroelectric development: who is listening?" *Global Environmental Change* 5(2):127–148.

Schneider, D. 1995–2001. Personal communications at the MIT Energy Laboratory. Cambridge, MA.

Tuxill, J. 1996. "Past dam disaster casts shadow over Three Gorges." *Environmental Intelligence*. WorldWatch Institute, Washington, DC (July/August)

US National Hydropower Association. 2001. Various reports.

Voith Hydro. 1999. "Industry Perspective on Hydro R&D." Presentation by R.K. Fisher, Jr., to the National Research Council. Washington, DC (July 22)

World Energy Council. 2001. "Hydropower," chapter 7 in *Survey of Energy Resources*. London, UK.

World Commission on Dams (WCD). 2002. "Water, development, and large dams," chapter 1 in *Dams and Development: a new framework for decision-making*.

Web Sites of Interest

www.eren.energy.gov/RE/hydropower
www.worldenergy.org/wec
hydropower.inel.gov
www.hydro.org/facts
www.energy.ca.gov/electricity/hydro.html
www.dams.org
www.ussdams.org

Problems

12.1 Estimate the required flow for a 10,000 MWe hydro installation as a function of the effective hydraulic head from 10–300 m.

12.2 Describe how you estimate the amount of land that will be innundated by a large-scale hydropower project like Three Gorges in China on the Yangtze River or the La Grande complex in St. James Bay, Canada.

12.3 Assume that the US wants to expand its hydropower capacity to its full technical potential without building any new dams as a means of offsetting carbon dioxide emissions from coal-fired plants. Estimate the impact on annual carbon dioxide emissions. What would be your estimate of capital investment needed to accomplish this?

12.4 You have been asked to finance US R&D for advanced hydropower technology for reducing the impact of current installations on fish migration and on developing economically and technically viable low-head, run-of-river technologies by allocating a portion of the current revenues generated from electricity sales from hydro resources. Estimate the revenue stream that would be produced annually as a function of the "sustainable R&D trust fund surcharge" amount in cents/kWh. You can assume that the average generating base price for hydropower in the US is $0.03/kWh.

12.5 A large mega-hydro project being considered will require approximately 1 million tons of concrete for the dam structure alone. The plant should produce about 2,000 MWe of power for its entire lifetime of about 100 years. How large might the impact of this plant on GHG emissions be?

Solar Energy 13

13.1 General Characteristics of Solar Energy

Throughout human history, solar energy has been utilized for domestic use in heating and cooking. The Anasazi Indian tribes of the American southwest were perhaps the first in North America to employ passive solar energy in their dwellings. The Greeks and Romans documented their use of solar power in Europe over two millennia ago. In the early 18th century, specific solar technologies were introduced to concentrate the sun's energy and put it to use in high-temperature processes. The development of a 1,700°C solar furnace by Lavoisier in the mid-1700s is an excellent early example of human progress in harnessing solar energy. In general, its ubiquitous nature and ability to be used effectively over a range of scales makes solar the popular choice among renewable energy enthusiasts today.

As discussed in Chapter 9, the sun's energy incident on the earth is the intrinsic source for many forms of renewable energy (including wind, ocean thermal, and bioenergy), and over a longer time scale, all of fossil energy. Technologies for solar energy capture have been actively pursued for over a century, with much engineering know-how and analysis developed during the last 50 years. As a result, the documentation available for solar energy is extensive. Those interested in more depth should examine the resources listed at the end of the chapter.

In this chapter, after discussing the characteristics of the solar energy resource, we will review capture and utilization technologies. Our treatment is limited to brief summaries of the main technical, economic, and policy issues for three modes of solar energy use.

- As thermal energy or heat collected passively or actively for space conditioning buildings
- As thermal energy that is collected and converted to electricity in solar concentrators
- By the direct conversion of solar energy to electricity using photovoltaic devices

13.2 Resource Assessment

Almost everyone qualitatively appreciates the variability of the sun's intensity during the day as the sun passes overhead and as its radiation encounters clouds in its path to the earth's surface. Seasonal variations are then superimposed on top of these diurnal changes. Fortunately, the daily and seasonal movement of the sun is both predictable and known in precise mathematical form. Changes in weather are less regular, but can be averaged for estimating the solar potential in different regions. The intermittent and variable characteristic of solar energy must be reckoned with to make effective use of it as a source of thermal or electrical energy. Passive or active storage in some form is almost always coupled to a solar energy system.

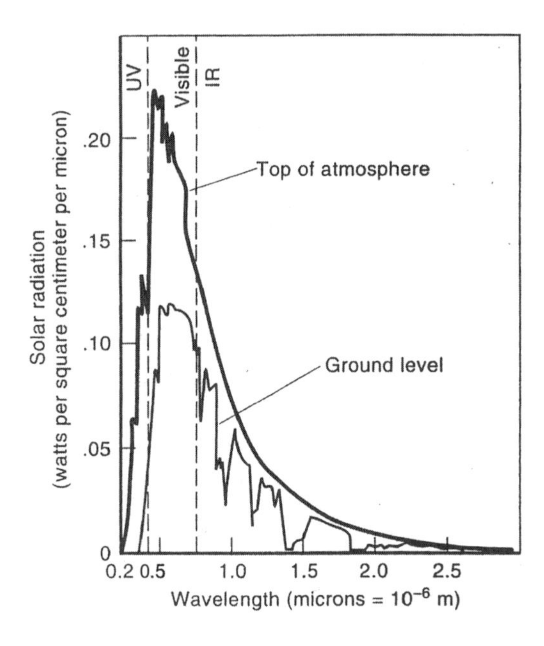

Figure 13.1. The solar spectrum, distribution of incident energy as a function of wavelength at ground level and the top of the earth's atmosphere. Adapted from Hinrichs (1996) and Kreith and Kreider (1978).

The intrinsic source of the sun's energy is a direct result of the thermonuclear fusion of hydrogen nuclei to form helium, which occurs at the phenomenally high rate of about 4×10^9 kg of mass conversion per second (Hinrichs, 1996). The solar fusion reaction results in temperatures of about 6,000°C at the sun's surface that, in turn, induces a large solar radiative flux that travels 93 million miles to the earth.

The distribution of the solar energy flux that intercepts the earth is a strong function of the wavelength of the incident light (as shown in the solar spectrum of Figure 13.1). Because of variations in the absorption and reflection characteristics of different molecules contained in earth's atmosphere, the distribution changes from the top of the atmosphere to the earth's surface. For example, most of the short wavelength ($\lambda < 0.4$ micron) ultraviolet radiation is absorbed by oxygen (O_2), ozone (O_3), and nitrogen (N_2) in the upper atmosphere, while H_2O and CO_2 capture a good portion of the longer wavelength radiation ($\lambda > 0.6$ micron) in the visible and infrared region.

The incident solar energy flux on the earth's surface is large. In general terms, the energy flux or total insolation that strikes the top of the earth's atmosphere is referred to as the *solar constant* and has a value of 1,354 W/m^2 (or equivalently 429 Btu/ft^{2}hr, 1.94 Langleys/min, or 4,870 kJ/m^{2}hr). Depending on the time of day and the month of the year, as well as the local weather and latitude of a particular location on the earth's surface, the amount of insolation that actually reaches the surface will vary from essentially 0 to about 1,050 W/m^2. On average, about half of the energy incident on the earth's upper atmosphere makes it to the surface, with the rest scattered, reflected, or absorbed and re-radiated into space. About 21% of the solar flux reaches the surface as direct radiation and about 29% as scattered or diffuse radiation.

Even with these losses, the amount of solar energy that reaches us is significant. For example, over 40,000 EJ of solar energy are incident on the US each year, which is more than 400 times the total primary energy consumed during all of 2002. Given this, why haven't we used solar energy more effectively to displace our demand for fossil fuels? Part of the problem has to do with intermittent and seasonal variability of solar. The highest solar fluxes are available during the period around noon, and they can vary by a factor of two or more from month to month depending on local cloudiness, humidity, distance to the equator, and seasonal changes in the position of the sun. Another problem has to do with the relatively low density of solar energy relative to chemically released energy due to combusting, for example, fossil fuels.

We can grossly characterize the solar flux or insolation into two forms: direct and diffuse light. The direct or unscattered portion can be focused or concentrated using mirrors or lenses, while the diffuse portion, which results from scattering by clouds and the atmosphere itself, cannot be focused. Scattering in the atmosphere can also be locally influenced by the presence of surface-emitted aerosols and particulates, especially in or near large urban areas. The summation of direct (adjusted by the cosine of its incident angle) plus diffuse radiation represents the total flux that hits the surface. Figures 13.2a–b provide overall annual averages for the US of total insolation incident on both horizontal and vertical surfaces, while Figure 13.2c gives the direct (unscattered) insolation incident on an east-west oriented concentrator that tracks the sun's path.

The sun's position in the sky varies from hour to hour, as well as from day to day, as the seasons pass. The overall motion of the sun has been characterized and correlated with various mathematical formulae and is available in graphical and tabular form (Michalsky, 1988). The orientation of the collector surface relative to the sun's pointing angle is a key element in determining the capture efficiency of the collector. To optimize the amount of energy recovered, the collector can be designed to track the sun's position or it can be oriented in a fixed position that is purposely selected to maximize recovery during certain times of the day or year. Systems that concentrate the solar flux by focusing energy on a fixed receiver usually track the sun's position continuously, while flat plate solar heating and photovoltaic (PV) systems usually have fixed orientations. (Details are provided in Sections 13.3–13.5 where specific technologies are discussed.)

In practical terms, the capture efficiency (η_{solar}) of any solar collector can be represented by the following equation.

$$\eta_{solar} \equiv \frac{\text{useful energy recovered}}{\text{total insolation incident on the collector}} \times 100\% \quad (13\text{-}1)$$

Recovered energy can be in the form of thermal energy (heat) or electrical energy (current × voltage). In thermal energy recovery applications, efficiencies can be high, ranging from 30– 60% or more. In photovoltaic systems, efficiencies are considerably lower with 8–15% being typical.

An operating variable that influences the capture efficiency of any solar collector is the pointing error, Ψ, which can be represented using the following nomenclature:

$$\Psi \equiv \text{pointingerror} = \xi - \beta \text{ in degrees}$$

and: (13-2)

$$\alpha \equiv \text{collector tilt relative to the latitude} = \beta - \varphi$$

where:

β = tilt angle of the collector in degrees from the horizontal
φ = latitude in degrees
ξ = pointing angle of the sun

Also important are two other orientation angles:

θ = altitude angle in degrees
ω = azimuth in degrees

The altitude angle θ and azimuth ω are defined as the angle of the sun above the horizon and the angle from true south, respectively, and are illustrated in Figure 13.3. Figure 13.4 illustrates the geometric relationship among these angles. The hourly variations of the sun's position are usually represented by the azimuth or hour angle, ω, that varies about 15° per hour and ranges from 0 to a maximum value that changes depending on the time of the year. ω is 0 at solar noon when the sun reaches its highest position in the sky for a specific location and reaches a maximum value when the sun sets below the horizon. The maximum value is <90° in fall and winter months and >90° during spring and summer months. Seasonal variations are usually given as a function of the declination angle, δ, which provides a quantitative measure of the tilted earth's position relative to the sun as the earth moves around the sun annually (see Figure 13.5). δ is 0 at the autumnal and vernal equinoxes, September 21 and March 21, respectively, and in northern latitudes, +23.5° at the summer solstice on June 21, and –23.5° at the winter solstice on December 21.

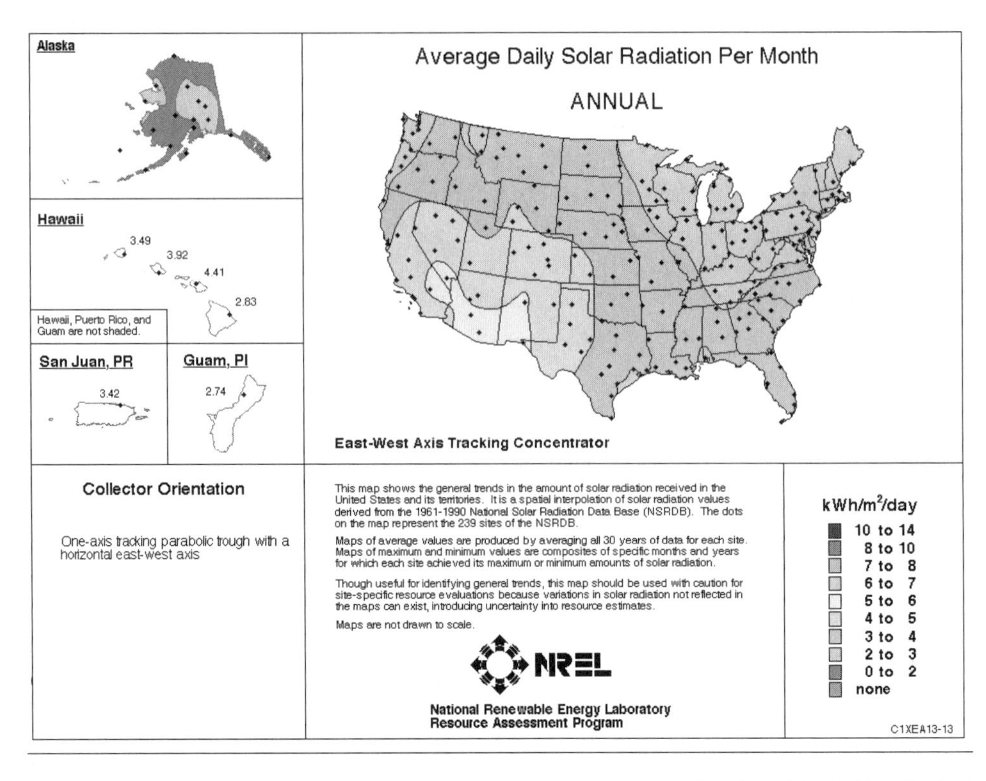
Alaska
Hawaii
3.49
3.92
4.41
2.83
Hawaii, Puerto Rico, and Guam are not shaded.
San Juan, PR
3.42
Guam, PI
2.74
Average Daily Solar Radiation Per Month
ANNUAL
East-West Axis Tracking Concentrator
Collector Orientation
One-axis tracking parabolic trough with a horizontal east-west axis
This map shows the general trends in the amount of solar radiation received in the United States and its territories. It is a spatial interpolation of solar radiation values derived from the 1961-1990 National Solar Radiation Data Base (NSRDB). The dots on the map represent the 239 sites of the NSRDB.
Maps of average values are produced by averaging all 30 years of data for each site. Maps of maximum and minimum values are composites of specific months and years for which each site achieved its maximum or minimum amounts of solar radiation.
Though useful for identifying general trends, this map should be used with caution for site-specific resource evaluations because variations in solar radiation not reflected in the maps can exist, introducing uncertainty into resource estimates.
Maps are not drawn to scale.
NREL
National Renewable Energy Laboratory
Resource Assessment Program
kWh/m²/day
10 to 14
8 to 10
7 to 8
6 to 7
5 to 6
4 to 5
3 to 4
2 to 3
0 to 2
none
C1XEA13-13

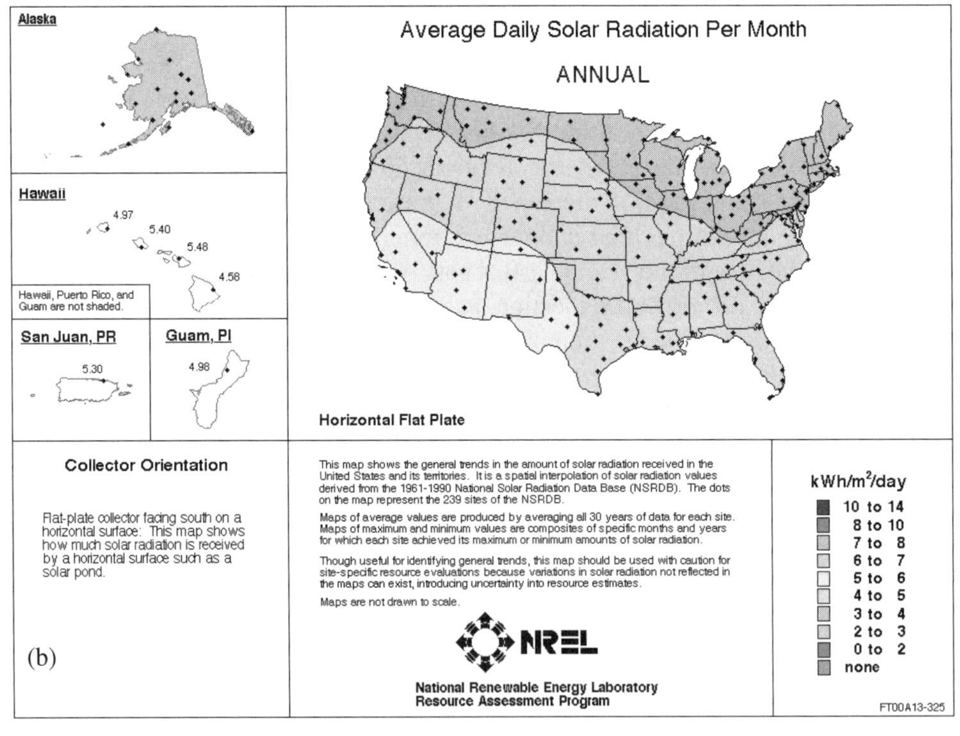
Alaska
Hawaii
4.97
5.40
5.48
4.58
Hawaii, Puerto Rico, and Guam are not shaded.
San Juan, PR
5.30
Guam, PI
4.98
Average Daily Solar Radiation Per Month
ANNUAL
Horizontal Flat Plate
Collector Orientation
Flat-plate collector facing south on a horizontal surface. This map shows how much solar radiation is received by a horizontal surface such as a solar pond.
This map shows the general trends in the amount of solar radiation received in the United States and its territories. It is a spatial interpolation of solar radiation values derived from the 1961-1990 National Solar Radiation Data Base (NSRDB). The dots on the map represent the 239 sites of the NSRDB.
Maps of average values are produced by averaging all 30 years of data for each site. Maps of maximum and minimum values are composites of specific months and years for which each site achieved its maximum or minimum amounts of solar radiation.
Though useful for identifying general trends, this map should be used with caution for site-specific resource evaluations because variations in solar radiation not reflected in the maps can exist, introducing uncertainty into resource estimates.
Maps are not drawn to scale.
NREL
National Renewable Energy Laboratory
Resource Assessment Program
kWh/m²/day
10 to 14
8 to 10
7 to 8
6 to 7
5 to 6
4 to 5
3 to 4
2 to 3
0 to 2
none
FT00A13-325

(b)

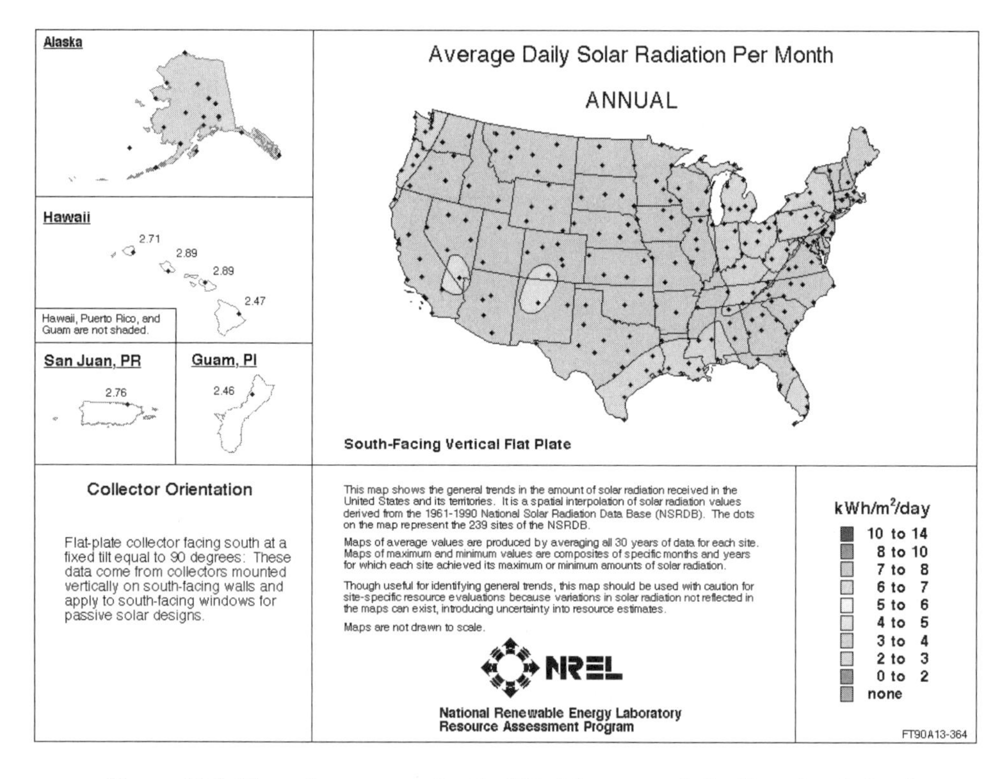

Figure 13.2. The solar resource for the US: (a) average daily direct insolation for an east-west tracking concentrator, (b) average daily total insolation flux for horizontal flat plate, and (c) average daily total insolation for a south-facing vertical flat plate. Source: Renne, Stovall et al. (2002).

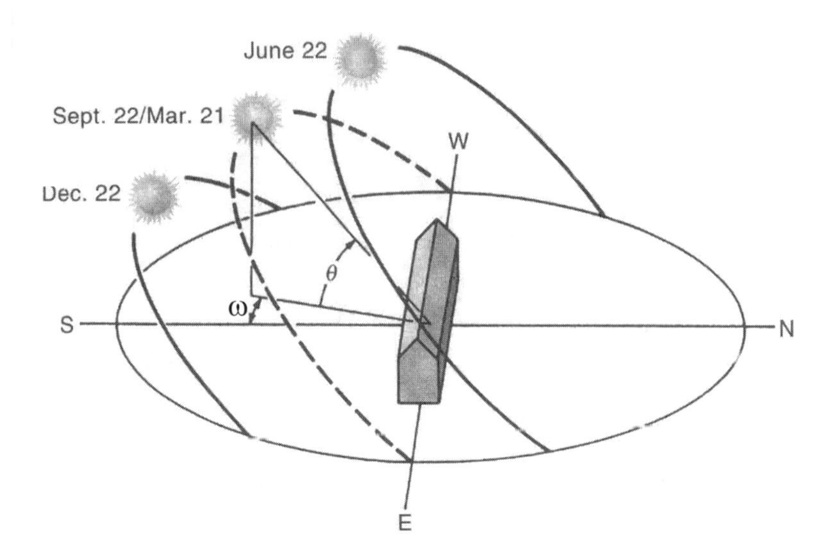

Figure 13.3. Schematic showing the definition of the hour angle (ω) and the solar altitude angle (θ) of the sun relative to an east-west oriented structure. Adapted from Hinrichs (1996) and Kreith and Kreider (1978).

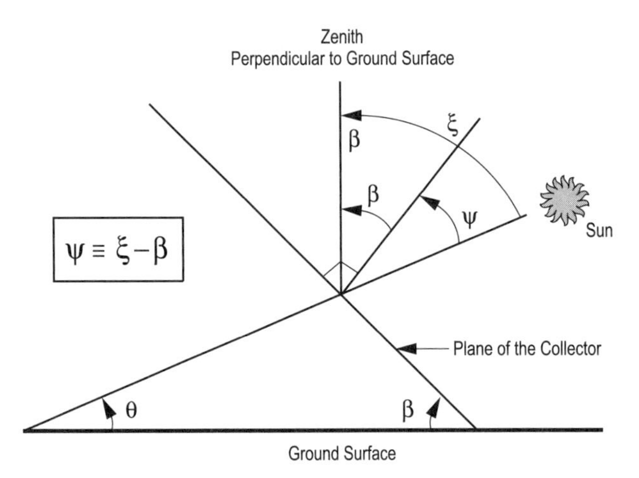

Figure 13.4. Geometric relationship among various angles characterizing the relative position of the sun to the solar collector surface.

For collectors facing the equator, $0 < \beta < 90°$; for collectors facing away from the equator, $90 < \beta < 180°$. Overall, a transcendental equation must be solved to determine how the pointing angle, ξ, varies as a function of both hourly and seasonal positional changes. In general terms, the relationship is given by,

$$\tan \frac{[90 - \varphi + \xi]}{2} = f(\delta,\omega) = \text{complex trigometric function} \tag{13-3}$$

Figures 13.6–13.8 illustrate these effects by showing how the sun's altitude angle varies as a function of the azimuth angle for three different northern latitudes (28°, 36°, and 44° north).

Here we introduce three examples of how collectors are normally placed to underscore the importance of orientation. For sites where the domestic heating and hot water loads are high during the winter months, solar thermal flat plate panels would be oriented facing south with a tilt angle greater than the latitude, typically at $\beta = \varphi + 15°$, to capture more of the sun's energy when it is lower in the sky nearer the horizon. For year-round solar heating, orienting flat plate collectors to face south at a tilt angle equal to the latitude ($\alpha = 0$ and $\beta = \varphi$) is commonly used. For sites where the air conditioning and electric loads are larger during the summer, flat PV panels would be placed facing south, but at a tilt angle less than the latitude, typically at $\beta = \varphi - 15°$ or $\alpha = -15°$ or so, given the sun's higher position in the sky. These choices are not optimal in a mathematical sense, as continuous, two-axis tracking would be needed to maximize efficiency, yet they do represent reasonable tradeoffs to improve performance while maintaining the simplicity of using a fixed orientation collector.

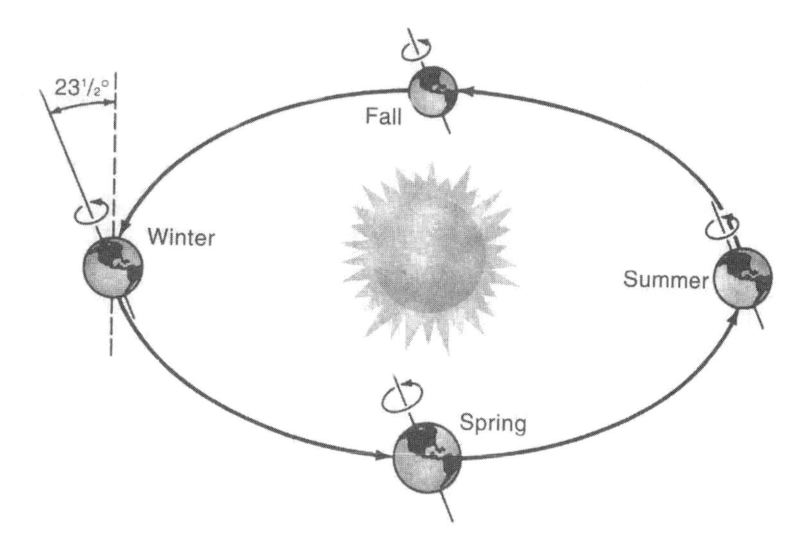

Figure 13.5. The yearly orbit of the earth around the sun. Adapted from Hinrichs (1996). From *Energy* second edition by Hinrichs. © 1996. Reprinted with permission of Brooks/Cold, a division of Thomson Learning: www.thomsonrights.com.

In addition to latitude, regional location is also important in determining the quality of the solar resource. Table 13.1 gives some representative annual values for total, direct, and diffuse insolation. For example, two sites, Seattle, Washington, and Bismarck, North Dakota, have similar latitudes of 47°N but far different levels of insolation.

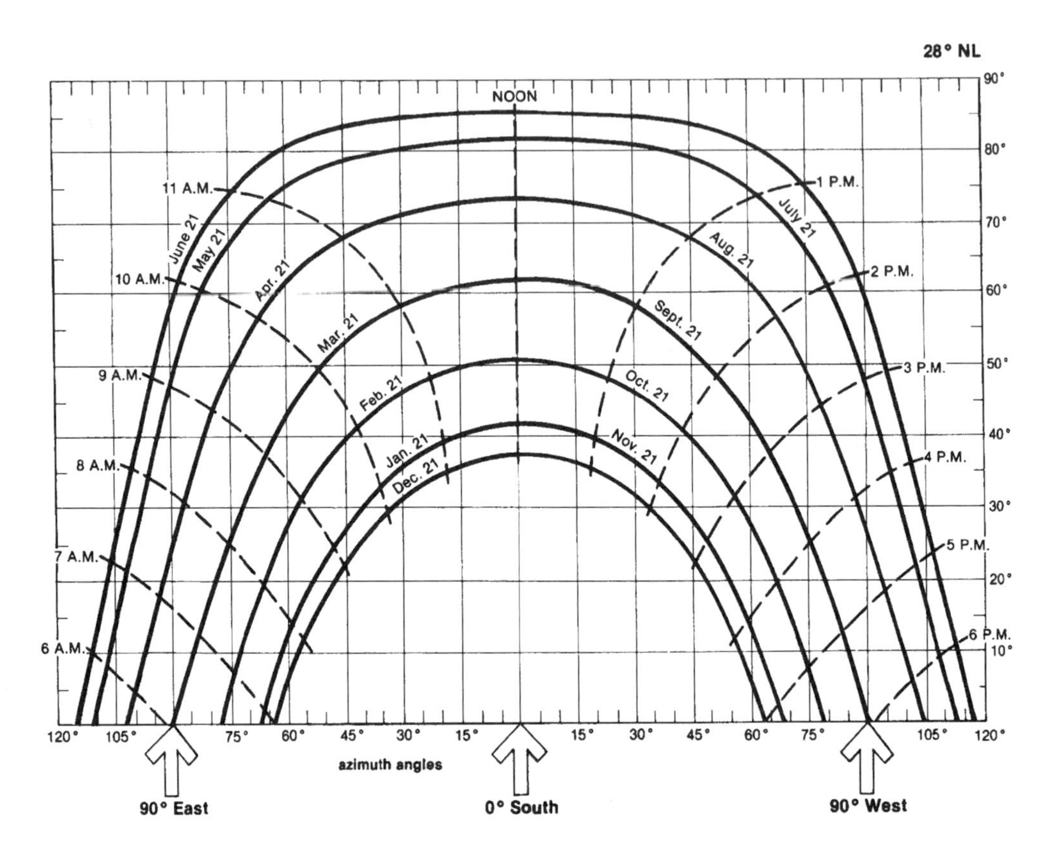

Figure 13.6. The variation of the sun's altitude angle as a function of azimuth or hour angle for a latitude 28° north of the equator. Adapted from Keisling (1983) and Kreith and Kreider (1978).

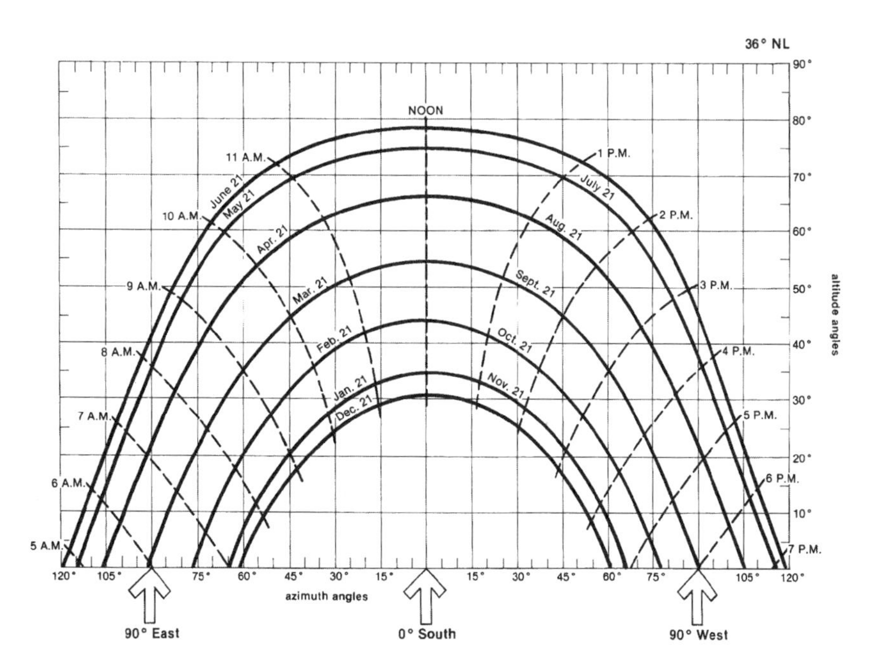

Figure 13.7. The variation of the sun's altitude angle as a function of azimuth or hour angle for a latitude 36° north of the equator. Adapted from Keisling (1983) and Kreith and Kreider (1978).

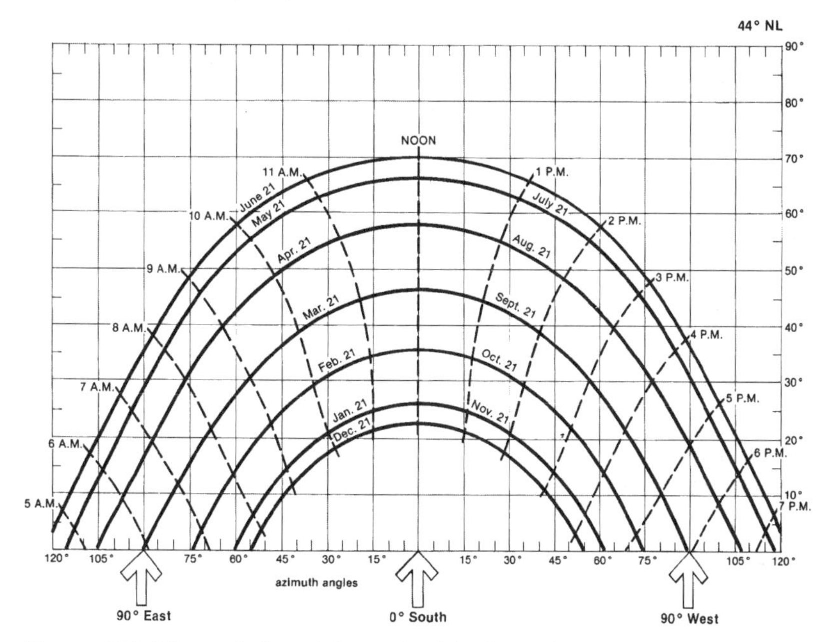

Figure 13.8. The variation of the sun's altitude angle as a function of azimuth or hour angle for a latitude 44° north of the equator. Adapted from Keisling (1983) and Kreith and Kreider (1978).

Table 13.1. Representative Monthly Values for Total (Direct Plus Diffuse) Insolation. Based on 30-Year Averages

Location	Latitude	Total (direct and diffuse) insolation on horizontal surface in Btu/ft^2 per day (Langleys per day)											
		Jan	Feb	Mar	Apr	May	Jun	Jul	Aug	Sep	Oct	Nov	Dec
Albuquerque, NM	35°	1151 (312)	1454 (394)	1925 (522)	2344 (635)	2560 (694)	2757 (747)	2561 (694)	2387 (647)	2120 (575)	1640 (444)	1274 (345)	1052 (285)
Atlanta, GA	33.5°	848 (230)	1080 (293)	1427 (387)	1807 (490)	2018 (547)	2103 (570)	2003 (543)	1898 (514)	1519 (412)	1291 (350)	998 (270)	752 (204)
Bismarck, ND	47°	587 (159)	934 (253)	1328 (360)	1668 (452)	2056 (557)	2174 (589)	2305 (625)	1929 (523)	1441 (391)	1018 (276)	600 (163)	464 (126)
Boston, MA	42°	505 (137)	738 (200)	1067 (289)	1355 (367)	1769 (479)	1864 (505)	1860 (504)	1570 (425)	1268 (344)	897 (243)	636 (172)	443 (120)
Boulder, CO	39°	750 (203)	1030 (279)	1390 (377)	1750 (630)	1960 (531)	2160 (585)	2120 (575)	1890 (512)	1580 (428)	1200 (325)	830 (225)	670 (182)
Ithaca, NY	42°	434 (118)	755 (205)	1074 (291)	1322 (358)	1779 (482)	2025 (549)	2031 (550)	1736 (470)	1320 (358)	918 (249)	466 (126)	370 (100)
Los Angeles, CA	25°	890 (241)	1150 (312)	1520 (412)	1920 (520)	2030 (550)	2090 (566)	2260 (612)	2070 (561)	1670 (453)	1320 (358)	1000 (271)	820 (222)
Miami, FL	25°	1292 (350)	1554 (421)	1828 (495)	2026 (549)	2068 (560)	1991 (540)	1992 (540)	1890 (512)	1646 (446)	1436 (389)	1321 (358)	1183 (321)
Washington, DC	39°	632 (171)	901 (244)	1255 (340)	1600 (434)	1846 (499)	2080 (564)	1929 (523)	1712 (464)	1446 (392)	1083 (293)	763 (207)	592 (161)

1 Langley = 3.69 Btu/ft^2 = 41.86 kJ/m^2 = 1 cal/m^2

Source: Renne, Stovall et al. (2002).

13.3 Passive and Active Solar Thermal Energy for Buildings

13.3.1 Motivation and general issues

About 1/3 of the energy we consume is used to heat, cool, and humidify/dehumidify the buildings we live and work in (see Chapter 20). In developed countries and megacities worldwide, people typically spend over 80% of their time inside such buildings. As such, indoor air quality can be a significant health issue that is strongly linked to energy use. The amount and type of energy required to condition buildings is dependent on climate conditions of the region where they are located.

Solar thermal energy utilization in buildings usually involves one or more of the following approaches:

1. Passive thermal gain and reuse
2. Active capture of solar heat using solar collectors
3. Direct or indirect daylighting

The first two require some type of thermal energy storage and a means for distributing the thermal energy. All require incorporation in the design of a building. In most instances, both direct and diffuse solar radiation are collected on a flat surface exposed to the sun's radiation where the absorber area is equal to the collector area. In some cases, a concentrating approach may be used to achieve higher storage temperatures where the collector area is larger than the absorber area.

In addition to capturing a portion of the solar spectrum for use, proper building design should strive for high performance by maximizing energy efficiency. This approach usually leads to increased building insulation (higher R values, reduced air infiltration and leakage) in the walls, floor, and roof and better window placement and materials. There are tradeoffs, of course, given the costs associated with reducing heat losses or heat gains, that must be balanced against the benefits of having lower energy demand. For example, indoor air quality can be compromised in a well-insulated building with low air-infiltration rates. In these cases, properly designed systems for air exchange with energy recovery are needed. Nonetheless, it is safe to assume that a building that has a passive or active solar thermal system is also designed for high energy efficiency. (Chapter 20 documents these sustainability aspects in more detail.)

13.3.2 Passive systems

The basic approach with passive systems is to utilize the building's structure to capture solar heat and transmit light, where appropriate, to reduce artificial lighting needs. The natural characteristics of certain building materials, such as stone, cement or concrete, and adobe clay, are ideally suited to capture and store heat. In a typical daily cycle, heat is collected and stored during the day and transferred by natural convection of air or water to condition the inside of the building over a period of time that extends into the evening.

Building location and orientation relative to the sun's movement is important in determining exactly what type of passive design will work best. In addition, the type of building poses different challenges. For instance, windowless or closed commercial office buildings that are loaded with people, lighting, fixtures, and their computer workstations represent a discrete set of small heat sources that introduce a substantial cooling load even in the winter months. Residential units with a lower density of people, greater opportunity for natural ventilation, and daylighting are better suited for classical passive designs.

Adobe and Trombe walls (as shown in Figure 13.9) represent popular options for certain locations. These options take advantage of the relatively high heat capacity and low thermal diffusivity of the solid stone or masonry material to store and transfer heat to the inside of the building. A combined mechanism of transient heat conduction through the wall material is coupled with natural convection of air in the building. Normally, the wall is placed on the south-facing side of the building and may be coated with a black or darkened surface to increase its solar absorptivity and covered with glass on the side facing outward with an air space between it and the solid wall. To reduce heat losses, the back and side surfaces may be insulated. A roof overhang is often used to limit the amount of solar gain during the hotter summer months. Alternatively, placing a set of windows or greenhouse on the building's south side to heat a stone, brick, or masonry floor of the room will accomplish a similar passive effect.

More recently, variations on the Trombe wall concept have made them more flexible and adaptable to a wider variety of building applications. The "transpiring wall" is one such idea. Introduced a number of years ago by engineering scientists at the National Renewable Energy Laboratory (NREL) (see Figure 13.10), transpiring walls have been effective for both passive heating and cooling applications.

13.3.3 Active systems

Active solar thermal systems are usually applied in residences and commercial buildings for providing hot water, heating, and air conditioning. What makes them different from passive systems is that they employ collectors that capture solar energy and rapidly transfer thermal energy to a circulating working fluid, which can be used immediately in the dwelling or stored for later use. Control systems are almost always employed to turn circulation pumps on and off and to divert fluid to storage vessels when collector temperatures reach specified levels.

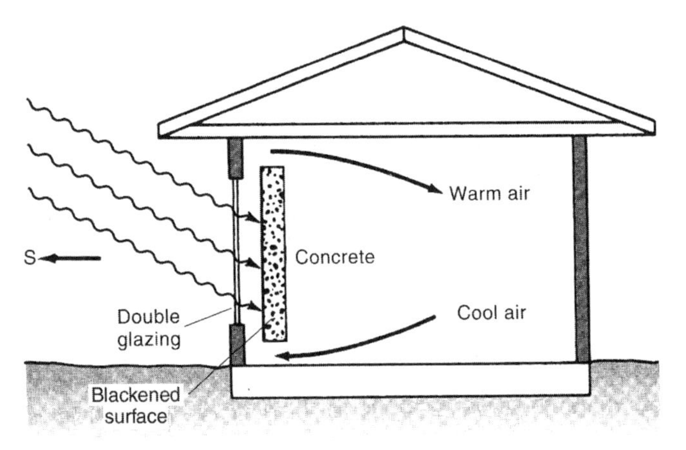

Figure 13.9. Schematic of a Trombe wall for passive capture of solar energy. Adapted from Hinrichs (1996). From *Energy* second edition by Hinrichs. © 1996. Reprinted with permission of Brooks/Cole, a division of Thomson Learning: www.thomsonrights.com.

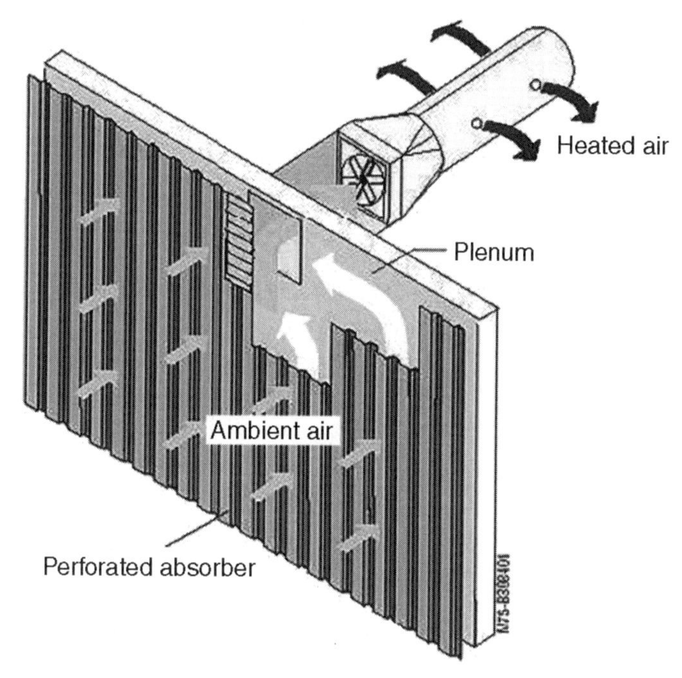

Figure 13.10. The NREL transpired passive solar collector configured as a vertical wall structure element. Source: Renne, Stovall et al. (2002).

Active systems have been in operation for over 80 years, mostly employed in homes for hot water or space heating. A typical design for supplying domestic hot water heating in a residence is shown in Figure 13.11. Here we see a flat plate collector that consists of a selectively coated metal plate with attached channels, as illustrated in Figure 13.12. A circulating fluid is heated as it is pumped through the channels on the collector and then passed to a coil contained inside of a hot water storage tank, where it transfers heat to the water in the tank. In the example shown in Figure 13.11, an antifreeze solution (typically a propylene glycol-water mixture) is used as the working fluid to avoid freezing and subsequent damage to the collector system during the winter. Alternatively, water could be employed with a gravity drain back loop to eliminate concerns about freezing.

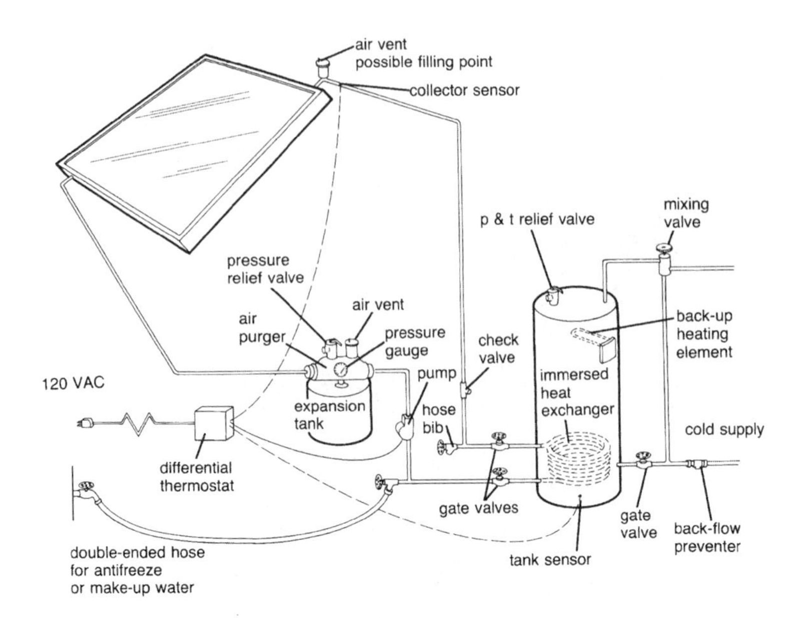

Figure 13.11. Typical active solar hot water system. Adapted from Keisling (1983).

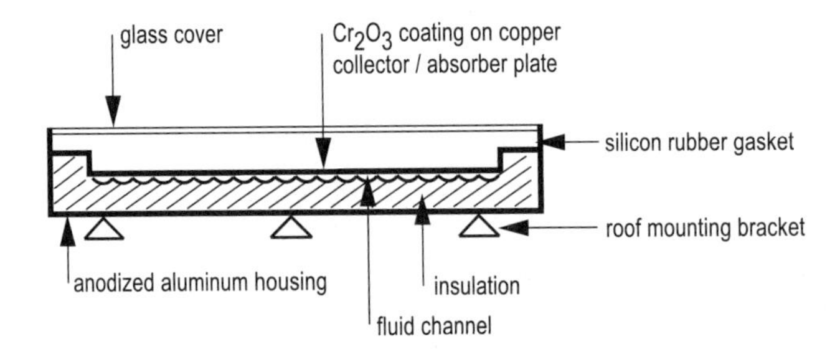

Figure 13.12. Schematic of a flat plate solar collector for application in a domestic hot water heating system.

Today, most flat plate collectors are modules that can be mounted on a roof or built into the roof structure. Each one contains a metal receiver that has been coated with special materials to produce a selective surface that has a high absorptivity for solar energy ($\alpha^* > 0.95$) in the visible and ultraviolet region at shorter wavelengths ($< 2\ \mu m$) and a low emissivity in the longer wavelength, thermal infrared region ($\varepsilon^* < 0.15$). This selectivity lowers the radiative heat loss from the collector surface. Although many materials have been used as selective surfaces, a favored material is a black chrome oxide, Cr_2O_3. To reduce heat losses from the collector, insulation surrounds the sides and back, and one or two transparent glass or plastic plates are positioned on the topside of the collector with an air gap of 1 cm or more. The choice of a transparent cover material is based on a number of factors, including its ability to transmit solar energy with small losses, durability to the weather (wind, rain, and ice), and cost. Tempered glass is often selected for solar hot-water heaters given its low cost and durability, even though it is opaque to radiation in the infrared region. An electronic control unit regulates the flow of working fluid and operates in response to a difference in temperature between the measured storage tank temperature and temperature of the collector surface on the roof.

Although the reliability of commercial solar hot water heaters was not universally good when they were extensively deployed in the 1970s and early 1980s, today's systems are very robust, carrying warranties of 20+ years (see also Section 13.6). According to the US Energy Information Agency (EIA), over 4.5 million solar hot water systems have been installed in the US alone.

Besides hot-water heating, solar flat plate systems can be used for space heating and cooling. In heating applications, air is often circulated through channels in the panels to capture the solar energy. It can then be used immediately for heating rooms by being forced through a set of room registers to distribute the heat or stored in a crushed rock bed for later night time use. Alternatively, water can be used as the heat-transfer fluid in a similar manner—the only difference is that a set of room radiators would be used to distribute the heat. Air has an advantage over water in that it does not freeze and/or cause corrosion problems, but it has lower heat capacity and higher parasitic losses in distribution and storage systems.

For cooling, both vapor compression refrigeration and absorption cycles can be used. In a vapor compression cycle, solar energy can be used as a heat source to power a turbine in a closed-loop Rankine cycle, which, in turn, drives the compressor of the refrigeration cycle. A disadvantage of these cycles is that they need to be fairly large to have reasonable operating efficiencies. For both large- and small-scale cooling loads, a lithium bromide (LiBr) absorption cycle, shown schematically in Figure 13.13, can be employed. Here, solar thermal energy at temperatures of 70–80°C is used to evaporate water from a low-pressure LiBr solution in the generator section of the cycle. Heat is rejected from the system as water is condensed while cooling occurs in the evaporator section, again operated under vacuum conditions at about 40°C. The cycle is completed as the concentrated LiBr solution reabsorbs the water vapor to complete the cycle.

13.3.4 Economic and policy issues

It is difficult to estimate costs for passive solar systems because they often become an integral part of a building's structure. For example, partial cost offsets result when a passive solar greenhouse, Trombe, or transpiring wall is incorporated into the design of a new building. In addition, guaranteeing trouble-free performance or other desirable attributes, such as enhanced daylighting, is as important as reducing heating costs in determining whether passive systems are deployed.

Although costs vary from location to location, a good average value to use for a roof-mounted solar hot water system is about $65/ft^2 (based on 2002 markets), which includes the costs for the collector panels, piping, controls, storage tank with electric or gas backup heating, and installation. Solar air-conditioning units are a factor or two more than the heating system in capital cost. Assuming that a typical family-sized home in the US might need anywhere from 60–120 ft^2 of collector surface, an investment of $4,000–$8,000 would be required. For comparison purposes for new construction, the installed cost of an electric hot water heater would be less than $500. Of course, the tradeoff is between higher capital costs for the solar system versus higher operating costs for the conventional hot water system.

Designs, and therefore costs, for solar space heating units vary substantially depending on load and whether the system was integrated into the building structure, was external to the structure, or was retrofitted. Furthermore, an inherent problem with solar space heating is that there is a seasonal mismatch between the demand for heat and the availability of solar energy. When the demand is highest in the winter, the insolation levels are at their lowest values (see Figures 13.6–13.8). Thus, having a means for seasonal storage of captured solar energy would enhance its value for space heating tremendously.

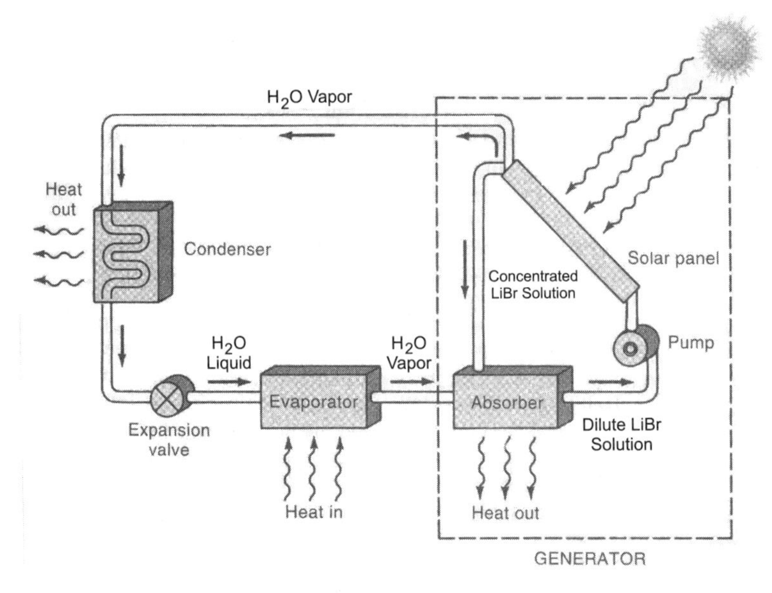

Figure 13.13. Flow sheet for a LiBr solar powered absorption refrigeration system for air conditioning applications. Adapted from Hinrichs (1996), p. 187, Figure 6-35. From *Energy* 2nd edition by Hinrichs. © 1996. Reprinted with permission of Brooks/Cole, a division of Thomson Learning: www.thomsonrights.com.

Several innovative concepts have been proposed for using the earth's subsurface in the form of water contained in a confined aquifer or as heated rock. While both of these concepts are technically possible, there are drawbacks. For example, additional costs are incurred to put such storage systems in place. Chapter 16 discusses some of the physical attributes of thermal storage that are relevant to implementing seasonal heat storage.

Given these limitations and constraints, deployment of existing passive and active solar heating and cooling technology for buildings has been severely limited by the high front-end capital costs that are incurred when a building is constructed. The potentially lower net lifecycle costs (amortized capital plus operating and maintenance costs) for a solar system may not be realized. The traditionally low prices of conventional fuels and baseload electricity, with the exception of occasional price shocks, is often the single most important factor that deflates interest in investing in energy efficiency and solar energy capture.

There are several ways of making solar heating systems more attractive. One is to achieve lower unit costs by improving and scaling up production levels, and the other is by introducing policy incentives. The high capital cost of solar hot-water systems is partly driven by limited production and lack of standardization of performance. For example, scaling up production to a million units per year in the US would have a substantial impact—reducing the current costs for these systems by 30–^40% or more to levels of $2,500–$5,000 or less, depending on the size of the system. Introducing incentives to homeowners or commercial building operators to install a solar system would also have an impact. Such incentives could be in the form of tax credits or lower interest rates on mortgages or loans. Alternatively, building codes and standards could be revised to encourage the use of solar energy.

13.4 Solar Thermal Electric Systems—Concentrating Solar Power

13.4.1 Fundamentals and options

Compared to the inherent simplicity of passive and active solar thermal systems for buildings, the use of concentrating solar energy for generating electricity is often perceived as a "high tech" option. Even though the scale of most concepts for concentrating solar power (CSP) is much larger than a single home would need, much of the technology required for CSP is relatively mature and readily available for deployment. For a number of reasons, the current menu of CSP options has been divided into three types: power towers, trough systems, and dish-engine systems.

A few features are common to all of these options. Since solar energy is concentrated, the collector area is larger than the absorber area of the system. This concentration of energy yields much higher operating temperatures than those for flat plate collectors. Steady-state temperatures range from about 400°C to above 1,000°C and provide an opportunity for generating electricity at reasonable thermal efficiencies or providing thermal energy for high-temperature industrial processes.

As only the direct, unscattered portion of the solar spectrum can be accurately focused on the absorber, high-grade areas for CSP applications usually exist where cloud cover and atmospheric moisture levels are normally low and year-round insolation levels are high. Almost exclusively, this limits high-quality CSP sites to arid desert or semi-arid regions at lower latitudes near the equator. For the US, the direct insolation resource is highest in the desert southwest states of Nevada, Arizona, New Mexico, and southern California. Regions of southern Spain and northern Africa are also particularly well suited for CSP deployment.

In almost all cases, CSP plants produce only electricity using some type of thermal energy to power the cycle. The incident solar flux is reflected and concentrated on a suitably designed absorber and is captured as thermal energy that would be used in a thermal cycle to heat a prime mover fluid, such as pressurized water or compressed air, which would be expanded to drive a turbo-generator unit for making electricity. Our coverage is limited to brief descriptions of each type of CSP technology in the subsections that follow. Those interested in pursuing CSP technology in more detail should consult the many available reviews and current Web sites for the latest information, for example DeLaquil et al. (1993) and www.eere.energy.gov/solar/.

13.4.2 Power tower—central receiver systems

A field of mirrors or heliostats is used to intercept and redirect the sun's energy. Each heliostat is focused on a centrally located receiver (as depicted in Figure 13.14). Typically, energy is absorbed into a high-temperature working fluid, such as a molten salt mixture, which is pumped through the absorber and stored for several hours at temperatures ranging from 500– 600°C. A steam Rankine cycle is then used to generate electricity with the working fluid providing the thermal energy needed to vaporize and superheat steam before expansion in a turbogenerator. Commercial power towers are large, base-load type installations capable of producing up to 100–200 MWe in a dispatchable mode when needed.

Several prototype, small-scale tower systems have been built and tested to demonstrate the concept. Notable demonstrations were conducted during the 1980s in Europe (e.g., Thémis at Targasonne, France; CESA and others at Tabernas, Spain) and Solar One and Two in the US, in the California desert at Daggett near Barstow. The Solar One and Two labels designate two separate phases of development in the US Department of Energy's (DOE) program. The Solar One experiment utilized about

71,000 m^2 of reflector surface with over 1,800 separate heliostats focused on a 55 m high, water/steam receiver with an outlet design temperature of 516°C at 105 bar. The capacity was rated at 10 MWe using an open steam Rankine cycle. Construction began in 1982 under joint sponsorship of the DOE, Southern California Edison, the Los Angeles Department of Water and Power, and the California Energy Commission. After two years of startup testing and upgrading, a four-year demonstration test was successfully conducted. The next phase led to the Solar Two demonstration that was focused on improving the operability and dispatchability of the system by modifying the receiver, working fluid, and heat storage system. In place of the once-through steam system of Solar One, a mixture of potassium and sodium nitrate salts was used as the prime mover fluid in the Solar Two demonstration (see Figure 13.14b). The change reduced the pressure of the receiver chamber because water was no longer used and allowed for thermal storage to be achieved at high temperatures (>500°C). In fact, the thermal storage capacity of molten salt systems of this type would permit continuous or dispatchable operations for periods of 24 hours or more, thereby increasing the value of the CSP-produced electricity.

Much has been learned about component performance for central receiver applications from the successes and failures that have occurred during the last 20 years of field testing, including improvements in the durability and ease of manufacture of mirrors, better heat transfer design and materials and control systems for the central receiver itself, and demonstrated dispatchability using molten salt storage. These improvements have increased reliability, lowered parasitic losses, and increased efficiency to a point where engineers are confident of scaling up tower designs to commercial levels.

Nonetheless, a large capital investment is needed to build a power tower system. Even with the best projections available for 100–200 MWe-sized systems, the lowest levelized busbar price for electricity produced in 2002 from central receiver power towers is about 8¢/kWh. In terms of capital investment, a range of $3,000–$4,000/kW is projected. Consider a 200 MWe power tower facility. Here an investor would have to come up with $600–$800 million to launch the project. In alternative energy markets, this magnitude of investment is not easy to realize and often requires national or international policies or subsidies including loans, tax incentives, guaranteed prices, and production credits.

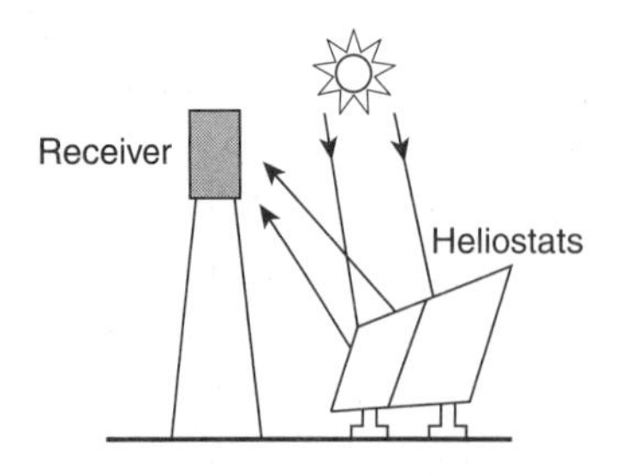

Large sun-tracking mirrors, called heliostats, focus the sun's energy on a receiver located atop a tall tower.

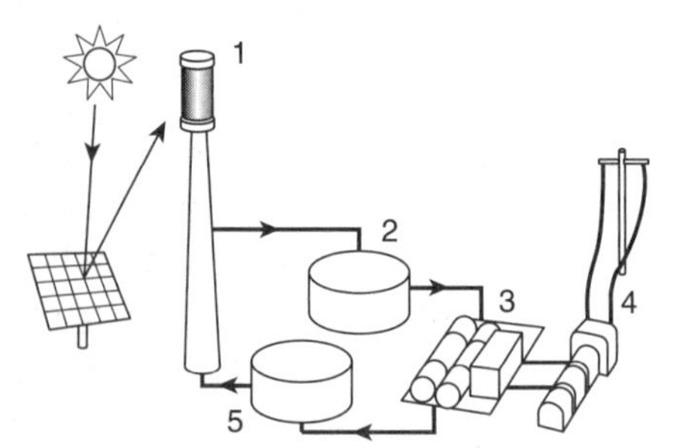

Schematic of electricity generation using molten-salt storage:
1) sun heats salt in receiver;
2) salt stored in hot storage tank;
3) hot salt pumped through steam generator;
4) steam drives turbine/generator to produce electricity;
5) salt returns to cold storage tank to be reheated in the receiver.

(a) schematic of power tower concept showing heliostats and receiver with storage and power generato

(b) Solar Two demonstration plant

Figure 13.14. Solar power tower concept and demonstration plant photographs. Sources: USDOE (1996); Renne, Stovall et al. (2002); and SunLab (1998).

13.4.3 Parabolic troughs

Trough concentrators reflect sunlight off a linear, parabolic mirror surface and focus it onto an absorber tube that is located along the focal line of the trough (as shown in Figure 13.15). A heat transfer fluid, usually water or oil, is pumped through the receiver tube to heat it to temperatures ranging from 100°C to about 400°C. Parabolic troughs have concentrating factors between 10 and 100 and usually employ a one-dimensional tracking system to maintain proper focus on the fixed receiver tube as the sun moves across the sky during the day. The recovered thermal energy can be used for high-temperature process heat applications or as an energy source in a thermal electric power plant. Commonly, a conventional steam Rankine cycle is employed to produce electric power. Today's designs have achieved solar energy to electricity efficiencies of about 12%, including all parasitic losses associated with operating the collector system.

Trough Systems

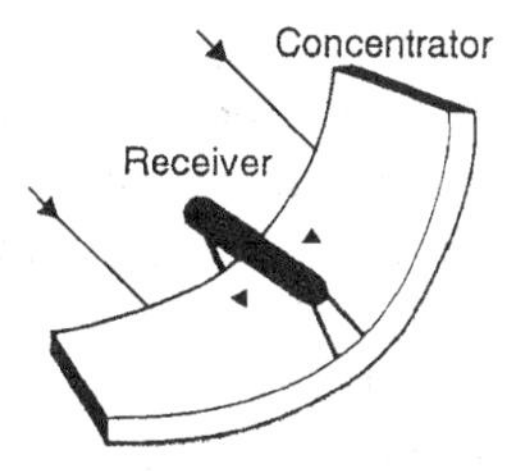

(a) Schematic of trough concentrator

(c) Photo of the Kramer Junction plant where nine trough power plants in California's Mojave Desert provide the world's largest generating capacity of solar electricity, with a combined output of 354 megawatts

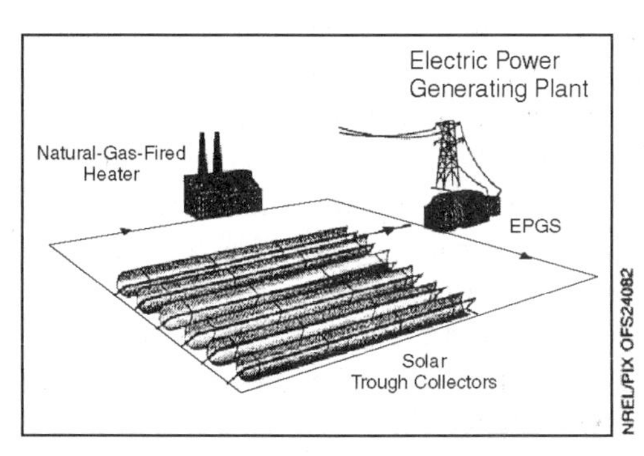

(b) Dispatchable hybrid solar-natural gas trough system

Figure 13.15. Parabolic solar trough concept and example. Sources: US DOE (1996) and SunLab (1998).

Trough systems have other desirable attributes that are worth mentioning. Each parabolic trough unit is modular and can be coupled to another for either series or parallel operation. In series, this increases the operating temperature, and, in parallel, this achieves higher energy flows to feed a process or power plant. For generating electric power, the generating capacity of a trough system for a single power plant ranges from about 10 MWe to a maximum value of 200 MWe. Given that the solar energy source itself has a fairly low energy density, the land area needed for a trough unit is substantial. For example, in desert regions, where the total and direct normal insolation percentage is the highest with about 2,500 kWh/m^2 of solar energy available annually, over 500,000 m^2 (0.5 square km) of collector surface area would be needed for a 100 MWe plant operating with a 12% solar to electric efficiency. Another resource that may be needed is cooling water for heat rejection in the power plant. If evaporative cooling is employed, annual water requirements can be very high. For example, for that same 100 MWe plant, about 1.5 million m^3 of water would be needed per year. Given that high-grade CSP resources are located in desert regions, the consumption of cooling water is both an economic and a sustainability issue.

A mismatch normally exists between the time the solar energy is collected and the time the energy is needed, thus storage may be used to permit the plant to operate in a dispatchable mode. Although some types of thermal storage (molten salts or heat transfer fluids) have been considered and tested, other types of storage, such as compressed air, pumped hydro, or magnetic energy storage, could be used if available. For hydropower, location might be a factor because high-quality pumped hydro sites are usually quite distant from large-scale CSP locations in arid, desert regions. Alternatively, the plant may be hybridized with natural gas or another combustible fuel to provide thermal energy when it is needed and not available from the sun. This would make the plant, even without any storage, fully dispatchable to supply power during periods of intermittent solar gain (e.g., when cloud cover exists) or during the night.

The receiver is designed to maximize temperature while minimizing heat losses. Typically, an evacuated glass tube with a high transmittance surrounds a metal absorber tube that has been selectively coated with a high-absorptivity, low-emissivity material. With the best available technologies today (e.g., Luz cermets or Solel UVAC coatings), absorptivities of 0.92– 0.96 and emissivitives of 0.14 or less are possible.

Other losses occur in capturing solar energy with trough systems. These include reflectivity and transmission losses in the trough mirrors, tracking errors and shading losses, losses in transferring energy from the heat transfer fluid used in the receiver tubes to the prime mover fluid used in the process heat or electric power generator, and losses in storing heat.

Given that the major capital items in a trough system are mirrors and absorber tubes, several issues are being addressed to reduce costs. The modularity of trough components makes them well suited to achieve significant reductions in manufacturing costs as production levels increase. Current emphasis is on the use of low-cost materials that have high-performance characteristics, are durable, and are easy to fabricate and maintain in the field. The operating experience gained by deploying a number of commercial scale systems in the 1980s and 1990s has been invaluable in moving the technology forward.

The history of troughs provides an interesting context for evaluating the potential of this technology form. Over 120 years ago, the first solar parabolic trough system was designed, fabricated, and tested by John Ericsson for supplying energy to an engine powered by a hot air cycle. Circa 1910, a 45 kW trough system was built to power a steam pumping plant in Egypt and operated until 1915 (DeLaquil et al., 1993). After that early period, it took some time before working units were constructed for capturing and utilizing solar energy in a useful manner. The concern over energy resources in the early 1970s renewed interest in trough technologies. In stark contrast to solar power tower and dish-engine methods, troughs were actually built and have been operated at a commercial scale for over 20 years.

For example, in Chandler, Arizona, process heat for a copper electroplating plant is supplied by a 5,500 m^2 solar trough system. At the Kramer Junction site in the Mojave Desert in southern California, there are nine separate solar trough electric plants operating with a total generating capacity of 354 MWe. These are often referred to as the Luz solar electric generating system (SEGS) plants after the company who designed and built all nine plants. The SEGS plants range in size from 14 to 80 MWe and have operating temperatures ranging from 325 to 410°C. The first plant (14 MWe), built in 1984, and the last one (80 MWe), built in 1990, incorporated many upgraded features that improved performance and lowered capital and operating costs. The first generation Luz plants had a $6,000/kW capital cost, compared to today's projected cost of about $3,000/kW or less.

Natural gas was used as the backup in all nine plants with no provision for thermal storage provided in SEGS plants 2 through 9. Public Utility Regulatory Policy Act (PURPA) regulations limited the maximum size of a plant to 30 MWe until 1987 when it was raised to 80 MWe and eventually eliminated altogether. The non-solar output of each plant was also restricted to no more than 25% of the total electricity generated. Without additional environmental offsets or green-power price credits, the existing level of investment tax credits granted and cost sharing development costs with the US DOE were still not enough to make trough-generated solar electric competitive with other alternatives, namely low-cost combined-cycles fired by natural gas. Unfortunately, even with all nine plants operating, the Luz Company went bankrupt in 1991. The SEGS plants, nonetheless, continued to operate under new ownership and are still running today. Useful operating data and upgrades have been introduced, particularly to reduce O&M costs. The SEGS experiment underscores the need for subsidies and real operating experience to help bring down the cost of a new technology while increasing reliability and reducing risks.

13.4.4 Dish engine systems

The third major category of CSP technology is the parabolic dish receiver-heat engine generator. Here, direct solar radiation is concentrated by a factor ranging from 600 to 3,000, with a small number of focusing parabolic mirror reflectors aimed at a single absorber target located at the focal point. Heat is absorbed and directly applied to a heat engine generator that is mounted near the focal point (as shown in Figure 13.16). A two-dimensional tracking system is used to point the dish to maximize the captured solar energy as the sun moves across the sky. Because of the high concentration ratios, absorber temperatures tend to be higher than troughs or towers, typically ranging from 600 to 1,500°C.

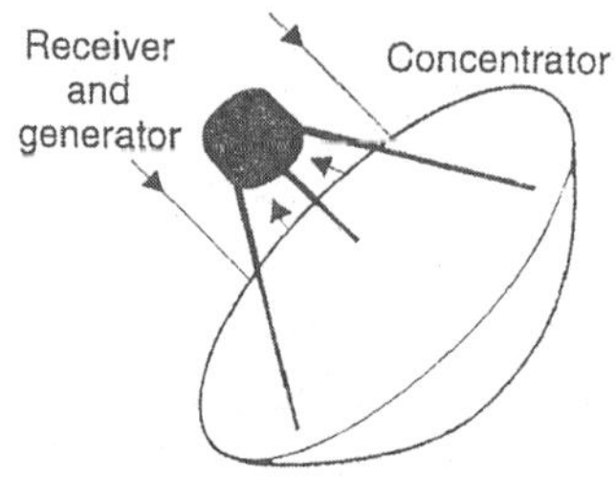

(a) dish-engine concentrator concept

(b) Boeing/Stirling Engine Systems 25 kW parabolic dish engine prototype (ca. 2002).

Figure 13.16. Solar dish-engine concentrator concept and prototype example. Sources: US DOE (1996) and SunLab (1998).

These higher collection temperatures, while potentially desirable for efficiently converting heat to electric power, introduce their own set of engineering challenges. One is developing a means for removing the thermal energy from the absorber. Oil or water heat transfer fluids have been used to accomplish this in several prototype systems. An alternative is to mount the heat engine at the focal point and to transfer the absorbed energy to the prime mover fluid without an intermediate step. Stirling cycle engines are good candidates for dish engine applications because they use air as working fluid and can achieve reasonably high efficiencies for high absorber temperatures. Early tests of prototype models have achieved solar to electric efficiencies approaching 30%, which is higher than either today's trough or tower technology. In addition, gas-fired backup systems can be used to provide heat for the Stirling engine when solar energy is not available, which makes dish systems dispatchable in distributed applications. However, there are challenges, as operating lifetimes needed for reliable power generation have not yet been achieved and installed capital costs are as high as the other CSP technologies. These issues pose significant barriers to commercialization. Trough and tower systems, which employ existing steam Rankine technology to generate electric power, can focus their cost-reduction development efforts on the solar end of the system (reflectors, absorbers, and thermal storage). In contrast, dish-engine systems require development of both solar and power converter components.

Dish-engine systems are sized in the 5–50 kWe range, making them ideally suited for remote, distributed applications. They can be deployed and operated in a larger integrated system in much the same way as a wind energy farm.

With an appropriate scale for remote power generation, potential applications in remote regions of developing countries are providing momentum to push dish-engine technology along. Here, the value of having dispatchable electricity at a kWe scale, albeit with associated higher unit costs per kWe, outweighs the economic gains of operating large, central station power plants. Smaller units require smaller investments and avoid having to build the infrastructure needed to distribute power over long distances. Some believe that early deployment of dish-engine units in these remote, distributed niche markets will bring down costs and reduce risks enough to lead to larger scale deployment in more competitive power markets.

13.4.5 Current status and future potential of CSP

The CSP technology portfolio has considerable diversity in both scale (from 5 kWe dish-engines to 100+ MWe power towers and troughs) and application (from on-grid, central station, baseload units, to moderate load, dispatchable plants, to remote, distributed units). Table 13.2 and Figure 13.17 summarize the characteristics and projected costs for the three types of CSP that were covered in this chapter.

Unlike other alternative energy technologies, there is almost 20 years of operating experience in providing dispatchable electric power at a commercial scale with solar-gas hybrid parabolic trough plants at Kramer Junction in California. The successful operation of the SEGS plants reduced O&M costs by introducing several innovations for replacing absorber tubes and for cleaning the parabolic mirrors. Nonetheless, like other renewable technologies, there continues to be a large barrier to further deployment of CSP due mostly to their high capital costs. With proper policy instruments that include consideration for a range of capacity scales and applications from large and small power markets, its environmental benefits, and its energy security aspects, solar concentrators would be attractive for providing power in high-grade, arid, and desert regions where sustained high levels of direct normal insolation can be counted on throughout the year.

Table 13.2. Summary of Operating Characteristics and Estimated Costs for Concentrating Solar Power Technologies

Operating Characteristics

CSP technology	Concentration ratio	Tracking requirement	Operating temperatures	Average solar to electric efficiency	Unit size range	Status
Power Towers	500–1,000	2-axis heliostats	400– 600°C	12–18%	30–200 MWe	Demonstration and testing at 10 MWe scale
Parabolic Troughs	10–100	1-axis reflector	100– 400+ °C	8–12%	30–100 MWe	20 years operating in California
Dish-Engines	600–3,000	2-axis	600–1,500°C	15–30%	5–50 kWe	Prototypes tested at 25 kWe or less for limited periods

Estimated Costs[a]

CSP technology	2002 capital costs \$/kW	2010+ capital costs \$/kW	O&M costs ¢/kWh	2002 levelized electricity cost ¢/kWh	2010+ levelized electricity cost ¢/kWh	Uncertainty
Power Towers	4,000	2,500	0.7	5–9	3.5–5[b]	moderate
Parabolic Troughs	4,000	2,500	1.0	7–11	6–9	low
Dish-Engines	3,500	1,500	2.0	9–13	4–6	moderate

(a) in 2002 US$; see www.eere.energy.gov/solar/ for current updates

(b) low estimate assumes municipal investment and high assumes private investment

Sources: SunLab (1998), US DOE (1996), Butler (1997, 2002), DeLaquil et al. (1993).

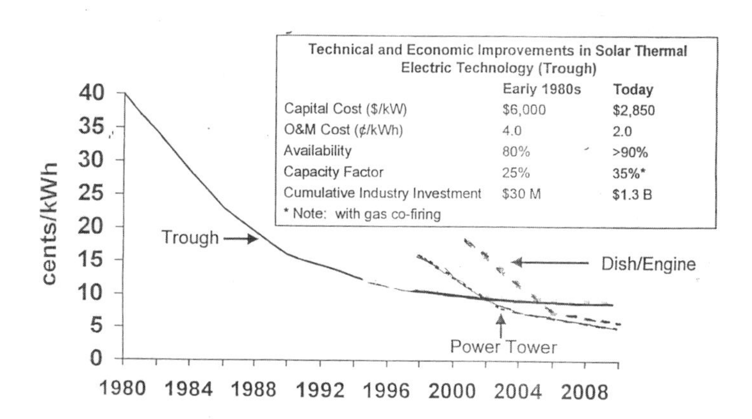

Figure 13.17. Actual and projected levelized electricity costs for concentrating solar power technologies in 2000 US$. Sources: SunLab (1998) and US DOE (1996).

Connecting demand growth to CSP resource quality will also drive markets. For example, the continued growth in population in the desert southwestern US creates a real opportunity for supplying a portion of its demand growth with CSP technologies. Further work on modular components, however, particularly heliostats and mirrors, absorber/receivers, and storage systems, is needed to take full advantage of the cost reductions that will appear as production levels rise. For example, the solar thermal manufacturing technology (SolMaT) initiative managed by the US DOE's SunLab is focused on cost-reduction methodologies in establishing partnerships with industry. The prospects for lowering costs for key components look good. For example, numerous studies indicate that, as production levels increase to 50,000 units with 150m^2 units per year, the cost of 2-axis tracking heliostats for power tower applications, including controls, can be reduced to about \$135/m^2 from a current level of about \$200–250/m^2 (Teagan, 2001). Validation of this hypothesis will require a commitment to build and deploy several large central receiver plants to achieve these heliostat production levels. International collaborations, such as SolarPACES, also provide considerable leverage to support key R&D and demonstration projects focused on common objectives for improving CSP technology and reducing costs.

Other ways of using concentrated sunlight for energy applications are also being considered. These include buildings integrated with trough technologies for providing heating and cooling needs with enhanced daylighting (Myles, 2002) and concentrating photovoltaics (see Section 13.5.5). A distinct advantage of these proposed applications is that they are small in scale (10 kW) and do not require being located in the highest-grade insolation regions.

13.5 Solar Photovoltaic (PV) Systems

Photovoltaic devices utilize the photoelectric effect in semiconductor materials to convert solar energy directly to electricity. Although the basis and proof of the concept were in place in the mid-1800s, actual PV devices were not developed until the 1960s, when they were motivated in large part by the US and Soviet space programs. The first practical PV cells were constructed of crystalline silicon and were very expensive, costing upwards of \$250/W (or \$250,000/kW). At this stage, PV technology was not suitable for large-scale terrestrial power applications. Following their use in space exploration, other applications appeared. As manufacturing technology and experience expanded, prices dropped into a range where PV units could be used to replace non-rechargeable batteries. Starting in the 1970s, PV devices were developed for watches, calculators, road signs, remote communication systems, and similar applications. Small scale (1–100 kW) power-generation applications began to surface in the late 1970s as well, with PV module collectors for building applications appearing in the marketplace. The growth in module production, primarily crystalline silicon, has been steady at about 15–20% per year for the last 20 years or so. By 2003, over 700 MWe of solar cells were manufactured at costs of \$2,500/kW or less, which represents more than a 100-fold reduction in module price. When other components that comprise the entire PV system are included, the total cost per kW is about twice the module cost. In 2002, entire PV systems that include polycrystalline silicon modules, power inverters to produce alternating current (AC) from direct current (DC), other power conditioning and control equipment, and a modest battery storage system could be installed at prices ranging from about \$600–900 per m^2 of active collector surface. Developing improved manufacturing methods for modules and associated PV equipment to bring costs down has become a major focal point of the PV industry. (We will say more about the costs and values of PV systems in Section 13.5.5.)

PV systems are attractive because of their simplicity. They have no moving parts. They are noise-free and potentially have low-maintenance requirements. Moreover, PV systems have built-in modularity and are scalable for applications ranging from watts to megawatts, and they can be integrated directly into the unit they are providing power for, whether it is a remote highway sign or residential dwelling.

The sections that follow provide a brief overview of PV technology. First, we review the basic concepts of photovoltaic cell operation and key performance and technology issues that are common to all types of cells. Next, we focus on the major types of systems being developed and deployed worldwide—crystalline and amorphous silicon, thin film copper indium diselenide (CIS), and cadmium telluride (CdTe), reviewing what is unique about each and the status of their development. In the final section, we discuss more advanced concepts and summarize the prospects for PV.

The popularity and scientific elements of engineering materials for photovoltaic applications have resulted in a rich field of technical literature on the subject. To provide more depth and details on PV technologies, readers are directed to a few archival references, as well as to the Web site resources listed at the end of the chapter. In particular, the reviews that appear in *Annual Reviews of Energy and the Environment* and in articles in the edited volume *Renewable Energy* (1993), authored by Kelly, Green, Carlson and Wagner, Boes and Lugue, and Zweibel and Barnett, as well as numerous publications by the National Center for Photovoltaics located at the *National Renewable Energy Laboratory* (NREL) in Golden, Colorado, are worth examining.

13.5.1 Solid state physical chemistry fundamentals

The photoelectric effect is the underlying phenomenon that controls how a photovoltaic device converts sunlight directly into electrical energy. When a photon strikes certain materials, such as conductive metals and semiconductors, the material's electrons are able to "capture" the photon's quantized energy. If that energy is of sufficient magnitude, exceeding the so-called band gap of that material, then electrons are released. The PV device actually works by creating a voltage difference within a nanoscale structural environment that directs the migration of electrons to produce a current. This is achieved by arranging certain semiconductor materials in a prescribed manner. Typically, p-type and n-type semiconductor materials are electrically connected, as illustrated in Figure 13.18.

The n-type or negative material is one that allows electrons to move freely within it. Thus, the current carriers in an n-type material are electrons. Silicon becomes an n-type semiconductor when small amounts of impurities, such as phosphorus, are added to it. In scientific terms, such n-type silicon is "doped" with phosphorus.

The p-type or positive material operates in an analogous but opposite manner. If a different impurity, such as boron, is added to silicon, a portion of the electrons associated with silicon are "tied up" or partially immobilized by the presence of boron atoms. This creates a network of "holes" within the silicon crystal that are locally positively charged. These "holes" act as if they were positive charges and serve as current carriers in analogous manner to the free electrons in an n-type material.

Placing an n-type material in contact with a p-type creates a p-n junction (as shown in Figure 13.18a), which has certain properties that are altered by the presence of light. First, without light present, the electrons of the n-type material tend to migrate across the junction to form a locally high concentration in the p-type material near the junction. Similarly, this creates an excess of "holes" or positive charges in the n-type material adjacent to the junction (see Figure 13.18a). These processes lead to a small voltage difference across the junction. For silicon-doped materials, this amounts to about 1/2 volt. The charge gradient creates a small current (I_d), which is exactly balanced by a current (I_v) that results from the thermally activated effects that release a small number of electrons and "holes."

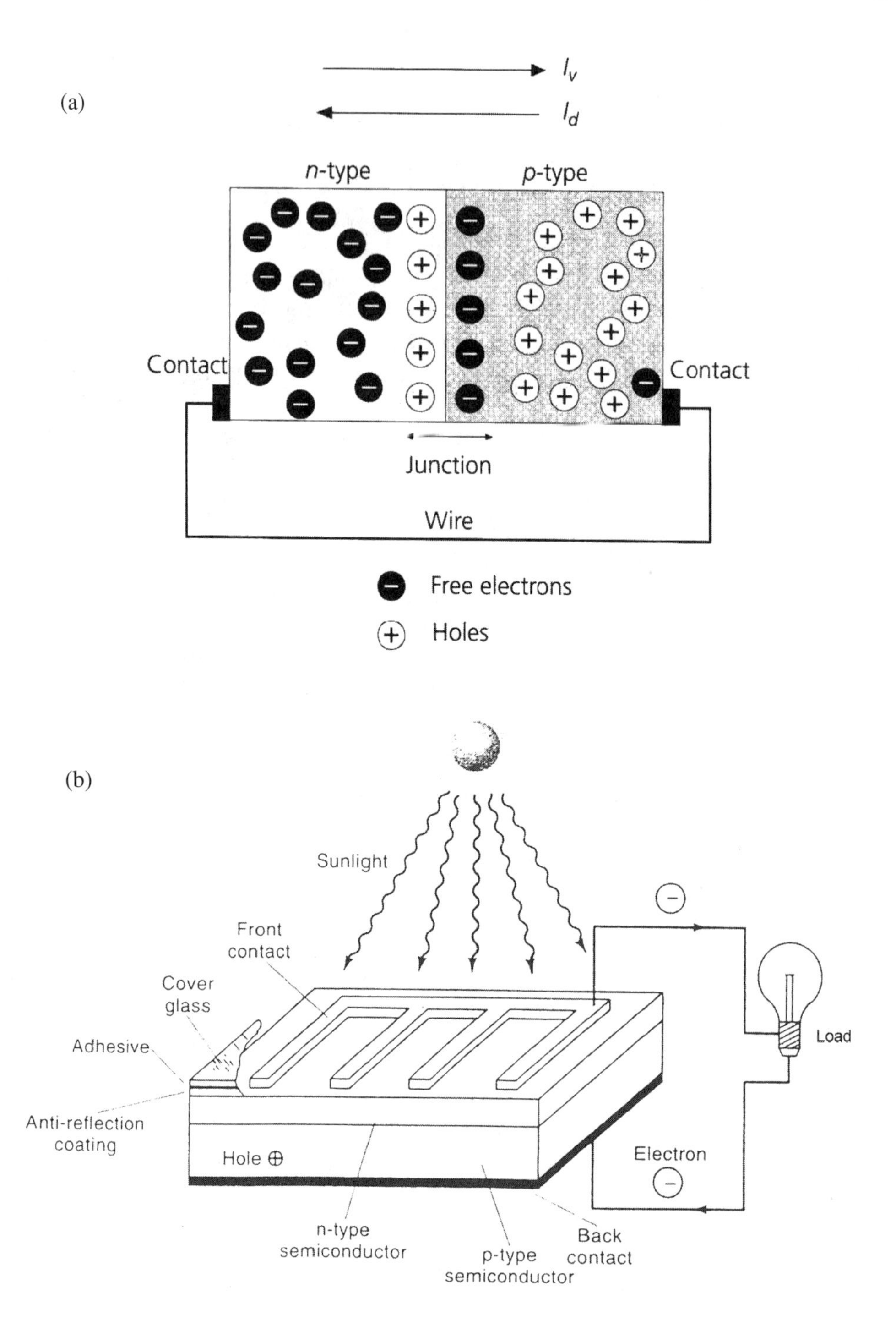

Figure 13.18. A p-n photovoltaic semiconductor device (a) p-n junction, contacts, and (b) circuit schematic. Sources: Kelly (1993) and Renne, Stovall et al. (2002).

Without light present, there is no net current flow through the p-n junction. When sunlight strikes the PV device (as shown in Figure 13.18b), the photoelectric effect causes the number of free electrons in the n-type material to increase and the number of "holes" in the p-type to increase as well. Connecting the two different materials across a load, using the electrical contacts shown in Figure 13.18b, causes electrons to flow from the n-type material through the load where they deposit measurable electrical power and then recombine with the excess of "holes" in the p-type material to close the circuit.

In Figure 13.18b, we see how things are arranged for a typical silicon solar cell. Metal contacts with a large fraction of open area for light to enter the module are attached to the front surface of a thin layer (about 1μm) of n-type material. The bottom layer of p-type silicon is normally thicker (typically from 100– 400 μm) and coated with a metal material that provides both structural rigidity and electrical contact. Although most PV devices have features similar to those found in silicon cells, there are differences (as will be discussed in the sections that follow).

13.5.2 Performance limits and design options

The key performance metric of any photovoltaic system is the net efficiency of converting sunlight to electricity. For a given amount of insolation, the required collector area to meet a given load decreases directly in proportion to an increase in efficiency. There are several measures of efficiency. The highest values reported are almost always obtained under well-controlled conditions in a laboratory with small cells, often using artificial sunlight as a calibrated energy source. The next level lower is measured for individual PV modules, again under controlled conditions. The final level is measured for complete PV systems installed in the field.

Under intense sunlight at peak periods, the solar flux to the surface of a cell is on the order of 1 kW/m^2, which produces a current of about 100 mA/cm^2 at a solar-to-electric energy efficiency ranging from a few percent to over 20% depending on a number of factors. Table 13.3 lists typical values of efficiency for a range of PV cell types. Important factors affecting efficiency include those listed below.

1. Reflective losses off the top surface of the PV collector
2. Loss of effective interception area by the presence of the metal contacts on the top surface
3. Ineffectiveness of photon-electron interactions in the layers of n- and p-type materials
4. Ohmic resistance losses in the circuit
5. Recombination of electrons and "holes" before reaching the p-n junction
6. Required band gap energies limiting the ability of photons at low energy levels to create free electrons and holes

Table 13.3 Representative photovoltaic cell efficiencies as a percent (%) of incident solar insolation to electricity.

Type	Field-deployed modules (%)	Prototype modules (%)	Lab-scale experimental cells (%)	Theoretical limit (%)
Flat Plate				
Single crystalline silicon (Si)	10–12	16–18	24+	30–33
Polycrystalline silicon	8–9	—	18.2	—
Single-junction amorphous silicon	3–5	5	6–8	27–28
Multi-junction stabilized amorphous silicon	8	10	12	27–28
Copper indium diselenide (CIS)	—	11	14.8	23.5
Cadmium telluride (CdTe)	—	10	15.8	27–28
Stacked multi-junction amorphous silicon and CIS	—	—	15.6	42
Concentrators				
Gallium arsenide (GaAs)	NA	22	28	—
Gallium arsenide and antimonide (GaAs and GaSb)	NA	NA	34	—
Two-junction, 2-terminal, monolithic	NA	NA	30	42+
Three-junction, 2-terminal, monolithic	NA	NA	32	42+

NA - not available
Sources: NCPV (2002) and Kelly (1993).

The durability of the module itself is also a key issue. Given that capital costs are high, PV systems should be able to run essentially maintenance-free for long periods—10 to 30 years are expected lifetimes for applications in competitive energy markets. As a result, the ability to maintain efficiency and resist the degrading effects of exposure to weather is essential for building deployed PV systems. While flat plate PV devices utilize both direct and diffuse sunlight, different PV devices have different capture and conversion efficiencies that vary with the intensity of light and its wavelength or frequency within the solar spectrum (see discussion below and Figure 13.1).

The performance of the PV module itself is centrally important. Nonetheless, there are several other components in the system that must perform well and not cost too much. These are often referred to as "balance-of-system components." They include power connections, power conditioning (e.g., inverters), control and interconnection equipment, and, in many cases, an electrical storage system usually consisting of batteries (see Chapter 16).

PV collectors fall into one of two categories, flat plates or concentrators. By far, flat plate configurations represent the largest fraction of manufactured and installed systems. Figure 13.19 shows the two common options for deploying flat plate PV systems—as roof-mounted units and as modular units within a separate structure. While flat plate systems utilize solar cell material over their entire exposed area, the interception area of a concentrator employs lower cost mirrors or lenses to focus the sun's energy on a smaller area of a PV device. This approach, in principle, leads to both increased performance of the PV device and lower costs because less semiconductor material is needed for each module. However, concentrators only utilize the direct, unscattered portion of the solar spectrum, which reduces their versatility over flat plate systems.

Regardless of which collector approach is used, the motivating principle of all PV technology development is to lower capital and operating costs per kWh generated. For flat plate technology, costs are lowered by increasing module performance (higher efficiency—smaller areas needed, and higher reliability—less O&M needed) and by decreasing module manufacturing costs. These costs are lowered by reducing the amount of highly refined semiconductor material needed by utilizing thin films or by increasing productivity per unit time. Most current R&D deals with lowering manufacturing costs without sacrificing efficiency and reliability. For concentrator systems, not only must the solar cell be optimized, but other critical components, such as the tracking system and reflectors, require careful attention in terms of performance and cost.

Figure 13.19. Photographs of typical PV installations. Sources: Renne, Stovall et al. (2002) and SunLab (1998).

Keep in mind that the module typically represents about 50–60% of the total capital cost. Other balance-of-system components, such as DC to AC inverters, interconnection devices, control and storage systems, and installation itself all are important and need to be considered in any cost reduction exercise. Balance-of-system elements alone currently add up to about $2,500/kW installed for roof-mounted PV units retrofitted on buildings.

13.5.3 Silica-based systems (crystalline and amorphous)

The main tradeoff between using single crystalline versus polycrystalline silicon is one of efficiency versus manufacturing cost. Single crystals of silicon have the highest efficiency, about 2–3% higher than refined polycrystalline material, but their manufacturing costs can be many times higher than polycrystalline modules. As a result, most non-space PV applications use polycrystalline silicon devices that are produced as thin wafers cut from cast ingots or drawn as a thin ribbon from molten silicon. The cost of producing the solar cell is in direct proportion to the amount of silicon used divided between the raw material cost and the cost of the energy needed to refine and process the silicon.

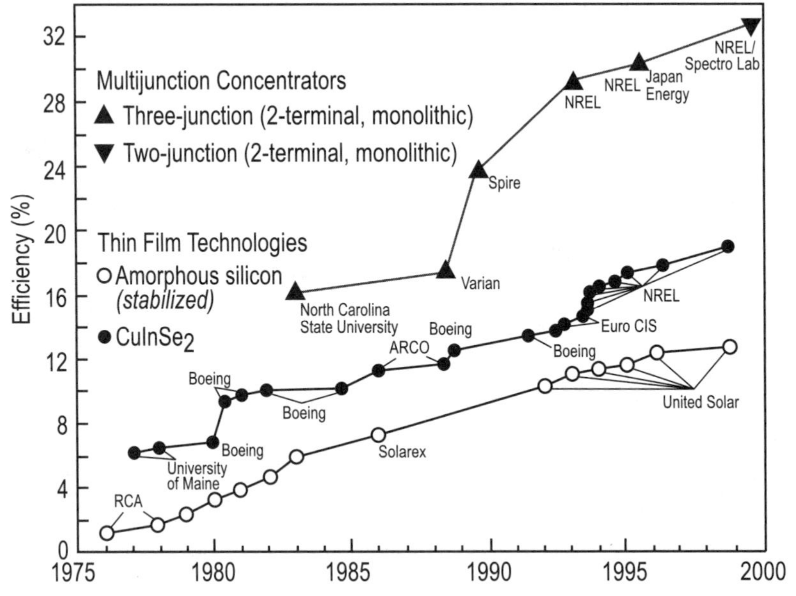

Figure 13.20. Thin film and multi-junction photovoltaic cell efficiencies. Source: NCPV (2002).

An alternative to using single or polycrystalline silicon was proposed in the 1970s. The basic idea was to deposit a thin film of non-crystalline, amorphous silicon directly on an inexpensive support material. Because of the excellent solar absorption characteristics of amorphous (α) silicon, films of only 1 μm or less thickness (compared to the 200–300+ μm needed for crystalline cells) would be required. Thus, manufacturing lines could be configured to produce solar cells by a continuous highly efficient process (see Carlson and Wagner, 1993, for details). There are a few drawbacks. The major one is that α-silicon cells have lower efficiencies (see Table 13.3), and they can be unstable, with decreases in efficiency appearing after a short period of operation. These issues are being addressed in a vigorous worldwide R&D effort and much progress has been made. Note in Figure 13.20 the progressive improvements in stabilized device efficiencies that have been achieved. This is remarkable given that the first α-silicon cells of 1974 vintage had efficiencies of only 1% (Carson and Wagner, 1993). By 2003, the major development effort was focused on reducing manufacturing costs.

13.5.4 Copper indium diselenide (CIS)

In addition to amorphous silicon PV devices, other thin film technologies are being pursued as possible replacements for crystalline silicon cells. Semiconductor PV devices consisting of 1–3 μm layers of copper indium diselenide ($CuInSe_2$ or CIS) along with comparable thin over-layers of silica (SiO_2) and cadmium zinc sulfide (CdZnS) or cadmium sulfide (CdS) have somewhat higher efficiencies than α-silicon devices (see Figure 13.20) and are predicted to have similar manufacturing costs (see Zweibel and Barnett, 1993, for details). Again much of the R&D effort is concentrated on simultaneously improving the electro-optical properties and durability of the films while streamlining the manufacturing process to reduce costs. This is particularly challenging because the process requires deposition of these films in a clean, high vacuum environment similar to that used to manufacture today's computer chips. The big difference with thin-film PV versus computer chip manufacture is that large volumes must be produced at very low cost.

CIS cell development on a large scale has sustainability and toxicity concerns that must be addressed. For example, workers involved with their manufacture need to be protected, and the impacts of extracting required elements (Zn, In, Mo, Se, Cd, and Cu) from natural mineral resources may create regional supply and demand constraints and a need to minimize the environmental damage associated with mining and refining operations, as well as with end-of-life disposal.

13.5.5 Cadmium telluride (CdTe)

Another popular thin-film candidate is CdTe. Its band gap energy of 1.5 eV is nicely matched to the solar spectrum with predicted theoretical efficiencies of 27–28% (see Table 13.3). Since their discovery in the 1950s, laboratory device efficiencies have improved from a few percent to over 16% (National Center for Photovoltaics (NCPV), 2002, Zweibel and Barnett, 1993). A number of manufacturing options exist for CdTe systems, including electrochemical deposition, which could lead to significantly lower module costs. Nonetheless, the same toxicity, resource consumption, and waste contamination concerns exist for CdTe as for CIS systems.

13.5.6 Current status and future potential of PV

There are a number of options for PV systems, each with its pluses and minuses. The performance and cost of the four major technologies, polycrystalline wafers and ribbon, and thin films consisting of α-silicon, CIS and CdTe, have been improving for the last 20 to 30 years as production has increased (see Figure 13.21). Since the mid-1970s, installed costs for rooftop flat plate PV systems have dropped from \$30,000/kW to less than \$7,000/kW, with reductions in manufactured PV module costs making the biggest contribution. Innovative collaborative work between government and industry has accelerated improvements to manufacturing processes for PV systems. For example, the PVMaT program managed by the NCPV and sponsored by the US DOE has been able to develop a practical roadmap for bringing down costs by focusing on critical manufacturing issues. (Figure 13.22 shows both achieved and predicted cost reductions that have and will result from a successful PVMaT campaign.)

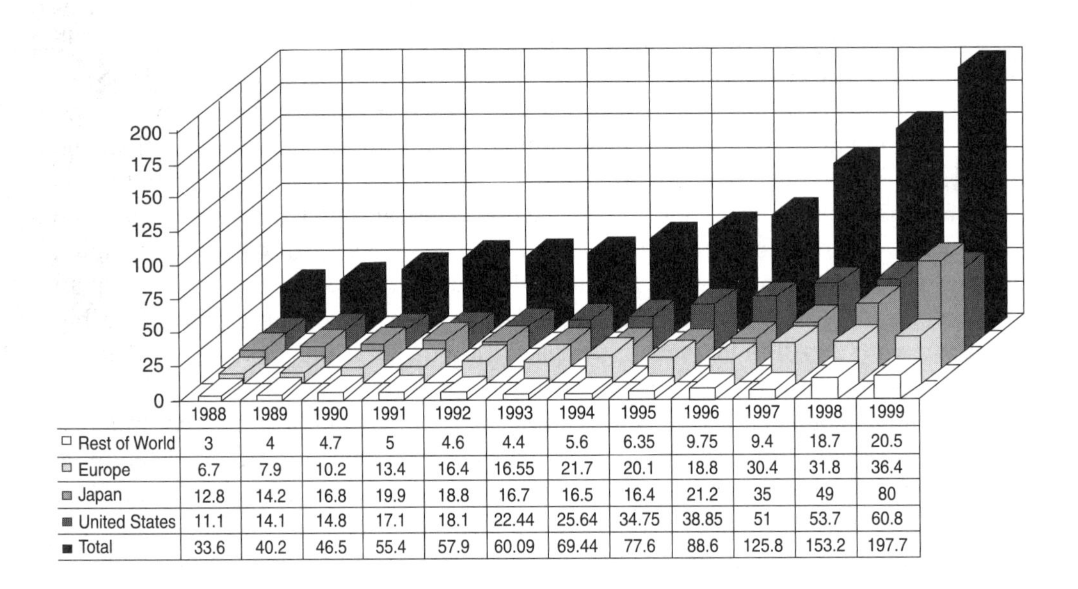

	1988	1989	1990	1991	1992	1993	1994	1995	1996	1997	1998	1999
Rest of World	3	4	4.7	5	4.6	4.4	5.6	6.35	9.75	9.4	18.7	20.5
Europe	6.7	7.9	10.2	13.4	16.4	16.55	21.7	20.1	18.8	30.4	31.8	36.4
Japan	12.8	14.2	16.8	19.9	18.8	16.7	16.5	16.4	21.2	35	49	80
United States	11.1	14.1	14.8	17.1	18.1	22.44	25.64	34.75	38.85	51	53.7	60.8
Total	33.6	40.2	46.5	55.4	57.9	60.09	69.44	77.6	88.6	125.8	153.2	197.7

Figure 13.21. History of PV module production worldwide from 1988–1999. Source: NCPV (2002).

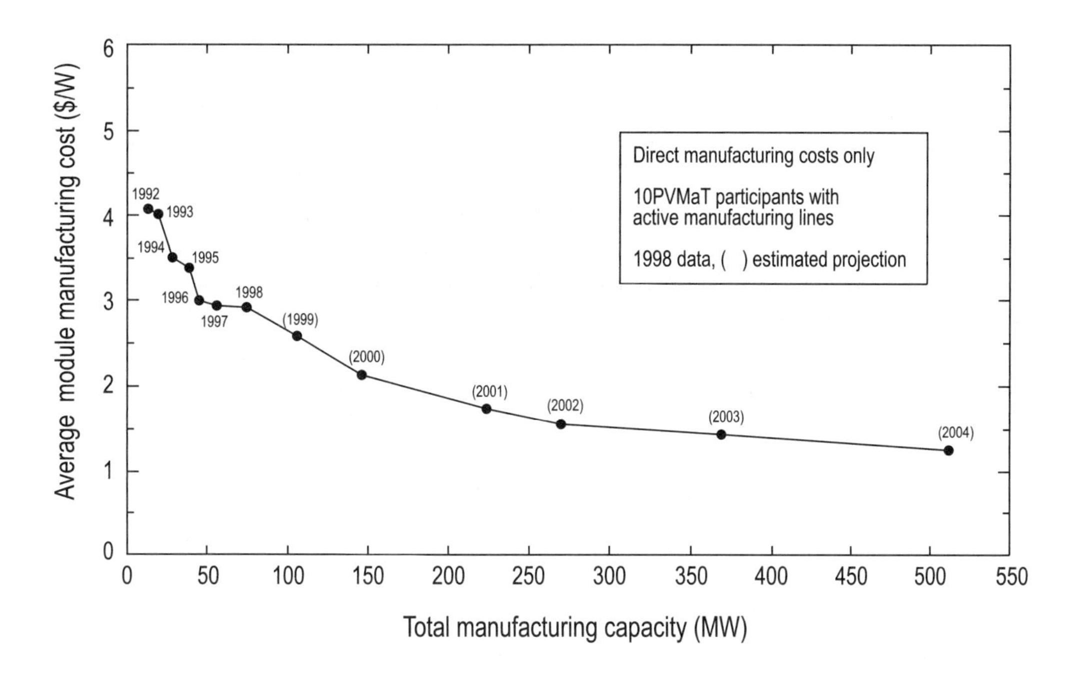

Figure 13.22. Cost reduction potential as a result of the US-led PVMaT initiative. Source: US DOE (1996).

A positive attribute of PV technologies in general is the opportunity for integrating the PV collector system into the building structure itself. Given that both direct and diffuse light can be converted to electricity, the external walls, windows, and roof of any structure could, in principle, be turned into a PV collector. Some policies in the US, Europe, and Japan are encouraging the development of building-integrated PV systems, as well as the installation of roof-mounted units, by offering subsides, tax incentives, or "green" energy production credits. In 1997, President Clinton announced a major deployment initiative for solar energy in the US, called "The Million Solar Roofs" initiative. The basic idea was to place one million PV or solar heating systems onto the rooftops of American residences and businesses by 2010. Federal buildings were to be involved in this initiative, as well.

In order for large gains to be achieved in deployment, more attention must be paid to critical balance-of-systems issues, especially electrical energy storage and interconnection standards and equipment. Reducing costs while increasing performance and reliability are key elements. While national policies will assist the process, a long-term commitment to R&D is required to develop new devices, encourage manufacturing process improvements, and create a set of uniform codes, standards, and testing procedures for components and complete systems to reduce risks and build a consumer base. On the research side, the most advanced tools of basic scientific research must be applied to the problems of producing low-cost, reliable PV systems. There are numerous examples that could describe basic research programs that could lead to significant gains in PV development. We have picked two to illustrate their value.

The first example involves thin-film, triple-junction semiconductors to engineer the PV device on a molecular level optimizing both the absorption and conversion of photon energy into electricity. Recent collaborative work by Spectrolab and the NCPV has produced efficiencies of >32%, the highest ever for a triple-junction PV. In this case, an optimized device results from using three connected layers of gallium indium phosphide, gallium arsenide, and germanium (see Staedter, 2002, for details).

The second example deals with developing a solar power window that produces PV electricity from the infrared and ultraviolet portion of the solar spectrum while transmitting visible light into the structure to achieve daylighting. Again, researchers at NREL (Frank et al., 2002) are pursuing this idea by utilizing a dye-sensitized PV cell—the so-called Gratzel cell, which was invented at the Swiss Federal Institute ETH in the early 1990s.

13.6 Sustainability Attributes

The desirable features of solar energy include its inherent flexibility and range of scales for many applications. Uses range from 1–10 kW-scale, passive and active heating and cooling systems (integrated into the designs of individual homes) up to 10–200 MW-scale, concentrating solar power towers and troughs that generate dispatchable electricity tied into a national grid.

Considering ways to capture and utilize a portion of the solar spectrum that intersects the walls, roof, and windows of a building for meeting a portion of heating, cooling, lighting, and other electricity needs is intriguing from a sustainability perspective. Currently, about 1/3 of our annual energy budget is used to condition and maintain the buildings we live and work in (see Chapter 20 for further discussion). Much of the time, we consume energy to offset the effects introduced by the conditions of the day or season (e.g., heating in winter and cooling in the summer, providing light when its dark). To a large extent, the sun's presence or absence is responsible for creating this demand for energy. Many argue that by capturing, storing, and converting solar energy into usable heat or electricity, we are achieving a sustainable energy system. Unfortunately, this rationale is too simplistic, as it does not consider the full lifecycle elements of performance, the environmental impacts, nor the costs involved in constructing and maintaining solar energy systems.

The hardware and space it takes to concentrate, convert, and store this intermittent and dilute energy resource carries its own environmental and economic burden. Consider the embedded energy content of a silicon PV module constructed with polycrystalline wafers. Rough estimates indicate that it takes about two years of operation for the silicon PV collector to recover the amount of energy that was needed to manufacture it. In fairness, the newer thin film PV concepts currently being pursued require much less energy to manufacture. Solar hot-water heaters and passive Trombe and transpiring walls also quickly recover their embedded energy content. In any case, the needs for energy, materials, and other natural resources in the manufacture and maintenance of each solar energy system should be evaluated.

Other environmental and health concerns for advanced PV systems arise from the types of materials being considered, such as CdTe and $CuInSe_2$ or GaAs. Many of these are toxic in several chemical forms, and the processes associated with recovering, purifying, and utilizing these materials carry their own impacts. Nonetheless, the use of thin-film technology for these PV applications has greatly reduced the amounts of semiconductor materials needed. Furthermore, the improved durability and reliability of today's PV systems, which have extended service lives to 25 years or more of operation, suggest that resource consumption and the associated contamination of the environment may be much less severe than originally thought.

Solar energy is a renewable resource that requires matching of the solar resource grade in a particular region to the demand for heat or electric power. While gaining access to the sun is relatively easy in remotely populated regions or for homes in suburban America or other locations, there are substantial challenges in using solar energy to provide a large fraction of the energy needs of a megacity like New York or Tokyo. In these cases, the transmission and distribution of solar energy from the point where it is captured to where it is used must be addressed. In addition, because it is inherently intermittent and variable on an hourly and seasonal time scale, storage of electricity and/or heat will be necessary in most solar energy applications.

The land required for large-scale solar electric installations is another environmental impact that must be dealt with. A few comparisons provide a basis for evaluation. Recall the amount of land associated with a typical large hydropower project (Chapter 12) or what is needed to cover the entire fuel cycle requirements of a bioenergy (Chapter 10) or strip-mined coal (Chapter 7) supply and conversion operation. A CSP plant in the US southwest desert would need about 20 square miles of land to produce roughly the same average amount of electricity as generated by Hoover Dam annually. The inundated portions of land behind Hoover Dam (forming Lake Mead) amount to about 250 square miles, over 10 times as much area. A key siting issue for CSP has to do with the height and brightness of large central receivers on power towers. For unit sizes over 100 MWe, the receiver heights will be 200 m or more.

During the ambitious deployment period of the 1970s and 1980s, reliability and durability problems plagued active solar systems. With tax incentives in place and growing customer demand for solar water heaters and PV panels for home use, many new companies were formed focusing on manufacturing components and assembling and deploying integrated products. Unfortunately, a large fraction of them were not able to produce a reliable product. In 1984, over 250 manufacturers in the US sold over 1.5 million m^2 of collectors, but by 1990, over 200 had gone out of business with the solar water heater market eroding to less than 0.5 million m^2 per year (Brower, 1992). To make things worse, poor choices of materials, inadequate control systems, and shoddy workmanship on-site led to frequent failures and distrust in solar technology. By 2000, with reliable solar products available, annual sales in the US increased again to a stable level about 1 million m^2 per year without strong policy or financial incentives. A similar trend has occurred in Europe with a few important exceptions. Israel, Australia, and Cyprus have seen growing markets and increased penetration of solar technology. This has been driven largely by two factors: a high-quality solar resource and favorable policies. At the turn of the century, national subsidy programs in several European countries and Japan have led to aggressive PV deployment programs.

13.7 Summary and Prognosis

To capture and efficiently utilize the sun's energy in today's energy markets requires multiscale integration of several factors, including proper matching of the solar resource and energy demands of a particular region, workable distributed energy systems with robust transmission and distribution and functional interconnection, a means for storing heat, and policy incentives that encourage the use of solar energy as a renewable resource. Given that the capital costs of solar energy systems are high relative to fossil fuels, financial restructuring of energy markets will be needed to place additional value on the attributes of solar energy over lower priced depletable fossil energy alternatives. These values include the absence of emissions and solar's frequent coupling to energy efficiency technologies that reduce consumption per unit of energy service provided, which are all linked to increasing sustainability indices.

In many ways, the inherent flexibility and simplicity of passive and active solar systems makes good sense for producing energy-efficient buildings that are affordable and desirable. The benefits of solar systems are substantially enhanced when coupled to advanced energy efficiency technologies, such as super insulation, low-E glass for windows, distributed combined heat and power, better natural ventilation, daylighting, and efficient appliances and lights. The ability to produce a portion of required hot water, air conditioning, and electricity using solar thermal and PV devices that are integrated into the building structure itself presents attractive opportunities.

Although progress continues to be made to lower manufacturing costs and increase collector performance, the relatively high capital costs of solar thermal and PV systems will continue to require incentives to encourage their widespread application. These incentives can reflect national or state initiatives (e.g., "The Million Solar Roofs" initiative or the Energy Star efficiency labelling program or LEED green buildings classifications in the US or the Massachusetts Renewable Energy Trust's Green Schools program). Or, they may be a result of a restructuring of state or local building codes requiring higher performance.

In addition to these more distributed, buildings-integrated applications, the use of concentrating solar energy to produce both dispatchable and distributed electric power with power towers, parabolic troughs, and dish-engine CSP technologies has some attractive features that could lead to deployment in many countries or regions that have a high-grade direct solar resource within their boundaries. By avoiding the use of fossil fuels for electric power generation and utilizing an indigenous energy source, CSP systems increase the diversity and security of the electric supply sector. Capital costs have been lowered to the point where, with modest incentives, CSP methods could be used to provide a renewable, carbon-free source of electric power at breakeven prices of 10¢/kWh or less on a large enough scale to make a difference. For example, with a major population shift in the US to the southwest, electric demand is increasing right where there is a high-grade resource for CSP. Advocates suggest that over 20,000 MWe of CSP capacity could be deployed in the US southwest within a decade if needed.

Further R&D support for both PV and CSP technologies will have an impact on increasing reliability and durability while lowering module manufacturing costs. To achieve these objectives, there is a need for improved standardization of components, and testing of components and complete systems, to increase product quality and consumer confidence. Centers such as the NCPV and SunLab's testing facilities are good examples of such assets to carry out these important functions for the US.

References

Boes, E.C. and A. Lague. 1993. "Photovoltaic concentrator technology," in *Renewable Energy*, ed. Johansson, H. Kelly, A.K.N. Reddy, and R.H. Williams, Chapter 8:361– 401. Washington, DC: Island Press.

Brower, M. 1992. *Cool Energy*. Cambridge, MA: The MIT Press

Butler, B. 1997, 2002. "Dish-engine commercialization." Los Angeles, CA: Science Applications International Corporation (SAIC).

Carlson, D.E. and S. Wagner. 1993. "Amorphous silicon photovoltaic systems," in *Renewable Energy*, ed. Johansson, T.B., H. Kelly, A.K.N. Reddy, and R.H. Williams. Washington, DC: Island Press.

DeLaquil, P., D. Kearney, M. Geyer and R. Diver. 1993. "Solar-thermal electric technology," in *Renewable Energy*, ed. Johansson, T.B., H. Kelly, A.K.N. Reddy, and R.H. Williams. Chapter 5:213–296. Washington, DC: Island Press.

Frank, A., B. Gregg, A. Nozik, D. Balcomb, et al. 2002. Personal communication. Golden, CO: National Renewable Energy Laboratory.

Green, M.A. 1993. "Crystalline- and polycrystalline-silicon solar cells," in *Renewable Energy*, ed. Johansson, T.B., H. Kelly, A.K.N. Reddy, and R.H. Williams. Chapter 7:337–360. Washington, DC: Island Press.

Hinrichs, R.A. 1996. "Electricity from solar energy," in *Energy*, Second Edition. Chapter 12:381–393. Fort Worth, TX: Saunders College Publishing.

Kelly, H. 1993. "Introduction to photovoltaic technology," in *Renewable Energy*, ed. Johansson, T.B., H. Kelly, A.K.N. Reddy, and R.H. Williams. Chapter 6:297–336. Washington, DC: Island Press.

Keisling, B. 1983. *The Homeowners Handbook of Solar Water Heating Systems.* Emmaus, PA: Rodale Press.

Kreith, F. and J.F. Kreider. 1978. *Principles of Solar Engineering.* New York: Hemisphere Publishing, McGraw-Hill.

Michalsky, J. 1988. "The astronomical almanac's algorithm for approximate solar position (1950-2050)." *Solar Energy*, 40(3):227–235.

Morse, F.H. 2000. "The commercial path forward for concentrating solar power technologies." *Report for Sandia National Laboratories.* Albuquerque, NM: US DOE.

Myles, J.F. 2002. "A brief overview of Duke Solar's programs." Presented to National Research Council, Washington, DC.

National Center for Photovoltaics (NCPV). 2002. L. Kazmerski, director, personal communication. Golden, CO: National Renewable Energy Laboratory.

Renne, D., Stovall, et al. 2002. Resource assessment program. Golden CO: National Renewable Energy Laboratory.

Staedter, T. 2002. "Triple-junction solar cells." *Technology Review*, 88–89.

SunLab. 1998. "Markets for concentrating solar power." Washington, DC: US DOE.

Teagan, W.P. 2001. *Review: Status of Markets for Solar Thermal Power Systems.* Cambridge, MA: Arthur D. Little.

USDOE, US Department of Energy. 1996. *A Strategic Plan for Solar Thermal Electricity: A Bright Path to the Future.* Washington, DC: US DOE.

Zweibel, K. and A.L. Barnett. 1993. "Polycrystalline thin-film photovoltaics," in *Renewable Energy*, ed. Johannson, T.B., H. Kelly, A.K.N. Reddy, and R.H. Williams. Chapter 10:437–481. Washington, DC: Island Press.

Web Sites of Interest

US-based
For solar resource characterization:
www.nrel.gov
www.eere.energy.gov/solar/

For CSP:
www.nrel.gov
www.energylan.sandia.gov/stdb.cfm

For PV:
www.eere.energy.gov/solar/

Australia
www.seia.com.au/
solar.anu.edu.au

International
www.solarpaces.org
www.wcre.org
www.eurosolar.org

Problems

13.1 Estimate how much collector area and storage capacity would be required for an active solar hot-water system designed to supply the total needs for two four-person families, one living in Manchester, New Hampshire, where the latitude is 44+° north and the other in Albuquerque, New Mexico, at 35+° North. The heat capacity of water is about 4,200 J/kg°C and the hot water supply temperature in both houses is 75°C (140°F). State and justify all additional assumptions made.

13.2 If we wanted to supply all the electricity needs for New York City (NYC), with a population of about 10 million people, using photovoltaics technology, how much land area would we need? What other elements would be required for such an idea to work for NYC? How large of an investment would this involve and how would the price of electricity in ¢ per kWh have to increase in order to recover that investment in 10 years at an interest rate of 8%? For comparison, the average New Yorker paid 12 ¢/kWh in 2002.

13.3 Estimate the amount of solar heating that is lost as a result of using one or two tempered 1/8 in thick glass plates covering a flat plate absorber. The transmission coefficient for radiation is 2% for wavelengths greater than 3 μm and 80% for wavelengths ranging from 0.2–3 μm.

13.4 The amount of solar energy captured in a given application is expressed in terms of how much thermal energy is delivered relative to the total demand or load. This amount is often referred to as the production function (Q_s). Using the Kreith and Kreider (1978) method of analysis, a simplified empirical model of a solar building in Boston can be developed to relate Q_s parametrically to the load in GJ (L), the amount of storage in m^3 (S), and the collector surface area in m^2 (A_c) *as:*

$$Q_s = L\left[0.8 + \ln\left[\left(\frac{A_c}{L}\right)^{1/3}\left(\frac{S}{L}\right)^{1/20}\right]\right] = 100 \text{ GJ}$$

If the collector area is increased by 10%, what happens to Q_s? If the amount of storage is increased by the same amount, what happens to Q_s? What do you conclude from this comparison?

13.5 CSP for cooking! Let's suppose that you want to cook 1 lb of hot dogs using a parabolic trough concentrator on a clear June day in Los Angeles. The hot dogs would be skewered on the stiff piece of wire and aligned along the axis of the trough in the same position as the absorber/receiver normally would be placed. The desired cooking time is 10 min with the surface temperature of the hot dog at a temperature of about 450°F. How would you design the reflector surface? Specifically, what would be the length and width of the parabolic cylinder?

13.6 Estimate how much hot water you would use for your personal needs for one week during a typical Christmas school break. If you wanted to use solar energy to provide the hot water, estimate what surface area of flat plate collectors would be required if you were located in Denver.

13.7 What are the incremental savings that result from increasing the efficiency of a PV collector by 1% in terms of collector area and costs? Assume that the PV module represents about 50% of the total PV system cost.

13.8 Let's assume a world of 10 billion people with an average demand for electricity of 1 kWe per person (720 kWh/month) for providing lighting and running refrigerators, other appliances, TVs, computers, etc. What surface area of polycrystalline silicon PV collectors would be required to provide the electrical needs for 10 billion people? How much electrical storage capacity would you estimate would be needed? If lead acid batteries were used to store the energy, what mass of batteries would be required? (Hint: See Chapter 16, Table 16.1.)

Ocean Waves, Tide, and Thermal Energy Conversion 14

14.1 Introduction

The awesome power of ocean waves and the relentless tides makes them obvious candidates for anyone's list of potential natural energy sources, while the more subtle observation that surface water is warmer than the briny deep tempts only those with some exposure to thermodynamics and the concept of heat engines. We can also point to a history of past development and recent applications in each area (see Table 14.1), and perhaps through them come to some understanding as to why more widespread deployment has not yet come about, and draw informed conclusions about whether they can eventually compete with other candidates for inclusion in future sustainable energy scenarios.

The three examples cited above—wave, tidal, and thermal—are not the only ocean-related energy prospects, as noted in the broader survey of Table 14.2 (which also provides references for finding more about each entry). But they are the most likely candidates. This chapter excludes energy sources at sea that are primarily straightforward extensions in locus of those that have a long history of prior terrestrial exploitation, such as oil, gas, coal, oil shale, geothermal, tar sands, or wind. (All of these energy sources are discussed in other chapters of this text.) Table 14.3 gives the magnitude of the potential and potentially exploitable resources of marine energy (also see Table 9.2 in Chapter 9).

14.2 Energy from the Tides

Wheel-type watermills, powered by tidal flow, were not rare in late medieval times, but were far outnumbered by their freshwater-driven counterparts. Our interest here is in more modern versions that generate electricity employing low-head, propeller-type turbines similar to those in run-of-the-river or other low-head hydro freshwater applications. It is important to note that only one large-scale complex of this kind has been constructed to date (at La Rance in France).

Tides arise from the gravitational interactions of the earth/moon (68%) and sun (32%) systems with successive high (and low) tides normally occurring twice daily, but only once daily at certain locations, such as Pensacola, Florida, and the Gulf of Tonkin (Cartwright, 1999, and Clancy, 1969). This extreme variation is the consequence of resonant interactions involving basin configuration, Coriolis forces, and the like, of sufficient complexity to require several dozen harmonics for accurate representation. In deep water far at sea, the rise or fall is only 0.5 m or so, which is not useful. However, tides vary greatly with location—for example, only a few cm at Nantucket Island, which is 150 km southeast of Boston, but up to a world-maximum of 16m between high and low water in the Bay of Fundy, which is 400 km northeast of Boston. It is these special and rare locations (several dozen sites worldwide: see Figure 14.1), with shelving shores or funnel-like inlets, which are candidate sites for practical tidal power stations. Estimates are that full exploitation of such resources (5 m rise) might contribute several tens of gigawatts to humankind's energy budget.

Table 14.1. Historical Timeline of Ocean Energy Exploitation

BC–AD	References to use of tides in classical Greece, possibly dating back to the time of Aristotle
960 AD	Reference to tide mills at Basra in southern Iraq
1041, 1078	First reference to European tide mills (around Venice)
1100–1900	Waterwheel-driven mills powered by tidal impoundments and currents operational in England, western Europe, and colonial Boston, among elsewhere
1135–WWII	Bromley by Bow Tidal Mill near London
1734	Slades Tidal Spice Mill, Chelsea, Massachusetts
1799	Gerard files first patent on wave energy device
1800–1900	25 tide mills cited in Britain
1871	Jules Verne's Captain Nemo posits thermoelectricity from ocean water in *20,000 Leagues Under the Sea*
1881	D'Arsonval proposes concept of ocean thermal energy conversion (OTEC)
1892	Stahl notes 19 wavepower concepts in American Society of Mechanical Engineers (ASME) transactions
1934	Claude tests open-cycle OTEC in Cuba
1935–1977	Succession of studies of Passamaquoddy/Bay of Fundy tidal power stations
1959	First of a number of small (<1MWe) tidal power plants reported in China
1966	Rance River tidal power plant operational in France
1969	Experimental tidal unit constructed in Kislaya Guba, Russia
1972–1984	US OTEC program
1976–1982	British launch, then quash, their wave power program
1977	Wells invents turbine which rotates in same direction when air flow is reversed
1978	Japanese install 125 kWe wave power unit off Honshu
1979	Mini-OTEC operated in Hawaii by the US and by Japanese off Shimane
1984	20 MW Annapolis tidal station operational in Nova Scotia
1985	KVAERNER wave energy converter deployed in Norway
1995	2 MWe OSPREY wave power station wrecked during installation
1997	La Rance scheduled to undergo renovation to enable operation far into the future
(1986–present)	Decline, on the average, of fossil energy prices in constant dollars saps motivation for vigorous pursuit of the more expensive categories of alternatives, e.g., anything out to sea

Table 14.2. A Broad Survey of Oceanic Energy Resources

Type (refs)	Characteristics of Interest
Thermal (Avery & Wu 1994), (Vega 1995)	About one-third of ocean surface water is (slightly) higher than 25°C and three-quarters of its volume is ≤ 4°C (deep water ≥ 500–1000m): sufficient to operate a low-efficiency heat engine
Waves (Ross 1995), (Hagerman 1995), (Edwards 1998)	Wave power averages ~ 10 kWe/m of coastline; mechanical energy can be extracted from their periodic motion
Tides (Charlier 1982), (Clark 1992, 1995, 1997)	The gravitational pull of the moon (68%) and sun (32%) causes 7m tides in several dozen coastal sites worldwide, which can be harnessed by low-head hydro
Current (Wick & Schmitt 1981), (Pearce 1998)	Steady flows of ~2.5 m/s at large volumetric rates are available at selected sites: gulfstream off Florida, Kuroshio off Japan; also from tidal currents; power can be extracted using large turbine rotors
Salinity (Wick & Schmitt 1981), (Levenspiel & de Nevers 1974), (Norman 1974)	Differences in salt content between freshwater (rivers) and the ocean can develop osmotic head of 240 m; Dead Sea and Great Salt Lake 3,000 m; more energy in salt of oil trap salt domes than in the oil
Biomass (Wilcox 1977), (Ryther 1979)	Giant kelp of the type growing off the California shore could be artificially farmed, harvested and processed to yield biogas
Unconventional Hydro	Flow via canals into the Dead Sea or Qattara depression or in a sea-level canal between the Atlantic and Pacific Oceans; or even through dams across the Bering Strait or Strait of Gibraltar!
Methane Hydrates (Suess et al. 1999)	Extensive deposits trapped in deep sea sediments
Offshore Geothermal	Geopressured reservoirs and hydrothermal vents
Mineral (Best & Driscoll 1980)	Extraction of uranium for fueling nuclear fission power plants from seawater or special sediments; extraction of deuterium and lithium for fusion reactor fueling

Table 14.3. Approximate Worldwide Ocean Energy Available (all values are Gigawatts, annual average)

Phenomenon	Maximum Potential	Practicable to Exploit
Tides	2,500 (total dissipation)	20
Temperature Difference (OTEC)	2×10^5 (based on total absorbed insolation in equatorial waters)	near or onshore: 40 offshore: 10,000
Waves	2,700 (on all the world's coastlines) 10,000 (open sea renewal rate)	500
Salinity Gradients	10^6	not presently viable
Marine Currents	5,000	50
Offshore near-surface winds	20,000	1,000
Compare world electric energy installed capacity.		in 1998: 2,800, of which US is: 750
Also compare world freshwater hydroelectric currently installed capacity at 550 GWe and total potential capacity of about 4,000 GWe.		

Notes:

1) The powers quoted are annual-average useful mechanical (electrical) values, converted from thermal, where applicable, by a representative thermodynamic efficiency: e.g., 2.7% for OTEC

2) Values are derived from similar estimates in the references at the end of this chapter or equivalent "back-of-the-envelope" calculations. Estimates in the literature vary widely, often by an order of magnitude (e.g., see Table 9.2 in Chapter 9).

3) Overall, the electric power grid operates at an average-to-peak load ratio of ~0.5, hence average power is about half of installed capacity. Individual sources (e.g., hydro) may have higher- or lower-than-average capacity factors.

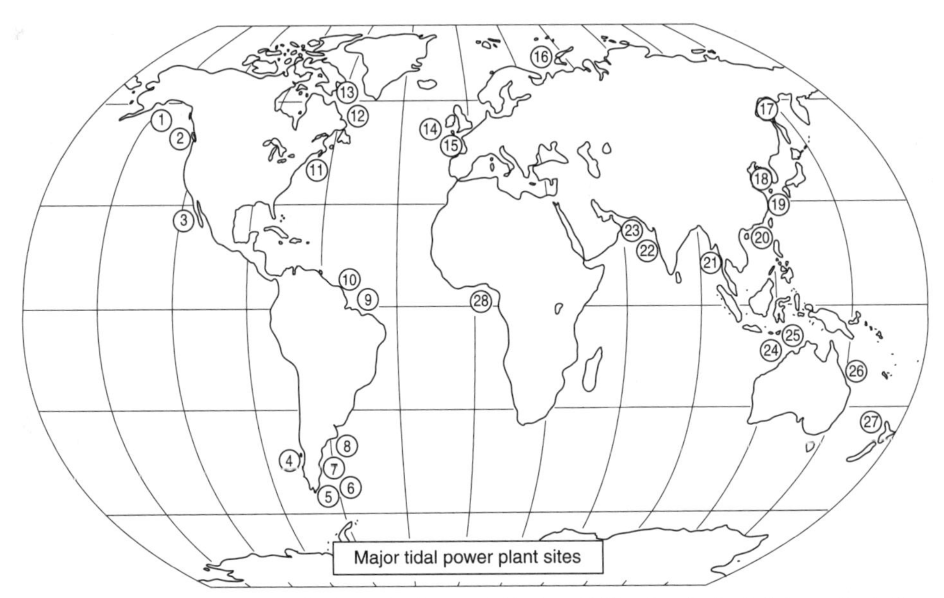

1. Cook Inlet; 2. British Columbia; 3. Baja California; 4. Chonos Archipelago; 5. Magellan Straits; 6. Gallegos/Santa Cruz: 7. Gulf of San Jorge; 8. San Jose Gulf; 9. Maranhão; 10. Araguaia; 11. Fundy/Quoddy; 12. Ungava Bay; 13. Frobisher Bay; 14. Severn/Solway; 15. Rance; 16. Mezen/Kislaya; 17. Okhotsk Sea; 18. Seoul River; 19. Shanghai; 20. Amoy; 21. Rangoon; 22. Cambay Bay; 23. Rann of Kutch; 24. Kimberleys; 25. Darwin; 26. Broad Sound; 27. Manukau; 28. Abidjan.

Figure 14.1. Sites with major potential for tidal power. Source: Borgese (1985).

Referring to Figure 14.2, calculating the maximum potential energy recoverable from a single tidal cycle is a straightforward exercise. The water impounded at peak high tide is, on average, a height $R/2$ (where R is the total range) above the minimum low water point, in which case:

$$\hat{E} = mg\, R/2 = (\rho AR)g\, R/2 \tag{14-1}$$

Substituting parameter values given in the appendix to this chapter and converting units appropriately:

$$\hat{E} = 1397\, R^2 A, \text{ kWhr per cycle} \tag{14-2}$$

for R in meters and A in square kilometers. The world mean period for tides is 12 hrs 24 min, hence there are 706 tidal cycles per year and, therefore, annual average generation is:

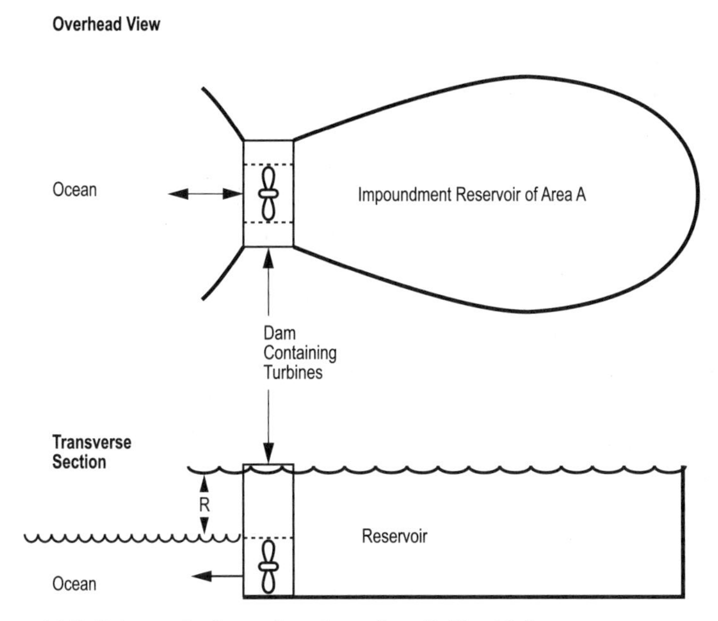

Figure 14.2. Schematic for estimation of available tidal energy.

$$\overline{E} = 0.986 \times 10^6 \text{ COP } R^2A, \text{ kWhr} \tag{14-3}$$

where we have introduced an overall coefficient of performance, COP (typically on the order of 0.2– 0.35) to account for a variety of factors.

1. Instead of single-basin one-way operation, as analyzed above, two basins could be employed and/or power could be generated in both the drain and fill phases (if reversible turbines are employed).
2. Using dual-purpose pump-turbines, by drawing power from the grid, the basin can be overfilled at the high water state and/or pumped down further at the low water mark, which allows a net energy gain later in the power generation phase.
3. Pumps/turbines/generators are not 100% efficient electromechanically, and a minimum head (~1.5 m) is needed to permit operation.
4. Instantaneous drain or fill is impractical, hence the difference in head between source and sink is less than the full tidal range.
5. Basin area (hence water volume per unit height) will decrease as a function of water depth if the reservoir has sloping sides.

Charlier (1992) describes La Rance station in France (see Table 14.4). Using R=8.5 m, A=22 km^2 and COP=0.33, Equation (14-3) predicts 517 GWhr/yr, about what is actually achieved.

Table 14.4 summarizes some characteristics of interest from this and a variety of other resources. It is by far the largest and longest operating of modern tidal-electric power plants. Even so, note that the mean power delivered of about 65 MWe is only about 1/3 that of a single combined cycle gas turbine module.

No discussion of tidal power would be complete without mention of the much-studied and debated (but never built) Passamaquoddy/Bay of Fundy projects bordering Maine, US, and New Brunswick/Nova Scotia, Canada (Baker, 1992). The Bay of Fundy has the world's largest tidal range and a topography that could support power stations totaling 2,000 MWe or more. Proposals for its exploitation date back to the 1930s, with periodic revivals thereafter. The principal forward progress has been the construction and operation, starting in 1984, of a 20 MWe demonstration unit near the mouth of the Annapolis River in Nova Scotia. It generates about 30 GWhr/yr (5% of La Rance) employing the largest straight-flow turbine in the world (7.6 m diameter). It is the only tidal station in North America.

Table 14.4. Characteristics of La Rance Tidal Power Station

Fully in service	1967
Length of dam	750 m
Area of basin	22 km^2
Usable basin volume	184×10^6 m^3 (max)
Tidal range	13.5 m (max) 8.5 m (avg.)
Nominal rating	240 MWe
Annual mean output	~65 MWe ~500– 600 GWe hr/yr
Range of output	Spring Tides 2,940 MWe hr/day Neap tides 738 MWe hr/day
Modes of operation	
capability	Double effect with pumping
practice	Not much reverse turbining or pumping
Turbine-generator	10 MWe each @ 3.5 KV 24 units, reversible flow and pumping capability; bulb configuration axial flow Kaplan type, useful range ≥ 1.2 m head
Cost of electricity	$0.037/ kW hr (1997)
Track record (1967–1997)	Hours of operation: 160,000 Generation: 16×10^9 kWh

To date, high projected capital cost and the need for Canadian-American coordination have stymied action on the Passamaquoddy project. Moreover, the recent move to exploit natural gas production offshore of Nova Scotia will create a local, tough-to-beat, low-cost alternative power source. Thus, tidal power appears to be, at best, a limited localized alternative and unlikely to make a near term impact of global significance on energy use sustainability.

14.3 Energy from the Waves: Overview

Waves are created by the interaction of the prevailing winds with long reaches of open water. Strong storm-force winds create a disorganized, chaotic, local wave field, which, through a process of interactive reinforcement and interference, leads to a more regular sequence of swells propagating away from the storm zone. As waves approach shallow water (depth approximately the same as wavelength), significant changes take place: wavelength shortens, speed is reduced, and the profile steepens—to produce the surfer's delight and breakers on the beach.

As discussed in Chapter 15, there is at least a weak correlation between wave power and wind power resources. For example, the western shores of Europe and the British Isles have high-grade resources of both. The Scottish coast is favored by waves having an average embodied linear power on the order of 50 kWe/m of wavefront. Therefore, it is not surprising that the British have played a leading role in R&D to exploit this potentially huge resource.

Many ingenious devices have been proposed for this application, based on exploitation of wave-induced pitching, heaving, or surging motion (Hagerman, 1995). Many of these devices have been tested under realistic in-service conditions. The simplest concepts induce waves to run up ramps to replenish a low-head hydro reservoir (e.g., The KVAERNER unit in Norway). Others employ floats with up and down motions that create mechanical work. In the Salter "nodding duck," a pear-shaped float cranks a hydraulic pump to create flow of a working fluid, which spins an electric generator via a hydraulic motor. Others use the oscillating wave motion to create hydraulic pistons, which force air through a turbine, again to power an electric generator. This latter scheme is the *modus operandi* for the only commercial units having a modest marketplace success to date: a Japanese buoy delivering on the order of 60 watts—enough to power navigational aids.

The lack of enthusiasm for wave power devices has largely been the result of their vulnerability to damage in storms, when linear power densities in excess of 200 kW/m may have to be withstood. This condition creates a costly design criterion.

Wave hydrodynamics are devilishly complicated, but a simplified model can give a rough estimate of wave-borne energy. Following Penner and Icerman (1984), consider a sinusoidal wave with crest-to-trough height, h (hence, mean sea level to crest height is $h/2$) as shown in Figure 14.3. Then the average height above the levelized sea is $2/\pi$ times this value, or h/π ; the center of gravity of the wave mass above sea level is $\pi h/16$. Thus, if the above sea level mass is (inverted and) dropped to fill in the below-level trough (which is its mirror image), potential energy is made available for extraction due to a total drop of $\pi h/8$. One has, per unit width of wavefront, and for wavelength, λ:

$$\Delta PE = mg\Delta h = \left[\rho\left(\frac{h}{\pi}\right)\left(\frac{\lambda}{2}\right)\right] g\left(\frac{\pi h}{8}\right) \tag{14-4a}$$

$$\text{Or } \Delta PE = \frac{1}{16}\rho\lambda g h^2 \tag{14-4b}$$

But power is energy per unit time (here, wave period, T) and wavelength is given by:

$$\lambda = \frac{gT^2}{2\pi} \tag{14-5}$$

Which all combine to yield:

$$P_{pe} = \frac{1}{32\pi}\rho g^2 h^2 T = 0.98h^2T, \text{ kW/m} \tag{14-6}$$

Where:

ρ = density of seawater (1,025 kg/m^3)

g = acceleration of gravity (9.807 m/s^2)

h = crest to trough wave height (e.g., 2 m)

T = wave period: time for successive crests to pass a fixed observer (e.g., 6 s)

We will not attempt to show it here, but the kinetic energy transported by the waves is, *mirabile dictu*, equal to the potential energy. Thus, total wave power is double that of Equation (14-6) (Rahman, 1995, and Elmore and Heald, 1969):

$$P = 1.96h^2T, \text{ kWe/m} \tag{14-7}$$

For the numerical values cited above, $P = 47$ kW/m, which is a high-quality resource several times the kWe per meter of rotor blade length for a representative wind turbine.

The various mechanical devices contrived to extract this energy are only of modest efficiency. Hence, energy actually delivered is significantly less.

The dependence on wave height squared in Equation (14-7) also highlights the vulnerability of wave machines to storm damage. Doubling wave height in our example would unleash power on the order of 200 kWe/m. Unlike wind turbines, few wave power devices have effective cutoff schemes to achieve a high degree of protection against excessive fluid velocities. Wave height spectra are sometimes fit to the same type of Rayleigh or Weibull functions as are employed for wind velocity characterization (see Chapter 15).

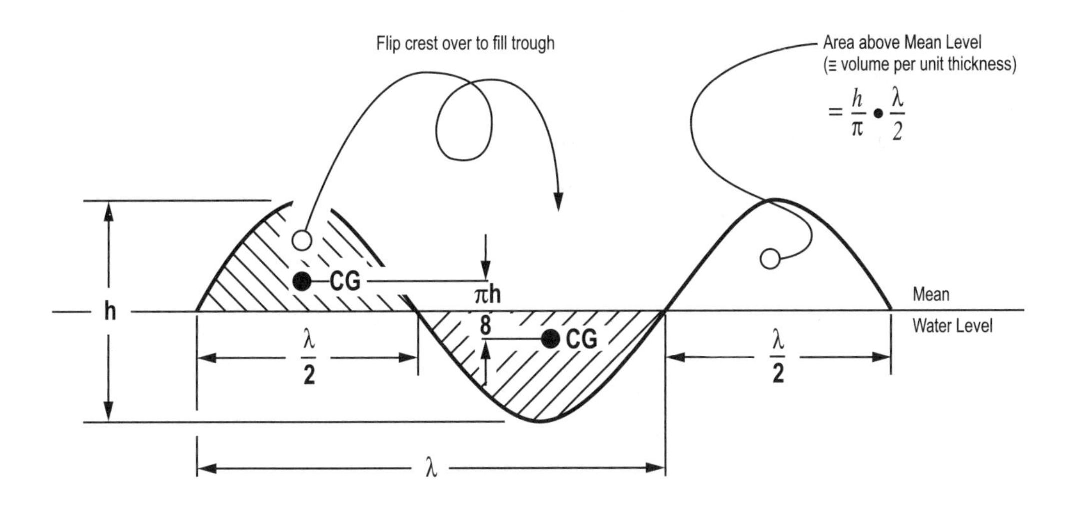

Figure 14.3. Schematic for estimation of wave potential energy.

14.4 Energy from Temperature Differences

14.4.1 Overview

As shown in Figure 14.4, there is a vast belt of warm ocean water girdling the globe between 20° north and 20° south of the equator, some 20–25°C hotter than the underlying deep water several hundred meters down. This is just barely enough to make heat engine operation technically feasible for the production of net electrical energy. As it turns out, ocean thermal energy conversion (OTEC) plant capital cost varies roughly as ΔT to the minus 2.5–3.0 power. Thus, only the highest-grade resources are worth consideration. Although the total resource is unfathomably immense—on the order of tens of thousands of Gigawatts—practical considerations limit, or at least defer, its exploitation. First, the existence of broad continental shelves force major OTEC plants to operate far offshore, which complicates transmission of energy back to large load centers on land. Production of hydrogen, ammonia, or methanol as carriers has been proposed (except for the limited number of situations where underwater power cables are practicable). This presupposes the existence of a viable synthetic-fuel-based economy—at best, a far future prospect. A secondary niche market involves the many islands scattered throughout the Pacific Ocean, where the cold water is close offshore. Here, the small scale can also be more than offset by the sale of co-products: distilled water, the products of mariculture and cool-climate agriculture, and air conditioning services. (For a discussion of a scenario for exploiting OTEC for future energy needs, see Savage, 1992.)

14.4.2 Performance limits

If hot and cold water are supplied to a heat engine at a rate sufficiently high that the energy processed by the engine is small compared to total throughput, then the source and sink temperatures (T_h and T_c, respectively) are constant, and Carnot's theorem gives the theoretical maximum efficiency of conversion of heat to work (Chapter 3):

$$\eta_c = \frac{W}{Q} = 1 - \frac{T_c}{T_h} \tag{14-8}$$

But pumping water through the hot and cold reservoirs at an excessively high rate would consume too much energy in the present instance. Several authors have optimized performance to maximize power when source and sink water are allowed to change temperature and when finite temperature differences across heat transfer surfaces are accounted for. This more practical Carnot cycle has the efficiency (Johnson, 1983):

$$\eta_{pc} = 1 - \sqrt{\frac{T_c}{T_h}}$$

$$\equiv 1 - \sqrt{1 - \left(\frac{T_h - T_c}{T_h}\right)} \tag{14-9}$$

And for small $T_h - T_c$, as is certainly the case here:

$$\eta_{pc} \cong 1 - \left[1 - \frac{1}{2}\left(\frac{T_h - T_c}{T_h}\right)\right] = \frac{1}{2}\eta_c \tag{14-10}$$

Under representative OTEC conditions, $T_h - T_c \approx 24°C$, $T_h \approx 300°K$ and:

$$\eta_c = 8\%$$

$$\eta_{pc} = 4\%$$

But lifting the cold water 1,000 m and other internal loads will typically consume 1% of the gross work output per unit of thermal energy, leaving a net:

$$\eta_{pc,net} \approx 3\%$$

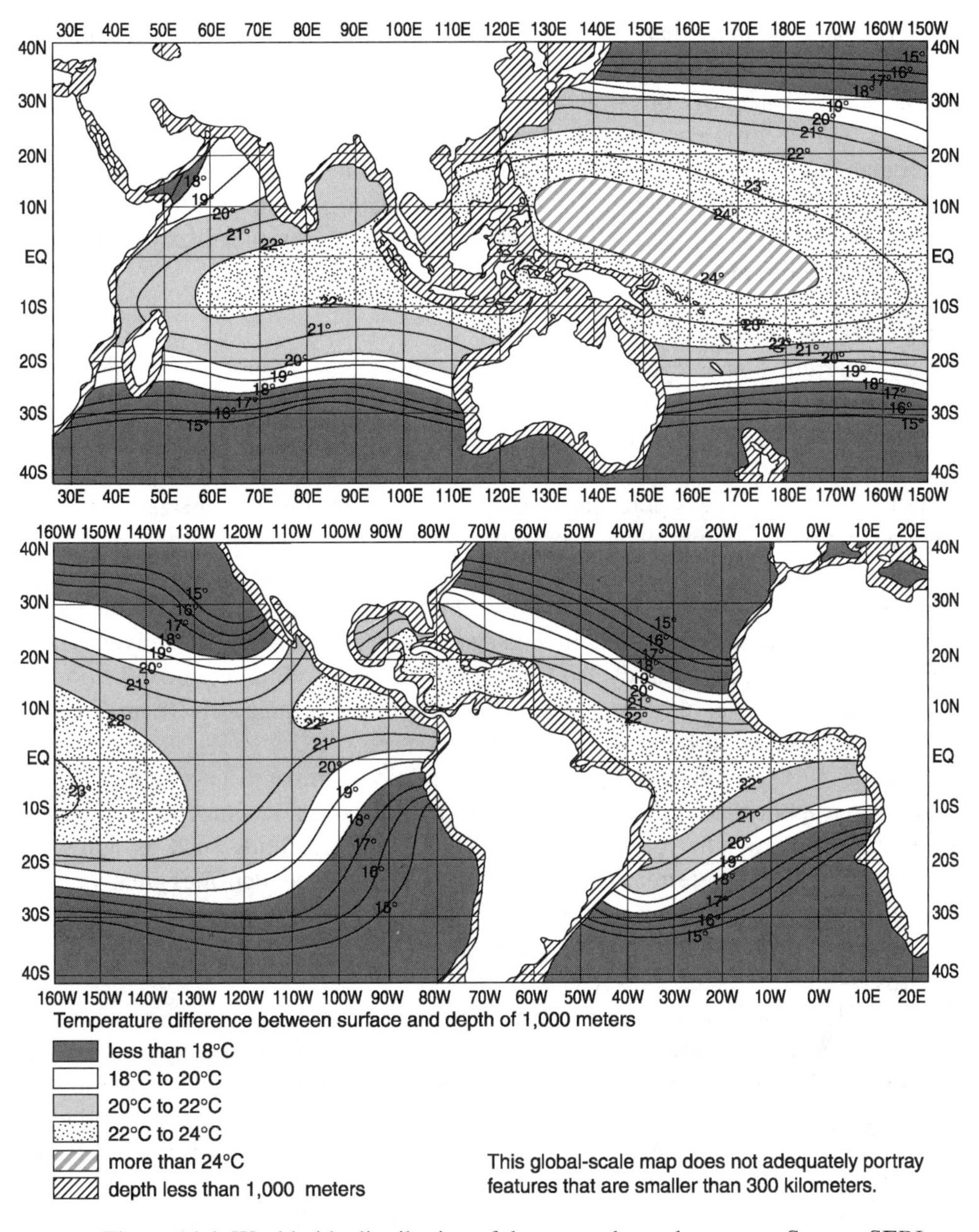

Figure 14.4. Worldwide distribution of the ocean thermal resource. Source: SERI (1990).

which is a factor of more than 10 less than achieved by nuclear or coal-fired plants and 20 times less than the best of the new gas turbine combined cycle units. Hence, while the fuel is "free," the large physical size of the OTEC unit per MWe of rating causes it to have a high specific cost, $/kWe. Operating and maintenance costs at sea are also higher than those on land. Together, these imposts require that fossil fuel costs undergo a significant, lasting escalation before OTEC once again appears as attractive as it did in the wake of the 1973 oil crisis. As noted in Chapter 2, this is an unlikely prospect for the foreseeable future.

14.4.3 OTEC technology

Two different power cycles—direct and indirect (see Figure 14.5)—have been pursued for OTEC service. In the direct cycle, pioneered by Claude in the 1930s, hot water is flashed to steam, passed through a low-pressure turbine, and the exhausted vapor is condensed in a direct contact condenser (or surface condenser if distilled water is a desired co-product). Two practical problems arise. Noncondensable gases must be removed by vacuum pumps, which consume an appreciable amount of the gross energy generated. More limiting is the low pressure drop and vapor density available to the turbine—conditions roughly similar to those in the last low-pressure stage of the largest commercial steam turbines used in nuclear power stations, which deliver only a megawatt or two. Thus, open-cycle OTEC units of higher rating require development of special large diameter turbines. A 100 MW unit could be as large as 100 m in diameter, hence several smaller turbines would probably be used.

Closed-cycle OTEC systems operate under conditions similar to air-conditioning and refrigeration applications, indirect cycle geothermal power stations, and bottoming cycles appended to other power sources. Thus, working fluids such as ammonia, propane, or Freon are preferred. Their higher pressure and vapor densities greatly reduce turbine size requirements, but with the tradeoff of adding two huge heat exchangers, each an order of magnitude larger in terms of m^2/MWe than the main condensers of large fossil-fired or nuclear power stations.

Both OTEC concepts share another unique feature: a long, large-diameter cold-water pipe and associated pump to bring cold sink water for the heat engine up from as deep as 1,000 m. Diameters as large as 7 m are required for a 40 MWe unit in order to keep pressure drop losses tolerably low. Shore-based OTEC plants may require a cold water pipe two or three times longer. Protection against damage by current-induced forces, especially during storms, is a challenge, because wave heights may exceed 15 m and wave energy increases as the square of height. The best OTEC sites are in the doldrums (regions of low average wind velocity). These zones are also spawning grounds of hurricanes. Incidentally, while hurricanes expend energy at an average rate of about 3,000 GW, these storms are more of a threat to energy facilities than they are candidates for useful energy extraction. This is because of their limited duration and changing location.

Table 14.5 summarizes key features of a representative OTEC power unit design.

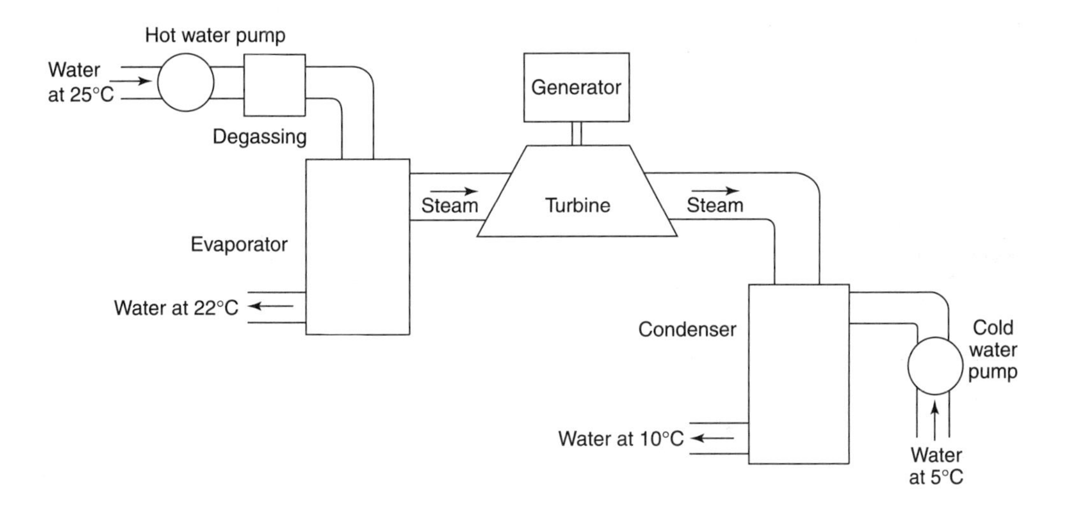

Figure 14.5a. Operating principle of an open-cycle OTEC plant. Adapted from: Brin (1981).

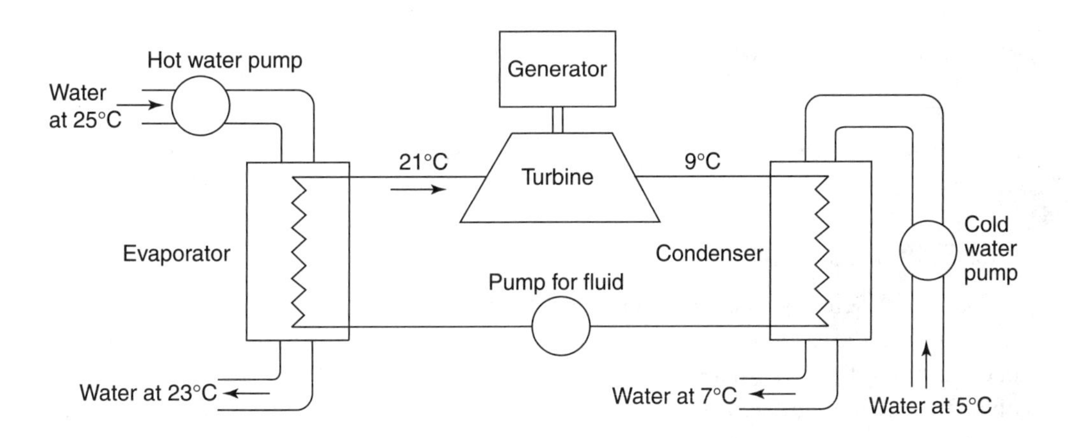

Figure 14.5b. Operating principle of a closed-cycle OTEC plant. Adapted from: Brin (1981).

Table 14.5. Characteristics of a Hypothetical 40 MWe OTEC Plant

Type	Closed Rankine Cycle Ammonia Working Fluid		
Nominal ΔT	23ºC ± 1º C seasonally		
Platform	Concrete Barge 150 m (1), 50 m (w), 30 m (d) 100,000 MT displacement multi-point moored plus dynamic positioning		
Cold-water pipe	segmented lightweight reinforced concrete 950 m long, 9 m diameter ball and socket joint with barge, flexible segment joints		
Hot water intake	15 m depth		
Effluent discharge	70m depth (below thermocline)		
Heat exchangers		Evaporator	Condenser
	flow rate, kg/s	130,000	100,000
	m^3/s per MWe	3.3	2.5
	m^2/kWe	6	6
	Material	Al	Al
Gross TG output	52 MWe underwater power cable to shore		
Net thermodynamic efficiency	2.7%		
Cost estimates	4,000 $/kW 10¢/kWh		

Note: This table is an amalgamation of design features from several sources, most notably Avery & Wu (1994).

14.5 Economic Prospects

The three energy technologies discussed in this chapter—tidal, OTEC, and wave—have the common burden of high capital-cost estimates (which increase in uncertainty in the order listed). The higher rates of return and imposed taxes required of the recently deregulated electric power industry in several Organisation for Economic Cooperation and Development (OECD) nations make it even more difficult for these capital-intensive options. Thus, plans for further R&D, much less deployment, have been deferred as a consequence of the low fossil-fuel prices during the 1990s. One exception is a modest new round of wavepower device tests, as described in Edwards (1998).

Tidal power, even with the benefit of the successful La Rance experience, and despite being technically proven, has not gone beyond successive rounds of paper studies (most notably, of late, for the Severn and Mersey estuaries in the UK). The limited number of premium sites worldwide also detracts from its status as a resource of truly global significance. In its favor, output, while cyclical, is predictable—unlike most other renewable sources.

OTEC applications, formerly of significant interest in the US and Japan, are also currently stymied by high capital cost and the limited number of good on- or near-shore sites. While global resources are immense, they are far offshore, which requires as a pre- or co-requisite the deployment of a synfuels infrastructure for transport of energy to users.

Wave power remains an enigma because none of the most promising universally applicable concepts have a successful, long-term track record of operation. This may change over the next several years (Edwards, 1998). If any of the dozen or so units live up to expectations, wave power could be cost competitive and capable of eventually supplying 10% or so of world electric energy needs—and even more in favored locations, such as the UK and Ireland.

As a rough generalization, good wavepower sites often correlate with better-grade, offshore windpower resources. Because of this, there may be useful synergism in co-location of these two types of units. Cost savings could arise from the sharing of power conditioning and transmission facilities, operating and maintenance infrastructure, and personnel. Wavepower units could also serve as protective breakwaters for the wind turbine foundations.

The ratio of lifetime energy generation to that consumed in construction and operation for OTEC is predicted to be a surprisingly large 10–15, achieved in large part by assuming a long lifetime (e.g., 50–100 years). Tidal installations come in at around 5—a bit lower than freshwater hydro. Early estimates for wavepower devices suggest 2:1 or lower (Spreng, 1998). Costs of electricity are projected to be higher than contemporary fossil competition: 50/100/150% for tidal/OTEC/wave, in that order. However, the lack of large-scale serial manufacturing, deployment, and operating experience introduces considerable uncertainty into all energy ratio and economics predictions.

Finally, shoreline sites are often expensive and strictly regulated because of competition for residential, recreational, and industrial uses.

14.6 Environmental and Sustainability Considerations

While the three major oceanic generator types of present concern are all rather benign, especially when it comes to regional or global impact, there are local effects to consider.

Tidal impoundment requires damming an estuary, thereby changing the regular cycle of wetting and dryout to which local flora and fauna have become accustomed. Change in the mix of species is inevitable, requiring that contentious value judgments be resolved. Non-migratory fish can be exposed, several times a day, to passage through the plant's turbines. Migratory waterfowl may be deprived of a useful way station. Here, as elsewhere, one must keep in mind that other alternatives may (and probably will) have different adverse effects.

OTEC units bring large amounts of cold, nutrient-rich seawater close to the surface. This is often cited as an advantage, as it would provide an opportunity for mariculture. Because the cold water also contains more dissolved CO_2 than the warm surface water, there is also the potential for CO_2 release to the atmosphere (but, perhaps, an order of magnitude less than combustion of fossil fuel to generate the same amount of electricity). Another hypothesized detriment is that close-to-shore units may bring up preserved pathogens accumulated during years of wastewater runoff.

Wavepower devices remove energy from the incoming waves and thereby alter sediment transport and suspension closer to shore. This breakwater feature may be credited as an advantage in some locations. However, they are also navigation hazards and may impede fishermen in their harvesting of bottom-dwelling fish and shellfish. For sites close to or onshore, aesthetic concerns must also be addressed.

In summary, the subject systems could all be credited with significant avoided emissions costs that may help defray the cost penalty associated with operations at sea. In general, their impacts are not serious, nor are they irreversible or global in nature.

14.7 The Ocean as an Externalities Sink

Another traditional *de facto* role of the ocean in energy generation has been as an ultimate natural sink for dispersed emissions, such as the carbon, sulfur, and nitrogen oxides created in fossil-fuel combustion. Coastally sited power stations also commonly reject waste heat to ocean water.

However, several more ambitious schemes for potential future implementation have also been proposed; the most notable are below.

- Seeding vast ocean tracts with trace element iron, or nitrogen-rich fertilizer, to promote the growth of biota—hence increasing the rate of CO_2 uptake and sequestration (Nadis, 1998, and Pearce, 2000)
- Capture of CO_2 at power plants and its injection into deep ocean waters, again to reduce its buildup in the atmosphere (Herzog et al., 2000)
- Disposal of encapsulated high-level radioactive waste from spent nuclear power plant fuel in ocean bottom sediments overlying subduction trenches. (Hollister and Nadis, 1998)

The technical and cost effectiveness of these options, and their environmental impact, are contentious. The ocean's legal status as a protected global commons ensures strong political challenges to the deployment of such practices.

14.8 Current Status and Future Prospects

In this category of options, only La Rance tidal station provides, at present, a sizeable commercial contribution to world energy needs. This situation is likely to continue for the foreseeable future. As always, it is a matter of choosing among alternatives. Of the alternatives having high capital and low (or no) fuel costs, conventional hydroelectric and nuclear fission reactors appear to have a significant advantage over tidal, OTEC, and wavepower installations. Thus, even if global warming becomes an overriding factor, ocean-derived power is not likely to be a highly ranked option. Moreover, experience with offshore oil and gas production offers a firm basis for estimating the cost-penalties of offshore operations. In addition, the rough geographic correspondence of premium wind and wave resources makes it likely that the former will be used in preference to the latter. Finally, near-shore OTEC and good tidal sites are too few in number and in size of exploitable capacity to merit the attention of those who are primarily interested in options of potential global significance.

From this perspective, freshwater hydro, with a currently installed capacity of more than 540 GWe, is a far more important source of water power. Compare, for example, the World Energy Conference prediction of ≤ 10 GWe from tidal energy plants by 2020 under their most optimistic scenario (Clark, 1997). This is an unlikely extrapolation from the current level of < 1 MWe. The prospects for power from the waves should become clearer with operation in late 2003 of the European Marine Energy Centre (EMEC) test facility in the Orkney Islands (Knott, 2003). It will be used to test modules of a wide variety of wavepower concepts at a commercially relevant scale under realistic environmental conditions.

References

Avery, W. H. and Chih Wu. 1994. *Renewable Energy from the Ocean: A Guide to OTEC*. New York: Oxford University Press.

Baker, G.C. 1992. "Current Status of Tidal Power in the Bay of Fundy," in *Tidal Power Trends and Developments: Proceedings Fourth Conference on Tidal Power*. London: Thomas Telford.

Best, F.R. and M.J. Driscoll. 1980. "The Prospects for Recovery of Uranium from Seawater." *Trans. Am. Nucl. Soc.*, 34 (June).

Borgese, E.M. 1985. *The Mines of Neptune*. New York: Harry N. Abrams.

Brin, A. 1981. *Energy and the Oceans*. Ann Arbor, MI: Ann Arbor Science.

Cartwright, D.E. 1999. *Tides: A Scientific History*. New York: Cambridge University Press.

Charlier, R. 1993. *Ocean Energies: Environmental, Economic, and Technological Aspects of Alternative Power Sources*. New York: Elsevier.

Charlier, R.H. 1992. *Tidal Energy*. New York: Van Nostrand Reinhold.

Clancy, E.P. 1969. *The Tides: Pulse of the Earth*. New York: Anchor Books.

Clark, R. 1997. "Tidal Power." Palo Alto, CA: *Annual Review of Energy* 2648–2674.

Clark, R.H. 1995. "Tidal Power," in *Wiley Encyclopedia of Energy Technology and the Environment*, Vol. 4. eds. A. Bisio and S. Boots. New York: John Wiley & Sons.

Clark, R., ed. 1992. *Tidal Power: Trends and Developments*. London: Thomas Telford.

Edwards, R. 1998. "The Big Break." *New Scientist* (October 3).

Elmore, W.C. and M.A. Heald. 1969. *Physics of Waves*. New York: Dover.

Hagerman, G. 1995. "Wave Power," in *Wiley Encyclopedia of Energy Technology and the Environment*, Vol. 4. eds. A. Bisio and S. Boots. New York: John Wiley & Sons.

Halmann, M.M. and M. Steinberg. 1999. *Greenhouse Gas Carbon Dioxide Mitigation*. Boca Raton, FL: Lewis Publishers.

Herzog, H.B., H.B. Eliasson, and O. Kaarstad. 2000. "Capturing Greenhouse Gases." *Scientific American*, 282(2).

Hollister, C.D. and S. Nadis. 1998. "Burial of Radioactive Waste Under the Seabed." *Scientific American,* 278(1).

Johnson, D.H. 1983. "The Exergy of the Ocean Thermal Resource and Analysis of Second-Law Efficiencies of Idealized Ocean Thermal Energy Conversion Power Cycles." *Energy,* 8(12).

Knott, M. 2003. "Power from the waves." *New Scientist*, 179(2473).

Levenspiel, O. and N. de Nevers. 1974. "The Osmotic Pump." *Science,* 183(4121): 157.

Nadis, S. 1998. "Fertilizing the Sea." *Scientific American*, 278(4).

Norman, R.S. 1974. "Water Salination: A Source of Energy." *Science*, 186.

Pearce, F. 2000 "A Cool Trick." *New Scientist,* 166(2233).

Pearce, F. 1998. "Catching the Tide." *New Scientist,* 158(2139).

Penner, S.S. and L. Icerman. 1984. *Energy: Volume II: Non-Nuclear Energy Technologies*, Second Edition. New York: Pergamon Press.

Rahman, M. 1995. *Water Waves: Relating Modern Theory to Advanced Engineering Applications*. Oxford, UK: Clarendon Press.

Ross, D. 1995. *Power From the Waves*. New York: Oxford University Press.

Ryther, J.H. 1979/80. "Fuels from Marine Biomass." *Oceanus,* 22(4).

Savage, M.T. 1992. *The Millennial Project*. Boston: Little Brown and Co.

SERI 1990. "The Potential of Renewable Energy." TP-260-3674. (March). US DOE.

Spreng, D.T. 1998. *Net-Energy Analysis and the Energy Requirements of Energy Systems*. New York: Praeger.

Suess, E., G. Bohrmann, J. Greinert, and E. Lausch. 1999. "Flammable Ice." *Scientific American,* 281(5).

Vega, L.A. 1995. "Ocean Thermal Energy Conversion," in *Wiley Encyclopedia of Energy Technology and the Environment*, Vol. 3, eds. A. Bisio and S. Boots. New York: John Wiley & Sons.

Wick, G.L. and W.R. Schmitt, eds. 1981. *Harvesting Ocean Energy*. Paris: The UNESCO Press.

Wilcox, H.A. 1977. "The US Navy's Ocean Food and Energy Farm Project," in *Ocean Energy Resources*, OED Vol. 4. ASME.

Web Sites of Interest

www.tidalelectric.com/

www.bluenergy.com

www.nrel.gov/otec/

Problems

14.1 a. Estimate the total flow rate in m^3/s of hot plus cold water for a fleet of closed-cycle OTEC units generating the same amount of electric power, 1,450 MWe, as a large nuclear power plant. The difference between warm surface water and cool deep water is 22ºC, and net OTEC thermodynamic efficiency is 3%. Express your answer as a fraction of the flow rate of the Mississippi River: 16,800 m^3/s. Seawater has a density of 1,025 kg/m^3 and a heat capacity of 1.16 kWh per metric ton per ºC.

b. Avery and Wu (1994) states that a medium size OTEC plant of 100 MW capacity and a temperature differential of 23.3ºC would require an intake of 300–500 m^3/s of warm surface water and an equal intake of cold deep water. Redo the estimate of part (a) using this estimate (e.g., 400 m^3/s) and explain any difference in results.

c. The nuclear reactor in part (a) requires 200 MT/yr of natural uranium to produce its required 30 MT/yr of low enrichment fuel. Seawater contains 3.3×10^{-9} U per kg seawater. If it were possible to remove 50% of this uranium from the water streams in part (b), what fraction of the reactor's annual requirement could be satisfied?

14.2 a. It has been proposed that large underwater single-stage multiblade turbines be installed in ocean currents such as the Gulf Stream off Florida and the Kuroshio near Japan. These units are entirely analogous in their fluid mechanics to the wind turbines discussed in Chapter 15. Using the relations in Section 15.4, evaluate the power rating of a current turbine having a rotor diameter of 170 m and an overall power coefficient, COP, of 0.30, driven by seawater of density 1,025 kg/m^3 flowing in a steady 2 m/s current.

b. Compare the pros and cons of this approach for renewable energy generation.

14.3 The winter air temperature over the floating ice sheets offshore of Antarctica (where the emperor penguin breeds) is as low as minus 62ºC, while the water beneath (home of the ice fish) is at –1ºC. Evaluate the prospects for siting OTEC units in this environment, using the sea as the "warm" reservoir and the air as the "cold" reservoir.

a. Estimate the achievable thermodynamic efficiency.

b. Discuss other important factors that would affect a decision to select this site for OTEC units compared to the usual proposal for their operation in equatorial seas.

14.4 Before experiential evidence suggesting otherwise, concern was expressed that an OTEC plant would release CO_2 to the atmosphere by transporting cold deep water containing 2.4 gram moles CO_2/1,000 kg seawater (as HCO_3^-) to the surface, at which CO_2 content is effectively 2.0 g mol/1,000 kg, and releasing the excess CO_2.

Carry out a worst-case evaluation of this scenario for an OTEC unit having a cold water flow rate of 5 m^3/sec per MWe. Compare the maximum hypothetical CO_2 release rate in kg CO_2/Kwhre to that of a modern high-efficiency combined cycle gas turbine (CCGT) (η=60%) unit burning natural gas and emitting 0.33 kg CO_2/Kwhre.

14.5 It has been suggested that wavepowered devices be towed by an OTEC barge to provide added power to help defray the large internal energy consumption by the cold and hot water pumps.

Estimate how long a string of wavepower converters would be required under the following conditions:

Average wave height = 2 m
Average wave period = 8 s
Efficiency of device = 50%
OTEC pumping power = 25 MWe (for a 100 MWe net unit)

14.6 A container ship having a displacement of 70,000 metric tons is raised 1 meter in 5 seconds by an ocean wave. Calculate the ratio of the average lift power to the ship's shaft horsepower of 50,000 hp.

Appendix

Constants and Conversion Factors

Quantity	**To Convert from units of**	→	**Into Units of**	→	**Multiply by**
Energy	Joules*		kWh		2.78×10^{-7}
Length	Feet		Meters		0.3048
	Miles		Kilometers		1.609
Area	Square miles		Square kilometers		2.59
Mass	Pounds		Kilograms		0.454
Flow rate	gpm		m^3/sec		6.309×10^{-5}
	Into	←	To convert from	←	Divide by

Density of seawater = 1,025 kg/m^3, Gravitational acceleration = g=9.807 m/sec^2

Temperature Conversion: $°C = \frac{5}{9}(°F - 32)$; °K = °C + 273.2; ºR = ºF + 459.7

*Note that 1 J = 1 Newton meter (Nm) and 1N = 1 kgm/s^2

Wind Energy 15

15.1 Introduction and Historical Notes

Extraction of energy from the wind dates back to antiquity, perhaps as early as 200 BC. Until the 19th century, windmills were surpassed only by waterwheels in duration and extent of use as mechanical generators of useful work. Reports of vertical, ducted windmills have been traced back to ancient Persia, but the familiar "Dutch windmill" rose to prominence after about 1100 in western Europe. (See Table 15.1 for an historical timeline.) Its design may have originated in England, perhaps with conceptual inspiration provide by returning crusaders (Kealey, 1987). Tens of thousands of such mills, of remarkably similar design, having rotor diameters of up to 30 meters and shaft power of five to as much as 30 kilowatts, were constructed in the 12th through the 19th centuries. A few dozen were built in the New England region in colonial times. The advent of steam power in the late 18th century and gasoline and electric motors in the late 19th century effectively ended this first windpower era, although small machines were widely used in the rural US, especially for pumping water (Righter, 1966).

Our interest here lies in the more recent use of large wind turbines and wind machines to generate electric power (see Part II of the Table 15.1 timeline). The first commercial machine of this genre in the US was constructed in Vermont starting in 1939 (Putnam, 1948). It had a rotor diameter of 53 m and a full power rating of 1.25 megawatts electric. Wartime exigencies and cheaper alternatives led to its demise in 1945. However, the oil supply crisis of 1973 ignited widespread interest in, and commitment to, wind turbines as a potential major component of the future electric energy generation system. Large wind farms were constructed, inspired in part by a generous tax policy. Perhaps the most notable wind farm was the Altamont Pass, California, array. Unfortunately, lagging interest followed, symbolized by the declaration of bankruptcy by the largest domestic manufacturer, Kenetech, in 1996. To some extent, these events led to reduced upkeep and a decline of the performance of installed turbines.

Internationally, the manufacture and deployment of wind turbines have remained more stable, with Germany, Denmark, and India being particularly active. The growing concern in the 1990s over CO_2-forced global warming has given new life to the prospects for greater use of wind turbines because of their credentials as non-polluting generators powered by winds created by solar energy, a renewable resource.

Readers interested in detailed information on certain topics addressed in this chapter should consult the references cited at the end of this chapter, as follows.

- Off- and on-the-grid contemporary small-scale use (Gipe, 1993)
- Historical US farm and ranch use (Righter, 1966, and Baker, 1985)

- History of European windmill use in the middle ages (Kealey, 1987)
- History of US use for electric power generation in the recent past (Berger, 1997, Gipe, 1995, and Sorenson, 1995)
- Vertical axis machines (e.g., Darrieus, Savonius) (Eldridge, 1980)

For in-depth coverage of the topics in this chapter, the books written by Gipe (1995) and edited by Spera (1994) are recommended, as is McGowan (1995).

Whatever the application or the era, wind machine performance is strongly dependent on the quality of the local wind resources, as explained in the following section.

Table 15.1. A Timeline of Wind Machine Milestones

Part I: The Historical Era

Dates	Events of Note
1	Hero (a.k.a. Heron) of Alexandria uses a wind machine to power an organ.
~ 400	Reference to wind-driven Buddhist prayer wheels
644, 915	References to pre-existing vertical axis windmills (panemones) in Persia
1137, 1274	References to pre-existing horizontal axis windmills in England and Holland, respectively
~ 1400	Smock (rotatable cap) mills originating in Holland supercede post mills (which turn on a vertical shaft).
1200–1850	Golden age of windmills in western Europe, totaling perhaps 10,000 in England, 18,000 in Germany, 9,000 in Holland, and 50,000 overall
1850s	Multiblade wind turbines for water pumping made and marketed in US
1769	James Watt granted a patent on his much-improved version of a steam engine
1877, 1893	Invention of the 4-stroke gasoline and the diesel internal combustion engines by Nikolaus Otto and Rudolf Diesel, respectively
1882	Thomas A. Edison commissions the first commercial electric generation stations in New York City and London
1900	Competition from alternative energy sources reduces windmill population to fewer than 10,000
1919, 1937	End of commercial operation of the last of the conventional old-style mills in Long Island (New York) and Denmark
1850–1930	Heyday of the small multiblade wind machine in the US midwest—as many as six million units installed
1936+	US Rural Electrification Administration extends the grid to most formerly isolated rural sites, which rapidly displaces wind machine use

Table 15.1 (continued). A Timeline of Wind Machine Milestones

Part II: The Modern (Electric) Era

Dates	Events of Note
1890–1893	LaCour in Denmark and Lewis Electric in New York state build wind machines to generate electricity
1933	Krasnovsky builds a 100 kWe wind machine in the Russian Crimea, near Yalta
1941–1945	The Smith-Putnam 1250 kWe wind turbine is installed and operated on Grandpa's Knob, Vermont
1973	The oil energy crisis inspires new interest in alternative energy sources
1974–1980	US Federal Large Wind Turbine Program
1976	US Energy Research and Development Administration (ERDA) small wind machine development program
1978	Public Utility Regulatory Policy Act (PURPA) requires that utilities purchase electricity from small producers at the utilities' "avoided cost"
1981–1993	Wind turbine boom times in California: more than 12,000 units installed, totaling some 1800 MWe
1982–1994	Record rating 4 MWe Hamilton Standard 4TS-4 unit operated in Wyoming
1985, 1986	US federal and California tax credits for wind projects expire, respectively
1988	Worldwide large wind turbine sales, R&D expenditures, market incentives bottom out due to decline in US
1991	First commercial offshore wind farm, Vindeby, Denmark
1992	National Energy Policy Act (NEPA) gives a 1.5 ¢/kWhr production tax credit for wind-generated electricity
1996	Kenetech Windpower (US Windpower), largest US and world manufacturer, declares bankruptcy. [assets sold to Enron Wind, then acquired by GE Wind]
1990–2000	Megawattage of installations in Europe grows at ~20%/year
1998–1999	European manufacturers open wind turbine factories in US and China
2001	US Department of Energy (DOE) announces goal and program to have 3¢/kWehr electricity from wind by 2012

15.2 Wind Resources

Winds are produced by uneven solar heating of the earth's land and sea surfaces. Thus, they are a form of "solar" energy. On average, the ratio of total wind power to incident solar power is on the order of two percent, reflecting a balance between input and dissipation by turbulence and drag on surfaces. Only a small fraction is close enough to the earth's surface to be practically accessible, and only selected locations have winds that are sufficiently strong and steady to be attractive for exploitation. Figure 15.1 shows a wind resource map for the US, and Figure 15.2 provides an overview for the world as a whole. As is evident from these resource maps, the best wind fields are generally near the coast, and there is commonly an overall decline in average quality in the central regions of large continental land masses. However, the great plains region of the US midwest has extensive resources. If fully exploited, those in North Dakota and South Dakota alone could be used to generate enough electricity to equal roughly half of current US consumption, and the totality of the US landscape could produce several times today's needs. A similar situation exists for the world as a whole, especially if offshore siting can be economically exploited. But again, there is wide variation by region and locale.

While the potential resources are immense, there are several constraints on use that limit near-term exploitation to perhaps 20% or so of total electric grid capacity:

- Winds vary in speed, hence incident energy flux, during the day and from season to season, and not necessarily in concert with demand for electricity.
- This non-dispatchable nature limits the portion of wind power in a utility's generator mix, with provision for spinning and standby reserve and grid stability being important concerns.
- Other than the fortuitous proximity of pumped storage hydro installations, there is no sufficiently inexpensive way, at present, to store energy for future use.
- The best wind fields (e.g., the Dakotas) may not be in reasonable proximity to large population centers, which necessitates the construction of expensive high-voltage transmission systems and results in large line losses of the input energy.

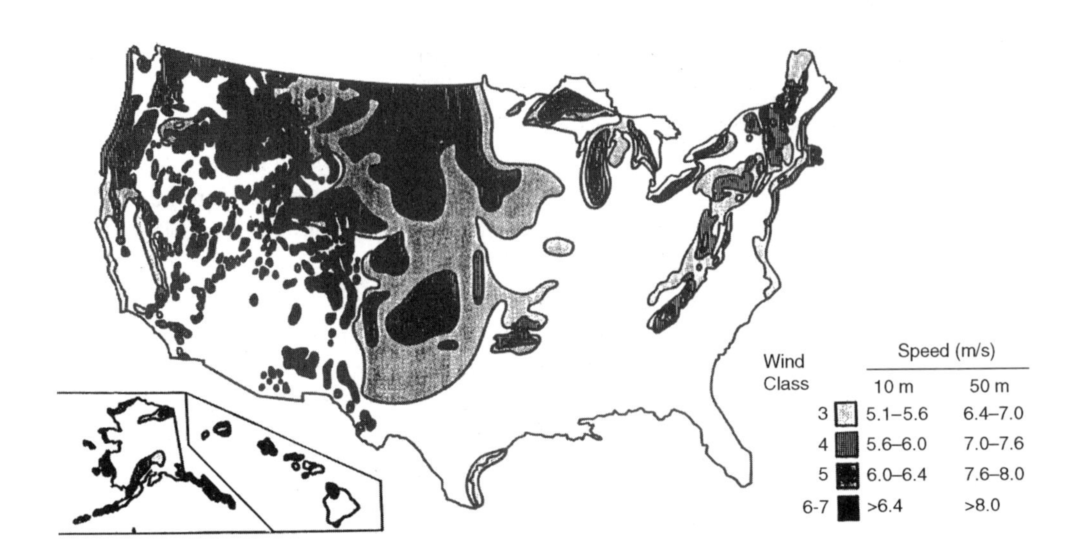

Figure 15.1. Annual average wind speed map of the US. Source: Pacific Northwest Laboratory (1986).

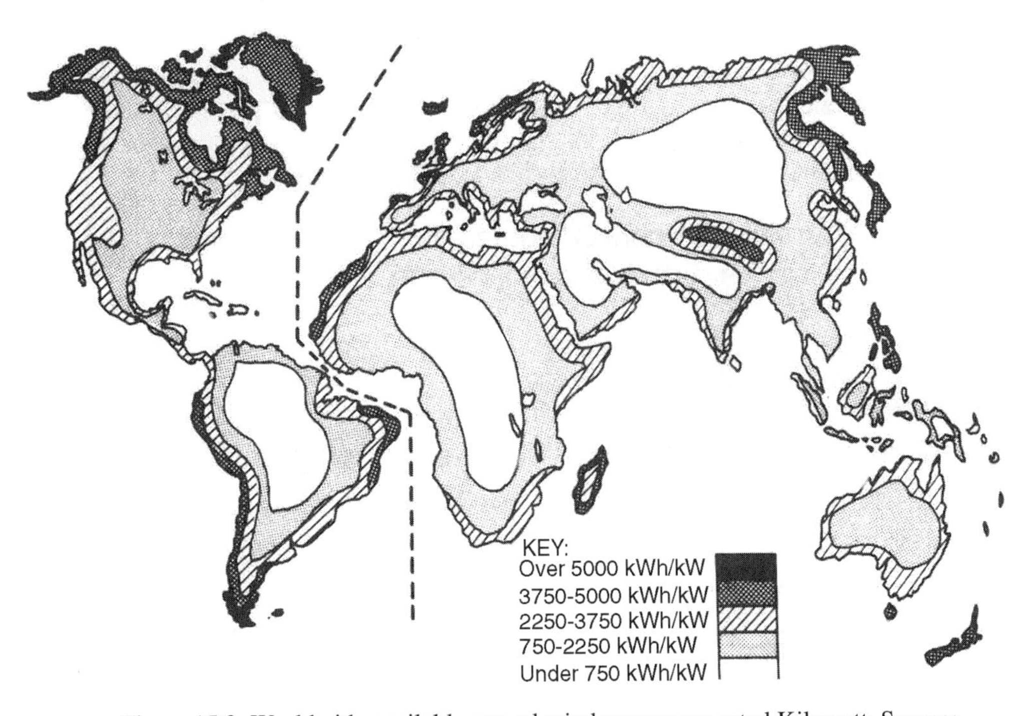

Figure 15.2. Worldwide available annual wind energy per rated Kilowatt. Source: Eldridge (1980). Reprinted with author's permission.

As with other solar-electric technologies, advances in storage technologies, whether centralized or dispersed, could greatly expand the prospects for market penetration by wind turbine generators. Compressed air energy storage (CAES), superconducting magnets, supercapacitors, advanced batteries, and flywheels are potential candidates (see Chapter 17). In addition, connecting spatially dispersed wind plants, together with an enhanced transmission system, could expand the wind-generated contribution beyond 20%.

The quality of wind resources is sufficiently variable and localized that accurate long-term site surveys are a prerequisite to deployment. As we will soon see, power in moving air is proportional to the cube of velocity; moreover, velocity spectra can differ to the extent that, for the same *average velocity*, *spectrum-averaged power* can differ by as much as 50%.

15.2.1 Wind quality

The power per unit area transported by a fluid system is proportional to the cube of the fluid velocity:

$$P''(v) = \frac{P(v)}{A} = \frac{\rho}{2} v^3, \ W/m^2 \qquad (15\text{-}1)$$

where ρ is the density of the fluid (kg/m^3), v its velocity (m/s), and A the area perpendicular to the direction of flow (m^2).

Because wind velocity varies over time, the spectrum-average power is of principal interest:

$$\overline{P}'' = \int_0^\infty P''(v)\ f(v)dv \qquad (15\text{-}2a)$$

where:

$$\int_0^\infty f(v)dv = 1.0 \qquad (15\text{-}2b)$$

in which $f(v)$ is the frequency spectrum, that is, the fraction of the time over the course of a year when the wind blows at velocity v.

It has become general practice to curve-fit measured velocities to a Weibull statistical function as defined in Table 15.2 and plotted in Figure 15.3, and, where simplicity suffices, to its more analytically tractable special case, the Rayleigh function (Weibull with shape factor $k = 2$), also defined in Table 15.2. Figure 15.4 illustrates normalized plots of Rayleigh velocity and power spectra. The dominant importance of wind in the higher velocity sector of the distribution when it comes to making power is obvious. Infrequent high velocity winds, which threaten turbine survival, should be independently characterized.

Table 15.2. Useful Properties of the Weibull and Rayleigh (k = 2) Wind Velocity Distribution

1. General Relations

Cumulative probability of velocity exceeding v:

$$F(> v) = e^{-\left(\frac{v}{c}\right)^k} \equiv 1 - F(<v) \tag{1}$$

where c is the scale parameter.

Frequency of v in dv about v:

$$f(v) = \frac{dF(<v)}{dv} = \left(\frac{k}{c}\right)\left(\frac{v}{c}\right)^{k-1} e^{-\left(\frac{v}{c}\right)^k} \tag{2}$$

Mean velocity and mean of nth power of velocity:

$$\bar{v} = c\Gamma\left(1 + \frac{1}{k}\right) \tag{3}$$

$$\overline{v^n} = c^n\Gamma\left(1 + \frac{n}{k}\right) \tag{4}$$

Most probable velocity (at max of $f(v)$):

$$\hat{v} = c\left(\frac{k-1}{k}\right)^{1/k} \tag{5}$$

and at maximum energy:

$$\hat{v}_e = c\left(\frac{k+2}{k}\right)^{1/k} \tag{6}$$

where Γ is the gamma function, having the properties:

$$\Gamma(z + 1) = z\Gamma(z) = z! \tag{7}$$

$$\Gamma(1) = \Gamma(2) = 1.0 \tag{8}$$

$$\Gamma(3/2) = \sqrt{\pi}/2 \tag{9}$$

In the interval, $1 \le z \le 2$, the following approximation is accurate within +0.15%:

$$\Gamma(z) \approx 2.3404 - 2.3040z + 1.1075z^2 - 0.1455z^3 \tag{10}$$

Ratio of average of velocity cubed divided by cube of average velocity. The energy pattern factor (EPF):

$$\text{EPF} = \frac{\overline{v^3}}{\bar{v}^3} \approx 1.03 + \frac{3.2465}{k^2} + \frac{0.8771}{k^4} + \frac{0.8750}{k^6} \; \textit{within } 0.2\% \textit{ for } 1 < k < 4 \tag{11}$$

2. Additional General Relations

Fraction of total power forgone below x (accurate to 1% of F for k between 1.0 and 4.0 for $x \le 0.75$):

$$F(<x) = \frac{kx^{k+3}}{(k+3)\Gamma(1+3/k)}\left[1 - \left(\frac{k+3}{2k+3}\right)x^k + \frac{k+3}{2(3k+3)}x^{2k} - \frac{k+3}{6(4k+3)}x^{3k} + \ldots\right] \quad (12)$$

Fraction of total power forgone above x (accurate to 1% of F for k between 1.0 and 4.0 for $x \geq 2$):

$$F(>x) = \frac{x^3 e^{-x^k}}{\Gamma(1+3/k)}\left[1 + \frac{3}{kx^k} + \frac{3(3-k)}{k^2 x^{2k}} + \frac{3(3-k)(3-2k)}{k^3 x^{3k}} + \ldots\right] \quad (13)$$

where the normalized velocity parameter is:

$$x = \Gamma\left(1 + \frac{1}{k}\right)\left(\frac{v}{\bar{v}}\right) = \frac{v}{c} \quad (14)$$

3. Properties of the Rayleigh Distribution ($k = 2$)

Velocity frequency distribution:

$$f(v) = \frac{\pi}{2}\frac{v}{\bar{v}^2} e^{-\frac{\pi}{4}\left(\frac{v}{\bar{v}}\right)^2} \quad (15)$$

$$\text{where } \bar{v} = \text{mean windspeed} = \int_0^\infty v f(v)\,dv \quad (16)$$

and:

$$\int_0^\infty f(v)\,dv = 1.0 \quad (17)$$

Most probable velocity (peak of $f(v)$ distribution):

$$\left(\frac{\hat{v}}{\bar{v}}\right) = \sqrt{\frac{2}{\pi}} = 0.80 \quad (18)$$

Mean of v^3 (power = 1/2 ρv^3)

$$EPF = \frac{\overline{v^3}}{\bar{v}^3} = \frac{6}{\pi} = 1.91 \quad (19)$$

Peak power at:

$$\frac{\hat{v}_p}{\bar{v}} = 2\sqrt{\frac{2}{\pi}} = 2\left(\frac{\hat{v}}{\bar{v}}\right) = 1.60 \quad (20)$$

Energy foregone below cut-in (fraction of total):

$$F(<x) \approx \frac{8}{15\sqrt{\pi}}x^5\left[1 - \frac{5}{7}x^2 + \frac{5}{18}x^4 - \frac{5}{66}x^6\right] \quad (21)$$

Energy foregone above cut-out (fraction of total):

$$F(>x) \approx \frac{4x^3 e^{-x^2}}{3\sqrt{\pi}}\left[1 + \frac{3}{2x^2} + \frac{3}{4x^4} - \frac{3}{8x^6}\right] \quad (22)$$

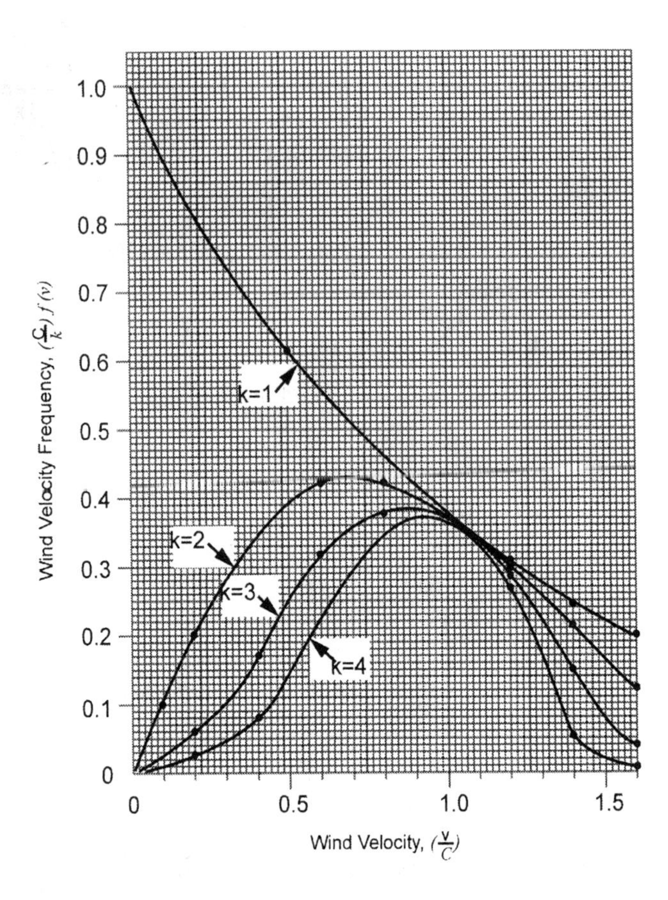

Figure 15.3. Weibull wind velocity frequency function $f(v)$ (for scale parameter c=1).

15.2.2 Variation of wind speed with elevation

For an idealized smooth plane surface, the average wind speed increases with height approximately as the 1/7th power:

$$\frac{v_2}{v_1} = \left(\frac{h_2}{h_1}\right)^{1/7} \equiv e^{1/7 \ln(h_2/h_1)} \tag{15-3}$$

Thus, a wind turbine with a hub elevation of 50 m will, relative to a height of 30 m, see an average wind speed some 7.6% higher. Because available power varies as velocity cubed, the higher position can increase turbine power by 24.5%, an appreciable improvement. As a result, selection of tower height is a major cost-benefit tradeoff. Other factors complicate this choice. For example:

- Topography and vegetation alter the wind field. Crests of treeless hills are advantageous; however, the flow above hills does not follow the 1/7th power rule.

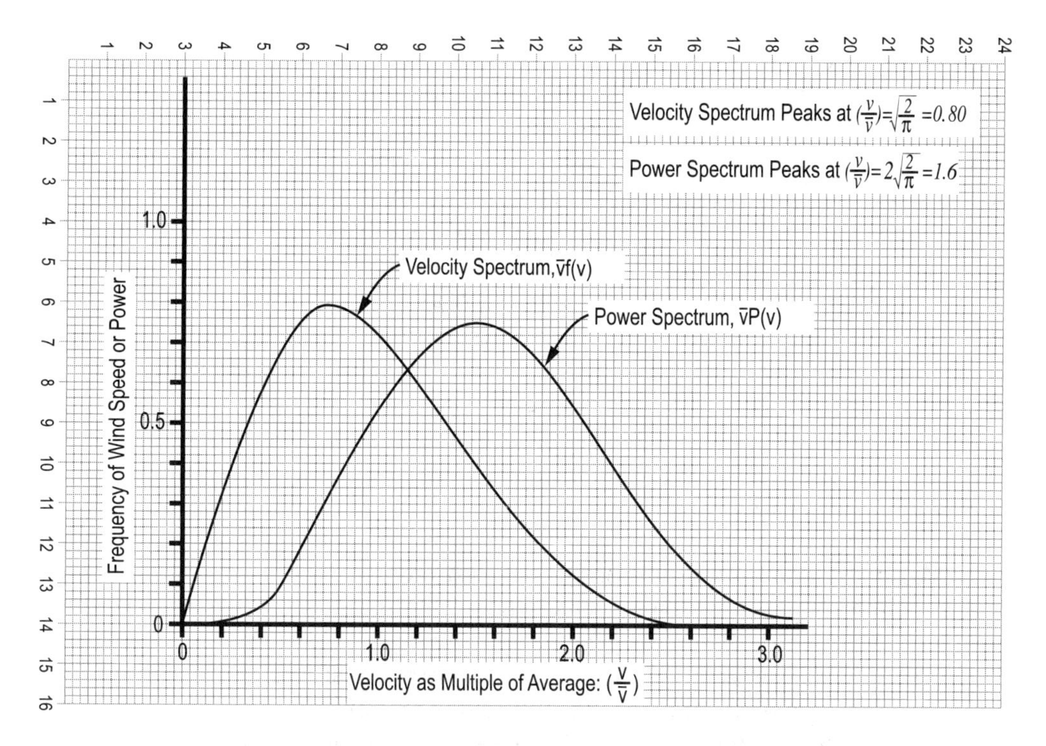

Figure 15.4. Wind velocity and power spectra for Rayleigh distribution.

- In general, there is less fluctuation in air flow at greater heights.
- Adjacent wind turbines interfere to reduce the energy flow field.

While grouping of units into large wind farms is advantageous for utility system operation and maintenance, too close a spacing leads to interference, which reduces the energy of downwind units. A rough approximation to the performance estimates plotted in Sorenson (1995) suggests that the relative output of a unit in the *Nr*th downwind row, with row spacing of Nd rotor diameters, is:

$$f \approx e^{-\frac{-2Nr}{Nd^2}} \tag{15-4}$$

Thus, if Nr = 10 and Nd = 10, a unit in that row will only generate 82% of the power of a stand-alone unit. Accordingly, spacings of several diameters across the wind front and ten or more diameters in the downward direction are advisable. The fewer rows deep, the better.

Finally, local wind speed can vary significantly with location within a given region; hence accurate measurements of the daily and seasonal wind field are essential to wind machine siting.

15.2.3 Air density

As is evident from Equation (15-1), the available power is also directly dependent on the fluid density. At 15°C and 1 atmosphere, dry air has a density of $\rho = 1.226$ kg/m^3; values at different temperatures and pressures are readily and accurately estimated using the ideal gas law. The most significant variation is with height above sea level, H_S; for H_S in kilometers, density is decreased by a factor of roughly $(1 - 0.1\ H_S)$. Thus, in "mile high" Denver (1.6 km above sea level), air density is only 0.84 that at sea level, which reduces available windborne energy by as much as a 6% decrease in average wind velocity. Water vapor in the air also decreases its density (at fixed temperature and total pressure) by a factor of roughly $(1 - 0.0005\ T$ (in °C)), for saturated air; hence, only on the order of a percent or so, which is negligible for present purposes.

We are now equipped to assay the energy available in a given wind field, and we can turn our attention to what fraction of this amount can be extracted by a wind turbine in the best of circumstances.

15.2.4 Maximum wind turbine efficiency: The Betz ratio

It is common practice to estimate maximum attainable efficiency for a wind turbine using an ideal, somewhat oversimplified, fluid flow model. Consider a unit mass of air (Figure 15.5) with upstream velocity v_u that intercepts a turbine disc of area A_t at a (lower) velocity v_t and moves downstream (at a still lower) velocity v_d.

The force on the disc (turbine blades) is the area times the rate of change of momentum between the upstream and downstream fluid, while the power extracted is force times velocity, hence:

$$F = \rho A_t v_t (v_u - v_d) \tag{15-5}$$

$$P = \rho A_t v_t^2 (v_u - v_d) \tag{15-6}$$

Now, we apply the Bernoulli relation (which, strictly speaking, applies to incompressible, irrotational flow—of somewhat dubious applicability here, one must admit).

Then, for a constant ambient pressure, p, far up and downstream, and constant air density everywhere:

$$p_t^+ + \frac{1}{2}\rho v_t^2 = p + \frac{1}{2}\rho v_u^2 \tag{15-7}$$

$$p_t^- + \frac{1}{2}\rho v_t^2 = p + \frac{1}{2}\rho v_d^2 \tag{15-8}$$

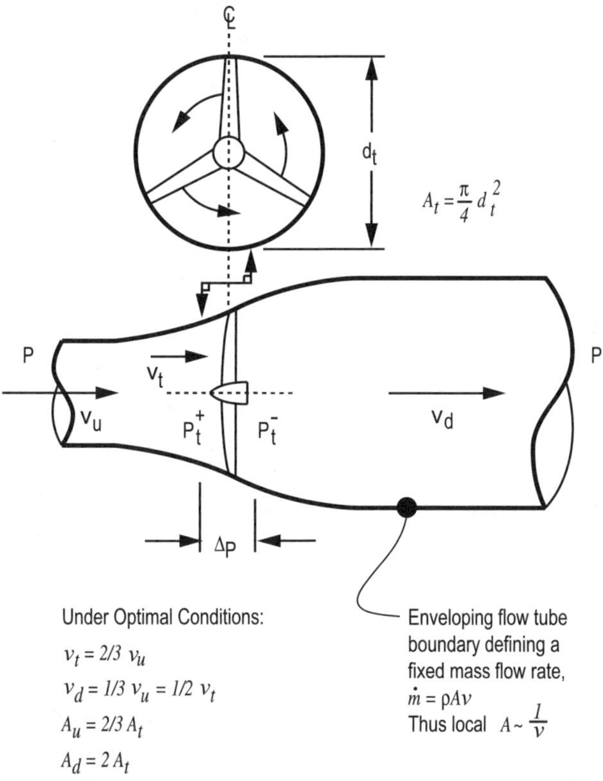

Figure 15.5. Schematic in support of derivation of Betz Efficiency Relation. Note that A_t = area swept by turbine disc, A_u = upstream stream tube cross-sectional area, and A_d = downstream stream tube cross-sectional area.

where the + and - superscripts on p_t indicate pressures *just* upstream and downstream of the turbine disc, respectively. Hence, the force on the disc (area times pressure drop) is:

$$F = A_t \Delta p = A_t(p_t^+ - p_t^-) = \frac{A_t}{2} \rho(v_u^2 - v_d^2) \tag{15-9}$$

or, factoring:

$$F = \rho A_t(v_u - v_d)\left(\frac{v_u + v_d}{2}\right) \tag{15-10}$$

Comparing the first and last equations for F, we see that:

$$v_t = \left(\frac{v_u + v_d}{2}\right) \tag{15-11}$$

Subtracting v_u from both sides and rearranging leads to:

$$(v_u - v_d) = 2(v_u - v_t) \tag{15-12}$$

So that

$$P = 2\rho A_t v_t^2 (v_u - v_t) \tag{15-13}$$

Dividing Equation (15-13) by the total power in the upstream airflow ($\rho A_t v_u^3/2$) gives the power coefficient:

$$\eta = 4\left(\frac{v_t}{v_u}\right)^2\left[1 - \left(\frac{v_t}{v_u}\right)\right] \tag{15-14}$$

Finally, to find the maximum value, set $\partial\eta/\partial v_t = 0$ which yields $v_t = 2/3 v_u$ and hence:

$$\eta_{max} = 4\left(\frac{2}{3}\right)^2\left[1 - \left(\frac{2}{3}\right)\right] = \frac{16}{27} = 0.593 \tag{15-15}$$

which is the so-called Betz value for maximum wind turbine power coefficient. A well-designed modern unit can achieve an aerodynamic η of about 50%, and a net value of 40% after gearbox and electrical losses.

The Betz limit is only as valid as the oversimplified model used in its derivation. There are effects that both degrade and augment the realized power coefficient. For example, in addition to vortex shedding at the blade tips, the downstream air flow is not irrotational but spins helically (opposite to the rotor spin). There is also beneficial air flow mixing across the downstream flow tube envelope. Thus, the Betz limit is not a hard constraint like Carnot's limit on heat engine performance, but a useful first order rule-of-thumb that captures the dominant physical phenomena.

Note that the number of rotor blades appears nowhere in the preceding analysis. Again, to first order, this is *not* a factor with respect to efficiency, although in practice three-bladed units appear to have a slight advantage over two- and one- (counterbalanced) bladed rotors. Multiblade rotors, as used for the small pumping units formerly common in the US midwest, are advantageous because more blades mean higher torque and functionality at lower wind speeds. Increasing the number of blades also decreases the rate of rotation under similarly optimized conditions of service, which offers other advantages such as lower aerodynamic noise. This is countered by increased cost. As a result, the tradeoffs involved appear to favor compromising on the use of three blades.

In the next section, practical aspects of actual wind turbine designs are presented. Following that, we will arrive at estimates of deliverable electrical energy.

15.3 Wind Machinery and Generating Systems

15.3.1 Overview

In this section we are concerned only with the intermediate (50 kWe–500 kWe) and large (500 kWe–5000 kWe) categories suitable for utility wind farm service. Furthermore, consideration is confined to the 2- or 3-bladed, horizontal axis machines that dominate today's market. Eldridge (1980) lays out a full display of the diverse taxonomy of wind turbine types. Vertical axis designs, such as the Darrieus and Savonius rotor types, are still being built, but in much fewer numbers than horizontal axis units. Table 15.3 summarizes the key features of a representative generic machine.

Figure 15.6 shows the principal subcomponents of a wind turbine unit.

- Rotor
- Nacelle containing gearing and electric generator
- Support tower and power conditioning equipment

Each deserves attention if a cost-effective synthesis is to be achieved. In what follows, the discussion is limited to the most common embodiments of current design practice. Gipe (1995) is a good source of further discussion of the many alternatives that have been tried.

15.3.2 Rotor blade assembly

Today's machines are much different from the frame-supported sails of their medieval predecessors. The new-style blades are more akin to airplane propellers or wings and helicopter rotors, and designers have made good use of prior theory and experience in these areas. It is perhaps more than just coincidence that the largest airplane wingspan (Howard Hughes's "Spruce Goose") and largest wind turbine rotor diameter were both approximately 100 m, and the widely deployed successful commercial units in both fields are each about two-thirds the span of these one-of-a-kind behemoths. Nevertheless, there are also important differences.

Two- and three-bladed designs are common. Three-bladed units are predominant today. Two-bladed turbines may find favor for very large offshore machines. Considerable sophistication has been introduced into the blade airfoil design process to optimize performance in the appropriate wind speed range and to satisfy other requirements, including:

- Intentionally stalling at storm-force wind velocities to protect the rotor, gears and generator against over-stress
- Avoiding loss of performance due to roughening by debris deposits (e.g., insect impingement)
- Tapering of hub-to-tip blade shape and twisting the angle of pitch in the horizontal plane to allow for the increase in circumferential speed as a function of radius
- Substituting lightweight composite materials for metal in blade design

Table 15.3. Generic Characteristics of a Representative Wind Turbine Unit and Its Site, Late 1990s

Machine Related	
Type	Horizontal Axis, Upwind Rotor
Rated power	500 kWe
Rotor diameter	40 m
Hence, swept area	1257 m^2
Wind-speed related specifications	
Cut-in	4 m/s
Rated	13 m/s
Cut-out	25 m/s
Survival	60 m/s
Rate of rotation	30 rpm (constant)
Tip speed	62.8 m/s
Tip speed/rated wind speed	5.2
Installed cost	1,000 $/kWe rated
Hence, specific cost	398 $/$m^2$
Rotor	3-bladed, fiberglass, variable pitch
Generator	Fixed speed
Type	4-pole induction, 660 V, 60 Hz
Rate of rotation	1,800 rpm
Gearbox	3-stage, planetary/parallel
Yaw control	Active, hydraulically driven
Over-power protection	
Reduce blade efficiency	Variable pitch blades
Mechanical brake	Disc and caliper
Tower	50 m, steel, tubular
Site Wind Field Related	
Weibull shape factor	k = 2.5
Average wind speed	7 m/s
EPF Ratio $\overline{v^2}/\overline{v}^3$ü;+çtvçhäKÇ_çB§22çt çæ	
Specific power available	300 W/m^2
Integrated Performance	
Overall power coefficient, COP	0.30
Availability	95%
Specific energy delivered	824 kWe hr/m^2 yr
Total energy	1.036 x 10^6 kWe hr/yr

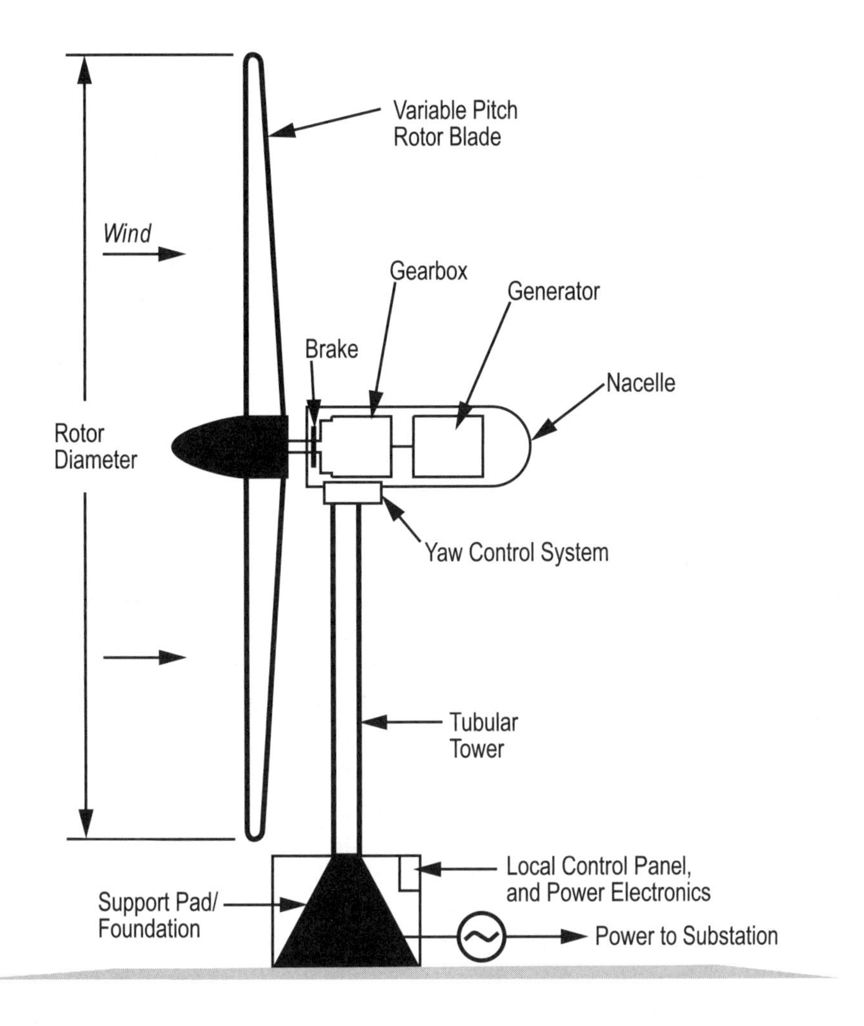

Figure 15.6. Principal elements of a wind turbine electric generator.

Good blade design is a challenge because efficient power extraction is only one goal. Wind gusts and changes in direction give rise to cyclic stresses in the blades, and through them, to the rest of the drive train. Blade and gear failures due to fatigue have been a continuing problem, even in next-generation machines. This undermines customer confidence. Nevertheless, there are enough unqualified success stories (machine cohorts with greater than 95% availability) to assure that technical feasibility should not be an issue in the long run.

15.3.3 Tower

We have previously noted that the wind *energy flux* increases approximately as the 3/7 power of height or 0.43% for every 1% increase in rotor hub height; hence the taller the tower, the better. However, cost increases with height, which leads to compromise. A rough rule of thumb is for hub heights to equal rotor diameter. (Obviously, height must exceed half the rotor diameter, plus a generous ground clearance.)

Two types of tower design are common: steel truss latticework and tubular pole. The latter appears to be gaining preference because of lower drag and downstream turbulence creation, and (to some) improved aesthetics. The elimination of perches for birds is also an advantage.

As rotor and nacelle pod components are improved, their weight, in general, decreases, which has a beneficial, synergistic effect on tower design, and minimizes overall unit cost of product.

15.3.4 Nacelle components

The nacelle houses the electric generator and associated gearing, control, and (optionally) some of the power conversion or conditioning components. The gear train is necessary to match rotor rotational speed (for example, 20 rpm) to that best suited for the electric generator. It must withstand highly variable loading and interface compatibly with a braking system.

The generator type is usually alternating current (AC) operated at constant shaft (hence rotor) speed to produce utility-grade 60 Hz AC power. Recently, variable-speed systems have been tested. They are attractive because they can operate more efficiently over a wider range of wind speeds and put less stress on the rotor components and gear train.

Because induction-type generators draw large amounts of reactive power from the utility grid, compensatory capacitors are now common. For a variable-speed generator, a power electronic converter is employed to maintain constant frequency line output. A transistor-based converter can rectify variable frequency AC to direct current (DC) then invert it back again into constant frequency, 60 Hz AC with low harmonic content. The overall cost-effectiveness of this approach is still at issue, but improvements in power electronics suggest this paradigm shift is inevitable.

15.3.5 Balance-of-station subsystems

Balance-of-station subsystems include power conditioning systems to interface the output of the windfarm with the utility transmission grid, a control and monitoring substation, and maintenance facilities. As noted in a previous section, variable-speed units can condition their power on the spot. Others may rely more on a central switchyard.

15.3.6 System design challenges

Although wind turbine engineering may, at first glance, appear prosaic, experience has highlighted the challenges of coping with the large number and severity of fatigue cycles created by fluctuating and asymmetric loading. These challenges are an order of magnitude beyond those associated with similar (e.g., aeronautical) applications. For example, the number of cycles over the wind turbine's lifetime, N_{cy}, is give by

$$N_{cy} = 60\, m \times w \times H \times t, \text{ cycles} \tag{15-16}$$

where

m = flexure cycles per revolution (typically 1)

w = rated (constant) rate of revolution (e.g., 30 rpm)

H = rotational hours per year (e.g., 4,000)

t = design life, years (e.g., 20)

For the numerical values cited, $N_{cy} \sim 1.4 \times 10^8$; this is a formidable prospect, particularly when long intervals between maintenance and downtime are essential to achieving cost effectiveness.

15.4 Wind Turbine Rating

As suggested in Sections 15.2.1 and 15.2.4, the nameplate power rating of a unit can be expressed as a fraction of the total energy of the wind intercepted by the spinning blades:

$$P_r = \text{COP}\,\rho\, A \frac{v_r^3}{2} \times 10^{-3}\ , \text{ kWe} \tag{15-17}$$

in which:

COP = overall coefficient of performance under rated conditions (for example, 0.30 in modern machines versus perhaps 0.05 in an old wooden Dutch windmill), which accounts for the velocity-dependent rotor blade aerodynamic efficiency and the smaller losses in the gearbox and generator

ρ = density of air under reference conditions at which the rating is quoted (e.g., 1.205 kg/m^3 for dry air at 20°C)

$A = \frac{\pi D^2}{4}$, where D is the tip-to-tip diameter of the circle swept out by the rotating blades (e.g., 40 m)

v_r = the reference wind velocity at which rated power is quoted (e.g., 13 m/s)

For the values quoted in parentheses, the rated power, P_r, is 500 kW.

In Equation (15-17), the manufacturer's quoted nameplate rating is somewhat arbitrary, as it strongly depends on the selection of "rated" wind velocity. For actual in-service performance, one must average over the product of the velocity-dependent COP(v_r) and the site wind frequency spectrum to define the overall average power coefficient to employ in Equation (15-17). Figure 15.7 shows how delivered power varies with wind velocity. Note that no power is extracted below a cut-in velocity sufficient to overcome drive train resistance or above a cut-out velocity selected to protect the drive train against excessive stress. Also note that (for a constant rotational speed) extracted power is constant over most of the operating range, whereas power available in the wind increases with velocity cubed.

15.5 Wind Power Economics

Of all solar technologies, wind turbines are the closest to being economically competitive with fossil-fired systems. Over the long term, this situation can only improve. To appreciate why, a bit of elementary analysis will suffice.

The busbar cost of wind power electricity in constant 1998 dollars is given by the conventional expression used to compare alternatives on a consistent, but relative, basis:

$$e,\ \text{¢}/kWe\ hr = \frac{100\varphi}{8766 \cdot L}\left(\frac{I}{K}\right) + \frac{100}{8766 \cdot L}\left(\frac{O}{K}\right) \tag{15-18}$$

where (representative values in parentheses):

φ = annualized carrying charge rate (0.1 per yr)

L = capacity factor: actual average kWe/hr per year delivered divided by rated non-stop kWe/hr per year (0.30)

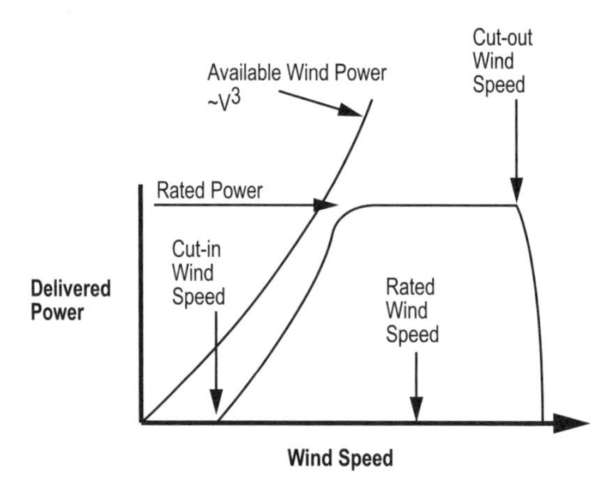

Figure 15.7. Generic wind turbine performance (variable pitch blades).

(I/K) = installed cost per *rated* kWe ($1,500/kWe), which for wind turbines is preferably given by the specific cost, $/m^2, divided by the specific loading, kWe / m^2.

(O/K) = maintenance costs per kWe ($20/kWe)

For the above-cited input, the predicted lifetime-levelized cost is (in current value dollars):

$$e = 5.7 + 0.8 = 6.5 \text{ cents / kWe/hr}$$

Under comparable assumptions, this is roughly the same as nuclear, on the order of 10% more expensive than coal, and perhaps 20% more costly than natural gas-fired combined cycle gas turbine (CCGT) generation, according to one recent comprehensive overview (OECD, 1998). Note, however, the admonition concerning comparing costs from different sources.

However, if one adds the externality costs of SO_x, NO_x, particulates and especially CO_2 emissions to the fossil options, wind underprices coal and draws even with CCGT. There are, moreover, additional, mostly favorable, considerations:

- Wind turbines can be installed under turnkey contracts within six months to a year of purchase, which significantly reduces the add-on of interest during construction.
- Its consequential reduction in associated uncertainty—the vulnerability of large, capital-intensive options involving protracted on-site construction, such as nuclear power and hydroelectric plants.
- Given a robust, steady market, economies of mass production can be realized; the cost of the Nth unit relative to that of the first is just

$$(I/K)_N = (I/K)_1 N^{\alpha} \tag{15-19}$$

  in which $\alpha = \ln(f/100) / \ln 2$, where f, the progress or learning factor, is typically on the order of 85%. Thus, the 100th unit will be cheaper than the 10th by a factor of 0.6.

- Fuel costs are zero, whereas for fossil units, they constitute a large fraction of the total busbar cost. A sizeable uncertainty is associated with projection of future fuel costs.
- There is considerable room for improvement in the operating and maintenance costs as experience accumulates in a field where most units are less than two decades in service.
- There is reason to believe that economies of scale can now be realized by gradually increasing unit rating. The classical prescription suggests that cost per kWe will decrease as:

$$\left(\frac{I}{K}\right)=\left(\frac{I}{K}\right)_0\left(\frac{K}{K_0}\right)^{n-1}, \$/\text{kWe} \tag{15-20}$$

where n, the scale exponent, is typically on the order of 2/3 for many systems. Thus, increasing size from 100 kWe to 1,000 kWe can potentially decrease cost per kW by a factor of 0.46. Faith in economy of scale was eroded due to the premature and ultimately short-lived attempts at instant giganticism (> 1 MWe) in the 1970s and 1980s, which culminated in the 4 MWe unit at Medicine Bow, Wyoming, noted in Table 15.1. In the 1990s, however, a more incremental evolution toward larger ratings was successful, and 1 MWe or larger units are again being marketed. Quarton (1998) shows an increase in kWe rating of the average new commercial unit by a factor of ten during the period 1980–1997.

- Siting costs can be kept low by dual use of the site and by leasing rather than purchasing. The clustering of wind turbine units in a windfarm also appears to be beneficial because it provides a critical size that permits proper staffing, an on-site spare parts inventory, and service equipment (for example, a high capacity crane). Maintenance delays were a major factor leading to low availability and capacity factors for early experimental and demonstration units.

Above all, one should note the dominant effect of capacity factor. This is, in turn, made up of several contributions:

$$L = L_0\left[1-\left(\frac{f+p}{100}\right)\right] \tag{15-21}$$

where:

L_0 = ratio of annual average power extracted in the ambient site wind spectrum to non-stop power at rated wind velocity

f = forced outage rate: percent of unplanned down time

p = planned outage rate: percent of time down for scheduled inspections and repairs

Thus, one must take into account how well the site measures up to the conditions implied in the nameplate rating. Moreover, because planned outages can be kept to less than 1%, the overriding importance of high reliability (small f) is evident. The term in brackets, also known as the "availability," now exceeds 95% for state-of-the-art, well-run units.

15.6 Measures of Sustainability

For our purposes, clearing the conventional hurdle of marketplace viability is an insufficient criterion for justification of deployment. To be sustainable, an energy system must also, at minimum, have a short energy payback time and a tolerable externality cost.

15.6.1 Net energy analysis

Common sense, as well as Public Law 93-577 (the Federal Nonnuclear Energy Research and Development Act of 1974), requires that a net energy analysis be performed on any new energy technology involved in the direct production, transmission, or utilization of energy. It may suffice to note that wind turbines measure up admirably in this regard, producing sufficient energy in about one year to pay back all energy consumed for their production (plus all other lifecycle events) (Gipe, 1995). (See Chapter 5 for other technologies.) A brief back-of-the-envelope calculation is instructive. The dominant energy input is in the form of fabricated steel for the turbine-generator and tower, amounting to about 50 kg/kWe. Taking automobiles as an analogous product, with their embedded energy of roughly 40 kWe hr/kg, and assuming that the wind turbine delivers 25% of rated power on an annual basis, the payback time is only:

$$50\,\frac{kg}{kWe,\,rated}\cdot\frac{kWe,\,rated}{0.25kWe,\,delivered}\cdot\frac{40kWhr}{kg}=8{,}000\,\frac{kWhr}{kWe}\sim 8{,}000\text{ hours}=0.9\text{ yr.}$$

Note the usual difficulty in that the wind turbine generates electric energy whereas the embedded energy is in thermal units—our calculation assumes (wastefully) that the electric energy is converted to thermal energy on a 1:1 basis (e.g., by resistance heating). If a heat engine were involved, it would be more appropriate to assume that 1 kWe~3kWth.

15.6.2 Cost of externalities

During operation, a wind turbine emits no noxious gases, and, unlike heat engine-type systems, requires no cooling water. Thus, most of the environmental costs not reflected in conventional accounting are associated with initial construction. Again, the steel components impose the lion's share of the burden. Building on the approximations of the preceding section, a rough estimate can be made.

We have to estimate the externalities associated with the 2,000 kWh of energy consumed to produce the steel components of the wind turbine per unit of rated power. Assume that all such energy has the same externality impact as a coal-fired fossil plant. Estimates range widely, but 2¢/kWhr is a plausible compromise value; at a generous 40% efficiency, this translates into 0.8¢/kWhr thermal.

If the wind turbine unit has a specific cost of $1,000 per rated kWe, which translates into a lifetime levelized capital cost component of 5¢/kWhr, then the corresponding levelized externality cost is:

$$\left[\frac{(2{,}000)(0.8)}{(1{,}000)(100)}\right] \cdot 5 = 0.08¢/\text{kWe hr}$$

This is vanishingly small and of a comparable order of magnitude to more sophisticated assessments.

The above estimate excludes costs and credits imputed to CO_2, both emitted and avoided during normal operation. For example, a coal-fired unit emits approximately 0.25 kg C/kWe hr. At a carbon tax (or avoidance credit) of $100 per metric ton of carbon (~27 $/mT CO_2), this adds approximately 2.5¢/kWe hr to the externality cost of the small amount of energy used to construct the windmill. This same assessment, however, can be taken as a production credit for displaced fossil-fuel combustion during wind system operation over the entire life of the facility. On a complete lifetime basis, a wind power station is responsible for about 2% as much CO_2 release as a comparable coal-fired unit.

15.6.3 Environmental impact of wind power

Wind turbines have low total externality costs—by widely shared consensus, less than a few tenths of a mil per kilowatt hour. As such, they outrank even other renewable/inexhaustible options. Most impacts are associated with the materials production and processing involved in original manufacture.

Undesirable features include:

- Aesthetic impact. The best wind farm sites are often in locations with a certain degree of wilderness cachet in the US and near the crests of horizon-defining topography. Moreover, many hundreds of individual units are required to constitute a GWe station comparable in output to a large central station fossil-fired or nuclear complex. The resulting visual impact is distressful to a non-negligible fraction of the public. In western Europe, proximity to a sizeable population is an even more stringent factor.
- Noise. Especially in western Europe, where open space is at a premium, the low-frequency blade noise is an important drawback. Machinery noise can be designed-out or muffled. Up to half a kilometer of buffer zone to the nearest habitations appears adequate, except perhaps for especially sensitive individuals.
- Raptor kill. In California, certain locations have proven hazardous to large, predatory birds such as hawks and eagles. If corrective measures prove ineffective, then this may become an additional constraint on siting.

- Interference with radio and TV transmission and radar. This has been, and still is, problematic in some instances, but appears to be diminishing as cable and satellite systems proliferate, as electronic technology improves, and as fiberglass blades are used in lieu of metal and siting guidelines evolve.
- Land use. Although wind energy fluxes through a horizontal plane are often several hundred watts per square meter, comparable to vertical solar energy fluxes, the large front-to-back and side-to-side spacings needed to avoid interference make wind power a dilute resource. Nevertheless, the actual footprint area of the tower is small, and dual land use (e.g., farming of all sorts, cattle grazing, or even solar photovoltaic or thermal units) is practical.
- Maintenance worker hazard. The dangers associated with windmill maintenance are similar to those faced by steeplejacks and high-rise-construction ironworkers. However, associated added costs are presumably reflected in the busbar price as higher workplace insurance premiums are paid by the wind farm owner/operator.

A recent instructive example of how these factors impact siting (and one close to home for the authors of this textbook) is the Cape Wind Associates' project, which seeks to build some 130 wind turbines in Nantucket Sound offshore of Cape Cod, Massachusetts. Important issues of aesthetics, environmental impact, licensing jurisdiction, and sociology (i.e., the NIMBY syndrome) are currently (as of August 2004) in a precedent-setting struggle to define the path forward for wind power projects of this type in the US—one already well-trod in Europe.

15.7 Current Status/Future Prospects

At the end of the 20th century, wind farm electric energy generation has come to a crucial crossroads. From an economic standpoint, it is the most deserving of all of the green supply options for more widespread deployment. However, like virtually all other candidates, it cannot compete with CCGTs unless there are significant taxes and/or credits assigned to CO_2 release/avoidance. In part, this helps to explain why demand is stagnant in the US, which (in California) once had (in the 1980s) over 90% of all wind power capacity, while it is growing at a moderate rate in western Europe, where it is currently responsible for approximately 20% of Denmark's electricity generation. Other impediments include:

- A lingering lack of confidence in reliability because of the high failure rate experienced by some product lines in the recent past
- The turmoil brought about by electric utility deregulation

- Historically low prices for coal, oil, gas, and uranium, which is lower than at any time in the past century
- The lack of a consistent, coordinated, stable, national energy policy and the resulting vacillation between acceptance and rejection of subsidies, tax credit allowances, and R&D support

Since the technological aspects of wind energy systems have attained near-maturity, we are unlikely to see further advances of the sort that improved rotor blade efficiency by 30% over the past two decades. Incremental improvements foreseeable in the near term include:

- Still better power electronics to improve wind generation interfacing with the grid
- Cheaper, better understood composite materials for blade construction.
- Power train simplification, for example, elimination of gear boxes (which can account for 30% of unit cost) by use of multipole generators (Lead units of this type are now entering service.)
- Intelligent online monitoring, diagnostic, and control systems to maximize power extraction and minimize the susceptibility to damage from off-normal conditions
- In the policy area, more widespread promotion of green power at a voluntary premium
- Improved modeling in all respects: aerodynamics and fluid mechanics, structural, power conditioning, grid integration, system planning, and environmental benefits

In the longer term, the single most beneficial development would be better, cheaper means of transmission and storage of energy, a need common to most other types of solar energy. At the top of the wish list, we see:

- Inexpensive room-temperature superconductors, which would enable loss-free long-distance transmission and both centralized and dispersed storage. At present, there is no theoretical assurance that this feat is even scientifically feasible.
- Hydrogen generation (both centralized and distributed) by electrolysis and pipeline transmission. A large end-use market will be needed to justify the enormous investment in infrastructure—perhaps fuel cells for cogeneration of electricity and thermal energy, which might offset the added cost of electrolysis at the production end.

- Greater exploitation of offshore siting could provide access to larger and higher quality wind fields. This has long been advocated, but until now, only implemented in shallow water, with towers embedded in the sea bottom. Some visionaries contemplate far-future scenarios in which floating or semi-submersible structures would operate farther at sea to generate hydrogen to supply a major fraction of world energy needs.

Even with no further technical advances, it is plausible to anticipate that wind power could generate 10–20% of the world's electricity by the year 2050—a value already achieved in Denmark.

One final caveat is especially appropriate in a chapter on wind power. Both technology and scale of deployment are advancing with sufficient rapidity that information soon becomes outdated. The reader is encouraged, indeed admonished, to consult the most recent literature, and especially the relevant Web sites, to become fully current.

Figure 15.8 tracks prior progress, and Table 15.4 shows a relevant country-by-country breakdown. In summary, this technology is attractive to already prosperous countries because of its excellent environmental credentials, and to developing nations because of its suitability for indigenous production and operation and its added avoided cost of long-distance transmission grid construction.

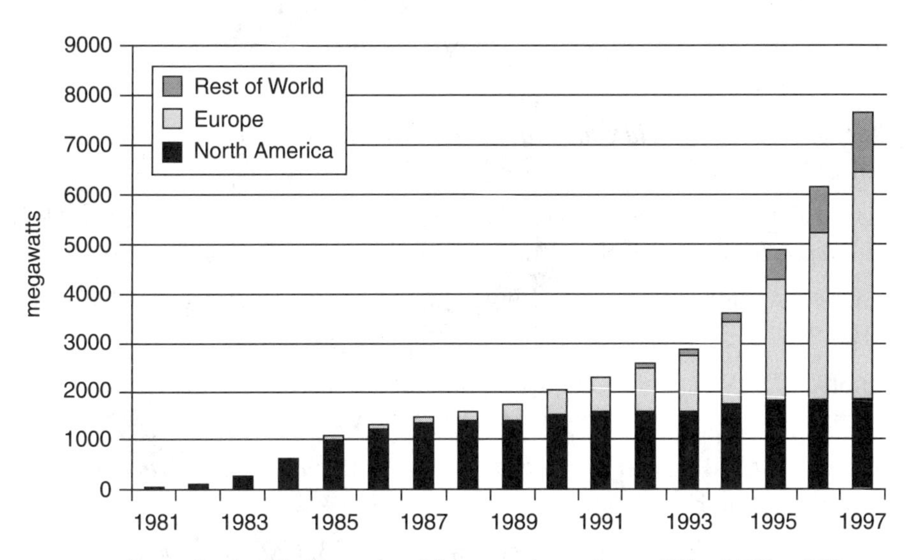

Figure 15.8. Worldwide installed wind-electric capacity. Sources: American Wind Energy Association (1997, 1998).

Table 15.4. Countrywide Use of Wind Power

Country	1998 MW Additions	1998 MW Year-end Total	1999 MW Additions	1999 MW Year-end Total
Germany	793	2,872	1,200	4,072
United States	193	1,770	732	2,502
Denmark	310	1,433	300	1,733
Spain	368	822	650	1,722
India	82	1,015	62	1,077
United Kingdom	10	334	18	534
Netherlands	50	375	53	428
China	55	224	76	300
Italy	94	199	50	249
Sweden	54	176	40	216

Source: Chambers (2000).

*Values for US, UK, China, India, and the Netherlands are final. Estimates for Germany, Spain, and Denmark are confirmed. Italy and Sweden are unconfirmed estimates. Additions only include projects that are installed and operating by the end of the calendar year. US figures are net of retired projects.

References

American Wind Energy Association (AWEA). 1997, 1998. Web site: www.awea.org.

Baker. T.L. 1985. *A Field Guide to American Windmills.* Norman, OK: University of Oklahoma Press.

Berger, J.J. 1997. *Charging Ahead: The Business of Renewable Energy and What it Means for America.* New York: Henry Holt & Co.

Chambers, A. 2000. "Wind Power Spins Into Contention." *Power Engineering* 104(2).

Connors, S.R. 1996. "Informing Decision Makers and Identifying Niche Opportunities for Windpower." *Energy Policy* 24(2).

DeBlieu, J. 1999. *Wind: How the Flow of Air Has Shaped Life, Myth and the Land.* Boston: Houghton Mifflin.

Eggleston, D.M. 1997. "Wind Power," in Chapter 23, *CRC Handbook of Energy Efficiency*, eds. F. Kreith and R.E. West, Boca Raton, FL: CRC Press.

Eldridge, F.R. 1980. *Wind Machines*, second edition. New York: Van Nostrand Reinhold.

Flavin, C. and N. Lenssen. 1994. *Power Surge: Guide to the Coming Energy Revolution.* New York: W.W. Norton & Co.

Gipe, P. 1995. *Wind Energy Comes of Age.* New York: John Wiley & Sons.

Gipe, P. 1993. *Wind Power for Home and Business.* White River Junction, VT: Chelsea Green Pub.

Heier, S. 1998. *Grid Integration of Wind Energy Conversion Systems.* New York: John Wiley & Sons.

Inglis, D.R. 1979. "A Windmill's Theoretical Maximum Extraction of Power from the Wind." *Am. J. Phys.* 47(3): 416.

Kealey, E.J. 1987. *Harvesting the Air, Windmill Pioneers in Twelfth-Century England.* Berkeley, CA: University of California Press.

McGowan, J.G. 1995. "Windpower," in *Wiley Encyclopedia of Energy Technology and the Environment,* Vol. 4, eds. A. Bisio and S. Boots. New York: John Wiley & Sons.

McGowan, J.G. and S.R. Connors. 2000. "Windpower—A Turn of the Century Review." *Annual Review of Energy and the Environment* 25:147–197. Palo Alto, CA: Annual Reviews.

Moore, T. 1999. "Windpower Gaining Momentum." *EPRI Journal* No. 4 (Winter).

Moretti, P.M. and L.V. Divone. 1986. "Modern Windmills." *Scientific American* 254(6).

Nakicenovic, N., A. Grubler, and A. McDonald, eds. 1998. *Global Energy Perspectives*, IIASA/WEC. Cambridge, UK: Cambridge Univ. Press.

National Renewable Energy Laboratory (NREL). Web site: www.nrel.gov.

Novitski, J. 1987. *Wind Star, The Building of a Sailship.* New York: Macmillan.

OECD (Organisation for Economic Co-operation and Development). 1998. "Projected Costs of Generating Electricity: Update 1998." Vienna.

Pacific Northwest Laboratory. 1986. *Wind Energy Resource Atlas of the United States.* Golden, CO: National Renewable Energy Laboratory.

Putnam, P.C. 1948. *Power from the Wind.* New York: Van Nostrand Reinhold.

Quarton, D.C. 1998. "The Evolution of Wind Turbine Design Analysis—A Twenty Year Progress Review." *Wind Energy* 1:5–24.

Ratto, C.F. and G. Solari, eds. 1998. *Wind Energy and Landscape.* Rotterdam: A.A. Balkema.

Righter, R.W. 1966. *Wind Energy in America, a History.* Norman, OK: University of Oklahoma Press.

Schmid, J. and H.P. Klein. 1991. *Performance of European Wind Turbines*. New York: Elsevier Applied Science.

Sorenson, B. 1995. "History of, and Recent Progress in, Wind Energy Utilization." *Ann. Rev. Energy Environ.* 20.

Spera, D.A., ed. 1994. *Wind Turbine Technology, Fundamental Concepts of Wind Turbine Engineering*. New York: ASME Press.

Wagner, S., R. Bareiss, and G. Guidati. 1996. *Wind Turbine Noise*. Berlin: Springer-Verlag.

Web Sites of Interest

www.windpower.org
www.eere.energy.gov
www.awea.org
www.nrel.gov

Problems

15.1 Consider a wind machine designed to deliver its rated power at the average velocity for a given wind spectrum. It will be marketed for sites having wind velocity spectra that have the same annual average velocity but with Weibull statistics ranging from k = 3/2 to 5/2.

a. Calculate the ratio of maximum to minimum annual average on-site wind powers available for exploitation for this range of wind fields.

b. What additional assumptions are necessary if the power actually produced is to be in the same ratio as in (a) above?

15.2 Estimate the total land requirements for wind farms generating an amount of electric energy equal to current US annual consumption (3.2×10^{10} kWe hr) if individual units are rated at 500 kWe, have a 50 meter diameter rotor, and are spaced 3 diameters side-to-side, 10 diameters front-to-rear in giant contiguous wind farms having an annual average capacity factor of 25%. Compare the required area to that of North and South Dakota (combined land area of 145,000 square miles), which have some of the best US wind resources.

Discuss the factors that would favor or impede the practicial implementation of a Grand Midwestern Windfarm plan to provide the bulk of US electric energy by the year 2050.

15.3 a. Identify and justify three current or proposed policy initiatives that could significantly increase the use of wind power.

b. Do likewise for three technological advances.

15.4 a. What fraction of the time will a wind turbine having a cut-in wind velocity of 4 m/s be non-rotational (and thus appear "broken" to an uninformed observer) in a wind field having a Rayleigh velocity distribution with a most probable speed of 6.4 m/s?

b. What percent of total possible windborne energy is forgone below this cut-in velocity?

c. If the same turbine has a cut-out velocity of 30 m/s, what percent of total possible wind-borne energy is forgone above this value?

15.5 a. A utility's existing wind farm has an average daily power output that varies as:

$$P(t) = \hat{P} \sin \frac{\pi t}{24},$$

where t is hours since midnight. It has located a comparable second site with a projected profile:

$$P(t) = \begin{cases} \hat{P} \cos \dfrac{\pi t}{24} & 0 \le t \le 12 \\ \hat{P} \sin \dfrac{\pi (t-12)}{24} & 12 \le t \le 24 \end{cases}$$

Compare the minimum-to-average, minimum-to-peak and average-to-peak power ratios for one- versus two-site wind systems, where $\hat{P}$ is the same for both sites.

b. Discuss the significance of one- vs. two-site windfarming with respect to overall utility system (generation, transmission, distribution) operations and planning.

15.6 Assume that wind turbines are designed such that irreparable damage should occur only for local storms, which occur, on average, once per hundred years. Calculate the annual insurance fee, \$/year, in constant dollars for a unit having an initial cost of \$300,000 if the worldwide windfarm insurance investment pool earns a net real inflation-free rate of return of 5%/year compounded continuously.

15.7 Sketch the outlines of a realistic and comprehensive policy with respect to birdstrikes, considering aspects such as regulations, monitoring, fines, mitigation, research, etc.

15.8 The US auto industry produces approximately 6 million vehicles a year, each containing roughly 1.5 metric tons of fabricated metal.

Consider a future wind-turbine industry having half of the auto industry's mass throughput of wind turbine units. Each unit has a specific weight of 75 kg/kWe rated (nacelle components plus tubular tower only).

a. At a wind-turbine capacity factor of 0.3, how long would it take to replicate the current US-installed average electric power rating of 600 GWe?

b Discuss the relevant factors that would determine whether a wind machine industry of this size could be successfully operated sustainably over periods on the order of several centuries.

15.9 Consider the crude energy payback time estimate sketched in Section 15.7.1. Discuss what other energy inputs should be considered in an overall lifecycle analysis of energy payback time for an offshore windfarm.

15.10 Consider the oversimplified externality cost estimate of Section 15.6.2. What, if any, other externality charges should be imputed to the subject unit on a complete lifecycle basis?

15.11 It is suggested that a wind turbine be developed to power a planetary probe being designed to land on the surface of Venus where atmospheric conditions are as follows:

pressure: 90 earth atm

temperature: 470°C

composition: 97% CO_2

atmosphere density: 64 kg/m^3

average wind speed: 1 m/s

Evaluate the practicality of this proposition if the wind turbine must provide 10 kWe to cool and power the proposed instrument package. Identify any important assumptions in your analysis and note changes in specifications compared to a unit designed for operation on planet earth.

15.12 Assume the following status for worldwide total and wind electric generation as of the year 2000:

	Total	**Wind**
Installed Capacity, GWe	2,000	10
Capacity Factor	50%	25%
Current Annual Growth Rate in Generation	3%/year	22%/year

By what year will wind generate 20% of all electricity if current trends continue?

Appendix

Windpower Relevant Conversion Factors

Quantity	**To convert ⟶ from units of**	**Into units of ⟶**	**Multiply by**
Velocity	Miles/hour (mph)	km/hr	1.61
		m/s	0.447
Power	Horsepower	Kilowatts, kWe	0.746
Rotational Speed	Revolutions per minute (rpm)	Rps	0.0167
		radian/s	0.105
Energy	Btu	Joules	1,055
	Btu	Watt Hour	0.293
Length	Feet	Meters	0.305
Mass/Weight	Tons	Kilograms	907
		Metric tons	0.907
	Pounds	Kilograms	0.454
Area	Hectares	Square meters	10,000
	Square miles	Hectares	259
	Acres	Hectares	0.405

Into ⟵ To convert from ⟵ Divide by

$$\text{Density of Dry Air} = 353 \frac{P(atm)}{T(K)}\ kg/m^3$$

Storage, Transportation, and Distribution of Energy 16

16.1 Overview of Energy Supply Infrastructure Needs

Several critical elements need to function properly in order to have a viable energy supply system. Requirements include a means for storing energy to provide it when needed and an infrastructure for supplying energy to locations where it is consumed. In earlier chapters, we concentrated on describing renewable and non-renewable energy technologies, focusing mostly on the capture, recovery, and conversion aspects of energy. In this chapter, we will consider three important elements of the system: energy storage, transmission, and distribution.

Energy storage permits decoupling of energy supply and demand periods, which is desirable for both economic and technical reasons. Economically, demand for energy and power has significant variation—both diurnally and seasonally. To maintain the capacity to meet peak demand periods without any storage would require more energy conversion equipment and capital investment. From a technical perspective, some types of renewable energy cannot be captured and converted when the demand requires energy. For example, solar and wind energy are intermittent on a daily and seasonal basis while biomass energy varies seasonally, requiring storage to function most effectively.

There are four major needs that storage addresses:

1. *Dispatchability*—responding to fluctuations in electricity demand
2. *Interruptibility*—reacting to intermittent energy supplies like wind and solar energy, the seasonal variations of hydropower and biomass, and the periodic instabilities that occur in fossil fuel supplies
3. *Efficiency*—recovering wasted energy
4. *Regulatory-driven needs*—meeting distribution and other transmission capacity expansion requirements

Electricity demand fluctuates on several cycles—daily and weekly due to variations in industrial and domestic loads, and seasonally due to weather and average temperature changes. As shown in Figure 16.1, these variations can be significant, placing strain on the electric supply system to vary its output on a regular basis.

The opportunity afforded by storage is illustrated in Figure 16.2, where the fraction of peak power is plotted as a function of the fraction of time or load duration curve. Jensen and Sorensen (1984) selected representative electric and heating loads over a one-year period. Note that the electric load never goes to zero, as some demand remains at night throughout the year, because of industrial users, while the heating load does go to zero during the mid-summer period. The cross-hatched areas on the electric curve illustrate how electric energy could be stored during periods of low demand and supplied from a suitable storage reservoir during peak demand periods. The nominal cycle time for shifting electric power to and from the storage system typically occurs during a 24-hour period. A similar scenario could be developed for using thermal storage to distribute the heating load but the cycle time now would be seasonal, or in a range from hundreds to thousands of hours. For these reasons, load management on energy grids using storage is important in both electric power and natural gas supply applications. Nonetheless, for smaller, more distributed energy systems, thermal storage is sometimes used, and it can be particularly effective when integrated with electric power storage in combined heat and power applications (see Section 16.5.2).

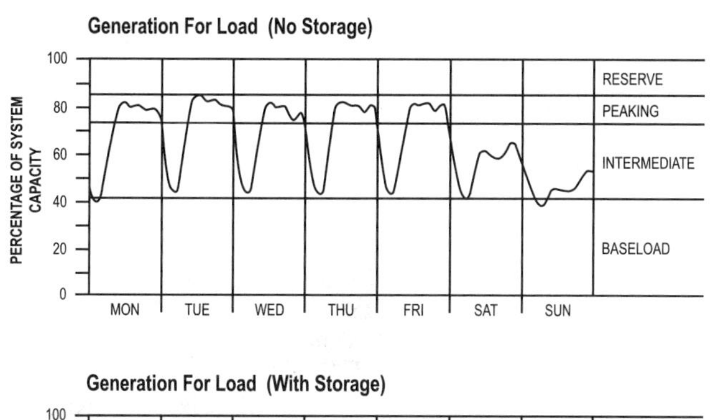

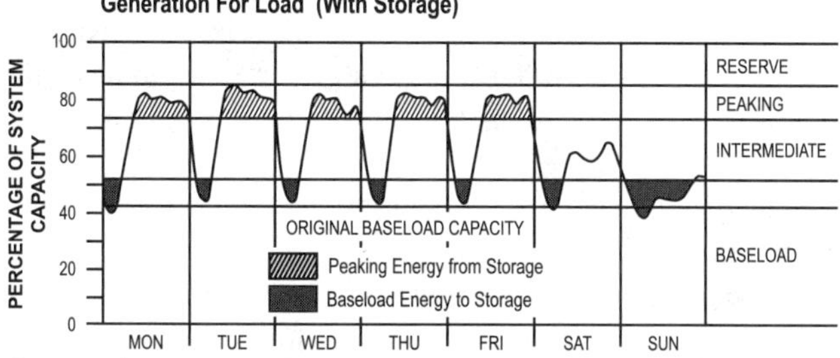

Figure 16.1. Typical weekly load curve of an electric utility. Source: National Academy of Sciences (1976).

Natural gas is used by providing both thermal energy and electricity. Demand variability depends on diurnal and seasonal variations—for example, a large heating demand exists in colder climates. In warm weather, natural gas can be fed to gas holding tanks, compressed or liquefied and stored for later peak sharing purposes, or perhaps, can be used to generate electricity to meet electric peaks when air-conditioning loads are high. Although natural gas and electric systems are managed separately, they are not really independent. When variable renewable energy sources and fuel cells and other distributed generators are in the mix, integrated energy planning is even more important.

For over a century, electric utilities have utilized conventional impounded and pumped storage hydropower systems (see Chapter 12) and rapidly dispatchable fossil-fired peaking power units to meet demand fluctuations and to achieve some measure of load leveling. Backup generators employing diesel fuel and natural gas, as well as electrochemical storage batteries, have also seen increasing use to ensure a non-interruptible supply. Recent deployment initiatives for renewable and distributed energy resources have reactivated interest in storage technologies. The value of both intermittent renewable energy sources and more distributed combined heat and power networks increases substantially if they can be dispatched when needed and if they have self-contained storage systems that make this possible.

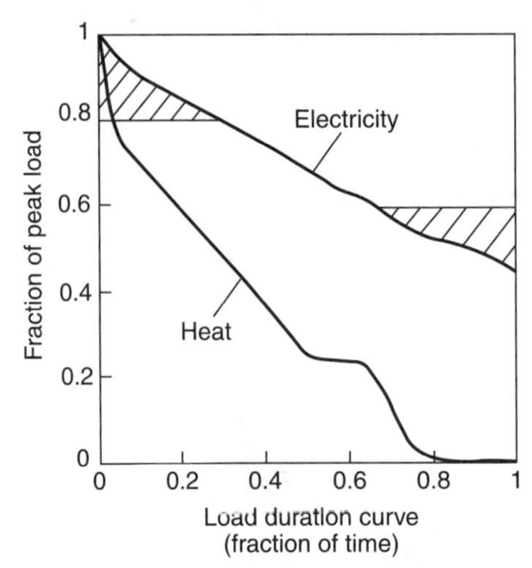

Figure 16.2. Opportunities for storage of electricity and heat based on a typical energy load or demand curve. The cross-hatched regions indicate the potential for storage to level out the demand. Adapted from Jensen and Sorensen (1984).

Recovery of wasted energy also provides an opportunity for energy storage technologies to increase efficiency (see Chapter 19 for more information). In transportation applications, electrochemical batteries are typically used to store electric energy for later use in engine starting and to meet high auxiliary loads during operation. Today, there is increasing emphasis on making cars and trucks more fuel efficient by conserving energy. Hybrid electric vehicles provide a means for doing this. They utilize an electric propulsion system that contains a generator powered by a small, fuel-fired engine. The smaller engine provides enough power to maintain normal cruising speeds on the highway. When more power is needed for acceleration, additional power is supplied to the electric motor from a robust battery storage system. As considerable energy is wasted by stop-and-go cycles in heavy traffic, regenerative braking systems are also being developed for hybrid electric vehicles to recover a fraction of dissipated energy as electric energy that can then be stored in the vehicle's battery system or as mechanical energy in a rotating flywheel storage system. All of these measures increase fuel efficiency.

The final motivation for developing and deploying energy storage systems comes from regulatory policies that promote more efficient, more secure energy supply systems. Recent arguments made in the US suggest that deregulation and restructuring of the electric power industry will lead to new opportunities for energy storage, particularly when it enhances power quality or reliability or eliminates a need to expand the transmission and distribution infrastructure (Schoenung et al., 1996, Gordon and Falcone, 1995, Yeager, 2002).

16.2 Connected Efficiencies and Energy Chains

An important aspect of supplying energy is that each step in the process carries some losses. For example, the transmission of electricity in the US national grid loses about 11% of the generated electricity before it ever reaches the consumer. If we look at the efficiency of the generation process, another 70%, on average, is lost. These effects are cumulative and lead us to think of the entire energy system as a chain of processes, each with a specified efficiency, η_i. Equation (16-1) captures this idea mathematically:

$$Overall\ efficiency = \eta_{overall} = \prod_{i=1}^{n} \eta_i \tag{16-1}$$

If we were evaluating a total energy system across its lifecycle, we would have to include all efficiencies along it path, from extraction to final utilization, to obtain a total efficiency. Consider the case of a gas-fired power station that supplies electricity to an electric motor located in a home in a large city on the east coast of the US located about 400 miles from the plant. The natural gas is supplied from a field in western Wyoming where it is produced, processed, and purged through a natural gas pipeline system to the power plant. Here, we might estimate the overall efficiency using Equation (16-1) as:

$$\eta_{overall} = \eta_{gas\ extraction}\eta_{gas\ processing}\ \eta_{gas\ transmission}\ \eta_{power\ plant}\ \eta_{electricity\ transmission}\ \eta_{distribution}\ \eta_{motor} \tag{16-2}$$

Another feature of specific component and overall efficiencies is that their magnitude depends on definitions that can vary considerably. Although common sense would suggest that the value of a particular η_i should never exceed 100% or 1.0, depending on specific definitions, this is not necessarily the case. Work or heat produced or power, the time derivative or rate of work or heat, are typical measures of output used. In some cases, the efficiency is defined as a ratio of power or work produced relative to the energy input to the system. In other cases, it is a ratio of power or work produced to the maximum amount of power or work that could be produced from the same input of energy. The first set commonly includes the so-called thermal cycle efficiency or fuel efficiency of a conversion process that produces electric or mechanical power by burning fossil fuel. In other words:

$$\eta_{thermal\ or\ fuel} \equiv \frac{\text{Work or Power Produced}}{\text{Energy Input}} = \frac{W_{net}}{m_{fuel}\,\Delta H_{combustion}}\quad or\quad \frac{W_{net}}{Q_{hot}} \tag{16-3}$$

Relevant performance metrics also include the efficiency of work exchange processes, such as pumped hydropower or compressed air energy storage where the cycle efficiency is given by:

$$\eta_{cycle} \equiv \frac{W_{recovered}}{W_{in}} \tag{16-4}$$

For these systems, η_{cycle} is given by the product of the input and output efficiencies of the power converter:

$$\eta_{cycle} = \eta_{in}\eta_{out} \tag{16-5}$$

In typical applications of pumped hydropower and compressed air storage, η_{cycle} can approach 0.6 or more as η_{in} and η_{out} generally are in the range of 0.8 to 0.9.

The second set of efficiencies are often called utilization or thermodynamic efficiencies, where:

$$\eta_{utilization} \equiv \frac{\text{Work or Power Produced}}{\text{Maximum Work or Power Produced}} = \frac{W_{net}}{W_{max}} \quad or \quad \frac{P_{net}}{P_{max}}$$

$$\eta_u = \frac{W_{net}}{\Delta \underline{B}} = \frac{W_{net}}{m\,(\Delta H - T_o \Delta S)} \tag{16-6}$$

where $\Delta \underline{B}$ is the availability or maximum work function. For most properly engineered applications involving heat-to-work conversions, η_u generally ranges from 0.5 to 0.7.

When only heat exchange is involved, such as in thermal energy storage and recovery, the efficiency of the overall process might be defined in terms of heat recovered to heat initially provided from the energy source or fuel. In this case:

$$\eta_{heat\ exchange} \equiv \frac{\text{Heat recovered}}{\text{Heat supplied}} = \frac{Q_{out}}{Q_{in}} \tag{16-7}$$

and efficiencies typically exceed 0.9 (or 90%).

For systems that involve the exchange and conversion of both heat and work, the situation is more complex. As we know from discussions in Chapter 3, heat and work are not equivalent, as the Second Law of Thermodynamics places an upper limit on the amount of heat that can be converted to useful work or power. We illustrate the effects of Second Law conversion efficiency in Figure 16.3 for the combined-cycle gas turbine (CCGT), which uses natural gas as the fuel source. The current state-of-the-art cycle efficiency for the CCGT is $\eta_{cycle} = 0.60$. In other words, for every 100 units of fuel energy fed to the combustion chamber, 60 units of electrical energy are produced and 40 units of thermal energy are rejected. In addition, Figure 16.3 shows the effect of definition on the magnitude of the overall efficiency for the entire system. In this case, the system is a fully integrated heat and power system connected to a thermal energy storage reservoir with a high-efficiency geothermal heat pump (COP = 4.0).

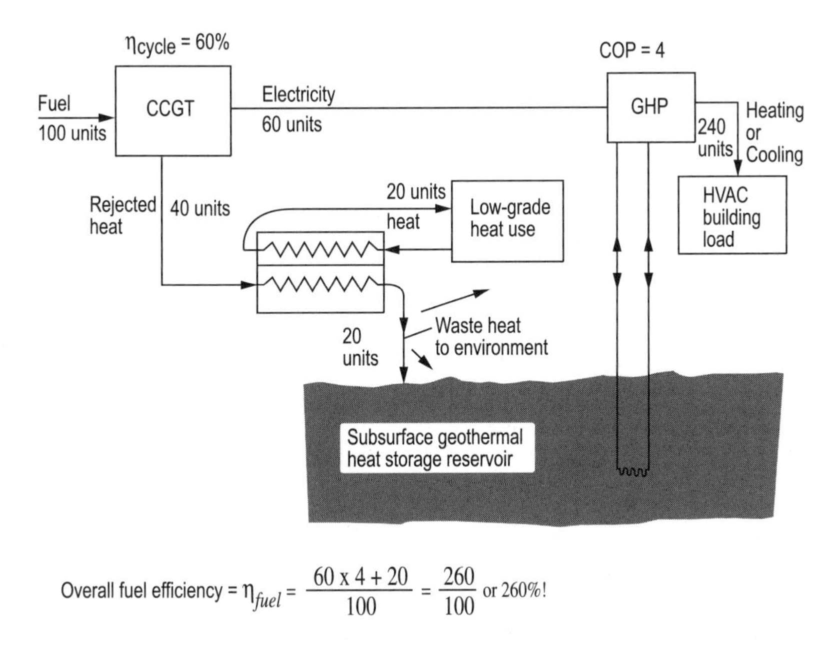

$$\text{Overall fuel efficiency} = \eta_{fuel} = \frac{60 \times 4 + 20}{100} = \frac{260}{100} \text{ or } 260\%!$$

Figure 16.3. Efficiency of a combined gas turbine-steam Rankine cycle (CCGT) coupled to a geothermal heat pump system with heat integration and waste heat recovery for supplying heating and cooling loads in a building application.

If we define the overall efficiency as the ratio of thermal energy utilized to the thermal energy content of the natural gas fuel, its value is an incredible 2.60 (or 260%). There are no violations of thermodynamic laws. The specific value of $\eta_{overall}$ often depends only on how it is defined. Clearly, one has to be careful in applying or interpreting efficiencies as measures of performance. Proper attention to definitions and conventions employed is essential to obtaining an unambiguous assessment.

16.3 Modes of Energy Storage

16.3.1 General characteristics

Energy can be stored in a variety of physical and chemical forms. In this chapter, we limit the discussion to energy stored in physical form as potential, kinetic, thermal, electrical, or magnetic energy. Chemical energy is usually associated with the molecular energy stored in chemical bonds. Energy is released for use as the chemical compounds that make up these fuels are transformed by combustion or oxidation and other means to more thermodynamically stable compounds. Important examples include the chemical energy released by the exothermic combustion of methane, petroleum liquids, coal, and wood to produce water and carbon dioxide.

The physical modes of energy storage may involve one or more mechanical, thermal, or electromagnetic forms. Table 16.1 lists common energy storage types and their characteristics. A particular energy storage technology may entail both a storage reservoir and a converter and transmission system for moving the power to and from the reservoir to its point of use. Compressed air and pumped hydropower systems, for example, contain both elements, whereas a molten-salt thermal storage system would typically involve only a reservoir for storing energy.

Energy storage modes are frequently determined by the particular end-use application. The magnitude of the energy or power load and the time scales involved are especially important (see Tables 16.1 and 16.2, and Figure 16.4). For example, to provide electric power for transportation or power outages requiring substantial power fluxes for only short periods of seconds to minutes, capacitors, batteries, and flywheels would be particularly well suited. When larger amounts of electrical or thermal energy are needed for periods of hours or longer, pumped hydropower or compressed air storage systems would be preferred for electric power, while hot water or molten salt storage reservoirs would be commonly employed to meet thermal loads.

Table 16.1. Conversion Energy Storage Modes

Mode	Primary Energy Type	Characteristic Energy Density kJ/kg	Primary Application Sector
Pumped Hydropower	Potential	1 (100 m head)	Electric
Compressed Air Energy Storage	Potential	15,000 in kJ/m^3	Electric
Flywheels	Kinetic	30–360	Transport
Thermal	Enthalpy (sensible + latent)	Water (100– 40°C)—250 Rock (250–50°C)—180 Salt (latent)—300	Buildings
Fossil Fuels	Reaction Enthalpy	Gas—47,000 Oil—42,000 Coal—32,000	Transport, Electric, Industrial, Buildings
Biomass	Reaction Enthalpy	Drywood—15,000	Transport, Electric, Industrial, Building
Batteries	Electrochemical	Lead acid—60–180 Nickel metal hydride—370 Li-ion—400– 600 Li-polymer ~ 1,400	Transport, Buildings
Superconducting Magnetic Energy Storage (SMES)	Electromagnetic	100–10,000	Electric
Supercapacitors	Electrostatic	18–36	Transport

Table 16.2. Energy Storage Technology Characteristics

Characteristic	Pumped Hydro	CAES[a]	Flywheels	Thermal	Batteries	Super-capacitors	SMES[b]
Energy Range	1.8×10^6 – 36×10^6 MJ	180,000– 18×10^6 MJ	1–18,000 MJ	1–100 MJ	1,800– 180,000 MJ	1–10 MJ	1,800 – 5.4×10^6
Power Range	100–1,000 MWe	100–100 MWe	1–10 MWe	0.1–10 MWe	0.1–10 MWe	0.1–10 MWe	10–1,000 MWe
Overall Cycle Efficiency[c]	64–80%	60–70%	~90%	~80–90%	~75%	~90%	~95%
Charge/ Discharge Time	Hours	Hours	Minutes	Hours	Hours	Seconds	Minutes to hours
Cycle Life	≥10,000	≥10,000	≤10,000	>10,000	≤2,000	>100,000	≥10,000
Footprint/ Unit Size	Large if above ground	Moderate if under ground	Small	Moderate	Small	Small	Large
Siting Ease	Difficult	Difficult to moderate	N/A	Easy	N/A	N/A	Unknown
Maturity	Mature	Early development	Early development	Mature	Lead acid mature, others under development	Available	Early R&D stage, under development

[a]CAES = Compressed Air Energy Storage

[b]SMES = Superconducting Magnetic Energy Storage

[c]For 1 full charge-discharge cycle

Sources: Jensen and Sorensen (1984), Schoenung et al. (1996), Boes et al. (2000).

The natural gas transmission and distribution system also requires storage as part of a system management program. Natural gas demand is variable on diurnal and seasonal time scales. Some gas can be stored in the pipeline system by raising the system pressure within design limits. Fifty years ago, many cities were dotted with large cylindrical "gas holders," which stored gas at slightly above-atmospheric pressure in a tank with a floating roof and a water seal. Later, higher pressure gas storage was developed, using inground caverns located near transmission lines. These formations are similar to those used for compressed air storage. Obviously, pressurization increases energy storage density.

In the future, if hydrogen is to be used as an energy carrier, safe and compact H_2 storage will become a critical issue. Hydrogen is a light gas (MW=2 versus 16 for methane) and, though it has a good energy density per unit weight of fuel, is difficult to store compactly. Storage tanks to handle high-pressure gas are heavy, and cryogenic systems to produce and hold liquid hydrogen (20 K [-253°C]) are cumbersome. At present, R&D on H_2 storage is receiving much attention with goals to design storage systems that will hold at least 5% hydrogen by weight. Even with these limitations on H_2 storage, the storage systems are still better than battery storage of electricity. This latter fact makes hydrogen interesting as a CO_2-free energy source that may be better than electricity for transportation applications (see Chapter 18).

Peak shaving in natural gas systems can also be provided by use of propane-air mixtures, blended to give a heating value that matches natural gas to be compatible with system appliance adjustments. Liquefaction of natural gas, by cooling it to around its boiling point of -260°F (-160°C), is another peak shaving possibility, and many gas utilities have built liquefaction plants that operate during low-demand periods to liquefy gas and store it in large insulated cryogenic storage tanks. The costs of such an operation are consistent with those of other peak shaving alternatives.

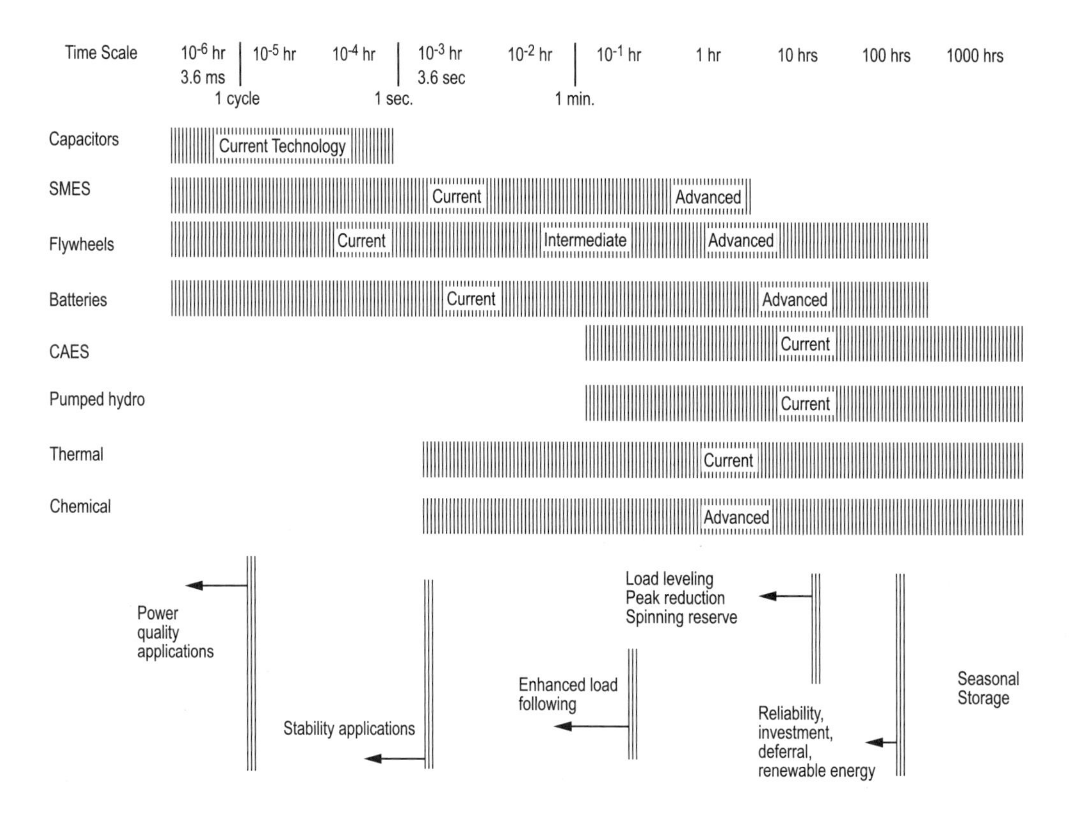

Figure 16.4. Characteristic times for energy storage. Adapted from Boes et al. (2000).

Some natural gas is located in remote areas, where it may be associated with remote oil production. Such stranded gas was either not produced, reinjected, vented, or flared. These latter two practices have significant environmental consequences and are rapidly being phased out. Where gas quantities are significant, liquefaction near a seaport allows transoceanic shipment in special cryogenic tankers. Japan generates a significant portion of its electricity using imported liquefied natural gas (LNG). In the US, several LNG importation projects were planned in the 1980s, but most were mothballed, not because of safety concerns, but because the availability of pipeline natural gas from Canada was more cost effective. (For more information on LNG, see Lom, 1974.)

A useful way of showing key performance characteristics for various energy storage technologies was originally developed by Ragone (1963), where energy density is plotted as a function of power density on a log-log scale. Figure 16.5 is a general Ragone plot with a range of storage technologies depicted to illustrate inherent differences between different storage modes. In the sections that follow, common storage technologies are discussed according to the inherent form of energy they utilize. Readers interested in exploring the full range of storage modes in detail should consult the monographs by Hassenzahl (1981) and Jensen and Sorensen (1984).

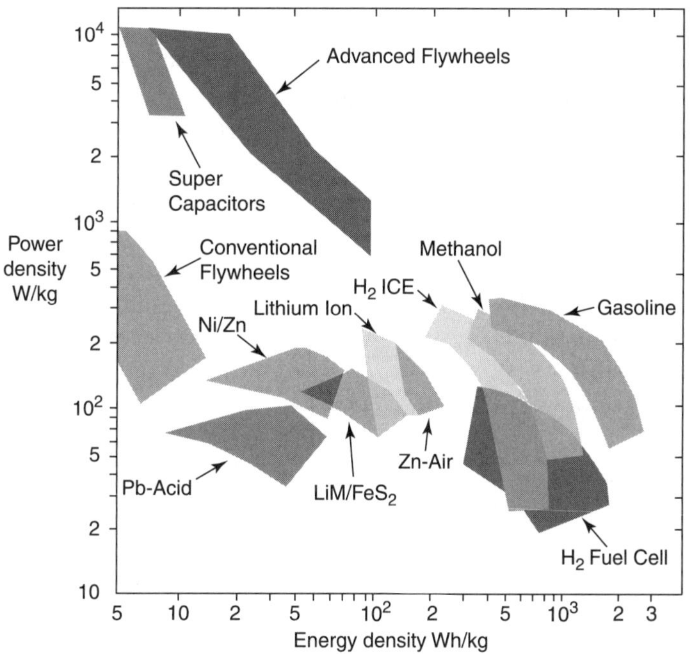

Figure 16.5. Ragone plot of power density (specific power/W per kg) versus energy density (specific energy/Wh/kg).

16.3.2 Energy storage technologies

Potential energy storage. Pumped hydropower and compressed air energy storage (CAES) are the two main storage systems that employ potential energy. In pumped hydropower systems (as illustrated in Figure 16.6a), a reservoir containing water at an elevated location represents stored potential energy that can be recovered by lowering it in the earth's gravitational field. In a typical cycle, when electric capacity exceeds demand during an off-peak period, the excess electrical power is used to pump water up to the storage reservoir. During high-demand periods, the water flows downward through a hydro turbine generator to produce additional power as needed. Pumped-hydro systems can employ separate units for pumping and power generation or may use a reversible, pump-turbine apparatus. Switching between the pumping and power generation phase can usually be accomplished rapidly. Pumped-hydro systems are in widespread use worldwide with over 300 systems operating with capacities ranging from about 20 MWe to more than 2,000 MWe (Schoenung et al., 1996).

The basic physics of power generation is the same as for normal hydropower (as described in Chapter 12). The energy density appropriate for pumped hydro corresponds to the energy content per unit mass of pumped fluid, which is proportional to the product of the gravitational acceleration (g), fluid density (ρ), and the total head (z) or elevation gain that occurs, or $g\rho z$. A well-designed pump or turbine expander or a combined pump-turbine apparatus can operate at efficiencies approaching 85% to 90% for each step. Thus, the overall efficiency of pumped hydro typically ranges from about $(0.85)(0.85) = 0.72$ or 72% to $(0.90)(0.90) = 0.81$ or 81%.

Commonly, pumped-hydro systems use naturally available sites where appropriate changes in topography (i.e., elevation) exist. The engineered system would be similar to a normal impoundment-type hydropower installation where a dam structure would be used to contain the pumped water. Land-use issues associated with pumped-hydropower systems are important and can result in significant public resistance to development. One way of avoiding the need for suitable topography and reducing the footprint of above ground pumped-hydro plants is to engineer an underground storage reservoir (as shown in Figure 16.6a).

CAES systems utilize the compressive energy associated with pressurized air contained in a closed underground reservoir consisting of natural or mined rock or salt cavities or in porous aquifers that are geologically contained. (Figure 16.6b schematically illustates these three types of CAES systems.) Finding a suitable site for CAES requires detailed geotechnical resource assessment and engineering and frequently, a phased development to achieve a full-scale system. Although there are currently only a few CAES installations operating worldwide, their potential is large, particularly if new lower-cost technologies appear for mining cavities in rock and salt. According to some experts, the most appropriate size range for CAES systems is between 10 and 1,000 MWe with storage capacity sufficient to operate 3–12 hours per day (Gordon and Falcone, 1995 and Thrasher and Lange, 1988). Nonetheless, there were only two CAES systems in operation in 2002, one at Huntorf in Germany, in a salt cavity rated at 290 MWe with an active storage volume of 300,000 m^3, and one in Alabama, also in a mined salt cavity, rated at 110 MWe for 6 hours of operation.

The large compressibility of air is utilized in CAES to produce power by expansion in a gas turbine at relatively high efficiency. Charging of the reservoir is accomplished by high-efficiency compression typically using the turbine expander operating in reverse. The energy transfer can be expressed as pressure-volume work that is represented by a simple equation:

$$\delta W = -PdV + d(PV) = +VdP \tag{16-8}$$

(a)

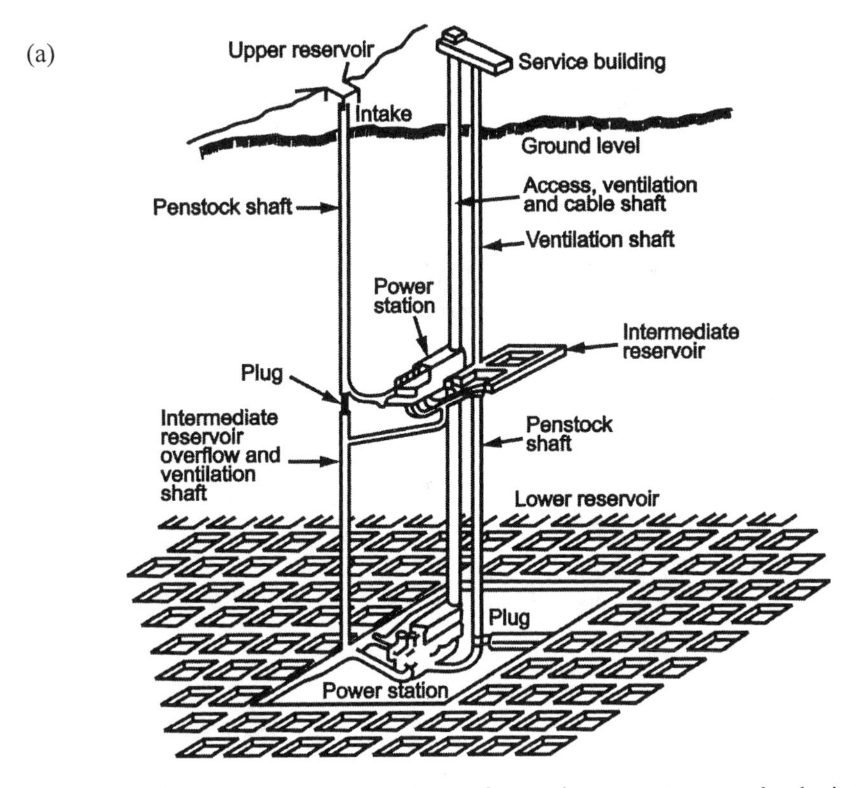

Figure 16.6a. Schematic illustrations of several energy storage technologies: *Aboveground and underground pumped hydro*. Source: Jensen and Sorensen (1984). Reprinted with permission of Elsevier.

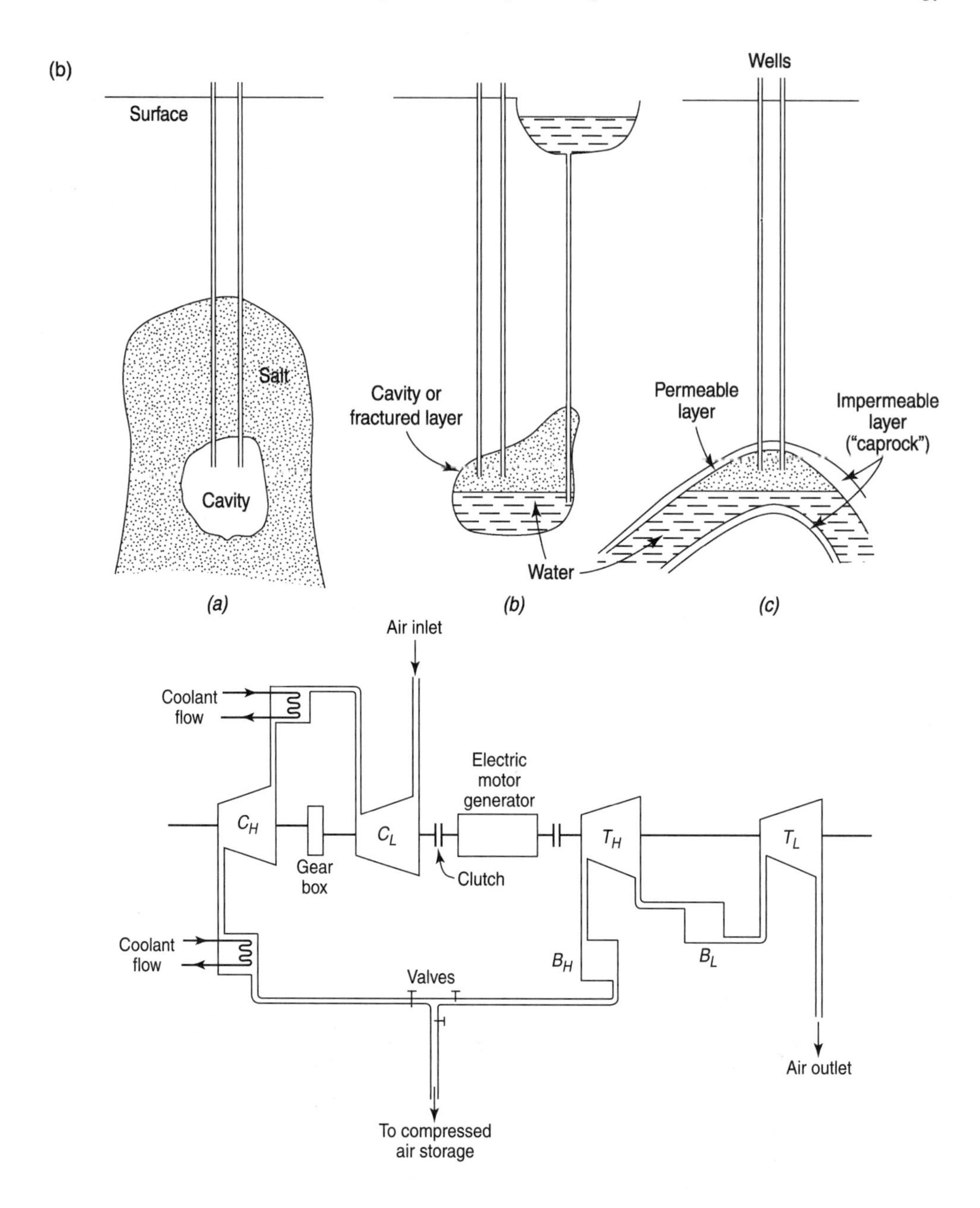

Figure 16.6b. Schematic illustrations of several energy storage technologies: *Compressed air energy storage*. Source: Jensen and Sorensen (1984). Reprinted with permission of Elsevier.

where the PdV term represents the total work done on or by a fluid element as it passes through the compressor or expander, and the $d(PV)$ accounts for the work associated with mass flow that is pushing the fluid element in and out of the compressor or expander. The sign convention used for work is as follows: $\delta W > 0$ means that work is done on the system or device. Therefore, $\delta W < 0$ if work is produced by the system or device, for example, by a turbine. Equation (16-8) can be easily integrated assuming that air obeys the ideal gas law—namely that $PV = nRT$. This is a good assumption for most pressure and temperature operating conditions currently envisioned for CAES. Assuming ideal gas behavior and using the mechanical efficiency of the turbine-compressor, the overall work or power levels for compression and expansion can be estimated:

$$net\ work = W_{net} = W_{turbine} - W_{compressor} \tag{16-9}$$

by integration of Equation (16-8) over appropriate pressure limits with defined reversible, quasi-static paths for both expansion and compression with $\eta_{turbine}$ and $\eta_{compressor}$ specified. The net work associated with one cycle of operation is given by:

$$W_{net} = \eta_{turbine} \int VdP \,|_{turbine} - \eta^{-1}_{compressor} \int VdP \,|_{compressor} \tag{16-10}$$

For the ideal gas case, closed-form integrals result from Equation (16-10) for isothermal, adiabatic, or polytropic paths. The net power for a complete cycle is easily obtained from Equation (16-10) by using the average mass flow rate of fluid that passes through the device. When the compression and expansion paths are roughly the same, the overall efficiency of the cycle can be defined in terms of the ratio of work released as gas is expanded from the pressurized CAES reservoir to the work needed to compress the gas as provided by excess power used to drive the compressor:

$$\eta_{overall} \equiv \frac{Work\ output}{Work\ input} = \frac{W_{turbine}}{W_{compressor}} = \eta_{turbine}\eta_{compressor} \tag{16-11}$$

Kinetic energy storage. The most common form of kinetic energy storage is achieved using mechanical flywheels with rotational energy being transferred to and from the device. Flywheels have been used for decades for a range of applications involving transportation propulsion, electric power generation, and other energy-intensive industrial processes. The most common functions for flywheels in the past have been for electric load leveling and pulse dampening, as well as providing short bursts of additional power. Specific applications include regenerative braking in large locomotives and some hybrid electric automobiles.

Because rotational kinetic energy can easily be stored and transformed to other mechanical or electrical energy forms, flywheels are well suited for energy storage. The basic physics is captured in a simple expression that relates the kinetic energy to the product of the moment of inertia and rotational speed of the device:

$$E_{rotational} = \frac{1}{2} I\omega^2 \tag{16-12}$$

where ω is the rotational speed or angular velocity in radians per unit time and I is defined as the moment of inertia in units of mass $\times$ length^2. For an axi-symmetric body like a solid or hollow cylinder, I is given by:

$$I = \int_0^R \rho(r) r^2 dr \tag{16-13}$$

where $\rho(r)$ is the mass density of the rotating body as a function of position from its rotational axis in units of kg/m. Thus, for a solid cylinder of uniform thickness t, $\rho(r)$ is given by:

$$\rho(r) = \rho_m 2\pi rt = \frac{m2\pi rt}{\pi R^2 t} = \frac{m2\pi r}{\pi R^2} \tag{16-14}$$

where R is the radius of the cylinder in meters, m is the total mass of the rotating cylinder in kg and ρ_m is the volumetric mass density in kg/m^3 or similar units. For the cylinder then,

$$I = \frac{1}{2} mR^2 \tag{16-15}$$

Thus, using Equation (16-12):

$$E_{rotational} = \frac{1}{4} mR^2 \omega^2 \tag{16-16}$$

Equation (16-16) illustrates the relative importance of mass, radius, and rotational speed to the amount of rotational energy that can be stored in a solid cylinder or disc. At first glance, it would appear that one would maximize the amount of energy stored in a flywheel by increasing its mass, particularly the amount that is contained at large radii, and by increasing its rotational speed. Ultimately, the strength of the material that the flywheel is constructed of will set the upper practical limit. Furthermore, other important issues are the frictional losses associated with the rotation of the flywheel itself and the need to add and extract power at high efficiency over a range of rotational speeds.

Although estimating the strength limit for a given flywheel requires careful engineering analysis, a simple expression has been developed to describe the effect quantitatively (Jensen and Sorensen, 1984). The energy density for the flywheel is given by:

$$\textit{Specific energy density} = \frac{E_{rotational,max}}{m} = \frac{k_m \sigma_{max}}{\rho_m} \tag{16-17}$$

where k_m is a shape factor that depends on the flywheel's geometry and σ_{max} is the maximum allowable tensile stress for the material. Here, high strength, large σ_{max}, and low density, small ρ_m, increase specific energy density. The shape factor k_m varies from about 1.0 for a well-designed constant stress disc to about 0.3 for a solid disc with a center hole. (Figure 16.6c shows a few representative designs and their shape factors.)

(c)

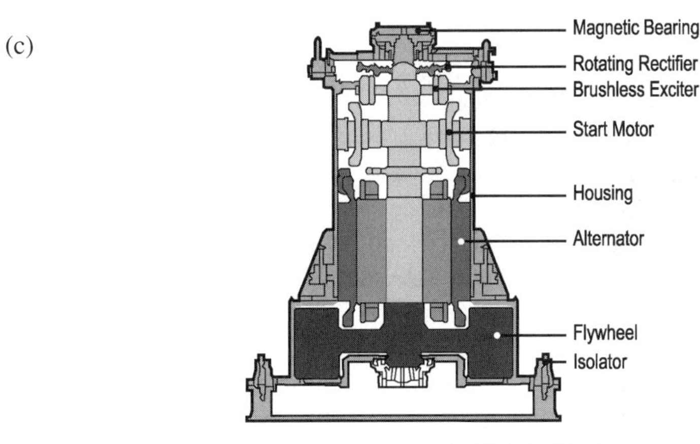

FLYWHEEL CONFIGURATIONS

DESCRIPTION	GEOMETRY	SHAPE FACTOR K
Flat unpierced disc		0.606
Flat pierced disk		0.303
Constand stress disc (typical)		0.931
Rim with web		0.400
Truncated conical disc		0.806
Thin rim		0.500
Bar		0.333
Shaped bar		0.500

Figure 16.6c. Schematic illustrations of several energy storage technologies: *Flywheels*. Adapted from Simpson, et al. (1975).

Steel and other metals are commonly used for flywheels because of their machinability, low cost, and high tensile strength. However, the generally high density of metals has limited practical devices to rotational speeds of about 3,600 rpm. Thus, in order to have a large power or energy capacity, massive steel flywheels are needed (see Equation (16-16)). The energy and power densities for first generation metallic flywheels range from about 5 to 10 Wh/kg for specific energy and from about 100 to 900 W/kg for specific power as shown in Figure 16.5.

Advanced flywheel concepts improve the energy density, power density, and overall efficiency, by using lower density, high-strength materials, such as fiber-reinforced, epoxy polymer composites, and advanced bearings that may employ magnetic levitation to reduce friction. Projections indicate that specific energy densities may range from about 5 to 100 Wh/kg and that specific power densities will exceed 1,000 and may be as high as 10,000 W/kg for advanced composite designs. Units with capacities as large as 1,000 kWh are becoming available for utility load management, while smaller units in the 1 to 5 kWh range are now used in hybrid electric vehicles.

Thermal energy storage. About 40% of our primary energy today is actually consumed as thermal, rather than electrical or mechanical, energy at temperatures of 250°C or less. Therefore, it makes sense to consider storing thermal energy. This is particularly true given daily and seasonal variations in demand and the intermittent availability of many renewable energy types, such as solar. The price fluctuations that exist for purchased bulk electricity during high- and low-demand periods also create an incentive for thermal energy storage.

There are two primary mediums used for thermal storage. One involves sensible heat stored in the heat capacity of single-phase materials by changing their temperature, while the other involves the latent heat associated with changing the phase of a material at a fixed temperature and pressure. Latent heat storage systems have some advantages in that thermal energy can be supplied or removed at essentially constant temperature, which is desirable for many applications.

In order to make thermal storage economically viable, low-cost stable materials need to be used. For sensible heat storage, water and solids like crushed rock, adobe, or concrete are commonly used. For latent heat storage, eutectic salt mixtures with phase transition temperature selected in a desirable range are commonly used for heating, whereas ice is often used for cooling. Table 16.1 shows that the range of energy densities possible for thermal storage systems is from 180 to 300 kJ/kg, which is relatively low in comparison to fossil fuels or biomass. Furthermore, the thermodynamic quality per kJ of thermal energy at temperatures less than 250°C is a factor of four or more, less than storing a comparable amount of electrical energy. Nonetheless, there are many end-use applications in which thermal storage is the most appropriate and economic option. Solar water heating or the making of ice for building air conditioning systems by refrigeration using off-peak electricity are good examples.

The inherent storage of thermal energy in a geothermal ground source or subsurface aquifer represents an attractive resource that can substantially increase the efficiency of an overall energy supply system (consider the combined heat and power cogeneration example discussed in Figure 16.3). In principle, applications for subterranean thermal energy storage could be greatly expanded to include seasonal energy storage as well. For example, solar energy as heat (or cold) would be captured during the summer (or winter) months and used months later when needed. Obviously, this would require an expanded infrastructure for capturing, storing, and distributing thermal energy. Taking such an aggressive energy conservation path has more to do with policy than the need for new technology to become available. By using water or ice as the storage medium, other benefits might arise with a distribution infrastructure in place, such as augmenting municipal water supplies during times of need.

Chemical energy storage. A considerable amount of energy is contained in the chemical bonds that hold atoms in place in molecules. Breaking these bonds selectively, such as during the oxidation of fossil or biomass fuels as they are combusted, can release a large amount of energy at high temperatures (see Table 16.1). On a time scale, such chemical energy is stored either for millennia, in the case of coal, oil, and other fossil fuels; for years, for wood and forest products; or for one season, for agricultural crop residuals, such as wheat chaff and rice straw.

In another proposed application of chemical energy storage, hydrogen would be produced from renewable or nuclear energy sources, stored, and later utilized for transportation needs by combining it with oxygen in a fuel cell, or for generating distributed electric power in heat-to-work cycles employing microturbines. A large set of issues for deploying hydrogen on a large scale, particularly for use in the transportation and building sectors, revolves around developing the needed infrastructure to produce, distribute, and store hydrogen in a safe, economic manner. For example, if hydrogen is to become a commodity transportation fuel like gasoline or diesel fuel, then both energy converters (fuel cells or internal combustion engines) and on-board storage technologies will have to be available at reasonable costs. In Chapter 18, some specific issues are discussed in more detail, but here it is appropriate to mention the storage of hydrogen as a transportation fuel.

One of the advantages of having gasoline and diesel as primary transportation fuels is their high energy density and storability onboard as liquids at ambient pressures and temperatures. The infrastructure for producing and distributing these fuels is highly developed. While hydrogen has a reasonable energy density on a mass basis of 120,000 kJ/kg compared to about 45,000 kJ/kg for gasoline or diesel, its low density as a gas at ambient temperature and pressure results in a volumetric energy density of only 10 MJ/m^3 compared to 35,000 MJ/m^3 for gasoline or diesel. The energy content of a full 20 gallon gasoline tank on an automobile is about 2.8 GJ. If we were to fill that same tank with hydrogen gas at 1 atmosphere, the energy content would only be 0.0008 GJ. One way around this problem is to pressurize the hydrogen and store it as a compressed gas, which introduces both infrastructure and safety challenges. Liquefaction and cryogenic storage is another option, which also has major infrastructure and safety issues. Yet another way would be to develop a suitable storage medium for hydrogen that has an acceptable energy density. Metal and chemical hydrides and carbon nanotube structures may provide such a medium, but more research and development is required before they will be competitive with conventional liquid fuel storage.

In other settings, chemical energy could be transferred in a reversible chemical reaction to augment the transfer and conversion of heat to work at high efficiency. Rapidly equilibrating chemical reactions have been proposed to manage energy in the carbon-hydrogen-oxygen system and for hydrogen storage in metal hydride systems. (For more details, see Jensen and Sorensen, 1984.)

Electrical energy storage. There are three major mechanisms for storing electrical energy: *electrochemical, electrostatic*, and *electromagnetic* energy. In this section, we focus on one technology for each storage mode— (1) batteries for electrochemical energy storage, (2) Supercapacitors for electrostatic energy storage, and (3) Superconducting magnetic energy storage (SMES) systems for electromagnetic energy.

(1) Batteries Electrical energy is stored and released in an electrochemical reaction cell that transports electrons to electrodes to carry out specific reduction/oxidation (redox) reactions. These reactions involve ionic conductors, which could be contained in liquid solutions, solid conductive polymers or gels, or ceramic host media. Frequently, catalysts are needed to accelerate reaction rates to acceptable levels. The important thermodynamic relationship that governs the amount of energy transfer that occurs per mole of reactant is the Gibbs-Faraday form of the Nernst equation in which the free energy change of the reaction is expressed in terms of a standard reference potential and activities of the reactants and products of the reaction. In mathematical terms, this is given by:

$$-\Delta G_{rx} = n_e F\,\varepsilon = n_e F \varepsilon^o - RT \ln\left[\prod_{i\ species} (a_i)^{\nu_i}\right] \qquad (16\text{-}18)$$

where ΔG_{rx} is the Gibbs free energy change for the electrochemical reaction, F is the Faraday constant = 96,500 coulombs per mole, ε is the cell voltage, ε^o the standard cell voltage with all reactants and products in their reference states, usually taken to be unit activity, a_i is the activity of species i, ν_i is the stoichiometric coefficient of species i in the electrochemical reaction, n_e is the number of electrons transferred in the reaction, R is the universal gas constant, and T is the absolute temperature. Although Equation (16-18) looks complicated, the key element to remember is that the ideal maximum work that can be produced from an electrochemical battery operating isothermally is equal to ΔG_{rx}. Most standard cell voltages are in the range of 0.5–2.5 volts DC, which require stacking and conversion to AC for most applications.

Electrical energy is stored and discharged in batteries by electrons that react at two different electrodes, a cathode and an anode, which are electrically connected by an ionic conductor or electrolyte. The energy produced is in DC form. For utility applications, it is normally converted to AC form using a suitable power inverter. During charging, energy is stored chemically by increasing the composition of charged ions contained in the electrolyte or conductor by selective redox reactions at the electrodes that consume or produce electrons. During discharge, energy is released by ion transport, causing redox reactions to occur in the reverse direction at the electrodes. Thus, the anode (oxidizing electrode) and cathode (reducing electrode) change position between charging and discharging.

The efficiency of energy transfer in batteries is a function of several linked effects. Frequently, it is expressed in the format of Equation (16-2) as a continuous product of specific efficiencies:

$$\eta_{battery} = \eta_{rev,max}\eta_{rx}\eta_{voltage\,losses} \tag{16-19}$$

where the maximum or reversible efficiency, $\eta_{rev,\,max}$, represents cell operation at zero current and yields the maximum theoretical energy for the device. The other two efficiencies, η_{rx} and $\eta_{voltage\ losses}$, capture non-idealities associated with operating the cell at practical current densities. Practical battery operation at finite currents places limits on energy and power densities that are considerably lower than their theoretical maximums—sometimes by a factor of 5 to 10 or more. Typical ranges are shown in the Ragone plot (Figure 16.5) for a number of battery types.

Mass and volumetric energy density for batteries become critically important for transportation applications. For utility or buildings applications, the size or weight of the battery storage system is usually not a limiting factor. Nonetheless, there are several common performance factors that apply for all applications. These include (1) lifetime (maximum number of charge and discharge cycles), (2) overall cycle efficiency, (3) depth of discharge per cycle, and (4) cost per unit of power or energy stored.

Although electrochemical batteries were discovered about 200 years ago, until recently, only one, the lead-acid battery, had been extensively commercialized. Since their initial release in 1859, lead-acid batteries for both stationary and mobile applications have been incrementally improved. Overall cycle efficiencies now exceed 80%, with cycle lifetimes up to 2,000, and relatively low costs (< \$200/kWh). The largest stationary lead-acid system for a utility application was installed by Southern California Edison in 1991 (Schoenung et al., 1996). It uses over 8,000 battery modules to deliver up to 10 MWe power for four hours of continuous discharge. Other units are planned or in operation as well.

Despite 150 years of slow, steady progress, nothing can be done to overcome the inherent limitations of lead-acid batteries in terms of energy and power density (see Figure 16.5). An average family-sized electric vehicle using the best lead-acid batteries available in 2002 would need about one ton of batteries to provide reasonable levels of performance in terms of speed and range.

In order to achieve the higher performance levels needed for totally electric or hybrid electric vehicles or other transportation applications, different electrochemical systems must be selected that preserve the efficiency and cycle lifetime advantages of lead-acid systems but have lower weight and volume requirements. During the past 25 years, new battery technologies have been aggressively pursued to find a replacement for lead-acid systems. Much progress has been made with the discovery and development of sodium sulfur, zinc air, nickel zinc, nickel metal hydride, and lithium ion systems. Figure 16.5 shows the position of several advanced battery systems that all perform at higher energy and power densities than lead-acid systems. Nonetheless, radical improvements that would revolutionize battery technology have not yet been achieved. A paper by Sadoway and Mayes (2002) suggests that the linkage of lithium ion chemistry with the modern tools for developing molecularly engineered materials may provide a platform to create a new generation of advanced Li ion solid polymer electrolyte batteries. (Figure 16.6d illustrates the main elements of this newly engineered battery.) Early testing of concepts at MIT suggests that they may meet specific energy and power densities needed for vehicles, namely 400 Wh/kg and 1,440 kJ/kg, respectively.

(2) Supercapacitors Capacitors store electrical energy in the form of confined electrostatic charges in a device consisting of two conductive plates separated by a dielectric medium. Recovery of the stored energy is achieved by connecting the conducting plates to a suitable load.

The governing equations for specific energy and power densities for capacitors can be represented as follows:

$$\textit{Energy density} = \frac{E}{Ad} = \frac{1}{2}\frac{A_{eff}}{(Ad)A}\varepsilon_m V^2 \tag{16-20}$$

$$\textit{Power density} = \frac{E/\tau}{Ad} = \frac{\frac{1}{2}\varepsilon_m V^2}{(Ad)RC}\frac{A_{eff}}{A} = \frac{\frac{1}{2}\varepsilon_m V^2 A_{eff}}{RCdA^2} \tag{16-21}$$

where τ is the characteristic time, = to RC, for the capacitor, A is the nominal surface area of the conducting plate or electrode, A_{eff} is the effective surface area of the capacitor, d is the spacing between plates, V is the applied voltage, R is the total effective resistance of the capacitor, and ε_m is the permittivity of the dielectric medium.

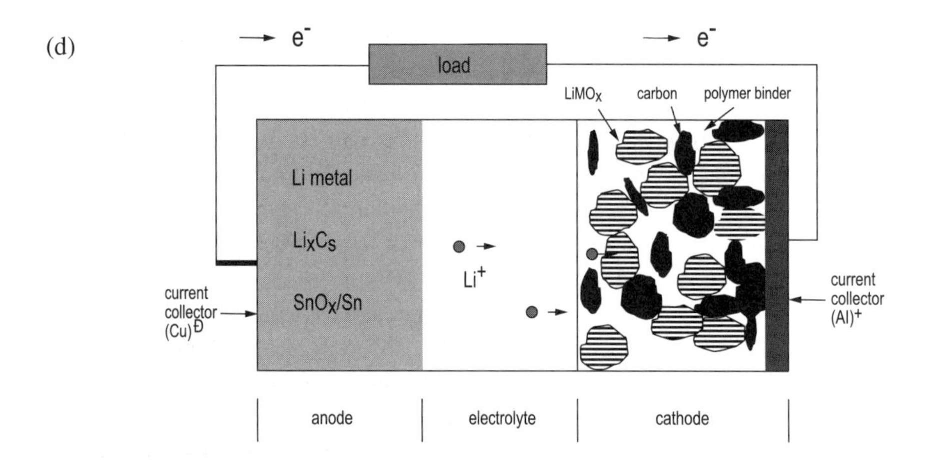

Figure 16.6(d). Schematic illustrations of several energy storage technologies: *Lithium ion solid polymer electrolye battery*. Source: Sadoway and Mayes (2002).

Since the overall capacitance C is given by:

$$C = \frac{\varepsilon_m A_{eff}}{d} \tag{16-22}$$

Equation (16-21) becomes:

$$Power\ density = \frac{E/\tau}{Ad} = \frac{\frac{1}{2}V^2}{RA^2} \tag{16-23}$$

From these operating equations, we can see that low-resistance systems will give high power densities and that the ratio of effective to nominal surface area A_{eff}/A could be important in influencing the magnitude of the energy density.

Capacitors have the ability to be charged and discharged quickly, on the order of seconds or less, which makes them useful for responding to power interrupts of short duration. Given that the power density of capacitors is inherently large, they have been used for mitigating power interrupts for many years in stationary utility applications over a range of scales. However, the specific energy densities of early generation capacitors were low, <<1 Wh/kg, making them unattractive for energy storage applications. The appearance of "supercapacitors" that utilize advanced materials that greatly increase the effective surface area of the capacitor's electrodes per unit mass has changed performance metrics in a manner that results in enormous gains in energy storage capacity. Typical ranges for energy and power densities are shown in Figure 16.5. Here we see that 3,000–10,000 W/kg at up to 10 Wh/kg are achievable. This makes supercapacitors attractive for regenerative braking and other power needs in electric and hybrid electric vehicles.

(3) *Superconducting magnetic energy storage (SMES)* In SMES systems, electromagnetic energy is stored and retrieved directly and with negligible losses using direct current (DC) flowing through superconducting coils to generate a magnetic field. The volumetric energy density is given by:

$$Volumetric\ energy\ density = \frac{1}{2}\mu_m H^2 \Omega \tag{16-24}$$

where H is the magnetic field intensity, μ_m is the magnetic permeability, and Ω is the characteristic volume of the SMES device. Resistive losses in the SMES unit are low, as are other losses associated with required AC to DC conversions and general power conditioning. To achieve superconductivity conditions will require some level of cryogenic refrigeration to maintain low temperatures. While initial designs required liquid helium temperature near 0 K, materials discovered in the 1980s have enabled superconductivity at liquid nitrogen temperatures of about 77 K, which reduces heat losses considerably and provides for more practical and economic SMES designs.

To sum up, SMES has many attractive features for utility-scale electrical power applications. In principle, large amounts of energy (up to 1,500 MWh) can be stored for long periods and utilized when needed at extremely high rates (10–1000 MWe) with overall cycle efficiencies of 95% or more. One of the disadvantages of such large units is their high total capital costs. To offset this, new technology is under development that could lead to micro-SMES units that would be less capital intensive and useful in distributed power applications from 1 to 30 MWe.

By 1985, SMES technology had been demonstrated at a small scale (30 MJ) in a project managed by the Bonneville Power Authority (BPA). A 30 MWe SMES demonstration unit has been designed by Babcock and Wilcox (Schoenung et al., 1996) for use in Alaska. Micro-SMES units at a capacity level of 1–2 MWe were becoming available in 2001. Nonetheless, even with these demonstrations, one should not assume that SMES is now ready for widespread deployment. Long-term performance testing and a number of economic issues need to be resolved.

In general, there is substantial latitude in choosing a specific energy storage technology for a particular application. For example, capacity and capital cost ranges for electrical storage applications are given in Table 16.3.

16.4 Energy Transmission

16.4.1 General characteristics of energy transmission systems

Energy can be safely and economically transported for long distances in a variety of modes. Pipelines are commonly used for oil and gas and have potential for coal transport in the form of coal-water slurries. Trucks, trains, barges, and tanker ships are heavily employed for distributing coal and oil. Biomass fuels can also be transported between their source and a power generator or chemical processing plant using these methods. Specialized refrigerated tankers haul LNG across the world's oceans. Electricity is primarily transported by wires, which are frequently interconnected in a complex distribution system or grid. Electric grids regulate and control currents and voltages at various levels from the generation site to the final use point. Even geothermal energy is transported from production wells by pipelines to electric power generation plants or to district heating distribution centers. Uranium ore is shipped to separation and processing plants by trucks or trains to be enriched into fissile power reactor fuels.

The efficiencies of most of the common energy transport modes are high—usually consuming only a small percentage of equivalent energy content of the delivered fuel. This is especially so for high energy density fuels like oil, gas, and coal. The equivalent energy efficiency for transporting lower energy density fuels, such as geothermal steam or hot water or residual biomass, is lower. The infrastructures for fossil energy transportation and electricity transmission systems are also highly developed, particularly in North America and Europe.

Table 16.3. Estimated Capital Costs for Representative Energy Storage Systems for Supplying Electric Power

System	Typical Size Range MWe	$/kWe	$/kWh
Pumped hydropower	100–1000	600–1000	10–15
Batteries			
Lead acid	0.5–100	100–200	
Nickel metal hydride	0.5–50	200– 400	150–300
Li-ion	0.5–50	200– 400	
Mechanical flywheels	1–10	200–500	100–800
Compressed-air energy storage (CAES)	50–1,000	500–1,000	10–15
Superconducting magnetic energy storage (SMES)	10–1,000	300–1,000	300–3,000
Supercapacitors	1–10	300	3,600

Sources: Turkenberg et al. (2000), Schoenung et al. (1996), and Boes et al. (2000).

Although there are many factors that control costs, a useful way to compare alternative energy transmission modes is to plot cost per unit of energy transported per unit distance as a function of distance. Hottel and Howard (1973) presented such a correlation in their energy monograph (updated in Figure 16.7). Readers should keep in mind that these estimates are rough—they indicate trends and are not site-specific. Local terrain, labor rates, and regulatory constraints vary widely from place to place. For example, compare the complexity of designing and constructing the Trans Alaska Oil Pipeline System in the 1970s with constructing a new gas pipeline in 2002 in eastern Wyoming. Careful attention to details must be given to develop accurate cost estimates.

16.4.2 Oil transport

Liquid petroleum is usually transported in one of two forms: crude oil or refined product. There are three primary modes of oil transport: pipelines, barge or tanker ship, and truck. Oil pipelines, ranging in diameter from about 10 cm to over 1 m, are common in developed countries. For example, in the US there are over 160,000 km (100,000 mi) of crude oil pipelines (Glasstone, 1983). Smaller, secondary pipelines are frequently used to transport crude oil from producing fields to major trunk lines, which bring the crude to centrally located refineries for processing into final products or to shipping ports where they are loaded onto suitably sized tankers.

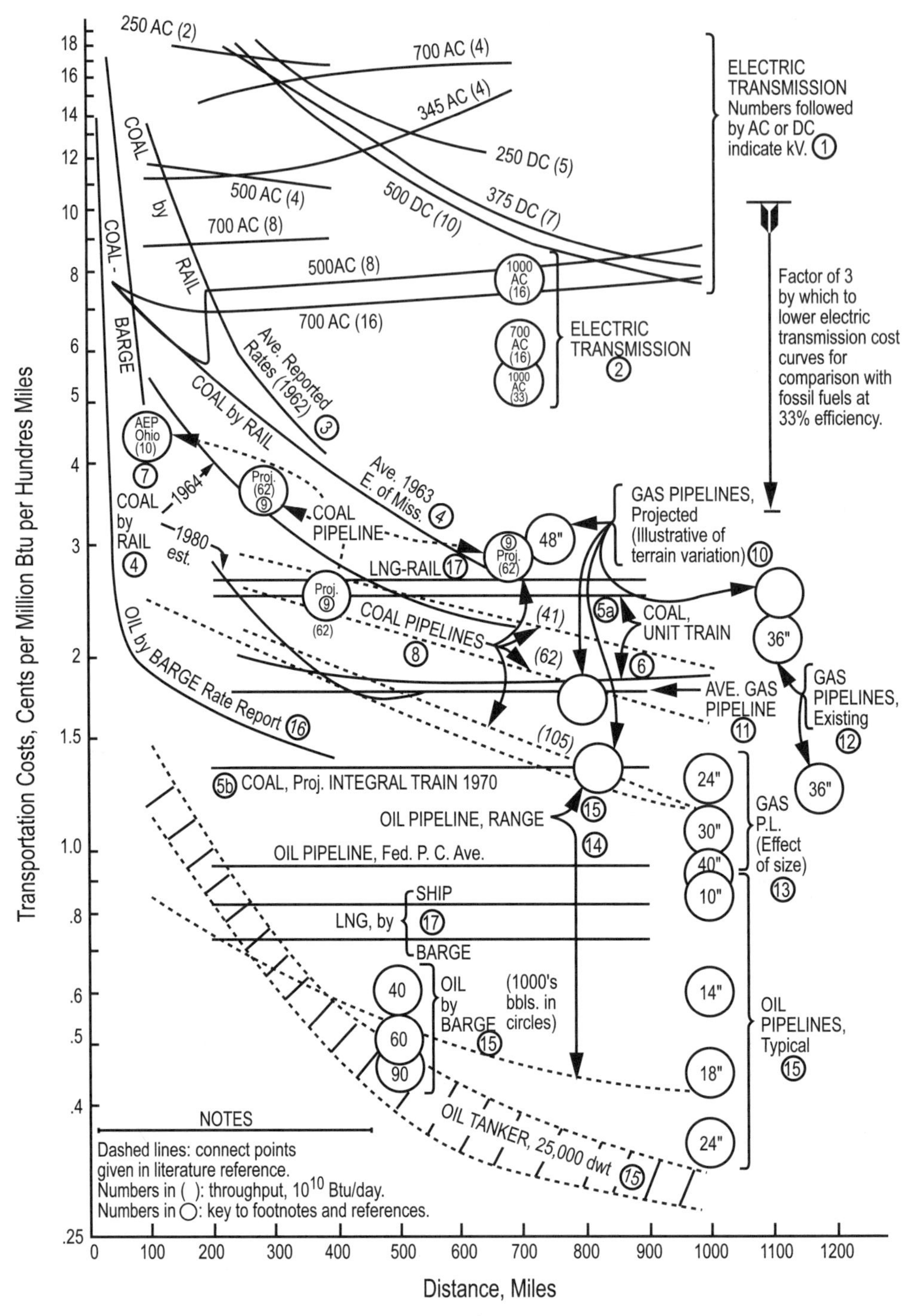

Figure 16.7. Estimated costs in 2000 US$ for energy transmission. Adapted and updated from Hottel and Howard (1973).

Footnotes and References for Figure 16.7

NB: Citations below refer to smaller circled numbers in Figure 16.7.

Electricity Transmission

1. Point-to-point transmission costs, 85% load factor (Federal Power Commission, 1964, pp. 193–203).
2. Point-to-point transmission costs for higher voltages, 90% load factor (Dillard, 1965).

Coal Transportation

3. Average reported rates by rail for 1962 (Committee on Interior and Insular Affairs, 1962, p. 46).
4. Average 1963 rail rates, 1964 volume rates, and 1964 estimate of 1980 volume rates (Federal Power Commission, 1964, p. 60).

5a. Average 1970 unit train (Federal Power Commission, 1970, p. III-3-77).

5b. Projected 1970 integral train (Federal Power Commission, 1970, p. III-3-77).

6. 1964 estimates for unit train (Energy R&D and National Progress, 1964, p. 129).
7. Actual costs, AEP Ohio pipeline (Energy R&D and National Progress, 1964, p. 131).
8. Projected costs as a function of distance for coal pipelines of various sizes (Energy R&D and National Progress, 1964, p. 131).
9. Projected costs for pipelines from (left to right) southern Illinois to Chicago, West Virginia to New York City, and Utah to Los Angeles, showing effect of distance and terrain on costs (Department of the Interior, 1962, Tables II-4 and II-7).

Gas Transmission

10. Projected pipeline costs, showing effect of terrain. From top: 48" pipeline from Prudhoe Bay to Valdez; Ft. Nelson, Canada, to Portland, Oregon; Portland to Los Angeles; (at right) Prudhoe Bay to Ft. Nelson. All except first case assume 100% load factor (Foster, 1970).
11. Average cost by pipeline (Federal Power Commission, 1970, p. III-3-77).
12. Existing pipelines in Canada and Canada to US West Coast, showing range of current costs (Foster, 1970).
13. 1954 estimates of costs for various pipeline diameters; absolute values are obsolete because of inflation, but relative values are of interest (*Gas Engineers Handbook*, 1965, p. 8–98).

Oil Transportation

14. Average oil pipeline cost (Federal Power Commission, 1970, p. III-3-77).
15. Ranges of cost for oil transportation in various quantities by various methods (*Oil and Gas Journal*, 1966).
16. Rates by barge in 1962 (Committee on Interior and Insular Affairs, 1962, p. 146).

LNG Transportation

17. Average costs by barge (Federal Power Commission, 1970, p. III-3-77).

There can be large distances between where oil is produced and where it is refined or used. For example, consider the 900-mile, 48-inch Trans Alaska Pipeline System (TAPS) that pumps oil from above the Arctic Circle at Prudhoe Bay over Alaska's Brooks mountain range to the all-season, deep sea port in Valdez where it is shipped to the lower 48 states. When it was completed in 1977, TAPS represented one of the world's largest and most complex civil construction projects, with total cost of $6+ billion. Given the environmental sensitivity of the arctic region, with permafrost and large herds of migrating caribou to deal with, the success of the Trans Alaska pipeline is an engineering marvel.

Large tanker ships are regularly employed for transporting oil over long distances. Transoceanic trips are common. For example, oil produced in the fields of the Middle East is transported for thousands of kilometers in supertankers to the US and Japan. These supertankers have capacities ranging from about 200,000 to 300,000 tons, which is the equivalent of 1.4 to 2.1 million barrels of oil. Special mooring in deep-water locations is needed to accommodate these supertankers. Smaller tankers and barges are used to ship oil over smaller distances using rivers and other smaller waterways. Trucks are commonly employed to deliver petroleum-based products such as gasoline and diesel jet fuel to their points of use (e.g., gasoline stations or airports).

16.4.3 Natural gas transport

Over its almost 200-year history, the infrastructure for transporting and distributing natural gas from a producing gas well to its final use point has become highly developed and sophisticated. Pipelines carry much of the load (75% or more) in North America and Europe. The US is the largest consumer of natural gas, produced both domestically and in Canada. It is delivered using 420,000+ km (250,000+ mi) of pipelines. (Major US gas pipelines are diagrammed in Figure 16.8.) In Europe, Norway now provides a substantial portion of the natural gas to Western Europe from its North Sea fields (Ager-Hanssen, 1990).

Compression of the gas to elevated pressures is used to drive gas through the pipeline system. Consequently, a premium is placed on operating gas compressors at high efficiencies. In addition, some storage of natural gas in pressurized tanks or natural aquifers or in caverns is frequently used for buffering supplies during peak demand periods.

A smaller, but important, portion of the natural gas market is assigned to refrigerated tanker ships that haul LNG from sites in Africa, such as Algeria, and Asia to load centers throughout the world. Although the energy requirements associated with liquefying natural gas are significant, the energy density of methane in its liquefied state (1 atm, -162°C) is about 600 times larger than as a gas at ambient conditions (1 atm, 25°C).

Advances in materials technology, automated welding, and non-destructive pipe inspection and testing have greatly enhanced the performance of natural gas pipeline systems. Larger diameter, higher pressure systems can reduce both frictional pressure losses and, with turbine expanders to recover compressed energy, can be quite efficient.

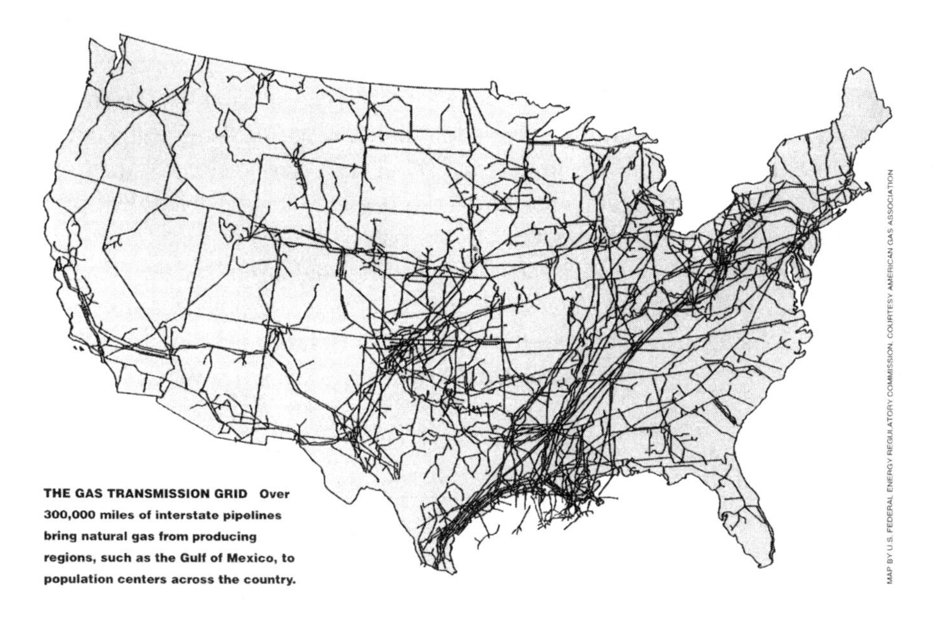

Figure 16.8. Illustration of the major gas transmission pipelines in North America. Source: http://www.platts.com

Mohitpour et al. (2001) predict that growth in the global production of natural gas will require an additional 120,000 km (72,000 mi) of gas pipelines globally by 2010, with North American expenditures exceeding $80 billion. Trends in 2001 have been directed toward larger diameter, high-strength steel pipes that significantly increase the throughput capacity of the pipeline. For instance, a major initiative is underway to bring existing gas from the North Slope of Alaska to the Edmonton hub for distribution to the lower 48 states using 52-inch-diameter, state-of-the-art, high-pressure, X-95 or better pipe. Comparable expansion in gas production and transmission is envisioned in the former Soviet Union to serve expanding European and other markets.

16.4.4 Coal transport

Most of the coal in the world is delivered from mines by rail or barge to power plants or other points of use. In contrast to liquid and gaseous fossil fuels, the transportation costs for delivering coal are usually a major portion of the fuel cost for electric power generation. For example (as shown in Table 16.4), in the last 20 years in the US, interstate coal transportation costs have increased from less than 25% to over 35% of the delivered coal price. Interestingly, the averaged delivered cost for coal in the US over this same period decreased from 28¢/ton mile to 13¢/ton mile (in constant 1996 US$) due to improved technology for mining operations as well as for transportation systems (Peters, 2001). (Table 16.4 provides other insights into the changing landscape of options that are used for transporting coal.) Barge shipments are typically limited to about 350 to 650 km (200– 400 mi), while rail transport distances have increased during the last 20 years from about 850 km (500 mi) to 1,400 km (800 mi). Multi-mode shipments involving a combination of rail, truck, and/or barge become important over long distances, while truck transport predominates for distances of 150 km (100 mi) or less. An alternative to these conventional modes for coal transport is the use of a coal-water slurry pipeline. The concept uses a mixture of about equal proportions of finely ground coal particles and water to create a slurry, which is pumped at near-ambient pressures in a steel pipeline. Although this idea was proposed about 110 years ago, and was operated at small scale in the UK in 1914 and at somewhat larger scale (108 mi) in the US in Ohio from 1956 to 1963 (Glasstone, 1983), only a few practical systems are in use today. A notable installation is the Black Mesa slurry pipeline, which started operation in 1970 between a mine in northern Arizona and a 1,500 MWe coal fired power plant in Nevada.

Table 16.4. Coal Transportation Statistics for the US

Year	US Average		Rail			Barge			Truck			Multi-mode			Other		
	% of total cost	¢/ton mile[a]	% of total shipped	average distance shipped	¢/ton mile[a]	% of total shipped	average distance shipped	¢/ton mile[a]	% of total shipped	average distance shipped	¢/ton milea	% of total shipped	average distance shipped	¢/ton mile[a]	% of total shipped	average distance shipped	¢/ton mile[a]
1980	24	28	59	550	2.8	10	310	1.6	7	50	12.3	10	–	5.8	14	60	3.0
1990	32	22	60	600	2.2	8	200	1.0	10	25	19.2	10	–	1.6	12	100	3.5
2000	37	13	73	850	1.2	10	300	0.9	5	40	15.0	6	–	1.1	5	100	2.4

[a]in 1996 U $

Source: Peters (2001).

16.4.5 Electric power transmission

The infrastructure for transmitting electricity from its points of generation to a multiple set of industrial and residential users has evolved for the last century to a sophisticated and interconnected level in most developed countries. In these developed areas, the great majority of customers are connected to a grid that provides them with electricity when needed. In remote locations, customers may not be on the grid, so that they must generate their own power in a distributed manner. The situation in undeveloped countries is far different. There is no national grid, and, in most cases, only major cities have operating electricity generation and distribution systems. In fact, it is estimated that over two billion of the world's people were without any electricity at the beginning of the 21st century.

Our focus in this section is on grid-based transmission and distribution. Transmission is usually characterized by overhead high-voltage lines carrying power long distances between central generation plants and main junction points on the grid. Distribution usually applies to lower voltage transmission over shorter distances from these nodal junctions to end users. Both AC and DC transmission systems are employed in practice. While there are advantages to each, AC is the dominant mode in most applications.

For transmission, the amount of power that can be transported increases in proportion to the voltage used. This is a result of Ohm's Law ($V = IR$), where power scales as:

$$Power = I_{eff}V_{eff} = \frac{V^2}{R} = I_{rms}V_{rms} \quad (in\ phase) \tag{16-25}$$

where the subscripts *eff* and *rms* refer to effective and root mean square levels of current (I) and voltage (V) carried by the wires with a total resistance (R). Note that *rms* values for normal sinusoidal AC power are equal to maximum values divided by 2. Equation (16-25), although somewhat simplified in the sense that inductance or capacitance effects are not explicitly included, shows that higher voltage and lower resistance will yield higher power-carrying capacity with lower losses. To account for inductance and capacitance effects, the so-called "*power factor*" is commonly introduced, where:

$$Power\ factor \equiv \frac{P_{actual}}{P_{apparent}} = \frac{P_{actual}}{I_{eff}V_{eff}} \tag{16-26}$$

Typically, copper or aluminum wires are used to transport electrical power in both AC and DC systems. Copper has the advantage of lower electrical resistance over aluminum, but aluminum has a lower mass density, which partially offsets its lower conductivity. In AC systems, three wires are needed, and the major portion of the energy is transmitted near the surface of the wires, which favors using stranded lines to increase the surface area to volume ratio of the transmission line. Long distance AC power transmission generally uses voltages between 250 and 765 kV, with a few systems at even higher voltages. The capacity of large AC transmission lines is over 3,000 MWe.

In DC systems, there is no "skin effect," as the entire cross-section of the wire is utilized for energy transport. Also, only two wires are needed. But where AC power is used by consumers, a rectifier must be employed to convert DC to AC. Even with this additional requirement, DC systems tend to have somewhat lower capital costs.

In the US, transmission and distribution losses average about 11% of the electrical energy supplied. These losses are attributed to Corona discharge, resistive (Ohmic) heating in the wires because of their finite electrical conductivity, and other irreversible losses in the power conditioning equipment that is required to step up and step down voltage levels using transformers. Transmission losses generally increase as the voltage level is lowered. According to Ilic (2001), over a distance of 1,000 miles, a high voltage AC (>500 kV) system loses about 3–5%, while a low voltage AC (<10 kV) distribution leg loses about 5–7% on the same basis. Without skin effects in the wires, DC transmission, on average for the same amount of energy transported, has lower losses.

Several environmental, safety, and health concerns directly related to high voltage power transmission strengthened public opposition towards the end of the 20th century. These included the manufacture and use of refractory and potentially carcinogenic PCBs (polychlorinated biphenyl compounds) as transformer fluids prior to 1970, fears about the health effects of electromagnetic fields (emf) generated near transmission lines, and safety concerns and the negative visual impact of bare wire and overhead transmission lines. Alternatives to existing AC and DC above-ground transmission systems do exist. For example, superconducting and conventional transmission carried out underground alleviate most of the major environmental and health concerns cited above.

Conventional underground transmission. In urban or highly populated regions, underground power transmission is usually required because of public safety concerns and the general lack of accessible, open space. Unfortunately, underground systems are more costly than overhead systems. Copper or aluminum cables placed underground need to be insulated to avoid charge transfer to the earth. This decreases their power carrying capacity to below 1,000 MWe and operating distances to about 100 km (60 mi) due to heat buildup resulting from Ohmic losses, dielectric losses, and capacitance effects. Improved insulation using gas blankets or active cooling with water or refrigerants can increase capacity at the expense of increased capital investment and somewhat higher operating and maintenance costs. Underground DC lines have lower resistive losses, but the need to rectify to AC introduces additional losses that often outweigh their Ohmic efficiency advantage. Similar technologies are also applied for underwater power transmission applications.

Superconducting transmission. This is a developmental area with great promise. As operating temperatures in a material approach its superconducting temperature limit, resistivity goes rapidly to zero, which virtually eliminates Ohmic losses and increases power capacity. The discovery of new compounds with superconducting properties near the normal boiling point of liquid nitrogen (77 K or –196°C) fueled much research and development work to manufacture practical superconducting wires and to integrate them into a complete transmission system that would be reliable and cost competitive. As of 2004, much remained to be done to achieve this goal.

16.5 Energy Distribution Systems

16.5.1 General characteristics of central versus distributed systems

Electric power in developed countries is usually produced by large central generating stations, then transmitted along high-voltage lines to local distribution systems that carry it to final consumers. Distributed generation (DG) and distributed energy resources (DER) concepts offer an alternative that has some attractive features. (Table 16.5a and Figure 16.9 compare the generation and distribution networks for central station and distributed generation systems.) As a practical matter, it is useful to define DG and DER as small-scale generation located near and connected to a load being served with or without grid interconnection. Both non-renewable and renewable energy resources could be involved, and there could be integration of electric power generation with heating and cooling loads.

Table 16.5a. Comparison of Central Station and Distributed Generation Systems

Characteristic	Distributed Generation	Central Station
Proximity	Variable, but typically, < 100 mi	Large Cities – 10 to 100 mi All others – 10 to 1,000 mi
Scale	1 kWe to 100^+ MWe	100 to 2,000 MWe
Energy form	Electricity or combined heat and power (CHP)	Electricity only
Fuel or energy source	Fossil, renewable (solar, wind, geothermal, biomass), or hybrid	Gas or coal, hydro, and possibly concentrating solar or geothermal
Grid connection	Optional	Required

Two key features that distinguish distributed networks from central station power are the shorter distances between the power generators and end users and the smaller scale of the generation plants within the system. Distributed generation technologies include internal combustion engines, small gas turbines, fuel cells, and photovoltaic and wind generation systems (see Table 16.5b and Figure 16.10). DG/DER can provide all or part of an individual's own needs, as well as produce excess power for sale. Excess power can be transmitted to a local distribution substation or distribution network.

DG systems are already in use in many countries, particularly for power quality or high-reliability applications, to provide emergency capacity or to avoid expansion of local transmission and distribution systems. Following the tragic events of September 11, 2001, the security of energy supplies and infrastructures received increased attention by local and national governments. DG/DER systems would de-localize and diversify how we generate and distribute energy, which should result in more secure energy services.

While preparing the final draft of this chapter, we were poignantly reminded of the vulnerability of centralized and interconnected grid systems. At 4:10 pm (EDT) on August 14, 2003, the US experienced its worst power outage when over 50 million customers lost power. Apparently, a large surge or outage developed somewhere in Ohio and quickly propagated to the remainder of the interconnected system in New York, Michigan, and parts of Connecticut and Vermont, as well as into the province of Ontario in Canada. For over 40 years, experts have warned of the weaknesses of the power grid system in North America. (An excellent reference that documents the vulnerability of the US power grid is Amory Lovins' 1982 book *Brittle Power*.)

Having additional spinning reserve capacity to meet peak demands on the hottest and coldest days of the year helps to reduce risks for outages. But fundamentally, a more robust control system with rapid diagnostics and response modes is needed to prevent a recurrence of blackouts on the scale of the 2003 event. Even with such improvements, centralized large grids inherently have vulnerabilities as a result of being regionally interconnected. A more distributed grid would avoid these systemic problems.

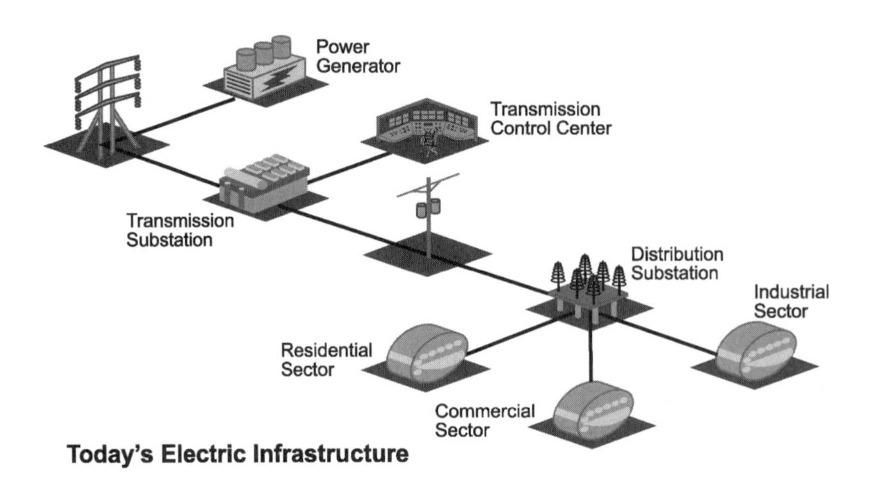

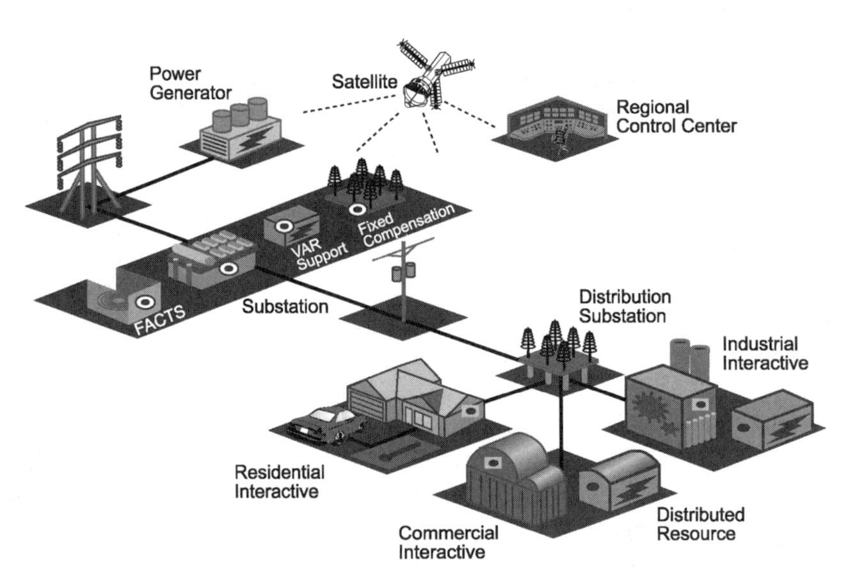

Figure 16.9. Base load, central station versus distributed electricity generation and distribution systems. Source: EPRI (2001).

Table 16.5b. Typical Ranges of Distributed Generator Unit Sizes

Type	Power Range
Internal combustion engines (ICEs)	1 kW to 10+ MW
Fuel cells	1 kW to 1 MW
Advanced CCGT	1+ MW to 200+ MW
Microturbines	1 kW to 1+ MW
Renewables	
wind	10 kW to 1MW
solar PV	1 to 10 kW
solar CSP	10 kW to 50+ MW
geothermal	1 to 100 MW
biomass	10 kW to 100 MW

CCGT – combined cycle combustion gas turbine-Rankine cycle
CHP – combined heat and power/cogeneration
CSP – concentrating solar power system
PV – photovoltaics
Source: DeBlasio (2003).

16.5.2 Combined heat and power opportunities

Most power distribution systems in developed countries are limited to providing electricity, not hot water or steam. Given that about 30– 40% of a country's total energy load is utilized in buildings or industrial processes as thermal energy or heat at temperatures below 250°C, it is unfortunate that we do not generally have thermal energy available from utilities for purchase. A few notable exceptions exist in European cities, in Reykjavik, Iceland, and in the US in New York City, where both electricity and thermal energy can be obtained from a utility company. As pointed out in Section 16.2, the combined use of heat and electric power (CHP) can result in substantial gains in efficiency. Thus, from a resource utilization perspective, CHP systems are more sustainable. Since distributed energy resources are close to end users, the infrastructure needed to install a CHP system is not as extensive as for centrally supplied grid power. Therefore, the capital investment needed to deploy CHP systems is more achievable for DG/DER applications.

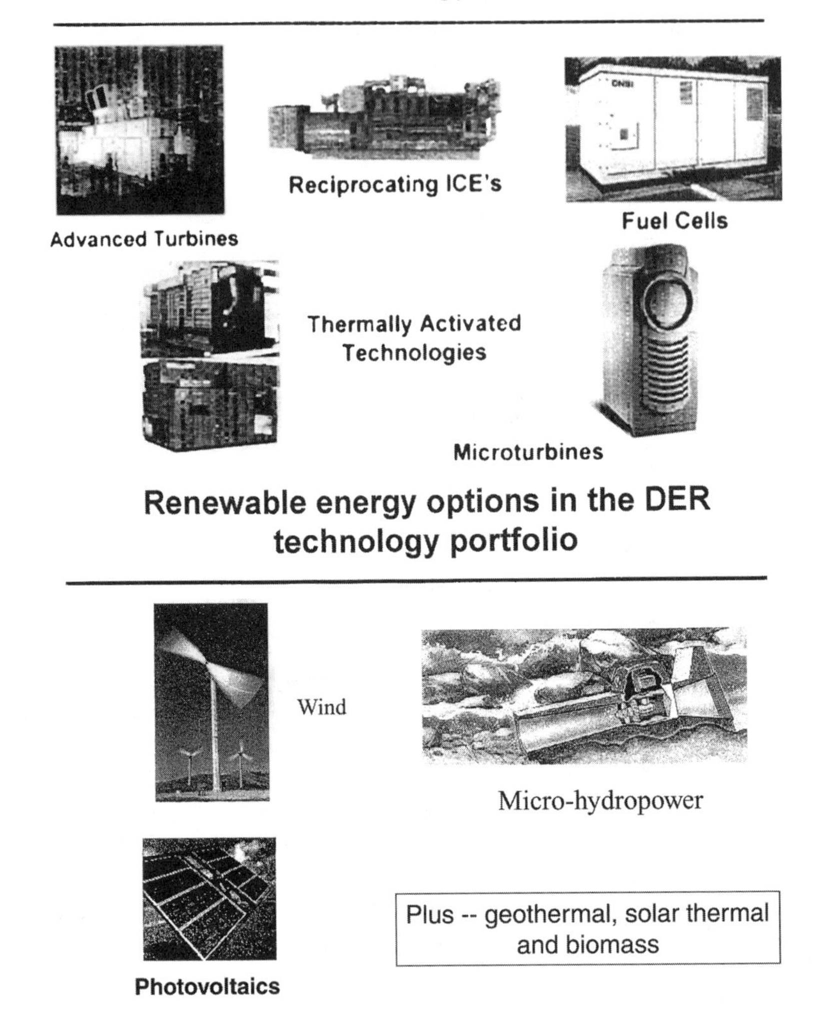

Figure 16.10. Examples of distributed energy technologies. Source: DeBlasio (2003).

16.5.3 Applications to renewable energy systems and hybrids

Given the smaller scale of DG/DER installations, their use of renewable energy resources, which tend to be more dispersed and less dense in terms of energy or power, is appropriate. Furthermore, given their localized characteristic with respect to resource grade, renewable energy is even more attractive. Because solar and wind energy are inherently intermittent on both daily and seasonal time scales, hybridized DG/DER systems that use stored fossil or biomass energy to provide dispatchable power are being considered for a range of remote off-grid, as well as some grid-connected, systems. The tradeoff here is a reduction in energy storage and a smaller renewable capture and generation system versus having two separate energy generators.

16.6 Sustainability Attributes

16.6.1 Improved resource utilization

Because energy storage generally increases system reliability and the ability to meet varying demands, one usually gets higher levels of resource utilization with well-designed storage in place. In particular, access to storage enables intermittent renewable technologies like wind, solar, and hydro to be more effective and can lead to higher energy capture and utilization efficiencies. For example, having thermal storage coupled to a solar system can avoid the use of backup systems to provide energy when demand is high at night or on a cloudy day. In another setting, having a flywheel, supercapacitor, or battery system installed on a hybrid vehicle reduces the size, fuel consumption, and emissions of the primary engine. As we have seen, the ability to use CHP systems with distributed energy generation can have an enormous positive impact on overall efficiency.

16.6.2 Environmental, safety, and health concerns

Energy storage, transmission, and distribution systems are part of the overall infrastructure, and, as such, they have an impact on the environment. Impacts can be associated with land use, in the case of a pumped hydropower storage installation, or the right of way required for an oil or gas pipeline or for a high-voltage electric transmission line. Impacts also come with the actual process of transporting the energy itself. One need only remember the fate of the Exxon Valdez in Prince William Sound in 1989 to understand the environmental risks associated with having millions of barrels of crude oil at sea in a large fleet of ocean-going supertankers.

There are a number of other potential hazards associated with the storage and transmission of natural gas and other energy forms. For instance, explosive mixtures of natural gas and air occur on a regular basis throughout the world. Given the potency of the methane molecule as a greenhouse gas—about 50 times greater than carbon dioxide—leaks from a natural gas pipeline and distribution system could have a significant impact that would offset its favorable hydrogen-to-carbon ratio over coal.

Concerns about the environmental, safety, and potential health risks of high-voltage overhead transmission of electric power are significant. Some US experts have suggested that it may be easier from a regulatory or permitting standpoint to site a smaller-scale, fossil-fired generation plant than it would be to build a new above-ground transmission line. This trend favors the development of distributed generation as a means to achieve capacity expansion on existing grids. Distributed systems also provide a more robust infrastructure that would be more secure against acts of sabotage or natural disasters. Furthermore, linkages between natural gas and electricity infrastructures allow better utilization of both systems when the peak period for one coincides with a low-usage period for the others.

16.6.3 Economic and operational attributes

The capital costs for some storage systems can be high, particularly for advanced flywheels, pumped hydro, CAES, and SMES. These costs need to be weighed against the benefits of the overall investment, which can be significant. For example, electricity and thermal energy storage provide several positive attributes that affect the economics of an electrical utility system. These include both capacity and operating benefits. By adding storage, expansions to existing generation and transmission systems can be avoided. As described in a review by Schoenung et al. (1996), adding capacity as gas-fired generation would involve $500– 600/kW and adding transmission lines about $200/kW or more. Thus, the investment required in the energy storage system has a built-in avoided cost that significantly improves the economics of the investment. Furthermore, there are other operational benefits that indirectly affect the economics (see Schoenung et al., 1996, for more details). They include:

1. Load following during dynamic shifts in power distribution that occur on the electrical grid
2. Spinning reserve that can supply needed generation capacity if sudden outages occur
3. Power quality improvements
4. Lower operating and maintenance costs
5. Reduced emissions by operating existing generation equipment at or near their peak performance conditions

16.7 Opportunities for Advancement of Sustainable Energy Infrastructures

The value of having workable energy storage systems is clear. Because they improve the utilization of natural resources and overall system efficiency, the development of storage technologies should be encouraged from a sustainability standpoint. However, storage is not often viewed as a focus area for R&D within compartmentalized scientific bureaucracies, and it suffers from a lack of support. One could make the case that storage, because of its crosscutting effects on energy system performance, should receive accelerated support. R&D programs need to be balanced among both small (e.g., batteries, flywheels) and large (e.g., pumped hydropower, CAES, SMES) storage technologies.

This chapter documents the necessity of being able to deliver energy reliably to customers with an operating transmission and distribution network for a variety of energy types. Without integrated, secure, and reliable infrastructures, the usefulness of all energy technologies would be severely compromised. In industrialized countries, these infrastructures are highly developed as a result of significant capital investment and governmental oversight.

From a sustainability perspective, it may be necessary to change existing infrastructures or develop completely new ones to achieve higher levels of performance. For example, if the US restructured its electricity supply system to a more distributed network of generation and transmission systems, it would facilitate the capture and storage of renewable energy, the development of hybrid systems, and an expansion of combined heat and power applications for residential buildings and industrial processes.

However, there is considerable inertia in many current systems with their electricity-only and central-station-generation focus. For example, North Dakota is a sparsely populated state with an extremely high-quality wind resource. In addition, given that the major industry in the state is farming, there is much open land available for siting wind turbines and an attractive social and economic climate that would embrace the development of wind power as another exportable product. In fact, the estimated wind power generation capacity is in the range of 100,000 MWe or more with acceptable capital investments of about $400–800/kW. However, there are few electric power transmission lines available for bringing power to large load centers located in other states. Furthermore, adding storage systems would greatly improve the attractiveness of wind as a substantial base load contributor to meet the nation's demand for electricity. Thus, to deploy wind at a level in North Dakota that would make a difference for the country would require policies and capital investments at a national level to develop the needed infrastructure.

The North Dakota example is easily transferred to other areas of the US and to other countries with established infrastructures. In developing countries, although there is no existing system to change, enlightened policies and approaches are still needed to reach a more sustainable destination. A strong force in making decisions today is the desire to have the initial capital cost, not the full lifecycle cost, as low as possible. Such attitudes usually do not favor developing energy storage technologies or the use of more distributed energy systems.

References

Ager-Hanssen, H. 1990. "European natural gas supplies and markets," in *Energy: Production, Consumption, and Consequences*. Washington, DC: National Academy of Engineering.

Boes, E.L., L. Goldstein, and G. Nix. 2000. "Energy Storage and Overview." Working paper. Golden, CO: National Renewable Energy Laboratory.

DeBlasio, R. 2003. National Renewable Energy Laboratories, personal communication, Golden, CO.

Energy Information Agency (EIA). 2002. "The Coal Transportation Rate Database," "National Trends in Coal Transportation," and related postings. Washington, DC: US Department of Energy.

Glasstone, S. 1983. *Energy Deskbook*. New York: Van Nostrand Reinhold Co.

Gordon, S.P. and P.K. Falcone, eds. 1995. "The Emerging Roles of Energy Storage in a Competitive Power Market: Summary of a DOE Workshop." Sandia Laboratory report SAND 95-8247. Albuquerque, NM: Sandia National Laboratories.

Hassenzahl, W.V., ed. 1981. *Electrochemical, Electrical, and Magnetic Storage of Energy*. Stroudsburg, PA: Hutchinson Ross.

Hottel, H. and J.B. Howard. 1973. *New Energy Technology*. Cambridge, MA: The MIT Press.

Huang, B., C.C. Cook, S. Mui, P.P. Soo, D.H. Staelin, A.M. Mayes, and D.R. Sadoway. 2001. "High Energy Density, Thin-Film, Rechargeable Lithium Batteries for Marine Field Operations." *Journal of Power Sources* Volumes 97 and 98: 674– 676.

Ilic, M. 2001. *Assessing Reliability as the Electric Power Industry Restructures*. Burlington, MA: Elsevier.

Jensen, J. and B. Sorensen. 1984. *Fundamentals of Energy Storage*. New York: Wiley.

Lom, W.L. 1974. *Liquefied Natural Gas. New York: J. Wiley.*

Lovins, A. 1982. *Brittle Power: Energy Strategy for National Security*. Rocky Mountain Institute, Snowmass, CO.

Mohitpour, M., A. Glover, and B. Trefanenko. 2001. "Key Worldwide Gas Pipeline Developments." *Oil and Gas Journal*, special report.

National Academy of Sciences. 1976. "Criteria for Energy Storage R&D." Report by the Committee on Advanced Energy Storage Systems, Energy Engineering Board, National Research Council, Washington, DC.

Peters, W.A. 2001. "Costs of US Central Station Coal, Gas, and Nuclear Electricity—Current Technology, Fuel Markets, and a Glimpse of the Future." Working paper. Cambridge, MA: Energy Laboratory, MIT.

Ragone, D.V. 1963. "Review of Battery Systems for Electrically Powered Vehicles." SAE paper #680453. Society of Automotive Engineers, Detroit, MI.

Sadoway, D.R. and A.M. Mayes. 2002. "Portable Power: Advanced Rechargeable Lithium Batteries." *Materials Research Society (MRS) Bulletin* 27(8).

Schoenung, S. 1998. "Flywheel Energy Storage for End-Use Power Equality." Electric Power Research Institute report TR-111831. Palo Alto, CA.

Schoenung, S., J.M. Eyer, J.J. Iannucci, and S.A. Horgan. 1996. "Energy Storage for a Competitive Power Market." *Annual Review of Energy and the Environment.* 21: 347–370.

Simpson, L.A., L.E. Oldaker, and J. Sterscheg. 1975. "Kinetic Energy Storage of Off-Peak Electricity." Atomic Energy of Canada report AECL-5116, Ottawa, Canada.

Stern, M.O. 1988. "Energy Storage in Magnetic Fields." *Energy* 1392:137–140.

Thrasher, J.E. and R.B. Lange. 1988. "Evaluation of Hard-Rock-Cavern Construction Methods for Compressed Energy Storage." Electric Power Research Institute report, EPRI AP-5717. Palo Alto, CA.

Trapa, P.E., B. Huang, Y.Y. Won, D.R. Sadoway, and A.M. Mayes. 2002. "Block Copolymer Electrolytes Synthesized by Atom Transfer Radical Polymerization for Solid-State, Thin-Film Lithium Batteries." *Electromechanical and Solid State Letters* 5(5).

Turkenburg, W.C. et al. 2000. "Renewable Energy Technologies" in *World Energy Assessment— Energy and the Challenge of Sustainability.* World Energy Council. Table 7.24: 263. New York, NY. United Nations Development Programme.

Yeager, K. 2002. "Electricity Infrastructure Technologies— Opportunities & Issues." *Forum on Energy Security and Restructuring of the Electric Industry–Technology Vision of the Electricity System.* Presentation delivered in Washington, DC.

Web Sites of Interest

electricity.doe.gov/
standards.ieee.org/resources/development/index.html
www.eere.energy.gov/
grouper.ieee.org/groups/scc21/1547/
grouper.ieee.org/groups/scc21/
www.eere.energy.gov/distributedpower/

Problems

16.1 You have been asked to design a rotating mechanical flywheel to supply 1 MWe of power for one minute. You have two choices for materials—steel and a proprietary composite (Rx-2002). The properties of both materials are given below. Key parameters to specify in your design are the volume and area (radius), mass, and maximum rotational speed. Discuss the issues that led to your final design. Are there other factors that need to be considered before your flywheel is practical?

Density – steel = 7.8 g/cm^3 and Rx-2002 = 2.0 g/cm^3

Maximum tensile strength – steel = 550 MPa and Rx-2002 = 2,100 MPa

16.2 A large pumped hydropower energy storage system is being considered for a site in Colorado near Denver where peak demand can be high at times. The plan calls for 1,000 MWe of dispatchable power for up to six hours. The proposed design would have two reservoirs, one located at Denver's mile-high elevation of 5,280 ft and the other in the foothills of the front range of the Rocky Mountains at an elevation of 8,000 ft. Estimate the minimum required volume of water that would need to be stored (state and justify all assumptions made). Assuming a margin of safety of 50% and an average water depth of 20 ft in both reservoirs, how large an area does this system impact? Are there other issues that may be important in assessing the sustainability attributes of this proposal?

16.3 Compressed air energy storage (CAES) is being considered as a means for storing energy during off-peak periods and for generating electric power during peak demand periods. The Alabama Electric Cooperative has adopted a novel scheme that pumps air underground into a huge cavern created deep in a solid salt deposit. The cavern is a cylinder about 300 m tall by 80 m in diameter and has an internal volume of about 1.5×10^6 m^3. Maximum storage pressures of 100 bar are possible using multistage compressors/turbine units that operate with equivalent efficiencies of $\eta_{compressor} = \eta_{turbine} = 0.85$. The ambient air temperature is 25°C, while the salt cavern's natural temperature is about 300°C as a result of prevailing geologic conditions. To keep the cavern from collapsing, a minimum pressurization level of 2 bar is required. Note that the thermal conductivity of the salt is high, and the thickness (radial and lateral extent) of the salt deposit is large relative to the dimensions of the cavity.

In the first part of the cycle, excess electrical power being produced during an off-peak period is used to operate the compressor to inject air into the cavern to pressurize it to 100 bar. In the final part of the cycle, power is produced during a peak demand period by expanding the cavern's pressurized air in a turbine (which is actually the compressor operating in reverse) to generate electric power.

Provide an expression for the net work produced during a single full cycle of energy storage and recovery and describe how you would estimate the net work produced. Please list all assumptions and simplifications used in your analysis. Your answer may be expressed as an integral equation involving the following variables: κ ($\equiv C_p/C_v$), cavern volume (V_{cavern}), T_{cavern}, $T_{ambient}$, $P_{cavern,\ max}$, $P_{cavern,\ min}$, $\eta_{compressor}$, $\eta_{turbine}$.

16.4 Electra and Caty Jones, distant cousins of Rocky and Rochelle, have a great idea that they feel will "revolutionize" the way electricity is produced by "avoiding the shortcomings of existing heat-to-power concepts that are constrained by the Second Law of Thermodynamics." In today's thermal cycles, fossil fuels are combusted in air to produce heat that is supplied to a practical power cycle to generate electricity. Typical fossil-fuel-fired power cycles include steam Rankine, gas turbine (Brayton type), or combined gas turbine-steam Rankine.

As an alternative to these cycles, Electra and Caty have invented an advanced fuel cell, which they call *Electrocat,* that isothermally converts natural gas, CH_4 (a common household fuel), directly to electricity using electrocatalyzed oxidation reactions in a proprietary electrolyte. Special proprietary catalysts promote the following redox reactions:

anode:	$CH_4 + 2H_2O \rightarrow CO_2 + 8e^- + 8H^+$
cathode:	$8H^+ + 8e^- + 2O_2 \rightarrow 4H_2O$
total reaction:	$CH_4 + 2O_2 \rightarrow CO_2 + 2\,H_2O$

According to their claims, the *Electrocat's* efficiency approaches 100% in recent tests using the conventional definition of cycle efficiency given in Chapter 14.

In these same tests, the *Electrocat* produced 0.75 MWe of power continuously with a feed rate of 1 mole of CH_4 per second.

Given the promise of obtaining electric power directly from this fuel cell, MITY Industries is considering making a large investment in *Electrocat* technology. However, a few of MITY's top executives are skeptical of the *Electrocat*, and they need your help to answer a few basic questions. In order to fully understand your answers, they have requested that you state and justify any assumptions you make.

a. What is the maximum power-producing potential of the *Electrocat* for a feed rate of 1 mole per second of CH_4 at ambient temperatures (25°C)?

b. What is the utilization efficiency (η_u) of the *Electrocat*?

c. Is their claim that η_{cycle} approaches 100% thermodynamically possible? Could η_{cycle} exceed 100%? Explain.

d. What are possible sources of loss of work producing potential in the *Electrocat*?

e. By making a First and/or Second Law analysis, determine whether it is possible to use the *Electrocat* for process cooling or heating as well as electric power production.

Property data for CH_4, CO_2, H_2O, and other compounds and ions are given below:

For pure water (H_2O): $\Delta G_{vaporization} = 8.58$ kJ/mol and $\Delta H_{vaporization} = 44.0$ kJ/mol.

Data: Standard Gibbs Energies and Enthalpies of Formation and Average Ideal-Gas State Heat Capacities at 298 K and 1 bar

Compound	ΔG_f^o kJ/mol	ΔH_f^o kJ/mol	ΔS_f^o J/mol K	$<C_p^o>$ J/mol K
$CO_2(g)$	-394.6	-393.8	2.68	36.8
$CO(g)$	-137.4	-110.6	89.93	29.3
$CH_4(g)$	-50.9	-74.9	-80.54	51.8
$N_2(g)$	0	0	0	29.3
$O_2(g)$	0	0	0	29.3
$H_2(g)$	0	0	0	22.15
$H_2O(g)$	-228.8	-242.0	-44.30	36.8
$H_2O(l)$	-237.1	-286.0	-164.09	75.3
$H^+_{(aq)}$	1517.0	-143.6	-5.57	—
$OH^-_{(aq)}$	-138.7	1536.2	5.62	—

16.5 Cornell University is considering a plan to utilize the cold temperatures available near the bottom of Lake Cayuga. The campus is located about 490 ft (150 m) above the surface of the lake. The average depth of the lake is about 210 ft (65 m). The Cornell plan would utilize the cool lake temperatures to assist central air conditioning needs on campus. You have been asked to evaluate a critical step in the proposed process. At a steady flow of 20 kg/s, warm water is discharged to the lake at about 60°F and cooler water at 40°F is pumped back to the central air conditioning plant on campus. Right now the warm water is released to the lake through an insulated expansion valve. You are to investigate the possibility of generating work to help offset the pumping power requirements. Consider using a hydraulic turbine to recover some of the PE from the elevation drop. MITY Industries has an adequate supply of turbines and pumps available with efficiencies of 90%,

i.e., (η_p $\eta_t = 0.90$ and $\dot{W}_{pump} = \dot{W}_{ideal}/\eta_p$ and $\dot{W}_{turbine} = \eta_t \dot{W}_{ideal}$)

Where $\dot{W}_{ideal}$ refers to reversible, isothermal operation. All sizes of ultra-high efficiency, reversible Carnot heat engines are also available. Is it possible to operate this system without external power? Explain your answer. In your analysis you can assume that the heat capacity and density of liquid water are constant, at 4,200 J/kgK and 1,000 kg/m^3, respectively.

16.6 Storage batteries that have small volumes and masses and high energy storage densities are in demand for both electric and hybrid vehicle applications. Based on your assessment of evolving technologies for advanced electrochemical batteries, estimate the smallest volume and weight battery system that might be available in the foreseeable future to provide full vehicle power for one hour of urban commuting in a family-sized five passenger sedan. You can assume that minimum power levels for such a vehicle are 20 hp at average speeds and 80 hp at peak acceleration. A dynamic flywheel or supercapacitor storage system integrated with a regenerative braking device can also be assumed to be on board to offset the peak loads by 50% or so.

16.7 Estimate the amount of coal delivered daily to the 1,500 MWe electric power plant in Nevada through the Black Mesa coal-water slurry pipeline. How many rail cars per day is this equivalent to, assuming each car has a capacity of 100 dry tons of coal? How much water is pumped and consumed each day? Does this pose a sustainability issue for this location? State and justify all other assumptions made.

16.8 Estimate the resistive losses over a 100-mile long, 100-MWe, DC transmission line constructed of two lines of copper conductor wire with an effective diameter of 2 cm at both the 100 kV and 250 kV levels. The electrical resistivity of copper is about 1.7 microhm-cm and its density is 0.321 lb/in^3.

Electric Power Sector 17

Let there be light.
—Genesis 1:3

17.1 Introduction and Historical Perspectives

The electric power industry evolved from a series of discoveries during the 18th and 19th centuries that opened possibilities for using electricity as a commercial energy carrier. Around 1750, Benjamin Franklin developed the concept of electricity as a fluid that flows when there is a difference in electrical charge levels. When Alessandro Volta invented an early battery around 1786, it was a scientific curiosity. However, it allowed further research on the nature of electricity. George Ohm later found that direct current flows according to the relationship: V(voltage) = I(current) × R(resistance) and that power scales as I × V (Ohm's Law, see Equation (16-25)).

Michael Faraday invented a crude motor around 1820, but it was not particularly useful because a source of power to run it was not readily available. Further studies by Faraday elucidated the inductive coupling between separate electrical conductors, which led to the theory of electromagnetic coupling. This means of translating mechanical motion to currents led to the development of practical motors and the wider understanding and use of alternating current.

By 1850, Gustav Kirchoff had extended Ohm's work to three-dimensional electrical systems and to networks of such systems. In the 1860s, James Maxwell elucidated the interrelationships between electrical and magnetic fields and advanced the understanding of the dynamic interactions between electricity, magnetism, and light. People began to imagine a future in which electricity brought new possibilities to the quality of everyday life and industrial operations. By 1880, Thomas Edison had developed a commercially viable light bulb, and pioneering power stations were beginning to come online. In 1882, a few power stations were opened in both the US and the UK. Another key invention at that time was the discovery of the transformer, which could be used to step up voltage. Almost simultaneously, the steam turbine was invented, facilitating the conversion of thermal energy to electricity using steam as the working fluid. Early plants ran off waterwheels in streams and hydropower from dammed rivers. Coal-fired plants expanded the technology further by generating steam that could be converted to electricity using steam turbines. The first successful gas turbine was produced in France in 1903.

By the end of the 19th century, power companies had started to meter electricity and charge for electrical services. The potential for electric heating and cooking was recognized and developed commercially. The electric washing machine appeared in 1918; the refrigerator, the following year. The early power system configuration evolved from these beginnings—a central generation plant with wires that brought services to local customers (see Figure 17.1).

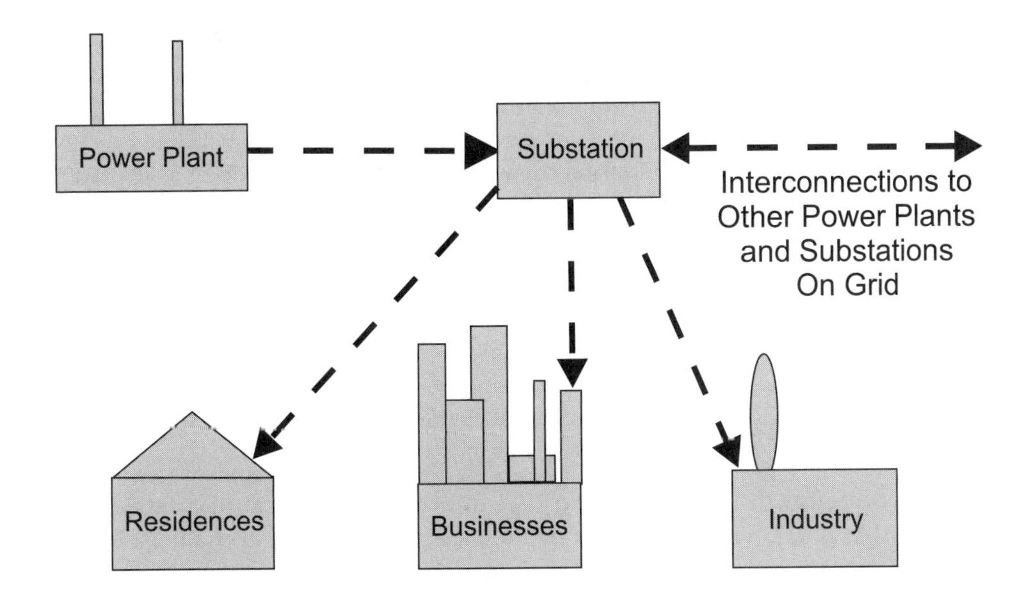

Figure 17.1. Traditional electric system configuration.

As discussed in Chapter 16, varying customer demands created problems for the generators who tried to match supply to demand. This led to the installation of generating equipment of differing sizes that could be brought on- and off-line, as well as units that could supplement the "base" power generation to meet "peak loads." In the early days of the industry, reliability of supply was shaky and outages were commonplace. As the industry grew, new power stations were added and redundancy was introduced to provide multiple feeds to major service areas. It seemed logical that this industry had to be a monopoly because of the complexity of interfaces between generation, transmission, and distribution. Also, companies needed the assurance of continuing revenues to provide a return on the large investments they had made in infrastructure. Local power authorities began to improve the reliability of their individual systems by interconnecting adjacent power companies and swapping power.

By 1930, most cities in the US and many in Europe had electrical transmission and distribution systems. Rural electrification followed. By mid-century, the advent of more sophisticated technologies that relied on electrical transmission frequencies for accurate performance made reliability of supply even more important. Utilities then developed sophisticated transmission and distribution networks to provide not only a high degree of service reliability, but precise control of frequency and phase. Diverse types of generators existed—running off different types of fuel (coal, oil, natural gas) or from nuclear fuel or hydroelectric power. As fuel prices fluctuated, or as price differentials between fuels varied, strategies for maximizing short-term profits while providing insurance against future price fluctuations in primary energy became complex.

The growth in demand for electrical services continued at a fast rate, further pressuring the industry to expand its generation and transmission/distribution capacity. Table 17.1 shows the historic growth rate for the main use sectors (Wattenberg, 1976).

Table 17.1. Growth in US Usage of Electrical Energy by Consumer Sector (millions of kilowatt-hours)

Year	Industrial	Residential	Commercial
1912	25,000	910	4,076
1932	100,353	11,875	12,106
1952	472,071	93,545	63,935
1970	1,631,731	453,015	295,057

Source: Wattenberg (1976).

More recent growth is seen in Table 17.2 from US Department of Energy (DOE) statistics (EIA, 2001), which only report retail sales of electricity to consumers. Because industry generates much of the power it consumes, the 1970 industrial consumption rates are significantly lower than those presented in Table 17.1.

Table 17.2. Recent Growth in US Usage of Retail Electrical Energy by Consumer Sector (millions of kilowatt-hours)

Year	Industrial	Residential	Commercial
1970	571,000	466,000	307,000
1980	815,000	717,000	488,000
1990	946,000	924,000	751,000
2000	1,064,000	1,192,000	1,055,000

Source: EIA (2001).

The electric industry, along with other utility monopolies, needed some regulatory structure to assure a functional national electric grid, to mediate disputes between parties, to monitor imports and exports of electricity, and to limit unfair pricing practices. By 1953, the Federal Power Commission was established in the US. About 20 years later, oversight and regulatory responsibilities were transferred to the DOE.

Meantime, the pollution associated with power plants was becoming more apparent as facilities proliferated. With the US National Environmental Policy Act of 1969, the US Environmental Protection Agency (EPA) was established and charged with the implementation of laws to protect public health and the environment. Similar environmental laws were enacted in European countries. Over the years, these regulations have become stricter, and air quality has generally improved (see Chapter 4 and Section 7.4.2).

As new facilities come online or as older facilities are upgraded, costs of environmental protection are passed along to consumers. However, most environmental regulations "grandfather" existing facilities. For power plants that have 40+ years of operating life, this means that pollution may continue until the technology becomes so outmoded that it is no longer economical to operate. Utilities, however, made major investments to meet environmental regulations and were allowed to incorporate charges for capital costs into their rate base.

Another major change came in 1978 with the passage of the Public Utility Regulatory Policy Act (PURPA) in the US. Market economists had long claimed that monopolies result in higher prices than would be obtained in a free market situation. Though it was difficult to unbundle the utility infrastructure, the strategy was to allow the entry of new generators into a competitive power market. Provisions of PURPA also allowed small generators to sell power back to the local utility company at a preassigned price. Although PURPA was ruled unconstitutional in 1981, it set the stage for more comprehensive deregulation of generators under the National Energy Policy Act of 1992 in the US.

As generating units entered free market competition, many utilities sold off some or all of their power plants. The remaining regional infrastructure was divided into two separate functions: the management of the power grid (under Independent System Operators [ISOs]) and the provision of electric services to consumers and local industry by local utility companies. Deregulation also spread to the UK and several other countries, with some differences in the structures that replaced the original monopolies. In the US, many consumers can now contract to purchase electricity from any generator (or through an electricity reseller), who then adds a markup to cover transmission costs between the generator location and the customer. The real situation is that electrons are injected into the grid at some remote location, and different electrons are extracted at another location. This makes the operation of the transmission grid complex—especially since it was originally designed for generating nodes that were under the control of the utility, not the free market.

The ISOs were faced with operating a complex system under new conditions and, not surprisingly, found that it was impossible to provide the required service and reliability without added investment in new equipment and technology. Advances in control and communications technology were becoming more available to help with the evolution of new grid architectures and monitoring/control configurations. However, when power generation, transmission, and distribution were unbundled, it was found that they were roughly equal in cost. Thus, deregulation provided some market advantages and reduced costs in the generation part, but that was only about 1/3 of the total cost of electricity to the consumer. Transmission costs increased somewhat because of the need for additional investment to operate the new system.

Putting generators into market competition also had some interesting transitional effects. Owners of existing power plants made decisions about selling off their plants based on the status of residual capital cost charges, as well as on operating costs. For example, some nuclear plants were sold at prices well below their original costs because most of the original cost had been recovered, and some companies attempted to cut possible future losses associated with ownership of these plants in a competitive market environment. Purchasers of older coal plants also found some bargains. New owners looked critically at the cost of power generation—and competition caused much leaner operation than had been common under the old utility structure. Non-competitive plants were closed, and, since all plants were competing on the cost of generation, plants with higher operating costs had to make deeper cuts to remain competitive. The generation industry has not made much new investment in capital-intensive technologies. Addition of gas turbine generators has been common. These have lower capital cost per unit of power installed and come in smaller units that allow operators to move generating units on- and off-line to meet variable demands at different cost levels.

Today's industry is still in a transitional phase. Problems in deregulation are being faced, and changes are being put into place. Gaming and unethical practices are being uncovered and addressed, but there still is a continuing evolution as new technology brings new challenges. Real-time pricing is a new possibility that will encourage consumers to use discretionary appliances when power demand is low and generators are willing to sell power at lower rates to shift the load away from peak use periods. Smart appliances are being developed that will turn themselves on or off depending on the spot price of electricity. The control and information technology needed to support these new possibilities are in an early stage of evolution, but they promise to provide more efficient and cost-effective electrical services in the future.

17.2 Power Generation

Chapters 7 through 15 describe a variety of primary energy sources and the conversion equipment that is needed to convert the primary energy source to electricity and other uses, such as work or thermal energy generation. Figure 17.2 (shown earlier in Chapter 1) displays the spectrum of energy sources and the conversion steps to societal end uses, including electrical power generation.

17.2.1 Electric energy

What is electricity? Electricity is energy transported via the motion of electrons. It is produced by an electric generator or electrochemical cell, which results in an electrical voltage difference being created on a conductor. In turn, this causes electrons to flow to a load. At the load, mechanical work can be performed by means of moving electromagnetically forced objects through a distance, or heat can be generated via ohmic (electrical resistance-based) heating. The generator requires a prime mover to cause its conducting coils to move through a magnetic field.

When a generator is employed, primary energy must be consumed in a heat engine in order to produce the prime mover's work. This is commonly done using a Rankine or Brayton thermodynamic cycle to convert a fraction of the heat injected into the heat engine into work. This work is transmitted as electrical energy, which can be recovered as either heat or work at the load. Thus, electricity is seen as an energy carrier moving energy released from the primary fuel to the point of consumption. It is converted, ultimately, into lower-availability heat by the load. Electricity is especially valuable because it permits a central prime mover to satisfy a distributed set of demands.

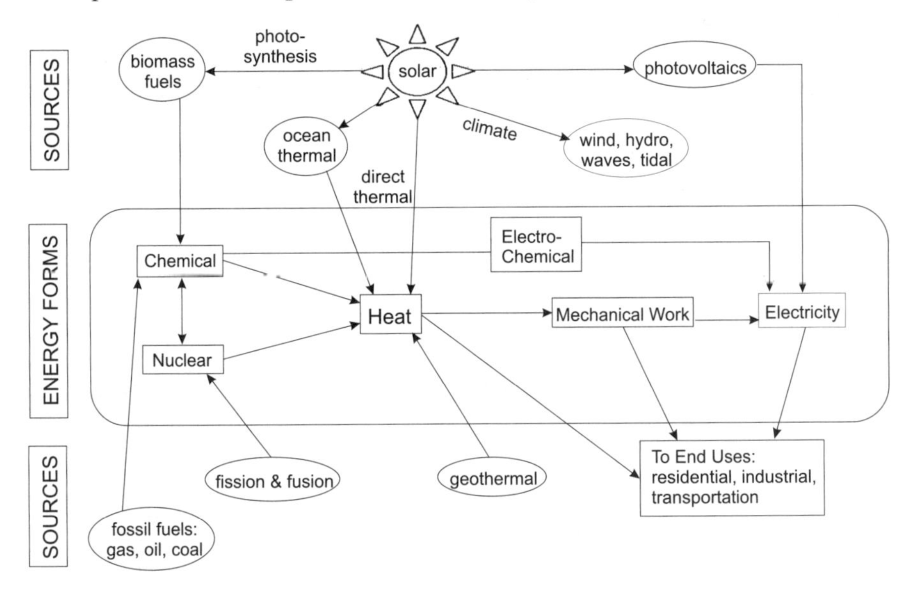

Figure 17.2. Energy sources and conversion processes.

Electricity is characterized by a number of attributes (e.g., voltage, current, frequency, phase angle, and their respective ranges of fluctuations). As such, electricity's suitability for satisfying a particular load depends upon how this combination of elements conforms to the load's needs. In other words, electricity supplied at the wrong voltage or current or frequency can be unsatisfactory, even though the values of the other attributes may be adequate. For some loads, departures from the satisfactory range of attribute combinations may be tolerated for a specified duration, while for others such a tolerable duration may be much different. For example, an electric heater can tolerate broad variations in voltage and frequency, or even current flow interruption for minutes or longer, and still provide satisfactory service. A credit card information data processing center, however, may be unable to tolerate a 5% voltage decrease for more than a millisecond. Thus, in determining the adequacy of an electrical system, it is necessary to specify the needs of the loads, as well as the capabilities of the system of generators and electrical transmission and distribution grid.

This distinction is important, as primary fuels are characterized only in terms of the energy released as kinetic energy by a single chemical or nuclear reaction. If that initially ordered kinetic energy can be captured in a work reservoir, the fraction of it available to perform tasks is typically much greater than if it is irreversibly converted to heat (i.e., disordered molecular motion). Except in small amounts (i.e., using capacitors or batteries), technologies for storing electricity are cumbersome and often costly, as discussed in Chapter 16. Rather, electricity can be stored via conversion to mechanical work, which then can be stored as potential energy, as is done with pumped storage hydroelectric facilities or stored compressed gas. Typically, the fraction of the energy consumed via the grid that can be stored is a small fraction of the total. Maintaining the essentially instantaneous balance between the amount of energy entering and that leaving the grid is a central problem in operating the grid stably and avoiding blackouts.

17.2.2 Centralized energy generation

The largest concentrated energy sources are those that come from the earth—fossil fuels, nuclear fuels, and some geothermal sources. Although these are depletable resources, the fuels are concentrated and permit generation of large quantities of energy from a relatively small power plant. Fossil fuels, because of the low cost of extraction and their high energy density, account for about 85% of global commercial energy usage. They also are the favored primary energy source of electrical power generation at present (about 64%, EIA, 2001). Coal and nuclear plants, in particular, benefit from economies of scale. Therefore, the formerly regulated and integrated utility industry designed its transmission systems around centralized electricity generation. In the recent past, deregulation has encouraged competition among generating units that are dispersed over the transmission grid. The existing grid system, operating under new requirements, must function over a much broader spectrum of conditions and closer to reliability limits. Large, longer-distance power flows strain the grid capabilities and create failure-prone conditions that can cause widespread blackouts (e.g., in 2003, in the northeast US and Canada, in Italy, and in London).

17.2.3 Electric power generation

Electricity generation accounts for about 25% of primary energy consumption in a typical industrialized country (the amounts produced in the US and worldwide, 3.7 and 14.6×10^9 kWhr, respectively, are shown in Table 17.3). In the US, this is done primarily by coal consumption (51%), nuclear fission (20%), natural gas (16%), and hydroelectric generation (10%). Worldwide electricity consumes about 18% of primary energy, reflecting the reduced amount of investment in electrical infrastructure in poor countries. As of 2001, other renewable energy sources (mostly biomass) comprised approximately 2.6% of the total. An overview of US electricity use, from generation through end use, is shown in Figure 1.15, including the losses involved at each state of system operations.

Until development of natural gas–fueled combined-cycle units, natural gas was too expensive and supplied only a small fraction of national electricity. Natural gas–fueled combined-cycle unit technology became practical about 12 years ago and achieved efficiencies of 55%—a huge increase over the 41% typical of large fossil-fueled steam Rankine cycle plants and 33% for light-water nuclear power plants. So, very quickly, the gas-turbine technology supplanted steam plants as the preferred technology for new electrical generation in the US and worldwide.

In the steam Rankine cycle, liquid water is turned to steam, which may then be heated further by a heat source (a flame or nuclear fission reactor) at constant pressure to a maximum temperature at which it enters a turbine. Within the turbine, the steam expands, thereby pressing the turbine blades in order to turn the turbine rotor (which, in turn, is connected to and turns the generator shaft). Low-pressure steam exiting the turbine flows into the condenser, where heat is rejected at constant pressure and temperature to the biosphere, thereby condensing the steam to a low-pressure liquid. This liquid is then pressurized by a pump and heated back to its turbine injection temperature. The thermal efficiency (the ratio of mechanical work performed to the heat injected to the steam) of the system depends approximately upon the mean temperature at which heat is added to the steam and at which it is rejected by the fluid in the condenser. The heat rejection temperature is determined by the heat sink (typically, the biosphere). The heat addition temperature depends on the structural strength characteristics of the heat source materials, typically 550°C for fossil-fueled units and 300°C for nuclear light-water reactor (LWR) units.

The Brayton cycle used in gas-turbine systems is similar to the Rankine cycle system except that the working fluid is gas, which remains in a single phase (rather than as a liquid, which boils and condenses). Heat is added to the gas until its maximum temperature is reached, after which it flows through a turbine. Following the turbine, the gas is cooled at constant pressure to a minimum temperature, following which it is depressurized and then heated again back to its maximum temperature. When this is done, the thermal efficiency may approach 30%. However, a lower temperature steam Rankine cycle system then captures the rejected heat. The combined system efficiency can be as high as 55%.

Electricity can also be generated by passing water through a hydro turbine in turning a generator. Typically, this is done in a hydroelectric facility, converting the stored potential energy into work as the elevation of the fluid is reduced. It can also be generated via photoelectric reactions in a solar cell or by collecting solar energy as a heat source that is then used to boil water or heat a gas. Electricity is also produced by the various geothermal and renewable technologies (e.g., photovoltaic, power tower, solar concentrator, solar geosynchronous satellite, biomass, wave, wind, ocean thermal gradient), which are discussed elsewhere in this book.

17.2.4 Environmental effects of electricity production

Electricity technologies differ considerably in the ways that they affect the environment (see Table 1.7), and none is free of serious effects. The plant site-related effects include thermal emissions, airborne and waterborne chemical emissions, and potential radioactive emissions. The greatest effect of operating a thermal electric power plant is heating of the local environment. The ratio of the energy rejected by a thermal power plant directly to the environment compared to the electricity produced is equal to the ratio $[(1 - \eta)/\eta]$, where η is the thermal efficiency of the power plant. Thus, this ratio for a light-water reactor (LWR) nuclear plant equals 2.0, versus 1.5 for a fossil-fueled Rankine cycle plant, and about 0.9 for a gas-fired combined-cycle plant.

Table 17.3. World Net Generation of Electricity by Region and Type, 2000. Billion kilowatt-hours (% of regional total)

Region	Fossil fuels	Nuclear	Hydroelectric	Total
North America	2,987 (65%)	830 (18%)	656 (17%)	4,574 (100%)
Central and South America	195 (25%)	11 (1%)	555 (74%)	779 (100%)
Western Europe	1,398 (49%)	849 (30%)	551 (21%)	2,875 (100%)
Eastern Europe and FSU	1,043 (66%)	266 (17%)	256 (17%)	1,569 (100%)
Middle East and Africa	759 (88%)	13 (2%)	86 (10%)	858 (100%)
Asia and Oceania	2,992 (74%)	465 (11%)	543 (15%)	4,043 (100%)
World	9,375 (64%)	2,434 (17%)	2,646 (18%)	14,697 (100%)

Source: EIA (2001).

Before 1970, most US plants were cooled using "once-through" surface discharge systems. These systems draw water from a nearby water body or river, heat it, and return it. Once returned, the warm water transfers its heat to the atmosphere and the parent water body. During the 1960s, criticism of such practices arose, organized around the slogan of preventing "thermal pollution." This was one of the earliest organized environmental protest issues in the US. The rise of this concern has been attributed to a combination of growth in the size of the largest power plants being built (by a factor of 10 during a decade) and of increased social activism arising from dissatisfaction with the Viet Nam War. Once concern over the environmental effects of such discharges had grown sufficiently and new environmental protection regulations were enacted, cooling plants with surface discharge systems largely stopped. They were typically (but not totally) replaced by requirements for cooling towers at river- and lake-sited plants and submerged diffuser discharge facilities for ocean-sited plants. As a result, there has been a reduction of the temperature increases caused by the discharges, an increase in human fatalities (typically about one per plant) and injuries in building these more elaborate facilities, and a diversion of funds from other uses to their construction.

The most airborne emissions come from coal-fueled plants that release about twice the CO_2 of a natural gas plant, as well as sulfur and particulates characteristic of the impurities in the original coal deposits. Gas-cleaning technologies can capture > 99% of the particulates, but have difficulty getting the sub-micron-sized ones, which are now believed to be responsible for much of the lung damage caused by particulates (Abbey et al., 1995). SO_2 is created in the burning of sulfur-bearing coal. It is typically captured in scrubbers (>90%), which results in a solid waste that requires disposal. All air-burning power plants create NO_x in amounts depending mainly on the temperature of combustion (which determines the rates at which O_2 and N_2 in the fuel and the air combine to form

NO and NO_2). Untreated plant emissions have been blamed for about 10 fatalities from lung disease nationwide each year, and for about an order of magnitude more treated instances of lung disease.

With oil-fired plants, emissions are lower than with coal, and particulates are not a problem. This is also true should the sulfur be removed prior to combustion. CO_2 emissions are lower, and NO_x emissions are about the same as with coal. With natural gas plants, the major emissions are CO_2 and NO_x. LWRs are free of all of these emissions, and pressurized water reactors (PWRs) have been shown to emit less routine radioactivity than coal-fired plants (due to activity in the escaping ash of the latter) (Eisenbud, 1964). Boiling water reactors (BWRs) routinely emit low levels of noble gas fission products.

Coal is a fossil fuel that creates serious environmental problems everywhere in its fuel cycle, although the record has improved in recent decades. Mining accidents (for surface mining <<1 per plant-year, and about 0.2 for underground mining) (MSHA, 2003), transport-related accidents and collisions (about 2 per plant-year, mostly for eastern plants using western coal) (FRA, 2003), and waste disposal (approximately 30,000–100,000 tons of ash and similar amounts of sulfur/calcium sludge per plant year) are the greatest problems caused by coal.

Oil spills, especially those polluting beaches and estuaries, are feared because they create dramatic television stories as well as damage to beaches and aquatic wildlife. More important, the scientific consensus on the proper response to most such spills is that benign neglect is usually the best policy after affected wildlife has been treated and oil is off-loaded from leaking tanks, because of the toxic effects of cleanup chemicals and disruption of the aquatic ecosystem by cleanup efforts. At most spill sites, evidence of the spill is usually gone after a decade. However, public reaction to spill stories usually leads to irresistible political pressure to do something about it, even if it is wrong. The chronic oil pollution of harbors due to mishaps in transferring oil between ships and land-based facilities appears to be the cause of much greater but less dramatic environmental disruption than that from infrequent near-shore shipwreck-based episodes. Better standards for construction, operation, and maintenance for all oil tankers also are reducing large oil spill risks.

Environmental concerns are growing about all of these major sources for generating power. Fossil fuel combustion releases greenhouse gases (GHGs). Nuclear plants create nuclear wastes and national security issues. Effects of large hydropower dams include changes in fisheries, land use, and river ecology. Many are looking toward larger use of renewable energy sources in the generation of electricity. However, though the renewable share of the supply is growing, the total impact is only a few percent because these sources are new, intermittent, and not fully cost competitive.

17.2.5 Power plant siting requirements

A problem for producing electricity is that suitable power plant sites may be hard to obtain, especially in populated regions. This is because of their physical requirements and because wealthy industrialized societies have become more averse to having large industrial facilities in their midst.

The physical requirements include those for connections to the transmission grid and transport systems for fuel and large components. The former need is best served when several independent connections are available, because this promotes the reliability of the overall support received by the plant from the electrical grid (this is especially important for the safety of nuclear plants). The latter needs can be served by road or rail, but are best served by waterborne transport, especially for transporting heavy components. Because of this factor and the need for cooling water, many plants are sited on the ocean and on large lakes.

The land area requirements for plants differ from about 200 ha for coal- or oil-fueled plants (note that the amount of land disrupted in surface mining of coal per plant-year can be substantially greater than this value), to 20 ha for natural gas-fueled facilities, to 1,000 ha (1×10^7 m^2) for plants using cooling ponds. The footprint of a LWR is about 20 ha, but the buffer "exclusion zone" surrounding the plant buildings can be substantially larger (Flavin and Lenssen, 1994).

Most plants need to be sited on a water source that is at least able to support the evaporative flow of water vapor and the cleansing "blowdown" flow out of the cooling towers (approximately 0.8 tonnes/s for a 1,000 MWe LWR). When once-through cooling is used, much larger water bodies (such as the Great Lakes or ocean) are required. Dry cooling using the atmosphere as the heat sink is also possible. Dry cooling is used at desert sites—but only rarely, because it is about a factor of 10 more expensive.

Since the recent focus on potential risks from terrorism directed toward infrastructure facilities, the amounts of land required for power plants, especially nuclear ones, may grow larger. Whether they need to grow in order to protect the public remains unclear. Still, the amount of land needed for power plants is likely to be a subject of debate for several years.

In addition to these factors, opposition to new power plants often arises when new facilities are proposed. Concerns about aesthetic and environmental degradation and health threats from the new facility usually motivate opposition. Such opposition is common in most industrialized countries and in some rapidly developing ones (e.g., South Korea, Taiwan). This opposition extends beyond energy facilities, also affecting new highways, airport runways (e.g., until a year ago the two largest airports in Japan, Tokyo and Osaka, were restricted to using a single runway each), and transmission facilities. This phenomenon can be explained as an institutional failure, where mechanisms have not been formulated to provide sufficient benefits to those affected by the facility so that they would be willing to accept it. Potential mechanisms include transfer payments from electricity consumers to those affected, reverse auctions among competing sites, tax payments to affected municipalities, and promises of jobs reserved for local populations.

In France, reduced electric prices are, with some success, used as such an incentive. In the US, generous municipal taxes are restricted to the town where a power plant is located. As a result, the plant is welcomed by the town, but often not by the neighboring towns. In other places (e.g., South Korea), no incentives are offered, and this results in predictable siting difficulties. In Japan, payments to towns and affected populations (e.g., fishermen) are common, but only after protracted discussions.

17.2.6 Electricity economics

The economics of electricity are affected by the costs of energy generation, transmission and distribution, consumer behavior, and the use of electricity as a vehicle of social engineering. Generation costs are determined in much the same way as the costs of owning an automobile. These costs include capital costs, operational costs, and fuel costs. Capital costs are strongly influenced by the amount of hardware in the plant and by the time duration required to build the plant. Fuel costs reflect market prices for the fuel used, and operational costs (primarily for maintenance) reflect the plant's complexity and material condition. Typical cost breakdowns for various electric technologies are shown in Table 17.4.

Of the first four technologies listed, each can be economically competitive under some circumstances, typically depending upon the costs of fossil fuels, the requirements for pollution abatement and other internalization costs, the time duration of construction, and the power capacity of the plant. Fuel costs depend on the market price of the fuel and the thermal efficiency of the plant. Operations costs reflect the amount of maintenance work required for the plant. Most geothermal plants and solar (except for wind) are not yet economically competitive.

Table 17.4. Cost Shares of Various Electric Generation Technologies

Technology	Capital (%)	Fuel (%)	Operations (%)	Areas of Special Expense
Coal	45	40	15	Coal mining releases of CO_2, Particulates, NO_x, SO_2
Natural gas	35	30	35	Releases of CO_2, NO_x
Oil	40	40	20	Releases of CO_2, NO_x, SO_2
Nuclear	60	15	25	Reactor accident risks High-level waste disposal
Solar	75	0	25	Land consumption
Geothermal	75	0	25	Water and air pollution

In a competitive power-generation market, the costs of generation are critical to the choice of source. Tables 7.5–7.7 show costs of electricity from different sources using available commercial technology. Table 17.5 summarizes some of the key information from those tables presented earlier in the text.

In the regulated utility environment, capital charges usually are approved by the regulatory authority and incorporated into the general rate base paid by customers. Therefore, capital-intensive technologies, such as nuclear plants, hydroelectric dams (where large capital costs are justified for other uses beyond power generation), and advanced coal facilities, can be justified if they are an essential component of the generation mix for a utility. Reasonable fluctuations in fuel costs were also reviewed and often passed on to consumers.

Capital-intensive technologies are generally not competitive in a deregulated power-generation industry, unless an existing plant can be acquired from a seller who has already recovered most initial capital costs and is willing to sell it well below the cost of a new plant. Table 17.6 shows the comparative costs if capital costs are pulled out of the charges. Of course, there will be some capital charges in most cases, but they will be more uniform across the different sources.

If capital costs are excluded, then nuclear power is economically competitive with natural gas (depending on current gas prices and the operating costs of the particular nuclear plant). An older coal plant remains the most cost-competitive option.

Table 17.5. Comparative Costs of Generating Electricity (see Tables 7.5 through 7.7)

Energy Source	Technology	Cost of Electricity $/MWh			
		Fuel	O&M	Capital	Total
Coal	Pulv. coal	8.5–13.8	3.5	18.4–19.6	31.1–35.9
Natural gas	Combined-cycle turbine	16.6–37.7	2.0	10.7	29.3–47.4
Nuclear	Fission reactor	~ 6	13–18	28.5–42.2	50–70

Table 17.6. Comparative Marginal Costs of Generating Electricity ex Capital Charges (see Tables 7.5 through 7.7)

Energy Source	Technology	Cost of Electricity (ex capital charges) $/MWh		
		Fuel	O&M	Total
Coal	Pulv. coal	8.5–13.8	3.5	12–17
Natural Gas	Combined-cycle turbine	16.6–37.7	2.0	19–40
Nuclear	Fission reactor	~ 6	13–18	19–24

The economies of scale influence the costs of all plants, so coal and nuclear plants are usually built in units of 500–1,400 MWe. While both types of plants can run at somewhat lower power outputs, they lose efficiency and become less economic. These base-load plants run best when demand is constant, and operators try to create demand by offering cheap electricity during low demand periods (as discussed in Chapter 16). Many industrial users can accept a cheap interruptible electricity supply from the grid and generate their own more expensive power when demand is high and they are disconnected from the grid. The net cost to the industrial user is less than if uninterruptible electric power were purchased.

The economic size for a gas turbine for power generation is around 100–200 MWe. Larger power stations are operated off multiple units running in parallel. This gives load-varying capability, because the smaller units can be brought on and off the grid as required.

As discussed in Chapter 16, the problem of providing power to meet peak demands is a real challenge to the generating industry. To maintain service that has been guaranteed to customers, generators work with the ISOs to meet demand fluctuations. Regional "power pools" are being established in many locations to help coordinate generation schedules and work with the ISOs to reduce bottleneck problems in the grid. More expensive peak shaving power technologies can be brought on-stream under these conditions—perhaps burning propane or other fuels in smaller generators. However, there is a real incentive to seek other ways, through pricing or incentives to customers to install more distributed sources or through more efficient equipment that can lessen peak demands.

17.2.7 Ways of organizing the electric economy

The electric industry has traditionally been organized as a regulated monopoly, where state-level public utilities commissions were empowered to determine whether new capital facilities will be built and to set electricity prices such that investors will receive adequate returns in order to attract future investments. Today, utilities are organized in different countries on scales ranging from municipal to national, and with different mixes of public and state ownership and oversight.

As efficiencies improved through the deregulation of other industries (e.g., airlines, telecommunications, water, and gas utilities), an international experiment began in the UK in 1990 with privatization of the nationally owned Central Electricity Generating Board. Many other countries have followed the UK's lead. The results have been mixed, depending on the details of how the new markets are designed and regulated. Some successes have been achieved, increasing competitive generation markets (e.g., US and Australia), but, in the areas of retail customer price deregulation and specification of allowable contracts, such markets have not worked well. The reluctance of the political system to permit voters to be exposed to the risks of potentially high prices and the inability to store electricity appear to create much of the difficulty. However, the degree of success of electricity deregulation is important to the future of electricity—in terms of the technologies used, the prices paid, and the practices of consumers.

17.2.8 Demand-side management (DSM) and distributed generation

As electricity use grew rapidly over the last half century, the problems of meeting peak demands grew more challenging and expensive for the utility companies. Most companies instituted a variety of programs to reduce peak demands:

- Offering much cheaper electricity to customers who were willing to be shut off during peak demand times
- Providing services to consumers to encourage more efficient electricity use (insulating buildings, energy efficient lighting, high-efficiency appliances, smarter management of time of use, etc.)
- Encouraging investment in distributed generation sources that would reduce peak demands

Under deregulation, many of the DSM programs were dropped as the generation sector became separated from direct interaction in providing consumer electrical services. New generation companies focused on producing electricity at the most competitive rates. Meeting supply and demand became a secondary interest since the main responsibility for providing continuous service lay elsewhere. In a competitive environment, these programs were easily eliminated from the investment portfolio. The ISOs were struggling with the matching of supply and demand across the complex grid system, and the consumer service providers were not profitable enough to make investments in such programs.

If customers could install some local power generation to meet a portion of their demand, this would reduce the need for transmission and distribution capacity, especially if there were a match between times of peak demand and the output of the local supply. Customers could pay about twice the operating costs for the distributed generation and still break even (since they are avoiding the transmission and distribution costs). Some of the renewable generation technologies (as discussed in Chapters 9–15) are well suited for distributed power generation. For example, where air conditioning peak loads are the problem, installation of photovoltaic panels might be an effective way to relieve the peak, since hot-sun and hot-weather periods tend to coincide. Other distributed technologies, such as local wind farms or natural gas–fed fuel cells or solar thermal water heaters, could be used to reduce overall electricity demand. If some energy storage capacity is provided, intermittent sources like solar and wind energy can also be used with the storage used to even out the supply and demand patterns on a daily basis. (A more detailed discussion of distributed technologies is provided in the previous chapter. Section 17.4 provides a case study that illustrates some of the trade-offs involved.)

17.2.9 Electricity transmission and distribution and economic deregulation

The system for electricity transmission is a set of high-current, high-voltage wires and associated hardware (e.g., transformers and circuit breakers) used for long-distance (hundreds of kilometers to 2,000 km or more) transport of bulk electricity. This system feeds energy to a more locally arranged (of the order of 100 km or less) distribution network of wires operating at low voltage and current. Together, they constitute the grid, linking generating units to electrical loads. The central problem in designing and operating the grid is balancing generation and loads such that the individual grid components do not become overloaded and fail. Failures can cascade via subsequent dependent failures, and large regions can lose electrical service. (This happened in the northeast and central US on August 14, 2003, the latest of an international succession of blackouts on approximately a decadal frequency. See Chapter 16.) Distribution failures are much more common. They are often caused by overloading a distribution transformer (e.g., caused by electrocution of a trespassing animal) or by tree limbs falling onto distribution lines. As deregulation of electricity sales has become more common over the past 12 years, the range of load and generation combinations has also become much greater, with a corresponding increase in the frequency of grid failures. Finally, weather-related grid failures occur regularly, particularly those associated with ice storms and high winds.

The infrastructure of the grid is land-intensive (although the land used may simultaneously be employed for agricultural or recreational purposes) and may be considered aesthetically intrusive. For these reasons, in wealthy countries opposition to new power lines has become so common that it has become very difficult to build new ones. As demands have grown, the result has been greater stress on the existing grid.

Concerns have also arisen that the electromagnetic fields associated with the grid's operation may cause diseases, particularly cancer. So far these fears appear to be unfounded, but their persistence may reflect the discomfort that many people have with large-scale technologies. In the absence of particular benefits for those living near power lines, these factors have often been sufficient to motivate strong opposition to new facilities. This has made it difficult to improve the grid in order to keep pace with growing demands and operating conditions.

Higher voltage transmission lines, some perhaps utilizing superconductivity to reduce energy losses, may make longer distance power transmission more attractive in the future. Smarter control systems and systems to facilitate real-time pricing of power may ease the inefficiencies associated with trying to match different supply and demand characteristics. Though the transmission and distribution grids have not yet been deregulated, many experts are trying to construct a new paradigm for the entire supply-to-end-use customer chain—where pricing is not based on electricity delivered, but on the electrical services (or even total energy services) delivered. The concept has many attractive attributes, but the reality of implementation is daunting. The evolution of the present system into tomorrow's system will likely entail a number of successful and unsuccessful experiments and a gradual learning path to a more sustainable electric industry future. Part of this challenge will involve the comparison of electricity and hydrogen as energy carriers and a resolution of the drawbacks and benefits of each for the spectrum of energy services that will be needed in the future.

17.3 An Example of Electric Industry Planning Using Multiattribute Assessment Tools

The preceding sections outline the complexity of choices facing a variety of participants in the electric industry, as well as their customers and regulators. With deregulation, choices become even more challenging. An ideal market operates efficiently to select the most cost-effective products and production processes among competitors. But the electric market is far from ideal. We do not know how to properly price things that are of common societal benefit, except through a variety of market interventions like taxes, fees, subsidies, and regulations. Where these are imperfect or do not exist, the societal benefits may not be addressed responsibly. Some of the major issues facing the electrical sector include:

- Assuring reliability of service
- Reducing GHG emissions
- Providing affordable basic energy services to the poor
- Siting facilities to reduce adverse land-use impacts

Such issues are of concern to the industry and to society. But the industry is fragmented and competitive. Various customers and regulatory groups are involved, as well as special interest groups that advocate for environmental responsibility and consumer rights. How can such a complex group of players associated with the electrical industry help it to evolve to a more sustainable state under the present ground rules?

Electrical services are provided by a multidimensional, complex system with many competing attributes and players. System complexity and options are beyond the possibility of productive decision modeling through assessing the utilities of each of the participants. The way that utilities actually dispatch power in real time is done using real-time algorithms that consider the spot prices of various fuels, as well as the operating costs and conditions of available generating units. These factors make equilibrium system models inappropriate.

Researchers (Connors, 1991) evolved the concept of modifying a real utility dispatch model to include the various attributes of potential interest to a broad group of stakeholders (e.g., the utilities, the regulators, the consumers, and various interested non-governmental organizations [NGOs]). They started with a set of knowledgeable representatives from the major local and regional stakeholder groups in New England who each agreed to participate in an interactive team effort to develop the model and guide analysis of options through a set of scenarios. In this project, the system model developers functioned as a technical support team to enable the work of the stakeholder steering group. Support for the project was initially provided by a group of utilities in the northeastern US.

The initial multi-stakeholder group suggested scenarios that they were interested in exploring. For instance, what happens if:

- The proportion of renewable energy on the system doubles.

- Grandfathered coal plants are phased out on a 10-year schedule.
- Certain nuclear plants are decommissioned.
- Primary energy prices vary over wide ranges.
- Aggressive DSM programs are pursued.
- More or less electricity is imported from Canada.
- Emissions regulations are tightened further

The basic methodology enabling the analysis follows the lines shown in Figure 17.3. The attributes (A_i, A_j, etc.) are chosen so that less is better—for example, lower costs or lower emissions. The group then decides on various options that they wish to explore relating to how the system will operate over the next 20 years. Options may include the types of new generation that are installed, the variability in primary fuel prices, and the DSM programs promoted. All the feasible combinations and permutations of options are identified as individual scenarios and the model computes the attributes of interest for each scenario.

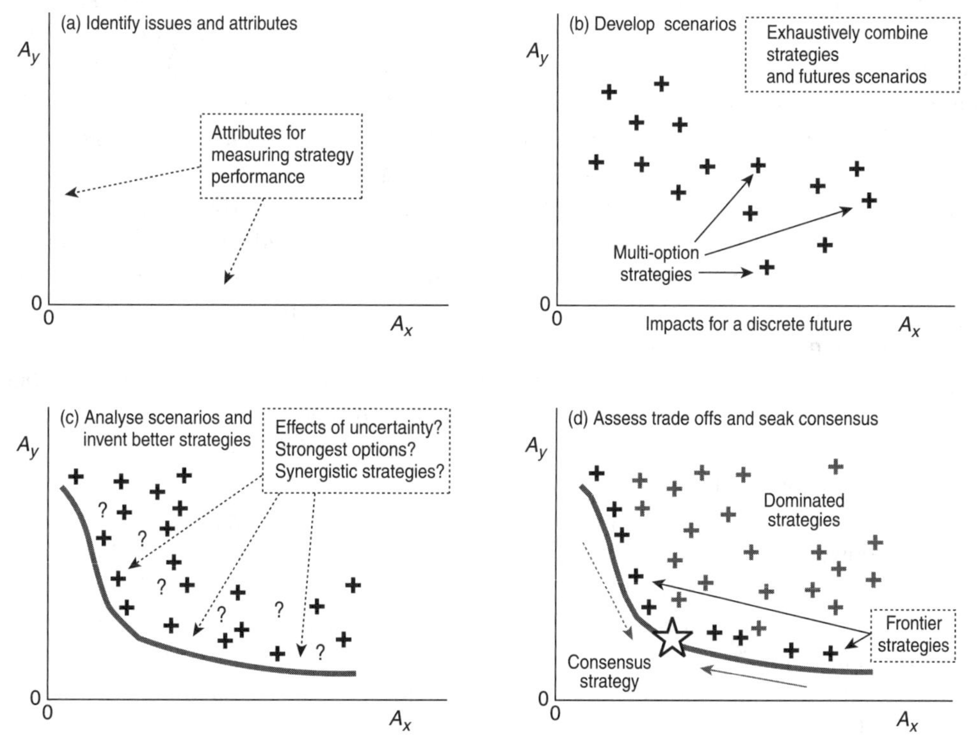

Figure 17.3. Basic elements in performing multiattribute trade-off analysis in a multi-stakeholder policy debate. Source: Connors (1991).

In Step (b) of Figure 17.3, the scenarios are plotted for two selected attributes. Similar plots can be made for other pairs of attributes as desired. There is a frontier of options that are cheaper and cleaner (in the case of a cost versus emissions plot), because they dominate other options that provide the same level of emissions at a higher cost or lower emissions at the same cost. As these frontier scenarios are identified in Step (C), the steering group discusses their attributes and gains ideas for new options or for synergistic strategies. With multiple attributes of interest, the group examines whether the best strategies for one set of attributes also provides good results for the other attributes of interest. When a final set of dominant strategies is identified, the group considers where they wish to be along the frontier line, which presents, in the case discussed here, costs for different levels of environmental performance.

The method is particularly powerful in getting the group to understand the positions and interests of the other stakeholder participants, in creating new ideas for options that may not have been identified initially, and in building a consensus among the participants. Participants can carry their new understandings back to their individual organizations and raise awareness among all stakeholders.

The Analysis Group for Regional Electricity Alternatives (AGREA) started by identifying the attributes of electricity generation that were of interest to some or all of the group members. Costs and emissions were of prime interest, although other attributes, such as system reliability, were included in other studies. In initial runs with the model, participants matched the model results with their own experiences—in some cases finding ways the model needed improvement and sometimes finding that their own understanding of the system behavior was flawed. As they built confidence in the reliability of the model, they were also able to learn and teach each other about some of the complexities of the system. Regulators learned that tightening emissions regulations on new plants would probably result in increased usage of the older coal plants that were grandfathered and worsen total emissions. Utilities were able to better understand how their dispatch protocols could be improved to provide better environmental performance at essentially no additional cost (AGREA, 1993). Some utilities were surprised to find that they were using gas turbines installed for peaking purposes as base-load power when gas prices were low. The basic techniques used have now been applied to many different scenarios and locations, including Switzerland (Connors, 1996) and Shandong Province in China (Connors et al., 2002).

As an illustration of the methodology, a summary of a study of niche opportunities for wind power in the New England region is presented here (Connors, 1996). This study was initiated because renewable technologies like wind power were usually evaluated on the basis of installation and operating costs without consideration of how they might actually improve the operational and flexibility characteristics of the system as a whole.

The steering group identified the following major attributes of interest:

Cost attributes:

> **PV (present value):** Total cost of electric service over 20-year study period in terms of present value (see discussion in Chapter 5) using utility cost of capital as the discount rate (standard economic analysis)

IA (intergenerationally adjusted): Same as above, except that future capital costs are not discounted but brought back to 1991 dollars and then summed ("sustainability economics" where intergenerational value is maintained)

RA (risk adjusted): Same as PV, but adjusted for financial risk by applying different discount rates to operational costs and capital costs (risk-adjusted discount rates) (See Awerbuch, 1995)

Environmental attributes:

NO_x: Cumulative nitrogen oxides stack emissions from power generation over 20-year study period

CO_2: Cumulative carbon dioxide stack emissions from power generation over 20-year study period

The dispatch model used in the study was based on the existing power generation facilities in place at the start of the study period, and the modelers added in various facility options for expanding power supply as appropriate to the scenarios studied. Other options included choices for primary energy sources, existing unit longevity, methods for new natural gas supply contracting, future levels of NO_x controls, and level of DSM.

Combinations and permutations of option sets formed the scenarios of interest. For each scenario, the attributes (costs and emissions, in this case) were computed.

To identify the individual cases, a five-letter name showed the option combinations:

First letter – Supply-side technology mix: G = gas/oil; H = gas/oil + clean coal; W = gas/oil + wind; D = gas/oil + wind + clean coal; K = gas/oil + wind + clean coal + biomass

Second letter – Existing unit longevity: I = life extension of existing units; O = retire/repower 10% by 2011; A = retire/repower 20% by 2011

Third letter – New unit natural gas contracting: S = all spot price gas (economic dispatch); M = 70% take or pay gas (must run status)

Fourth letter – CAAA Title I – NO_x controls: A = Phase I 1995—reasonably available control technology (RACT); E = Firm Phase II NO_x controls (target 60%); I = Hard Phase II NO_x controls (target 80%)

Fifth letter – Level of DSM: R = 1992 level of utility DSM programs; D = double the 1992 level of utility programs; C = Triple commercial and industrial conservation.

Thus, a scenario named WISER would be (W) gas/oil + wind; (I) life extension; (S) spot gas pricing; (E) firm Phase II NO_x controls; (R) 1992 DSM levels.

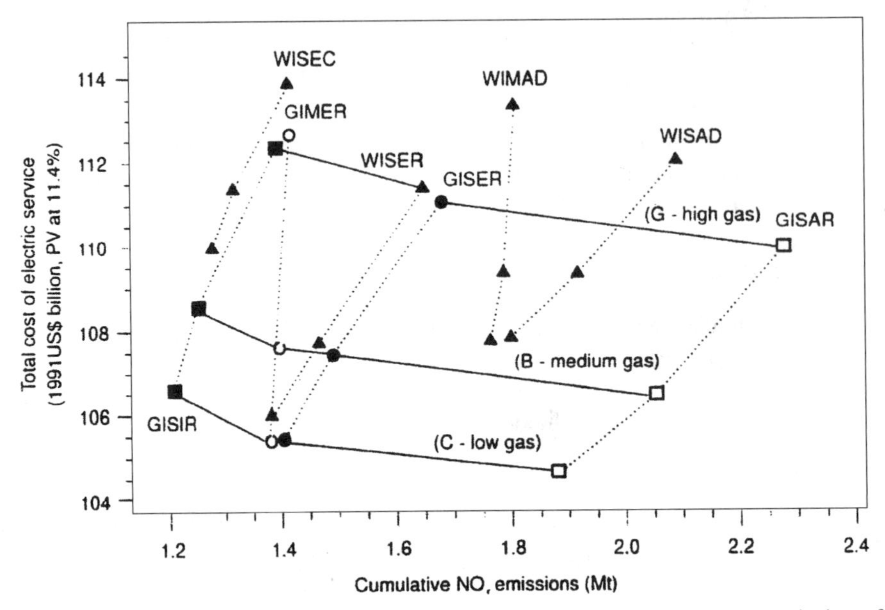

Figure 17.4. Scenario analysis for utility present value costs and NO_x emissions for selected multi-option strategies for three natural gas pricing options.

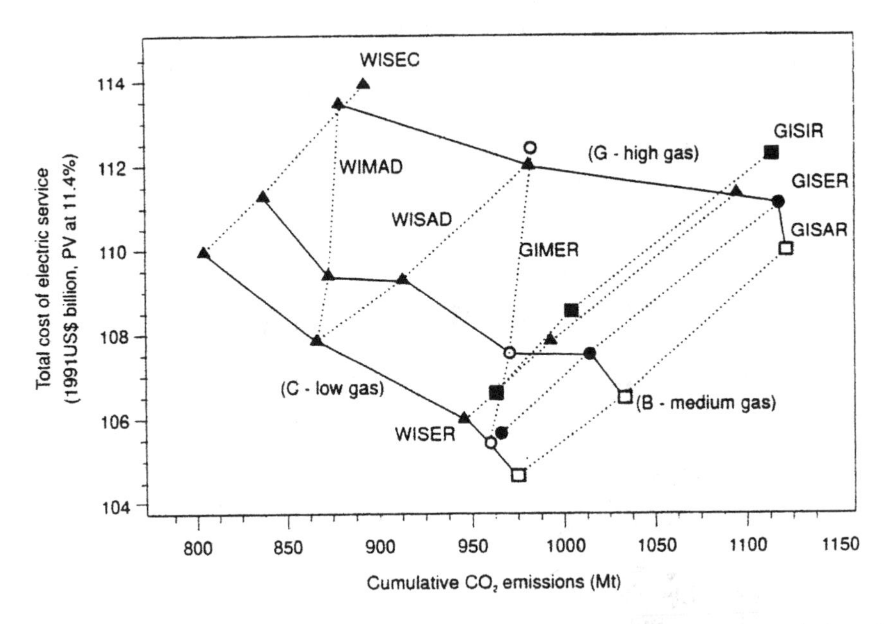

Figure 17.5. Scenario analysis for utility present value costs and CO_2 emissions for selected multi-option strategies for three natural gas pricing options.

The full group of scenarios are run for the 20-year future period. The results are plotted to show cost versus environmental attributes. For example, Figure 17.4 shows the utility present value cost of electrical service over the next 20 years (1991 US$ billion, PV at 11.4%) as a function of cumulative NO_x emissions (million tons). Figure 17.5 shows the same cost basis as a function of cumulative CO_2 emissions (million tons).

Figure 17.4 shows that the wind strategies, along with those with higher levels of DSM, are not cost competitive (on a PV basis) with the gas/oil strategies for NO_x reduction even when gas prices range considerably. However, they are much more attractive for CO_2 abatement, as shown in Figure 17.5. For CO_2 abatement, wind strategies and DSM also provide protection against increases in CO_2 emissions if gas prices increase substantially (encouraging more use of older coal plants). Connors (1996) also explores the effects of other cost bases and of variation in the future capital costs for wind turbines. In his conclusion, he states:

> ...As transition issues such as stranded investments become clarified, topics such as who will be responsible for demonstrating overall environmental compliance in a decentralized industry will need to be addressed. Trade-off analysis will again be useful in evaluating such stranded emissions risks, and how to avoid them. As that happens, the ability to use wind, solar, and DSM resources to both "buy down" pollutant emissions and reduce fuel cost and environmental compliance risks will become apparent.

He ends with the following challenge:

> While the multiattribute framework can be used to evaluate the performance and risks associated with resource choice, a question must be asked. If the development and broad-based application of learning tools like multiattribute trade-off analysis did not occur in a vertically integrated utility industry, who—if anyone—will promote their development and use once the planning function has become decentralized? While a somewhat rhetorical comment, it raises doubts about who will be responsible for longer-term issues of market coordination, and identifies the need to develop an institutional niche in the restructured electric industry where balanced policies—which give wind and other resources their due—can be evaluated and implemented.

17.4 Energy Market Impacts on Electricity Generation Options

The low cost of coal has made it the backbone of the electric power industry. However, in the past decade or so, costs of coal-fired generation have increased because of strict emissions regulations. Also, demand growth slowed, so that the major capital costs associated with large coal plants were hard to justify. Adding smaller and cheaper gas-fired units became the industry's new capacity choice.

The major source of uncertainty in the COE from natural-gas-fired electric power plants is fuel costs. Assuming stable coal prices, $2.70/MMBtu ($2.56/GJ), gas is an apparent threshold above which coal-based generation is economically favored (Table

7.5). An understanding of natural gas markets, and in particular of gas deliverability at various prices is important to gauging the profitability and sustainability of power generation options and of projects to supply new gas. Utilitarian analysis of prospective energy markets is challenging, owing in part to appreciable uncertainty regarding future economic progress, social and political behavior, and advances in technology. Nevertheless, careful exegesis of publicly accessible data can provide valuable insight into short- and long-run planning.

Demand Behavior From 1980 to 1993, there was a significant excess of gas in the US. This "gas bubble" had several causes. Wellhead price controls were removed, creating a competitive market that was more responsive to prices than demand. Anticipated growth in demand was delayed by repowering of coal-fired electric generating stations and fewer than expected shutdowns of nuclear power plants. Recently, several forces have conspired to eradicate this surplus and push demand and prices upward. A major driver has been increasing demand from electric power generators spurred on by economic expansion and population growth, especially in California. Recent cold winters and higher than normal drawdowns of hydro reservoirs in the Pacific Northwest (Lynch, 2001) have exacerbated the situation. Despite the US economic slowdown, electric sector gas demand growth will likely persist, because gas-fired plants offer electricity generators lower capital costs and the option of staging capacity expansions in smaller increments reducing the risk of overbuilding. Further, there is considerable consumer and political pressure to correct for under-building of electric generating capacity in the 1990s.

California is up to 10 GWe short on in-state supplies. Consistent with these projections, the US EIA forecasts that US gas consumption will expand over the next 20 years. They project 24% and 54% higher US gas demand in 2010 and 2020, respectively, primarily from increased gas use in electric power generation. It is no surprise that US gas demand in 2000 exceeded 1972 demand for the first time in 28 years. Happily, North American gas reserves are plentiful—about 2,000 trillion cubic feet (tcf), or about 80 years supply at recent production rates. Canadian domestic use will grow but this will still leave ample stocks for export (see below). But how much gas will be available in the Lower 48, when, and at what price? The answers will be determined by complex interactions among various forces that govern what Jensen (2000) describes as a "new North American natural gas economy."

Jensen's New Gas Economy In May 2001, unlike the early- to mid-1970s, a market-place economy, not government regulations, determined North American gas prices and the value of gas transmission assets. The previous 18 months witnessed substantial increases in gas prices, punctuated by near stratospheric spikes. For example, Henry Hub (Louisiana) Bid Week Spots were $9.70/MMBtu in December of 2000 versus $4.00/MMBtu in December of 1997. (The December, 2000 spot corresponds to $56.26/bbl oil on a Btu equivalent basis.) These high prices resuscitated dormant supply projects and stimulated investor interest in new initiatives to provide more gas to the

Lower 48. Jensen (2001) teaches that fresh thinking about natural gas market behavior under competition provides a more consistent understanding of prospective gas availability and the profit potential of long-term gas supply projects.

Pitfalls in Market Forecasting To evaluate the economic feasibility of capital intensive (up to several billion dollars each) long-term gas supply ventures, it is essential to understand long-term gas prices (Jensen, 2001). Short-run prices (see below) are just too volatile to illuminate profitability prospects. A major pitfall is the misconception that futures markets forecast long-term prices. To illustrate, Jensen (2001) calculated pre-tax returns of 9, 30, and 90% for a hypothetical project importing Nigerian LNG to Cove Point, Maryland, based on the March 1999, 2000, and 2001 New York Mercantile Exchange (NYMEX) 12-month strip prices of 2.12, 2.97, and 5.44 ($/MMBtu), respectively. The NYMEX strip is a 12-month prospective price, or a "future." If this were a reliable index, the March 1999 strip would presumably attract little capital other than from "mattress stuffers," whereas the March 2001 strip would create a stampede of investors anxious to cash in on hedge-fund-like returns. Clearly, neither return on investment (ROI) matches industry experience in capitalizing such projects.

Short-Run Market Behavior - Coupling with Oil Prices Real-world, short-term gas supply/demand relationships are far more complex than the behavior implicit in the simple "X-like" supply/demand curves found in introductory economics textbooks. Jensen (2001) argues that it is not the absolute price of gas but rather the gas-to-oil price ratio (GOPR) that is important in determining gas demand. Under gas surpluses, gas prices become decoupled from oil prices and there can be gas-to-gas competition, as occurred in the US for much of the 1990s. Gas prices then decline substantially. However, when there is oil-to-gas competition, price buffering is provided by electricity generators' ability to replace gas with petroleum-derived fuels. Gas prices triggering switching are often expressed as a percentage of refiners' acquisition cost (RAC) for crude oil of the same heating value. For example, there was once a "ten-to-one" thumb rule, meaning that $25.00/bbl oil resulted in a switch from gas to residual fuel oil when gas reached $2.50/MMBtu. $25.00/bbl oil is equivalent to $4.31/MMBtu (divide $/bbl by 5.8 million Btu/bbl) and 2.50/4.31= 58% so that switching elasticity set in when gas prices rose to 58% (interestingly not 100%) of the RAC. More recently, owing to tightening environmental standards and lower availability of resid, the switchover point has risen to about 90% RAC. Under oil-to-gas competition, developers of long-term projects need to be concerned that a collapse in oil prices will drag down gas prices and interrupt or even extinguish the high ROIs needed for project profitability.

Another petroleum-derived liquid fuel, middle distillate oil (MDO), provides a higher price buffer against unbounded increases in gas prices. MDO, a petroleum fraction boiling in the 350–700°F range (versus 700^{+}°F for resid), consists primarily of No. 2 residential heating oil, diesel fuel, jet fuel, and kerosene (NRC, 1980). The kerosene and jet fuel components are much cleaner than resid and more volatile. They are well suited to direct firing in stationary gas turbines for electric power generation. MDO has typically been available at 140% of RAC, implying a $6.00/MMBtu natural gas price bench with oil at $25.00/bbl. Gas prices above this value have, so far, not been sustainable, which Jensen (2001) attributes to buyers' embracing the purchase of lower cost MDO instead of gas after a modest learning period. Gas prices become inelastic if their relationship to resid or MDO prices change (e.g., owing to resid or MDO scarcity). Connors (2001) notes a complication in MDO fuel switching. Some state regulatory bodies (e.g., Massachusetts) sharply limit the amount of MDO an electricity generator can substitute for natural gas.

Long-Term Gas Supplies What really matters over the long haul is what amount of gas will markets absorb at various prices, and if the market will sustain prices high enough for long-term supply projects to compete for needed capital. Natural gas prices are typically forecast from economic models. Gas supplies are projected from models that combine detailed estimates of the resource base in various regions with information on drilling productivity and costs of production to predict how much gas will be available at what price. Several models have projected substantial expansion of long-term supplies with little or no change in prices (Jensen, 2001). These models have trouble in forecasting the time details of availability, especially whether cost relationships established for gas in surplus will remain valid under pressures for accelerated rates of discovery. The ultimate measure of gas shortage or excess is neither the resource base nor the rate at which developers can find gas within these deposits, but rather the deliverability of the gas or the rate at which discovered gas can be produced.

To expand gas production rates, reserve additions must increase, which means discovery rates must accelerate. However, between about 1985 and 2000, surplus gas in the US and Canada provided little incentive for intense exploration. Today, both countries seem to have left their days of excess gas behind. The US gas bubble has ended. After 1985, Canada began working off its large inventory of gas development prospects. Between 1978 and 1985, production from these wells had been stultified by export limits and domestic set asides. The US and Canada may even be approaching "technical" limits on deliverability (yearly production rate as a percentage of proved reserves), which are set by geology and technology. Economics will provide an ultimate limit on deliverability (i.e., when sales from incremental production of gas can no longer provide acceptable rates of return on the cost to produce that gas).

Jensen (2001) believes that, owing to environmental concerns, neither nuclear nor coal-based electric power generation will significantly moderate electric sector natural gas demand in the US. This contrasts with Japan, where coal and nuclear are combined to confine expensive liquefied natural gas (LNG) imports to about a 30% market share, largely intermediate load dispatch. Others, including Ellerman (2001a,b), envisage a greater market share for coal in base-load US electricity generation and, thus, in tempering at least some of the demand growth for natural gas in this sector. Interestingly, the fear or reality of sustained upward movement of gas prices may have begun to challenge the long-run stability of coal prices. Ellerman (2001a) points out that recent increases in coal prices from major US provinces show a strong correlation with high gas prices in the regions serviced by that coal. For example, Powder River Basin coal, which serves Texas, Oklahoma, and Arkansas, has tripled in price, and southern Appalachian coal has doubled. Prices of northern Appalachian coal, which serves regions in less competition with gas-fired electric power production, have gone up only 50%.

Summary Although contrarian views exist, there is now evidence that Lower-48 gas markets are exhibiting non-traditional behavior that has important implications for future US electricity costs and for the market shares of coal, gas, and nuclear electricity generation. Jensen (2001) proposes that long-term North American gas prices must exceed their historical \$2.00–\$2.50/MMBtu range to attract the multibillion dollar investments necessary to provide enough new supply to meet anticipated demand growth from the electric power sector. How high gas prices may need to rise is open to debate. When gas is not in surplus, short-run prices can be buffered by electricity generators' substituting residual fuel oil or middle distillate oil for gas when gas prices (per MMBtu), respectively, rise to about 90% and 140% of refiners' acquisition cost (RAC) for crude oil (140% of RAC corresponds to \$6.00/MMBtu gas for \$25/bbl oil). These price benches would be extinguished by scarcity of resid or MDO. LNG importation is another means of expanding gas supply at these higher price levels. Long-range gas supply projects could be undermined by declines in drilling yields with increasing drilling activity (now in evidence) or by weakened gas prices induced by a collapse in oil prices (very possible) or a return to a long-run gas surplus (plausible but probably unlikely).

To avoid service interruptions and unpopular rate increases, gas-dependent wholesalers and retailers of electricity must understand Lower-48 gas markets. A diverse national and regional portfolio of electricity supply options will hedge against vicissitudes in fuel markets and public acceptance, while retaining the nation's ability to provide affordable electricity to all consumers. Central hydro, nuclear, coal, and gas, distributed generation, and various renewables, when economical, will contribute.

17.5 Sustainability Issues

Over the last century, the electric industry has gone through a long period of rapid expansion and now provides fairly inexpensive and reliable service throughout most developed countries. The industry has partially moved from a centrally controlled and regulated architecture into a new configuration in which the generation section is deregulated and operates competitively. The intent has been to further reduce electricity prices by introducing the benefits of market competition. The transition has been difficult because the reconfiguration of a part of the system has had some major unanticipated repercussions. These are being dealt with in a variety of ways, but today's electric industry will have to evolve much further to achieve efficient, not to mention sustainable, operations in the future.

Meanwhile, the developing world is struggling with bringing the benefits of electrification to both its urban and rural populations. Their growing pains mirror those of the earlier development of electric systems in the developed world, although they have access to the technological developments that have made power generation cleaner and more efficient. However, some developing countries need power so desperately and have such limited resources that they are using cheaper and dirtier technologies in the interim. Since electricity is key to industrialization, these developing countries will address problems of pollution and reliability of service when they can afford to do so—or if they receive foreign assistance. Meanwhile, electrification is likely to bring benefits of improved health care, better education, and falling birth rates.

In the process, developing nations are likely to contribute significantly to both global warming and regional air pollution (e.g., China and India rely heavily on coal).

The electrical system worldwide is in an expansion mode as new customers are acquired and as existing customers demand new services. The industry is a major emitter of GHGs—notably CO_2 from fossil fuel combustion, but also CH_4 from gas leakage during transmission to gas-fired power generators. Expansion in service for the same primary fuel mix necessarily means increases in GHG emissions. Further, when bulk sources of GHG emissions are identified, power plants are at the top of the list (as opposed to transportation GHG emissions, which are distributed among the tailpipes of millions of individual vehicles). Until there are regulations limiting GHG emissions or economic penalties associated with emitting GHGs, it is unlikely that improvements in efficiency and the introduction of new technologies that are cost-competitive under present pricing rules will be able to slow the future increases in GHG emissions associated with growth.

What are the other options available in a future where GHG emissions from the electric sector have to be reduced? Possibilities include (see Figure 4.18):

- Increased use of renewable energy sources
- Increased use of nuclear power
- Introduction of carbon capture and storage at fossil fuel power plants
- Reduction in consumer and industrial usage of power (but not necessarily in services)

In Chapter 6, we learned that it is important to examine the consequences of each of these options on a lifecycle basis and to understand the impact of the changes on society as a whole. Increased use of renewable energy sources can reduce GHG emissions, but this will probably come at some higher cost and will have land-area impacts because the renewable sources are more distributed. Flavin and Lenssen (1994) present estimates in Table 17.7.

Table 17.7. Land Requirements for Electric Generation Technologies

Technology	Land Requirements [km^2 per exajoule-year]
Renewables	
Dedicated biomass plantation	125,000–250,000
Large hydro	8,300–250,000
Small hydro	170–17,000
Wind	300*–17,000
Photovoltaic (PV) central station	1,700–3,300
Solar thermal trough	700–3,000
Fossil fuels (including mining/production)	
Bituminous coal	670–3,300
Lignite coal	6,700
Natural gas-fired turbines	200–670

*Land actually occupied by turbines and service roads

Source: Flavin and Lenssen (1994).

Biomass production (see Chapter 10) is land intensive because most flora have solar energy conversion efficiencies of 1–2%. In addition, harvesting and transport of biomass may require use of fuels and will further reduce the net amount of biomass energy gained. Solar PV systems have conversion efficiencies of 10–15%. The peak solar energy flux varies with geographical location. (Solar-based projects are more attractive in areas nearer the equator.) Solar flux is intermittent because of diurnal cycles, seasonal cycles, and cloud cover. Wind energy is also variable by location and intermittent. Further, the best sites for wind turbines are often in water bodies. Wind power also has fairly large land requirements when the area for generating the wind is considered, although the land between wind turbines can be used for other purposes. The intermittent nature of solar and wind energy usually means that another backup energy supply or energy storage system is needed. Large hydropower also has major land requirements and disrupts natural flows of water with associated perturbations of the aquatic ecosystem. In addition to costs, the land-intensive nature of the renewable sources will be a limitation on their use as bulk power generation sources, although they will certainly have a small but growing role in reducing GHG emissions.

The nuclear power option (see Chapter 8) also has some other impacts. On a lifecycle basis, the large amounts of concrete and other materials needed in power plant construction add to GHG emissions. Safety concerns are widespread, and the pressure on today's nuclear plants to compete with fossil plants in the generation sector may cause premature shutdowns. Nuclear safety may not be priced properly in today's market structures, although the industry pays high insurance premiums and the government provides further backing through the Price-Anderson Act. New generations of safer nuclear plants may be developed, but safety and proliferation concerns will need to be addressed.

Carbon capture and storage may be an interim strategy to facilitate a transition from major dependence on fossil fuels to other energy sources. Table 17.8 shows some of the technologies being evaluated for use in the power industry.

Herzog (2002) examines the increases in electricity production costs that would be associated with capture and sequestration from a pulverized coal plant (PC) and from an integrated gasifier combined-cycle (IGCC) coal plant. The third case evaluated is for capture and sequestration of CO_2 from a natural gas combined-cycle (NGCC) plant. Figure 17.6 shows the impact of sequestration on the costs of electricity for those three technologies. The triangles show typical costs of production without capture and sequestration; the circles and error bars show the estimates for the same net power generation with 90% of the emitted CO_2 captured.

The costs of capture vary by technology and might influence the choice of fossil generation technology if it were to be used for a few decades or more. However, carbon taxes in the range of \$100–200 per tonne CO_2 could make the above technologies economic. This option also would have infrastructure impacts as CO_2 would have to be transported to sequestration locations—probably in depleted oil and gas fields or in deep aquifers, or even in deep ocean disposal. All these potential impacts will need to be carefully weighed against those for other mitigation options.

Reduction in energy consumption sounds like a positive goal, but expansion in markets has produced the capital to replace or repower existing generation facilities. A new costing model will be needed if our sustainable future results in a stabilized or dwindling demand for power. There will be effects on the electric industry, and consequently, on national and global economies during a transitional period. In a society that values economic growth, this will mean a major value shift.

Table 17.8. Approaches to CO_2 Capture from Power Plants

Approach	Coal Technology	Gas Technology
Combustion with flue gas clean-up (Ambient pressure)	Pulverized coal or conventional coal combustion with steam turbine	Flue gas scrubbing with amine solution to remove CO_2 (amine regenerated and CO_2 sequestered)
Combustion in oxygen (Slightly above ambient pressure)	Oxygen plus recycled flue gas instead of air in combustor with steam turbine	Flue gas (CO_2 and water) passed through modified combined-cycle turbine, dried, and CO_2 sequestered
Gasification (High pressure)	Coal gasification	Syngas passed through shift reactor, CO_2 separated and H_2 burned in combined cycle turbine

Source: Herzog (2002).

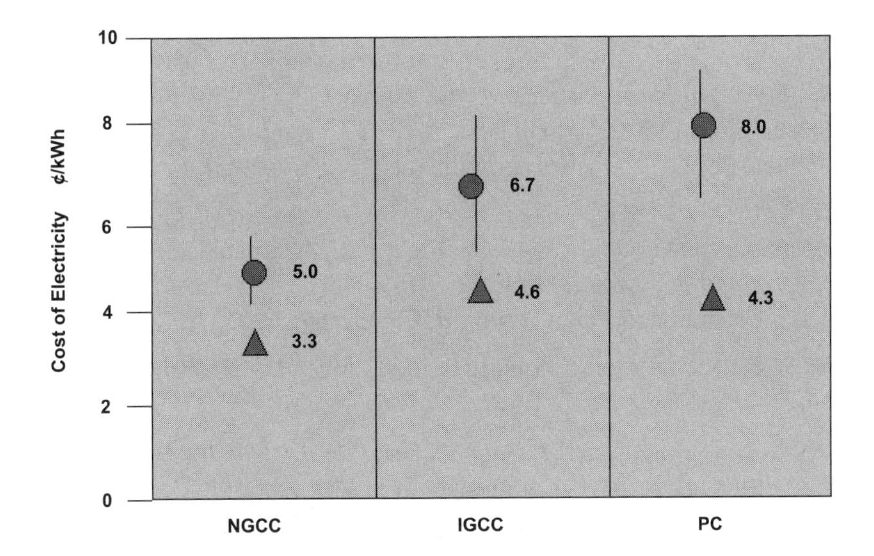

Figure 17.6. Estimated costs of electricity with capture. Source: Herzog (2002).

Perhaps the hydrogen economy will emerge. Hydrogen is an energy carrier like electricity—not a primary source of available energy. If hydrogen is to be made from electrolysis rather than from fossil fuel sources, the role of the electric industry might even expand to include transportation energy.

However, this would mean that even large amounts of electricity would be needed and that such electricity would have to meet limitations on total GHG emissions. If hydrogen were to be made from natural gas (the most economic pathway) there would be significant additional demand on natural gas, having impacts both on short-term markets and rates of depletion. GHG emissions would still occur, but at a lower level.

The future is always unclear. Electricity offers many advantages to a sustainable future if used carefully, but perhaps we will find better ways to provide the services we value today with less power consumption. Perhaps we will change our attitudes toward what we need to live good lives, or we will stumble our way into the future learning from our mistakes as humanity has been doing for centuries. Perhaps our mistakes will be correctable without disaster to ourselves, future generations, and the ecosystems that coexist on this planet.

References

Abbey, D.E., Ostro, B.E., Petersen, F., and Burchette, R.J. 1995. "Chronic respiratory symptoms associated with estimated long-term ambient concentrations of fine particulates less than 2.5 microns in aerodynamic diameter (PM2.5) and other air pollutants." *J. Expo. Anal. Environ. Epidemiology*, 5:137–159.

AGREA (Analysis Group for Regional Electricity Alternatives). 1993. *Background Information for the 1992/1993 Scenario Set: Second Tier*. MIT Energy Laboratory Technical Report. Cambridge, MA.

Awerbuch, S. 1995. "Market-based IRP: it's easy!!!" *The Electricity Journal,* 8(3): 50–67.

Connors, S.R., comments at *A Forum On Enabling Distributed Generation*, Energy Laboratory, MIT, Cambridge, MA, May 18 (2001).

Connors, S.R. 1996. "Informing Decision Makers and Identifying Niche Opportunities for Wind Power." *Energy Policy*, 24(2).

Connors, S.R. (Analysis Group for Regional Electricity Alternatives). 1991. *Externality Valuation versus Electric Power Systems Analysis*. MIT Energy Laboratory Working Paper EL 91-002. Cambridge, MA.

Connors, S.R. and W.W. Schenler. 1998. "Climate Change and Competition—On a Collision Course?" Proceedings of the 60th American Power Conference, April 14–16, Chicago, IL.

Connors, S., C. Cheng, C. Hansen, and C. Barker. 2002. *Shandong; China Electric Sector Simulation Assumptions Book (CETP)*. MIT Laboratory for Energy and the Environment, Report LFEE 2002-002. Cambridge, MA.

EIA (US Department of Energy, Energy Information Administration). 2001. *Annual Energy Review 2001*. Washington, DC: US DOE.

Eisenbud, 1964. "Radioactivity in the Atmospheric Effluents of Power Plants that Use Fossil Fuels," *Science,* 144:288.

Ellerman, A.D., comments during discussion period, *Energy and Environmental Policy Workshop*. MIT Center for Energy and Environmental Policy Research, MIT, Cambridge, MA, May 4 (2001a).

Ellerman, A.D., Center for Energy and Environmental Policy Research, MIT, Cambridge, MA, personal communications to W.A. Peters, Feb 7, (2001b).

Flavin, C. and N. Lenssen. 1994. *Power Surge— Guide to the Coming Energy Revolution*. Worldwatch Institute. New York: W.W. Norton & Co.

FRA (Federal Railroad Administration). 2003. "Railway Safety Statistics Annual Report 2001." US DOT Fed. Railroad Administration report. Washington, DC: US Department of Ttransportation.

Herzog, H. 2002. *Understanding the Economics of Carbon Capture and Storage. Proceedings of the Third MIT Carbon Sequestration Forum*, November 13–14, Cambridge, MA: The MIT Press.

Jensen, J.T., "North American Natural Gas Markets—The Revenge of the Old Economy?", oral presentation and written handout materials, *Energy and Environmental Policy Workshop*, MIT Center for Energy and Environmental Policy Reseach, MIT, Cambridge, MA, May 4, (2001).

Lynch, M.C., WEFA, personal communications to W. A. Peters, Feb. 26, (2001).

MSHA. 2003. "Roof/Rib Fatalities." Mine Safety & Health Administration report. Washington, DC: US Dept. of Labor.

Meyer, H.W. 1972. *A History of Electricity and Magnetism*, MIT Press, Cambridge, MA.

NRC, "Refining Synthetic Liquids from Coal and Shale," Panel on R&D Needs in Refining of Coal and Shale Liquids, National Research Council, National Academy Press, Washington, DC (1980).

Nye, D.E. 1998. *Consuming Power—A Social History of American Energies*. Cambridge, MA: MIT Press.

Wattenberg, B. 1976. *Statistical History of the United States*. New York: Basic Books.

Web Sites of Interest

www.eia.doe.gov/fuelelectric.html

www.epri.com

www.eei.org

Problems

17.1 Recently, a spokesman for a power plant that is being built near the US border in Mexico to supply power primarily to US markets (California area) said that the new plant would have the best emissions performance of all similar plants now operating in Mexico, and would exceed the performance of many power plants now operating in California. Was he telling the truth? Explain your answer in a few sentences and present the basis for your conclusion.

17.2 Compare the land requirements given by Flavin in Table 17.6 with the data presented in Table 9.2. Are the values reasonably consistent? If not, what do you think the reasons are for discrepancies?

17.3 Great Britain has become largely dependent on natural gas for power generation, as a result of their finds of gas in the North Sea. Most coal plants have been phased out, and new nuclear plants are not being built. As the North Sea gas field becomes depleted, Great Britain will have to increase its gas imports—primarily from Russia and Libya—or move to new power generation from coal. List the pros and cons of these future options and make your policy recommendation to the British government.

Transportation Services 18

An agreeable companion on a journey is as good as a carriage.
—Publius Syrus, 42 B.C.

18.1 Introduction and Historical Perspectives

Urban planners use systems models to develop transportation systems that maximize "mobility" at reasonable cost. Transport needs depend strongly on population distributions and on people's commuting, shopping, recreational, educational, and other travel patterns, as well as freight-distribution patterns. Theoretically, cities might be designed to co-locate housing, jobs, and recreation to minimize travel needs. But in developed countries, society is fairly affluent and mobile, so the use of the personal automobile to travel significant distances on a routine basis is pervasive.

However, urban transportation has historically been plagued with problems of congestion, pollution and safety (*Scientific American*, 1997). Improvements (like new roadways, efficiency improvements, or cleaner vehicles) intended to address these problems usually end up creating more vehicle usage and a recurrence of the situation addressed. In the US, in about 10 years after interstate highways are completed, they are jammed at peak hours. Progress in reducing hydrocarbon emissions per vehicle has been offset by increases in vehicle-miles traveled and emissions while idling in traffic jams. Improved safety systems seem to encourage drivers to travel at higher speeds and follow other vehicles more closely.

Megacities in many developing countries tend to follow this pattern. Congestion problems are shifting as bicycles are increasingly competing with motorized bikes, motorcycles and cars for limited road space. Increased diesel truck and bus transport are also adding to pollution problems. And safety is often not a priority.

From an historical perspective, people spend about 45–90 minutes per day travelling, whether they live in an African village or Europe or Asia or the US. They also spend about 10–15% of their income on transportation. Once per capita income exceeds about $1,000 (1985 US dollars), people begin to demand motorized mobility. At the lower income levels, bicycles, motorbikes, or local, low-speed public transit dominates. As income rises above about $5,000, the automobile takes an increasing share of travel miles, with a leveling occurring at an income of about $10,000. Air travel and high-speed trains come into the picture at still higher per capita income levels.

In developed countries, the transportation sector constitutes about 1/3 of the total national energy usage. Most of the energy is used in personal road vehicles, followed by over-the-road trucking. Figure 18.1 shows the 1999 US energy consumption (EIA, 2000) and the distribution within the transportation sector (ORNL, 2000).

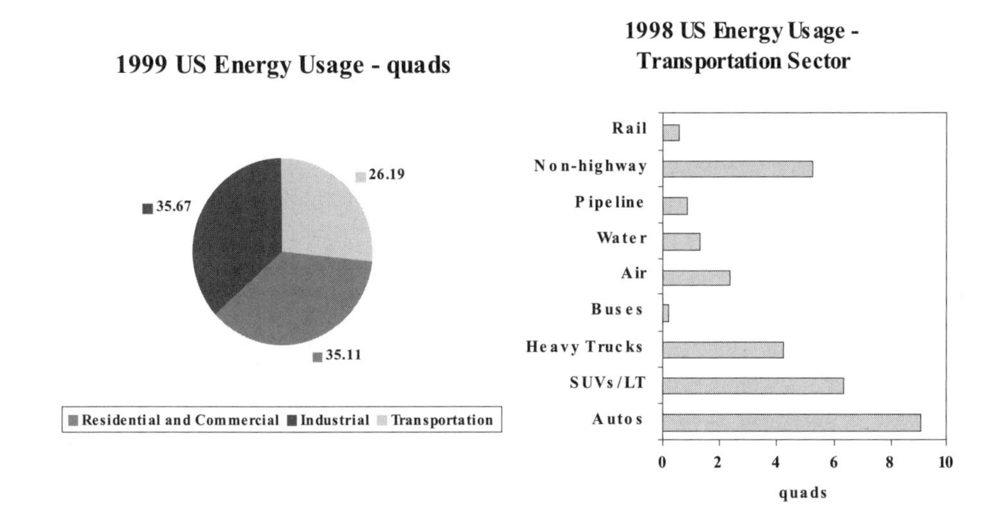

Figure 18.1. Energy usage in the US and in the transportation sector. Source: EIA, 2000.

In urban areas, population densities and transportation needs lead to road congestion, pollution, the waste of time and energy, and lack of parking space availability. With high-enough population densities, public mass transit systems become desirable, both for speed of intracity transport and for reducing the adverse effects of road congestion. Mass transit systems are usually designed to meet peak commuting loads and are often underutilized. Unless a mass transit system is heavily utilized, it may be less energy efficient per person-trip than personal transportation. Moreover, almost all mass transit systems require some subsidies, and many people prefer the flexibility of personal vehicle travel to mass transit. In suburban areas, transportation may be designed so that people drive and park at a mass transit terminus for commuting purposes.

Between major population centers, rail and air transport offer some advantages. For distances of about 300 miles, high-speed rail passenger transport can be more efficient and faster than air transport, when delays associated with travel to and from airports, check-in delays, and other factors are considered. Rail transport is a desirable option for heavy freight between major populous cities and industrial centers. Air transport offers a speed advantage for longer distances, but a higher cost. Some 1996 data on energy intensity of different modes of transport in the US per passenger mile are shown in Table 18.1. For freight transportation in the US, 1996 data are shown in Table 18.2.

Table 18.1. Energy Intensities of Passenger Modes—1996

Mode	Btu/vehicle-mile	Btu/passenger-mile
Automobiles	5,874	3,671
Light trucks/SUVs	7,247	
Transit buses	39,307	4,512
Intercity buses		816
School buses	13,680	
Certified air carriers		4,081
General aviation		10,481
Rail transit		3,444
Intercity Amtrak		2,389

Source: ORNL (1998).

Table 18.2. Energy Intensities of Freight Modes – 1996

	Energy Intensity – 1996 data
Heavy trucks (Btu per vehicle-mile)	21,964
Class I freight RR (Btu per freight car-mile)	15,747
Class I freight RR (Btu per ton-mile)	368
Domestic waterborne commerce (Btu per ton-mile)	412

Source: ORNL (1998).

In 1999, essentially all highway and off-highway vehicles were fueled with petroleum products. About 2.5% of *all* transportation energy use was from natural gas (for compressors in pipeline transport of natural gas), and about 1.2% was from electricity (a mix of rail and pipeline transport). The combustion of petroleum products produces some air pollutants that are of concern from a human health and pollution perspective. Table 18.3 presents the total 1996 US annual emissions for the "Criteria Air Pollutants" as identified by the Environmental Protection Agency (EPA).

The EPA also regulates allowable emission limits from motor vehicles, as discussed in Chapter 4. These are based on standard driving cycles (a combination of urban and highway driving) and specify the maximum amount of the regulated pollutant that can be emitted per distance traveled. Over the years, these standards have gotten more stringent, influenced by health concerns and international trends in standards. In the US, the EPA phased in standards in a series of levels or "tiers." For NO_x, for example, EPA Tier 1 standards in 1994 were 0.373 g/km for gasoline engines and 0.777 g/km for diesels. Japanese standards for 2002 were 0.80 g/km for gasoline; 0.280 g/km for diesels. The EPA Tier 2 standards, effective in 2004, are 0.043 g/km for both gasoline and diesel engines (NRC, 2003).

Annual highway fatalities in the US have been of the order of 40,000 per year recently (down from around 50,000 in the 1960s), plus about 2,000 more in boat, plane, and train accidents (*Statistical Abstract*, 2000). Traffic accidents kill about 885,000 people annually worldwide (O'Meara, 1999).

What can be done about the pollution, safety, and congestion problems? Pollution and safety regulations continue to push toward cleaner and safer technologies. But, with fairly low fuel and vehicle costs, there is little self-regulation on personal travel until congestion approaches deadlock. Traffic bans, automated toll highways, limited urban parking, and other solutions offer some promise, but usually have the effect of differentially penalizing the low-income part of the population.

There are a few success stories. For example, Curatiba, Brazil, has integrated high-population-density buildings with radial busways and a 150-km network of bike paths. While about 1/3 of the population use cars, 2/3 of all trips in the city are by bus. Car traffic has declined by 30% since the plan was instituted in the 1970s even though the urban population has doubled. This suggests the potential for decoupling car ownership from car use in densely populated areas (*Scientific American*, 1997).

These integrated planning measures may be effective in a rapidly growing "new city," but they are harder to implement in existing urban areas. Some creative initiatives are beginning to be tried in existing regions. For example, in the US, the state Department of Transportation in Maryland has launched a "Smart Growth" program in which state and local officials target areas for revitalization and/or growth. These designated locations are the only (or principal) areas that are eligible to receive state investment in transportation infrastructure. The focus includes major investment in maintaining or upgrading existing infrastructure rather than adding infrastructure that will foster more "urban sprawl." The program links all transportation modes and looks for operational improvements and coordination, as well as physical improvements (MDOT, 1999). Other cities have tried experiments such as banning of cars on certain days or times, but congestion still presents major challenges, especially in the growing megacities in developing countries (O'Meara, 1999).

Automated highways and smarter cars are evolving technologies. New automated toll collection technology allows for congestion pricing of highways, which could discourage travel at peak times and increase automotive safety. In addition, the communications revolution may impact the transportation sector by substituting some telecommuting for physical travel.

Table 18.3. Total National Emission of the Criteria Air Pollutants by Sector, 1996

CO=carbon monoxide, NO_x=nitrogen oxides, VOC=volatile organic compounds,

PM-10=particulate matter smaller than 10 microns, SO_2=sulfur dioxide

Sector	Millions of short tons / percentage				
	CO	**NO_x**	**VOC**	**PM-10**	**SO_2**
Transportation	70.0 / 79% *	11.8 / 50%*	7.9 / 41.5%	0.9 / 3%	0.7 / 3.5%
Highway	53.0 / 60% *	7.2 / 31% *	5.5 / 29% *	0.3 / 1%	0.3 / 1.5%
Aircraft	1.0 / 1%	0.2 / 0.7%	0.2 / 0.9%	0.04 / 0.1%	0.01 / 2%
Railroads	0.1 / 0.1%	0.9 / 4%	0.05 / 0.3%	0.03 / 0.1%	0.24 / 1%
Vessels	0.1 / 0.1%	0.2 / 1%	0.05 / 0.3%	0.03 / 0.1%	0.1 / 0.6%
Other off-highway	15.9 / 18%	3.3 / 14%	2.2 / 11%	0.5 / 1.6%	0.01 / 0.0%
Stationary source fuel combustion	6.0 / 7%	10.5 / 45%*	1.1 / 6%	1.2 / 4%	17 / 88% *
Industrial processes	4.6 / 5%	0.8 / 3%	9 / 47% *	0.9 / 3%	1.6 / 8.5%
Waste disposal/recycle	1.2 / 1.4%	0.1 / 0.4%	0.4 / 2.3%	0.3 / 0.9%	0.05 / 0.3%
Miscellaneous	7 / 8%	0.2 / 1%	0.6 / 3%	28 / 90% *	0.01 / 0.0%

* Major contributor

Note: PM-10 does not differentiate between potentially harmful and more benign particles.

While lead is not listed for the US, it contributes to major health problems in many countries that still use leaded gasoline. Over 60% of the children in cities like New Delhi and Singapore have blood levels of lead high enough to cause adverse health effects. Ozone levels are a frequent respiratory health problem in major cities, such as Santiago and Mexico City, and are due to atmospheric chemical reactions between hydrocarbon emissions and NO_x (O'Meara, 1999).

Source: ORNL, 1998.

As concerns about the impacts of greenhouse gas emissions grow and international accords to limit emissions develop, the transportation sector has a new set of issues to face. With its consumption of about 28% of US total energy use, and with almost all of the energy coming from petroleum sources, the transportation sector emitted about 580 million metric tons of carbon in 1998 (ORNL, 2000). This constitutes a little more than 30% of all the US carbon emissions distributed from the tailpipes of more than 200 million vehicles. Thus, the general options available to the transportation sector, if major reductions in greenhouse gas emissions are required, are:

- Major improvements in energy efficiency per vehicle-mile
- Decarbonization of the fuel supply
- Changes in consumers' usage of transportation services

Though concerns about rising greenhouse gas emissions continue, there is still enough uncertainty about cause and effect and the possible magnitude and timing of impacts that most people are perpetuating bad energy-use habits. Because of the availability, suitability, and affordability of petroleum fuels for the transportation sector, any *rapid* requirement to reduce transportation sector emissions would entail significant costs and disruptions. However, recent investment in R&D on more efficient transportation vehicles and systems and on alternative fuels has identified some new road transportation possibilities that might emerge in a greenhouse gas-constrained world. These investments are likely to make transitioning to a suitable pathway less costly and disruptive to global society. (A report by the World Business Council for Sustainable Development (2001) provides a comprehensive overview of the issues discussed in this chapter.)

18.2 Elements of the Transportation System

About 3/4 of transportation energy use is by road vehicles and rail transport. These are all wheeled vehicles with drive trains and propulsion systems. Automobiles and light trucks constitute the largest group of these vehicles, especially since many light trucks (sport utility vehicles or SUVs) are also largely used for personal transportation, especially in the US. At present, there are emerging technologies that may offer opportunities to reduce the environmental impacts of road transportation while meeting the transportation needs of consumers. However, to assess the potential improvements, along with any barriers and uncertainties, it is important to compare options on a full system ("well-to-wheels" for automotive technology) basis, with consistent assumptions and boundary conditions. Lifecycle analysis is one useful framework for assessment, but it must be used judiciously with regard to temporal and spatial locations of various impacts. One might use a "snapshot in time," where the fleet of vehicles at different lifecycle stages is evaluated. One might compare one technology—say a fuel-cell-powered hybrid vehicle manufactured in 2010 with an evolved gasoline-spark-ignition-powered vehicle also manufactured in 2010. In this latter comparison, one should also account for equivalence in vehicle performance to meet consumer demands—size, carrying capacity, driving range, acceleration, fueling convenience, serviceability, durability, cost, etc.

For today's typical gasoline, internal combustion engine car, about 10–15% of the total source energy is consumed in the production and processing of the fuel, about 2–5% in the fuel distribution, and the remainder in the operation of the vehicle. Energy is also used in the production of materials used to make the vehicle and in recycling/scrapping operations, but these are small quantities in comparison to the energy used over the multi-year vehicle operational lifecycle. Figures 18.2 and 18.3 show the major "stakeholder" groups and key elements in the lifecycle of a typical automobile.

These private stakeholders represent clusters of industries that have their own competitive aspects and are the subject of specialized regulations and constraints. Their purpose is to profit through satisfying consumer demands for their products and services—and they help create some consumer demands through marketing and advertising. The government invests some national funds in supporting infrastructure development but also benefits from the overall contribution of the transportation system activities to national GDP and to national economic and social health.

In order to compare different technology options on a consistent basis, both the fuel and vehicle cycles need to be considered as a system. A lifecycle analysis of the system should track the cumulative materials use (virgin and recycled, including energy), cumulative costs, cumulative emissions of different types, and cumulative wastes. For consistency, the two systems should offer similar performance and reliability characteristics to the user. Additional factors may also be involved: if one system requires a major investment in new infrastructure and the other one doesn't, the transitional impacts should be weighed. For this reason, inertia favors evolutionary, rather than revolutionary, changes.

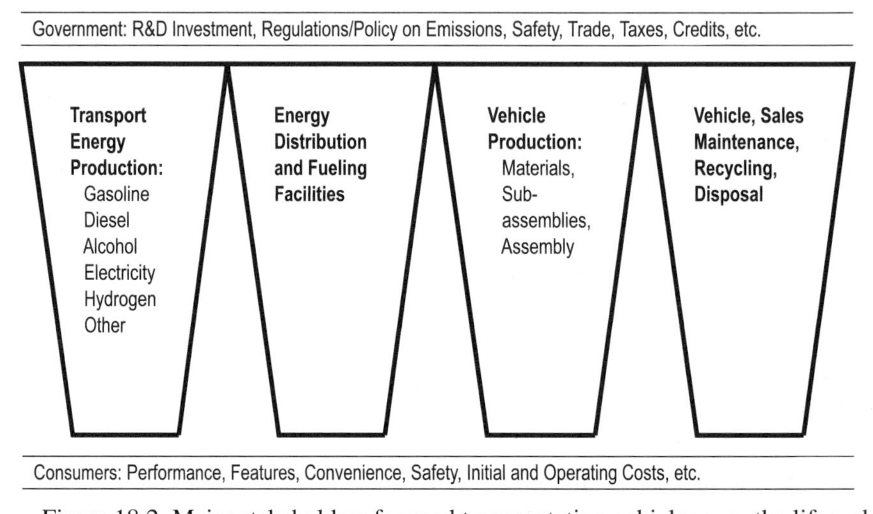

Figure 18.2. Major stakeholders for road transportation vehicles over the lifecycle.

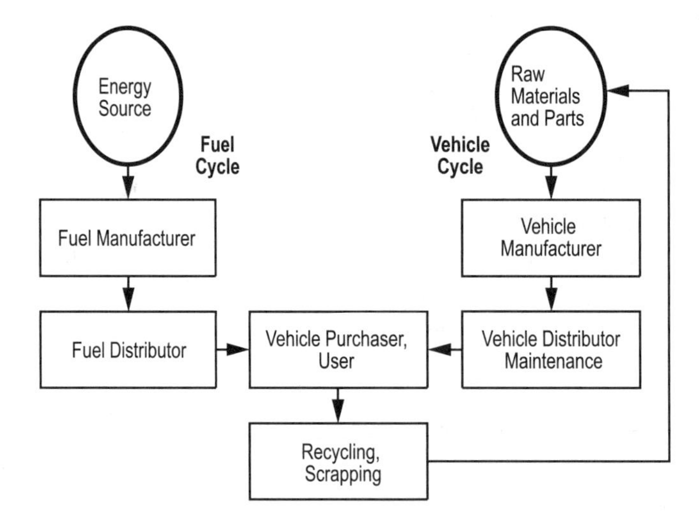

Figure 18.3. Key elements of a road transportation vehicle lifecycle.

Over the past decades, auto efficiency has improved by about 1% per year on a fairly consistent basis, because of various combinations of incremental improvements. Auto emissions continued to decrease as regulations gradually become more stringent. However, if there is a major sudden change in performance requirements, such as some restriction on greenhouse gas emissions, then a revolutionary technology may become competitive—if it can meet the new requirement at the same or lower cost than the modified existing technology. Some new technologies appear interesting. For example, hydrogen and electricity are CO_2-free on-board energy sources, and fuel cells offer higher efficiencies than fossil fuel combustion engines, which are limited by thermodynamic constraints. However, these and other options need to be assessed in a full-system, lifecycle context to evaluate whether they have the potential to make a significant system improvement. Sometimes, improvements in one aspect, like GHG reduction, will have to be traded off against another aspect, such as total system cost.

18.3 Transportation Fuels and the Fuel Cycle

Petroleum-derived gasoline and diesel are the predominant fuels for today's road transportation sector. The popularity of petroleum-based fuels began in 1859 when "Colonel" Drake discovered oil in Pennsylvania. This liquid fuel found a number of early uses. When gasoline demand developed from the growing fleet of cars being built in the early 1900s, the oil industry grew rapidly and developed a major infrastructure of pipelines and refineries that support transportation fuel needs.

Liquid fuels can also be made from coal as demonstrated in South Africa by the SASOIL plant and later implemented by the Germans during World War II when their access to petroleum was cut off. However, the processes are more costly than use of petroleum.

Gasoline and diesel fuels today are cleaner and perform better than earlier fuels; they seem ideally suited to supplying energy for transportation. Energy supplies appear adequate to support demand for a number of decades into the future (see Chapter 2). However, growing concerns about GHG emissions are making the transportation industry look at other possibilities for fuels.

Less carbon-intensive alternatives include fuels derived from natural gas (mainly methane). These include compressed natural gas (CNG), liquefied natural gas (LNG), methanol, and hydrogen. New Zealand, in a desire to reduce its dependence on petroleum imports around 1980, invested in a plant to convert natural gas first to methanol and then to gasoline using a patented Mobil catalytic process. The gasoline plant was never cost-competitive and was idled in 1997. More recently, Shell has been developing gas-to-liquids synfuel plants in Malaysia and Qatar. Fuels can also be made from biomass sources—ethanol and methanol are the most common examples—which do not produce net CO_2 emissions except for fossil energy used in fertilizer and harvesting/transportation/conversion activities. Electric cars have been promoted as a way to avoid local combustion emissions, but the fuels used to generate electricity ordinarily have associated emissions. Depending on the mix of generation fuels and the individual conversion efficiencies to electricity, GHG emissions vary between those associated with older coal-powered plants and low emissions associated with nuclear or hydroelectric generation. Hydrogen is a fuel that can be used in fuel cells to produce electricity, or in combustion engines, with water as the final emission product. However, hydrogen does not exist in significant quantities in nature because of its high reactivity. It is most economically made from methane (with substantial carbon emissions associated with the production). It can also be produced by electrolysis of water (with more electricity used in its production than can be recovered from its subsequent use, because of inefficiencies and irreversibilities) or from fossil sources with carbon capture and sequestration. (See Chapter 7 for further discussion of research on thermal processes and Chapter 10 for bioproduction of hydrogen.)

In evaluating the fuel cycle, two important factors are the scale of the production and distribution activity and the ease of storage of the energy at distribution stations and onboard the vehicle. In 1998, 17.8 million barrels of oil were consumed per day in the US, mostly for transportation fuel purposes (BP, 1999). The production of gasoline and diesel fuel products at such a large rate has required the development of a large infrastructure of refineries and fuel transportation facilities that have planned operating lives of decades. These liquid fuels have good energy storage density, so that today a 15-gallon fuel tank (about 2 ft^3 volume) gives a typical travel distance of 300 miles.

What are candidate fuels for the future? Let us look forward to 2020 and examine the most likely options. Table 18.4 shows typical auto fuel properties.

Gasoline from petroleum. Assuming that trends toward cleaner fuel continue, the gasoline of the future will move toward low sulfur content, with possible changes in volatility, aromatics, or other specifications driven by increasingly stringent environmental emissions standards. Additives to gasoline to improve performance have had a somewhat checkered history. Lead was used in the mid-1900s as an octane enhancer, but was found to cause environmental health problems and has since been phased out (although there is still some use in developing countries). More recently, MTBE (methyl tertiary-butyl ether) was mandated as a component in reformulated gasoline. Now, groundwater contamination concerns have shown that this is not a desirable additive. Whether or not oxygenate additives to gasoline significantly improve combustion performance is still being debated, and ethanol is being evaluated as a possible substitute.

Costs will not be materially different from the present when adjusted for inflation because the oil industry has been creative in finding technologies to produce cleaner fuels at little or no additional cost. There may be some special grade of higher purity gasoline that is designed for fuel cell vehicles fitted with gasoline reformers, but it may not be radically different from the cleaner gasoline that will evolve for use in internal combustion engine cars. Prices, on the other hand, may be influenced in the future by tax policies, by market manipulations, or other exogenous factors. However, costs are an important factor in examining the competitive introduction of alternative fuels.

Table 18.4. Properties of Established and Suggested Fuels for Automobile Internal Combustion Engines

Fuel	Property			
	Density	**LHV/mass***	**LHV/vol.****	**LHV/vol. of stoich. mixture*** (at 1 atm, 300K)**
	(kg/m^3)	(MJ/kg)	(MJ/m^3)	(MJ/m^3)
Gasoline	750	44	3.3×10^4	3.45
Diesel	810	42	3.4×10^4	3.37
Natural Gas				
gas @ 1 bar	0.72	45	3.2×10^1 (*X*)	3.25
gas @ 100 bar	71		3.2×10^3	
LNG @ 180K, 30 bar	270		1.22×10^4	
Methanol	792	20	1.58×10^4	3.16
Hydrogen				
gas @ 1 bar	0.082	120	0.984×10^1(*X*)	2.86
gas @ 100 bar	8.2		0.984×10^3	
liquid @ 20K, 5 bar	71		8.52×10^3	

Source: Weiss et al., 2000.
*detemines mass of fuel to carry
**determines size of fuel tank
***determines power density of engine
(*X*) completely not feasible

Diesel from petroleum. Diesel fuel will evolve from current properties toward low sulfur content, with possible changes in volatility, polynuclear aromatics, cetane, and other specifications. While there is general agreement that lower sulfur will reduce particulate emissions, it is not yet clear whether changes in other properties can substantially reduce harmful emissions from current heavy-duty diesel engines. However, changed fuel specifications would be important in new diesel engine designs or exhaust gas treatment, or in light-duty diesel engines for passenger cars. Modifications to reduce emissions to low levels may reduce the relative efficiency advantage of the diesel engine over a spark ignition internal combustion engine. Fuel costs will not change significantly, although a large increase in diesel demand will require some modifications to existing refineries and their operations.

Diesel from natural gas conversion. Diesel fuel from Fischer-Tropsch (F-T) synthesis or another gas-to-liquid (GTL) process will have high cetane content and zero sulfur. Low temperature properties may be somewhat compromised. Interest in F-T diesel stems from the hope that ultra-clean fuel will make a significant difference in particulate and other emissions associated with the use of conventional diesel fuel. While F-T diesel is more expensive than conventional diesel, prices can be reduced if a cheap source of plentiful natural gas is available. There are many remote gas locations worldwide, but major investments in new plants to produce significant amounts of F-T diesel will be needed.

Methanol from natural gas conversion. If methanol is produced as a commodity fuel, it will be manufactured at remote gas sites. An investment of about $12–15 billion is estimated to install remote methanol capacity to meet 5% of the present US gasoline demand. Smaller-scale investments will be required for new or converted facilities to store, transport, and dispense methanol fuel. Methanol has about half the energy storage density of gasoline, but still has the advantages of a liquid fuel. Methanol is toxic through skin contact or ingestion, so facilities need to be designed to minimize spills.

Compressed natural gas (CNG). Natural gas (primarily methane) can be used as a transportation fuel and is typically stored in vehicles as a pressurized gas (around 200 atm pressure). It is a clean fuel (NO_x being the primary emission of concern) with about 25% lower carbon emissions than gasoline—but only if no methane is leaked. Methane is a more powerful greenhouse gas than carbon dioxide (by a factor of about 21, averaged over a 100-year period), so 1–2% leakage can offset its lower carbon advantage. High-pressure gas is less dense than gasoline (and requires a pressurized cylindrical storage vessel), so fuel storage volume impinges on vehicle interior space.

Liquefied natural gas (LNG). LNG is another alternative, involving liquid storage at about –260°F and a pressure just a little above atmospheric. LNG has lower density than gasoline, requires about a 10% energy penalty for liquefaction, and requires a sophisticated insulated container to minimize boil-off when the vehicle is not in use, so it is probably not as attractive an option as high pressure gas storage for passenger vehicles. LNG is more attractive for fleet use, where vehicles operate continuously from a central location.

Liquefied petroleum gas (LPG). LPG, mostly a mixture of propane and butane, has been used for years as a convenient "bottled gas" for remote locations, for camping, and for vehicle fuel. At 5–10 atm pressure, saturated liquid can be stored at typical ambient temperatures. LPG is usually produced along with oil or natural gas, but it is not as plentiful as either.

Biofuels. Liquid transportation fuels, such as ethanol and methanol, can be made from agricultural wastes or from farmed energy crops like corn, grain, or fast-growing cellulosic materials. Basically, biomass is a solar energy conversion process that operates at about 1% efficiency, so the energy content is relatively low per unit planted area. Energy is needed to harvest and transport biomass to a processing location, and, if the transport distances reach about 30 km, the transportation energy starts to exceed the energy in the harvested crop. Biomass, by nature, is a dispersed resource that fits smaller dispersed uses or can supplement other energy sources. While biomass takes carbon dioxide from the atmosphere as it grows, it is not GHG-neutral to the extent that fossil fuels are used in cultivation, fertilization, irrigation, harvesting, and processing.

Hydrogen from natural gas reforming. For the near term (the next two decades), if hydrogen is developed as a widespread transportation fuel, it will likely be manufactured most economically by reforming natural gas at "service stations" located on the natural gas pipeline network. It probably will be stored and dispensed as a gas at about 350–400 atm pressure. Other more expensive options include generating hydrogen by electrolysis of water, reforming natural gas in large centralized facilities and piping compressed hydrogen or trucking liquid hydrogen to service stations. In all cases, large investments in infrastructure will be required. A further barrier is the low energy storage density of hydrogen fuel on-board the vehicle, relative to gasoline. Research goals for hydrogen storage are typically 5–10% hydrogen per unit weight of storage system (fuel plus tank).

Electric Power. While electric cars are considered "zero emissions vehicles," there are still emissions associated with the original production of the electricity. Electricity from nuclear power and renewable energy (including hydropower) can have low GHG emissions, but fossil fuel combustion is presently the predominant source of electricity generation. Because large power plants can be more efficient than individual vehicle engines, some GHG reduction may be achieved for the same fuel even if electricity comes from fossil sources. The main barrier to use of electric energy in transportation vehicles is the difficulty in storing electricity. Batteries are heavy, cumbersome, and expensive. Much research is ongoing in search of a more efficient battery storage system, but a major breakthrough is needed if electric cars are to compete in price, convenience, and range with today's gasoline vehicles.

18.4 Personal Vehicles

18.4.1 Historical perspectives

In 1860, Lenoir invented an operable two-cycle internal combustion engine (ICE), and, by 1865, Marcus had built and tested a primitive road vehicle using this type of engine. In Germany, Daimler and Benz developed the four-cycle Otto engine, which has evolved into the most common type of engine used today. By 1889, automobiles were being manufactured in France. Over the next few decades, a number of vehicles were developed, using a variety of engine technologies (e.g., steam, electric) and fuels (e.g., wood and gasoline bought from hardware stores). By 1908, when Ford manufactured the first Model T, there were about 200,000 vehicles in the US (1/3 of those were new that year), driving on about 2 million miles of mostly rutty roads. The population of the US was a little over 85 million, more than 6 million new telephones were in use, and carriage manufacturers like Buick and Olds were starting to make motorized vehicles.

When Henry Ford began building commercial automobiles in 1903, he chose an 8-hp, 2-cylinder, $850 car as his Model A. He experimented over the next few years with different engines, bodies, and types of body steel. By 1908, he had built a large factory to mass-produce the Model T, also sold at $850, which he decided would have a large popular market. In a test drive of an early production model from Detroit to Chicago, the car averaged 21 mph, with about 20 miles per gallon of gas. After experimenting with assembly line production, he cut the assembly time for a vehicle from over 12 hours to about 1 1/2 hours. Profits boomed, and he doubled the pay of his factory workers to a minimum of $5 per day—putting them into a wage bracket that could afford a car.

This phase of automobile development is of interest because it contains lessons about introduction of a new technology that moves into dominance. This early car had controls on the left for driving on the right side of the road, had the engine, the transmission, a flywheel with a built-in magneto, and the universal joint all inside an oil-lubricated case, a single block four-cylinder engine, and semi-elliptic springs mounted across the two axles. A lead-acid battery was optional as a back-up to the magneto. Ford manufactured this model with few changes for the next 19 years. In the early 1920s, these cars sold at a rate of about 2 million per year. His competitors quickly moved to capitalize on his successful experiment and offered variations of what was to become the mainstream technology.

With the growing car market, road infrastructure grew rapidly, and improved roads facilitated improvements in tire technology. Likewise, the petroleum industry developed an infrastructure to support the growing demand for gasoline. Many of the technologies introduced in the Model T became dominant and remained dominant, with continuing improvements in all of the components over the 20th century. The hand crank was replaced with an electric starter, tires and suspensions were improved to give a better ride, and travel speed capabilities gradually increased.

In the past 20 years, to meet the changing expectations of the market, cars have become sophisticated, reliable, quiet, and comfortable means of transportation. The interstate highway system, originally designed for alleged national security reasons, provided the high-quality infrastructure that made the auto the preferred transportation mode for most people—as well as for much freight transport. Different types of combustion cycles were developed and tested (see Chapter 3), evolving to the predominance of spark ignition Otto cycle and diesel engines in today's road fleet. Pollution problems associated with auto exhaust emissions were largely addressed through regulations that led to development of catalytic converters to treat exhaust. Leaded gasoline, which would poison the catalysts, was already being phased out because of toxicity problems. Electronics and computer controls have fine-tuned performance. Seat belts, air bags, antilock braking systems, and traction control have enhanced safety.

The Arab oil embargo of 1973 aroused concern about US dependence on imported petroleum for transportation, and legislation followed that drove the industry to pay more attention to the fuel economy of cars. Better aerodynamic design and smaller cars became attractive to many consumers in the aftermath of the oil embargo, but as memories of gas shortages faded, consumers returned to larger vehicles once again. In the mid-70s to the mid-80s, the US instituted a program to encourage auto makers to improve the efficiency of motor vehicles through a Corporate Average Fuel Economy

(CAFE) program. By 1985, the average mileage per gallon for all the new passenger cars made by a particular manufacturer each year was required to be at least 27.5 miles per gallon (8.55 liters per 100 km). A similar standard for light-duty trucks, instituted in 1996, is 20.7 miles per gallon. Further, a "gas-guzzler" tax was imposed on those cars with the poorest fuel economy. These standards have remained unchanged, but the percentage of light truck sales (e.g., SUVs) has risen from about 30% of new personal sales in 1990 to nearly 50% in 2000. Although the US auto fleet size is stabilizing and vehicles in each size category are getting more fuel efficient, more vehicle miles traveled per person and the preference for large vehicles continues to increase annual US petroleum consumption.

18.4.2 Looking forward

How might growing concerns over GHG emissions influence the development of the car in the next decades? The US government and the US auto industry are vitally interested in this question and are searching for solutions to the rising GHG emissions from the road transport sector. In 1993, the Partnership for a New Generation of Vehicles (PNGV) was initiated by the US government in cooperation with the "Big 3" US automakers. The goal of this program was to develop technologies that would allow a family size sedan (similar to a Ford Taurus) to achieve fuel economies of up to 80 mpg while meeting stringent emission standards at a comparable market price. Concept cars (demonstrating the technical capability) were to be built by 2000, with production prototypes to be available by 2004.

To achieve this major improvement in fuel economy, about a factor of three, it was necessary to examine all the aspects of each component (or alternative new component) of the vehicle system to search for any opportunity. Concept cars are now available, but these do not fully meet the original goals of the PNGV program, especially with respect to market price. However, investment in new technology (about a billion a year by the industry, and well over a $100 million per year by the government) has sparked the development of many improved and new technologies. Major program areas of focus include combustion engines, fuel cells, electrochemical energy storage, power electronics and electrical systems, structural materials and safety, and hybrid designs with features like regenerative braking.

More recently, the focus has shifted under the second Bush administration, to research on hydrogen-fueled vehicles in the "Freedom Car" initiative.

Europe and Japan are also investing heavily in new automotive technologies and are making significant progress in the development of highly fuel-efficient cars. With higher fuel costs, more urban driving, and shorter trip distances than in the US, the overseas markets are focused on smaller vehicles.

Candidate technologies for the future are in various stages of development, and predicting "winners" for the future is difficult. The present mainstream technology has been improving continuously in the past—at perhaps 1–1½% per year—and can be expected to do so in the future. Common wisdom that mature technologies stagnate does not apply to complex systems like an automobile that are composed of many components that are in various stages of maturity and provide many different opportunities for improvement and innovation. Further, the presence of major infrastructure to support the mainstream technology provides significant inertia to change because accelerated infrastructure changes will require increased investments relative to those needed for infrastructure maintenance.

Emerging alternative technologies may offer considerable promise, but research breakthroughs are unpredictable, and often a new technology requires many ancillary developments to allow its integration into a vehicle system. Therefore, more uncertainty exists about the development pathways for new technologies. Finally, if alternative technologies are being compared for impacts, it is important to look at the full lifecycle of production, use, and disposal to learn the net advantages and disadvantages of making a change.

Some of the technologies that might contribute to reductions in fossil fuel use and are being considered for the future include:

- Lightweighting. Use of high-strength steels, aluminum, plastics, and plastic composites can all reduce the weight of the vehicle and reduce overall fuel consumption. Other options include further streamlining of the vehicles and improvements in tires to reduce friction losses. These changes may have some differential impacts on cost and on vehicle safety. If consumers begin to find smaller vehicles more attractive for their needs, this is another opportunity for fuel saving, although most US consumers still are choosing larger vehicles.
- Electric vehicles. Electric cars are sometimes considered "zero emission" vehicles, but emissions are produced with electric power generation. In addition, affordable battery technology requires major weight and volume penalties for energy storage. Lead acid batteries have a specific energy of 25–55 Wh/kg; nickel metal hydride batteries, of 50–80 Wh/kg; lithium polymer batteries of 100–200 Wh/kg (Lankey et al., 2000). Gasoline, on an equivalent energy delivered basis, stores about 3,000 Wh/kg (assuming a 25% engine efficiency). Thus, electric cars are limited in performance and distance between recharges by the size of batteries required and the increasing weight of the vehicle associated with additional battery power.
- Hybrid designs. Hybrid cars typically combine a smaller engine (diesel or ICE) with a larger rechargeable battery backup power capacity. The engine runs continuously at peak efficiency, and the battery is used to provide extra power for acceleration or hill climbing. When the car is idling or has excess power, the engine generates electricity to charge the batteries. In some designs, energy recovered in braking is also used to recharge batteries (regenerative braking). Full hybrid design requires that both mechanical and electrical power be supplied to the car wheels.

Thus, a separate electrical drive and motors on the vehicle wheels are required along with a control system to coordinate the delivery of mechanical and electrical power. Hybrids offer the most efficiency improvement potential in urban (stop-and-go) driving cycles. For extended acceleration or long hill climbing, the battery may be drained and power capability reduced. This is a problem in countries like the US where drivers want performance for both urban and long-distance driving. At present, Honda and Toyota are both manufacturing hybrid cars for markets, primarily in the OECD countries. Some partial hybrids are also available.

Hybrid cars are generally more expensive than conventional cars because of their complex transmissions and additional batteries. Hybrids cover a whole range of options—small engines with large batteries, a larger engine with smaller batteries, parallel or series (the combustion engine generates electricity) drives to the wheels, etc. As with electric cars, the weight and space penalties associated with battery storage limit the electrical energy delivery. Hybrids have less stringent battery requirements than do electric cars because the batteries provide short supplemental power bursts. Nevertheless, additional battery power still imposes considerable extra weight. Highly efficient hybrids for urban driving can be designed, but these have poorer performance for driving through mountainous areas or for high-speed highway passing.

- Fuel cells. Fuel cells generate electricity from electrochemical reaction of oxygen and hydrogen. Proton Exchange Membrane (PEM) fuel cells are best suited for use in cars. Fuel cells have highest efficiency at low power outputs, but also have a wide range of power output with only minor losses in efficiency—and with stack efficiencies in the 60% and above range (Thomas et al., 1998). However, this is for pure hydrogen fuel. Like electricity, hydrogen is not a primary fuel but an energy carrier that must be produced from other fuels or by electrolysis of water. The most cost-effective means to produce bulk hydrogen fuel today is from steam reforming of natural gas, but some energy losses are associated with the reforming process, and the carbon in the natural gas is lost to the atmosphere. Hydrogen is also difficult to store onboard a vehicle. Present technology (either high-pressure gas, liquid, or metal hydride storage systems) allows about 5% H_2 per unit storage system weight. Hydrogen has a specific energy about three times that of gasoline, but there is the storage system weight penalty of a factor of 20 (or about 10 if hydrogen storage research goals for the future are met). Suppose that a fuel cell that has twice the efficiency of a combustion engine is available—then the net equivalent specific energy for hydrogen storage is about 800 (or

1,600 future) Wh/kg. This compares favorably with battery energy density storage but is still about four times bulkier than gasoline.

Another option is to make the hydrogen onboard the vehicle by "reforming," or the chemical conversion of liquid fuels to hydrogen gas. This takes advantage of the onboard energy density of liquid fuels. There are energy losses associated with the reforming process, and the carbon from the liquid fuel is discharged to the atmosphere. Depending on the developments that may occur in reformer and fuel cell technologies, there is the promise of some system energy efficiency gain from such fuel cell vehicles. The fuel cell vehicle can also be a series hybrid, as supplemental batteries can be used to provide extra power for rapid acceleration or hill climbing.

18.5 A Lifecycle Comparison of Road Transport Alternatives for 2020

A comprehensive comparative lifecycle assessment (LCA) of a number of reasonable technology options for personal use road vehicles in 2020 was recently completed and provides some interesting insights for the future (Weiss et al., 2000). This report looks at comparative system performance for 13 representative fuel/vehicle systems: today's "typical" passenger car (similar to a 1996 Toyota Camry), an "evolved" car in 2020 with similar space and performance attributes to the 1996 Camry (the baseline car for the study), and 11 similar alternative cars that employ new fuel or vehicle technologies. The LCA considers energy use, costs, and GHG emissions for a lifecycle that contains the elements shown in Figure 18.3: fuel extraction, manufacturing, and use; vehicle materials extraction, production and manufacturing; vehicle use; and final scrapping or recycling of the vehicle. Table 18.5 shows the system combinations evaluated. (Similar lifecycle assessments have been performed by OTA, 1995, Singh et al., 1998, Hoehlein, 1998, and others.)

Of course, when one predicts technology improvements 20 years in the future, there is considerable uncertainty involved. Experience has shown that technological systems like the automobile are subject to continuous improvement because they are composed of multiple technological parts that are at different stages of development. So, it is not unreasonable to expect that cars, in each size category, will get more efficient at a continuing rate of about 1% per year into the future. For the evolutionary vehicle in 2020, we judge the uncertainty range to be on the order of plus or minus 10%. When it comes to predicting improvements in emerging technologies, the uncertainties become greater. Hybrid vehicles have a more complex system of delivering power to the wheels because both electrical power and mechanical power are transmitted in varying combinations. While hybrids are now appearing on the market, they are still in an early stage of development, and our estimates for hybrids are in the range of plus or minus 20%.

Table 18.5. Vehicle Technologies Assessed

Year and Technology	Fuel	Engine(s)	Transmission
1996 (Reference)	Gasoline	SI	Auto
2020 Evolutionary (Base Case)	Gasoline	DI SI	Auto-Clutch
2020 Advanced Vehicle ICE	Gasoline	DI SI	Auto-Clutch
	Diesel	DI CI	
	F-T Diesel		
2020 Advanced Vehicle ICE Hybrids	Gasoline	DI SI + Battery	CVT
	Diesel	DI CI + Battery	CVT
	F-T Diesel		
	CNG	DI SI + Battery	CVT
2020 Advanced Vehicle Fuel Cell Hybrids	Gasoline	Reformer-FC + Battery	Direct
	Methanol	Reformer-FC + Battery	Direct
	Hydrogen	FC + Battery	Direct
2020 Advanced Vehicle Electric	Electricity	Battery	Direct

Source: Weiss et al., 2000.
Abbreviations:
Advanced Vehicle = Lightweight body and efficient design
CI – Compression ignition
CNG – Compressed natural gas
CVT– Continuously variable transmission
DI – Direct injection
FC – Fuel cell
F-T – Synthetic Fischer-Tropsch diesel from natural gas
ICE – Internal combustion engine
SI – Spark ignition

For fuel cell systems and electric cars (where the battery technology today is a major limitation), continuing R&D may lead to some breakthroughs. But here, we judge the uncertainty bounds to be on the order of plus or minus 30%.

The costs of owning a new passenger car fall into two general categories: variable costs (such as fuel cost, which depends on how much the vehicle is driven) and fixed costs (such as finance costs, which are independent of vehicle use). Based on assumptions (e.g., present US tax rates, 20% annual capital costs) that are clearly stated in the full report, the comparative operating costs are shown in Table 18.6.

Differences in fuel costs worldwide are primarily due to differences in tax policy. Some countries have used gasoline taxes as a major way to raise additional revenues for the government. France has had higher taxes on gasoline than on diesel fuel to encourage wider use of more efficient diesel cars. Other countries have subsidized fuels, sometimes differentially, to encourage use. For example, developing countries often subsidize fuel for use in the agricultural sector or for a particular industry that they wish to develop.

Low energy taxes have had a role in spurring US development and are now considered a basic right by many American consumers and industries. There are a few important insights that can be gained from examination of Table 18.6.

- Fixed costs are much larger than variable costs (mostly fuel costs) for a new vehicle. New car buyers will be more sensitive to initial cost of the vehicle.
- Encouraging fuel efficiency through fuel taxes is a weak policy lever unless major changes in fuel tax are made. (UK tax rates of $3.5 US per gallon raise the fraction of variable costs for the baseline car to about 20% of total costs.)
- For older used cars, fuel costs become a larger part of the cost, but vehicle fleet efficiency is a legacy of the past choices of new car buyers.
- We are only comparing cars of a fixed size and performance. Past experience shows that car buyers are willing to pay more to obtain features they want (e.g., size, luxury features).

Figure 18.4 shows point estimates for energy use (MJ/km) for each of the fuel/vehicle systems. It estimates lifecycle energy use (based on 15-year vehicle life and annual travel of 20,000 km) for each combination and further shows the contribution to total energy from "embodied energy" (that used in manufacture, distribution, and disposal of the vehicle), from the "fuel cycle" (that used in making the fuel), and from the vehicle operation cycle.

Table 18.6. Operating Costs for New Passenger Cars in 2020, ¢ (1997)/km

2020 Technology (Vehicle Price)	Total	Fixed	Variable
Baseline gasoline ICE ($18,000)	30.6	25.0	5.6
Advanced gasoline ICE ($19,400)	32.1	26.8	5.3
Advanced diesel ICE ($20,500)	32.8	28.1	4.7
Hybrid gasoline ICE ($21,200)	34.1	29.2	4.9
Hybrid diesel ICE ($22,200)	34.8	30.4	4.4
Hybrid CNG ICE ($21,700)	34.6	29.7	4.9
Hybrid gasoline FC ($23,400)	37.3	31.7	5.6
Hybrid methanol FC ($23,200)	36.5	31.5	5.0
Hybrid hydrogen FC ($22,100)	35.7	30.3	5.4
Battery electric ($27,000)	40.8	36.3	4.5

Source: Weiss et al., 2000.

Only about 0.25 MJ/km of total energy is associated with the manufacture of the vehicle, assuming a reasonable level of recycling. This is a small fraction of total energy for today's vehicle, but becomes a larger relative portion as cars get smaller, more efficient, and use more exotic energy-intensive materials. Fuel cycle energy is also a relatively small fraction of total energy use, except when fuel conversion is required (e.g., Fischer-Tropsch, methanol, hydrogen, electricity). It is evident that the most energy-efficient choices in this study are the diesel hybrid and the CNG hybrid, closely followed by the gasoline hybrid and the hydrogen-fueled fuel-cell hybrid. When the uncertainty bands are considered, most of these future system choices overlap as shown in Figure 18.5.

The charts in Figure 18.5 show a dot for the best estimate values and also the projected uncertainty bars. The values are given as a ratio to the performance of the 2020 evolutionary baseline car (Value = 100).

Here are some observations:

Evolutionary changes between now and 2020 could result in the typical passenger car being lighter (about 1,240 kg versus 1,440 kg for a 1996 equivalent vehicle) and more efficient than today's car. About a 35% reduction in total energy consumption per km and carbon emissions over present vehicle usage for comparable size and performance could also occur. In the absence of major regulatory interventions to the contrary, cars will probably use petroleum-based gasoline. However, this gasoline will be of an improved nature to meet more stringent pollution limitations.

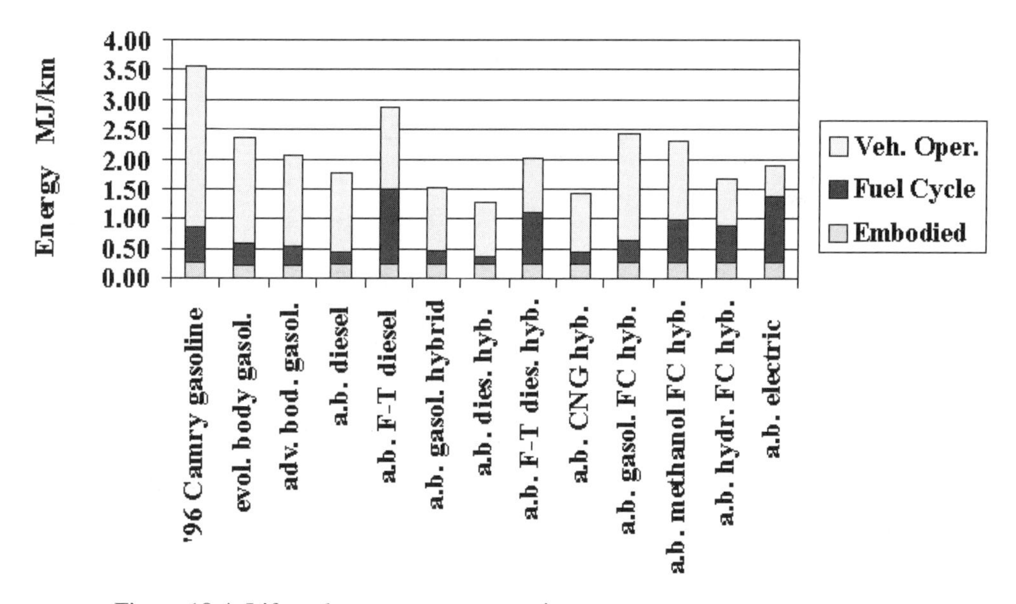

Figure 18.4. Lifecycle energy use comparisons.

All cars are 2020 technology except for 1996 "Reference" car
- ICE = Internal Combustion Engine, FC = Fuel Cell
- 100 = 2020 evolutionary "baseline" gasoline ICE car
- Bars show estimated uncertainty

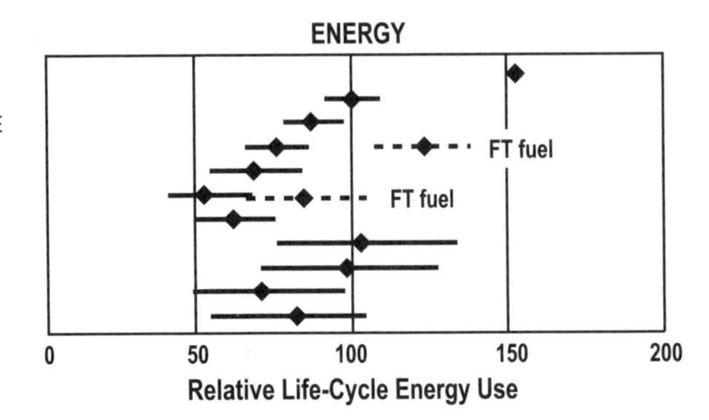

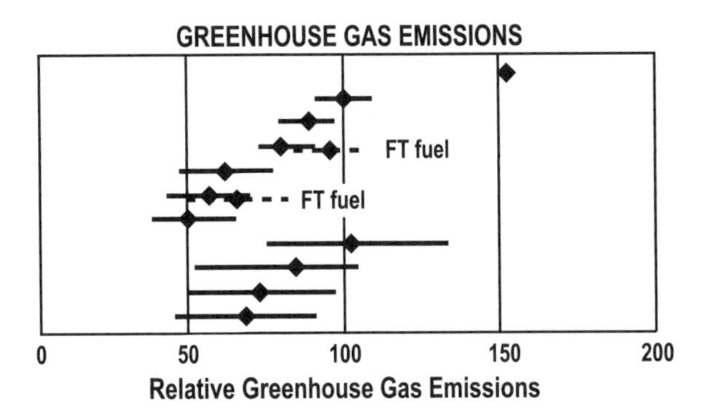

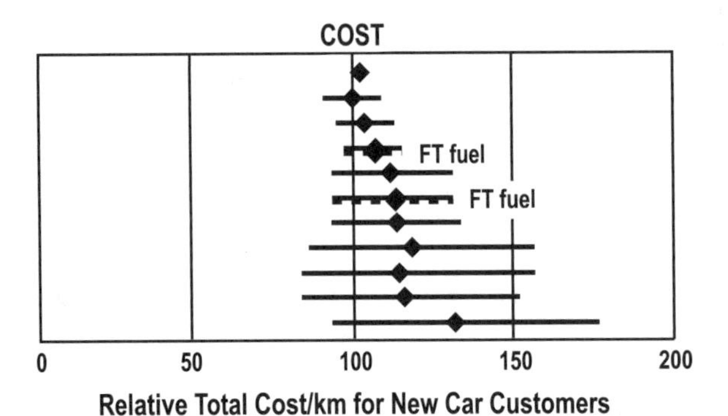

Figure 18.5. Lifecycle comparisons of technologies for new mid-sized passenger cars in 2020. Source: Weiss et al., 2000.

Advanced body design involving the substitution of lighter-weight structural materials (e.g., aluminum, plastics), reductions in drag and rolling resistance, and other improvements could cut fuel use by about an additional 11% at about an 8% increase in cost over the baseline car. GHG and local emissions will be cut proportionally. Because this is a sound way to reduce energy needs of the vehicle, it is likely that all the advanced propulsion system designs of the future will be configured with an advanced body design, although lighter weights raise some safety concerns.

Liquid fuel hybrid propulsion systems are particularly suited to urban driving cycles. The gasoline engine hybrid we examined would require about 25% less total energy consumption per km than is needed for the equivalent standard ICE gasoline car with a similar advanced body. Vehicle cost is higher by 8% for the advanced body and about an additional 5% for the hybrid configuration, which includes dual drives (mechanical and electrical) and special battery energy storage capacity. Carbon emissions are reduced by about 30% relative to the advanced body ICE, and 40% relative to the baseline vehicle (down about 60% from those of the typical 1996 vehicle).

Advanced diesel engines offer, by our calculations, about a 12% energy efficiency improvement over the gasoline engine with about a 2% operating cost increase. Combined with an advanced body, the energy efficiency improvement is about 24% over the baseline gasoline ICE. Diesel technology has improved over the recent past and could be further improved with modest investment. Many countries have or are tightening diesel emission standards, and attention is also turning toward requiring cleaner diesel trucks. In a hybrid configuration, diesel system fuel use is about 45% less than that of the baseline car (compared to about 30% less than the baseline for the hydrogen fuel cell car). Lifecycle GHG emissions are about 45% lower than the baseline (compared to about 30% lower for the hydrogen fuel cell car).

Diesel emissions, especially nitrogen oxides and particulates, have given this technology a bad image in the past. In the US, the EPA is tightening emissions regulations for diesels, and much research is underway to ascertain the connections between fine particulates in diesel exhaust and various respiratory ailments. To help reduce diesel emissions, the US and Europe are requiring reductions of sulfur content in diesel fuels to 15 and 10 ppm, respectively, over the next 10 years.

It is also unclear what level of NO_x reduction is possible with diesel technology. Additional R&D on fuel and engine design and on exhaust gas cleanup to make the diesel acceptably clean with respect to future regulations is needed. Meeting these emission requirements will take additional energy and will reduce the efficiency benefit that makes the diesel technology attractive in the first place. Some allowances for these factors were made in this assessment.

CNG hybrid vehicles provide vehicle energy-efficient performance between the gasoline and diesel hybrids but provide a substantial reduction in carbon emissions because of the switch to lower carbon intensity natural gas as the fuel. This option gives the greatest carbon emission reduction of all the options included in our study—about 50% relative to the baseline and 70% lower than today's equivalent car. The heavier and larger fuel-pressurized storage tank for high-pressure gas is bulky and infringes on available trunk space, and fueling times will be longer than for liquid fuel transfers. Leakage of methane, a potent GHG, can also offset benefits.

Fuel cell vehicles offer an alternative energy conversion system that has several advantages over internal combustion engines. They avoid high combustion-generated gas temperatures and, when fueled with hydrogen, do not produce harmful gaseous or particulate emissions. The fuel cell unit itself operates at higher efficiency than internal combustion engines. However, they do require hydrogen as fuel, and the emissions and inefficiencies associated with the production and distribution of hydrogen are comparatively large. Thus, comparisons on a "well-to-wheels" basis are important if net system improvements are to be assessed. The fuel cell vehicles with a liquid fuel reformer onboard, by our estimates, do not appear to offer any energy use benefits by 2020 over the advanced body gasoline vehicle and are inferior in performance to the similar fuel ICE hybrid options considered. If hydrogen fuel is stored onboard, a fuel plus vehicle system reduction in energy use of 30% over the baseline car is estimated. GHG emissions would then be reduced by about 30%, and local emissions would be virtually eliminated. In this case, the hydrogen would be manufactured at distribution/filling stations from reforming natural gas. The state of the art for hydrogen storage is still a limitation on this technology. The authors assumed a storage density of 5% in their analyses.

Electric cars are considered zero emission vehicles, but on a "well-to-wheels" basis, the emissions depend on the emissions associated with electricity and vehicle production. Large power plants can be more efficient than individual vehicle engines, so some GHG reduction is gained for the same fuel. Nuclear power and electricity from renewable energy can reduce associated GHG emissions to near zero; however, fossil fuels are projected to remain the predominant fuel for power generation in the 2020 timeframe. Using the electric sector supply mix for 2020 as forecast by the EIA, lifecycle carbon emissions for the electric vehicle option are 30% lower than those for the baseline gasoline vehicle, and the lifecycle energy consumption is lower by about 20%. Costs for electric vehicles are high, mainly because of the cost of batteries. A major limitation in performance is due to the energy storage limitations of batteries, so the electric car in this study has a shorter driving range than all the other vehicles.

Behavioral changes in auto buyers are likely to be slow and will require that new options are at least as attractive as the old. Although behavioral changes were not quantitatively addressed in this assessment, they may come about in response to, or perhaps independently of, the new technologies. Increased use of public transportation, more carpooling, car sharing, and less use of cars for short trips can be encouraged through pricing mechanisms, convenience, and education. If major reductions in GHG emissions are needed, more attention will have to be paid to land use planning, integrated with efficient transportation systems. The present market structure and the interests of car manufacturers encourage the acquisitive instincts of many auto buyers to upgrade to larger, more luxurious vehicles that are large consumers of energy. Behavioral changes seem unlikely without government intervention or a major change in the environmental consciousness of customers.

The full report (Weiss et al., 2000) draws the conclusions below.

- A valid comparison of future technologies for passenger cars must include lifecycle analysis for the total system, which includes assessment of fuel and vehicle manufacture and distribution in addition to assessment of vehicle performance on the road.
- Successful development and penetration of new technologies requires acceptance by all major stakeholder groups: private-sector fuel and vehicle suppliers, government bodies at many levels, and ultimate customers for the products and services. Therefore, the economic, environmental, and other characteristics of each technology must be assessed for their potential impacts on each of the stakeholder groups.
- Continued evolution of the traditional gasoline car technology could result in 2020 vehicles that reduce energy consumption and GHG emissions by about 1/3 from comparable current vehicles and at a roughly 5% increase in car cost. This evolved "baseline" vehicle system is the one against which new 2020 technologies should be compared.
- The more promising advanced technologies for propulsion systems and other vehicle components could yield additional reductions in lifecycle GHG emissions (up to about 50% lower than the evolved baseline vehicle) at increased vehicle costs (up to about 20% greater than the evolved baseline vehicle).
- Vehicles with hybrid propulsion systems using either ICE or fuel cell power plants are the most efficient and lowest emitting technologies assessed. In general, ICE hybrids have advantages over fuel cell hybrids with respect to lifecycle GHG emissions, energy efficiency, and vehicle cost, but the differences are within the uncertainties of the results and depend on the source of fuel energy.

If automobile systems with drastically lower GHG emissions are required in the very long run future (perhaps in 30–50 years or more), hydrogen and electrical energy are the only identified options for "fuels," but only if both are produced from non-fossil sources of primary energy (such as nuclear or solar) or from fossil primary energy with carbon sequestration.

18.6 Freight Vehicles

Today, global freight transportation consumes over 40% of all transportation energy. Freight transportation is crucial to most human activities, ranging from the movement of foodstuffs, water, and energy to raw materials, finished commercial products, and wastes. With economic development and globalization, freight transportation has grown both in efficiency and extent and has become an indispensable part of economic activity. While the benefits of freight transportation are crucial to most economic activities, there are some associated side effects of concern. Large trucks compete with personal transport vehicles for road space and contribute to congestion and pollution problems. Because the diesel engine is the propulsion system of choice for moving heavy loads, and because truck emission regulations have typically lagged those imposed on cars, trucking is a significant contributor to air pollution in both developed and developing countries. New emissions regulations, however, will impose stricter limits on trucks and will require a combination of cleaner fuels and after-treatment technologies that will result in some higher costs. In addition, efficient and affordable global freight capabilities could allow some developing countries to sell cheap labor, poor environmental practices, and undervalued raw materials into world markets to accelerate their own economic growth.

Freight is moved by many modes, including rail, trucks, pipelines, ships, and air. These global systems are interconnected and interact for bulk and smaller shipments, over long and short distances. Much attention has been given to bulk, long-distance shipments, and the introduction of containerization has made a major improvement in freight handling, allowing easy and largely automated connections from ships to rail or trucks at port facilities. Within cities, smaller delivery trucks distribute goods to residents and businesses. With the advent of mass catalog marketing and, more recently, of online marketing, the rate of increase of small package deliveries is rising.

Increased demand in freight transportation creates more congestion, pollution, safety problems, and GHG emissions. Congestion is usually handled by imposing limitations on freight transport in urban areas that in turn have some adverse impact on business. Pollution problems are being addressed retroactively through more stringent emission standards. Safety problems are also addressed through a combination of regulations and design of traffic systems that provide some separation of cars and larger trucks. To date, the impacts of increased energy usage by the freight sector have not been directly addressed. This is a currently unpriced environmental externality.

Similarly, the SUV, which has been subject to light truck, rather than car, emission standards, has been growing in popularity. These vehicles are less efficient than the smaller passenger cars and have contributed to increased air emissions. As of 2003, regulators are bringing light truck emission standards into line with those for smaller cars and are considering fleet efficiency standards intended to discourage future consumers from opting for even larger vehicles.

18.7 Interurban and Intercontinental Transport

Although many developing countries have rail systems that are used for interurban freight and passenger transportation, air travel is becoming the preferred mode of travel for longer interurban trips and the predominant mode for intercontinental transport. Air travel passenger-kilometers worldwide grew from 0.1×10^{12} in 1960, to 2×10^{12} in 1990, to projections of 14×10^{12} in 2020 (Schafer and Victor, 2000). In 1960, global air travel was 2% of global total travel; in 1990, 9%; in 2020, the air travel share is projected at 24%.

Airports around the world are already feeling limitations in capacity as demand rises. Plans are in place for more sophisticated air traffic control and for new runways or satellite airports. In heavily congested urban areas, siting new airports and runways is becoming increasingly difficult. As demand for air transportation grows, operational limitations may impose some constraints on the projected growth rates indicated above.

Jet fuel contributions to GHG emissions are leveraged in importance by the fact that they are injected into the upper troposphere and lower stratosphere (in the range of 9–13 km in altitude). Aircraft engines emit not only CO_2 but also NO_x, SO_x, and particulates. The impact of releasing these emissions at altitude is estimated to create about double the effect of burning the same fuels at ground level (Lee et al., 2001).

18.8 Motorization Trends

The US development of road transportation and road infrastructure took place over the first half of the 20th century; today there is almost one car per registered driver in the US, and much of our lifestyle incorporates the automobile. China is starting on its pathway to motorization. In 2001, there were 18 million road vehicles in China (excepting the ubiquitous bicycle), of which 5 million were passenger cars. With a population of over a billion, the car is still a rare personal possession. But with rapid growth in the Chinese economy, more and more people, particularly in urban areas, are becoming able to buy cars. The motorization of China is happening over a couple of decades, and the Chinese are eager to emulate the lifestyles of the developed world. The question is whether they can find development pathways that avoid the mistakes that were made in the developed countries. The Chinese government in 1999 requested that experts from the Chinese Academy of Engineering partner with the US National Academy of Engineering to prepare a report containing information and advice that might be helpful to the government of China in the management of Chinese motorization (NRC, 2003).

The Chinese government, in its eighth five-year plan (1991–1995), designated the Chinese auto industry as one of eight "pillar industries" that would drive China's 21st century economy. In the tenth five-year plan (2001–2005), plans for a massive restructuring of the fragmented domestic auto industry are described, along with a goal of producing an affordable Chinese family economy car that could be purchased for less than 80,000 RMB ($9,800). Major changes in the demography of large cities are already underway to accommodate a motorized future. The old large cities of China were too densely populated to have room for the addition of cars. For example, in the densest part of Shanghai, typical per capita living space was about 2 m^2 (20 ft^2)—about the footprint of a typical car. Most Chinese cities have large (and relatively inexpensive) taxi fleets that provide mobility to many people. However, congestion and pollution problems are growing in these cities.

Today, major Chinese cities are being ringed with new satellite cities, and per capita living space is increasing to 20– 60 m^2 in new high-rise apartment buildings. Inner city occupants are moving to these new living areas. New highways and parking facilities are being built, as well as new public transit lines. The bicycle is still the major transportation mode for most, but accident rates are increasing as autos are encroaching on road space. In some congested areas, bicycles are banned, or complicated lane arrangements for separating bicycle and auto traffic are provided. Motor scooter use is also increasing to further add to the mix of vehicles trying to use the same roads.

China is proactively adopting (and enforcing) new emission standards for vehicles to address the serious pollution problems. These are being phased in, and the intention is to have China's emissions standards at world quality levels by 2010. Public transportation is also growing as the government is aware that most of their citizens will not have access to personal cars for many years. Roads are being built at double-digit increase rates in the main cities that are motorizing. Businesses and industries are relocating from the inner cities to the new areas, making travel distances for consumers and commuters much longer than they had been. The new demographics are following patterns in developed country cities like Tokyo and New York, and will make the auto an integral part of the future of China.

China has limited petroleum reserves, and it is meeting the growing demand for gasoline by increasing imports. This is of some concern to the government. Also, with China's recent entry into the World Trade Organization (WTO), it must remove many protective tariffs that allowed its domestic auto industry to sell non-competitive cars within the country. Thus, the Chinese auto industry has to reorganize and come up rapidly to world-competitive levels. There are a number of joint-partnership auto manufacturing plants within China that produce world-class cars. However, the foreign partners have been careful to protect the trade secrets and know-how that make them competitive. Thus although many Chinese work in these plants and are increasingly involved in their planning and management, there remains a large gap between today's domestic industry and the government's hope for a domestic industry that will be a "pillar" of the future economy.

Nevertheless, the Chinese are aware of these issues and are rapidly catching up with the world in many other areas of technology—and starting to lead in some. They are dealing with a difficult transition in which Chinese auto companies and suppliers consolidate and streamline production operations and search for special opportunities that will allow them to be more competitive. Perhaps the small economy car will be one, although that will be up to Chinese consumers. At present, the sales of cars in China are around 1 million per year—about the minimum sales volume needed to support just one major car manufacturer—but the rate of growth is high. Joint-venture foreign design cars are a major part of sales, so it will be a challenge to capture a viable domestic market for a Chinese auto manufacturer. Already the Chinese are looking for export opportunities, perhaps in Russia or India.

As motorization progresses in the developing countries, the demand for gasoline will increase accordingly, and this in turn will lead to increases in GHG emissions, even though countries like China place major emphasis on fuel economy. While the use of hybrid technology might slow the increase, the added cost will deter sales to consumers who are barely able to afford their own vehicle. It can be argued that car ownership can create a larger gap between the rich and the poor, as car owners have greater job mobility and better employment opportunities. The balances between the benefits of enhanced mobility and the associated negative impacts are part of the tradeoffs to be addressed in a transition to a more sustainable world.

18.9 Sustainability Issues

The major issues relating to sustainability and the transportation sector involve complex tradeoffs between economic, societal, and environmental sustainability. The major challenges are outlined below:

- Transportation and mobility are vital parts of our global economic system. They are making major contributions to sustainable development of developing countries while providing services that are in increasing demand in the OECD world. Population growth and economic development increase demand for mobility services. If global car use increases to current OECD levels, sustainability goals for GHG emissions will be seriously compromised.
- Almost all of the transportation sector is dependent on petroleum-based fuel. While it appears that supplies are adequate for a number of decades into the future, at some point society will have to transition to some alternative transportation energy supply. The faster the growth in global mobility demand, the sooner and more challenging the transition will be.
- The transportation sector is growing more rapidly in GHG emissions than other sectors, and it is harder to control because of the dependence on fossil fuel and the dispersed nature of the emission sources.
- Road and rail infrastructures in developed countries have already impinged on natural habitats and have resulted in isolating species into pockets of reserves that are vulnerable because natural migration options have been removed. This pattern is being repeated as infrastructure is growing in developing nations.

- Automobiles have a major impact on urban land use, leading to decentralization and redistribution of living, work, and shopping patterns. Cars now compete successfully with subsidized public transportation systems.
- People love mobility and see it as a major enhancer of quality of life. Cars provide easy mobility for the more affluent, but lack of access to personal cars limits the mobility of the poor, the aged, etc.
- Subsidized energy prices (not including societal costs of some of the externalities) provide people with little incentive to conserve.

What are some of the future opportunities?

- Encouraging stewardship of transportation and associated energy services so that economic and societal benefits can accrue without unnecessary mobility demand (and resultant congestion and pollution problems) and wasteful use of energy (through pricing, policies, and/or education).
- Avoiding energy subsidies as a tool for accelerating economic development, unless other measures are in place to prevent inefficiencies and waste.
- Reducing the gaps in per capita energy consumption to allow the poorest to enjoy at least a basic quality of life.
- Utilizing communications technology to replace actual business trips with virtual trips.
- Developing longer-term alternative fuels and policies that will encourage a responsible transition from essentially total petroleum dependence.
    - Natural gas as a transitional possibility
    - Low GHG-emission alternatives: electricity or hydrogen generated from coal with carbon sequestration; biofuels (with careful consideration to land use competition with agriculture and natural habitats)
- Using integrated land use planning to modify present settlement patterns and build new settlements that include energy-efficient transportation options for the residents. These options should consider placement of residences, work locations, shopping locations, and recreational locations along with the mobility needs of the population. Improved provision of energy-efficient and convenient public transportation, both local and longer-distance, can significantly reduce reliance on personal cars.
- Encouraging the use of more locally-produced products to decrease the amount of long distance air freight transportation.

- Examining of all short-term strategies for important sustainability impacts on other activities and goals worldwide and, especially, on future generations.

References

BP (British Petroleum). 1999. *BP Statistical Review of World Energy*. London (see www.bp.com/centres/energy/).

EIA (Energy Information Administration). 2000. *Monthly Energy Review*. Washington, DC: US Department of Energy. (October)

EIA (Energy Information Administration). 1997. *Emissions of Greenhouse Gases in the United States*, 1996. Washington, DC: US Department of Energy. (October)

Hoehlein, B. 1998. *Brennstoffzellen-Studie* (fuel cell studies); Ganzheitliche Systemuntersuchung zur Energiewandlung durch Brennstoffzellen. Frankfurt am Main: Forschungsvereinigung Verbrennungskraftmaschinen e V.

Lankey, R., F. McMichael, L. Love, H. MacLean, and S. Joshi. 2000. "Alternative Vehicle Power Sources: Towards a Life Cycle Inventory." *Proceedings of the Total Life Cycle Conference*, Society of Automotive Engineers. 2000-01-1516: Detroit, MI.

Lee. J.J., S.P. Lukachko, I. Waitz, and A Schafer. 2001. "Historical and Future Trends in Aircraft Performance, Cost, and Emissions." *Ann. Rev. of Energy and the Env*. 26: 167–200.

MDOT (Maryland Department of Transportation). 1999. "Facing Urbanization, The Engineering Challenges." Keynote address, National Academy of Engineering Symposium. John Porcari, Secretary, MDOT. (October)

NRC (National Research Council and Chinese Academy of Engineering). 2003. *Personal Cars and China*. Washington, DC: National Academies Press.

O'Meara, M. 1999. "Reinventing Cities for People and the Planet." Paper 147. (June). Washington, DC: Worldwatch Institute.

ORNL (Oak Ridge National Laboratory). 2000. *Transportation Energy Data Book*, Edition 20. Oak Ridge, TN: US Department of Energy.

ORNL (Oak Ridge National Laboratory). 1998. *Transportation Energy Data Book*, Edition 18. Oak Ridge, TN: US Department of Energy.

OTA (Office of Technology Assessment). 1995. "Advanced Automotive Technologies: Visions of a Super-Efficient Family Car." OTA-ETI-638. Washington, DC: US Government Printing Office.

Rabinovitch, J. and J. Leitman. 1996. "Urban Planning in Curitiba." *Scientific American*.

Schafer, A. and D. Victor. 2000. "The Future Mobility of the World Population." *Transptn. Res. A*. 34(3):171–205.

Scientific American. 1997. "The Future of Transportation," Special Issue. (October)

Singh, G. 1998. "Total Energy Cycle Assessment of Electric and Conventional Vehicles: An Energy and Environmental Analysis." ANL/ES/RP-96387. Argonne, IL: Argonne National Laboratory.

Statistical Abstract of the United States, National Data Book. 2000. Washington, DC: US Government Printing Office.

Thomas, C.E., B. James, F. Lomax, and I. Kuhn. 1998. *Societal Impacts of Fuel Cell Vehicles*. Society of Automotive Engineers. Paper No. 982496. Washington, DC: SAE.

Weiss, M., J. Heywood, E. Drake, A. Schafer, and F. AuYeung. 2000. "On the Road in 2020." MIT Energy Laboratory Report No. 00-03. Cambridge, MA.

World Business Council for Sustainable Development. 2001. *Mobility 2001*. Switzerland: Atar Roto Presse (online at: www.wbcsd.org).

Web Sites of Interest

www.dot.gov

www.eere.energy.gov/EE/transportation-doe.html

www.wbesdmobility.org/

Problems

18.1 World production of crude oil in 2000 was about 3.7 billion tonnes. Assume that we wish to replace 10% of this supply with a transportation fuel derived from biomass. Assume that the energy content of biomass is about 1% of incident solar energy during the growing season. How many hectares of arable land worldwide would be needed to supply this quantity of biomass fuel? (Hint: Look at Chapter 13 for solar energy resource information and at Chapter 10 for information on productivity of energy crops.) What fraction of arable land area worldwide does this represent?

18.2 Demographia (http://www.demographia.com/db-intlua-data.htm) gives 1990 population densities for major cities:

Tokyo: 7,100 people/km^2

Vienna: 6,800 people/km^2

Paris: 4,600 people/km^2

New York: 2,100 people/km^2

And in the central core of Shanghai, population densities reach 50,000 people/km^2. What are the implications of increasing car ownership in such an area? Suppose 10% of the population owns cars. What parking space area is needed to accommodate these cars?

18.3 Calculate the specific energy (Wh/kg), the energy density (Wh/L), and the fueling rate (Wh/s) for gasoline, methanol, ethanol, natural gas stored at 200 atm, hydrogen stored at 350 atm, and electricity. State your assumptions about the fueling conditions at the fueling location. Estimate and discuss the lifecycle GHG emissions associated with each of these fuel options.

18.4 What is the impact of a carbon tax of $50/tonne on the price of a gallon of gas? How would you decide whether such a tax is justified if you were a lawmaker?

Industrial Energy Usage 19

Engineering is the science of economy, of conserving the energy, kinetic and potential, provided and stored up by nature for the use of man. It is the business of engineering to utilize this energy to the best advantage, so that there may be the least possible waste.
—William A. Smith, 1908

19.1 Introduction and Historical Perspectives

Following the Industrial Revolution, our society entered into a period of continual technological change, which brought many advantages and problems. Initially, progress was slow, and small industrial activities were located near suitable sources of materials and energy. Pollution resulting from the activities was also localized, and accidents associated with early machinery and chemical processes were usually on a fairly small scale. As industries grew, they learned how to avoid the most severe problems and to improve competitiveness and efficiency. As noted in Chapter 6, along with industrialization came a better standard of living, improved human health, urbanization, and population increases. In 1700, world population was only 600 million. Three hundred years later, population is over 6 billion people, and it is still rising.

Not surprisingly, the expansion of industrial activities to meet the needs of a rapidly growing and more affluent population has magnified the drain on resources for production and the impacts of pollution associated with industrial wastes. In the mid-20th century, governmental agencies began to institute regulations designed to reduce industry's more harmful environmental impacts—release of species affecting human health into the air, pollution of lakes and rivers from waste disposal, serious impacts on land from careless mining and forestry practices—and to protect worker and consumer safety.

In the early 21st century, environmental regulations in many countries in the developed world are being expanded and made more stringent. Developing countries are already beginning to deal with some of the pollution problems of early industrialization and are implementing environmental and safety regulations that will balance their desire to gain economic ground with needs to protect the health of their populations and the environment. However, most of these efforts are instituted in response to obvious problems and do not anticipate consequences of future growth in industrial demand generated by growing populations that are able to and want to consume more goods. The associated increase in fossil fuel use is also of growing concern, as the world faces the possibility of near-term climate change as a result of greenhouse gas (GHG) emissions.

The industrial sector in developed countries typically accounts for about 1/3 of total commercial energy consumption (when both thermal and electricity use are included). Heavy industries require large amounts of continuous thermal and electrical energy and rely on bulk energy sources like fossil fuels. Table 19.1 presents some data on the carbon equivalent emissions from fossil fuel use in US industries.

Table 19.1. Greenhouse Gas Emissions from Major US Industry Sectors (million tonnes of carbon equivalent)

Industry (excluding electric use)	**1990**	**1995**	**2000**
Iron and Steel	23.0	20.0	18.0
Cement and Lime	13.5	15.0	17.0
Ammonia	5.0	5.0	5.0
Aluminum	2.0	1.5	1.5
Agriculture*	142.0	153.0	152.0

*Agriculture impacts are dominated by nitrous oxide and methane emissions.
Source: EPA (2000).

If energy required for electricity production were included, the total GHG emissions for the iron and steel industry would almost double, and the emissions for the aluminum industry would more than double.

The petroleum and petrochemical industries use fossil fuels both as feedstock and energy sources. Some industries, like aluminum production, require large amounts of electricity, and plants are often sited near major hydroelectric facilities to utilize cheaper power. The production of concrete, another major commodity, also generates significant quantities of carbon dioxide as part of the manufacturing process. Wood products and the pulp and paper industries are increasingly using waste biomass as a process heat source. Just these few examples show the diversity of issues and opportunities for GHG reduction that abound in the industrial sector. This diversity also means that there are no simple sets of rules that will apply to all industries, especially because the economics of energy use varies depending on the particular application and on the location of a facility. In a global economy, location of production plants is based on economic motives and may take advantage of less stringent environmental or labor regulations in developing countries.

In this chapter, we will present some general principles and concepts that are likely to be employed by a variety of industries in addressing their own transitions to more sustainable production. Then we will discuss several major industries that are energy-intensive and/or emit GHGs as part of their processing activities. Processes that generate CO_2, for example, might be attractive candidates for capture and sequestration of that GHG. Finally, the opportunities and barriers to a transition to sustainable industrial production will be explored.

19.2 Lifecycle Analysis and Design for Sustainability

Chapter 6 introduced the importance of using a life cycle approach in comparing product or process alternatives to achieve desirable outcomes. In the automotive technology example discussed in Chapter 18, it was evident that looking at the improved

GHG performance of a vehicle technology without considering the emissions associated with energy production (e.g., for the electric car) can lead to a sub-optimal choice. Yet the auto industry tries to improve its performance without any control over the fuels industry. Only recently are these two industries getting together to explore some of the important lifecycle issues associated with the combination of new fuels and new automotive technologies for the future. It is challenging, however, to get industries whose products are related but whose members are in competition to collaborate in planning for the future. In addition, the industries that provide the raw materials for auto manufacturing, the industries that support highway infrastructure, and the industries involved in the recycling/scrapping of vehicles at the end-of-life are all connected to the lifecycle impacts of auto transportation.

When a lifecycle analysis is used, attributes usually include material and energy flows, costs, and associated waste streams. Here are some of the common options available to move toward a more sustainable product or process (NAE, 1999):

- Materials. Reduce use of virgin depletable materials; make use of recycled or renewable materials; choose materials that minimize production of harmful waste products; design for efficient use of materials (includes the idea of making products that are durable and reusable); facilitate reuse or recycling of product components at the end of a product's useful life
- Energy. Improve process efficiency; employ heat and work integration; recover waste energy; reduce use of fossil fuel sources; reduce need for transportation of raw materials and final products; choose efficient transportation modes where needed
- Wastes. Choose processes that minimize harmful waste production; find uses for waste materials that will convert them to a useful purpose; treat harmful wastes so that they do not cause human health or environmental impacts

All of these measures can be used to differing extents and as appropriate to the specific conditions associated with a particular industry. Unfortunately, we have not included the overriding issue of cost in the above list. Industry has found that many of the changes that lead to eco-efficiency also provide cost savings or are cost neutral. Most responsible industries are motivated to implement such changes. However, when there are significant costs involved in implementing a more sustainable process or product, most industry today is reluctant to take this route and hurt their competitive position.

There are different aspects of industrial sustainability. Some may set reduction in GHG emissions as a top priority (Riemer et al., 1998). For a text examining more sustainable energy use, this is certainly a major concern. And if we look at energy needed to produce materials of various sorts, a reduction in material consumption also will reduce energy demands and any associated GHG emissions. Depletion of non-renewable materials is another issue of concern (Gardner and Sampat, 1998), although many economists subscribe to the idea that markets will continue to drive technologies to substitute different materials as depletion begins to drive prices up. Other drivers toward sustainability are problems caused by waste and pollution. The drivers of concern will be highly location- and culture-dependent (Graham and Hartwell, 1997).

Progress will be slow until pricing structures change, consumers modify their purchasing habits, and/or the need for a transition to sustainability becomes a high priority. Under present pricing in the US, some metals recycling is cost effective; paper, plastics, and glass recycling is growing through environmental awareness (even though it is marginally profitable); and much waste is still landfilled at fairly low cost. In Europe and Japan, where land is much more expensive, landfilling is not a real option, and more costly waste incineration is the main method of waste disposal. It is not surprising that the Japanese are implementing strong regulations to minimize waste by establishing recycling regulations, requiring that manufacturers take back major products at their end-of-life, and charging the consumer for each bag of waste disposed.

Today's industry depends on a growth model for success: sell more products by creating new customers, developing new desires, or through engineered obsolescence of existing products. In a future sustainable business world, population and markets would be stabilized and businesses would make profits through selling fewer sustainability-designed goods and services to a relatively constant group of customers. Consumers would purchase goods based solely on durability and service. However, because technology changes allow improvements in performance and efficiency, provisions are needed for replacing older technology with better, newer technology. Perhaps products will be designed in a modular fashion so upgrades can be done in a simple manner and the older technology portions can be recycled. The transition to a sustainable future involves major changes both in industrial and economic practices.

Some companies are anticipating such a transition by experimenting with new ways to conduct their businesses. Some are implementing carbon-emission accounting practices that allow emission patterns to be understood and actions undertaken to reduce emissions where most feasible. Some are implementing principles of sustainable design into products and anticipating that design for disassembly and recycling of used product components will provide new materials for future products. McDonough and Braungart (2002) present one idealistic view of a future sustainable world that eliminates waste, and where materials are perpetually recycled, producing a "cradle-to-cradle" approach to design.

Cradle-to-cradle design is a theoretical concept, because we usually use unrecoverable energy in making products, and, by using things, we degrade their integrity. Chapter 3 discusses the thermodynamic laws of conservation and of entropy that limit our ability to get something for nothing. However, we are able to do much better than we do now. The ability to reuse waste products, to recycle their materials back to the same use, to recycle back to a lower integrity use, to recover energy from waste streams, and to utilize waste heat are all tools that can be used. Each industry has its own feedstocks, processes, products, and wastes, and each, therefore, has a different slate of opportunities to move toward more sustainable lifecycle operations. The following sections briefly outline some of the characteristics of major energy-intensive industries and some of the opportunities for them to move to more sustainable operations.

19.3 Metals Industries

The metals industries are major users of commercial energy and are connected to a variety of other impacts, including land impacts of mining, disposal of mining wastes, and carbon emissions. This sector produces over 10% of all manufacturing carbon emissions in the US. Figure 19.1 shows EIA (1994, 1998) statistics on such emissions and indicates that the iron and steel industry emits over 60% of the carbon produced by the US metals industries.

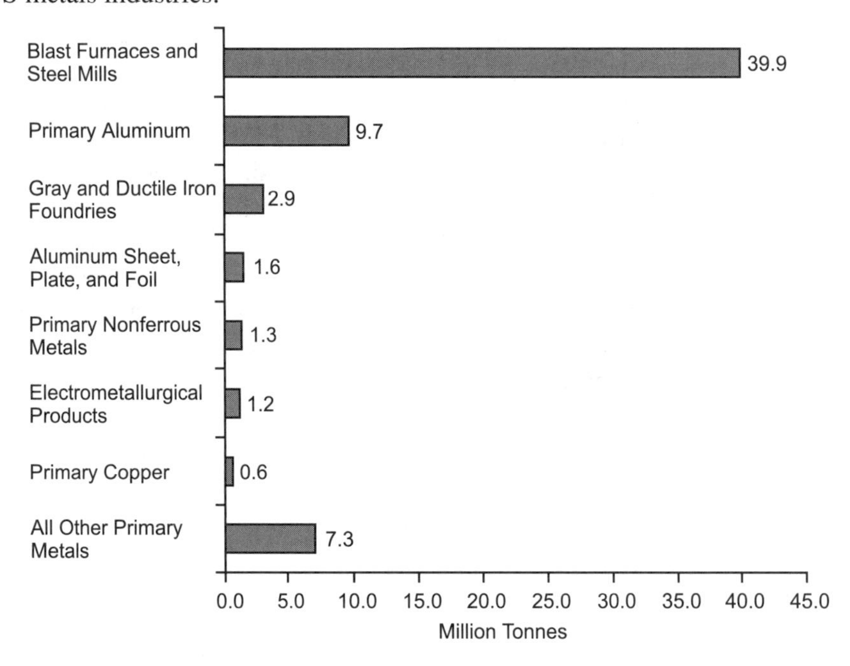

Figure 19.1. Carbon emissions from the metals industry. Source: EIA (1994, 1998).

Location of iron ore deposits and their mineral forms (hematite, magnetite, goethite, taconite, etc.) vary widely. The ores all are in some oxidized state, and many steps are required between the mine and the final metal product.

High-quality deposits are located in the crustal layers of the earth, and mining techniques vary from open pit mining to underground mining. Mining involves major energy-intensive movements of earth and ore materials and generates large quantities of wastes. Water is often used in separation and washing processes that occur at the mine, and wastewater is discharged into the environment. Ore is extracted, separated from extraneous deposits, and pelletized at the mine for shipment to processing plants. Iron ore is reduced to pig iron in a blast furnace fueled by coke, a carbonized form of coal that is made in ovens where the coal is heated to distill off coal gases. Both processes are energy intensive. Subsequent steps involve the further refinement of pig iron to steel in a basic oxygen furnace or an electric-arc furnace. The electric-arc furnace is widely used for recycling scrap metal, although some scrap can be co-fed to a basic oxygen furnace. Further processing, like rolling and stamping or blending to make specialized alloys, also requires some energy.

Opportunities for carbon reduction in the ferrous materials industry include higher levels of recycling, increased use of electric-arc technology (which is somewhat more efficient than the basic oxygen furnace), and co-siting with other industrial activities, such as combined cycle power plants, that might utilize some of the waste energy.

Primary aluminum production is also energy-intensive. Metallic aluminum is prepared by electrolysis of alumina (Al_2O_3) in a bath of molten cryolite ($AlF_3 \cdot 3NaF$). The alumina reacts with the carbon electrodes in the electrolytic cell to produce metallic aluminum and carbon dioxide. The carbon consumed is about 60% of the weight of metal produced. Molten aluminum is tapped from the bottom of the cell, and the carbon electrodes are continually replaced as the carbon reacts.

Because this is an electrical process, aluminum plants are often located near major hydroelectric facilities. In fact, the co-siting of aluminum plants often facilitates the development of remote hydroelectric power plants.

The energy required for electrolysis of aluminum is many times greater than the energy for melting aluminum scrap, so there are major incentives to recycle aluminum. The offgases from the process are a concentrated mixture of CO_2 and CO and could be suitable for capture and sequestration.

A major challenge to the recycling of metals is the widespread use of specialized alloys to achieve particular physical properties for different applications. Mixed alloys, when recycled, are likely to have compositions that do not match requirements for applications similar to their original uses. Careful attention to design and the ability to sort out different alloys for recycling offer the promise of more successful reuse of metals streams.

19.4 Cement and Lime Industries

Production of materials to support the construction industry is another major activity in terms of resource use and GHG emissions. The steady increase in US cement and lime production is shown in Table 19.1. Although the US cement industry has made steady improvements in energy efficiency over the past three decades, it is still energy-intensive, and coal and coke are still the primary fuels used (Martin et al., 1999). Each kilogram of cement produced in the US results in about 0.5 kg of carbon dioxide emissions. The global ratio is about 0.8 kg CO_2/kg cement. Globally, cement production is growing at a rate close to 4% per year, especially as developing countries build transportation infrastructure and upgrade the housing provided for their populations to meet urban apartment living standards in more developed countries. Cement production consumes about 2% of global primary commercial energy production, or almost 5% of total industrial energy consumption (BMI, 2002).

Concrete is made from a mixture of cement with water, gravel, and sand added. Cement is manufactured from calcining of limestone (calcium carbonate) along with smaller quantities of clay and sand (sources of silica, aluminum, and iron) to make a composite material. The limestone is quarried, crushed, and blended with clay and sand before introduction to a calciner—usually a rotary kiln furnace that has maximum temperatures approaching 1,870°C. Flow is countercurrent, with the hottest part of the kiln in the flame zone near the exit. Figure 19.2 shows a schematic of a typical rotary kiln cement process.

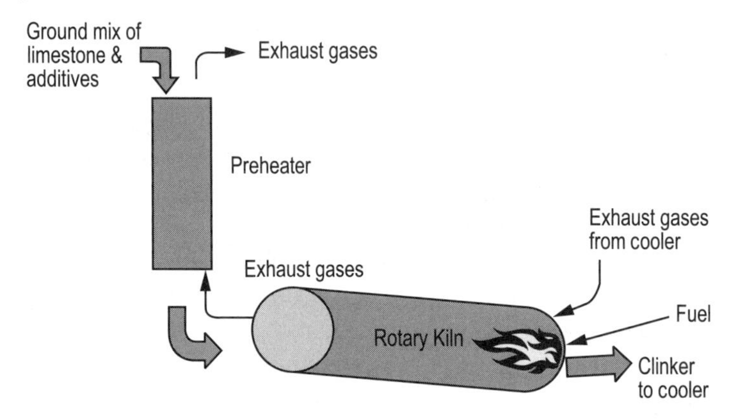

Figure 19.2. Rotary kiln for cement manufacture.

The reactions in the furnace result in the production of a material called "clinker," which is partially molten at the exit of the furnace. The clinker is cooled and then passed through a series of grinding processes to produce cement. Typically, half of the CO_2 emissions come from the reduction of the limestone and half come from the fuel used in the calcining process. There are both "wet" and "dry" processes for cement manufacture because water facilitates the initial crushing and grinding process. However, this water requires additional energy as it is evaporated in the calciner.

There are a number of ways to reduce fossil fuel use and CO_2 emissions in the overall process. Several global cement companies have cooperated under the aegis of the World Business Council on Sustainable Development to evaluate the present status of the global cement industry and to formulate a "Cement Sustainability Initiative" (CSI). The CSI includes the development of action plans by 2006, reporting of emissions, tracking of progress, and a commitment to invest in the rehabilitation of quarry and plant technologies and practices.

Some possibilities for GHG reduction in the cement industry are explored by Martin et al. (1999) and include switching to a dry process (10–15%), using waste fuels (30%), adjusting the cement additives (30–35%), and using alternative cements (up to 80%).

Concrete formulations also can incorporate other waste materials and serve as a convenient use for other process residual materials. Research is underway on cements made from geopolymers and mineral polymers (e.g., kaolin) instead of limestone. Of course, energy efficiency improvements also can be made, especially in smaller outmoded plants in developing countries.

Cement plants usually operate for four or five decades, so there are many older plants with opportunities for upgrading, though this will require substantial capital investment. A more expensive option for CO_2 emission reduction might be to heat the calciner with a fuel/oxygen flame so that the combustion gases do not contain much nitrogen. In that case, a nearly pure CO_2 and water combustion product could produce a CO_2 stream that could be captured and sequestered.

19.5 Chemical Industries

Because the chemical industries cover a wide range of products and processes, there are a diverse set of process-specific actions that are involved in a transition to more sustainable operation. The chemical industry category includes over 60,000 different products, including major categories of basic organic and inorganic commodity chemicals that serve as feedstocks for a variety of manufacturing processes to make plastics, fertilizers, paints, detergents, textiles, and other products. The plastics industry also uses large amounts of petroleum and natural gas as feedstocks for making widely used organic polymers, such as polyethylene, polypropylene, and polyvinyl chloride. The chemical industry uses about three quads of thermal energy annually and an equal amount of energy sources as feedstock.

Opportunities for efficiency improvements in the chemical industry exist, but they are usually not economically viable enough to implement. Chemical process design has become sophisticated, and powerful software tools facilitate economic optimization (see www.aspentech.com/). These design tools include properties databases, thermodynamic relationships, and energy and material balances, as well as more sophisticated design techniques like "pinch technology"—a method for integrating heat and work flows throughout a process system to minimize net energy requirements. However, these techniques require economic data as inputs and produce an optimization based on overall costs. Energy costs today do not include costs for GHG reduction, so this will not be a priority in today's design activities.

Today, several major energy-intensive chemical industries produce relatively pure CO_2 streams that might be of future interest for capture and sequestration projects. The fertilizer industry and the ethylene industry are two candidates.

In the manufacture of ammonia, the building block of the fertilizer industry and a major industrial chemical, nitrogen is reacted with hydrogen in a type of Haber-Bosch or similar process at elevated temperatures and pressures. The main reactions of interest are shown below, but the process requires multiple stages and the use of special catalysts to enhance reactivity.

$$N_2 + 3H_2 \rightarrow 2\,NH_3 \tag{19-1}$$

The hydrogen is usually made from steam reforming of methane (or other hydrocarbon, depending on economics).

$$CH_4 + H_2O \rightarrow CO + 3H_2 \tag{19-2}$$

The reaction products are mixed with enough air to provide the nitrogen for ammonia synthesis, and the product gas is passed into a "shift reactor" where the CO is further reacted with steam.

$$CO + H_2O \rightarrow CO_2 + H_2 \tag{19-3}$$

The CO_2 is removed prior to ammonia synthesis, and it could be recovered for sequestration.

The above steam reforming and shift reactions are key steps in many processes because they are routes to hydrogen. Steam cracking of ethane in a tube furnace is now the primary method for making ethylene—with CO_2 as a relatively pure byproduct. Ethylene is a major commodity chemical, being the feedstock for polyethylene manufacture as well as for a host of other chemicals, such as ethylene glycol.

Hydrogen demand for refining operations is also growing as feedstocks get heavier and the demand grows for lighter fuel products. If the source of the hydrogen is from hydrocarbons, a byproduct will be waste carbon, probably in the form of CO_2. Hydrogen transportation fuel also bears the same penalty in carbon emissions from manufacture, but this could be another opportunity for carbon sequestration.

Much work is in progress on new processes that will reduce both energy intensity and CO_2 emissions from major industrial processes. Some of these involve government-industry partnerships because the economic incentives are not yet in place to encourage major industrial investments in new technology.

19.6 Forest Products and Agriculture

The forest products industries include industries that produce wood goods, as well as the pulp and paper industry. These industries do not show up as major sources of CO_2 production because of limited fossil fuel consumption. In papermaking, either mechanical or chemical processes can be used to convert wood or other fibers to pulp. About half the energy used in papermaking is associated with forming and drying paper products. Over the past few years, the industry has replaced much of its use of fossil energy with energy derived from waste wood combustion.

However, because the global carbon cycle involves the biosphere and, consequently, the management of forests, these industries are involved in issues beyond those affecting the conventional chemical industry. In the US, improved forest management, use of advanced timber-harvesting techniques, and reforestation practices have resulted in a modest net annual increase in carbon uptake (DOE, 1999).

The impacts of the agricultural industry are most evident when the greenhouse forcing effects from methane and nitrous oxide emissions are weighted along with CO_2 emissions. Table 19.2 shows data from 1997 (DOE, 1999).

There are many opportunities to reduce emissions from agricultural activities through better management of livestock, fertilization, irrigation, and waste products. Improvements in food processing and transportation energy use are also possible. However, agriculture is a diverse activity with many different and competing players, so change must be driven by cooperation between governments and major industrial participants. New policies and practices will need to be demonstrated successfully to encourage any change from long traditions.

Table 19.2. Human-Caused GHG Emissions from Global, US, and US Agricultural Sources for Three Major Greenhouse Gases, 1997

	Carbon dioxide	Methane	Nitrous oxide	All three gases
Radiative forcing per unit mass relative to CO_2	1	58	206	
Mean atmospheric lifetime (years)	125	12	120	
GWP*, 100 years relative to CO_2	1	21	310	
Emissions of equivalent gigatons carbon (GTCE)				
All global sources	7.10	2.15	0.507	9.76
All US sources	1.50	0.180	0.109	1.79
All US agricultural sources	0.044	0.054	0.077	0.175
Emissions, % total global GTCE				
All global sources	72.8	22.0	5.20	100.
All US sources	15.4	1.84	1.12	18.4
All US agricultural sources	0.45	0.55	0.79	1.79

* GWP is the Global Warming Potential, a numerical index developed by the Intergovernmental Panel on Climate Change (IPCC) to evaluate the relative radiative effects of different GHGs based on their forcing effect and on their residence time in the atmosphere. Integration of net forcing effects relative to CO_2 over a 100-year period is taken as the basis. Source: DOE (1999).

19.7 Waste Management Industries

As we follow through the lifecycle of industrial products, we eventually come to the issue of waste disposal. Gardner and Sampat (1998) estimate that the US uses 10 billion tons of materials every year—excluding food, fuel, and water. That amounts to about 100 kg per person per day. Much of this quantity does not enter into our direct use. For example, most of the volume of ore extracted for metal production is discarded at the mine site. Eventually, even the materials that we use in our daily lives will end up in the waste stream.

We all are aware of household and commercial "garbage" that constitutes municipal solid waste (MSW). On the average, each US citizen generates about 2 kg per person per day (excluding water content). These quantities do not include construction and demolition debris or agricultural and industrial waste (much of which enters the disposal stream separately). Garbage is usually collected weekly by an army of fossil-fueled trucks and taken to disposal sites. At present, in the US, 55% is buried in landfills, 17% is incinerated, and about 28% is recovered either in recycling facilities or through composting (EPA, 2000).

The US has much larger amounts of land per capita than do Japan and Europe. Consequently, we make use of large landfills and have a fairly inexpensive way of disposing of waste (clean incineration is about twice as expensive as landfilling). In Europe and Japan, incineration is the main option for dealing with waste, and both those regions have much stricter constraints on waste disposal than exist in the US. Typical constraints include recycling goals, waste disposal charges, and initiatives to encourage industry to design for maximum reuse and recycling of their products. In Germany, for example, autos and large appliances must be returned to the manufacturer at the end of their normal use life.

However, wastewater and landfills both emit methane as a product of gradual fermentation. In 1990, these emissions accounted for about 4% of the US GHG emissions, and about 90% of the effect was due to methane emissions. Incineration results in some CO_2 and nitrous oxide emissions, but these are relatively small compared to methane emission effects.

The US EPA has been active in promoting better MSW management. Large landfills are designed to capture emitted methane and use it for energy that displaces fossil fuel use. In addition, the EPA has developed a Waste Reduction Model (WARM) that supports voluntary municipality reporting efforts. This model facilitates the computation of equivalent GHG emissions from different waste treatment sources. (Details of the model are available at: yosemite.epa.gov/oar/globalwarming.nsf/content/ActionsWaste.html)

This model also includes a feature that allows a credit for carbon sequestered in the landfill— one of the controversial carbon credits that are being debated internationally. If credit is allowed for landfill carbon sequestration, the EPA WARM model provides the estimates shown in Table 19.3.

Table 19.3. Net GHG Emissions in Million Tons of Carbon Equivalent from US Municipal Waste Streams (Results including carbon sequestration)

Scenario	Estimate of Carbon Sequestered	Estimate of GHG Emissions	Total Net GHG Emissions
1974	-22	36	13
2000	-48	10	-38
2000 with 1974 technology	-40	51	11

The 1974 technologies assumed in the above tabulation include a much larger use of landfilling, as well as landfill technologies that did not delay decomposition or provide for methane emissions capture.

19.8 Sustainability Issues

Industry is the primary producer of the many products that enhance our quality of life. And industry, operating in a competitive free market economy, is proficient at finding the production methods that minimize costs and produce products that will attract consumers. Society imposes some constraints on industry that protect values of importance (e.g., clean environment, fairness, stable institutions) through regulations or policy or economic measures. Often these regulations or policies emerge in response to public concerns about specific effects of a particular industrial activity. In developed countries, such regulatory and policy measures are extensive and vary according to local cultural, economic, and environmental needs. In developing countries, priorities for providing basic services to large populations are usually much higher than policies for environmental protection, which later evolve as the country's economic base improves to the point where "luxury problems" begin to be addressed.

Because sustainability is a longer-term issue and the effects of non-sustainable behavior are often deferred, industry and governments generally focus on near-term issues and defer spending to address problems that are not imminent. However, many larger companies, especially those with visionary leadership, are becoming increasingly aware of the need to make significant changes in their future practices. Investments are being made in research on new technological possibilities, and production practices are being reviewed to better understand lifecycle aspects, including net emissions of GHGs. Further, there is a growing population of concerned consumers who are willing to pay some small premium for "green products." But the competitive market also contains players who are willing to cut corners or make false claims. And the CEO who moves a company to costlier products that are non-competitive in the market will soon be out of a job.

Industry will be an important player in any transition to more sustainable behavior, but much of the driving force will have to come from a new breed of consumers who want less materials and more performance and from citizens who will elect government officials who are willing to divert some part of current economic gain to invest in the future. Educated consumers and voters will be key to initiating change.

The technological tools to facilitate change have been discussed at length in this text: lifecycle analysis, efficiency, heat and work integration, waste minimization, reduction in use of virgin materials where possible through reuse and recycling, and smart design. However, as long as we have energy that is "cheap enough to waste," we cannot expect industry to treat it as a limited resource that should be used sparingly.

If we recognize that energy today is not priced to include the potential costs associated with GHG emissions (and some other factors, such as land impacts, social and political issues, depletion), the resulting process designs will not include provisions for reducing GHG emissions. GHG control is likely to require additional costs. Remember that past environmental legislation and standards initially imposed additional costs, but industrial creativity found ways to incorporate changes so that new results were not too onerous. Nevertheless, more energy-intensive industries will have a bigger challenge than other industries in making a transition.

References

BMI (Battelle Memorial Institute). 2002. *Toward a Sustainable Cement Industry.* Colombus, OH: BMI.

DOE (Department of Energy). 1999. "Emission Reduction of Greenhouse Gases from Agriculture and Food Manufacturing." Office of Industrial Technologies, DOE/GO-10099-646. Washington, DC: US DOE.

EIA (Energy Information Administration) 1994. *Annual Review of Energy 1993.* Washington, DC: US DOE.

EIA (Energy Information Administration) 1998. *Annual Review of Energy 1997.* Washington, DC: US DOE.

EPA (Environmental Protection Agency). 2000. *Municipal Solid Waste Generation, Recycling, and Disposal in the United States: Facts and Figures for 1998.* EPA530-F-00-024. Washington, DC: US EPA.

Gardner, G. and P. Sampat. 1998. "Mind Over Matter: Recasting the Role of Materials in Our Lives." Worldwatch Paper 144. Washington, DC: Worldwatch Institute.

Graham, J.D. and J.K. Hartwell. 1997. *The Greening of Industry: A Risk Management Approach.* Cambridge, MA: Harvard University Press.

Halwell, B. 2002. "Home Grown: The case for local food in a global market." Worldwatch Paper 163. Danvers, MA: Worldwatch Institute.

Martin, N., E. Worrell, and L. Price. 1999. "Energy Efficiency and Carbon Dioxide Emissions Reduction Opportunities in the U.S. Cement Industry." Lawrence Berkeley National Laboratory Report 44182. Berkeley, CA: US DOE.

McDonough, W. and M. Braungart. 2002. *Cradle to Cradle: Remaking the Way We Make Things*. New York: North Point Press.

NAE (National Academy of Engineering). 1999. *Industrial Environmental Performance Metrics: Challenges and Opportunities*. Washington, DC: National Academy Press.

ORNL (Oak Ridge National Laboratory). 1997. "Energy Technology R&D: What Could Make a Difference?" ORNL-6921. Oak Ridge, TN: ORNL.

Riemer, P., A. Smith, and K. Thambimuthu, eds. 1998. *Greenhouse Gas Mitigation: Technologies for Activities Implemented Jointly*. Oxford, UK: Elsevier Science Ltd.

World Business Council for Sustainable Development. 2003. www.wbcsdcement.org/pdf/agenda_summary.pdf

Web Sites of Interest

yosemite.epa.gov/oar/globalwarming.nsf

www.wbcsdcement.org

www.aspentech.com/

Problems

19.1 A process designer decides to set fossil fuel prices at a factor of ten above market prices so that the resulting process will be highly energy efficient. His boss rejects the design, citing five major problem areas. List the likely concerns his boss raised and the rationale (a couple of sentences) used by the boss in explaining her concerns to the process designer.

19.2 Write two paragraphs containing your ideas (more than one) on how industry can be proactive in promoting sustainable energy (and materials) use without jeopardizing their bottom line in a significant manner.

19.3 Should chemical companies replace petroleum-based plastic production with biopolymer plastics?

- Are there equivalent performance biopolymers available for most common plastics?
- What are the land use and emissions impacts of producing enough biopolymers to satisfy today's US markets for plastics?
- Which would be cheaper at today's petroleum prices? With a $10/MTC tax?
- Is this a practical idea today? Twenty years from now? In a developing country? (on a lifecycle basis, of course.)

19.4 Table 19.3 shows the carbon sequestration that is associated with landfilling wastes as a carbon credit for the US net GHG emissions calculation. Describe two industrial processes that generate carbon-containing waste, and use lifecycle arguments to indicate whether you agree or disagree with the EPA proposal.

Commercial and Residential Buildings 20

There's no place like home.
—Dorothy, *The Wizard of Oz.*

20.1 Introduction and Historical Perspectives

After water and food, shelter is the most important human need. In many parts of the world today, humans live in caves or simple mud or thatch huts, heating and cooking on open fires with fumes exhausted through a roof hole. In contrast, the developed world population is accustomed to comfortable homes and workplaces with climate conditioning and a variety of electrical devices that provide light, comfort, task assistance, and entertainment. In Organisation for Economic Cooperation and Development (OECD) countries, about 1/3 of total energy and more than 1/2 of electricity is consumed by the building sector. Peak electrical loads are often determined by cyclical demands for lighting and cooling within buildings.

Given the variety of cultures, business activities, and associated economic development, people have developed lifestyles that involve different patterns of energy use.

- **Rural**. Low population density and low energy consumption per unit land area. In developing countries, clusters of people living in small villages of simple family dwellings are likely to practice labor-intensive farming. In developed countries, farming has become more mechanized, so population densities can be even lower, but energy use per land area is increased. Typical rural population densities are in the range of 5–20 people per square kilometer.
- **Urban**. High population density, with high energy consumption per unit land area. In developing countries, population densities today reach 50,000 people per square kilometer in central areas of megacities like Shanghai. Here, people live in crowded multistory buildings and shop and work locally. Commuting is largely by bicycle or public transportation because there is no room for many private cars. In developed world cities, the downtown area tends to be high-rise commercial buildings mixed with some high-density apartments. Typical population densities in OECD cities are in the range of 2,000–7,000 people per square kilometer. Developing country cities are rapidly transitioning to the lower density core city model, with population distributed more widely in suburban areas. In all cases, cities are faced with providing some minimal living conditions for the urban poor, since they are squeezed by the rising costs of city living and cannot access affordable transportation. Urban slums contribute to health, crime, and drug problems.

- **Suburban**. Automobiles have allowed redistribution of population from crowded central areas to much less dense suburbs. Suburban land is typically less expensive and people can afford larger homes and property. Businesses and shopping locate along highways to accommodate suburbanites. Some commute by public transportation to core city areas for work and entertainment. The range of suburban living spans from satellite cities with high-rise apartments to more rural suburbs. Population densities range from 20 to 2,000 people per square kilometer.
- **Undeveloped**. Considerable terrain globally is still unpopulated because of the ruggedness of climate and/or terrain. In developed areas, nature preserves or national parks are sometimes designated to preserve local habitats and species. However, growing global populations are placing increasing pressure on undeveloped land.

Over the centuries, housing and commercial buildings have become increasingly sophisticated. In hot climates, builders learned to design for high ceilings to trap the warmest air, to place windows away from direct afternoon sun, to build shaded verandas, and to use trees for shading. In cold climates, window areas were designed to face the afternoon sun, and living spaces were smaller to require less heat. In climates with large diurnal temperature variations, thick walls were used to provide thermal mass to stabilize interior temperatures. These early lessons have been rediscovered by today's building designers, and they are being integrated into more sophisticated, energy-efficient building designs.

Energy for climate conditioning in buildings and for the electrical appliances within buildings has been a relatively inexpensive part of the lifecycle costs of buildings. Heating and air-conditioning of living and working spaces are common and are not much of a limitation on the decisions about floor space per occupant. Per capita floor area has been increasing steadily with growth in the standard of living. Likewise, energy use in buildings has grown, especially with the introduction of new appliances and office equipment. Because utility costs have been small relative to other costs (e.g., mortgages, maintenance, taxes), designers and consumers have not focused on paying extra for more energy-efficient products or conserving energy through careful use. Recent increases in energy prices and possible future increases in energy prices because of carbon taxes or other greenhouse gas (GHG) mitigation measures may change this behavior and lead to the use of more energy-efficient building components. Windows and doors can be made tighter so energy-intensive climate conditioned air is not lost—but there is a trade-off because indoor air quality in poorly ventilated spaces may cause adverse health effects. Europe is ahead of the US in designing efficient buildings because of higher energy prices, more awareness of "green" issues, and fewer litigation threats associated with the introduction of innovative technologies.

The life of most buildings is measured in decades. Architects who design buildings are primarily interested in appearance, location, uses, and attractiveness to buyers. Building details and installation of utilities are usually left to consulting building engineers and contractors after the major design elements are fixed. Few architects think about the long-term maintenance and efficiency of the structure. Old ideas of siting and designing buildings for passive climate control are largely forgotten in the US.

The construction industry is a major consumer of materials and generates a major fraction of all waste materials. The construction industry is also a fragmented industry, as it contains multiple suppliers of specialty products like lumber, windows, flooring and roofing materials, wall board, electrical systems, plumbing, heating and cooling, and ventilation, and an even larger number of builders. Building contractors employ workers from many trades and prefer to minimize construction costs through use of standard components and practices. Building codes set minimum standards for construction, and zoning requirements may set additional limitations, but the industry is primarily focused on cutting initial capital costs, not on the long-term energy efficiency of the building and the health of its occupants. However, a way to introduce new, efficient building components could be to package them as prefabricated parts with associated reductions in labor costs and construction details. Just a 1% decrease in labor costs could pay for most differential material costs in new sustainable buildings.

Because energy efficiency has not been a priority, there are opportunities for improvements in energy efficiency in the building sector. However, these opportunities are distributed over thousands of different items and practices. Because investments in energy efficiency today are not cost effective— or, if they are, they require capital or changes in well-established practices—there is little incentive to change. However, in countries like China, where electricity demand is growing rapidly, there is an emerging awareness that investment in energy efficiency and better construction quality in new buildings can be a cheaper energy option than the construction of more power plants.

20.2 Lifecycle Analysis

A fragmented construction industry is unlikely to think about the lifecycle analysis for the buildings it helps construct. An architect will focus on the attractiveness of the design and the use of space. A builder will be concerned about products that are easy to procure and use and provide acceptable performance (usually set by building codes) at the lowest price. Often, insulating materials with higher energy efficiency, energy-efficient lighting, and high-efficiency space conditioning systems and appliances are somewhat more expensive than less-efficient alternatives. Builders usually choose on the basis of price, since most home and commercial building buyers are more concerned with the purchase price than with the operating costs for the building. However, these operating costs, when summed over 40+ years of service, are significant, and energy costs are a major portion of such services. In Europe and Japan, where energy costs are much higher than in the US, the operating costs of buildings are of more concern to buyers. Typically, European and Japanese buildings provide smaller space per capita, and their building codes emphasize higher levels of energy efficiency.

Only in the last decade have government and academic researchers begun to evaluate the lifecycle aspects of buildings. (Lifecycle aspects of a typical building are shown in Figure 20.1. More information on lifecycle analysis is presented in Chapter 6.) The lifecycle also needs to consider unusual events that might happen over the life of the building—fires, floods, winds, and earthquakes. Building codes for each locality are usually a source of guidance for this part of design and construction.

The building sector in most developed countries consumes about 40% of the annual demand for raw stone, gravel, and sand, and comparable shares of processed materials such as steel. It uses about 25% of virgin wood, and about 10% of total lifecycle energy use occurs in the construction process itself. (Construction energy is usually apportioned to the "industry sector"—the energy consumed by all building sector operations makes up about a third of total industrial energy use. Table 20.1 shows energy embodied in producing typical construction materials.) Different materials have significantly different levels of embodied energy. Also, a variable component in the embodied energy is the amount of transportation energy required in bringing the materials from source locations to the particular building site. However, as a rule, the embodied energy in a building is much less than the energy consumed over the operational life of the structure.

Figure 20.1. Lifecycle of a building.

Table 20.1. Energy Used in Virgin Production and in Recycling Typical Building Materials

Material	Typical Energy Used – Gigajoules per ton	
	Virgin Production	Recycling
Concrete	0.5–1.5	0.5–1.5
Brick	2.5– 6.1	(reuse?)
Wood (domestic)	4–5	(reuse, reprocess, heat?)
Glass	13–25	10–20
Plastics	80–220	50–160
Steel	25– 45	9–15
Copper	70–170	10– 80
Aluminum	150–220	10– 15

Source: Roodman and Lenssen (1995).

Metals are energy intensive and expensive enough so that recycling is often justifiable. However, where metals are alloyed, mixed recycled metals may have different properties from virgin materials and may not be able to be recycled back to original uses. In recycling, there are tradeoffs between the complexities of separating different composition materials and the suitability for primary recycling (back to original use) or secondary recycling (back to lower quality use) or for recovery of thermal energy in the case of combustible materials like wood, paper, and plastics. Recycling also has the advantage of reducing waste. In Europe and Japan, where waste disposal is much more expensive than in the US (due to availability of US land for landfills), use of recycled materials is strongly encouraged. In fact, if design for recycling is considered in the original product design, there can be considerable savings in materials and energy use on a lifecycle basis—and sometimes cost savings, if the costs of waste disposal are properly valued.

During the operating life of a building, energy and water are used on a regular basis. Sanitary wastes along with waste products from purchases of food and other goods are continually generated for disposal. Let us focus on energy use. The amount of energy needed for space conditioning a building depends not only on the local climate and the energy efficiency of the structure but also on the siting of the building. Positioning of windows relative to sun locations during the year, shading by trees and other structures, and annual wind patterns around the building all contribute to the heating and cooling loads. Indoor comfort standards also vary considerably. Before air conditioning, people accepted higher temperatures in the summer and modified their activity levels accordingly. Likewise, people could dress a little more warmly in winter and accept a slightly cooler indoor temperature. With fairly inexpensive energy available, US residents have come to expect high levels of indoor comfort and expect malls and other public spaces to provide similar comfort levels.

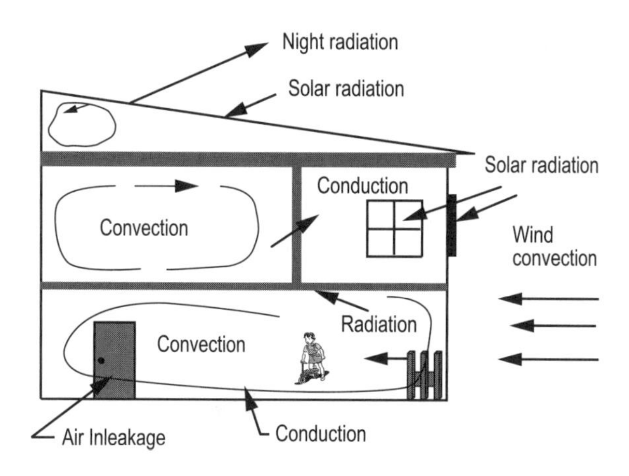

Figure 20.2. Some heat transfer aspects of building design.

Figure 20.2 shows some of the heat transfer aspects of building design. (The theoretical explanations of different forms of heat transfer are presented in Section 3.5 of this text.) Solar energy (see Chapter 13) warms the outside of the building during the day and penetrates through windows. At night, radiational cooling causes heat losses. Shielding by clouds, awnings, trees, or adjacent structures reduces these radiational transfers. Natural convection driven by temperature differentials and forced convection driven by wind also exchange heat between the outside air and the building shell. Heating and cooling systems inside the building maintain indoor climate control, while discharging some waste heat to the outside environment, along with products of combustion. (If electricity is used, the products of combustion are discharged at the power plant.) Lights and appliances in the building also discharge some waste heat to the interior, helping to heat the building in winter and adding to the cooling load in summer. Air also leaks between the inside and the outside through small cracks, ventilation systems, and open doors and windows. Air leakage puts an additional burden on the indoor climate control systems. However, air-to-air heat exchangers between intake and exhaust air can help reduce energy losses for commercial and larger residential buildings.

Renovation and demolition of buildings generates a substantial waste stream. US construction wastes are about equal to the total of all municipal solid waste. On a lifecycle basis, much of this material could be reprocessed or reused, but present economics do not justify much effort spent in sorting out materials from old buildings that never were designed for recycling.

The long life of most buildings means that change in this sector is slow, and changes in present design and construction practices will take decades to penetrate the building sector. Thus, it is important to look for ways to remodel or modify existing buildings to improve energy efficiency. Recycling of building materials is another way to reduce the energy input into construction materials, and better long-range land use planning practices can avoid the common need to demolish buildings before the end of their useful lives. Such long-range planning will ultimately benefit from being combined with transportation planning to find energy efficiencies from siting residences and commercial properties to minimize travel needs.

Several groups are starting to develop tools that will assist the building industry in transitioning to higher levels of sustainability on a lifecycle basis. In Europe, eight groups in five countries are collaborating in a major project (Kohler and Klingele, 1996, and Polster et al., 1996) called the REGENER project. Its objectives are listed below.

- To define a common methodology for applying lifecycle analysis in the building sector
- To develop a design toolbox
- To perform first applications of the methods, concerning, for example, integration of renewable energy sources

This European team includes research institutes involved in lifecycle analysis, as well as professionals from the building sector, consultants, and a regional agency for environment and renewable energy.

The US Department of Energy (DOE) is also developing a number of tools and software to support the US building industry in "greener" building design and the effective use of renewable energy as part of buildings' energy systems. However, most DOE tools are useful later in the design process, after the important initial design parameters have been set.

Guidelines and aids to help designers consider passive design options for enhanced energy efficiency and healthy indoor air quality early in the siting and initial design concept process are also needed.

20.3 Residential Building Design

Buildings are designed to meet the needs of the people who use them. Designs usually reflect a compromise between the wants and the pocketbooks of the users. In central Shanghai, for example, typical living space per person was recently about 2.5 m^2 or about 24 ft^2. Such limited space has to have multiple uses (e.g., sleeping mats are rolled up to make room for daytime activities like eating and household chores). Workers usually occupy separate additional space outside the home to perform their jobs.

As new satellite cities are being developed around Shanghai, the per capita living space is being increased by factors of 4–10 times (and provision is being made to accommodate larger numbers of cars, each having a parked footprint of around 3–5 m^2). Germans typically have per capita living areas of around 30 m^2; typical Americans expect around 60 m^2 per capita (Schipper and Meyers, 1992). As per capita space increases, residents allocate separate spaces to different activities—living rooms, dining rooms, bedrooms, kitchens, studies, etc. Permanent furniture is acquired along with lighting and other accoutrements. All the added space requires proportionally larger quantities of materials and more energy to maintain the indoor environment. Cleaning and maintaining larger living spaces requires more income, and more family members work to support a more affluent lifestyle. With less time available for household chores, a variety of appliances and convenience foods have appeared to save time—and most of these involve the use of additional energy. Workspaces have also become larger and much more energy intensive with space conditioning, high-quality lighting, computers, communications equipment, and mechanized manufacturing systems.

Given these factors, there are some major opportunities for designing more energy efficient buildings.

> **Efficient use of living space**. How much space is really needed for the activities of working and living? Most people today believe that "more is better." Larger spaces encourage accumulation of more "things," and many of these acquisitions are used infrequently, if at all. If energy and materials were priced to include the societal and environmental costs of use, then more consideration would be given to their conservation. However, people may choose to live a simpler lifestyle to reduce the stress in their lives. This will lead to smaller and more highly utilized living and working spaces.
>
> **Passive thermal design**. Depending on the local climate and the site surroundings, a variety of opportunities exist for reducing energy demands for space conditioning (Givoni, 1994). Where summer heat requires large air-conditioning loads, trees, awnings, or overhangs can shade windows. Taller ceilings can trap the warmest air. Windows can be located on the shady side of the building, reflective coatings can be used on the sunny exposures, and thick walls can help even out diurnal temperature swings. Likewise in cold climates, window areas can be reduced on the sides of the building away from sunlight, and good windows can be used on the sunny side to capture limited sunlight, wall and roof insulation can minimize heat losses, trees can be used as windbreaks to reduce convective cooling by cold air currents, and enclosed porches can limit heat losses through entries. Wind patterns around nearby buildings and terrain can be utilized to promote selective heating and cooling. Natural ventilation can be incorporated into passive thermal design with placement of windows to facilitate nighttime cooling as shown in Figure 20.3.
>
> **Active thermal design**. Heating and cooling systems can be chosen on the basis of lifecycle energy efficiency and performance. There are a number of primary energy options for supplying space conditioning to buildings. The choice depends on local climate, the type and use of the building, and the cost and availability of alternative types of energy. Building designers today need to be conscious that energy costs may change in the future to reflect environmental and societal values or that purchasers of buildings may be attracted to "greener" attributes for buildings. These factors lead to a goal of designing climate-conditioning systems that are highly energy efficient.

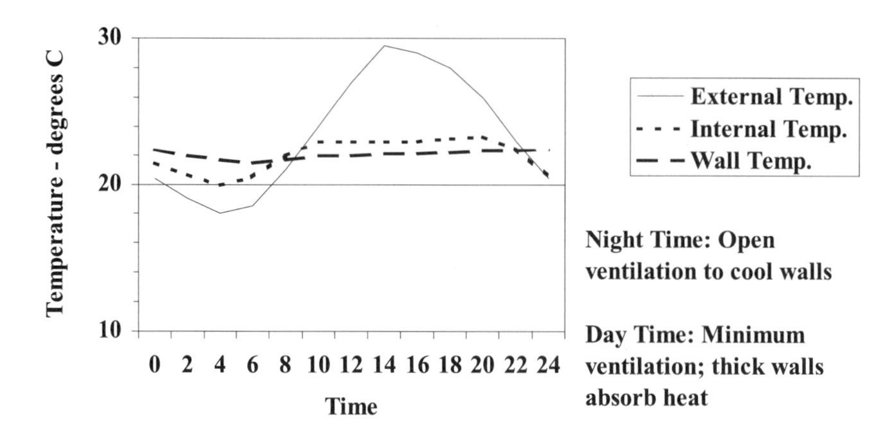

Figure 20.3. Night cooling ventilation strategy.

Energy efficiency pertains not only to the conversion of the primary energy source but also to the integrated use of complementary energy supplies and reuse of waste energy. For example, waste heat from furnaces may be used for water heating along with solar thermal converters. Geothermal heat pumps can be used to warm or cool indoor spaces by using average subsurface ground temperatures as an energy source or sink. Solar energy may also be used to power air conditioning devices in regions where summers are hot. A backup energy supply is usually needed to supplement such renewable energy sources, but commercial energy demand can often be reduced substantially. Heat from electrical appliances needs to be considered in designing the building space conditioning systems. Ventilation systems can be designed to minimize the leakage of energy to the outside environment while maintaining healthy interior air quality (see section 20.5).

Adapting "comfort" standards. Inexpensive energy has led modern society to expect shirtsleeve warmth indoors in winter and enough cooling in summer to allow (or necessitate) wearing of a light jacket or sweater. Large spaces, like theaters and shopping malls, are heated or cooled to these standards. But people can be comfortable with less vigorous space conditioning. Use of fans to cause air circulation and humidity adjustment can provide comfort with less use of energy. Before the advent of air conditioning, people modified their dress and activity levels in summer to adapt to the heat. Before central heating, people dressed more warmly indoors in winter and clustered around a heat source like a stove or fireplace.

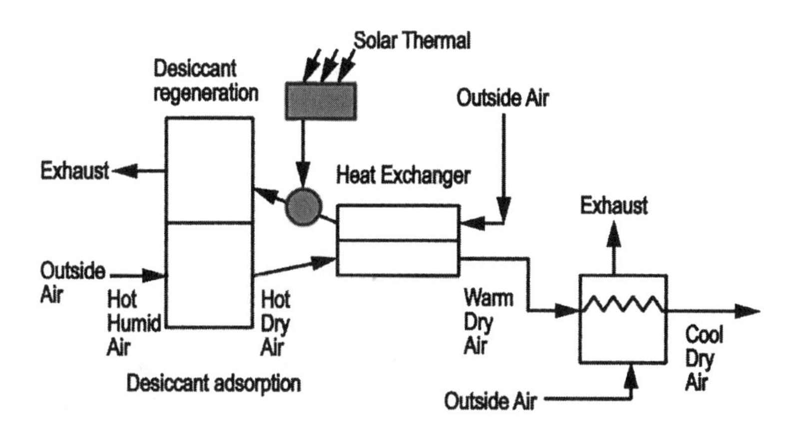

Figure 20.4. Conceptual diagram of a desiccant system.

International thermal comfort standards (ASHRAE, 1997, and ISO, 1994) identify 26°C as the upper bound of thermal comfort when humidity is high. If the air is moving, convection and evaporative cooling from the skin allow comfort at higher temperatures. Lechner (1991) indicates that an air speed of 1 m/s compensates for about a 3.3°C increase in temperature at the same comfort level.

Humidity control also can be used to enhance comfort. The rate of evaporative cooling from the skin depends on humidity levels. In winter, when outdoor absolute humidity levels are low due to the low temperature of the air, indoor humidity levels can be raised to enhance comfort. With reduced evaporative cooling, building inhabitants can be comfortable at lower indoor temperatures and also benefit from reduced drying of skin and respiratory passages. In the summer, reducing humidity indoors can promote comfort at higher indoor temperatures. Figure 20.4 shows a conceptual diagram of a desiccant and cooling system that runs on fan and solar energy. Two desiccant beds are provided so that one can capture moisture while the other is being regenerated.

Thermal comfort standards used in developed countries are largely Western in origin and may not apply in all parts of the world. Recent research has shown that comfort standards are different in naturally ventilated buildings where people are comfortable at higher temperatures. A study in Thailand (Khedari et al., 2000) showed comfort at 28°C in air-conditioned offices and at 31°C for naturally ventilated offices. A number of other studies show that many people from more tropical climates are comfortable in conditions outside the traditional comfort limits established by ASHRAE.

Creating "smart" buildings. With the advent of sophisticated electronic controls, more options are available to use energy in buildings more selectively. For example, buildings could be kept at energy-efficient temperature levels outside of normal comfort zones and then brought up to comfort levels upon receipt of a timed or occupant-generated signal. Today, thermostats that control daytime and nighttime temperatures at different levels are widely available. Within the house, control systems could be developed to climate condition only those spaces that are actually being used at a particular time. Looking to the future, one can imagine systems that only condition climate around the immediate space occupied by an individual.

With the electrical sector starting to use real-time pricing as an economic tool to level out power demand fluctuations as discussed in Chapter 17, even more sophisticated automated energy-control systems may come into use. "Smart windows" may have internally actuated shades or coating layers that could be controlled for emissivity using sensors or timed programming. Appliances within the home could be controlled to operate during periods of low electric demand to get favorable rates, and the waste energy could be treated accordingly in the overall design. Individual lighting controls based on daylight levels, as well as occupancy, could also save substantial energy. In a situation where there were strong incentives to reduce peak electric demands during midday heat, a system, coupled with some of the techniques discussed above under "passive design," could overcool at night and use thermal mass in the walls to maintain comfort during the day.

The opportunities for advances in energy efficient building design are great, but they are often not implemented because of resistance to change and fear of litigation in the fragmented building industry and because of the priorities of the purchasers of most buildings. However, if energy is to be used more sustainably, changes will be required.

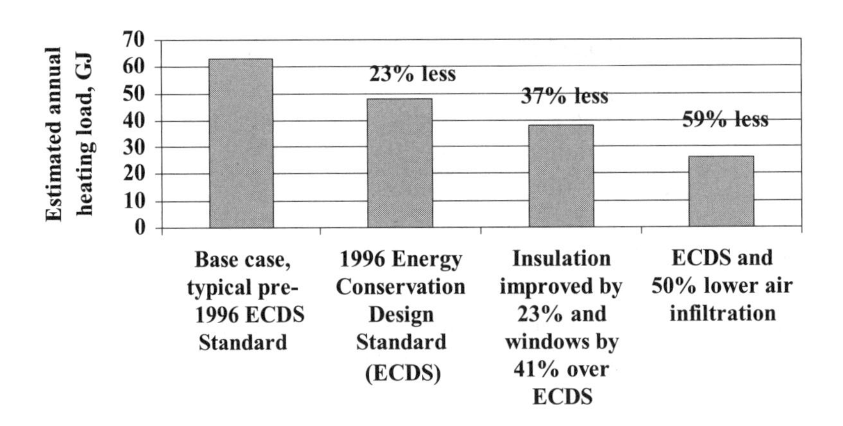

Figure 20.5. Model results: heating load for a typical Beijing housing unit. Source: Glicksman et al. (2001).

Some leadership in this area may come from activities in developing countries. One example, for new residential housing developments in China, is described in detail in a comprehensive review article by Glicksman et al. (2001). Increases in personal wealth in China for large population groups are creating a rapid demand for new housing with more living space (though much less than is common in OECD countries). Energy supplies for China are becoming more adequate, but there are strong incentives to not waste energy. It can be shown that investment in energy efficient buildings is more cost-effective than the construction of new power plants to provide the energy wasted in inefficiencies. Figure 20.5, taken from the Glicksman et al. (2001) paper, shows that almost a 60% improvement in efficiency is possible with good cost-effective design relative to pre-1996 Chinese construction standards.

20.4 Commercial Buildings

Multiple-unit buildings, such as commercial and manufacturing buildings or multi-unit residential developments, offer more opportunities for energy management than individual residences. Where small amounts of waste heat are available, the costs of equipment to recover the heat for other use may not be justified. But at a larger scale, the costs of energy recovery equipment may be feasible. Also, waste heat from power generation (fossil fuel or nuclear) can be used to generate steam, which can then be used for heating. Older US cities often had steam distribution systems that provided thermal energy to buildings and commercial activities in central urban areas. So-called "district heating" is also used in a number of European cities and townships. New towns are often designed to incorporate thermal energy recovery as a way to lower overall greenhouse gas emissions from fossil fuel use.

The options for reducing energy demand in the commercial building sector start with most of the same design possibilities as discussed in the previous section for residential buildings.

> **Efficient use of space**. Just as people enjoy having spacious homes, they like having good office space. So there is a trade-off between the costs and environmental/social impacts associated with building space and with the desires and needs of the occupants. In designing office and other commercial buildings, thought needs to be given to enhancing worker productivity through provision of space adequate to perform the work and provide the intangible qualities of a pleasant working environment. Clean air, proper lighting, noise management, and provisions for working alone and/or in teams are all important. Office and manufacturing equipment require space and maintenance—and may release heat, solvents, or other species into the air, as discussed in Section 20.5 on indoor air quality.
>
> Although most office space is used only during normal working hours, energy is still needed to maintain the space when it is not fully utilized. Even allowing temperatures to drop at night or over weekends results in having to reheat the space

in the morning (or vice versa for air cooling). If the space could be used for more hours, for example by working two shifts or by finding an alternative use for the off hours, there should be a net saving of energy per worker.

Passive thermal design. As with residential buildings, commercial buildings can be designed with insulating materials, walls with thermal mass, and appropriate placement of windows. In core cities with concentrations of tall buildings, a "heat island" effect is noted because the energy used within the building dissipates to the outdoors in quantities large enough to affect the local climate. Likewise, buildings in a cluster can shade each other or may block natural light. Tall buildings may channel wind and cause local regions where natural winds are either abated or intensified, depending on the relative alignments. Many architects and building designers are developing tools to help with understanding these phenomena and using them to aid in energy efficient passive thermal design of high-rise building clusters.

Active thermal design. In recent times, many taller buildings were designed with sealed fixed windows. This meant that the entire climate conditioning system for the building was dependent on machinery and distribution duct designs. Many of the buildings constructed in this manner developed problems with indoor air quality. Tall buildings today may incorporate options for some natural ventilation along with improved air circulation and cleaning strategies.

Commercial buildings can incorporate more efficient space conditioning systems and have more opportunities for recovery and reuse of waste energy than single residences. With time-dependent electricity and natural gas pricing, there are opportunities for designers to take advantage of off-peak energy pricing in developing an integrated climate conditioning strategy for the building. Use of advanced facing systems designed to facilitate natural ventilation in high-rise buildings can make the interiors more comfortable with less energy use.

Adapting "comfort" standards. As discussed for residential buildings, there are standards for comfort in the working environment that depend on the nature of the work and local standards of "comfort." There also are opportunities for reducing overall space conditioning energy needs through judicious use of dehumidification and air circulation.

Creating "smart" buildings. Commercial and manufacturing buildings typically require large amounts of electricity for office and other types of equipment. Switching large machines on and off creates power surges, and control systems are available that not only monitor the electricity flow into the building but analyze the patterns of the different demands on the system. Controls can manage the sequence of starting and shutting down equipment to prevent transients that might affect other equipment. These types of control systems might be extended to managing the interactions between climate conditioning systems and major types of equipment in the building. In the long term, local space conditioning around each person in the building may offer promise of still more efficient energy use. However, there will be tradeoffs between the complexity of the control systems and the value of energy savings.

Larger-scale sustainable design. Just as energy can be saved by integration of utility services within commercial buildings, there are opportunities for even more integration in groups of buildings. Siting and design to facilitate energy conservation through modifying insolation patterns to the building group or by changing airflow around and through it are useful techniques. If there are power plants or industrial operations located nearby, there may be opportunities for utilization of waste steam or other energy or even for integration of natural gas and electric services that would level out peak demands. There are also opportunities to co-locate work opportunities with housing to minimize the need for additional transportation energy.

20.5 Indoor Air Quality

Buildings require less energy for space conditioning if there is a minimum amount of outdoor air infiltrating the building. However, in tighter buildings, any pollutants that are emitted inside the building keep recirculating and accumulating. If levels build enough, health impacts may occur. Even before serious health impacts occur, building inhabitants may feel some sleepiness, mild headaches, or general malaise. In office or manufacturing buildings, worker productivity may suffer and absenteeism may rise. This "sick building syndrome" has led to considerable research on the importance of balancing air renewal in a building with the associated energy inefficiencies (EPA, 1991).

What are the sources of pollution in buildings? They vary considerably, but here are some major sources:

Materials and furnishings. A first source is the offgassing of materials and furnishings in the buildings. Many insulating materials are made from gas-blown foams, and the gas trapped inside the insulation diffuses slowly into the building. (Modern building codes ban the use of toxic foaming agents.) Fabrics may be treated with coatings. Electrical equipment may produce some ozone or other pollutants. There are a variety of volatile organic compounds (VOCs) that are slowly released from plastics, glues, paints, and composite materials. With poor ventilation, such pollutants can accumulate indoors to levels that are hundreds of times higher than outdoor concentrations. Dust and mold spores can also accumulate indoors.

Occupants. People, pets, and their activities also create a variety of emissions indoors. Our bodies generate gases, sweat that evaporates, dandruff, and other wastes that may cause some emissions during disposal. Our activities include food preparation, eating, doing laundry, and various types of work and hobbies that may cause emissions. In the workplace, inks, solvents, toners, and other equipment emissions are common. Indoor heating and ventilation systems may produce leakage emissions of their own. Uncleaned air filters can be a source of dust and molds.

> Outdoor air is introduced through open windows and entries and exits. In developing countries or rural areas, more basic heating and cooking may be done indoors with air drawn from the interior. Exhaust gases can accumulate in a poorly ventilated house and cause major oxygen depletion in addition to producing pollutants like carbon monoxide, acid gases, and soot. Smoking and illnesses may also lead to the accumulation of harmful species in the indoor air.

Over the past few decades, as the effects of poor indoor air quality have become more apparent, society has recognized that losses in worker productivity and increased medical costs warrant serious attention. Research on the effects of poor indoor air quality has led to major changes in how buildings are designed and climate is conditioned. Natural ventilation will help reduce pollutant levels in buildings and reduce "sick building" problems. Occupants in naturally ventilated buildings will have a greater sense of well being relative to people occupying sealed buildings.

Outdoor air is intentionally brought into buildings, and indoor air is exhausted to prevent unhealthy buildup of pollutants. In large buildings where this might involve a significant energy penalty, heat exchange between intake and exhaust air may be warranted. In residences, the energy penalty associated with enough air leakage to maintain health is considered a wise investment.

New research suggests improved ventilation strategies for buildings. Present practice is to mix indoor air as it recirculates so pollutants are distributed evenly throughout the indoor environment. Yuan et al. (1999) have described the concept of displacement ventilation and its application in the US. For air conditioning in summer, inlet cooled air is introduced along the floor level and rises by displacement as new cooled air is introduced. Warmer air is removed at ceiling level and recooled with the addition of fresh air and/or with pollutant removal. The process is reversed in a heating cycle. The advantage of this system is that the continuous plug flow of air substantially reduces pollutant accumulation and horizontal mixing of pollutants. (This ventilation concept will soon be tested in several campus buildings at MIT.)

With rising concerns about potential terrorist activities, the Department of Health and Human Services has released new guidelines for protecting ventilation systems in commercial and government buildings from chemical, biological, and radiological attacks. The guidelines provide recommendations that address the physical security of ventilation systems, airflow and filtration, systems maintenance, program administration, and maintenance staff training. (More information is given in the Center for Disease Control [CDC] Web site listed in Web Sites of Interest.)

20.6 Sustainability Issues

The building sector today can take advantage of many opportunities for more sustainable use of energy. However, these opportunities are difficult to implement because the industry is decentralized and because the price structure for energy and materials at present encourages more people to acquire ever-larger living spaces.

Because many can easily afford energy services, most do not think about the value of energy conservation. This is true both in affluent economies and in developing economies, where energy prices are often subsidized to encourage economic development. In developing economies, people are seeking better living space with more amenities consistent with their rising incomes. However, we all enjoy as many building services as are readily affordable at whatever income level we have.

Pollution and waste generation can be managed through regulatory standards that limit emissions or set standards for waste treatment and disposal. Broader sustainability issues, such as the potential effects of today's GHG emissions on future generations, the consequences of more land development on natural habitats and biodiversity, eventual depletion of non-renewable materials on resources needed for future generations, or growing quality-of-life gaps between the rich and the poor, are not given much thought by many. This is because these issues do not always produce immediate and visible local consequences.

These longer term issues are not yet well incorporated into our present economic pricing schemes (although some countries tax certain GHG emissions and many more are considering economic options for encouraging GHG emission reduction consistent with the Framework Convention on Climate Change). As a result, present market structures do little to encourage change. But a growing number of people are becoming more environmentally and socially aware. There may be market pull for "green buildings." Building designers and contractors are beginning to think about sustainability concepts and incorporate energy-efficient or other environmentally-attractive concepts into their designs. Utility companies are also moving toward the marketing of energy-efficient appliances and lighting as part of an overall management policy. Some consumers are opting for simple, but comfortable, lifestyles.

Because the life of buildings is so long, society will have to live with the inefficiencies of the past for many decades. The industry is not in a position to make a coordinated effort to implement sustainability standards into its operations without being forced to do so by regulation or the market. Therefore, change in this sector will have to come from the people who purchase building space and make decisions on the contents and use of the space. As more people become knowledgeable about impacts of the building sector and commit to change, more action will occur.

Most of us could easily reduce our home energy use by 10–15% without causing ourselves any hardship. If we were willing to give up some of the space we don't often utilize, even larger energy savings would result. As people become more aware of the true value of energy, we will become committed to buying greener products and living and working in more sustainable buildings.

References

ASHRAE (American Soc. Htg. Refrig. Air Cond.). 1997. "Fundamentals," p.8.1–8.28. *ASHRAE Handbook*. Atlanta, GA.

EPA (US Environmental Protection Agency). 1991. "Indoor Air Facts No. 4: Sick Building Syndrome." Washington, DC: EPA Office of Air Research.

ISO (Intl. Stds. Org.). 1994. "Moderate thermal environments— determination of the PMV and PPD indices and specifications for thermal comfort." Standard 7730. Geneva: ISO.

Givoni, B. 1994. *Passive and low energy cooling of buildings*. New York: John Wiley & Sons.

Glicksman, L., L. Norford, and L. Greden. 2001. "Energy Conservation in Chinese Residential Buildings: Progress and Opportunities in Design and Policy." *Ann. Dev. Energy Environ.* 26:83–115.

Khedari, J., et al. 2000. "Thailand ventilation comfort chart." *Energy Build.* 32:245–249.

Kohler, N. and M. Klingele. 1996. "Simulation von Energie- und Stofflussen von Gebauden wahrend ihrer Lebendsdauer." International Symposium of CIB W67, Vienna.

Lechner, N. 1991. *Heating, Cooling, Lighting: Design Methods for Architects*. New York: John Wiley & Sons.

Polster, B., B. Peuportier, I. Sommereux, P. Pedregal, C. Gobin, and E. Durand. 1996. "Evaluation of the environmental quality of buildings—a step toward a more environmentally-conscious design." *Solar Energy* 57(3).

Roodman, D. and N. Lenssen. 1995. "A Building Revolution: How Ecology and Health Systems are Transforming Construction." Paper 124. Washington, DC: Worldwatch Institute.

Schipper, L. and S. Meyers. 1992. *Energy Efficiency and Human Activity: Past Trends and Future Prospects*. Cambridge, UK: Cambridge Univ. Press.

Yuan, X., Q. Chen and L.R. Glicksman. 1999. "Models for Prediction of Temperature Difference and Ventilation Effectiveness with Displacement Ventilation." *ASHRAE Transactions* 105(1):353–367.

Web Sites of Interest

www.cdc.gov/niosh/hhs-ventrel.html
www.eere.energy.gov/femp/program/newconstruction.cfm
www.eere.energy.gov/EE/building.html
www.oikos.com
www.sustainable.doe.gov/greendev/tools.shtml
www.cenerg.ensmp.fr/english/themes/cycle/html/bcycle.htm
www.ubka.uni-karlsruhe.de/vvv/1997/architektur/3/3.pdf
www.umich.edu/~nppcpub/resources/compendia/architecture.html

Problems

20.1 Estimate your personal annual energy use (GJ/year) associated with building space that you use: your home, your workspace, space rented for temporary accommodations, etc. How does your energy use compare with that of an average American? An average Japanese?

20.2 Estimate the percentage energy savings in the US if we kept the thermostat at 67°F in the winter and only used air conditioning when outdoor temperatures were above 80°F. State your assumptions clearly.

Synergistic Complex Systems 21

The engineer is the key figure in the material progress of the world. It is his engineering that makes a reality of the potential value of science by translating scientific knowledge into tools, resources, energy and labor to bring them into the service of man ... To make contributions of this kind the engineer requires the imagination to visualize the needs of society and to appreciate what is possible as well as the technological and broad social understanding to bring his vision to reality.
— Sir Eric Ashby

21.1 Introduction and Historical Notes

The choices made about energy are guided not only by technological superiority but by economics, local availability, customs, and cultural preferences. We each make choices when we purchase or use almost anything, and the companies that provide our utility services also make choices on the supply side. Figure 21.1 shows the complexity of the energy picture.

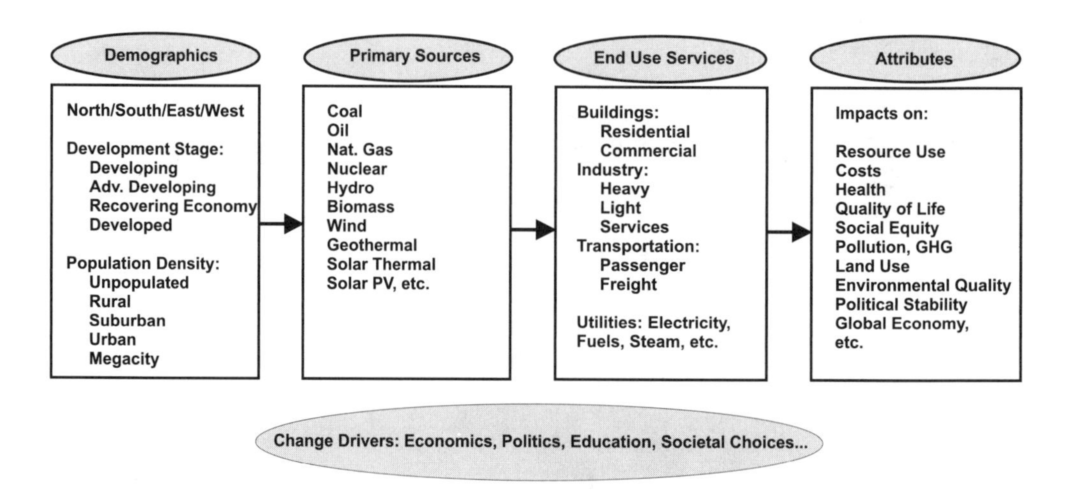

Figure 21.1. Aspects of global energy uses.

Historically, our society has developed energy choices based on local conditions and economics. The availability of new and inexpensive energy has been an engine of industrialization. The steam engine was critical to the early stages of development, and the availability of oil and gas fueled spurts in development during the past century. In developed countries, markets contain produce from around the world thanks to inexpensive energy in the transportation sector; our homes are loaded with various appliances for convenience and amusement; our businesses manufacture all sorts of products that are sold to consumers to create profitability and add to Gross Domestic Product (GDP). The free market provides a framework that encourages creativity in finding technologies and services that maximize benefit per unit cost.

But what happens when pricing does not include some values that may be important to society as a whole or to future generations? The wealthy usually do not have undernourished poor on their radar screen— or perhaps they contribute a little to some charity organizations to assuage any feelings of guilt. Wealthy nations also contribute foreign aid to poorer countries. For example, the US contributes a little more than 0.1% of GDP to overseas development. At the Earth Summit in Rio de Janeiro in 1992, a goal was set for wealthy nations to give 0.7% GDP to foreign aid. Ten years later, at the Rio + 10 Summit in Johannesburg, only Denmark, Norway, Luxembourg, Sweden, and the Netherlands had met this commitment. In fact, foreign aid fell from $69 billion in 1992 to $53 billion in 2000 (of which $18 billion came from the US). Developing-country debt increased 34% during the same period.

Also of concern at the Rio Earth Summit were issues such as health, biodiversity, gender roles, human rights, and the potential for global climate changes as a result of greenhouse gas (GHG) emissions. This last concern is coupled with fossil fuel use for energy supply and is an important consideration in sustainable use of energy. In the past decade, global CO_2 emissions have risen 9% (and US emissions have risen 18%), in spite of the Rio goal of stabilizing CO_2 emissions of Organisation for Economic Cooperation and Development (OECD) countries near 1990 levels. However, 10 years later at the summit in Johannesburg, the US opposed any binding constraints on GHG emissions and withdrew from the Kyoto Protocol. The Summit ended by endorsing only some voluntary goals.

The problem seems to be that most of us who are comfortable do not wish to disrupt our way of life for some possible adverse consequence that may impact others who live far away or who will live in future generations. When there is so much uncertainty, we have problems understanding the tradeoffs. Unfortunately, most people are unaware that inaction is actually a decision.

In developing countries, the problems are different but still challenging. Where large segments of the population do not have enough to provide a basic minimum standard of living, people are usually dependent on local power structures for their survival. Often the governmental structures are corrupt and exploitive and work for the enrichment of those in power. Foreign aid, from groups such as the World Bank and the US Agency for International Development, usually passes through the local governments. In addition to the potential influences of local government corruption and misuse of funds, there may also be political and special interest agendas attached to the funding by the donor.

Expanding trade is another means of helping poorer countries. The World Bank has estimated that elimination of all trade barriers could result in about $300 billion being transferred to poorer countries—many times the level of foreign aid. However, in this free-trade scenario, there is also exploitation of cheap labor and resources in the poorer countries and a likelihood of accelerated environmental degradation.

Where most of the population is struggling for survival, and where education is minimal or biased, the voice of the public and the common interests of society as a whole are often lost. In the midst of competing self-interests for power and wealth, any concerns about long-term sustainability of natural habitats, of water and resources, and of improved social equity are usually ignored.

The more quantitative tools and frameworks presented in Chapter 6 for evaluating the pros and cons of specific energy technologies or policies within a larger system are not designed to look at the whole system nor to project impacts out to future generations. This chapter addresses more complex system issues. How will GHG emissions affect future generations? How will poverty and social inequity today impact energy development? Will energy development lessen the inequities between the rich and the poor or increase them? How do different cultures maintain their heritages in a society with access to the Internet and television? Does it matter if we sacrifice some economic growth for societal and environmental values? Who wins and who loses?

Today we have incomplete knowledge about the complex systems humans have constructed on planet Earth and how they interact with natural systems. Nevertheless, decisions are made daily on policies that impact the system as a whole as well as its subsystems and its future. Some of these decisions are made at national levels, some at local levels, and some without our being conscious that daily choices are really decisions. Therefore, this chapter begins with a discussion of some of the ways that society makes choices within imperfectly understood complex systems. Then it describes some of the techniques that people have used to better understand the evolution of complex systems with decisions being made by groups of diverse "stakeholders," each with special priorities and interests. Projecting our complex systems out to the future, even with great uncertainty, may provide some insights on how a transition to a more sustainable future might begin, and it may identify some of the major challenges that need to be addressed to facilitate change.

21.2 The Complex Systems View

Scholars from a variety of fields have worked hard to develop models of complex systems. Most model subcomponents first and then link them to simulate a larger system. But there are major problems with this approach. First, the modeler has imperfect knowledge and may miss important elements or interactions. Second, the analytical tools themselves introduce uncertainties. Third, computational capabilities are limited, and trade-offs between solution time and cost make approximations necessary. Fourth, most models cannot anticipate the transience and inherent variability that are associated with natural and human aspects of the global community.

Humans and organizations base their energy choices on many factors. Usually, economics is most important, but we are also influenced by advertising, convenience, information availability and many other inputs to each individual decision. Societal choices are a summation of individual decisions worldwide. Moreover, many choices are based on major capital investments in infrastructure that lock us into patterns that will persist for decades.

Although all models are fraught with levels of uncertainty, even our imperfect models can give insight into trends and outcomes. In Chapter 6, we suggested that the early work of Malthus (1798) was correct in its analysis but underestimated our capability to increase crop yield through technology. However, his warning about the impact of population growth on the depletion of resources was sound even if his scenario of global starvation did not ensue on the predicted schedule. Fortunately for the human race, we are clever at using technology and finding solutions to problems. But with over six billion people on this planet, all striving for a better way of life, we are probably pushing some of the limits. This chapter gives an overview of some of the most common modeling and decision tools in use today. Section 21.3 will present examples of different approaches taken by researchers toward understanding the complexities of the future.

21.2.1 Expert panels

A widely used aid to decisionmaking is the use of a "panel of experts." This approach attempts to get around complexity and uncertainty by assembling a group of people with varied expertise and backgrounds to develop answers to a set of questions posed by decision makers. Sometimes the experts are used to review and assess a decision that has already been made.

Most organizations have mechanisms for expert review and advice. Some organizations will hire specialist consultants to give advice, but if the issues are controversial, consultants are often viewed as "hired guns" who may not be objective. Companies have boards of directors that often authorize special expert studies if decisions require it or if they wish to have independent approval of a decision. Independent review reduces the liability associated with a possible "wrong" decision, although it offers no protection against negligence. University departments and centers have "visiting committees" or "advisory boards."

Government agencies often form advisory groups to get expert opinions on planned activities or programs. Sometimes government agencies will bring special issues to the National Research Council, which then draws on the membership of the National Academies of Science and Engineering, the Institute of Medicine, and other outside experts for advice. The advice is provided through peer-reviewed reports. (Examples of such reports are found at: http://www.nationalacademies.org/publications/.)

For government advisory boards, a key objective is to represent the major interests in a way that balances the potential biases of the group. For example, if the US Department of Energy wishes outside advice on its R&D priorities, it is hard to find anyone who is totally unbiased. A solar expert is optimistic about the future of solar; a nuclear expert, about nuclear; a university researcher thinks that more spending on university research is important (especially at his or her own university); an industry representative may think that the government does not know how to pick winner technologies the way that industry R&D managers do. An expert panel will give consensus advice except when the advice involves a conflict of vested interests within the panel. However, some panels present majority and minority opinions that contain information useful to the decision makers. The advice given by expert panels is certainly not binding, and often the panel report is only accepted in part or is left to gather dust on a bookshelf. Nevertheless, the process of reaching a consensus on a difficult topic leads to active and informed discussions of many strategies, problems, and options, and it can generate some useful insights that would be difficult to achieve through strictly analytical methods. However, these expert panels also suffer from the dangers of "group think," where positions may be taken to impress the group or to posture. They often display overconfidence and couch their advice so that they can avoid future criticism if they are wrong or have overlooked an important factor.

21.2.2 Decision analysis techniques

Suppose a decision is to be made by a group of people who have widely divergent goals and opinions. Decision aids provide frameworks that help the group reach a decision in a rational manner. Such models necessarily have to consider individual preferences (termed "utility"), trade-off strategies, and methods for addressing uncertainties. Of course, decision making frameworks, which are known in advance, set the stage for "gaming" to manipulate the system to gain an edge.

Suppose you are a government administrator or a corporate executive. You have several competing business or focus units reporting to you, and they compete for budgets. Are there techniques for sorting through conflicting priorities to optimize the distribution of resources to meet centralized objectives?

Decision analysis consists of a set of tools that aid a *single* decision maker in sorting out differences between several fairly complex and distinct courses of action *in a rational manner*. To do this, the decision maker needs some sort of *model* for characterizing and describing a useful abstraction of the decision problem. The model would at least enumerate the sequences of immediate and later choices available to the decision maker, and describe the possible outcomes of these choices in terms of an attribute of interest (often cost). It is also important to describe the *uncertainties* associated with the characterization and to revise this characterization as additional information becomes available.

The decision maker may have some preferences that contribute to the decision, so that decision outcomes can be sorted for relative desirability. The decision maker's preferences are described by a *utility function,* which can be generated following axioms of rational behavior. Some decision makers may be *risk seeking*, some *risk neutral*, and some *risk averse*. This behavior can be measured and described by asking the decision maker to choose between sets of lotteries with consistency checks built in to see if his or her preferences are consistent. For example, do you buy lottery tickets? If the lottery were designed to be completely breakeven, would you buy lottery tickets? If the lottery were designed to cost the average player net money but offer the possibility of winning, would you still play?

There are various methods and techniques that are used for making decisions when there is a group of vested interests or stakeholders.

Voting methods. Each stakeholder gets to vote according to an established rule. For example, one rule could be that each votes for his or her favorite and the choice getting the most votes wins. Another rule might be that each votes for his or her first, second, and third preferences. If first place votes don't show a majority, the second place votes of people who did not vote for the leading choice are distributed. Note that this allows for the possibility of gaming.

Weighted-scoring methods. Alternatives are scored on several attributes, which are then weighted to reflect the relative importance to the decisionmaker. Associated techniques divide attributes into "musts" and "wants" and sort options accordingly. Still others attempt to "minimize negatives" or "maximize positives."

Cost-benefit analysis. For each option, costs are tallied and benefits are described in terms of monetary value. This is straightforward for strictly economic decisions, but gets complicated when there are effects on persons or entities other than the group benefitting from the activity (i.e., health risks, job losses, environmental impacts, etc.). Such *externalities* may be valued differently by different stakeholders, so resulting estimates are usually highly uncertain.

Mathematical programming. This analytical method is useful when alternatives can be expressed as linear functions of a few decision variables. An *objective function* is also described, which presents the decision maker's objective (i.e., maximize profit, maximize long-term yield, etc.). Constraints are also quantified (i.e., total budget, rate of spending, emissions limits). The tools of linear algebra are then used to find the values of the decision variables that maximize the objective function subject to the constraints.

Payoff matrix analysis. This technique incorporates uncertainty by assessing the probabilities of possible outcomes for a set of alternatives. Such analysis is useful for limited sets of alternatives and outcomes.

Decision analysis. This is a methodology for structuring series of sequential decisions involving a variety of options into a logical framework that fully defines the decision space. The "utility" of the decision maker for each possible outcome is assessed. The utility function describes the strength of preference for each outcome—the higher the value the more desirable the outcome. Then "axioms of rational choice" are applied to find the decision sequence that optimizes the decision maker's utility. Where there is uncertainty in the utility function or where there are multiple decision makers, an "expected utility" function is used, which represents a consensus value.

Multiattribute utility analysis. This is an extension of decision analysis in which there are competing multiple objectives and utilities. A decision hierarchy is established that allows mathematical solution of the decision system to find the best decision strategy.

Simulation models. Simulation models try to replicate the differential spatial details of the particular system being studied at a point in time and to step forward in time and space differentially to predict the behavior of the system. Most climate models are based on large simulation models. To allow computational results in finite time, sub-models of the atmosphere, oceans, and biosphere may be run individually and then linked through other models. Uncertainties are usually treated by exercising the models for different "scenarios" that are thought to span the ranges of likely future pathways. The Intergovernmental Panel on Climate Change (IPCC, 2001) based its assessments of the likelihood of climate change over the next century on these types of models.

Economic models. Microeconomists have developed complex models based on the assumption that prices govern choices and that the economy can be modeled as a system that, when perturbed by price changes in certain portions of the system, will find a new "general equilibrium." These models were used to support Adam Smith's theory of the "invisible hand" (Smith, 1776). Smith claimed that individuals working for self-interest in a market economy would collectively produce benefits for the society as a whole. Subsequently, economists have learned that these microeconomic models are not necessarily stable, nor do they necessarily have unique equilibrium points. Their vulnerability comes from the reduction of human choices and behavior to pure economics (an interesting discussion of economic models is presented by Ackerman [1999]).

Systems dynamics modeling. A systems dynamics model links the important aspects of a complex system through a system of feedback loops that represent the individual relationships and influences between different parts of the system. The feedback influence can be any type of mathematical relationship, so the solution of large problems can be computationally demanding. An example of this type of modeling is described in Section 21.3.

Sometimes the values of the decision makers can be so diverse that expected utilities are averages of totally divergent values. Sometimes the choice of one technique results in a different answer than would another technique. Sometimes, the knowledge base of some of the decision makers is much greater in one area of the decision than another. How can we find a solution that will satisfy all the decision participants that the answer is a sound compromise? This leads us to the next methodology.

There is extensive literature on decision analytical methods, including the classic work by Raiffa (1968), Keeney (1992), and a book that gives interesting case studies from a behaviorist viewpoint (Kahneman et al., 1982). The American Institute of Chemical Engineers, Center for Chemical Process Safety, also has published a book on decision techniques oriented toward chemical industry practitioners (CCPS, 1994).

21.2.3 Negotiation

A different approach to dispute resolution involves mediation or negotiation. Many of us are familiar with these tools as ways to resolve legal disputes without going to trial. Others are aware of the diplomatic negotiations that occur to resolve conflicts between nations or other political entities. More recently, these tools have been used to facilitate decision making over a broad spectrum of controversial issues, including environmental and human rights issues.

The key to successful negotiation is to apply the talents of a skilled and impartial mediator or negotiator to guide the opposing parties through a series of possibilities and outcomes that will allow continuing compromise until a final accord is reached. Usually, the parties in conflict start out as adversaries and are forced into negotiation either because the parties each wish to break an unwanted deadlock or because they are forced to find a resolution by a higher authority. The success of negotiation involves a common acceptance of the role and rules of the negotiator, the parties' knowledge of the factors of importance, and an exploration of the consequences of various actions in a broader context.

Bazerman and Neale (1992) presents an overview of rational rules of negotiation based on both decision analytic and behavioral aspects of dispute resolution. Three excellent books for further reading on environmental negotiation and politics are by Susskind (1994), Porter and Brown (2000), and Young (2002).

21.2.4 How are decisions really made?

The techniques discussed in the prior sections are often employed when a major decision needs to be made. However, most of us make little decisions every day without much thought to the "big picture." Here are some motivations that guide the different groups of people who make our daily energy decisions:

> **Consumers.** Consumers base decisions on cost, convenience, reliability, and availability (may also include some value for "greenness"), as well as imitation, habit, and social status.
>
> **Regulators.** Most operate on the basis of political ideology, consensus technology information, and a desire to protect themselves from litigation and criticism.

Congress. Individuals in Congress operate in a highly political climate and keep a strong eye on protecting re-election chances. Trading or gaming to gain consensus on special issues is common. Decisions may be influenced by interests of financial backers and information provided by lobbyists of various sorts, though reform measures are reducing abuses in these areas. Most committees thrive on negotiation.

Industry. Industry's main focus is on short-term competitive positions and shareholder value. Some companies have farsighted leadership and may play an important role in the transition to sustainability, if they are willing to take some short-term risks.

Environmental groups. These smaller groups do not have the large resources needed to take on major industries or the government, so they employ litigation to delay or improve environmental performance of projects perceived as harmful, utilize public relations to raise awareness on special issues, lobby key government agencies and committees, and introduce shareholder initiatives to influence corporate practices.

Nations. Most are motivated by issues of national security, growth, and economic health.

UN. The UN engages in lots of negotiation reflecting national and regional self-interests while working on global issues—but implementation of policies is usually voluntary among individual nations.

Investors. Investors make decisions with emphasis on return on investment (ROI) and windfalls.

Who will lead us to the sustainable future? Many people may be concerned, but history indicates that we usually delay behavioral changes until a disaster provides the motivation. Some scholars have tried to alert society to potential disasters in order to encourage us to anticipate and respond to emerging problems rather than wait for disasters that force us to change.

21.3 Some Case Studies

In this section, we provide brief overviews of three approaches taken by different researchers that use models and analysis to better understand what might happen to global sustainability in the future. The first is an overview of the system dynamics modeling approach used by Meadows et al. (1992) in recent studies exploring the sustainability of humankind. Although the Meadows' work is controversial and is not presented by its authors as much more than an exploration of possible futures in a complex and uncertain world, it does serve as an example of an attempt to look into the interactions that drive human activities, and their impacts on the economy, society, and the environment.

The second example is more qualitative. Basically, it is a look at three potential pathways that might be followed by global society. It is more representative of the output that might be gained from an expert panel addressing the issues of sustainability.

The third example is a complex simulation model that integrates global human activities (including resource use, land use, economics, and industrial activities) with climate models that represent interactions between the atmosphere, the oceans, and the biosphere. A number of climate models are being used to represent potential changes in global natural circulations because of human-caused perturbations, and there are many models that address economic futures, including the role of changing technology. For our example, we have selected an integrated model that interconnects these two types of predictive models and provides a tool for scenario analysis based on a close interactive working relationship between climate scientists and economists. This example also contains broad stakeholder comment.

21.3.1 *Beyond the Limits* (Meadows et al., 1992)

In 1972, Donella and Dennis Meadows, in collaboration with Jay Forrester, published the results of a system dynamics model of the world developed under a commission from the Club of Rome and with funding from the *Volkswagen* Foundation. That book, *The Limits to Growth* (Meadows et al., 1972), showed that then-present growth trends in world population, industrialization, food production, and resource depletion would likely reach some planetary limits within the next century. The most probable effect of such limits would be to precipitate a sudden decline in both population and industrial capacity. Most important, they concluded that such decline was not inevitable—that a sustainable society could be developed, and that the sooner work toward that goal began, the easier the transition would be. This work was extremely controversial at the time, but was widely viewed as a wake-up call that our globally expansionist way of life was not sustainable.

Twenty years later, the Meadows made refinements in both the world model and their analyses (Meadows et al., 1992). Their results were published as *Beyond the Limits: Confronting Global Collapse—Envisioning a Sustainable Future*. This book provided both an introduction to system dynamics modeling and a set of valuable insights that were derived from scenarios studied using the model.

Their World3 model is based on aggregated stocks, sources and sinks, and methods for tracking their interactions through feedback loops. Positive feedback loops generate growth, in this case in population, economic output, and technology, until constrained by various limits. Negative feedback loops generate depletion of stocks, such as the earth's resources, pollution absorption capacity, and life-support systems. Below certain levels, the depletion results in impacts on health, well being, and ultimately, human survival. Stocks are aggregated at a high enough level that substitution of alternatives is possible within broad general categories. Figure 21.2 is taken from the Meadows' 1992 book and shows how the feedbacks between main stocks (population, cultivated land, and industrial capital) interact to change levels of pollution. As stated in their book,

"Each arrow indicates a causal relationship, which may be immediate or delayed, large or small, positive or negative, depending on the assumptions included in each model run."

The model is exercised for a series of scenarios that indicate potential alternate futures. The "standard run" is the same case that was used in the 1972 *Limits to Growth* assessment: the world follows the present pathways as long as possible without any major policy changes. Figure 21.3 shows the Scenario 1 (standard run) results from the 1992 analysis. The curves shown are global values, and the variability between the wealthy and the poor is hidden in the composite.

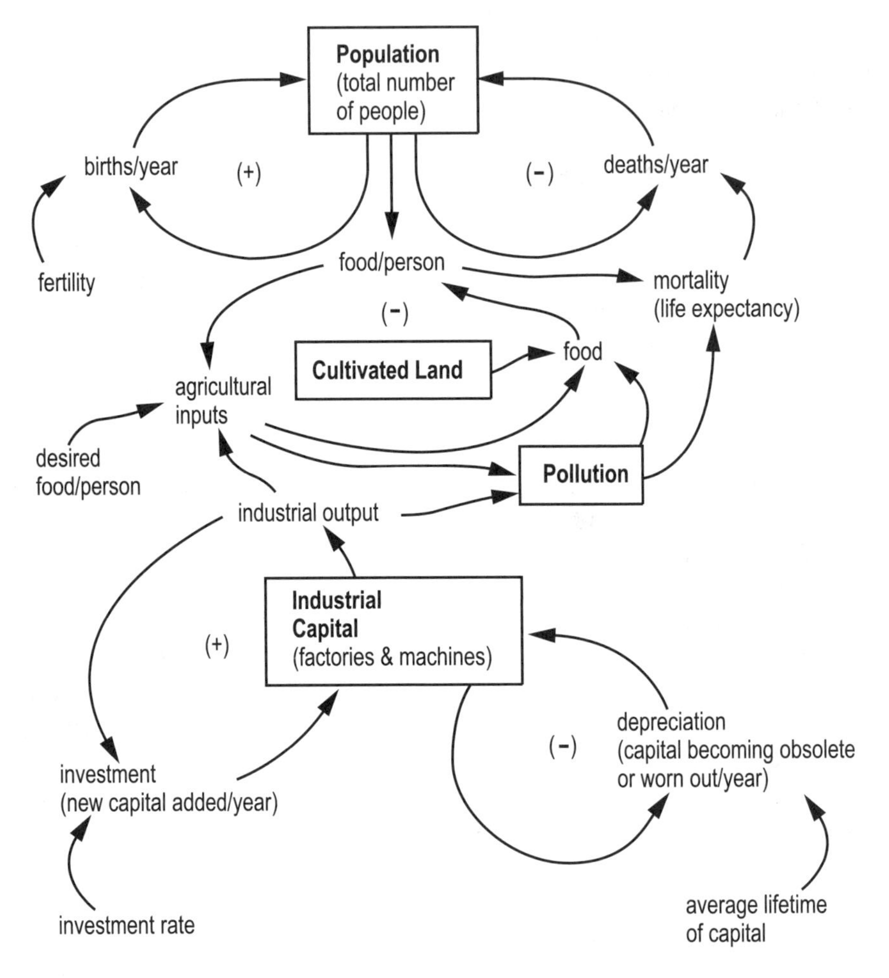

Figure 21.2. Feedback loops of population, capital, agriculture, and pollution. Source: Meadows et al. (1992). Reprinted with permission of Chelsea Green Publishing Co., Post Mills, VT.

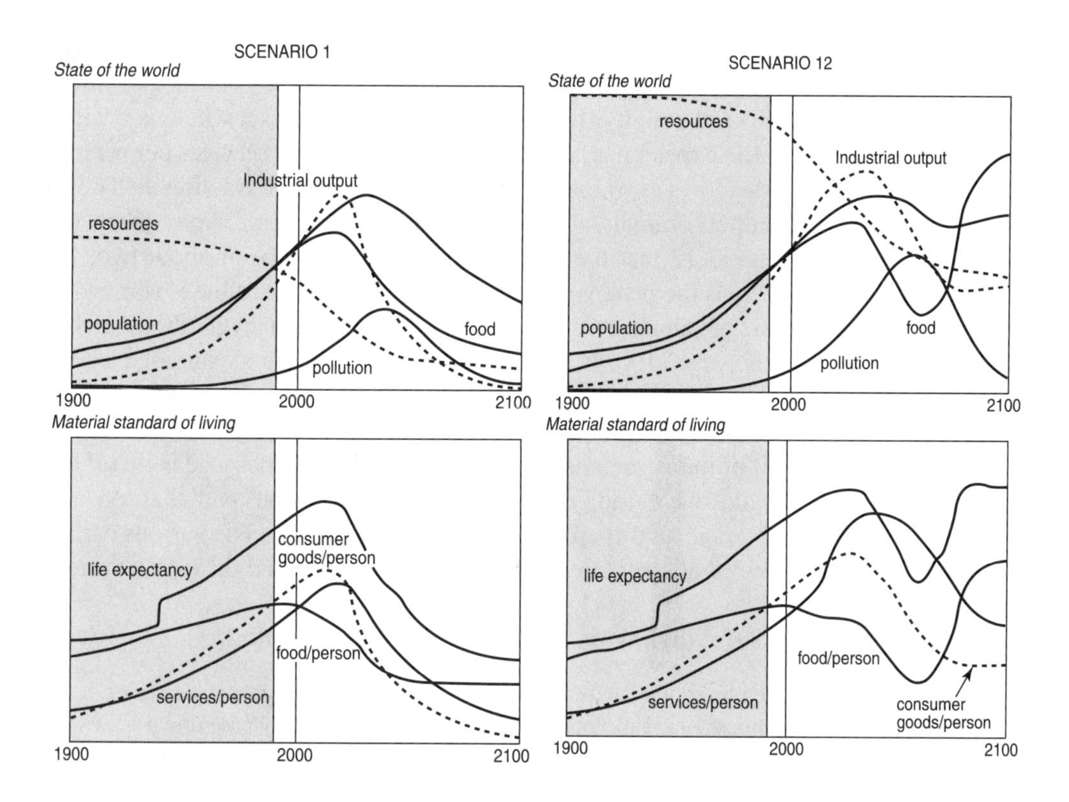

Figure 21.3. World3 system dynamics model results for Scenario 1—"standard run" (continuing present trends without major policy changes)—and Scenario 12 (stabilized population and industry with technologies to reduce emissions, erosion, and resource use adopted in 2015). Source: Meadows et al. (1992). Reprinted with permission of Chelsea Green Publishing Co., Post Mills, VT.

However, there are increasing indications that society may have to make some significant changes. Already, pollution control measures are becoming more severe, and investment in more sustainable technologies is increasing. The scenarios studied by Meadows et al. (1992) look at many of these possibilities individually and in combination. Figure 21.3 presents their results of Scenario 12 (stabilized population and industry with technologies to reduce emissions, land erosion, and resource use adopted in 2015).

Scenario 1 leads to continuing growth in population, industrial output, and food production until resources are seriously depleted, causing food and industrial output to decline rapidly. As the decline starts, pollution rises to a peak level and then falls. At that point, general decline sets in as the economy collapses and society approaches a new equilibrium state. By 2100, industrial output is a few percent of peak levels, population has declined (because of health problems and likely political conflicts), most major resources are seriously depleted, and pollution again declines because of lower population and less industrial activity. In this scenario, the world as a whole is back to or below 1900 levels of life expectancy, consumer goods, food, and services per person.

In Scenario 12, the Meadows examine a world that takes corrective action in 2015 to downsize industrial output dramatically as population stabilizes. Note from the resources curve in Scenario 12 that these are expanded by a factor of about two. By slowing industrial growth as the peak is being reached, the rate of decline is somewhat abated, and the indicators of standard of living in Scenario 12 reach levels about double those in Scenario 1 in the year 2100. In these and the other scenarios examined by the Meadows, the shape of the response curves is more meaningful than the actual numbers predicted by the model. Several scenarios show the advantages of earlier action and more severe curtailment of industrial and population growth. The message is that if the real world is something like the World3 model, we are on a dangerous path that can lead to a major decrease in average quality of life on this planet by 2100. The various earlier preventive actions analyzed all mitigate the rate of collapse and end up with a variety of better end points.

The Meadows summarize their work in *Beyond the Limits* as follows:

> Is any change we have advocated in this book, from more resource efficiency to more human compassion, really possible? Can the world actually ease down below the limits and avoid collapse? Is there enough time? Is there enough money, technology, freedom, vision, community, responsibility, foresight, discipline, and love, on a global scale?
>
> Of all the hypothetical questions we have posed in this book, those are the ones that are unanswerable, though many people will pretend to know the answers. The ritual cheerfulness of many uninformed people, especially many world leaders, would say that the questions are not even relevant; there are no meaningful limits. Many of those who are informed and who worry about the problem of overshoot are infected with the deep public cynicism that lies just under the ritual cheerfulness. They would say that there are severe problems already, with worse ones ahead, and there's not a chance of solving them.
>
> Both those answers are based, of course, on mental models. The truth is that no one knows.
>
> We have said many times in this book that the world faces not a preordained future, but a choice. The choice is between models. One model says that this finite world for all practical purposes has no limits. Choosing that model will take us even further beyond the limits and, we believe, to collapse.

> Another model says that the limits are real and close, and that there is not enough time, and that people cannot be moderate or responsible or compassionate. That model is self-fulfilling. If the world chooses to believe it, the world will get to be right, and the result will also be collapse.
>
> A third model says that the limits are real and close, and that there is just exactly enough time, with no time to waste. There is just exactly enough energy, enough material, enough money, enough environmental resilience, and enough human virtue to bring about a revolution to a better world.
>
> That model might be wrong. All the evidence we have seen, from the world data to the global computer models, suggests that it might be right. There is no way of knowing for sure, other than to try it.[1]

21.3.2 *Which World?* (Hammond, 1998)

Another look at the future is contained in a set of scenarios developed by Hammond (1998) in his book *Which World?* He examines three different scenarios, which each hold elements of plausibility and largely parallel the World3 cases discussed in the previous section. The quotations below are taken directly from the book.

- **Market World.** A future of expanding globalization with rapid economic growth that brings developing nations into the economic mainstream, provides new technologies to solve social and environmental problems, and results in "widespread prosperity, peace, and stability." This world is the hope of many economists and corporate leaders.
- **Fortress World**. A pessimistic scenario in which market forces are unsuccessful in promoting societal equity and preventing environmental disasters. "The scenario describes the dark side of global capitalism, a future in which enclaves of wealth and prosperity coexist with widening misery and growing desperation, a future of violence, conflict and instability."
- **Transformed World.** A "visionary scenario in which fundamental social and political change—and perhaps even changed values and cultural norms—give rise to enlightened policies and voluntary actions that direct or supplement market forces. Transformed World envisions a society in which power is more widely shared and in which new social coalitions work from the grass roots up to shape what institutions and

1 For more information on the book and the World3 model, see: http://www.rpi.edu/~simonk/ESP/BeyondTheLimits.html

governments do. Although markets become effective tools for economic progress, they do not substitute for deliberate social choices; economic competition exists but does not outweigh the larger needs for cooperation and solidarity among the world's peoples and for the fulfillment of basic human needs. In effect, this optimistic scenario asserts the possibility of fundamental change for the better—in politics, in social institutions, in the environment."

Hammond goes on to analyze global and regional trends that support or are counter to each of these three scenarios, but he has chosen these extreme futures to show the potential consequences of our present choices. He, like the Meadows, leads us to the conclusion that we need to make some changes as soon as possible.

21.3.3 MIT Joint Program on the Science and Policy of Global Change: Integrated Global Climate Model

The MIT Integrated Global System Model (IGSM) is designed to simulate the global environmental changes that may arise as a result of natural and anthropogenic causes, the uncertainties associated with the projected changes, and the effect of proposed policies on such changes (Prinn et al., 1998). The current IGSM formulation includes an economic model for analysis of greenhouse and aerosol precursor gas emissions and mitigation proposals, a coupled model of atmospheric chemistry and climate, and models of the oceans and terrestrial ecosystems.

Figure 21.4 illustrates the framework and components of the IGSM. Feedbacks between the component models that are currently included or under development for future inclusion are shown as solid and dashed lines, respectively.

In the integrated model, the combined anthropogenic and natural-emissions model outputs are driving forces for the coupled atmospheric chemistry and climate model, the essential components of which are chemistry, atmospheric circulation, and ocean circulation. The climate model outputs drive a terrestrial ecosystems model that predicts land vegetation changes, land CO_2 fluxes, and soil composition, which feed back to the coupled chemistry/climate and natural emissions models.

The model computes anthropogenic emissions of the key GHGs from 12 economic regions and converts them into distributions by latitude where needed. Special provision is made for analysis of uncertainty in key influences, such as the growth of population and economic activity and the pace and direction of technical change. The model also supports analysis of emissions control policies, providing estimates of the magnitude and distribution among nations of the costs, and clarifying the ways that changes are mediated through international trade. The model also supports the examination of potential feedbacks of climate change onto predicted emission rates.

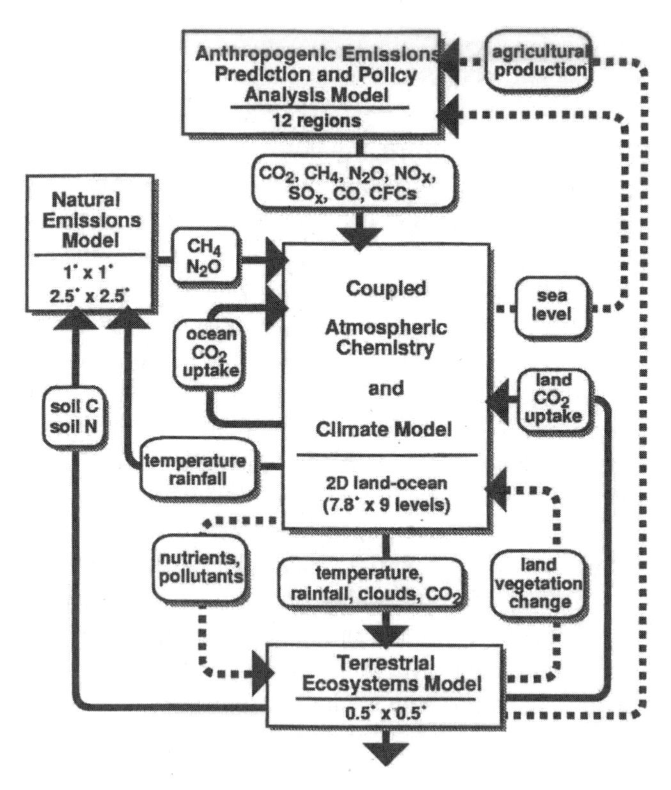

Figure 21.4. Schematic of MIT Integrated Global System Model. Source: Prinn et al. (1998).

The exercise of the IGSM provides an understanding of our future paths and of how today's decisions influence those paths. For example, Figure 21.5 shows some average projections of carbon-equivalent emissions per capita for the next century (this would have to be multiplied by world population to get total projected emissions). The assumption for the upper curve (EPPA Global Reference) is that we continue without major changes, that population increases to about 10 billion by 2100, and that we continue to make technological progress. It is evident that this scenario leads to more than a doubling of global CO_2 equivalent emissions by 2100. The curve representing constraints applied to limit emissions to a doubling is significantly lower than the normal behavior curve.

The dots on the vertical axis representing the present are where we are today: the OECD countries are the uppermost dot, the recovering economies the next lower, and the two lowest dots represent the developing countries, divided into richer and poorer groups. These developing countries are aspiring to a better standard of living, which will require more energy. Yet we see that the world average for the "doubling" scenario is only slightly above the present usage of the more successful developing countries and below the present OECD average. If the developing countries try to reach OECD standards (and the consequent GHG emissions), the world is likely to create a major climate problem over the next century.

What is the message? Countries that are asked to limit their development paths to protect the environment are almost certainly going to expect that the OECD countries will make substantial progress in reducing emissions. This means that we all face changes in our behavior—behavior that has been based on unconstrained growth. We all have to address the issue of how to get energy to provide our standard of living while greatly reducing GHG emissions, and similar sustainability challenges will be faced in other spheres of human impacts (e.g., water, biodiversity, institutional systems).

The climate component of the IGSM currently includes a simplified, two-dimensional land- and ocean-resolving (LO) model of the atmosphere, coupled to a three-dimensional general circulation model (GCM). A submodel is incorporated to include the coupling between chemistry and climate (through GHGs and aerosols), and between climate and chemistry (through temperature, humidity, cloudiness, rainfall, and surface fluxes). Together, the interacting components provide predictions of climate and air composition over both land and ocean as a function of latitude. In addition, the IGSM includes mass-balance models of the Greenland and Antarctic ice sheets. A mass balance model of the world's mountain glaciers is under development. These models, when combined with the calculation of the thermal expansion of seawater already included in the ocean models, will make it possible to simulate changes in sea level.

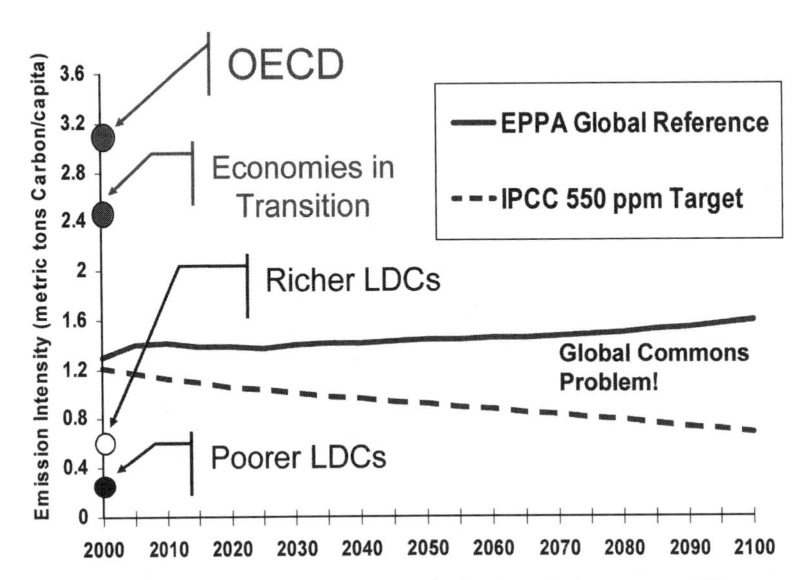

Figure 21.5. Global per capita carbon emissions needed to meet a 550 concentration target (Jacoby, 2004).

The terrestrial ecosystems model component, developed by the Marine Biological Laboratory, is used to predict global ecosystem states and support the investigation of feedbacks in the climate system. It also provides a crucial step to studies of the impact of climate on natural ecosystems and agriculture. A model component is used to simulate the natural emissions from the terrestrial biosphere to the atmosphere.

The linked IGSM components provide an emissions/chemistry/climate framework that is applied to questions relevant to policymaking. For example, predictions of GHG emission levels can be linked to uncertainties in future influences, such as population growth and economic performance. Further, the effect of various emissions control policies on climate can be pursued, allowing for analysis of the sensitivity of policy-relevant outcomes to critical assumptions in the various components.

The IGSM has evolved in complexity and content since it was originally conceived in 1991. Each of the submodels has been improved and tested against evolving data and knowledge; the original CO_2 focus for GHG emissions has been expanded to include all the GHG constituents recognized in Framework Convention on Climate Change (FCCC) circles, as well as other species that may be of significance in climate forcing. An abstract of a recent report by Webster et al. (2001) gives a flavor of how the model is being used today to guide policymaking and to recognize uncertainties in both our understanding of the economy and the ecosphere—as well as the uncertainties in projecting behaviors into the future.

> Future global climate projections are subject to large uncertainties. Major sources of this uncertainty are projections of anthropogenic emissions. We evaluate the uncertainty in future anthropogenic emissions using a computable general equilibrium model of the world economy. Results are simulated through 2100 for carbon dioxide (CO_2), methane (CH_4), nitrous oxide (N_2O), hydrofluorocarbons (HFCs), perfluorocarbons (PFCs) and sulfur hexafluoride (SF_6), sulfur dioxide (SO_2), black carbon (BC) and organic carbon (OC), nitrogen oxides (NO_x), carbon monoxide (CO), ammonia (NH_3) and non-methane volatile organic compounds (NMVOCs). We construct mean and upper and lower 95% emissions scenarios (available from the authors at 1° x 1° latitude-longitude grid). Using the MIT Integrated Global System Model (IGSM), we find a temperature change range in 2100 of 0.9 to 4.0°C, compared with the Intergovernmental Panel on Climate Change emissions scenarios that result in a range of 1.3 to 3.6°C when simulated through MIT IGSM. (Webster et al., 2001)

Another feature of the Joint Program activity is the use of sponsors and experts from a broad spectrum of industrial, governmental, and non-governmental organizations who convene on a regular basis to discuss results of the program in the context of global trends and politics. These discussions are "not-for-attribution," so attendees can speak freely. These forums provide guidance to the researchers and inform those in attendance about the issues involved in making progress toward a more sustainable future. (Extensive details about the Joint Program research and the IGSM are available at http://web.mit.edu/globalchange/www/.)

21.4 Transitional Pathways

The cases discussed in the previous section illustrate the breadth of future choices available to humanity and the potential outcomes of those choices, which range from global catastrophe to various transitions to some sustainable new equilibrium level where each average global citizen has less economic wealth and material goods than are enjoyed by citizens in OECD countries today. Does this mean that future generations will live less happily and well than we do today?

Not necessarily. One can imagine a transition to a new set of societal values in which personal achievement is not measured primarily by wealth and stature in the power structure of the world. The market economy is still likely to be the most efficient way of transferring goods and services among people, but it could include charges to cover the costs of environmental impact reduction and a reduction in the huge gaps between the richest and the poorest. The very poor today will improve their living standard in this new world and will have their basic needs well met—a luxury not available to about 1/3 of the population today. The wealthier, as discussed in Chapter 6, might enjoy a life with less financial pressures and fewer possessions—but select possessions that can provide satisfaction and service in a more productive manner. Technology can help in this regard, along with new industrial product and process design standards that optimize performance over the service life and beyond in recycled reincarnations.

There are many activities that bring pleasure with little need for materials and/or commercial energy: looking at a beautiful sunset, listening to music, interacting with friends, expressing and receiving love and friendship, sharing past experiences with each other, reading, sitting quietly and feeling peace, taking a walk alone or with a friend, practicing whatever spiritual program we individually may find rewarding, balancing our individual competitiveness with a sense of sharing and community, and so forth. A transition to a new set of values will benefit both those who suffer from isolation and those who are overstressed from hectic and over-programmed lives.

Unfortunately, humans have the potential for greed, selfishness, deceit, and cruelty, as well as the potential for good. In the US, many shareholders in the late 1990s and 2000 were delighted to see the stock market rise beyond their past experiences and they expanded their expectations about future living standards. When the market reversed direction and fell, upon news of corporate misdeeds that had artificially inflated the market and of concerns about terrorism and unrest, many people were seriously troubled and had to revise their expectations. We are resilient beings and survivors, however, and most of us will find ways to make our lives rewarding in spite of decreased expected affluence.

While most of us try to anticipate our own futures and adjust our behavior accordingly, it is harder for us to do the same on a community, regional, or global level. Yet the world today is becoming rapidly globalized through trade and communications. Once they are beyond poverty levels, people in developing countries are purchasing televisions and mobile phones, and gaining Internet access. We can now see and learn how others live and function. Business is globalized, and goods are transported worldwide. The affluent have access to global travel. On the international level, the United Nations provides a system for leadership and decision making. We enjoy international sports competitions like the World Cup and the Olympics. We also have global problems that range from GHG emission effects to the spread of infectious diseases. And the threats of global terrorism and conflicts are of concern to all of us. Can we start acting as responsible global citizens?

The key to a transition to a more sustainable society is individual and collective reinforcement of the positive values we share throughout all cultures and lifestyles:

- Change our values from overemphasis on material wealth to a balance between wealth and cultural, social, environmental, and spiritual riches.
- Provide lifelong education and cultivation of independent thinking; facilitate constructive discussions in an atmosphere of tolerance, respect, and wisdom.
- Actively participate as members of families, communities, and the society at large—including giving feedback to government and businesses.

21.5 The Challenge to Society

The United Nations is providing leadership to address some of the issues critical to a transition to sustainability. Concerns about GHG emissions have been formalized in the FCCC, and the Kyoto Protocol—a first step—might be ratified even without US participation. Although population control is a highly sensitive issue, many countries are recognizing the global problem and taking steps to stabilize their own populations in the future.

In a meeting of the UN Millennium Summit in September 2000, participants committed to a number of goals related to social equity, human rights, health, and international political stability to be accomplished by 2015. A total of 84 delegations (of which 59 were at the level of head-of-state and government) signed or deposited instruments of ratification or accession to 40 multilateral agreements, deposited with the secretary-general of the United Nations. A total of 273 treaty actions took place during the three days (187 signatures and 86 ratifications or accessions).

These included:

- Reduce the number of people living in extreme poverty by 50% (1.6 billion to 0.8 billion).
- Reduce the number of people without access to clean drinking water (now 1.3 billion) by 50%.
- Reduce women's deaths in childbirth (now 600,000 per year) by 75%.
- Reduce child mortality (now 58 deaths per 1,000 births) by 67%.
- Reverse the spread of AIDS.

- Ensure the implementation of treaties in the areas of arms control and disarmament.
- Ensure the implementation of international humanitarian and human rights laws.
- Take action against international terrorism.
- Call on states to accede to the international landmines ban and the Rome Statute authorizing establishment of the International Criminal Court.

But what about energy? At the Rio + 10 Summit in Johannesburg in September 2002, the principles of the Millennium Declaration were generally reaffirmed, and it was agreed to encourage and promote the development of a 10-year framework of programs to accelerate the shift toward sustainable consumption and production. But after much debate about energy issues, the main conclusions of the Summit in the energy area were:

- **Renewable energy**. Diversify energy supply and substantially increase the global share of renewable energy sources to increase its contribution to total energy supply.
- **Access to energy**. Improve access to reliable, affordable, economically viable, socially acceptable and environmentally sound energy services and resources, sufficient to achieve the Millennium Development Goals, including the goal of halving the proportion of people in poverty by 2015.
- **Energy markets**. Remove market distortions, including the restructuring of taxes and the phasing out of harmful subsidies. Support efforts to improve the functioning, transparency, and information about energy markets with respect to both supply and demand with the aim of achieving greater stability and ensuring consumer access to energy services.
- **Energy efficiency.** Establish domestic programs for energy efficiency with the support of the international community. Accelerate the development and dissemination of energy efficiency and energy conservation technologies, including the promotion of research and development.

Missing in the agreements are specific targets or any binding commitments, though these were discussed.

The challenge we face in a transition to energy sustainability is that we enjoy the benefits of inexpensive and reliable energy both as individuals and as a society. If the world is to stabilize GHG emissions while achieving a better balance between the richest and poorest inhabitants of this planet, OECD countries, and especially the US, will need to change their energy habits.

This book has presented a variety of energy choices available to us, along with the trade-offs associated with use of each. Reducing emissions through efficiency and reasonable conservation are obvious first steps, but then we also have to face the reality that energy is under-priced today, because it is subsidized in subtle ways, and we overlook some critical externalities (e.g., GHG emissions). Expensive renewable and lower-GHG-emitting technologies (e.g., nuclear, carbon sequestration) find strong competition from low-price fossil fuels. Until mechanisms of price rationalization are implemented, we are unlikely to make major progress. Even then, we will be forced to address the broader issue of the land impacts associated with much wider use of renewable energy. If we continue to live a high-consumption, urban lifestyle, we may find that renewable sources of energy are not enough. If we also want some level of social equity among world populations, we will have to choose a mix of renewable energy and nuclear energy sources or other options like fossil fuel use with carbon capture and sequestration.

As humans, we tend to resist proactive change. We do, however, respond well to change when disaster strikes. In dealing with global sustainability issues like climate change (not to forget other issues such as water supply, biodiversity, social equity, etc.), we can start to take some early actions. These actions, which are not painful, will begin a transition to a more sustainable way of life. Strong leadership will be an important element in guiding the transition. But who chooses our leaders? Who are the consumers who determine the course of industrial production? Where is the community that wants to protect "the Common"?

"Let it begin with me!" —Charles Spurgeon.

References

Ackerman, F. 1999. "Still Dead after All These Years: Interpreting the Failure of General Equilibrium Theory." Working Paper 00-01. Medford, MA: Tufts University Global Development and Environment Institute.

Bazerman, M. and M. Neale. 1992. *Negotiating Rationally*. New York: Free Press.

CCPS (Center for Chemical Process Safety). 1994. *Tools for Making Acute Risk Decisions*. New York: American Institute of Chemical Engineers.

Hammond, A. 1998. *Which World? Scenarios for the 21st Century*. Washington, DC: Island Press.

IPCC (Intergovernmental Panel on Climate Change). 2001. *Third Assessment Report. Working Group I: Climate Change 2001: The Scientific Basis*. Geneva: IPCC Secretariat.

Jacoby, H. 2004. From presentation to MIT Sustainable Energy Course, 10.391J. Feb. 26, 2004. Cambridge, MA: MIT.

Kahneman, D., P. Slovic, and A. Tversky. 1982. *Judgment Under Uncertainty: Heuristics and Biases*. Cambridge, UK: Cambridge University Press.

Keeney, R. 1992. *Value-Focused Thinking: A Path to Creative Decision Making*. Cambridge, MA: Harvard University Press.

Malthus, T.R. 1798. "Essay on the Principle of Population." Reprinted by Oxford University Press (1993).

Meadows, D., D. Meadows, J. Randers, and W. Behrens. 1972. *The Limits to Growth*. New York: Universe Books.

Meadows, D.H., D.L. Meadows, and J. Randers. 1992. *Beyond the Limits*. Post Mills, VT: Chelsea Green Publishing Company.

Porter, G. and J. Brown. 2000. *Global Environmental Politics*, Third Edition. Boulder, CO: Westview Press.

Prinn, R., H. Jacoby, A. Sokolov, C. Wang, X. Xiao, Z. Yang, R. Eckaus, et al. 1998. "Integrated Global System Model for Climate Policy Assessment: Feedbacks and Sensitivity Studies." *MIT Joint Program on the Science and Policy of Global Change, Technical Report No. 36*. Cambridge, MA: MIT.

Raiffa, H. 1968. *Decision Analysis*. Reading, MA: Addison-Wesley.

Smith, A. 1776. *The Wealth of Nations: An Inquiry into the Nature and Causes*. Reprinted by Modern Library (1994).

Susskind, L. 1994. *Environmental Diplomacy*. Oxford, UK: Oxford University Press.

Webster, M.D., M.H. Babiker, M. Mayer, J.M. Reilly, J. Harnisch, R. Hyman, M.C. Sarofim, and C. Wang. 2001. "Uncertainty in Emissions Projections for Climate Models." *MIT Joint Program on the Science and Policy of Global Change, Technical Report No. 79*. Cambridge, MA: MIT.

Young, O. 2002. *The Institutional Dimensions of Environmental Change: Fit, Interplay, and Scale*. Cambridge, MA: MIT Press.

Web Sites of Interest

web.mit.edu/globalchange/www
www.ipcc.ch/
www.nap.edu
www.sustainableliving.org/
unfccc.int/
www.un.org/millennium/declaration/ares552e.htm
www.johannesburgsummit.org/
www.rpi.edu/~simonk/ESP/BeyondTheLimits.html

Problems

21.1 At the Millennium Summit, nations debated whether it was reasonable to include a goal that 10% of energy come from renewable sources by 2015. This was not passed. List arguments that might be made for and against this proposal from the following viewpoints:

a. United States
b. Switzerland
c. Costa Rica
d. Japan
e. Russia
f. China
g. India

21.2 How would the arguments be made by each of the countries in Problem 1 for the inclusion or exclusion of the following as renewable sources:

a. Hydropower
b. Biomass

21.3 What levels of carbon taxes are considered high enough to drive a significant reduction in GHG emissions? (Norway already has some carbon taxes.)

21.4 Assume a carbon tax of $100/tonne carbon emitted. If this were implemented in the US, what fraction would it constitute of the US GDP?

21.5 Write a brief essay on what actions you can take to help a transition to sustainable energy use in the world. Are you likely to implement the actions you describe?

Choosing Among Options 22

This chapter brings to a close our journey to investigate and understand sustainable energy. Along the way, we examined renewable energy sources, conservation, and efficiency. We also looked at non-renewable energy sources (fossil fuel and nuclear energy, for example), even though some view these as unsustainable.

This book provides no simple road map to guide the reader to a sustainable global energy economy, nor does it advocate a particular suite of energy technologies or policies. Instead, we have tried to help readers discover the core tenets of sustainable practices, and we have tried to encourage readers to use these tenets to formulate their own ideas for realizing more sustainable energy futures. Given this approach, we subtitled our book "Choosing Among Options."

Many experts agree that choosing appropriate sustainable energy options has much to do with harmonizing the supply and utilization of energy with the near- and long-term protection of natural and social ecosystems. Many also believe that sustainable energy practices will emerge and improve as humans discover, implement, and update enabling technologies and public policies. However, these experts also believe that opinions will diverge as societies attempt to translate ideas into practice. These diverging opinions will present important questions for the future of sustainable energy, including:

- Over the next 50 years, what will various nations determine to be the appropriate percentages of renewable and non-renewable energy sources?
- What is the right balance between government regulation and voluntary initiative?
- What are the fair energy-related responsibilities for developed, emerging, and less-advanced economies?

Answering these questions and attaining viable, long-term agreements will involve the cooperation of governments, industries, societies, and individuals around the world. The answers to these questions will also depend on an understanding of critical time scales, e.g.:

- The onset of environmental jeopardy from abusive energy practices
- Technological innovation
- Transformation of institutions
- Economic dislocation
- Consensus building within and across political jurisdictions

Unfortunately, most of these time scales are marked by uncertainty. For example, in understanding the environmental impacts of greenhouse gases (GHGs), recall the substantial variations in the times for a given increase in global mean temperature predicted by the Global Circulation Model for a doubling of atmospheric CO_2 concentrations (Figure 4.15). Likewise, consider the time scales for multi-national treaties concerning economic development (e.g., the General Agreement on Tariffs and Trade [GATT]). These treaties can take years or even decades to be negotiated and approved. Because of these uncertainties, many experts believe that sustainable energy practices must be achieved by evolutionary transitions rather than by abrupt modifications in the status quo.

The size of the energy sector also affects the time scales for change. The energy sector vastly exceeds all other industries in infrastructure size, capitalization, money transfer rates, and annual throughputs of raw materials and products. Consequently, time scales for major technological innovations in the global energy economy are typically decades rather than years. Breakthroughs in research or transitioning may accelerate rates of technological transformation, but the extraordinary complexity of the energy sector presents formidable challenges to rapid change.

Complex challenges, uncertain time scales, and diverging opinions are central to the examination of sustainable energy practices. Indeed, our preface described this text as a technology book concerned with a dilemma—the dilemma of providing humankind with energy-derived advantages without damaging the environment, affecting societal and economic stability, or threatening the well-being of future generations. Will we solve this dilemma? Will we move from our largely reactive and narrow approach that addresses major problems one-by-one only after they each become severe, to an anticipatory and broader approach that finds less damaging pathways? Will sustainability become the preferred instrument to solve the dilemma? Or will sustainability end up as merely a fad in humankind's search for answers?

Throughout this book, we teach that sustainable energy will be the preferred approach to solving our energy problems. Given the need to reconcile environmental, economic, technological, and socio-political requirements, we argue that the sustainability approach is the only viable means of gaining and retaining public support to address our energy concerns. We conclude that sustainability will, over time, provide better and better solutions to the dilemma of reconciling energy needs with ecosystems protection and economic progress.

The dilemma will be solved gradually, and solutions will depend on the outcomes of technological and non-technological instruments. Many of these instruments will affect one another. For example, some governments may mandate phased rates of decarbonization of select energy sectors. These mandates may, in turn, give rise to modifications in fuel-manufacturing technologies or the accelerated adoption of low-carbon energy alternatives or a change in energy consumption patterns.

The challenge for current and future architects, builders, and proponents of sustainable energy practices is to discover and implement the right enabling measures—both technical and non-technical—at the right time. This book will have contributed to progress in sustainable energy if it helps readers to:

- Discover and understand the tradeoffs, difficulties, and opportunities involved in harmonizing economic progress and societal needs with the protection of ecosystems
- Understand the requirements for assessing and selecting technologies and policies for energy supply, utilization, conservation, environmental protection, and the responsible management of earth-human interactions
- Identify and quantify the technical, environmental, and geopolitical time scales essential to implementing the sustainable stewardship of energy and other natural resources
- Practice the principles of sustainable energy and communicate to others the importance of more sustainable energy options in our workplaces and communities

Conversion Factors

Quantity	To convert from	Into	Multiply by
Length	Feet	Meters	0.3048
	Centimeters	Inches	0.394
	Meters	Yards	1.09
	Miles	Feet	5,280
	Kilometers	Miles	0.62
Area	Square meters	Square feet	10.8
	Hectares	Square meters	10^4
	Square miles	Acres	640
	Acres	Hectares	0.4047
	Square miles	Square kilometers	2.59
Mass	Pounds	Kilograms	0.454
	Tons (US)	Pounds	2,000
	Tons (US)	Kilograms	907
	Tonnes (metric)	Kilograms	10^3
	Tonnes (metric)	Pounds	2,205
Volume	Liters	Quarts	1.057
	Liters	US gallons	0.264
	Liters	Imperial gallons	0.22
	Barrel (oil)	Gallons	42
Flow rate	Gallons per minute	Cubic meters/second	6.31×10^{-5}
Energy	Btus	Calories	252
	Btus	Joules	1,055
	Joules	Newton-meters or 1 kg m^2/s^2	1
	Btus	Foot-pounds force	778
	Kilowatt-hours	Btu	3,412*
	Quads (10^{15} Btus)	Exajoules (10^{18} joules)	1.055
Power	hp or horsepower (US = 550 $lb_f ft/s$)	Kilowatts	0.746

*When converting thermal energy to electricity, it is important to recognize the inefficiencies associated with power generation. A power plant with 33% thermal efficiency needs three times as much primary energy (e.g., coal) to produce an equivalent amount of electric energy.

Metric Prefixes:

Exa-	E	10^{18}	deci-	d	10^{-1}
Peta-	P	10^{15}	centi-	c	10^{-2}
Tera-	T	10^{12}	milli-	m	10^{-3}
Giga-	G	10^{9}	micro-	μ	10^{-6}
Mega-	M	10^{6}	nano-	n	10^{-9}
kilo-	k	10^{3}	pico-	p	10^{-12}
			femto-	f	10^{-15}

Temperature Conversion:

°C = (5/9)(°F - 32)
°F = 1.8 °C + 32
°K = °C + 273.2
°R = °F + 459.7

Constants:

Gravitational acceleration = 9.807 m/s^2 = 32.2 ft/s^2
Density of water = 1,000 kg/m^3 = 62.4 lbs/ft^3 (standard conditions)
Density of seawater = 1025 kg/m^3 (standard conditions)
Density of dry air = 1.226 kg/m^3 (at 15°C and 1 atm)

Nominal calorific values:

Coal = 7,000 cal/g
Oil = 10,180 cal/g
Gas = 1,000 Btu/ft^3 at standard condition

Financial:

1 mil = 10^{-3} US$ or 0.1 US cent
1 ECU ~ (0.8–1.2) US$ (varies with exchange rate)

Constants and Conversion Factors

Quantity	To Convert from units of ⟶	Into Units of ⟶	Multiply by
Energy	Joules*	kWh	2.78×10^{-7}
Length	Feet	Meters	0.3048
	Miles	Kilometers	1.609
Area	Square miles	Square kilometers	2.59
Mass	Pounds	Kilograms	0.454
Flow rate	gpm	m^3/sec	6.309×10^{-5}
	Into ⟵	To convert from ⟵	Divide by

Density of seawater = 1,025 kg/m^3, Gravitational acceleration = g=9.807 m/sec^2

Temperature Conversion: $°C = \frac{5}{9}(°F - 32)$; $°K = °C + 273.2$; $°R = °F + 459.7$

*Note that 1 J = 1 Newton meter (Nm) and 1N = 1 kgm/s^2

Energy Conversion Factors Chart

To convert from the first column units to other units, multiply by the factors shown in the appropriate row (e.g., 1 Btu = 252 calories)

Key: MWy = megawatt-year; bbls = barrels; tonnes = metric tons = 1,000 kg = 2204.6 lb; MCF = thousand cubic feet; EJ = exajoule = 10^{18}J. Nominal calorific values assumed for coal, oil, and gas.

	Btus	quads	calories	kWh	MWy
Btus	1	10^{-15}	252	2.93×10^{-4}	3.35×10^{-11}
quads	10^{15}	1	2.52×10^{17}	2.93×10^{11}	3.35×10^{4}
calories	3.97×10^{-3}	3.97×10^{-18}	1	1.16×10^{-6}	1.33×10^{-13}
kWh	3413	3.41×10^{-12}	8.60×10^{5}	1	1.14×10^{-7}
MWy	2.99×10^{10}	2.99×10^{-5}	7.53×10^{12}	8.76×10^{6}	1
bbls oil	5.50×10^{6}	5.50×10^{-9}	1.38×10^{9}	1612	1.84×10^{-4}
tonnes oil	4.04×10^{7}	4.04×10^{-8}	1.02×10^{10}	1.18×10^{4}	1.35×10^{-3}
kg coal	2.78×10^{4}	2.78×10^{-11}	7×10^{6}	8.14	9.29×10^{-7}
tonnes coal	2.78×10^{7}	2.78×10^{-8}	7×10^{9}	8139	9.29×10^{-4}
MCF gas	10^{6}	10^{-9}	2.52×10^{8}	293	3.35×10^{-5}
joules	9.48×10^{-4}	9.48×10^{-19}	0.239	2.78×10^{-7}	3.17×10^{-14}
EJ	9.48×10^{14}	0.948	2.39×10^{17}	2.78×10^{11}	3.17×10^{4}

	bbls oil equivalent	tonnes oil equivalent	kg coal equivalent	tonnes coal equivalent	MCF gas equivalent	joules	EJ
Btus	1.82×10^{-7}	2.48×10^{-8}	3.6×10^{-5}	3.6×10^{-8}	10^{-6}	1055	1.06×10^{-15}
quads	1.82×10^{8}	2.48×10^{7}	3.6×10^{10}	3.6×10^{7}	10^{9}	1.06×10^{18}	1.06
calories	7.21×10^{-10}	9.82×10^{-11}	1.43×10^{-7}	1.43×10^{-10}	3.97×10^{-9}	4.19	4.19×10^{-18}
kWh	6.20×10^{-4}	8.45×10^{-5}	0.123	1.23×10^{-4}	3.41×10^{-3}	3.6×10^{6}	3.6×10^{-12}
MWy	5435	740	1.08×10^{6}	1076	2.99×10^{4}	3.15×10^{13}	3.15×10^{-5}
bbls oil	1	0.136	198	0.198	5.50	5.80×10^{9}	5.80×10^{-9}
tonnes oil	7.35	1	1455	1.45	40.4	4.26×10^{10}	4.26×10^{-8}
kg coal	5.05×10^{-3}	6.88×10^{-4}	1	0.001	0.0278	2.93×10^{7}	2.93×10^{-11}
tonnes coal	5.05	0.688	1000	1	27.8	2.93×10^{10}	2.93×10^{-8}
MCF gas	0.182	0.0248	36	0.036	1	1.06×10^{9}	1.06×10^{-9}
joules	1.72×10^{-10}	2.35×10^{-11}	3.41×10^{-8}	3.41×10^{-11}	9.48×10^{-10}	1	10^{-18}
EJ	1.72×10^{8}	2.35×10^{7}	3.41×10^{10}	3.41×10^{7}	9.48×10^{8}	10^{18}	1

List of Acronyms

A&G	Administrative and general
AC	alternating current
AGR	advanced gas reactor
AIChE	American Institute of Chemical Engineers
ARL-NREC	Alden Research Lab–Northern Research & Engineering Corporation
BACT	best available control technology
BAU	business as usual
BCE	before current era
bTOE	billion tonnes of oil equivalent
Btu	British thermal unit
BWR	Boiling water reactor
CAA	Clean Air Act
CAES	compressed air energy storage
CANDU	Canadian design heavy water reactor
CCFD	complementary distribution function
CCGT	combined cycle gas turbine
CCPS	Center for Chemical Process Safety (AIChE)
CCS	carbon capture and sequestration (storage)
CDF	cumulative distribution function
CDF	computational fluid dynamics
CESA	Central Electro Solar de Almería

CFC	chlorofluorocarbon
CH_4	methane
CHP	Combined heat and electric power, or cogeneration
CI	compression ignition
CIS	copper indium selenide
COE	cost of electricity
CO_2	carbon dioxide
COP	Conference of Parties, coefficient of performance
CSP	concentrating solar power
DC	direct current
DOE	Department of Energy
DPU	Department of Public Utilities
E	primary energy
EEC	European Economic Community
EIA	Energy Information Administration (US DOE)
ECCS	emergency core coolant system
EGS	enhanced geothermal system
EJ	exajoules
em	electromagnetic
emr	electromagnetic radiation
EOS	equation of state
EPRI	Electric Power Research Institute
ERDA	Energy Research and Development Administration
ESE	earth system ecology
FCCC	Framework Convention on Climate Change
GCM	general circulation model
GCR	gas-cooled reactor
GDP	gross domestic product

GHG	greenhouse gas
GHP	geothermal heat pump
GHPC	Geothermal Heat Pump Consortium
GNP	gross national product
GTSTCC	gas-turbine steam-turbine combined-cycle
HAP	hazardous air pollutant
HC	hydrocarbon
HDI	human development index
HDR	hot dry rock
HEU	highly-enriched uranium
HHV	high heating value
HLW	high level waste
HTGR	high temperature gas-cooled reactor
IAQ	Indoor air quality
IE	industrial ecology
IGSM	integrated global system model
ILW	intermediate level waste
INPO	Institute of Nuclear Power Operations
IPCC	Intergovernmental Panel on Climate Change
ISO	International Standards Organisation
ITER	International Tokamak Experimental Reactor
J	joule
J-T	Joule-Thompson
KE	kinetic energy
kWe	kilowatt of electrical power
kWh	kilowatt-hour
kWth	kilowatt of thermal power
LCA	life cycle analysis

LDC	less developed country
LEU	low-enriched uranium
LHV	low heating value
LLW	low level waste
LMFBR	liquid metal– cooled fast breeder reactor
LNG	liquefied natural gas
LWR	light water reactor
MACT	maximum achievable control technology
MCF	thousand cubic feet
MEU	medium-enriched uranium
MTBE	methyl tertiary-butyl ether
MTOE	million tonnes of oil equivalent
MW	megawatt (million watts)
MWe	megawatt of electrical power
MWt	megawatt of thermal power
NAAQS	national ambient air quality standards
NCPV	National Center for Photovoltaics
NEMS	national energy modeling system (US EIA)
NEPA	National Energy Policy Act
NNP	net national product
NOAA	National Oceanic and Atmospheric Administration
NO_x	nitrogen oxides
NREL	National Renewable Energy Laboratory
NSPS	new source performance standards
O_2	oxygen
O_3	ozone
OECD	Organisation for Economic Co-operation and Development

OMB	Office of Management and Budget
O&M	operation and maintenance
OPEC	Organization of Petroleum Exporting Countries
OTEC	ocean thermal energy conversion
P	population
PC	personal computer
PCAST	President's Council of Advisors on Science and Technology
PDF	probability density function
PDL	person-days lost
PE	potential energy
PHWR	pressurized heavy water reactor
PIC	product of incomplete combustion
PM	particulate matter
PPP	purchasing power parity
PSC	Public Service Commission
PSD	prevention of serious degradation
PURPA	Public Utility Regulatory Policy Act
PV	photovoltaic
PVMaT	Photovoltaic Manufacturing Technology Project
PWR	pressurized water reactor
Q	quadrillion Btu (quad)
RACT	reasonably available control technology
R&D	research and development
RHS	right hand side
RBMK	Soviet designed large tube-type reactor
RPS	Renewable Portfolio Standard
SCF	standard cubic feet
SEGS	solar electric generating system

SI	spark ignition
SMES	superconducting magnetic energy storage
SNM	special nuclear material
SO_2	sulfur dioxide
SO_x	sulfur oxides
SPR	strategic petroleum reserve
T	tonnes (metric tonne = 1000 kg)
TAR	Third Assessment Report (IPCC)
TDS	total dissolved solids
TOE	tonnes of oil equivalent
TQM	total quality management
TRU	transuranic waste
TWh	terawatt-hour
UNAT	natural uranium
USGS	US Geological Survey
UVAC	ultraviolet absorptive coating
VOC	volatile organic compound

Index

H

I

J

K

L

M

N

Q

R

S

U

W

Y